Conjugated Polymers
A Practical Guide to Synthesis

RSC Polymer Chemistry Series

Series Editors:
Professor Ben Zhong Tang (Editor-in-Chief), *The Hong Kong University of Science and Technology, Hong Kong, China*
Professor Alaa S. Abd-El-Aziz, *University of Prince Edward Island, Canada*
Professor Stephen L. Craig, *Duke University, USA*
Professor Jianhua Dong, *National Natural Science Foundation of China, China*
Professor Toshio Masuda, *Fukui University of Technology, Japan*
Professor Christoph Weder, *University of Fribourg, Switzerland*

Titles in the Series:
1: Renewable Resources for Functional Polymers and Biomaterials
2: Molecular Design and Applications of Photofunctional Polymers and Materials
3: Functional Polymers for Nanomedicine
4: Fundamentals of Controlled/Living Radical Polymerization
5: Healable Polymer Systems
6: Thiol-X Chemistries in Polymer and Materials Science
7: Natural Rubber Materials: Volume 1: Blends and IPNs
8: Natural Rubber Materials: Volume 2: Composites and Nanocomposites
9: Conjugated Polymers: A Practical Guide to Synthesis

How to obtain future titles on publication:
A standing order plan is available for this series. A standing order will bring delivery of each new volume immediately on publication.

For further information please contact:
Book Sales Department, Royal Society of Chemistry, Thomas Graham House, Science Park, Milton Road, Cambridge, CB4 0WF, UK
Telephone: +44 (0)1223 420066, Fax: +44 (0)1223 420247
Email: booksales@rsc.org
Visit our website at www.rsc.org/books

Conjugated Polymers
A Practical Guide to Synthesis

Edited by

Klaus Müllen
Max Planck Institute for Polymer Research, Germany
Email: muellen@mpip-mainz.mpg.de

John R. Reynolds
Georgia Institute of Technology, USA
Email: reynolds@chemistry.gatech.edu

and

Toshio Masuda
Fukui University of Technology, Japan
Email: masuda@fukui-ut.ac.jp

RSC Publishing

RSC Polymer Chemistry Series No. 9

ISBN: 978-1-84973-799-9
ISSN: 2044-0790

A catalogue record for this book is available from the British Library

Published by The Royal Society of Chemistry,
Thomas Graham House, Science Park, Milton Road,
Cambridge CB4 0WF, UK

Registered Charity Number 207890

For further information see our web site at www.rsc.org

This book is dedicated to our spouses, Renate Müllen, Dianne Reynolds and Keiko Masuda, for the understanding and support they have given us in all that we do.

Preface

Why conjugated polymers? A convincing answer to this question can be given in three parts: conjugated polymers are fascinating species from i) a structural point of view due to the many ways of establishing an extended pi-conjugation; ii) a functional point of view due to their electronic and optical properties, which qualify them as active components of organic electronics; and iii) a research point of view due to their potential of fostering cross-disciplinary research.

Since we live in a material world there cannot be novel technology without new or improved materials, and they must be synthesized. Synthesis is both a matter of creative design and experimental skill, and no target structure can prove this claim better than conjugated polymers. The crucial role of synthesis is what has led to the concept of this book; not so much as a theoretical exercise "on-paper", but as a practical, "hands-on" approach. It is intended to convince the reader of the beauty of conjugated polymer design and synthesis, but not set aside the technical difficulties and experimental pitfalls.

A good case can be made by first taking a look at the device and then work backwards to the synthesis of the functional molecules. Technologies and also the underlying materials stand in competition with each other. Organic electronics face the fact that inorganic semiconductors, such as silicon, have beautiful properties and are thus strong contenders. One may think of the high charge carrier mobilities and the vanishing exciton binding energy. What stands against that is the "designable" structural and functional versatility of their organic counterparts, highlighted by the ease of their synthesis and processing. Even in the organic field, conjugated polymers must compete with small organic molecules, which can be made structurally perfect and can be processed by vacuum techniques.

Whatever material is considered, organic synthesis is generally performed in solution. The active function of an organic electronic device, however, cannot

RSC Polymer Chemistry Series No. 9
Conjugated Polymers: A Practical Guide to Synthesis
Edited by Klaus Müllen, John R. Reynolds and Toshio Masuda

Published by the Royal Society of Chemistry, www.rsc.org

be understood or even tailored by restricting one's attention to the dilute solution. It is rather a matter of solid thin films and their interfaces. An important criterion, when judging the function of an organic field effect transistor (OFET) as a switch, is high charge carrier mobility. This depends upon the solid state packing and thus the supramolecular order of the semiconductor. Likewise, "bright" emission from an organic light emitting diode (OLED) requires different layers with defined interfaces to keep the charge recombination zone away from the metal electrode; and this, next to high film quality, depends upon the orthogonal solubility of the different conjugated polymers used. The situation in an organic photovoltaic device (OPV) is, from a morphological point of view, even more complex. What is needed in such a device is a nanophase separation of electron donor and electron acceptor components to ensure, after charge separation, efficient percolation of holes and electrons. It is the complexity of the underlying functional processes that explains why physics, engineering and materials science are so important – or at least have the last say when it comes to device performance and stability.

Again, the whole endeavor starts with synthesis and this shall be considered in greater detail now. Even the more distant reader will be aware of "classical" conjugated polymer structures, such as poly(*p*-phenylene) (PPP), poly(phenylenevinylene) (PPV) and polythiophene (PT). Here the "usual suspects" are incorporated as building blocks: benzene, thiophene or a C=C double bond. It does not need much imagination, however, to dream up structural modifications, which brings us to the unbelievable manifold of design opportunities. When we take a closer look at the nature of the building blocks, benzene can be replaced by larger polycyclic aromatic hydrocarbons (PAHs) that also leave us with different ring positions as coupling points. Hydrocarbons can be replaced by their heterocyclic analogues, which is particularly important for the electron donor or acceptor character of the final macromolecule. The search for even more repeat units can have different driving forces, empirical or theoretical, and is often coupled to device results. This can also explain why the community sometimes follows fashion. Presently, donor units, such as bridged dithiophenes, and acceptor units, such as benzothiadiazole or diketopyrrolopyrrole, seem to be particularly popular. When different building blocks are incorporated there are many ways of doing this, statistically or in a defined sequence. Conjugated polymers with an alternating array of donor and acceptor repeat units have been studied as low band gap materials for efficient light harvesting in OPV. Another important structural modification becomes possible by an increase in the dimensionality of the macromolecules when going from a (linear) chain to a step ladder or even full ladder polymer. A logical extension then is a true two-dimensional pi-system, such as a graphene sheet, which has recently attracted immense attention. Finally, all these polymer architectures possess inherent rigidity that can severely limit solubility and thus solution processability. Alkyl substitution is therefore a mandatory step. There is of course, the danger of compromising extended conjugation when the additional substituent causes torsion of the building blocks about the interring bonds, and substituents may be considered to "dilute" electronic properties. Nevertheless, alkyl substituents can also have beneficial effects by directing supramolecular order during film formation. This is also true when more complex topologies are targeted, such as diblock or multiblock copolymers made from rod or coil segments. The latter aspect is not only relevant for electronic reasons – bringing electron rich and electron poor domains into close proximity to one another – but also for morphological reasons, where phase separation can be enforced at the desired length scale.

All these design opportunities concern the macromolecular structure – prior to processing – and they are, indeed, decisive for key electronic and optical features, such as, for example, the wavelength of absorption and emission or their function as n-type *vs.* p-type semiconductors. One might go as far as to claim that, based on these subtle structure–property relationships, (opto)electronic properties can be "synthesized". Or one might even go a step further and encode desired supramolecular characteristics to introduce weak intermolecular forces during molecular design, although in this case structure–property relationships are much less well developed. Here is, however, a very critical point: all these design opportunities are easily sketched on paper but do not necessarily translate easily into a glass flask. This brings us to the key issue of this book: the synthesis of conjugated polymers.

When talking about conjugated polymer synthesis, an initial question concerns the availability and purity of the suitably functionalized building blocks. Thereby, the required synthetic efforts can be quite different and demands by the community can lead to the commercial availability of starting compounds. Closely connected to this is the question of the polymerization reaction, that is to say how the building blocks are subjected to a repetitive coupling. Remarkably enough, while step-growth methods have long been the "classical" polymerization mode, chain-growth methods have recently been realized as well. This of course holds promise for establishing "living" end-groups and offers new opportunities for block copolymer synthesis. Transition metal-catalyzed reactions developed in organic and organometallic chemistry have had a huge impact on conjugated polymer synthesis, and this holds true for methods not only of carbon-carbon, but also carbon-heteroatom (*e.g.* nitrogen) connections. Many of these commonly used pathways are addressed as "named reactions" honoring their inventors and, indeed, these inventions can hardly be overestimated. These reactions allow the connection of aromatic ring systems to a polymer chain *via* homo- or hetero-atom couplings, and there the reader will readily envisage how important the above mentioned strictly alternating incorporation of donor and acceptor moieties is. Apart from aromatic building blocks, vinylene and ethynylene elements can be introduced as well. Here it is rewarding to compare transition-metal catalysis with older procedures, such as connective double-bond formation.

Is it thus sufficient, when targeting a particular polymer structure, to check the literature for the right building blocks and the best mode of coupling? The answer is definitely no, and there are at least two reasons for that. The first one, even if somewhat discouraging for a physicist who feels that the synthetic plan looks quite easy, concerns the experimental conditions. A typical case is rigorous exclusion of oxygen and water since these can affect catalyst activity and trigger unwanted side-reactions. The choice of reaction temperature and time should be mentioned here as well. The second reason is that the nature of the building block and the chosen synthetic method must be carefully adjusted. Thus, steric hindrance may slow down the growth of the macromolecule so much that side reactions come into play, which can either stop further growth or induce side reactions. The issue of side reactions, even if occurring as minor processes, is critical. A typical case is the so-called Wessling Zimmermann synthesis of PPV, which played an important role for the development of light emitting devices based upon conjugated polymers in the early nineties. This advance in the synthetic procedure was achieved *via* a precursor polymer, which was finally transformed into the conjugated chain *via* a 1,2-elimination process. Failure to accomplish this quantitatively will, of course, lead to an interrupted pi-conjugation. Even more severe is the fact that side reactions can lead to (ketonic) defects that act as traps for the excited states of the light emitting device. Let us come back to the question of the mechanism of the polymer forming reaction. In a step-growth polycondensation, high molecular weights can only be achieved for high degrees of conversion and thus the coupling of intermediate oligomers. Even if limitations imposed by solubility can be excluded, mismatch of the stoichiometries must be avoided, and this relates to the absence of side reactions.

Conjugated polymer synthesis, in order to be meaningful, requires an even higher level of sophistication, and this raises the question of how we describe its molecular structure. We commonly denote the prevailing repeat units, written in parenthesis, and then add "*n*" for the number of repeat units, even if it is often not specified. There are ample cases, particularly in recent literature, where the molecular weight and the polydispersity of an individual polymer play a decisive role in the device performance, such as charge carrier mobility. This, by the way, can readily be understood when considering the above mentioned ordered packing of polymer chains. Then, it is clear that further features come into play, such as the mode of end-capping. Impurities, as has been mentioned, can serve as traps not only for excitons, but also for charges, and structural irregularities do not only interrupt pi-conjugation, but also hamper regular packing. It follows that the above way of denoting a polymer is in urgent need of careful specification. Otherwise, the whole research field is in danger of criticism, such as "*they measure very precisely, but they do not precisely know what they measure*".

This book is intended to show the beauty of conjugated polymer synthesis without ignoring the difficulties and obstacles. While organic electronics are often tempted to follow the fascination of device fabrication, this book takes the opposite direction in a synthesis-first approach. This approach requires imagination, but also scrupulous optimization of experimental conditions combined with careful proof of structural perfection.

Considering the above, a strong plea is made herein to activate the good old virtues of organic synthesis, and these will certainly prove their value when, for example, pushing the limits of molecular weight or handling a low band gap polymer with its inherent chemical instability. There are, however, still many open territories for conjugated polymer synthesis. The new challenges could concern a complex combination of properties, such as making a charge transporting polymer also responsive to stimulus-driven switching or by connecting a conjugated polymer segment to a biomacromolecule for recognition and self-assembly processes. Furthermore, while we have so far focused on the power of solution synthesis, performing synthesis in the (bulk) solid state or on a surface could create significant advantages as one could not only build in the supramolecular order, but also avoid the additional difficulties of controlling solution processing. Whether synthesis is structure or method oriented, knowledge or application driven, or whether it targets new polymer structures or up-scales established ones, it finds room everywhere. But it is not always easy. However, we believe that this book will help to make it easier – and more successful.

Last but not least, we would like to thank all the authors who actively do research in the forefront of the conjugated polymer field for their excellent contributions. Thanks are also due to Professor Ben Zhong Tang, Editor-in-Chief of this book series for his continuous support and to Ms. Leanne Marle at the RSC office for her clerical assistance.

Klaus Müllen
On behalf of the Editors

Contents

RSC Polymer Chemistry Series No. 9
Conjugated Polymers: A Practical Guide to Synthesis
Edited by Klaus Müllen, John R. Reynolds and Toshio Masuda

Published by the Royal Society of Chemistry, www.rsc.org

CHAPTER 1

Pi-Conjugated Polymers: The Importance of Polymer Synthesis

JOHN R. REYNOLDS

School of Chemistry and Biochemistry, School of Materials Science and Engineering, Center for Organic Photonics and Electronics, Georgia Institute of Technology, 901 Atlantic Drive, Atlanta, GA 30332-0400, USA
Email: reynolds@chemistry.gatech.edu

1.1 Historical Perspective

When one considers the early days of conjugated and conducting polymer synthesis, the early work by Letheby on the oxidation of aniline, presumably forming polyaniline (**1**),[1] and Dall'Olio *et al.* on polypyrrole (**2**)[2] are often referred to as landmark developments in the field. To gain an important historical view, the reader is directed to the work of Rasmussen who provides a perspective on the field where the work of Weiss on polypyrrole, as well as that of Buvet and Jozefowicz on polyaniline, are highlighted.[3] These materials, while completely insoluble and infusible as formed from oxidative polymerization from the parent monomers, served as the basis for inducing electroactivity into polymer systems. Over the years, numerous review articles, book chapters and reviews have issued around the field, most often directed to a class of polymers or type of property they impart, with the 1986, 1998, and 2007 editions of the "Handbook of Conducting Polymers" providing a deep and complete scientific overview[4–6] (Chart 1.1).

The 1950s saw the Nobel Prize winning work of Karl Ziegler and Giulio Natta on coordination polymerization of unsaturated molecules, which

RSC Polymer Chemistry Series No. 9
Conjugated Polymers: A Practical Guide to Synthesis
Edited by Klaus Müllen, John R. Reynolds and Toshio Masuda

Published by the Royal Society of Chemistry, www.rsc.org

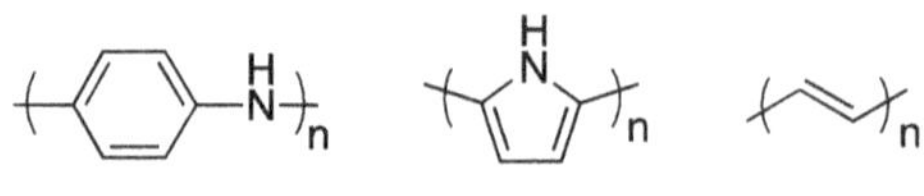

1 2 3

Chart 1.1

provided a route to structurally well-defined polymers.[7] Most known for the development of commercially important polyolefins (*e.g.* polyethylene and polypropylene), the work of Natta demonstrated that acetylene polymerization could yield the conjugated polymer polyacetylene $(CH)_X$ (**3**) as an infusible grey powder.[8] While the 1960s saw the development of many aromatic units containing polymers, it also became understood that pi-conjugation led to semiconducting material properties. In one fascinating study, Berets and Smith[9] examined the vapor phase treatment of $(CH)_X$ powders with various Lewis acids and bases. In fact, when using BF_3 as the reacting dopant, they measured conductivity enhancements by a factor of 1000 on pressed pellets. In this work, they also treated $(CH)_X$ with Cl_2, yet only saw small conductivity increases by a factor of 5. Interestingly for the discussion that will follow, this work did not use iodine or bromine as an oxidizing system.

Serendipity and careful observation go hand-in-hand in science, and this has been especially important for many discoveries in the conjugated and conducting polymer field. An especially important discovery was the formation of free-standing films of $(CH)_X$ by Ito *et al.* in 1974 caused by rapid polymerization of acetylene at a quiescent Ziegler–Natta polymerization initiator system solution.[10] While the Shirakawa research group was mainly dedicated to understanding structural properties (*e.g. cis–trans* ratios), the fact that these films were mechanically flexible, with a metallic silver luster, suggested important underlying electronic properties.

In a landmark series of experiments, Heeger, MacDiarmid, and Shirakawa combined efforts to study the electronic properties and gas-phase doping of polyacetylene films. Their discovery that treatment of these films with controlled amounts of Cl_2, Br_2, I_2 and AsF_5 could yield conductivity enhancements greater than 10^7, and ultimately yield electronic conductivities in excess of 500 S/cm, demonstrated unprecedented conductivity in an organic polymer.[11] While charge transfer salts, such as those based on tetrathiafulvalene and tetracyanoquinodimethane were known to exhibit high conductivity and metallic properties,[12] the fact that such properties were accessible in the more disordered and flexible polymer films was especially stunning. Researchers around the world quickly picked up on this, and it was demonstrated that the high level of conjugation in the polymer, along with pi-stacking and interchain interactions, all played an important role in the electronic properties. Chemists quickly realized that many polymer structures could be prepared that were fully conjugated; thus, the stage was set for a major synthetic effort. This work has now spanned 30 years and has led from insoluble, infusible, materials that were highly unstable conductors to well-characterized, solution processable polymers with fine structural control that are finding utility across a broad number of applications.

With this background in mind, this book seeks to teach the details of synthetic preparative polymer chemistry in all of the major classes of pi-conjugated polymers that have been developed to researchers in the field. The authors of each chapter have carefully overviewed the various polymer types employed in the field with a special focus on experimental details that yield reproducible and high-quality materials. Prior to moving to those specific chapters, let's take some time to review the general concepts in polymerization that are important for the development of such materials.

1.2 Considerations in Polymerizations

Fundamental polymer chemistry teaches us two main relevant mechanisms for polymerization; specifically step-growth and chain-growth methods.[13] These methods provide polymers with distinctly different structures in terms of repeat unit functionality, molecular weight, and dispersity. As these molecular structures relate to higher-level macromolecular considerations, such as chain–chain interactions and the development of material morphology, it is important that the mechanism be understood for any system under study. Step-growth polymerization sees the step-wise buildup of molecular weight as a function of the extent of conversion of reactive monomer functional groups. As taught by Carother's equation, high molecular weight polymers are obtained at high extents of conversion requiring especially high degrees of monomer purity. The excess of any one monomer type (more formally the excess of any one functional group in polymerization) limits the molecular weight considerably where oligomers can provide non-optimal properties.

Chain-growth polymerizations to form addition polymers are most often accomplished using monomers with multiple bonds and loss of unsaturation. In this mechanism, a reactive intermediate is first created in an initiation step and subsequently propagates *via* repeated monomer addition to provide a macromolecule. When the reactive intermediate is ionic, impurity termination or quenching processes can kill the reactive intermediate, while in the case of radical polymerization, coupling termination can lead to an overall doubling of the average molecular weight. Many early attempts at forming conjugated polyarylenes and polyheterocycles attempted to use step-growth polymerization under non-optimized conditions, thus yielding low molecular weight polymers. Significant efforts detailed in this book demonstrate how careful control of the reagents and polymerization conditions now lead to quality polymers as high molecular weight, well-defined chemical systems. In fact, in some instances where it was believed that step-growth couplings were occurring, detailed studies show that indeed chain-growth (and in some instances living) polymerizations were in fact underway. While one of the benefits of a chain-growth polymerization can be the formation of high molecular weight polymers at a low degree of monomer conversion, the fact that unsaturation is lost tends to limit simple chain-growth polymerizations to directly form conjugated polymers to alkyne derivatives.

1.3 Side Chains, Processability and Molecular Weight

One of the most important physical limitations that have been addressed by synthetic chemists over the years is the inherent insolubility of pi-conjugated polymer chains. With a tendency towards rigidity and strong interchain pi-stacking interactions, the inherent systems tend to be completely insoluble and infusible, as illustrated by the structures of unsubstituted (R=H) polythiophenes (**4**), poly(*p*-phenylenes) (**5**), and poly(*p*-phenylene vinylene) (**6**). A major success of synthetic efforts over the years has been to create highly soluble pi-conjugated polymers that can be processed into thin-film and fiber forms for potential applications. Overcoming these solubility issues was one of the most important early contributions the synthetic community made to the field. The introduction of pendant flexible side chains (R = alkyl and alkoxy in **4**, **5**, and **6**) on conjugated polymers provides conformational entropy that induces solubility into the polymer product. As a generality, alkyl groups on the order of 6–8 carbons in length (hexyl to octyl), provide sufficient conformational disorder to induce solubility in the high molecular weight polymers with simple single aryl ring repeat units. This method is illustrated throughout this text, as it has become the main approach for preparing usefully processable conjugated polymers (Chart 1.2).

Considering the high molecular weights possible with chain-growth polymerizations, the synthesis of soluble and processable polymer precursors to fully conjugated materials has proven to be an excellent route for preparing useful materials. An early example of this is the synthesis of poly(*p*-phenylene vinylene) (PPV) *via* the polymerization of bis-sulfonium salts of bis-dichloromethylbenzene.[14,15] Basic treatment of the bis-sulfonium salt leads to *in situ* formation of a quinoidal structured intermediate (not isolated), which subsequently polymerizes to form a nonconjugated polyelectrolyte that is soluble in alcoholic media. This soluble precursor polymer solution can be stored for quite some time, and subsequently processed into thin films by any number of solution processing methods. Thermal treatment of the solid material leads to elimination of HCl and dialkylsulfide or tetrahydrothiophene yielding the final conjugated PPV derivative. This general concept of soluble precursor polymer synthesis has found use in the preparation of various polyacetylenes,[16] poly(*p*-phenylenes),[17] and poly(thienylene vinylenes)[18] along with numerous other

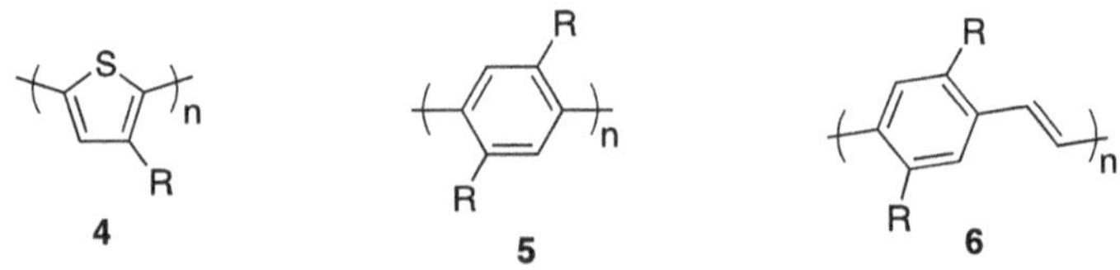

Chart 1.2

poly(arylene vinylene) systems in general. A major benefit of this methodology is that high molecular weight polymers can be obtained, even at low monomer conversion, with the precursor polymers tending to solution process well. A major drawback of the polymer precursor route is the chemical purity of the final conjugated polymer. As with any reaction on a macromolecule, complete conversion is not possible. In addition, many of the conversion reactions are thermally driven eliminations where side reactions are induced.

A standard question asked in any new polymer study is, what is the magnitude of the molecular weight that is required to provide the limiting properties for a particular application? This will be an important concept addressed throughout the many chapters in this book, as the polymerization chemistry used to attain conjugated systems can be quite varied. As just one example, consider the evolution of the optical absorption spectra of conjugated polymers as a function of chain length and the effect on the resultant color (important when considering electrochromic applications) transmitted or reflected by the final polymer films. In a combined size exclusion chromatography/UV-Vis spectroscopy experiment, it was found that a series of cyanovinylene-linked dioxythiophene polymers attained their limiting spectra at a GPC estimated number average molecular weight of approximately 10 kg/mol.[19] Simultaneously, this molecular weight also provided materials with sufficient film-forming properties for stable and reproducible electrochemical switching and, as such, this molecular weight is adequate for this specific electrochromic polymer application. In general, many step-growth polymerization methodologies can provide conjugated polymers of sufficient molecular weight for the application at hand where the materials are used as thin, electrode-supported, films. Standard equilibrium controlled step-growth polymerizations have degrees of polymerizations controlled by Carother's equation. The necessity for a high degree of functional-group conversion for molecular weight creates the situation in which the synthetic chemist must be especially careful about monomer purity and functionality. At the same time, it has been demonstrated that higher molecular weights, beyond which there is no visible change in the spectroscopic signature of a conjugated polymer, can provide elevated power conversion efficiencies in solar cell and field effect transistor applications.[20] These considerations of molecular weight are subtle from polymer to polymer, and application to application, and must be addressed separately for each system. These concepts are illustrated nicely throughout this book.

1.4 Structural Control *via* Repeat Unit and Functionality

When one considers how synthetic chemistry has impacted the development of conjugated polymers, there is no better example than the poly(3-alkylthiophenes) (P3ATs, **4**). Early work focused on oxidative polymerization methods as a means of preparing soluble forms of this polymer.[21] Subsequently, Grignard coupling reactions were able to prepare the polymer directly in the reduced state, such that there were no residual charge carriers in the materials (this would ultimately prove useful in the concepts of using P3ATs as semiconducting and charge-transporting organic electronic materials).[22] Disorder through the formation of head/head and tail/tail defects led researchers to develop controlled polymerizations that provided regioregular P3ATs with a high degree of order.[23] Even finer control has been brought through the utilization of Grignard metathesis reactions and the examination of polymerization catalytic processes, such that the polymerization can be carried out under living conditions.[24] It is just these considerations that are the major driving force that led us to edit this book. It is crucial that synthetic chemists obtain polymers with high repeat unit purity, backbones with no branching or crosslinking, high molecular weights with low dispersity, and overall high purity in the removal of residual chemical species formed during polymerization, such as entrapment of metallic catalyst impurities. Further, the controlled introduction of end groups on the conjugated polymer chains provides another degree of purity, and depth of structural understanding.

This book is designed to provide the reader with a comprehensive view of how the various classes of conjugated polymers are synthesized. Contained within these 20 chapters are overviews of the reactions, structures, and synthetic conditions required for effective polymer formation, along with experimental details. Throughout the text, the evolution of structural build up is a focus; moving from simple polymer repeat units, to highly functionalized polymers, to more complicated structures with specific property design in mind. Building on the fundamental conjugated polymer systems introduced above, a number of chapters are directed to various forms of polyarylenes such as the poly(phenylene ethynylenes) (**7**, Chapter 8), polyfluorenes (**8**, Chapter 5), and polycarbazoles (**9** and **10**, Chapter 6), to name just a few. Property modification becomes evident through the extent of conjugation provided in these types of polymers. For example, comparison of the 2,7- and 3,6-linked polycarbazoles allow examination of the effects of full compared to broken conjugation, where the latter structure leads to discrete chromophores. Ultimately, the ability to form these specific linkages in polymers plays an important role in determining ultimate properties as the 2,7-linked carbazole units are found to be useful in high-performance solar polymers,[25] while the electron-rich 3,6-linked carbazoles find use in easily switchable, redox-active electrochromic polymers[26] (Chart 1.3).

As noted earlier, polythiophene has served as an easily functionalized system where the nature of the side chains, and their regio-orientation, provides a

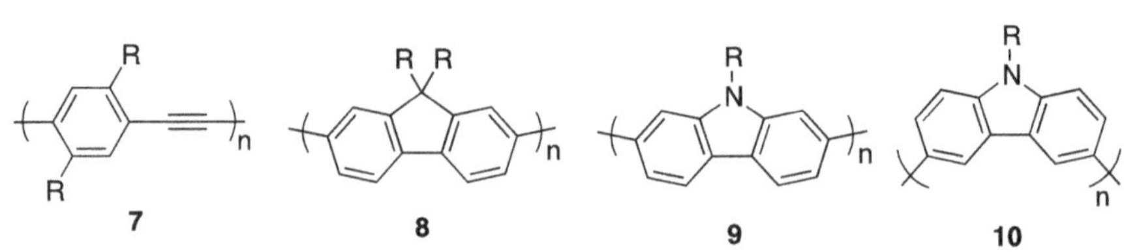

Chart 1.3

broad range of controllable solution and solid-state order properties. Many functionalized conjugated polymers are illustrated throughout this collection of chapters where the side chains bring added functionality. Using the oligoether and naphthylene containing side-group-substituted polythiophenes **11** and **12**, respectively, (Chapter 9) as examples, polar ion coordinating and liquid-crystalline behavior can be introduced into the resultant materials. Dioxythiophene chemistry, led by poly(3,4-ethylenedioxythiophene) (PEDOT) (Chapter 10),[27] provides a class of polymers that are easily oxidized, thus providing highly stable conducting materials. The poly(3,4-propylenedioxythiophene) (PProDOT, **13**) family of polymers can be prepared using oxidative, Grignard metathesis, and direct arylation conditions to yield a family of polymers that are especially vibrantly colored in their neutral states and transmissive in their oxidized forms, as desired for electrochromic applications[28] (Chart 1.4).

The range of properties introduced by side chains on conjugated polymers is quite broad and can include redox activity, charge transporting capabilities, optical absorption and emission, and chemical reactivity. This is illustrated by the two polyfluorenes, **14** and **15** (Chapter 5), which are functionalized with electron-rich and hole-transporting groups. In these polymers, light emission is provided by the polymer backbone, while the charge-carrying properties are dominated by the pendant side chains. Synthetic chemistry employed in preparing conjugated polymers with functional side chains must take their potential reactivity (such as ease of oxidation) into account, an aspect that is nicely illustrated throughout this book (Chart 1.5).

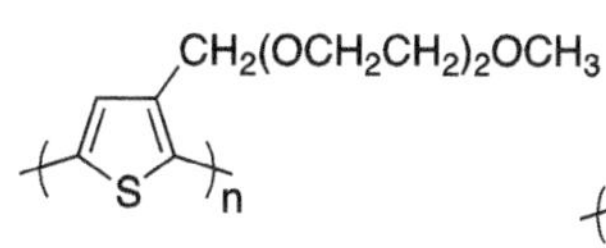

11

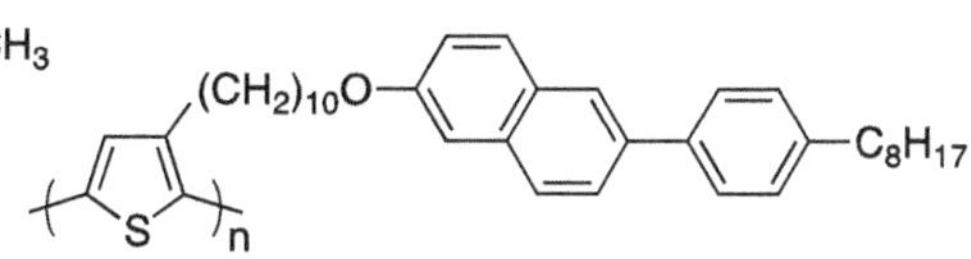

12

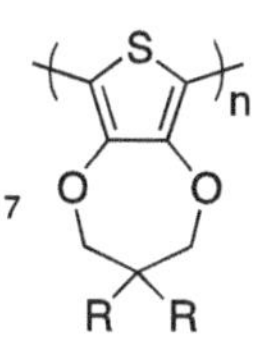

13

Chart 1.4

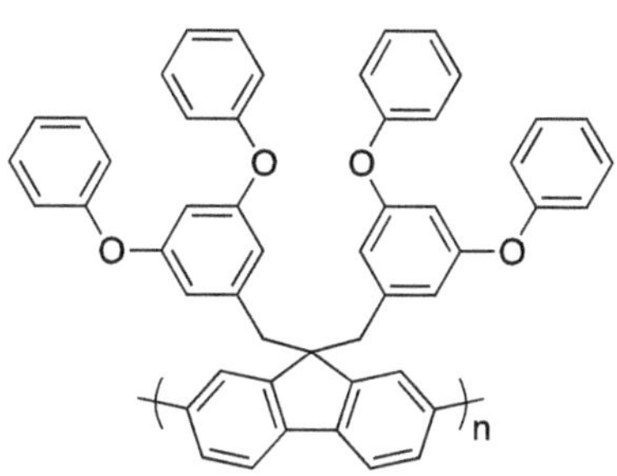

14

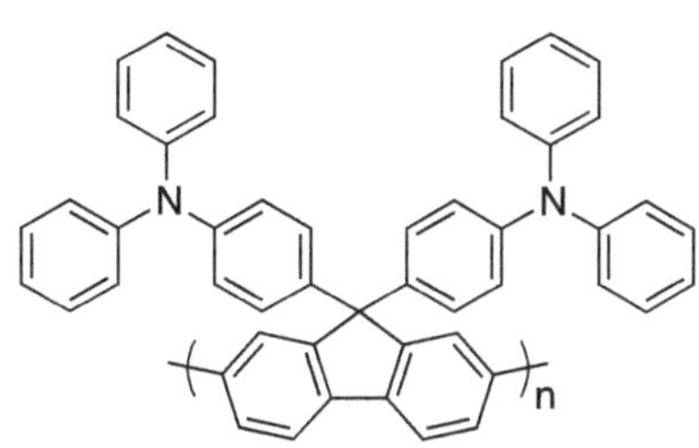

15

Chart 1.5

As synthetic chemists desired to tune the optoelectronic and redox properties of conjugated polymers in a fine manner, more complicated conjugated systems were required. The two fused heterocycles-substituted polythiophenes **16** and **17** (Chapter 18) illustrate this as electron-poor imine functionality in **16** brings donor–acceptor character to the material, while the more electron-rich thio-based system **17** provides for especially easy oxidation. Polymerization of these complicated bis-2-thienyl monomers by electrochemical methods paves the way for fundamental structure–property relationships to be understood, ultimately directing the synthetic chemist towards soluble polymers (Chart 1.6).

Initiated around concepts of self-doping in which an anion is covalently bound to a pi-conjugated redox-active polymer and provided charge balance during oxidative doping in **19** and **20**,[29,30] the synthesis of ion-containing conjugated polymers has required a unique set of synthetic capabilities. While early work focused on controlling the dominant ion transport during redox switching, many derivatives, such as that shown in the poly(*p*-phenylene vinylene) (**21**) and poly(*p*-phenylene) (**22**) derivatives, led to water-soluble polymers. Due to their amphiphilic nature, a number of these ionic polymers have been processed *via* solution methods (*e.g.* layer-by-layer film formation) and are used as active materials for sensing applications, exemplified by the highly fluorescent poly(*p*-phenylene ethynylene) derivative **23**. The use of organic solvent soluble precursor polymers that could be purified prior to conversion to their ionic forms gave a synthetic route to more structurally defined and pure conjugated polyelectrolytes (Chapter 16)[31] (Chart 1.7).

Polyheterocycle synthesis has been especially prevalent in the synthesis of new polymers for organic electronic and photovoltaic applications.[32,33] Revolving around a series of metal-mediated coupling reactions (Heck, Suzuki, Kosugi–Migita–Stille, direct arylation, *etc.*) electron-rich donor (D) and electron-poor acceptor (A) monomer units are combined in DA polymer motifs that allow fine control over the redox and electronic states of the pi-conjugated system. Examination of structures **24** and **25** shows the subtle synthetic control chemists have used in providing new and optimal structures in polymers designed for bulk heterojunction solar cells.[34,35] For example, the use of

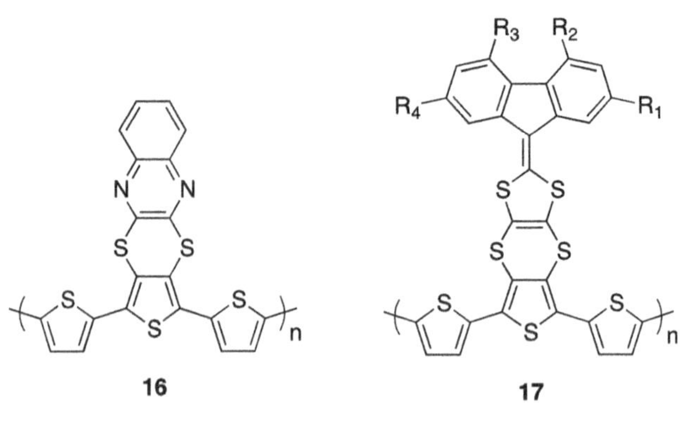

Chart 1.6

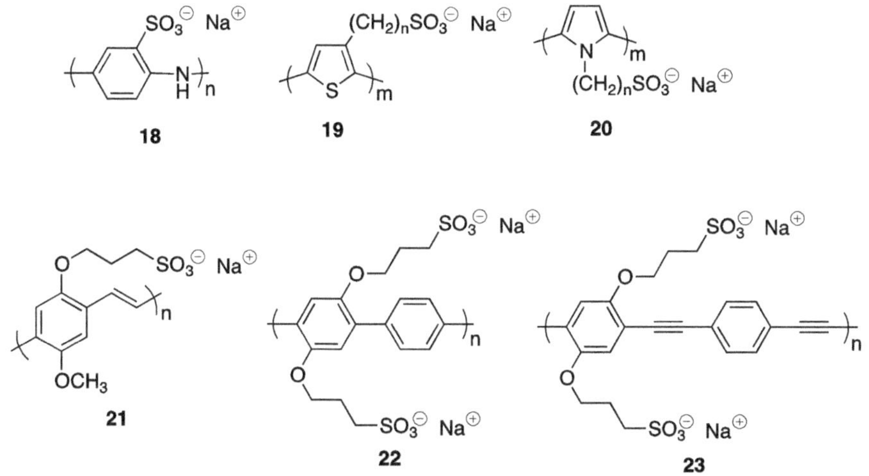

Chart 1.7

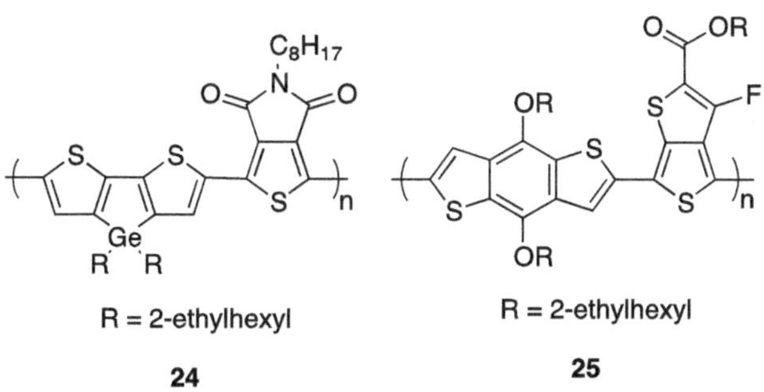

Chart 1.8

germanium (replacing carbon and silicon) in **24** controls bond length and stacking, while the incorporation of fluorine in **25** tunes electronic properties, such that organic solar cells constructed using both of these polymers provide high AM 1.5 power conversion efficiencies in excess of 8% (Chart 1.8).

1.5 Summary

It is evident from this collection of repeat-unit structures that synthetic chemistry, both at the molecular and macromolecular levels, is the enabling science for the preparation of a host of new materials with a broad array of properties. The collection of chapters assembled in this book address the structural and experimental details that are required for the preparation of the main classes of pi-conjugated polymers in high quality. It is hoped that both those having a general interest in the field, and those that are actively involved in the laboratory synthesizing these materials, will find this text useful and enjoyable.

Acknowledgements

Technical assistance from Mr. James Ponder during the assembly of this manuscript is greatly appreciated.

References

1. H. J. Letheby, *Chem. Soc.*, 1862, **15**, 161–163.
2. A. Dall'Olio, G. Dascola, V. Varacca and V. Bocche, *C. R. Acad. Sci.*, 1968, **433**, 267C.
3. E. T. Strom and S. C. Rasmussen, ed., *"Electrically Conducting Plastics: Revising the History of Conjugated Organic Polymers" In 100+ Years of Plastics. Leo Baekeland and Beyond*, Strom, ACS Symposium Series #1080, American Chemical Society, Washington, DC, 2011.
4. T. A. Skotheim, ed., *Handbook of Conducting Polymers*, 1st edn, Marcel Dekker, Inc., New York, NY, 1986.
5. T. A. Skotheim, R. L. Elsenbaumer and J. R. Reynolds, ed., *Handbook of Conducting Polymers*, 2nd edn, Marcel Dekker, Inc., New York, NY, 1998.
6. T. A. Skotheim and J. R. Reynolds, ed., *Handbook of Conducting Polymers*, 3rd edn, CRC Press, Boca Raton, FL, 2007.
7. J. J. Eisch, *Organometallics*, 2012, **31**, 4917–4932.
8. H. Shirakawa, *Angew. Chem. Int. Ed.*, 2001, **40**, 2574–2580.
9. D. J. Berets and D. S. Smith, *Trans. Faraday Soc.*, 1968, **64**, 823–828.
10. T. Ito, H. Shirakawa and S. Ikeda, *J. Polym Sci.*, 1974, **12**, 11–20.
11. C. K. Chiang, C. B. Fincher, Jr., Y. W. Park, A. J. Heeger, H. Shirakawa, E. J. Louis, S. C. Gau and A. G. MacDiarmid, *Phys. Rev. Lett.*, 1977, **39**, 1098–1101.
12. J. Ferraris, D. O. Cowan, V. Walatka Jr. and J. H. Perlstein, *J. Am. Chem. Soc.*, 1973, **95**(3), 948–949.
13. G. Odian, *Principles of Polymerization*, 4th edn, Wiley-Interscience, Hoboken, NJ 2004.
14. R. A. Wessling and R. G. Zimmerman, U.S. Patent, 3,401,152, Sept. 10, 1968.
15. D. R. Gagnon, J. D. Capistran, F. E. Karasz, R. W. Lenz and S. Antount, *Polymer*, 1987, **28**, 567–573.
16. T. M. Swager, D. A. Dougherty and R. H. Grubbs, *J. Am. Chem. Soc.*, 1988, **110**(9), 2973–2974.
17. D. G. H. Ballard, A. Courtis, I. M. Shirley and S. C. Taylor, *Chem. Commun.*, 1983, 954–955.
18. K. Jen, M. Maxfield, L. W. Shacklette and R. L. Elsenbaumer, *Chem. Commun.*, 1987, 309–311.
19. B. C. Thompson, Y. G. Kim, T. D. McCarley and J. R. Reynolds, *J. Am. Chem. Soc.*, 2006, **128**, 12714–12725.
20. R. J. Kline, M. D. McGehee, E. N. Kadnikova, J. Liu, J. M. J. Fréchet and M. F. Toney, *Macromolecules*, 2005, **38**, 3312–3319.
21. M. Sato, S. Tanaka and K. Kaeriyama, *Synth. Met.*, 1987, **18**, 229–232.

22. K. Jen, G. G. Miller and R. L. Elsenbaumer, *Chem. Commun.*, 1986, 1346–1347.
23. T. Chen, X. Wu and R. D. Rieke, *J. Am. Chem. Soc.*, 1995, **117**, 233–244.
24. M. C. Iovu, E. E. Sheina, R. R. Gil and R. D. McCullough, *Macromolecules*, 2005, **38**, 8649–8656.
25. N. Blouin and M. Leclerc, *Acc. Chem. Res.*, 2008, **41**(9), 1110–1119.
26. C. L. Gaupp and J. R. Reynolds, *Macromolecules*, 2003, **36**, 6305–6315.
27. A. Elschner, S. Kirchmeyer, W. Lövenich, U. Merker, and K. Reuter, *PEDOT: Principles and Applications of an Intrinsically Conductive Polymer*, 1st edn, CRC Press, Boca Raton, FL, 2011.
28. A. L. Dyer, E. J. Thompson and J. R. Reynolds, *Appl. Mater. Inter.*, 2011, **3**, 1787–1795.
29. A. O. Patil, Y. Ikenoue., N. Basescu, N. Colaneri, J. Chen, F. Wudl and A. J. Heeger, *Synth. Met.*, 1987, **20**, 151–159.
30. N. S. Sundaresan, S. Basak, M. Pomerantz and J. R. Reynolds, *J. Chem. Soc., Chem. Commun.*, 1987, 621–622.
31. H. Jiang, P. Taranekar, J. R. Reynolds and K. S. Schanze, *Angew. Chem.*, 2009, **48**, 4300–4316.
32. C. Nielsen and I. McCulloch, *Prog. Polym. Sci.*, 2013, Published online, 10.1016/j.progpolymsci.2013.05.003.
33. B. C. Thompson and J. M. J. Fréchet, *Angew. Chem. Int. Ed.*, 2008, **47**, 58–77.
34. C. M. Amb, S. Chen, J. Subbiah, K. R. Graham, C. E. Small, F. So and J. R. Reynolds, *J. Am. Chem. Soc.*, 2011, **133**, 10062–10065.
35. Y. Liang and L. Yu, *Acc. Chem. Res.*, 2010, **43**(9), 1227–1236.

CHAPTER 2

Polyacetylenes

KAZUO AKAGI

Kyoto University, Department of Polymer Chemistry, Katsura Campus, Kyoto 615-8510, Japan
Email: akagi@fps.polym.kyoto-u.ac.jp

2.1 Introduction

Since the synthesis of polyacetylene (PA) thin film and the discovery of chemical doping disclosed the uncultivated field of conductive polymers,[1–4] conductive polymers have been extensively investigated and widely used in such products as electrolytic capacitors and secondary batteries.[5–7] Today, polymers have become essential for lightweight, high-performance batteries used in notebook computers, cellular phones, and other portable equipment. Much research and development has also been conducted on polymer light-emitting diodes, organic solar cells that are anticipated for use in next-generation displays and energy sources.[7] Conductive polymers are also being studied for their use as materials in molecular devices, called the ultimate electronic devices.[8] Thus, although many conductive polymers have been developed for various applications, PA is still the highest conductive polymer, showing an electrical conductivity of 10^5 S/cm after iodine doping.[9,10]

Helical PA (H-PA) is a unique conductive polymer because it has a super-hierarchical helical structure forming a spiral morphology, and it is synthesized in an asymmetric reaction field consisting of a chiral nematic liquid crystal, even though an acetylene monomer has no chiral moiety.[11] Despite the helical structure, the relatively high conductivity of 10^3 S/cm allowed us to anticipate that H-PA might be a prototype exhibiting novel electromagnetic properties such as a nanosize polymer solenoid. It is intriguing to elucidate H-PA from an

RSC Polymer Chemistry Series No. 9
Conjugated Polymers: A Practical Guide to Synthesis
Edited by Klaus Müllen, John R. Reynolds and Toshio Masuda

Published by the Royal Society of Chemistry, www.rsc.org

interdisciplinary viewpoint between polymers, liquid crystals and synthetic metals.

2.1.1 Polyacetylene (PA)

Acetylene was first polymerized by Natta and coworkers using a Ziegler–Natta catalyst, Ti(O-*n*-Bu)$_4$ – AlEt$_3$.[12] The polymer obtained was a gray, infusible, and powdery material, which was not soluble in any common solvents. It was not amenable to most characterization methods and did not possess the interesting electrical properties that were anticipated. Several other catalyst systems were shown to polymerize acetylene, yielding similar products.[13–16] Pristine PA is a typical semiconductor, but its electrical conductivity of 10^{-9}–10^{-5} S/cm can be varied by over 14 orders of magnitude through chemical doping.[17,18] The maximum conductivity reported to date for mechanically stretched and highly aligned PA is more than 10^5 S/cm after iodine doping,[9,10] which is comparable to that of copper and gold.

2.1.1.1 Shirakawa-Type PA

Using the same Ziegler–Natta catalyst employed by Natta *et al.*, but with different experimental conditions, Shirakawa and coworkers succeeded in preparing free-standing polyacetylene (PA) films having metallic luster (Figure 2.1).[19,20] The polymerization at low temperature yields *cis* PA. The *cis* content of the polymer decreases with increasing polymerization temperature. Thus, a polymer synthesized and purified carefully at –78 °C contains 98% *cis* form, whereas a polymer obtained at room temperature is about 60% *cis* (40% *trans*).[21] Since the rate of isomerization of *cis*-rich polymer in the solid state is too slow at room temperature, heat treatment at 75 °C for 300 min is necessary to obtain a polymer containing the same *trans* content as the polymer synthesized at room temperature. The *trans* PA film is an intense black material with a metallic luster (Figure 2.1), whereas the *cis* PA film has a copper-like

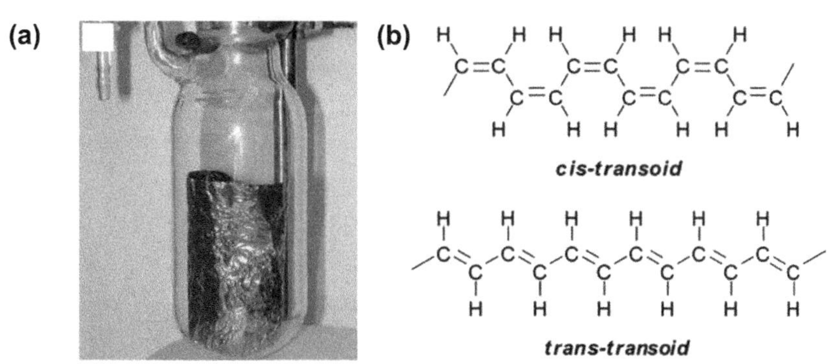

Figure 2.1 Polyacetylene film with metallic luster (a) and molecular structures (b).

luster. Very thin films of *trans* and *cis* polymers show deep blue and clear red colors, respectively.

PA has a planar structure, irrespective of *cis* and *trans* forms, due to strong π-conjugation between the sp^2-hybridized carbon atoms in the PA polymer chain.[22,23] Strong interchain interactions give rise to a fibrillar crystal consisting of rigid, π-stacked polymer chains.[19,20,23] This makes PA infusible and insoluble in any kind of solvent. Thus, the solid-state structure and morphology of PA are determined during acetylene polymerization. The fibril morphology of PA films is randomly oriented, as is typical for ordinary polymers, depressing the inherent one-dimensionality of this polymer (Figure 2.2). Hence, several types of procedures and polymerization methods for macroscopic alignment of the polymer have been developed to achieve higher electrical conductivity with an anisotropic nature.[24–34]

2.1.1.2 *Mechanically Stretchable PA*

PA films can be stretch-aligned to a limited degree of orientation, resulting in anisotropic electrical and optical properties.[35] However, it is unclear to what extent the orientation of the polymer chain contributes to these properties. Other morphological parameters are unknown, such as the distribution of crystalline and amorphous domains in the PA films, as well as their influence on physical properties of the films. Therefore, a novel synthetic method for controlling the morphology of PA during acetylene polymerization was carried out. Heeger and coworkers synthesized a partially oriented PA film by polymerizing acetylene on the surface of a biphenyl crystal.[36,37] Despite the incompleteness of this work in obtaining macroscopically aligned PA, this experimental trial yielded an important idea for how one can control the morphology of PA *via* the polymerization of acetylene.

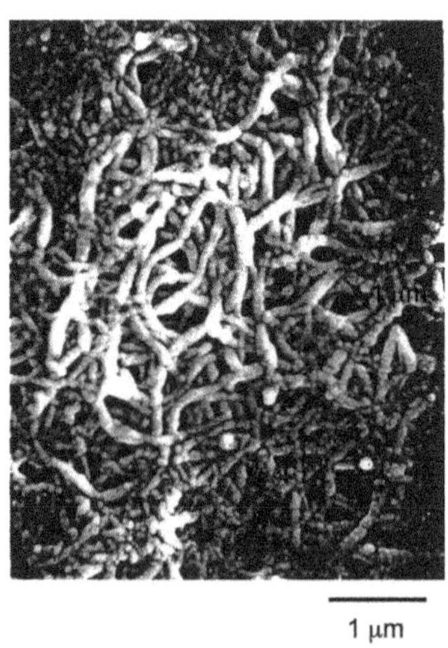

Figure 2.2 Random fibril morphology of polyacetylene film.

One approach is the mechanical stretching of an as-grown PA film.[9,10,26–31] The conductivity of the film is enhanced through a uniaxial alignment of the fibrils associated with the mechanical stretching of the film. It is therefore desirable to develop highly stretchable PA films with high-modulus and high-tensile properties that might prevent the cleavage of fibrils and/or of PA chains during mechanical stretching. In other words, profound mechanical strengths such as Young's modulus and tensile strength are responsible for the uniaxial alignment and hence the electrical conductivity of the stretched film.

2.1.1.2.1 Naarmann and Theophilou-Type PA. Naarmann and Theophilou[9] synthesized highly stretchable PA films using a high-temperature aged Ziegler–Natta catalyst, Ti(O-*n*-Bu)$_4$ and $AlEt_3$, dissolved in viscous silicone oil or toluene, and reported high conductivities of 10^5 S/cm after iodine doping in CCl_4 solution. Their polymerization method is characterized by use of silicone oil as a solvent, aging of the catalyst at 120 °C and addition of reducing agents (ARA) such as *n*-butyllithium or Mg-derivative before polymerization.[3,9,38] An acetylene polymerization using the aged catalyst in a silicone oil yields a homogeneous, defect-free PA film that can be stretched mechanically by up to 550%, corresponding to stretching rates of 6.5. When doped with iodine in CCl_4, the film displays an electrical conductivity of $\sim 1.6 \times 10^4$ S/cm. An aged catalyst in a silicone oil that has been mixed with various quantities of *n*-butyllithium and allowed to react with acetylene yields PA that is highly regular, compact and crystalline in well-defined parallel planes. In special cases, the transparent films can be stretched by up to 400% ($l/l_o = 5$) and gives a conductivity of higher than 10^5 S/cm after doping in CCl_4. Bulk densities of the films are 0.85–0.9 and 1.12–1.15 before and after doping. The stability studies show that the *cis* fraction (70–80%), conductivity and morphology of the films are unchanged after several months. Here, it should be noted that Naarmann's method including the ARA technique lacks for reproducibility of the conductivity higher than 10^5 S/cm.[38] The transmission electron microscope (TEM) results indicate inhomogeneous doping within the fibrils, showing highly doped regions separated by non-doped regions, which are typically 1–3 nm wide. The temperature dependence of the dc conductivity of iodine-doped PA film can be fitted with the Sheng formula within a model of a heterogeneously conducting material consisting of highly conducting regions that are separated by insulating barriers.[39]

2.1.1.2.2 Tsukamoto-Type PA. Tsukamoto *et al.*[10] modified Naarmann's method using a higher-temperature aged catalyst dissolved in decaline and employing acetylene gas under low pressure. An SEM micrograph of unstretched PA film shows that the film is composed of a densely packed globular structure and is contrasted with the fibril structures of conventional PA. Such densely packed structure indicates a high volume density. The bulk density of PA is around 1.1 g/cm^3. The *cis* content is estimated to be 96% and the mechanical stretchability is 8–10. The mechanically stretched films show electrical conductivities of higher than 10^5 S/cm after iodine doping.

However, all the samples do not always yield conductivity higher than 10^5 S/cm. The reproducibility is 70 to 80%.[40] As the thickness of highly conducting PA is around 1 μm, accurate thickness measurement is one of the most crucial points in determination of conductivity. Temperature dependences of conductivities of fully iodine-doped PA were measured in the temperature range between 12 K and 300 K. The temperature dependence shows that activation-type conduction is still dominant in this temperature range, just as does that of doped conventional PA. Namely, the temperature dependence of conductivity is not metallic, suggesting that interdomain and interfibrillar processes are still dominant in the conductivity.

2.1.1.2.3 High Mechanical Modulus and Strength PA. Akagi *et al.*[24,25] developed two solvent-free acetylene polymerization methods: a solvent-evacuation (SE) method and an intrinsic nonsolvent (INS) one. In the former, cumene used as a solvent is evacuated after the high-temperature aging of the catalyst. In the latter, no solvent is used from the preparation and high-temperature aging of the catalyst to the polymerization. PA films synthesized by these methods exhibit extremely high mechanical strengths characterized by a Young's modulus of 40–100 GPa and a tensile strength of 0.9–2.1 GPa,[24,41] comparable to those (132 and 3.9 GPa) of well-known engineering plastics such as Kevlar, in addition to a high bulk density of more than 1.0 g/cm^3. The films show well-reproducible and high conductivities of $2.2–4.3 \times 10^4$ S/cm, which are enhanced by the uniaxial alignment of polymer chains associated with a mechanical stretching of 8–9 times. It can be noted from these results that the high mechanical strength is a prerequisite for the achievement of high electrical conductivity, and that in the INS polymerization, the cocatalyst of AlR_3, by virtue of its liquidity, plays the role of a solvent toward catalytically active complexes formed between $Ti(OR)_4$ and AlR_3, which guarantees the homogeneity of the catalyst system in spite of the absence of solvent.[42]

The ESR studies[43,44] show that high-temperature aging at 150–200 °C on the Ti(O-*n*-Bu)$_4$ and $AlEt_3$ catalyst yields a bulky aggregate of Ti^{3+} complex with a bridged structure through butoxy and/or ethyl groups. Such a bulky complex should lead to a decrease of apparent catalytic activity and hence a depression of the exothermic heat associated with the acetylene polymerization. This allows a moderate condition that guarantees production of a *cis*-rich and highly homogeneous PA film.

The numerical evaluations based on the analytical formulae[45] indicates that the inplane alignment contributes to enhance the conductivity of the aligned film in a cooperative manner with the mechanically forced fibril alignment (parallel alignment). In particular, the inplane alignment becomes crucially dominant when the film thickness is less than 1.0 μm. On the other hand, although the parallel alignment expectedly decreases the conductivity perpendicular to the elongation direction along with the mechanical stretching, the decrease is partly compensated by an increase due to the inplane alignment.

2.1.1.3 Directly Aligned PA

Another approach is a direct alignment of the film using a liquid crystal as an anisotropic solvent for acetylene polymerization under an external perturbation such as gravity flow and magnetic field.[32–34] This approach affords an alignment of even ultrathin films with semitransparency that is suitable for measurements of nonlinear optics.[46] A nematic LC (N-LC) has long-range orientational order. Furthermore, the orientational order of a N-LC can be easily controllable using external perturbations, such as an electric field, magnetic field, sheer force, or gravity flow. For this reason, N-LC is useful for an anisotropic reaction solvent for controlling the morphology of PA during acetylene polymerization. N-LCs that could be used as solvents for the Ziegler–Natta catalyst are limited, because they can act as reactive substituents with the active species and/or components of the catalyst. Among the available N-LCs, a phenylcyclohexane (PCH) derivative is stable in the presence of Ziegler–Natta catalyst.[32–34]

Acetylene polymerization was carried out under a gravity flow of the LC reaction field, in which the Ziegler–Natta catalyst was dissolved. Macroscopically aligned fibrils were found along the gravity flow direction of the N-LC catalyst solution. The direction of fibril growth appeared to coincide with that of N-LC molecules under the gravity flow of the solution. In fact, PA chains propagate along the director (average direction of LC molecules within a domain) of the N-LC. This novel and simple method produces highly oriented PA films, making it possible to investigate various anisotropic properties of PA.[32–34,47]

An N-LC molecule can be aligned using an external magnetic field by virtue of its anisotropy in diamagnetic susceptibility. In particular, LC molecules with positive or negative anisotropies in diamagnetic susceptibility were aligned parallel or perpendicular to the magnetic field direction, respectively. Akagi *et al.*[34] constructed a monodomain structured LC reaction field from a multidomain one, using an external magnetic field 2–14 kG. Owing to the positive anisotropy in diamagnetic susceptibility of the phenyl moiety in the LCs, the multidomain structure of N-LCs tends to be aligned in parallel with the magnetic field direction, making a monodomain structured N-LC. Highly aligned PA was synthesized in a monodomain structured LC reaction field (Figure 2.3).[34,48,49]

The macroscopically aligned PA films prepared in N-LCs under magnetic field show high conductivities of 10^4 S/cm after iodine doping, and anisotropies of ca. 5 defined as a ratio of parallel to perpendicular to the aligned fibril direction. The unexpectedly small anisotropy is attributed to the relatively high conductivity in the perpendicular direction. This is because the aligned film has highly condensed fibril morphology and hence the interchain and even interfibril hopping of the charged carrier are largely enhanced. The degree of alignment of the film is governed by that of the liquid crystal used as the solvent. That is, the morphology and therefore the macroscopic properties of the film are crucially dependent on the anisotropic polymerization field.[48,49]

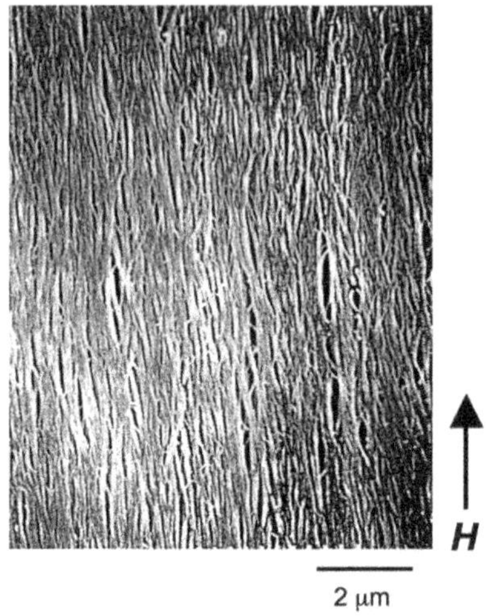

Figure 2.3 Macroscopically aligned polyacetylene film synthesized in nematic LC (N-LC) reaction field under magnetic field.

This situation is common to an epitaxial polymerization, where acetylene polymerization is carried out on the substrate crystal with a homologous isomorphic compound such as naphthalene, anthracene, biphenyl, or terphenyl.[50] The fibril orientation is strongly dependent on the lattice matching between PA and the substrate crystals.

It is worth noting here that a PA bearing an LC moiety in the side group was reported in 1993.[51,52] The LC-substituted PA is not only soluble in organic solvents, but also easily aligned by spontaneous orientation of the LC group. In addition, it was macroscopically aligned by an external perturbation, such as shear stress, electric field, and magnetic field.[51–56] This situation means that a monodomain structure of the LC phase can be constructed on a macroscopic level. Thus, the polymer should have a higher electrical conductivity, compared with the case of random orientation. At the same time, one can control the molecular orientation and, hence, the electrical conductivity of the polymers with an external force. As the macroscopic alignment was first achieved under magnetic field,[56] it is straightforward to study many kinds of LC-substituted PA derivatives.[54,56–63] However, the electrical conductivity of LC-substituted PA is significantly lower than that of nonsubstituted PA. This is due to a lower coplanarity of the main chain, which arises from steric repulsion between substituents, a higher ionization potential, and a lower electron affinity.

2.1.2 Helical Polyacetylene (H-PA)

It has been generally accepted that PA has a planar structure, irrespective of *cis* and *trans* forms. This is due to π-conjugation between the sp^2-hybridized carbon atoms in the polymer chain.[1,19–22] If it were possible to modify this planar structure of PA into a helical one, novel magnetic and optical properties might be expected.[64,65] Here, we present the polymerization of acetylene in an asymmetric reaction field, which is constructed with chiral nematic LCs (N*-LCs), and show that PA films formed from helical chains and fibrils can be synthesized.[11,66–80]

2.1.2.1 Asymmetric Liquid Crystal Reaction Field

The N*-LC to be used as an asymmetric solvent is prepared by adding a small amount of chiral compound, as a chiral dopant, into nematic LC (Figure 2.4). The formation of N*-LC is recognized when a Schlieren texture characteristic of nematic LC changes into a striated Schlieren or a fingerprint texture in a polarized optical microscope (POM). The distance between the striae corresponds to a half-helical pitch of the N*-LC. Note that as the degree of twist in the N*-LC is larger, the helical pitch observed in POM is shorter.

The helical pitch of the N*-LC can be adjusted by two methods: changing the concentration or changing the twisting power of the chiral dopant. However, the mesophase temperature region of the N*-LC is affected by changing the concentration of the chiral dopant. Namely, it becomes narrow as the concentration increases, and finally the mesophase is destroyed when the concentration is close to a critical value. Herein, owing to the limitation of the concentration method, an alternative approach of utilizing the chiral compound with large twisting power is adopted. Axially chiral binaphthyl derivatives are used as chiral dopants,[74] since they have been reported to possess larger twisting powers (HTPs) than asymmetric carbon-containing chiral compounds.[67]

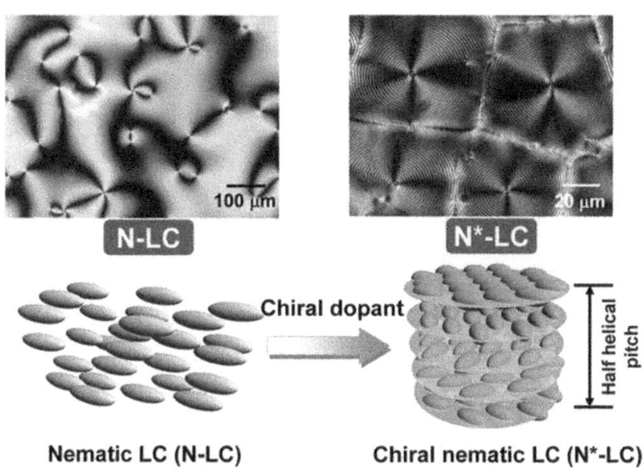

Figure 2.4 Chiral nematic liquid crystal (N*-LC) induced by the addition of a chiral dopant into N-LC. Schlieren texture (left) and fingerprint texture (right) are observed for N-LC and N*-LCs, respectively, with a polarized optical microscope.

The axially chiral binaphthyl derivatives, (*R*)- and (*S*)-1,1-binaphthyl-2,2-di-[*p*-(*trans*-4-*n*-pentylcyclohexyl) phenoxy-1-hexyl]ether, are synthesized using Williamson etherification reactions of chiroptical (*R*)- and (*S*)-1,1-bi-2-naphthols, respectively, with phenylcyclohexyl derivatives. The products are abbreviated as (*R*)- and (*S*)-D-1 (Figure 2.5). To prepare an induced N*-LC, approximately 5 to 14 wt% of (*R*)- or (*S*)-D-1 is added as a chiral dopant to an equimolar mixture of the N-LCs 4-(*trans*-4-*n*-propylcyclohexyl)ethoxybenzene (PCH302) and 4-(*trans*-4-*n*-propylcyclohexyl)butoxybenzene (PCH304). The LC substituent group in (*R*)- and (*S*)-D-1 enhances miscibility between the N-LC mixture and the binaphthyl derivative used as the chiral dopant. Note that usage of similar substituents with a shorter methylene spacer such as PCH503 or normal alkyl substituent gave insufficient miscibility, yielding no chiral nematic phase. In polarizing optical micrographs of the mixture of PCH302, PCH304, and (*R*)-PCH506-binaphthyl and that of PCH302, PCH304, and (*S*)-PCH506-binaphthyl, a striated Schlieren or finger printed texture characteristic of N*-LC phase is observed.

Although each component (PCH302 or PCH304) shows a LC phase, the LC temperature region is very narrow, being less than 1 to 2 °C. This is not suitable for acetylene polymerization in a N-LC or N*-LC reaction field, because the exothermal heat evolved during the acetylene polymerization would raise the temperature inside a Schlenk flask. This would easily destroy the LC phase, making it isotropic. Hence, the LC mixture is prepared by mixing two equimolar LC components. In the LC mixture, the nematic – isotropic temperature, $T_N - I$, and the crystalline – nematic temperature, $T_C - N$, are raised and

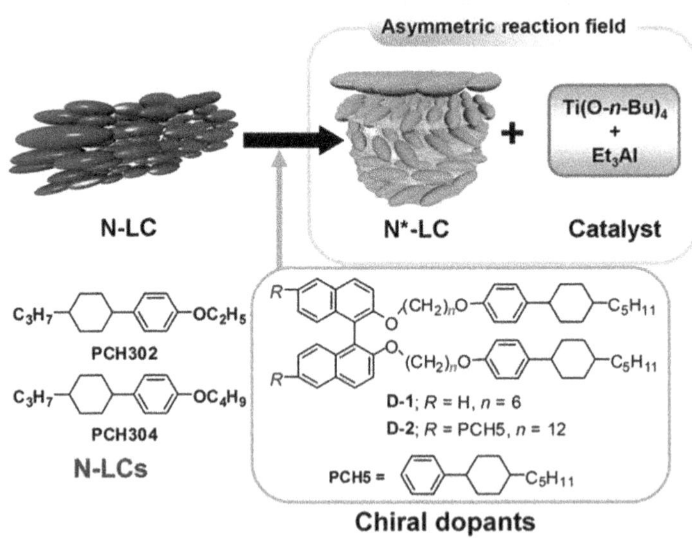

Figure 2.5 Construction of asymmetric reaction field for acetylene polymerization by dissolving Ziegler–Natta catalyst, $Ti(O\text{-}n\text{-}Bu)_4$–$AlEt_3$, into the N*-LC. The N*-LC includes an axially chiral binaphthyl derivatives, D-1 or D-2.

lowered, respectively. In fact, the mixture exhibits the LC phase in the region from 20 to 35 °C. Subsequently, the change of $T_N - I$ upon addition of Ti(O-*n*-Bu)$_4$ − AlEt$_3$ catalyst is examined using differential scanning calorimetry (DSC). Taking into account the effect of supercooling for LCs, a catalyst solution that consists of a LC mixture and chiral dopant is obtained for room-temperature polymerization ranging from 5 to 25 °C. Note that supercooling is the process of lowering the temperature of a liquid or a LC below its freezing point without it becoming a solid. This sufficiently wide temperature region enables the performance of acetylene polymerization in the N*-LC phase. A Ziegler – Natta catalyst, consisting of Ti(O-*n*-Bu)$_4$ and AlEt$_3$, is prepared using (*R*)- or (*S*)-N*-LC as a solvent (see, Figure 2.5). The concentration of Ti(O-*n*-Bu)$_4$ is from 15 to 50 mM, and the mole ratio of the cocatalyst to catalyst, [AlEt$_3$] / [Ti(O-*n*-Bu)$_4$], is 4.0. The catalyst solution is aged for 0.5 h at room temperature. During the aging, the N*-LC containing the catalyst shows no noticeable change in optical texture and only a slight lowering of the transition temperature by 2 to 5 °C. The transition temperature between the solid and N*-LC phases is 16 to 17 °C. The transition temperature between the N*-LC and isotropic phases is 30 to 31 °C. No solidification is observed down to −7 °C, as a result of supercooling. Thus, the (*R*)- and (*S*)-N*-LCs are confirmed to be chemically stable in the presence of the catalyst. It is therefore possible to employ these LCs as an asymmetric solvent for acetylene polymerization.

2.1.2.2 *Characterization of H-PA*

SEM images of H-PA films show that multidomains of spiral morphology are formed (Figure 2.6a), and each domain is composed of a helical structure of a bundle of fibrils with a one-handed screwed direction (Figure 2.6b). The multidomain-type fibril morphology of H-PA seems to replicate that of the N*-LC during the interfacial acetylene polymerization. A closer observation of

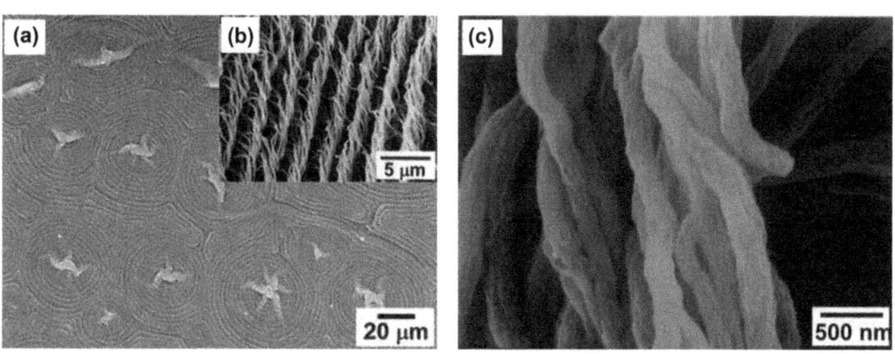

Figure 2.6 SEM micrographs of helical polyacetylene (H-PA) film synthesized in the right-handed (*R*)-System-1 {PCH302 : PCH304 : (*R*)-D-1 = 100 : 100 : 2 (mole ratio)} with a helical pitch of 5 μm. The photograph of (b) shows the magnified one of (a).

SEM images indicates that H-PAs synthesized in the (*R*)- and (*S*)-N*-LCs form the screwed bundles of fibrils and even screwed fibrils with left-handed and right-handed directions, respectively. This result implies that the screw direction of H-PA is controllable by choosing the helicity, *i.e.* optical configuration of the chiral dopant, so far as the N*-LC induced by the chiral dopant is employed as an asymmetric polymerization solvent. In addition, it is of keen interest that the screw directions of bundle and fibrils are opposite to those of the (*R*)- and (*S*)-N*-LCs used as solvents.

It has been elucidated so far that the PA chains propagate along the director (an averaged direction for the LC molecules within a domain) of the N*-LC. Since the helical axis of PA is parallel to the PA chain, and the director of the N*-LC is perpendicular to the helical axis of N*-LC, the helical axis of PA is perpendicular to that of N*-LC. Taking these aspects into account, one can describe a plausible mechanism for interfacial acetylene polymerization in the N*-LC, as shown in Figure 2.7. In the case of a right-handed N*-LC, for instance, the PA chain would propagate with a left-handed manner, starting from the catalytic species, but not with a right-handed one. This is because the PA chains with the opposite screw direction to that of the N*-LC could propagate along the LC molecules, but those with the same direction as that of the N*-LC would encounter LC molecules, making the propagation stereospecifically impossible. The detailed mechanism of acetylene polymerization in N*-LC has been elucidated.[77,78]

In circular dichroism (CD) spectra of the PA thin films synthesized with (*R*)- and (*S*)-N*-LCs, positive and negative Cotton effects are observed, respectively, in the region from 450 to 800 nm. This corresponds to a $\pi \rightarrow \pi^*$ transition in the PA chain, despite the absence of a chiroptical substituent in the side chains. This indicates that the PA chain itself is helically screwed.[69] It is evident that this Cotton effect is not due to the chiral dopant, because the Cotton effect of the chiral dopant is only observed at shorter wavelengths (*i.e.* 240 to 340 nm). Left-handed (counterclockwise) and right-handed (clockwise) H-PA chains are

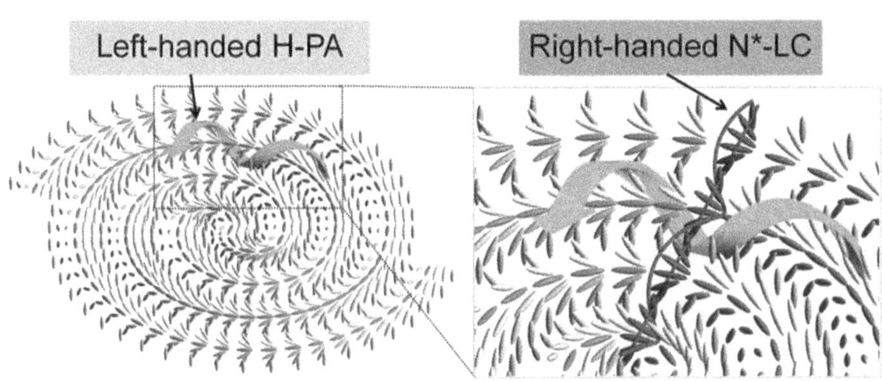

Figure 2.7 Schematic representation of mechanism for acetylene polymerization in the N*-LC. The H-PA with left-handed screw direction grows starting from the catalytic species in the right-handed N*-LC.

formed in (*R*)- and (*S*)-N*-LCs, respectively. These helical chains are bundled through van der Waals interactions to form helical fibrils.[66] The bundles of fibrils form a spiral morphology with various sizes of domains. From the results mentioned above, it is worth noting that, by using the N*-LC as an asymmetric polymerization solvent, a hierarchical helical structure having primary to higher orders in H-PA is synthesized (Figure 2.8).[66]

The dihedral angle between neighboring unit cells, (–CH=CH–), of the H-PA is estimated to be from 0.02° to 0.23°.[11] Although such a very small dihedral angle may allow us to regard the present PA as an approximately planar structure, the polymer is rigorously screwed by a one-handed direction with the nonzero dihedral angle. The present H-PA films have high *trans* contents of 90% and become highly conductive upon iodine doping. In fact, the electrical conductivities of the doped films are $1.5-1.8\times10^3$ S/cm at room temperature, which are comparable to those of metals. The iodine-doped H-PA showed the same Cotton effect as that of nondoped PA. This indicates that the helical structure is preserved even after iodine doping. Furthermore, CD and X-ray diffraction measurements show that the helical structure is also preserved after heating to 150 °C (which corresponds to the isomerization temperature from *cis* to *trans* form). The most stable structure of PA is the planar one. However, since the PA is actually insoluble and infusible, the helical structure

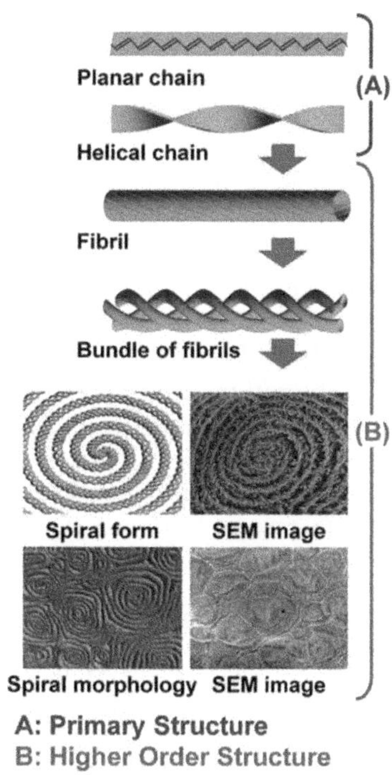

Figure 2.8 Hierarchical helical structures from primary to higher order in helical polyacetylene. SEM = scanning electron microscope.

formed during the polymerization can be preserved even if it is washed by toluene or thermally heated below the isomerization temperature. In other words, the insolubility and infusibility of PA are indispensable for preserving the metastable helical structure.

2.1.3 H-PA with Bundle-Free Fibril Morphology

H-PA is anticipated to act as a prototype of a molecular solenoid, by virtue of its helical structure and high electrical conductivity.[64,65] It is therefore desirable to synthesize a sufficiently screwed H-PA and prevent the individual fibrils from forming bundle morphology. Toward this aim, it is essential to construct a more highly twisted LC reaction field. Here we show that a H-PA film, consisting of only single fibrils and not a fibril bundle, can be synthesized using a highly twisted N*-LC reaction field.

2.1.3.1 Powerful Helicity Inducers

It is known that binaphthyl derivatives substituted with LC groups at the 2,2′,6,6′ positions of binaphthyl rings exhibit good miscibility towards the host N-LC, due to their liquid crystallinities.[69] They also have HTPs based on axial chirality. However, despite using binaphthyl derivatives with large HTPs, it is difficult to induce the N*-LCs with nano-ordered helical pitches. Thus, both miscibility and large HTP are required for the effective transfer of axial chirality from binaphthyl derivatives to the host N-LCs. To investigate the amplification of HTPs in the axially chiral binaphthyl derivatives, several rigid substituent groups are introduced into the 2,2′,6,6′ positions of the binaphthyl rings.[74] Among them, a *tetra*-substituted binaphthyl derivative, D-2, which has a direct linkage between the mesogenic core of phenylcyclohexyl (PCH) moieties and the 6,6′ position of the binaphthyl rings, shows an extremely large HTP of 449 μm^{-1} when added as a chiral dopant into N-LC (Figure 2.5, see also Scheme 2.2). The HTPs of D-1 and D-2 are 171 μm^{-1} and 449 μm^{-1}, respectively. It is clear that the HTP of the chiral dopant D-2 is *ca.* 2.6 times larger than that of D-1. This may be rationalized with a difference in the number of substituents. Namely, the axially twisting torque of D-2 is more effectively transferred to environmental N-LC molecules, by virtue of intermolecular interactions between the four PCH substituents of D-2 and the PCH moieties of LC molecules. In contrast, D-1 bears only two PCH substituents.

2.1.3.2 Highly Twisted N-LC Reaction Field*

The N*-LCs, including D-1 at 1 mol% {PCH302:PCH304:D-1 = 100 : 100 : 2 (mole ratio)} and D-2 at 1.5 mol% {PCH302:PCH304:D-2 = 100 : 100 : 3 (mole ratio)}, abbreviated as System-1 and System-2, respectively, are prepared as the asymmetric reaction field for acetylene polymerization. The helical pitches of System-1 and System-2 are 5 μm and 270 nm, respectively. Figure 2.9 shows POMs of the N*-LCs, System-1 and System-2. System-1 gives a fingerprint

Figure 2.9 Polarizing optical micrographs of N*-LCs at 27 °C. (a) (*S*)-System-1 containing 1.0 mol% of chiral dopant, (*S*)-D-1. (b) (*R*)-System-2 containing 1.5 mol% of chiral dopant, (*R*)-D-2.

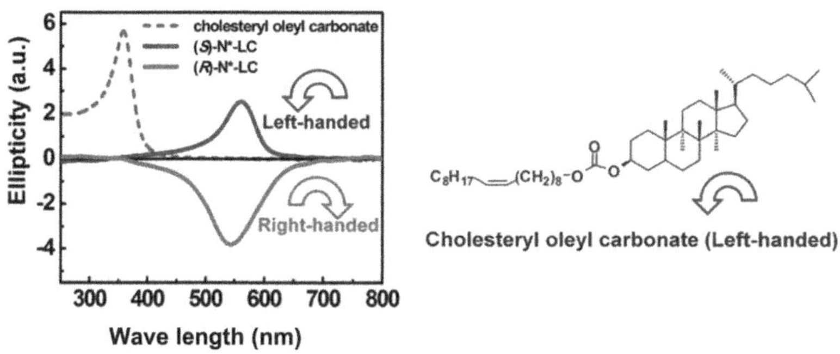

Figure 2.10 Reflection CD spectra of the cholesteryl oleyl carbonate and the N*-LCs induced by (*R*)-D-2 and (*S*)-D-2. (*R*)-N*-LC; PCH302 : PCH304 : (*R*)-D-2 = 100 : 100 : 2 (mole ratio). (*S*)-N*-LC; PCH302 : PCH304 : (*S*)-D-2 = 100 : 100 : 2 (mole ratio).

texture with striae (Figure 2.9a). The distance between the striae (2.5 μm) corresponds to one half of the helical pitch in N*-LC. Meanwhile, the POM of System-2 gives a fan-shaped texture, but no striae is observed (Figure 2.9b). This is due to the fact that the distance between the striae formed in System-2 is too small to be detected in the POM microscope, which has a resolution limit of *ca.* 1 μm.

The helical sense of System-2 is examined through a selective light reflection in circular dichroism (CD) spectra.[66] It is known that cholesteryl oleyl carbonate is a left-handed cholesteric LC that shows a selective light reflection in the visible region. Therefore, cholesteryl oleyl carbonate is used as a reference for determining the handedness of the N*-LC. As shown in Figure 2.10, the N*-LC inducing (*R*)-D-2 shows a negative sign, while the N*-LC including (*S*)-D-2 and cholesteryl oleyl carbonate shows peaks having a positive sign in the CD spectra. These results indicate that the helical senses of the (*R*)- and (*S*)-N*-LCs are right and left directions, respectively.

2.1.3.3 Synthesis of H-PA with Bundle-Free Fibril Morphology

Figure 2.11 show scanning electron microscope (SEM) photographs of the H-PA films synthesized in a right-handed N*-LC with a helical pitch of 270 nm {(*S*)-System-2}. Hierarchical helical structures are observed in H-PA film. It is confirmed that the screw direction of the fibril, including even the fibril bundle, is opposite to the helical sense of the N*-LC. For instance, in the right-handed N*-LC of (*R*)-System-1, the fibrils are screwed left to form the fibril bundle. Similarly, in the left-handed N*-LC of (*S*)-System-2, the fibrils are screwed right, although no bundle is formed.[69] It is of particular note that the highly twisted N*-LC (System-2) gives the fibrils but not the fibril bundle (Figure 2.11). This is in distinct contrast to the morphology of H-PA that is synthesized in the moderately twisted N*-LC (System-1). It is evident from Figures 2.11b and c that the PA fibrils synthesized in (*S*)-System-2 are more highly twisted than those in (*R*)-System-1 (see also Figure 2.6).

To elucidate the relationship between the helical pitch degree of the N*-LC and the morphology of H-PA, various N*-LCs are prepared with helical pitches between 5 μm and 270 nm (*i.e.* 2.3 μm, 850 nm, and 470 nm), by changing the mole percent of the D-2 chiral dopant. The N*-LC with a helical pitch of 2.3 μm gives the fibril bundle in H-PA morphology. In the case of the N*-LC with a helical pitch of 850 nm, bundles consisting of several fibrils are observed. The distance between bundles is less than 550 nm. When the N*-LC having a helical pitch of 470 nm is used, almost only single fibrils are observed; however, portions of the fibrils are overlapped. These results indicate that the morphology of H-PA is dominated by the helical pitch degree of the N*-LC, but not by the species of the chiral dopant.

Figure 2.12 shows a schematic representation of the relationship between the twisting degree of the N*-LC and the hierarchical morphology of H-PA. In the weakly twisted N*-LC, PA fibrils (diameters from 70 to 120 nm) are gathered to form fibril bundles (diameters up to 1 μm). Interestingly, the distance between

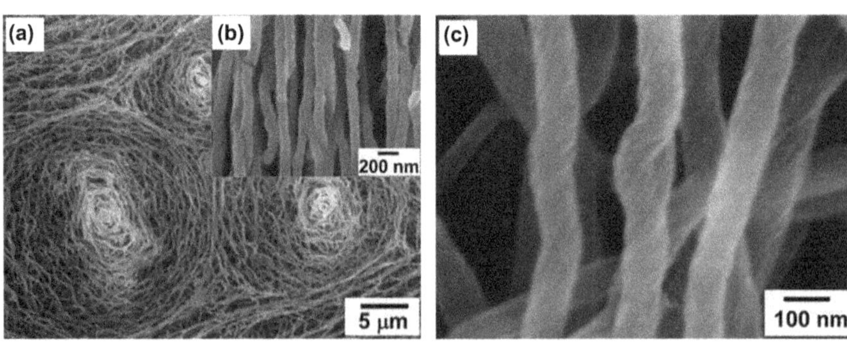

Figure 2.11 SEM micrographs of helical polyacetylene film synthesized in the left-handed (*S*)-System-2 {PCH302 : PCH304 : (*S*)-D-2 = 100 : 100 : 3 (mole ratio)} with a helical pitch of 270 nm. The photograph of (b) shows the magnified one of (a).

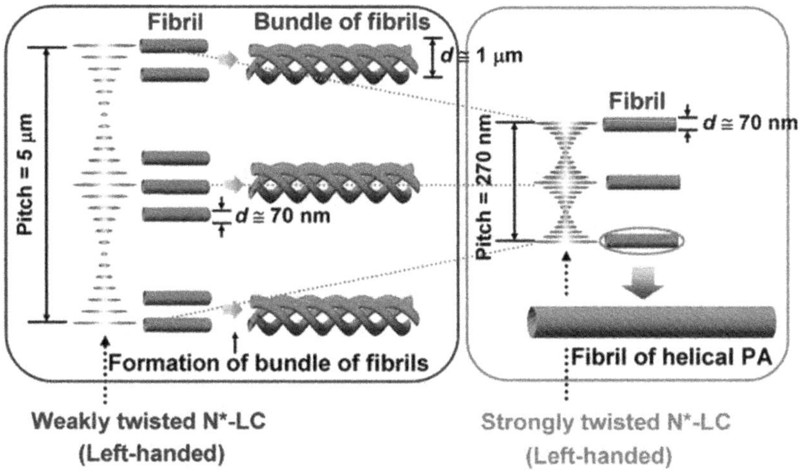

Figure 2.12 Relationship between the twisting degree of N*-LC and the morphology of helical polyacetylene.

fibril bundles is close to one half of the helical pitch. Fibril bundles can be formed in N*-LCs with helical pitches up to 1 μm. However, in the highly twisted N*-LC (helical pitch narrower than 1 μm), the PA has highly screwed fibrils, but not a bundle of fibrils. Therein, the diameters of the fibrils are in the range from 70 to 120 nm. This may be due to the fact that the helical pitch of 270 nm in the strongly twisted N*-LC is smaller than the diameter (*ca.* 1 μm) of the fibril bundle. This situation might prevent the formation of fibril bundles. This type of morphology that is free from fibril bundles should make it much easier to evaluate the electromagnetic properties of the screwed fibril.

H-PA films having a morphology free from fibril bundles have high *trans* contents of 90%. This is mainly due to the polymerization temperature of 20 – 21 °C. It is known that the *cis* form of the PA segment is a kinetically favorable product, because of the so-called *cis* opening mechanism of acetylene polymerization. On the other hand, the *trans* form is a thermodynamically stable product. The *cis* form is actually transformed into the *trans* form, depending on the degree of exothermal heat produced during acetylene polymerization. As exothermal heat is evolved in the polymerization at 20 °C, the *cis* – *trans* isomerization is enhanced to give high *trans* content in the present H-PA.[76] H-PA shows high electrical conductivities, such as around 1.8×10^3 to 2.0×10^3 S/cm, at room temperature after iodine doping. Meanwhile, the PA films synthesized in Systems-1 and 2 have bulk densities of *ca.* 0.5 and 0.96 g/cm^2, respectively. This indicates that single fibrils give a more closed morphological packing than the fibril bundle structure.

By virtue of the high helical twisting power and liquid crystallinity of the novel *tetra*-substituted binaphthyl derivative (D-2), we obtain a N*-LC with a helical pitch in the nanoregime. This is performed by adding a high mole percentage of chiral dopant to the N-LC without destroying the LC phase.

The helical pitches (270 to 850 nm) are smaller than the radius (1 μm) of the fibril bundle of H-PA. In particular, the highly screwed N*-LC (helical pitch of 270 nm) depresses the formation of the fibril bundle, resulting in a bundle-free fibril morphology consisting of single fibrils. Thus, we find that the degree of screwing in the N*-LC reaction field is a key factor in controlling the bundle formation and/or depression in PA fibril morphology. It is expected that a highly twisted H-PA without the fibril bundle might be feasible for the evaluation of unprecedented electromagnetic properties of a single fibril of conducting polymer. It is worth emphasizing that by using the N*-LC as an asymmetric polymerization solvent, helix formation is possible not only for PA but also for aromatic π-conjugated polymers without chiral substituents on the side chains.[66,79,80]

2.1.4 Morphology-Retaining Carbonization of H-PA

Lastly, it is of particular interest that the H-PA film is useful as a precursor for helical carbon and graphite film when the morphology-retaining carbonization *via* an iodine doping is used.[81,82]

It has been considered difficult to prepare a freestanding carbon thin film through the carbonization of an organic polymer film. This is because carbonization at high temperatures causes thermal decomposition and volatilization of hydrocarbon gases, destroying the morphology of the original film. However, the iodine-doped PA film was almost completely carbonized at temperatures above 800 °C. No indication of thermal decomposition was observed in differential thermal analysis curve for the doped PA film. Surprisingly, the fibrous morphology of the original structure remained unchanged, even after carbonization and heat treatment. In addition, the weight loss of the films due to carbonization at 800 °C was very small, amounting to only 20% of the film weight before iodine doping. Thus, the nanofibril-fabricated carbon film was obtained through carbonization of the doped PA film. It is elucidated that iodine doping prevents the PA film from thermally decomposing at high temperatures. According to the structural model of the iodine-doped PA film, polyiodide ions (*e.g.*, I_3^- and I_5^-) are situated between the PA chains, forming a charge-transfer complex. Iodine tends to react with hydrogen at high temperatures. Actually, outgassing of hydrogen iodide was detected by gas chromatography mass spectrometry during heating of the doped PA film. It can be assumed that hydrogen contained in the doped PA is removed to some extent as hydrogen iodide from the PA chains, and that the PA chains partially crosslink between the neighboring chains. Furthermore, most of the hydrogens are removed with increasing temperature. As a result, networks of sp^2 hexagonal carbon bonds are formed during carbonization. It should be emphasized that the carbon film heat treated at 2600 °C had almost the same helical structure as those of the original H-PA film and the carbon film that was prepared at 800 °C. We found a new aspect of the iodine doping by disclosing its indispensable role in maintaining the nanostructure and morphology of precursors during carbonization.

2.2 Experimental Procedures

2.2.1 Synthesis of Shirakawa-Type PA[19–22]

PA is prepared in the form of film by introducing acetylene gas to an unstirred concentrated solution of the tetra-*n*-butoxytitanium $Ti(O\text{-}n\text{-}Bu)_4$ and triethylaluminium $AlEt_3$ ($Al/Ti = 4$) catalyst system or a reaction vessel in which a concentrated solution of the catalyst is spread over the glass wall to form a polymer film on the surface of the solution. To control the thickness of films, the polymerization is interrupted by evacuating the system. Polymerization is carried out at several temperatures over a wide range of *ca.* –100 and 180 °C. At polymerization temperatures lower than 80 °C, toluene is used as solvent, whereas at temperatures higher than 80 °C *n*-hexadecane is used. The film is washed repeatedly by toluene, until the solution becomes colorless. Purification of the films is carried out at the same temperature as the polymerization temperature to prevent *cis–trans* isomerization of the polymers. The purified films are spread on a copper mesh and dried by blowing nitrogen gas on them at room temperature for a few minutes. All the processes are carried out under an atmosphere of nitrogen to prevent the oxidation of the polymers. Very thin films of PA obtained at low temperature are red in color and have a *cis* structure. High-temperature polymerization leads to blue-colored *trans* polymer films.

2.2.2 Synthesis of Naarmann and Theophilou-Type PA[9,38,39]

2.2.2.1 Standard Method

The silicone oil is stirred and degassed for 20 min at 0.05 mbar. $AlEt_3$ is added in a countercurrent stream of argon, and $Ti(O\text{-}n\text{-}Bu)_4$ is run in, drop by drop, through the inactivated dropping funnel over a period of one hour at 38–42 °C. Afterwards, the mixture is degassed for one hour at room temperature and subsequently stirred for two hours at 120 °C in a weak current of argon. This is followed by stirring and degassing for another hour at 0.1 mbar. An even, homogeneous layer of standard catalyst is applied to a flat carrier consisting of a stretchable polymer-supporting material, *e.g.*, high-density polyethylene or polypropylene film. The amount of catalyst required to obtain a PA film of 15 μm in thickness is 7 ml. The carrier, coated with the catalyst, is hermetically sealed in the glove box by means of a hood fitted with a gas inlet valve. First, the hood is evacuated; then 600 ml of acetylene is passed into it over a period of 15 min. (Note that the acetylene is previously purified and a separate trap for $AlEt_3$, sulfuric acid and molecular sieve.) As it is being admitted, the acetylene polymerizes at the surface of the catalyst on the carrier to form PA, which is a black, homogeneous film. The PA film is extremely porous and contains all the catalyst. The standard PA and transparent films are washed under the following conditions: 3 h in toluene; 16 h in methanol containing 6% of hydrochloric acid and two periods of 1–1.5 h with methanol. The films are dried under argon in such a way that they do not contract in volume during the drying stage.

2.2.2.2 ARA (Addition of Reducing Reagents) Method

The catalyst Ti(O-*n*-Bu)$_4$ - AlEt$_3$ is mixed with the different soluble reducing agents. An even, homogeneous layer of catalyst that is obtained by mixing 10 ml of catalyst, *e.g.*, with 1 ml of *n*-BuLi, is applied to a flat carrier consisting of a polymer-supporting material (high-density polyethylene or polypropylene films). Acetylene is then introduced and allowed to react with this modified catalyst to give a PA film. Different experiments are performed by mixing 10 ml of the standard catalyst with various quantities of *n*-BuLi. Using 3 ml of *n*-BuLi, it is allowed to synthesize a PA film of the highest conductivity. The PA films prepared by the ARA method are stretched with the supporting film by up to 400%, corresponding to $l/l_o = 5$, and then removed from the film. The PA films prepared by the ARA method are washed in toluene.

2.2.3 Synthesis of Tsukamoto-Type PA[10,40]

The catalyst used in the acetylene polymerization is Ti(O-*n*-Bu)$_4$ - AlEt$_3$ with an Al/Ti ratio of 2.0. The catalyst is dissolved in an organic solvent with a high boiling point, such as decaline (b.p. 196 °C), with Ti(O-*n*-Bu)$_4$ concentration of 1.0 mol/l. The aging of the catalyst solution is carried out by refluxing at 220 °C for 30 min in an argon gas atmosphere. Aging followed by quick quenching to room temperature is repeated twice. Upon aging, there appear in the catalyst solution numerous minute aggregates, which are removed by filtering before polymerization. An acetylene monomer of six-nines grade is used without further purification. The polymerization procedure basically follows the conventional Shirakawa method, although the monomer gas is diluted with argon gas to 5 vol%. The polymerization is allowed to progress for several hours at the temperature of –60 °C. After polymerization, the PA film is washed in toluene and a hydrochloric acid/methanol solution. The prepared PA film has an average thickness of 10 to 20 μm and a shiny silvery luster.

2.2.4 Synthesis of High Mechanical Modulus and Strength PA

2.2.4.1 Solvent Evacuation Method[24]

All experimental procedures are carried out under inert argon atmosphere unless otherwise stated. Acetylene gas of six-nines grade and AlEt$_3$ are used without further purification. Ti(O-*n*-Bu)$_4$ and *n*-hexane, toluene and cumene used as polymerization solvents are distilled under argon gas prior to use. The concentration of catalyst is 0.5 mol/l of Ti(O-*n*-Bu)$_4$ and the ratio of AlEt$_3$ to Ti(O-*n*-Bu)$_4$ is 2.0–4.0. After preaging at room temperature for 0.5–1 h, the catalyst solution is aged for 1–5 h at temperature near boiling point of each solvent, *i.e.* 70–80 °C, 110–120 °C and 140–150 °C, for *n*-hexane (b.p. 68.7 °C), toluene (110.6 °C) and cumene (152.4 °C), respectively. Such high-temperature aging is carried out by refluxing the catalyst solution with stirring in a weak current of argon gas. Afterwards, a Schlenk flask of the thermally treated catalyst solution is connected to vacuum line *via* a flexible joint. The solvent in the flask is dynamically pumped off with simultaneously rotating the flask kept horizontal until it is completely evaporated (practically below 0.01 Torr). This procedure gives rise to a catalyst layer homogeneously coated over the inner glass wall of the flask. At this stage, the concentration of the catalyst is increased by 2–3 times compared with the initial concentration before the solvent evaporation, and hence the catalyst becomes viscous enough not to be flowed down along the glass wall of the flask. Polymerization is carried out at −78 °C by introducing acetylene gas onto the catalyst. The initial acetylene pressure is 680–750 Torr, and polymerization time is 2–26 h. After the polymerization, PA films are washed several times with toluene cooled by dry-ice and ethanol solution (−78 °C) under argon gas. The PA films prepared, unlike usual films, are found to be not shiny but dull and/or almost black, and also hard and tough irrespective of thickness. The *cis* contents are 80–90% through the films.

2.2.4.2 Intrinsic Nonsolvent (INS) Method[25,41,42]

All experimental procedures are carried out under inert argon atmosphere unless otherwise stated. Neat $AlEt_3$ is added dropwise to neat Ti(O-*n*-Bu)$_4$ in a Schlenk flask. The flask is precooled on an ice bath to avoid an evolution of heat due to a rapid exothermic complex formation between Ti(O-*n*-Bu)$_4$ and $AlEt_3$. The catalyst is aged at room temperature for 1 h, followed by a high-temperature aging (150 °C) for 1 h. The flask is then degassed and the catalyst is coated on the inner wall of the flask by rotating it. After cooling the flask by dry-ice and ethanol mixture (−78 °C), an acetylene gas is introduced in it. The polymerization is carried out for 0.5–1 h. PA film prepared is washed several times by toluene cooled at −78 °C and dried through vacuum pumping. Note that all procedures from the catalyst preparation to the end of polymerization are free from solvent.

2.2.5 Synthesis of Directly Aligned PA[32–34]

Acetylene, Ti(O-*n*-Bu)$_4$ and $AlEt_3$ are purified before use. As a liquid-crystal solvent is used an equimolar mixture of 4-(*trans*-4-*n*-propylcyclohexyl)ethoxybenzene and 4-(*trans*-4-*n*-propylcyclohexyl)butoxybenzene. This solvent is used without further purification except degassing prior to use. The concentration of catalyst is 0.01 mol/l of Ti(O-*n*-Bu)$_4$ and ratio of $AlEt_3$ to Ti(O-*n*-Bu)$_4$ is 4.0. The catalyst solution is aged for 30 min at room temperature. Polymerization of acetylene is carried out along the following procedure. By using a microsyringe, the catalyst solution is transferred into a flat-bottomed container placed in a Schlenk flask. The flask is connected *via* flexible joints to a vacuum line and then sufficiently degassed. After the flask is fixed between both poles of a magnet so that the fluid surface of the catalyst solution in the container can be kept horizontal, acetylene gas is introduced into the flask. The initial pressure of acetylene was 600–760 tort and polymerization time 4–60 min. The magnetic force is 2–14 kG, and the temperature is kept at 10–15 °C during the polymerization in order to maintain the nematic phase of the liquid crystals. After the polymerization, the PA film (thickness of 2–50 μm) is carefully stripped off from the container and washed with toluene several times under inert gas at room temperature. The film is dried in vacuum on a Teflon sheet and stored in a freezer at –20 °C.

2.2.6 Synthesis of Helical PA[11,66–68]

Acetylene gas of six-nines grade (99.9999% purity) is used without further purification. The polymerization temperature is kept between 17 and 18 °C to maintain the chiral nematic LC (N*-LC) phase, by circulating cooled ethanol through an outer flask enveloping the Schlenk flask. The initial acetylene pressure is 11.6–22.6 Torr and the polymerization time is 10–43 min. After polymerization, PA films are carefully stripped from the container and washed with toluene several times under argon gas at room temperature. The films are dried through vacuum pumping on a Teflon sheet and stored in a freezer at –20 °C.

2.2.7 Synthesis of Nematic Liquid Crystals and Chiral Dopants

The mesogenic compounds, 4-(*trans*-4-*n*-pentylcyclohexyl)phenol [PCH500] and 4-(*trans*-4-*n*-pentylcyclohexyl)bromobenzene [PCH5Br], are also commercially available. N-LCs of phenylcyclohexane derivatives, 4-(*trans*-4-*n*-propylcyclohexyl)ethoxybenzene (PCH302) and 4-(*trans*-4-*n*-propylcyclohexyl)butoxybenzene (PCH304), shown in Chart 2.1, are synthesized according to previous reports.[11,32,34]

The chemical compounds, (*R*)- and (*S*)-2,2-dihydroxy-1,1-binaphthyl (optical purity, 0.99), are purchased from commercially available sources. The di-substituted binaphthyl derivatives, (*R*)- and (*S*)-1,1'-binaphthyl-2,2′-di-[*para*-(*trans*-4-*n*-pentylcyclohexyl)phenoxy-1-hexyl]ether, are synthesized through the Williamson etherification reactions of chiroptical (*R*)-(+)- and (*S*)-(-)-1,1′-bi-2-naphthyl, respectively, with phenylcyclohexyl derivatives. The products are abbreviated as (*R*)-, (*S*)-D-1 (Scheme 2.1). The substituent is composed of phenylcyclohexyl (PCH) moiety, *n*-pentyl group (the number of carbon of 5), and hexamethylene chain linked with an ether-type oxygen atom, [-$(CH_2)_6O$-, 06], and thus abbreviated as PCH506.

Host Nematic LCs

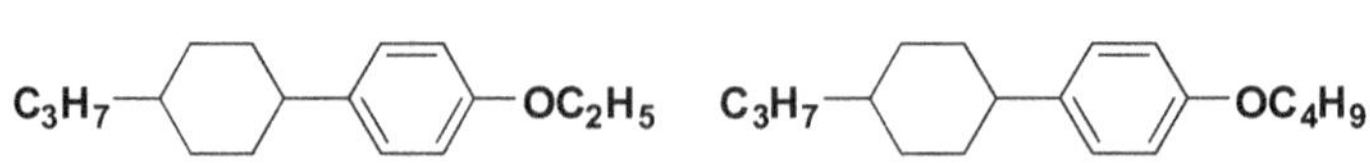

PCH302 PCH304

Chart 2.1 Molecular structures of host nematic LCs, PCH302 and PCH304.

Scheme 2.1 Acetylene polymerization using Ziegler–Natta catalyst.

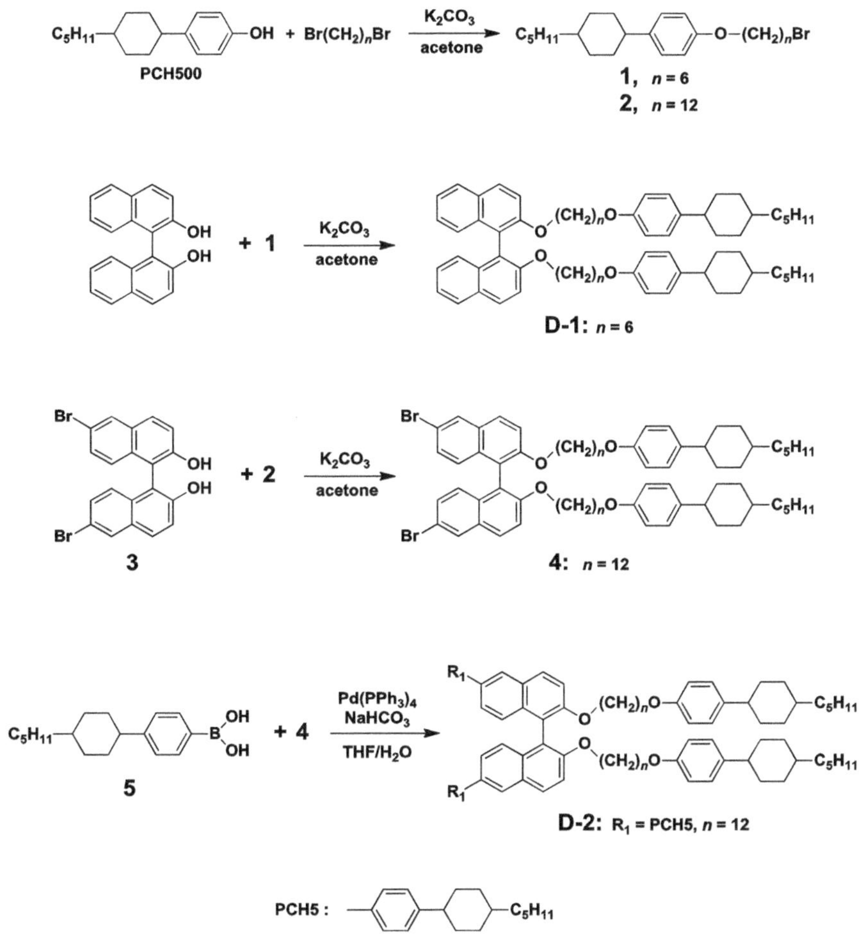

Scheme 2.2 Synthetic routes for di- and tetra-substituted axially chiral binaphthyl derivatives.

The *tetra*-substituted binaphthyl derivatives, D-2 are synthesized by substituting aromatic (PCH5) moieties into the 6,6′ positions of the binaphthyl rings without methylene spacers. Synthetic routes for the *tetra*-substituted binaphthyl derivatives are also shown in Scheme 2.1. It should be mentioned that to improve a miscibility of chiral dopant to the host N-LCs, PCH moieties are introduced into the 2,2′ positions of binaphthyl rings (D-1) or the 2,2′,6,6′ positions of binaphthyl rings (D-2).

References

1. J. C. W. Chien, *Polyacetylene - Chemistry, Physics and Material Science*, Academic Press, Orlando, FL, 1984.
2. H. Shirakawa, *Angew. Chem. Int. Ed.*, 2001, **40**, 2574.

3. A. J. Heeger, *Angew. Chem. Int. Ed.*, 2001, **40**, 2591.
4. A. G. MacDiarmid, *Angew. Chem. Int. Ed.*, 2001, **40**, 2581.
5. *Handbook of Conducting Polymers*, T. A. Skotheim, ed. Marcel Dekker, New York, 1986.
6. *Handbook of Organic Conductive Molecules and Polymers*, H. S. Nalwa, ed. Wiley, New York, 1997.
7. *Handbook of Conducting Polymers, Conjugated Polymers,* 3rd edn, T. A. Skotheim, J. R. Reynolds, ed. CRC Press, New York, 2007.
8. *Electrical and Optical Polymer Systems: Fundamentals, Methods, and Applications*, D. L. Wise, G. E. Wnek, D. J. Trantolo, T. M. Cooper, and J. D. Gresser, ed. Marcel Dekker, New York, 1998.
9. H. Naarmann and N. Theophilou, *Synth. Met.*, 1987, **22**, 1.
10. J. Tsukamoto, A. Takahashi and K. Kawasaki, *Jpn. J. Appl. Phys.*, 1990, **29**, 125.
11. K. Akagi, G. Piao, S. Kaneko, K. Sakamaki, H. Shirakawa and M. Kyotani, *Science*, 1998, **282**, 1683.
12. G. Natta, G. Mazzanti and P. Corradini, *Atti Accad. Naz. Lincei Rend. Cl. Sci. Fis., Mat. Nat.*, 1958, **25**, 3.
13. W. H. Watson, W. C. McMordie and L. G. Lands, *J. Polym. Sci.*, 1961, **55**, 137.
14. L. B. Luttinger, *J. Org. Chem.*, 1962, **27**, 1591.
15. I. V. Nicolescu and E. M. Angelescu, *J. Polym. Sci., Part A-1*, 1966, **4**, 2963.
16. J. H. Edwards, W. J. Feast and D. C. Bott, *Polymer*, 1984, **25**, 395.
17. H. Shirakawa, E. Louis, A. G. MacDiarmid, C. K. Chiang and A. J. Heeger, *J. Chem. Soc. Chem. Commun.*, 1977, 578.
18. C. K. Chiang, C. R. Fincher, Y. W. Park, A. J. Heeger, H. Shirakawa, E. J. Louis, S. C. Gau and A. G. MacDiarmid, *Phys. Rev. Lett.*, 1977, **39**, 1098.
19. H. Shirakawa and S. Ikeda, *Polym. J.*, 1971, **2**, 231.
20. T. Ito, H. Shirakawa and S. Ikeda, *J. Polym. Sci., Polym. Chem. Ed.*, 1974, **12**, 11.
21. T. Ito, H. Shirakawa and S. Ikeda, *J. Polym. Sci., Polym. Chem. Ed.*, 1975, **13**, 1943.
22. H. Shirakawa and S. Ikeda, *Polym. J.*, 1973, **4**, 460.
23. Y. Tanabe, H. Kyotani, K. Akagi and H. Shirakawa, *Macromolecules*, 1995, **28**, 4173.
24. K. Akagi, M. Suezaki, H. Shirakawa, H. Kyotani, M. Shimomura and Y. Tanabe, *Synth. Met.*, 1989, **28**, D1.
25. K. Akagi, K. Sakamaki and H. Shirakawa, *Macromolecules*, 1992, **25**, 6725.
26. Y. Cao, P. Smith and A. J. Heeger, *Polymer*, 1991, **32**, 1210.
27. S. Shibahara, M. Yamane, K. Ishikawa and H. Takezoe, *Macromolecules*, 1998, **31**, 3756.
28. O. A. Scherman and R. H. Grubbs, *Synth. Met.*, 2001, **124**, 431.
29. O. A. Scherman, I. M. Rutenberg and R. H. Grubbs, *J. Am. Chem. Soc.*, 2003, **125**, 8515.

30. D. E. Schuehler, J. E. Williams and M. B. Sponsler, *Macromolecules*, 2004, **37**, 6255.
31. H. Gu, R. Zheng, X. Zhang and B. Xu, *Adv. Mater.*, 2004, **16**, 1356.
32. K. Araya, A. Mukoh, T. Narahara and H. Shirakawa, *Synth. Met.*, 1986, **4**, 199.
33. K. Akagi, H. Shirakawa, K. Araya, A. Mukoh and T. Narahara, *Polym. J.*, 1987, **19**, 185.
34. K. Akagi, S. Katayama, H. Shirakawa, K. Araya, A. Mukoh and T. Narahara, *Synth. Met.*, 1987, **17**, 241.
35. Y. W. Park, M. A. Druy, C. K. Chiang, A. J. Heeger, A. G. MacDiarmid, H. Shirakawa and S. Ikeda, *J. Polym. Sci., Polym. Lett. Ed.*, 1979, **17**, 195.
36. T. Woerner, A. G. MacDiarmid and A. J. Heeger, *J. Polym. Sci., Polym. Lett.*, 1982, **20**, 305.
37. T. Woerner, A. G. MacDiarmid, A. Feldblum and A. J. Heeger, *J. Polym. Sci., Polym. Lett.*, 1984, **22**, 119.
38. H. Naarmann, *Synth. Met.*, 1991, **41**, 1.
39. T. H. Schimmel, D. Claser, M. Schwoerer and H. Naarmann, *Synth. Met.*, 1991, **41**, 19.
40. J. Tsukamoto, *Adv. Phys.*, 1992, **41**, 509.
41. K. Akagi, K. Sakamaki and H. Shirakawa, *Synth. Met.*, 1993, **55**, 779.
42. K. Akagi, K. Sakamaki, H. Shirakawa and H. Kyotani, *Synth. Met.*, 1995, **69**, 29.
43. K. Akagi, D. Hashimoto, H. Shirakawa and J. Isoya, *Polym. Commun.*, 1990, **31**, 411.
44. K. Akagi, D. Hashimoto, H. Shirakawa and J. Isoya, *Synth. Met.*, 1991, **40**, 197.
45. K. Akagi and H. Shirakawa, *Synth. Met.*, 1993, **60**, 85.
46. M. Sinclair, D. Moses, K. Akagi and A. J. Heeger, *Phys. Rev. B*, 1988, **38**, 10724.
47. M. Kyotani, S. Matsushita, M. Goh, T. Nagai, Y. Matsui and K. Akagi, *Nanoscale*, 2010, **2**, 509.
48. A. Montaner, M. Rolland, J. L. Sauvajol, M. Galtier, R. Almairac and J. L. Ribet, *Polymer*, 1988, **29**, 1101.
49. N. Coustel, N. Foxonet, J. L. Ribet, P. Bernier and J. E. Fischer, *Macromolecules*, 1991, **24**, 5867.
50. Y. Yamashita, S. Nishimura, K. Shimamura and K. Monobe, *Makromol. Chem.*, 1986, **187**, 1757.
51. S.-Y. Oh, R. Ezaki, K. Akagi and H. Shirakawa, *J. Polym. Sci., Part A: Polym. Chem.*, 1993, **31**, 2977.
52. S.-Y. Oh, K. Akagi, H. Shirakawa and K. Araya, *Macromolecules*, 1993, **26**, 6203.
53. S. K. Choi, Y. S. Gal, S. H. Jin and H. K. Kim, *Chem. Rev.*, 2000, **100**, 1645.
54. J. Liu, J. W. Y. Lam and B. Z. Tang, *Chem. Rev.*, 2009, **109**, 5799.
55. S. Koltzenburg, D. Wolff, F. Stelzer, J. Springer and O. Nuyken, *Macromolecules*, 1998, **31**, 9166.

56. K. Akagi, H. Goto, Y. Kadokura, H. Shirakawa, S.-Y. Oh and K. Araya, *Synth. Met.*, 1995, **69**, 13.
57. S.-Y. Oh, F. Oguri, K. Akagi and H. Shirakawa, *J. Polym. Sci., Part A: Polym. Chem.*, 1993, **31**, 781.
58. K. Akagi and H. Shirakawa, *Macrmol. Sympos.*, 1996, **104**, 137.
59. K. Akagi, H. Goto, H. Shirakawa, T. Nishizawa and K. Masuda, *Synth. Met.*, 1995, **69**, 33.
60. B. Z. Tang, X. Kong, X. Wan, X.-D. Feng and H. S. Kwok, *Macromolecuels*, 1998, **31**, 2419.
61. H. Kuroda, H. Goto, K. Akagi and A. Kawaguchi, *Macromolecules*, 2002, **35**, 1307.
62. C. Ye, W. Y. Lam, Z.-F. Liu, S. Z. D. Cheng, E.-Q. Chen and B. Z. Tang, *J. Am. Chem. Soc.*, 2005, **127**, 7668.
63. K. Akagi, *J. Polym. Sci., Part A: Polym. Chem.*, 2009, **47**, 2463.
64. H. J. Lee, Z. X. Jin, A. N. Aleshin, J. Y. Lee, M. Goh, K. Akagi, Y. S. Kim, D. W. Kim, D. W. Kim and Y. W. Park, *J. Am. Chem. Soc.*, 2004, **126**, 16722.
65. A. N. Aleshin, H. J. Lee, Y. W. Park and K. Akagi, *Phys. Rev. Lett.*, 2004, **93**, 196601.
66. K. Akagi, *Chem. Rev.*, 2009, **109**, 5354.
67. K. Akagi, I. Higuchi, G. Piao, H. Shirakawa and M. Kyotani, *Mol. Cryst. Liq. Cryst.*, 1999, **332**, 463.
68. K. Akagi, S. Guo, T. Mori, M. Goh, G. Piao and M. Kyotani, *J. Am. Chem. Soc.*, 2005, **127**, 14647.
69. M. Goh, M. Kyotani and K. Akagi, *J. Am. Chem. Soc.*, 2007, **129**, 8519.
70. M. Goh, M. Takayuki, M. Kyotani and K. Akagi, *Macromolecules*, 2007, **40**, 4762.
71. K. Akagi, *Polym. Int.*, 2007, **56**, 1192.
72. T. Mori, M. Kyotani and K. Akagi, *Macromolecules*, 2008, **41**, 607.
73. K. Akagi and T. Mori, *Chem. Record*, 2008, **8**, 395.
74. M. Goh and K. Akagi, *Liq. Cryst.*, 2008, **35**, 953.
75. T. Mori, T. Sato, M. Kyotani and K. Akagi, *Macromolecules*, 2009, **42**, 1817.
76. M. Goh, G. Piao, M. Kyotani and K. Akagi, *Macromolecules*, 2009, **42**, 8590.
77. T. Mori, M. Kyotani and K. Akagi, *Macromolecules*, 2010, **43**, 8363.
78. T. Mori, M. Kyotani and K. Akagi, *Chem. Sci.*, 2011, **2**, 1389.
79. M. Goh, S. Matsushita and K. Akagi, *Chem. Soc. Rev.*, 2010, **39**, 2466.
80. S. Matsushita and K. Akagi, *Isr. J. Chem.*, 2011, **51**, 1075.
81. M. Kyotani, S. Matsushita, T. Nagai, Y. Matsui, M. Shimomura, A. Kaito and K. Akagi, *J. Am. Chem. Soc.*, 2008, **130**, 10880.
82. S. Matsushita, M. Kyotani and K. Akagi, *J. Am. Chem. Soc.*, 2011, **133**, 17977.

CHAPTER 3

Substituted Polyacetylenes

FUMIO SANDA,[a] MASASHI SHIOTSUKI[b] AND TOSHIO MASUDA*[c]

[a] Kyoto University, Department of Polymer Chemistry, Katsura Campus, Kyoto 615-8510, Japan; [b] Kinki University, Molecular Engineering Institute, Kayanomori 11-6, Iizuka 820-8555, Japan; [c] Fukui University of Technology, Department of Environmental and Biological Chemistry, Gakuen 3-6-1, Fukui 910-8505, Japan
*Email: masuda@fukui-ut.ac.jp

3.1 Introduction

It is now well known that the polymerization of substituted acetylenes using transition-metal catalysts provides substituted polyacetylenes (Scheme 3.1). In the 1970s, it was found that the group 6 transition metal chlorides, WCl_6 and $MoCl_5$, work as active catalysts for the polymerization of phenylacetylene. Subsequently, many classic metathesis catalysts were developed, which turned out to be highly effective for various substituted acetylenes, including monomers that are sterically crowded. More recently, $MoOCl_4$-based ternary catalysts and so-called Schrock's metal carbenes have been shown to induce living polymerization of substituted acetylenes. In addition, Rh catalysts were found to polymerize monosubstituted acetylenes to give *cis*-stereoregular polymers, with some also effecting the living polymerization of phenylacetylene. Development of these numerous catalysts has enabled the synthesis of a variety of substituted polyacetylenes. The formed polymers possess carbon–carbon alternating double bonds along the main chain with various side groups, and consequently, exhibit unique and interesting properties based on the particular conjugated structure. The present chapter deals with the polymerization of

RSC Polymer Chemistry Series No. 9
Conjugated Polymers: A Practical Guide to Synthesis
Edited by Klaus Müllen, John R. Reynolds and Toshio Masuda

Published by the Royal Society of Chemistry, www.rsc.org

substituted acetylenes and characterization of the formed polymers, focusing on the experimental details of their synthesis. For details of the synthesis, properties, and functions of substituted polyacetylenes, the reader is referred to the review papers and monographs in References 1–10.

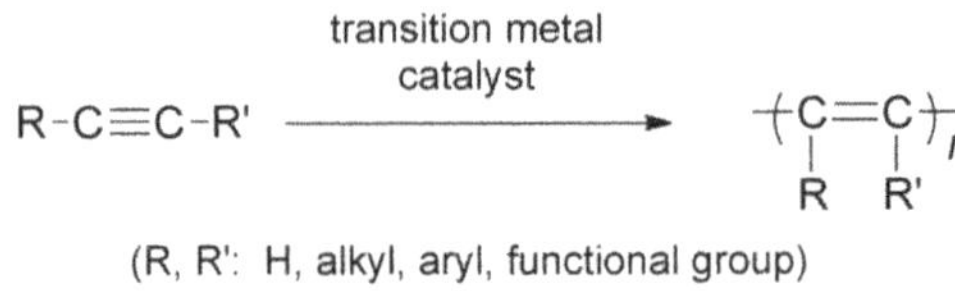

(3.1)

3.1.1 Polymers of Aromatic Monosubstituted Acetylenes

Aromatic monosubstituted acetylenes, represented by phenylacetylene (**1** in Chart 3.1) and its ring-substituted derivatives (**2**) can be efficiently polymerized using transition-metal catalysts to give high molecular weight polymers. Various aromatic groups can be introduced in the side chains by polymerization of the corresponding monomers, giving rise to air stability, solubility, and processability, in addition to numerous unique properties. Generally, Rh, Mo, and W catalysts in combination with suitable cocatalysts are most effective for the polymerization of aromatic monosubstituted acetylenes.

Rh catalysts used in the polymerization of acetylenic monomers are classified into two categories: classic and well-defined catalysts. The former have the advantage of being able to be used directly as they are easily accessible and/or commercially available Rh complexes such as $[(nbd)RhCl]_2$ (nbd = 2,5-norbornadiene)[11,12] and $[(nbd)Rh(\eta^6\text{-Ph})(BPh_3)]$.[13] The well-defined catalysts usually require several steps for their preparation, but can often give precisely controlled polymers in terms of molecular weight, polydispersity index (M_w/M_n), and stereoregularity of the main polyene chain. Representative catalysts of this type include $[(nbd)Rh(\text{-C}{\equiv}\text{C-Ph})(PPh_3)_2]$[14] and $[(nbd)Rh(\text{-CPh}{=}CPh_2)(PPh_3)]$.[15] Substituted polyacetylenes formed using Rh catalysts usually have the *cis* form of the double bond in the main polyene chain, in contrast to the case of group 6 transition-metal catalysts.

Group 6 transition-metal halide catalysts, represented by WCl_6 and $MoCl_5$ here, are the most convenient for the polymerization of monosubstituted acetylenes because of their accessibility. Sterically crowded monosubstituted acetylenes such as ortho-substituted phenylacetylenes can be efficiently polymerized using these catalysts. Addition of cocatalysts such as n-Bu_4Sn, Ph_4Sn, Et_3SiH, Ph_3Sb, and Ph_3Bi enhances catalytic activity.[16,17] Group 6 carbonyl

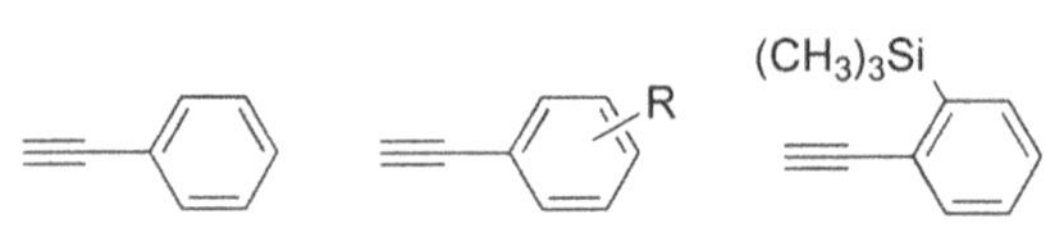

1 2 3

Chart 3.1

complexes polymerize monosubstituted acetylenes when the catalyst solution is irradiated with UV light. Their catalytic activity is usually lower than those of the halide catalysts, but the carbonyl compounds have the advantage of being stable in air.[18] Mo and W alkylidene complexes, represented by the Schrock carbenes, have explosively evolved the polymerization chemistry of substituted acetylenes. Although they need several steps for their preparation, many are able to achieve precisely controlled polymerization of monosubstituted acetylenes (*e.g.*, **3**).[19]

3.1.2 Polymers of Aliphatic Monosubstituted Acetylenes

t-Butylacetylene (**4** in Chart 3.2) is a typical aliphatic monosubstituted acetylene, polymerizable using Mo and W catalysts.[20] Aliphatic monosubstituted acetylenes, including those with heteroatoms, undergo polymerization with Rh catalysts to give *cis*-stereoregular polyacetylenes. When appropriate chiral groups are substituted, the polymers show extremely large optical rotations due to the formation of helical structures with predominantly one-handed screw sense. In the cases of poly(*N*-alkynylamides) (**5**)[21] and poly(*N*-propargylcarbamates) (**6**),[22] the key importance of helix formation is intramolecular hydrogen bonding between the amide and carbamate moieties in the side chains, as well as steric repulsion. On the other hand, 1-methylpropargyl alcohol (**7**) and 1-methylpropargyl alkanoates (**8**) provide helical polymers solely based on steric repulsion.[23] The remarkable ability of such a small chiral moiety to induce helicity is probably due to the location of a stereogenic center adjacent to the main chain.

3.1.3 Polymers of Aromatic Disubstituted Acetylenes

Diphenylacetylene (**9** in Chart 3.3) itself forms a polymer in the presence of $TaCl_5$–n-Bu_4Sn catalyst.[24] The polymer possesses a very high thermal stability,

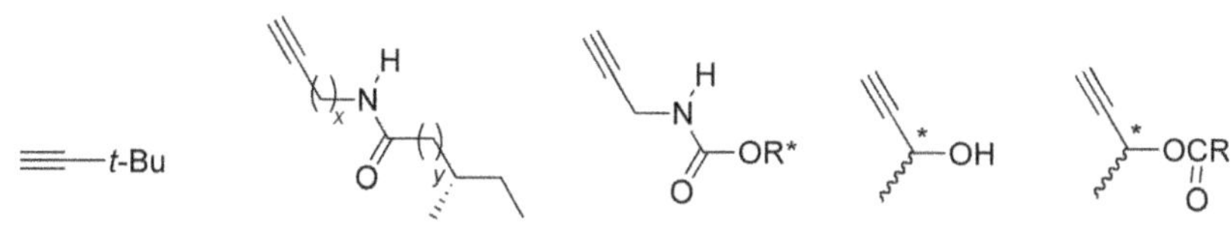

x = 1–3, *y* = 0–3 R* = *s*-Bu*, etc

4 **5** **6** **7** **8**

Chart 3.2

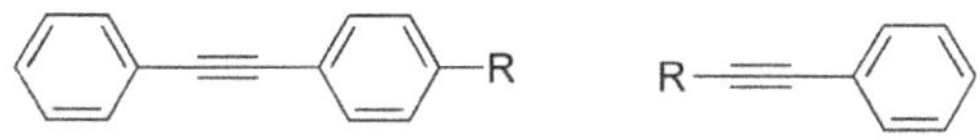

9: R = H
10: R = $SiMe_3$
11: R = $OSiMe_2$-*t*-Bu

12: R = Me
13: R = Cl

Chart 3.3

but it is not soluble in any solvent. This lack of solubility resembles a trend for polyacetylenes that have two identical alkyl groups in the repeat unit, whereas polyacetylenes with a methyl and a long alkyl group are soluble in various solvents. Thus, one can hypothesize that *para*- or *meta*-substituted monomers will provide soluble polymers. In fact, soluble high molecular weight polymers have been obtained from many derivatives with bulky substituents. For instance, 1-phenyl-2-[(*p*-trimethylsilyl)phenyl]acetylene (**10**) polymerizes using $TaCl_5$–cocatalyst systems in high yield.[25] The polymer thus obtained is totally soluble in toluene and $CHCl_3$, and its molecular weight can reach as high as approximately two million. In contrast, $TaCl_5$ alone or a $NbCl_5$–cocatalyst system is ineffective for this particular monomer. The derivatives with *m*-Me_3Si, *m*-Me_3Ge, *p*-*t*-Bu, and *p*-*n*-Bu groups also polymerize, leading to totally soluble, high molecular weight polymers.

Since only Ta and Nb catalysts, which are not tolerant to polar groups, are available for the polymerization of sterically crowded disubstituted acetylenes, it is generally difficult to directly synthesize disubstituted acetylene polymers that have a highly polar substituent such as a hydroxy group. Recently, the synthesis of poly[1-phenyl-2-(*p*-hydroxyphenyl)acetylene] has been achieved by the polymerization of 1-phenyl-2-(*p*-*t*-butyldimethylsiloxyphenyl)acetylene (**11**) and the subsequent acid-catalyzed deprotection reaction.[26]

Since 1-phenyl-1-alkynes [*e.g.*, 1-phenyl-1-propyne (**12**)] exhibit relatively large steric effects, Nb and Ta catalysts produce polymers with molecular weights of 1×10^5–1×10^6.[27] In contrast, W catalysts yield only oligomers with molecular weights lower than 1×10^4, and Mo catalysts are inactive.

Chlorine-containing aromatic disubstituted acetylene monomers, including 1-chloro-2-phenylacetylene (**13**) afford high molecular weight polymers.[28,29] For instance, the polymerization of 1-chloro-2-phenylacetylene and 1-chloro-2-naphthylacetylene is catalyzed by $MoCl_5$–*n*-Bu_4Sn to give polymers whose molecular weight exceeds 1×10^5. It appears that the electron-withdrawing Cl atom plays some role in the inertness of these monomers to Nb, Ta, and W catalysts.

3.1.4 Polymers of Aliphatic Disubstituted Acetylenes

2-Alkynes [*e.g.*, 2-octyne (**14** in Chart 3.4)], which are sterically not very crowded, polymerize using Mo catalysts to give polymers with molecular weights over one million.[30] For these monomers, W and Nb catalysts are less active, and Ta catalysts yield only cyclotrimers. Symmetrical dialkylacetylenes (*e.g.*, 4-octyne) are slightly more crowded, and consequently, Nb, Ta, and W catalysts exert high activity, while Mo catalysts are inactive.

14: R = n-C_5H_{11}
15: R = $SiMe_3$

16

Chart 3.4

Heteroatom-containing aliphatic disubstituted acetylenes with polymerizability include silicon- and halogen-containing monomers. 1-Trimethylsilyl-1-propyne (**15**), which is a sterically crowded Si-containing acetylene, polymerizes with Nb and Ta catalysts but not with Mo or W catalysts.[31] The molecular weight of the polymer obtained using $TaCl_5$–Ph_3Bi reaches four million, which is among the highest for all the substituted polyacetylenes.[32] 1-(Trimethylgermyl)-1-propyne polymerizes in a similar way to monomer **15**. As chlorine-containing acetylenes, 1-chloro-1-alkynes (*e.g.*, 1-chloro-1-octyne **16**) are known to polymerize with Mo-based catalysts.[33] Both $MoCl_5$–*n*-Bu_4Sn and $Mo(CO)_6$–CCl_4–*hν* catalysts are useful for the polymerization of such monomers as 1-chloro-2-phenylacetylene (**13**).

3.1.5 Functions of Substituted Polyacetylenes

Recent examples of functions of substituted polyacetylenes are introduced here. Poly(phenylacetylene) and its ring-substituted derivatives have often been examined with respect to their functions (Chart 3.5). Tang and coworkers reported a soluble poly(phenylacetylene)/CdS hybrid **17** that was solution cast to give a thin membrane that demonstrated excellent photoconductivity.[34] A ferrocene-containing polymer was shown to be able to wrap around carbon nanotubes (CNTs) to give a processable nanohybrid **18**, which could be ceramized into a magnetically susceptible material.[35] Advincula and coworkers synthesized double conjugated polymer **19**, which is expected to act as a crosslinked LED material.[36] Poly(phenylacetylene) containing many TEMPO radicals **20** has been prepared, which is useful as a positive electrode material for organic radical batteries.[37]

The functions of polymers formed from disubstituted acetylenes have also been studied extensively (Chart 3.6). Kwak *et al.* found that poly(diphenylacetylene) derivative **21** showed strong blue–green emission.[38] Hsu and coworkers fabricated an LED composed of ITO/PEDOT/diphenylacetylene copolymer **22**/Ca/Al and observed very high device performance with a maximum luminescence of 4230 cd/m^2 at 14 V and a maximum current efficiency of 3.37 cd/A at 7 V.[39] Imidazole-containing polymer **23** can be prepared from a chlorine-bearing polymer. The fluorescence of **23** is quenched by a trace amount of Cu^{2+}, but not when also

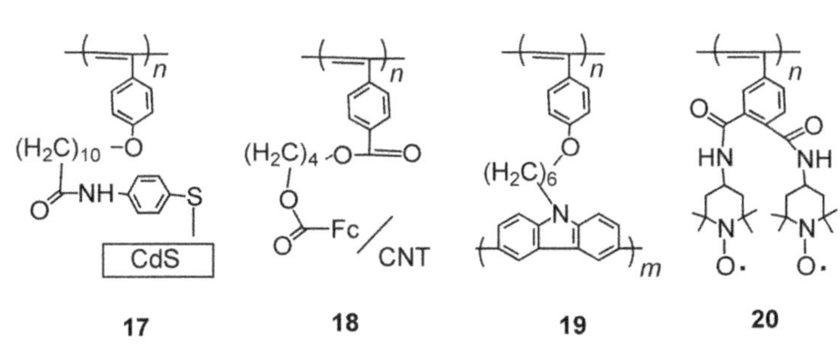

Chart 3.5

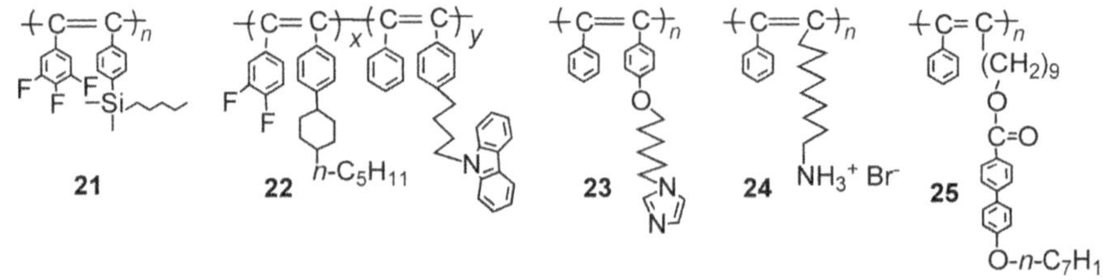

Chart 3.6

in the presence of a small quantity of amino acids. Thus, polymer **23** is effective as a sensor of Cu^{2+} and amino acids.[40] Tang and coworkers clarified that a functional perovskite nanohybrid prepared from **24**, which possesses ammonium bromide moieties and $PbBr_2$, shows a higher photoconductivity than its parent polymer **24**.[41] Polymer **25** bearing mesogenic pendants with a biphenyl core displays liquid crystallinity featuring a smectic A phase in a temperature range of 158–172 °C.[42]

3.2 Experimental Procedures

3.2.1 Materials and General Polymerization Procedures

Many monomers with simple structures, including phenylacetylene, *t*-butylacetylene, 1-phenyl-1-propyne, 2-octyne, and 1-trimethylsilyl-1-propyne, are commercially available. These monomers are usually purified by distillation in the presence of suitable drying agents prior to use. On the other hand, monomers that are more complex, such as *ortho*-substituted phenylacetylenes, *N*-propargylcarbamates, ring-substituted diphenylacetylenes, and 1-chloro-1-alkynes, must be synthesized. Derivatization of simple alkynes rather than formation of the acetylenic moiety, is frequently applied to synthesize such monomers. These are then purified by vacuum distillation or column chromatography.

Components of classic metathesis catalysts, *i.e.* group 5 and 6 transition-metal chlorides and organometallic cocatalysts, are more or less sensitive to moisture and air, and so should be handled in a dry, inert gas atmosphere. Although a few Schrock carbenes are commercially available, they are very sensitive to moisture and air, and so must always be handled under a strictly dried inert gas. Rh catalysts are relatively stable to air in the solid state but decompose readily in solution.

Toluene is a popular solvent for carrying out polymerization using metathesis catalysts, because it dissolves both catalysts and formed polymers well and does not inactivate the catalysts. Toluene is usually used after distillation from calcium hydride or sodium metal. On the other hand, protic solvents like ethanol can also be used in the case of Rh catalysts, which are insensitive to polar groups.

In most cases, the polymerization of substituted acetylenes is performed using a flask equipped with a three-way stopcock under a dry, inert gas (nitrogen or argon) atmosphere. The reaction flask should be dried and charged with the inert gas before use. Solid materials may be handled in a drybox, and solutions transferred with syringes. In more sensitive cases where stricter handling is needed, a vacuum line is employed.

To quench a polymerization reaction catalyzed by a metathesis catalyst, addition of a methanol/toluene mixture (*e.g.*, 1:4 v/v) is effective. Pouring the polymerization mixture into a large volume of methanol is useful to precipitate the formed polymer and to remove oligomers and catalyst residues that are soluble. In the case of Rh catalysts, acetic acid is added to quench the polymerization, giving rise to the cleavage of the metal–polymer chain bond.

Many of the following examples of polymerization provide conventional procedures for the synthesis of substituted polyacetylenes. If more information is required, the reader is referred to the original papers cited. Living polymerization has also progressed significantly for substituted acetylenes, which enables the control of molecular weight and its distribution, and the construction of various polymer structures. Thus, several examples of living polymerization are given in Examples 3, 4, 10, and 25.

3.2.2 Polymers of Aromatic Monosubstituted Acetylenes

Example 1: Polymerization of Phenylacetylene with $[(nbd)RhCl]_2$–Et_3N Catalyst

A toluene solution (2.0 mL) of phenylacetylene (1.0 mmol, 102 mg, 110 μL) is added to a toluene solution (3.0 mL) of $[(nbd)RhCl]_2$ (10 μmol, 4.61 mg) and the reaction solution is kept at 30 °C for 24 h. The polymers formed are isolated by precipitation in a large volume of methanol, filtered, and dried under vacuum to constant weight.[11] Detailed results are shown in Table 3.1. Triethylamine can also be used as the solvent for the polymerization.[12]

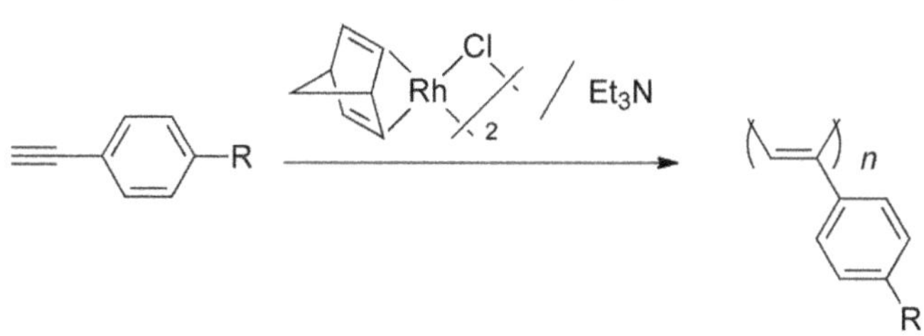

(3.2)

Table 3.1 Polymerization of phenylacetylenes catalyzed by $[(nbd)RhCl]_2$.[a]

para-substituent	*solvent*	*yield, %*	*molecular weight*	*cis %*	*ref.*
H	toluene[a]	100	255 000[c]	92	11
H	Et_3N[b]	83	40 000[d]	100	12
OMe	Et_3N[b]	67	15 000[d]	100	12
Cl	Et_3N[b]	96	260 000[d]	100	12

[a]Conditions: $[monomer]_0 = 0.20$ M, [catalyst] = 2.0 mM, $[Et_3N]$/[catalyst] = 10, 30 °C, 20 h.
[b]Conditions: $[monomer]_0 = 0.65$ M, [catalyst] = 5.0 mM, 30 °C, 20 h.
[c]Number-average molecular weight.
[d]Weight-average molecular weight.

Example 2: Polymerization of Phenylacetylene with (nbd)$Rh^+[(\eta^6\text{-}Ph)B^-Ph_3]$

The reaction is carried out under an argon atmosphere in a prebaked Schlenk tube. The polymerization is initiated by adding a monomer solution (0.375 M, 4 mL) to Rh complex solution (30 mM, 1 mL). The reaction is subsequently quenched by adding triphenylphosphine (0.18 mmol, 47 mg), and the mixture is poured into methanol (100 mL). Methanol-insoluble material is collected by filtration, washed with methanol, and then dried under vacuum at room temperature for several hours. Phenylacetylene derivatives bearing alkyl, alkoxy, and ester groups as ring substituents are also polymerizable with the same catalyst under similar reaction conditions (see Table 3.2 for the results).[13]

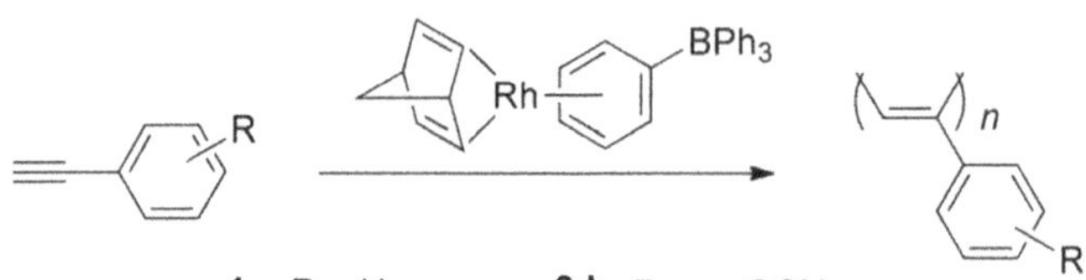

(3.3)

1: R = H
2a: R = *o*-CH_3
2b: R = *o*-OCH_3
2c: R = *o*-CF_3
2d: R = *m*-OCH_3
2e: R = *m*-$COOCH_3$
2f: R = *p*-OCH_3
2g: R = *p*-$COOCH_3$

Example 3: Living Polymerization of Phenylacetylene with [(nbd)Rh(CPh=CPh_2)(PPh_3)]/Ph_3P

Polymerization is initiated by adding a monomer solution (2 mL) to a solution of the vinyl catalyst and phosphine ligand (3 mL) under dry nitrogen (Scheme 3.4), and quenched with a drop of acetic acid. The polymer obtained is precipitated in 50 mL of methanol, filtered, and dried to constant weight. The smallest M_w/M_n value is 1.05 with M_n of 25 000 for the formed polymer, which is accomplished under the following conditions: in toluene, 30 °C, 1 h, [phenylacetylene]$_0$ = 0.50 M, [catalyst] = 2.0 mM, [$(4\text{-}FC_6H_4)_3P$] = 10.0 mM.[15]

Table 3.2 Polymerization of phenylacetylenes catalyzed by [(nbd)Rh(η^6-C_6H_5)B(C_6H_5)$_3$].[a]

Monomer	*time, min*	*conv, %*	*yield, %*	$M_n/10^4$	M_w/M_n
1	270	97	90	3.8	3.02
2a	30	49	35	insoluble	
2b	30	100	100	insoluble	
2c	30	53	26	insoluble	
2d	1	100	100	12.4	2.77
2e	1	99	100	13.3	2.66
2f	1	100	100	15.7	1.79
2g	1	100	100	21.8	2.58

[a]Conditions: [monomer]$_0$ = 300 mM, [catalyst] = 6 mM, in THF at 19–21 °C.
[b]Determined by GPC using polystyrene standards.

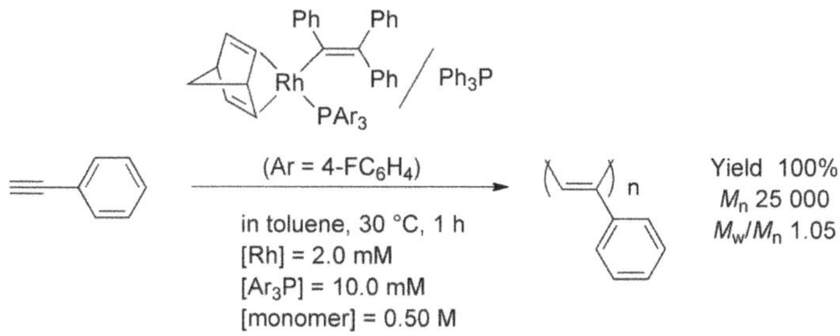

(3.4)

Example 4: Synthesis of High Molecular Weight Living Poly(phenylacetylene) with [(tfb)Rh(CPh=CPh$_2$)(PPh$_3$)]/Ph$_3$P

Under a nitrogen atmosphere, a solution of well-defined vinylrhodium complex [(tfb)Rh-{-C(Ph)=CPh$_2$}(Ph$_3$P)] (0.625 μmol, 0.53 mg; tfb: tetrafluorobenzobarrelene) and Ph$_3$P (3.13 μmol, 0.82 mg) in toluene (3.0 mL) is prepared. A toluene solution (2.0 mL) of phenylacetylene (2.5 mmol, 255 mg, 274 μL) is added to this catalyst solution, and the polymerization is carried out at 30 °C for 1 h. The formed polymer is isolated by precipitation in a large volume of methanol, filtered with sintered glass, and dried under vacuum to constant weight. Polymer yield = 100%, M_n = 401 000, M_w/M_n = 1.12, [P*]/[Rh] = 0.98.[43]

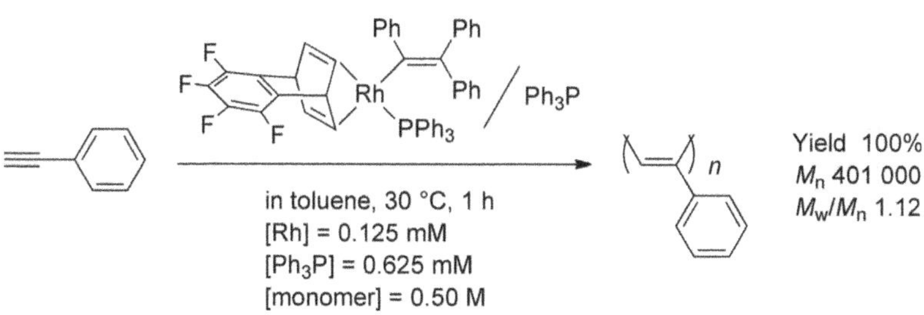

(3.5)

Example 5: Synthesis of Helical Poly(phenylacetylene)s with [(nbd)Rh(CPh=CPh$_2$)(PPh$_3$)]/Ph$_3$P

A portion (1 mL) of a solution of [(nbd)Rh(C(Ph)=CPh$_2$)(PPh$_3$)] (5.3 mM) and PPh$_3$ (27 mM) in DMF/dichloromethane (DCM) (3:1 v/v) is added to a solution of monomer **2h** (127 mg, 0.40 mmol) in DMF/DCM (1 mL, 3:1 v/v) in a glass tube equipped with a three-way stopcock under argon. After 24 h at 30 °C, the polymerization mixture is poured into methanol to afford the formed polymer as precipitate, which is filtered and finally dried under vacuum to give high molecular weight poly(**2h**) (M_n 128 800) in good yield (66%). In the case of monomer **2i**, the highest molecular weight poly(**2i**) (M_n 70 300, yield 82%) is obtained by polymerization under the following conditions: [**2i**] = 0.20 M, [**2i**]/[catalyst] = 150, and [Ph$_3$P]/[catalyst] = 5. The Kuhn dissymmetry factor, g ($\Delta\varepsilon/\varepsilon$), which gives quantitative information on the degree of preferential screw sense, takes constant large values when the polymer molecular weight exceeds

2×10^4. This means that the present polymers assume stable helical structures when their M_n is above 2×10^4.[44]

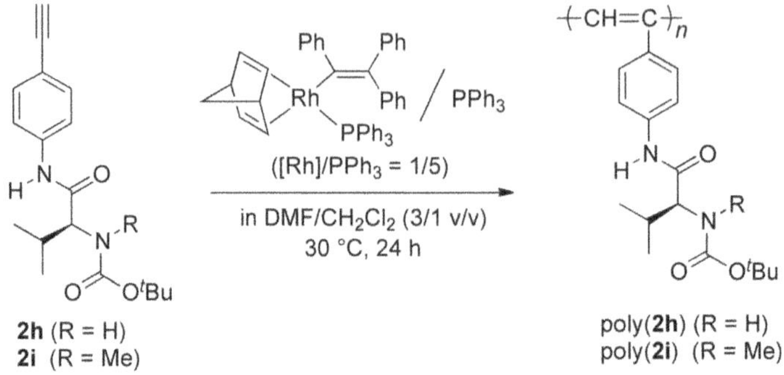

(3.6)

Example 6: Synthesis of End-Functionalized Poly(phenylacetylene) with Cationic Pd(II) Catalysts

To a solution of Pd complex (0.020 mmol, 16.9 mg) in CH_2Cl_2/CH_3CN (3:1 v/v, 2.85 mL), 2.85 mL of an AgOTf solution in CH_2Cl_2/CH_3CN (3:1 v/v, 8.4 mM; AgOTf, 0.024 mmol) is added to form a cationic Pd derivative with AgBr as a white precipitate. After 5 min at room temperature, the silver salt is removed by filtration. Ten minutes after the addition of the AgOTf solution, phenylacetylene (1.0 mmol, 102 mg, 0.110 mL) is added to the yellow filtrate to start the polymerization. The reaction mixture is kept at 30 °C for 24 h for the polymerization to proceed. Subsequently, all the volatiles are evaporated under reduced pressure to quench the polymerization. The residue is dissolved in DCM (0.3 mL), and then, the mixture is poured into a large volume of methanol to isolate the produced polymer. Polymer yield = 94%, $M_n = 3700$. High molecular weight polymer is formed ($M_n = 26\ 000$, yield 69%), at [phenylacetylene]/[Pd catalyst] = 500. It is confirmed by mass and IR spectra that the polymer carries a *p*-cyanophenyl group at its initiating chain end. Apart from this group, a variety of functional groups, such as methyl, *o*-tolyl, 2-mesityl, triphenylethynyl, *o*-hydroxymethylphenyl, and *p*-nitrophenyl, can be introduced to the initiating chain end by employing the corresponding Pd catalysts. The hydroxy-group-bearing polymer can be used as a macroinitiator to synthesize a block copolymer containing the β-propiolactone block.[45]

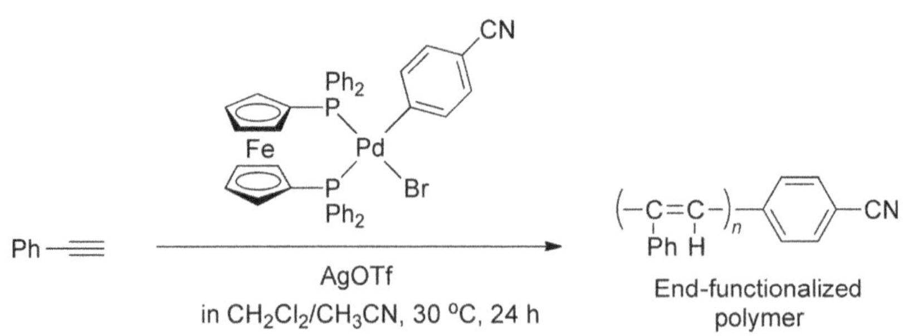

(3.7)

Example 7: Polymerization of Phenylacetylene with a WCl_6-Based Catalyst

A Schlenk tube equipped with a three-way stopcock is baked and dried under nitrogen and then charged with WCl_6 (0.10 mmol, 39.7 mg) and toluene (5.0 mL) at 30 °C. A solution of phenylacetylene (10 mmol, 1.02 g, 1.10 mL) and tetralin (0.5 mL; for GC) in toluene (3.4 mL) is then added to this catalyst solution at 30 °C. Polymerization is quenched after 1 h by adding methanol (1 mL). After determining the monomer consumption by GC (Silicone DC 550, 2 m, 100 °C), the polymerization mixture is poured into methanol (500 mL) to precipitate the formed polymer, which is then separated by filtration and dried. Monomer conversion = 90%, polymer yield = 80%, and M_n = 12 000 (vapor pressure osmometry, VPO). It appears as a *trans*-rich auburn powdery polymer, which is gradually oxidized and degraded. $MoCl_5$ provides a cis-rich yellow poly(phenylacetylene) (M_n = 7000).[16,17]

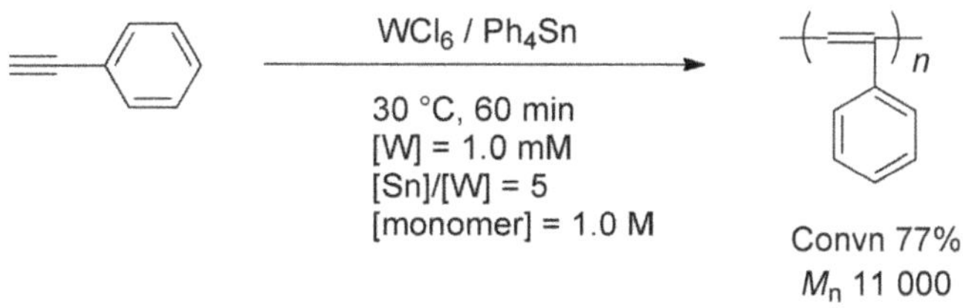

(3.8)

Example 8: Polymerization of Phenylacetylene with a $W(CO)_6$-Based Catalyst

The catalyst solutions and polymerization solutions are kept in the dark whenever UV irradiation is not applied. A general polymerization procedure is as follows: A CCl_4 solution (5 mL) of $W(CO)_6$ (0.10 mmol) is irradiated with UV light (200 W) at 30 °C for 1 h. A solution of phenylacetylene (10 mmol, 1.02 g 1.10 mL) is then added with chlorobenzene (0.50 mL, as internal standard for GC analysis) in CCl_4 (3.4 mL). The mixture is subsequently kept in the dark at a constant temperature for a given time. Polymerization is quenched with methanol, and the polymer formed is precipitated in methanol, separated by filtration off, washed with methanol, and dried to constant weight.[18]

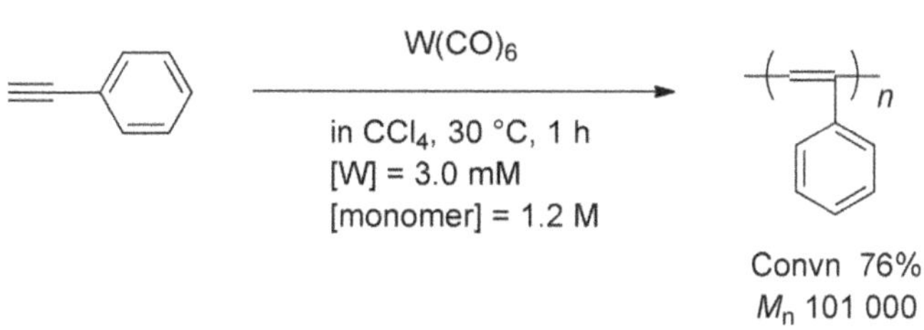

(3.9)

Example 9: Polymerization of *o*-(Trimethylsilyl)phenylacetylene with a WCl_6-Based Catalyst

A Schlenk tube equipped with a three-way stopcock is baked and dried under nitrogen and then charged with the monomer (10.5 mmol, 1.83 g, 2.07 mL), bromobenzene (for GC, 0.50 mL), and toluene (1.63 mL). In another Schlenk tube, WCl_6 (0.10 mmol, 38.6 mg), Et_3SiH (0.10 mmol, 11.6 mg, 15.9 μL), and toluene (6.0 mL) are mixed and the solution is aged at 30 °C for 15 min. Polymerization is initiated by the addition of the monomer solution (4 mL) to the catalyst solution at 30 °C. Polymerization is then quenched after 24 h by adding a toluene/methanol mixture (4:1 v/v; 2 mL). The polymerization mixture is diluted with toluene (30 mL), and the monomer conversion is determined by GC (PEG 6000, 3 m, 145 °C). The polymerization mixture is further diluted with toluene (120 mL), and then poured into methanol (2 L) under stirring to precipitate the formed polymer. The polymer is separated by filtration and dried to constant weight. Monomer conversion = 100%, polymer yield = 100%, $M_w = 1.6 \times 10^6$, $M_n = 67 \times 10^4$. The polymer appears as a purple fibrous solid, which is stable in air at room temperature for a long period.[46] o-(Trifluoromethyl)phenylacetylene polymerizes in a similar way.[47]

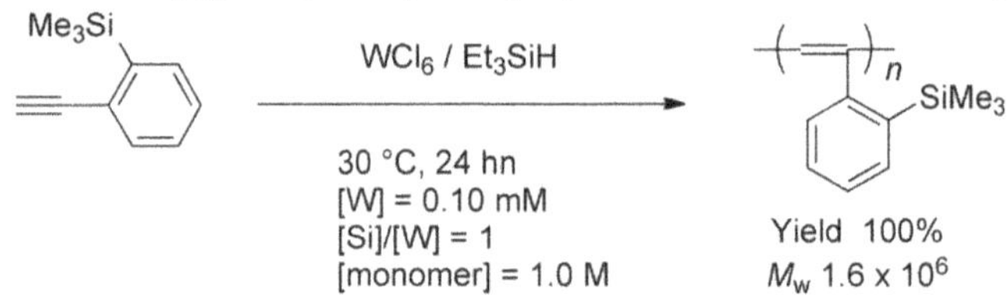

(3.10)

Example 10: Living Polymerization of *o*-(Trimethylsilyl)phenylacetylene with a Mo Carbene Catalyst

Stock toluene solutions are prepared for both the catalyst (13.9 mM, Scheme 3.11) and the monomer, *o*-(trimethylsilyl)phenylacetylene (0.23 M). The polymerization reaction is carried out under nitrogen in a dry box with quick addition of the monomer solution to a vigorously stirred catalyst solution at room temperature. In a typical experiment, the monomer solution (0.24 mL; monomer 55.2 μmol) is quickly injected into a toluene solution of the catalyst (50 μL bulk solution diluted to 3 mL). These conditions correspond to [M]/[cat] = 80. The solution turns purple within 30 min, and the color does not change significantly thereafter. After 24 h, the reaction is brought into the air, and methanol is quickly added to induce the precipitation of polymer as a dark purple solid. The polymer is collected by filtration, washed with methanol, and dried in vacuo (yield = 9.0 mg, 93%; M_n = 11 650, $M_w/M_n = 1.05$). The molecular weight of the polymer increases in proportion to the monomer/catalyst ratio ([M]/[cat]): *e.g.*, $M_n = 6930$ and $M_w/M_n = 1.06$ at [M]/[cat] = 50; and M_n = 13 900 and $M_w/M_n = 1.04$ at [M]/[cat] = 100.[19]

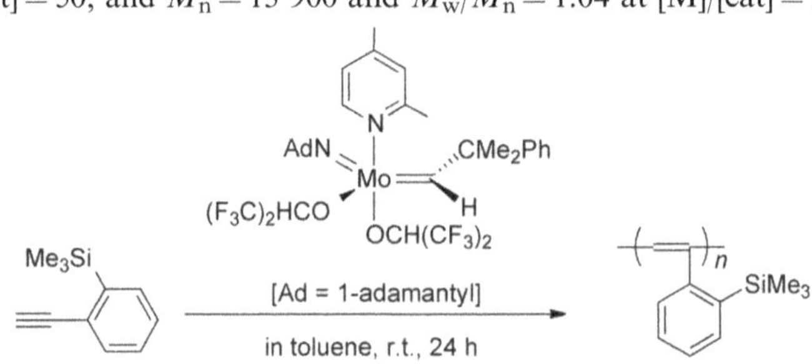

(3.11)

3.2.3 Polymers of Aliphatic Monosubstituted Acetylenes

Example 11: Polymerization of *t*-Butylacetylene with a $MoCl_5$ Catalyst

A Schlenk tube equipped with a three-way stopcock is baked and dried under nitrogen, and then charged with $MoCl_5$ (0.10 mmol, 27.3 mg), followed by the addition of toluene (5.0 mL) to dissolve the catalyst with stirring at 30 °C. A toluene solution (5 mL) of *t*-butylacetylene (10 mmol, 0.822 g, 1.23 mL) is then added at 0 °C. Polymerization is quenched after 24 h by adding a toluene/methanol mixture (4:1 v/v; 2 mL). The polymerization mixture is subsequently diluted with toluene (40 mL), and then poured into methanol (1 L) to precipitate the formed polymer. The polymer is separated by filtration and dried to constant weight; appearing as a white powder that is stable in air at room temperature for an extended period. Polymer yield = 96%, $M_n = 38 \times 10^4$.[20]

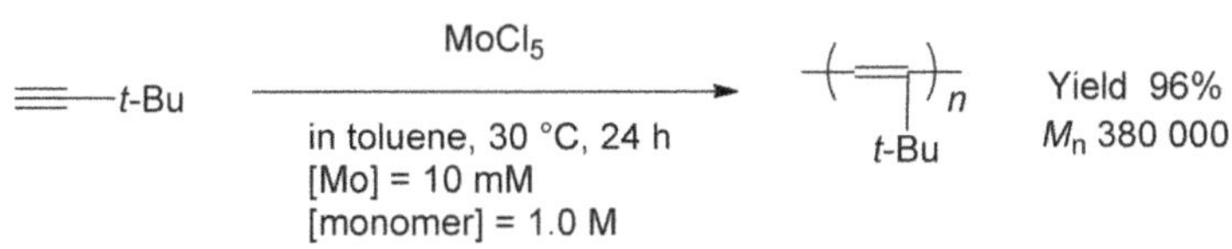

(3.12)

Example 12: Polymerization of *N*-Alkynylamides with $(nbd)Rh^+[(\eta^6\text{-Ph})B^-Ph_3]$

A solution of monomer in $CHCl_3$ ($[M]_0 = 1.0$ M) is added to a solution of $(nbd)Rh^+[(\eta^6\text{-Ph})B^-Ph_3]$ ($[M]_0/[Rh^+] = 100$) in $CHCl_3$ under nitrogen, and the solution is kept at 30 °C for 24 h. The reaction mixture is poured into a large volume of hexane to precipitate the resulting polymer, which is then separated by filtration and dried under reduced pressure (see Scheme 3.13 and Table 3.3 for the results).[21]

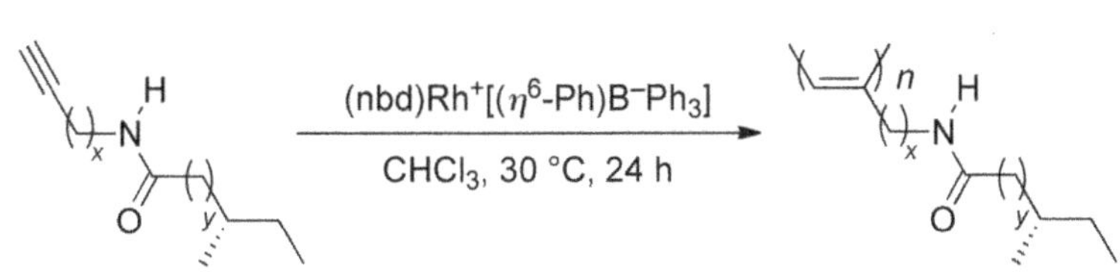

$x = 1–3$, $y = 0–3$

(3.13)

Example 13: Polymerization of *N*-Propargylcarbamates with $[(nbd)RhCl]_2/Et_3N$

A solution of $[(nbd)RhCl]_2$ and Et_3N in $CHCl_3$ is added to a solution of monomer ($[M]_0 = 0.50$ M) in $CHCl_3$ under nitrogen, and the resulting mixture is kept at 30 °C for 1 h. The mixture is poured into a large volume of methanol to precipitate the polymer, which is then separated by filtration and dried under reduced pressure (Scheme 3.14 and Table 3.4).[22]

Table 3.3 Polymerization of *N*-alkynylamides.[a]

x	*y*	*yield*,[b] %	M_n [c]	M_w/M_n [c]	*cis* % [d]	$[\alpha]_D$ [e]
1	0	99	16 000	2.55	100	+1 610
1	1	98	21 000	1.99	79	+876
1	2	67	13 000	1.71	100	+320
1	3	73	11 000	1.85	96	−13
2	0	79	–[f]	–[f]	–[f]	–[f]
2	1	85	14 000	1.53	82	+268
3	0	98	11 000	3.46	90	+2

[a]Polymerized with $(nbd)Rh^+[(C_6H_5)B^-(C_6H_5)_3]$ in $CHCl_3$ at 30 °C for 24 h. $[monomer]_0 = 1.0$ M, [catalyst] = 10 mM.
[b]Hexane-insoluble part.
[c]Estimated by GPC (eluent $CHCl_3$, polystyrene calibration).
[d]Determined by ^1H NMR.
[e]Determined by polarimetry in $CHCl_3$ ($c = 0.051$–0.083 g/dL) at room temperature.
[f]Not determined because the polymer is insoluble in $CHCl_3$.

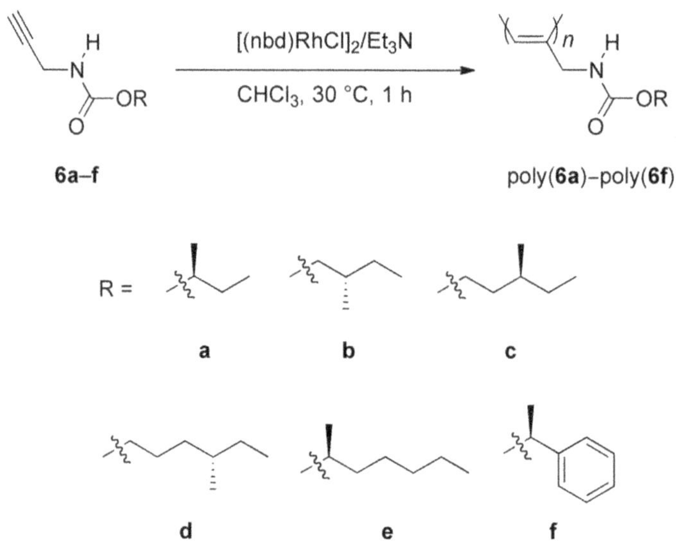

(3.14)

Example 14: Polymerization of 1-Methylpropargyl Alcohols and 1-Methylpropargyl Hexanoates with $(nbd)Rh^+[(\eta^6\text{-}Ph)B^-Ph_3]$

A dimethylformamide (DMF) or tetrahydrofuran (THF) solution of monomer ($[M]_0 = 3.0$ or 1.0 M) is added to a DMF or THF solution of $(nbd)Rh^+[(\eta^6\text{-}Ph)B^-Ph_3]$ ($[M]_0/[Rh^+] = 100$) under nitrogen, and the solution is kept at 30 °C for 24 h. The reaction mixture is poured into a large volume of diethyl ether or methanol to precipitate the polymer, which is then separated by filtration and dried under reduced pressure (Scheme 3.14).[23,48] See Table 3.5 for the polymerization results. A macromonomer bearing the chiral 1-methylpropargyl ester group also provides a helical polymer.[49] It is possible to introduce photoluminescent and redox active groups by esterification of

Table 3.4 Polymerization of *N*-propargylcarbamates.[a]

monomer	*yield,*[b] *%*	M_n [c]	M_w/M_n [c]	*cis %* [d]	$[\alpha]_D$ [e]
6a	95	–[f]	–[f]	–[f]	–[f]
6b	82	52 000	17.2	97	+777
6c	38	46 000	1.8	85	–230
6d	59	35 000	2.0	83	+231
6e	91	24 000	6.1	100	–711
6f	97	30 000	5.5	96	–553

[a]Conditions: $[monomer]_0 = 0.50$ M, $[(nbd)RhCl]_2] = 5.0$ mM, $[Et_3N] = 2.5$ mM in $CHCl_3$, 30 °C, 1 h.
[b]MeOH-insoluble part.
[c]Estimated by GPC (eluent $CHCl_3$, polystyrene calibration).
[d]Determined by ^1H NMR.
[e]Determined by polarimetry in $CHCl_3$ ($c = 0.44$–0.45 g/dL) at room temperature.
[f]Not determined because the polymer is insoluble in $CHCl_3$.

Table 3.5 Polymerization of 1-methylpropargyl alcohols and 1-methylpropargyl hexanoates.[a]

					$[\alpha]_D$ [d]	
monomer	*solvent*	*yield,*[b] *%*	M_n [c]	M_w/M_n [c]	*MeOH*	*toluene*
(*S*)-**7**	DMF	79	19 000	2.03	+1187	–[e]
(*R*)-**7**	DMF	77	21 000	1.70	– 1214	–[e]
(*S*)-**8a**	THF	88	34 000	3.38	–[e]	+840
(*R*)-**8a**	THF	79	32 000	3.16	–[e]	−873

[a]Polymerized with $(nbd)Rh^+[(C_6H_5)B^-(C_6H_5)_3]$, $[monomer]_0$ in DMF (3.0 M) or THF (1.0 M), $[monomer]_0/[Rh] = 100$, 30 °C, 24 h.
[b]Insoluble part in Et_2O {poly[(*S*)-**7**] and poly[(*R*)-**7**]} or MeOH {poly[(*S*)-**8a**] and poly[(*R*)-**8a**]}.
[c]Determined by GPC eluted with DMF {poly[(*S*)-**7**] and poly[(*R*)-**7**]} and $CHCl_3$ {poly[(*S*)-**8a**] and poly[(*R*)-**8a**]}.
[d]Determined by polarimetry at room temperature, $c = 0.063$–0.604 g/dL.
[e]Not determined because the polymer was insoluble in the solvent.

1-methylpropargyl alcohols, leading to the preparation of photofunctional and electrically functional helical polymers.[50,51]

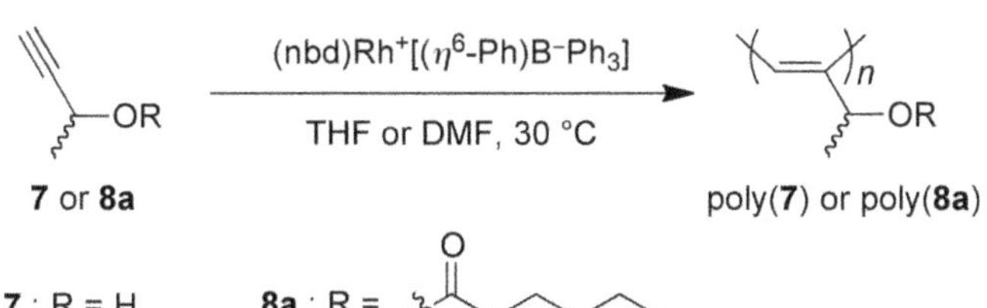

(3.15)

3.2.4 Polymers of Aromatic Disubstituted Acetylenes

Example 15: Polymerization of Diphenylacetylene with a $TaCl_5$-Based Catalyst

Polymerization of diphenylacetylene is carried out under dry nitrogen. Polymerization conditions are as follows: in toluene, 80 °C, 24 h,

$[monomer]_0 = 0.50$ M, $[TaCl_5] = [n\text{-}Bu_4Sn] = 20$ mM. The mixed catalyst is used after aging in toluene solution at 80 °C for 15 min. The monomer conversion is measured by GC to be 87% and the yield of methanol-insoluble polymer, as determined by gravimetry, is 77%. The polymer appears as a yellow solid that is insoluble in all solvents.[24]

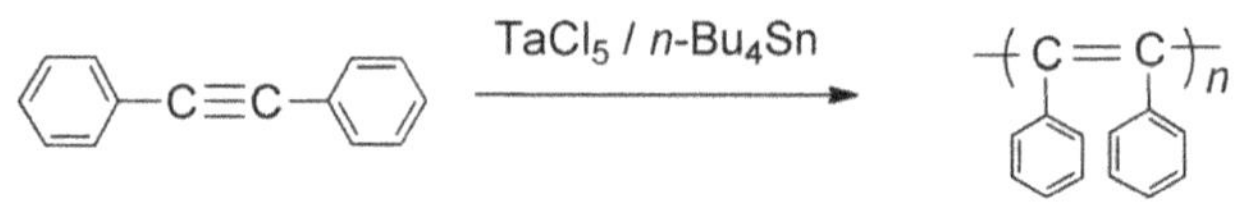

(3.16)

Example 16: Polymerization of 1-Phenyl-2-[(*p*-trimethylsilyl)phenyl]acetylene with a $TaCl_5$-Based Catalyst

The manipulation of reagents and the polymerization reaction of 1-phenyl-2-[(*p*-trimethylsilyl)phenyl]acetylene are carried out under dry nitrogen. A monomer solution (2.4 mL) is prepared in a Schlenk tube by mixing the monomer (3.0 mmol, 0.75 g, 0.71 mL), eicosane (0.50 mL; internal standard of GC), and toluene (1.19 mL). Another Schlenk tube is charged with $TaCl_5$ (0.10 mmol, 35.8 mg), *n*-Bu_4Sn (0.20 mmol, 69.4 mg, 65.6 μL), and toluene (3.0 mL) and the solution is aged at 80 °C for 15 min. Following this, monomer solution (2.0 mL) is added to the catalyst solution. Polymerization is carried out at 80 °C for 24 h, and then quenched with a mixture of toluene and methanol (4:1 v/v; 1 mL). The monomer conversion, as determined by GC, (Silicon DC 550, 0.5 m, 180 °C) is 95%. The polymerization mixture is diluted with toluene (300 mL) and poured into methanol (4 L) under stirring to precipitate the formed polymer, which is separated by filtration and dried. The polymer yield, as determined by gravimetry, is 85%. $M_w = 2.2 \times 10^6$, $M_n = 0.75 \times 10^6$, as measured using gel permeation chromatography (GPC). The polymer appears as a yellow solid and is highly stable in air at room temperature, with the onset temperature of weight loss measured by TGA in air reaching up to 400 °C.[25]

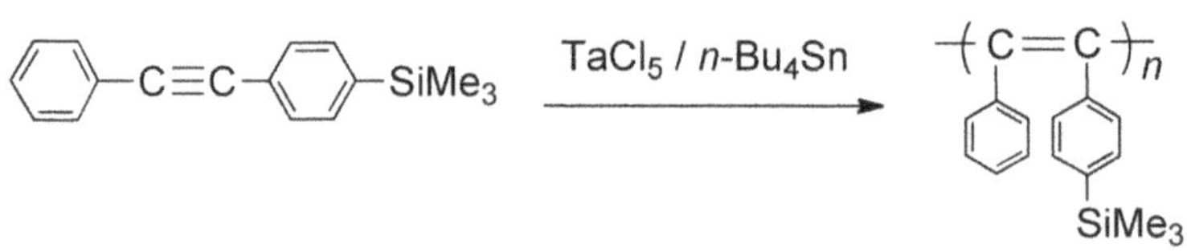

(3.17)

Example 17: Syntheses of Poly[1-phenyl-2-(*p*-*t*-butyldimethylsiloxyphenyl)-acetylene] and Poly[1-phenyl-2-(*p*-hydroxyphenyl)acetylene]

The polymerization of 1-phenyl-2-(*p*-*t*-butyldimethylsiloxyphenyl)acetylene is accomplished using a 1:2 mixture of $TaCl_5$ and *n*-Bu_4Sn as catalyst in toluene solution at 80 °C. The polymerization conditions are as follows: 24 h, $[TaCl_5] = 20$ mM, $[n\text{-}Bu_4Sn] = 40$ mM. The formed polymers are isolated by precipitation into a large volume of methanol, and the polymer yields are determined by gravimetry. The molecular weights of the polymers are estimated using GPC ($CHCl_3$ as eluent, polystyrene calibration).

Poly[1-phenyl-2-(*p*-*t*-butyldimethylsiloxyphenyl)acetylene] is solution cast to produce a membrane that is subsequently transformed into a durable membrane *via* desilylation using a mixture of trifluoroacetic acid (TFA)/water. A detailed method of desilylation of membranes is as follows: the membrane of polymer is immersed in a mixture of TFA and water (9:1 v/v) at room temperature for 24 h. To neutralize the remaining acid in the polymer matrix, the membrane is then immersed in aqueous sodium carbonate solution at room temperature for 24 h, and then dried in air at room temperature. Finally, the membrane is immersed in hexane for 24 h to remove residual impurities and dried to constant weight at room temperature for 24 h. Complete desilylation is confirmed by IR spectroscopy (no absorptions are seen at 1250 (d_s, Si–CH), 855 (ν_{as}, Si–CH_3), and 812 cm^{-1} (ν_{as}, Si–CH_3), which are observed in the siloxy polymer). Desilylation is also confirmed by measuring the weight of the membrane before and after desilylation, thereby verifying loss of the silyl group (*e.g.*, mass of siloxy polymer = 53.6 mg, mass of hydroxy polymer = 33.5 mg (measured), 33.5 mg (calculated).[26]

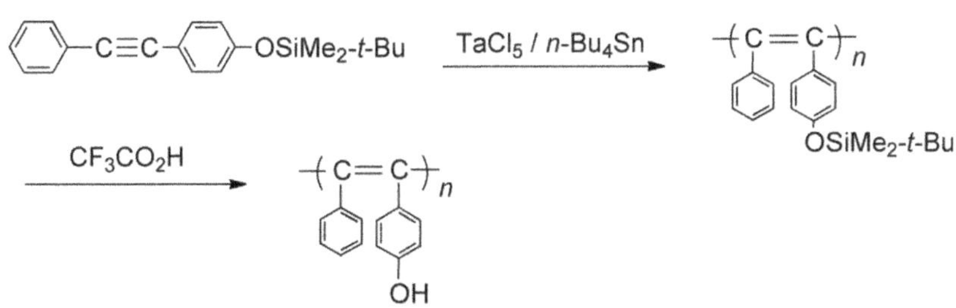

(3.18)

Example 18: Polymerization of 1-Phenyl-1-propyne with a $TaCl_5$-Based Catalyst

Polymerization is carried out under dry nitrogen. A monomer solution is prepared by adding 1-phenyl-1-propyne (12.0 mmol, 1.39 g, 1.50 mL) and bromobenzene (0.40 mL, as internal standard for GC) to toluene (4.10 mL). A catalyst solution is prepared by mixing $TaCl_5$, (0.20 mmol, 71.6 mg) and *n*-Bu_4Sn (0.20 mmol, 69.4 mg) in toluene (5.0 mL) and aged at 80 °C for 15 min. Following this, monomer solution (5 mL) is added to the catalyst solution at 80 °C. After 24 h, the reaction is quenched by adding a mixture of toluene (5 mL) and methanol (0.5 mL). The monomer conversion, as determined by measuring the initial and final monomer concentrations using GC, (silicone DC 1 m, 140 °C) is 100%. The polymerization solution is diluted with toluene (100 mL) and poured into acetone (1 L). In this case, methanol, which is often used as a polymer precipitant, cannot be used because cyclotrimers of 1-phenyl-1-propyne (byproducts) also precipitate in this particular solvent. The polymer is separated by filtration, washed with acetone, and dried to constant weight under vacuum. The polymer appears as a white solid, which is stable in air at room temperature. The polymer yield, as determined by gravimetry, is 76%. The M_w, as determined by GPC, is 1.5×10^6.[27]

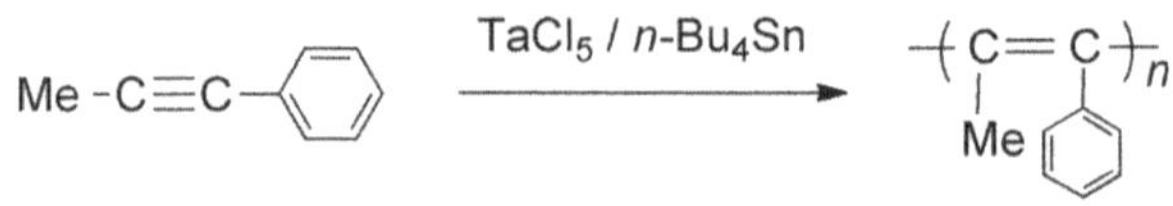

(3.19)

Example 19: Polymerization of 1-Chloro-2-phenylacetylene with a $Mo(CO)_6$-Based Catalyst

A solution of $Mo(CO)_6$ (0.10 mmol, 26 mg) in CCl_4 (5.0 mL) is irradiated with UV light at 30 °C for 1 h. To this solution is added, *via* a syringe, a mixture of 1-chloro-2-phenylacetylene (10 mmol, 1.4 g, 1.3 mL) and CCl_4 (3.7 mL), and polymerization is allowed to proceed in the dark at 30 °C for 24 h. Polymer yield = 80% and $M_w = 2.0 \times 10^6$.[28]

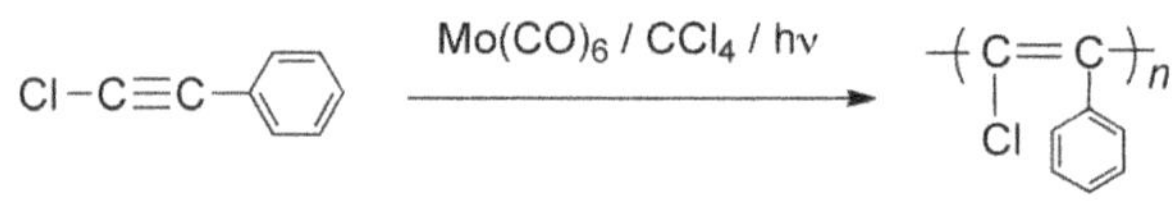

(3.20)

Example 20: Polymerization of 1-Chloro-2-phenylacetylene with a $MoCl_5$-Based Catalyst

The $MoCl_5$–*n*-Bu_4Sn catalyst is prepared by mixing the two components at a 1:1 molar ratio in toluene solution and subsequently aging the mixture at 30 °C for 15 min. Polymerization is carried out under nitrogen in toluene at $[\text{monomer}]_o = 1.0$ M, $[MoCl_5] = [n\text{-}Bu_4Sn] = 20$ mM, 30 °C for 24 h. Monomer conversion is determined by GC (PEG 6000, 3 m; 135 °C; bromobenzene as internal standard). The formed polymer is isolated by pouring the polymerization mixture into a large volume of methanol and subsequently filtering the polymer. Polymer yield = 90% and $M_w = 690 \times 10^3$.[29]

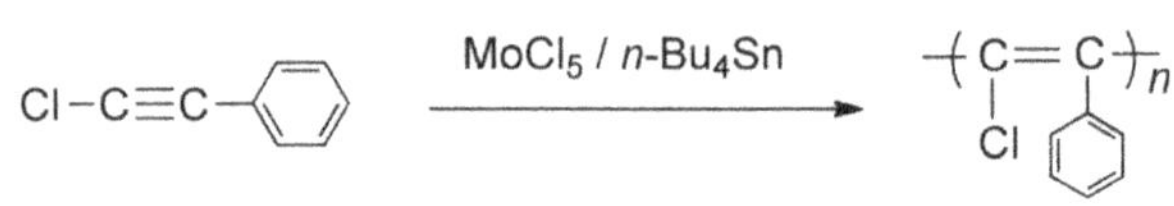

(3.21)

3.2.5 Polymers of Aliphatic Disubstituted Acetylenes

Example 21: Polymerization of 2-Octyne (16b) with a $MoCl_5$-Based Catalyst

A Schlenk tube equipped with a three-way stopcock is baked and dried under nitrogen, and then charged with $MoCl_5$ (0.30 mmol, 81.9 mg), Ph_4Sn (0.30 mmol, 128 mg), and toluene (8.6 mL). The mixture is aged at 30 °C for 30 min and then 2-octyne (10 mmol, 1.1 g, 1.5 mL) is added at 30 °C. Polymerization is terminated after 24 h by adding a toluene/methanol mixture (4:1 v/v; 2 mL). Following this, the polymerization mixture is diluted with toluene (100 mL), and poured into methanol (2 L) to precipitate the formed polymer, which is then separated by filtration and dried to constant weight. It appears as a white powder that slowly decomposes in air at room temperature. Polymer yield = 63% and $M_w = 1.5 \times 10^6$.[30]

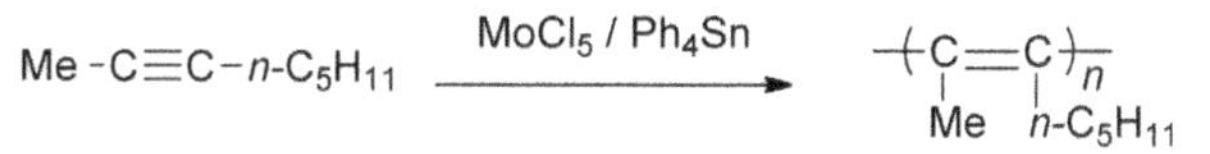

(3.22)

Example 22: Polymerization of 1-(Trimethylsilyl)-1-propyne with a $TaCl_5$ Catalyst

The polymerization reaction is carried out under dry nitrogen. A 10-mL flask equipped with a three-way stopcock is flushed with dry nitrogen, and charged with 1-(trimethylsilyl)-1-propyne (12 mmol, 1.34 g, 1.77 mL), chlorobenzene (0.43 mL; for GC), and toluene (3.8 mL) *via* syringes. This monomer solution is maintained at 80 °C until use. $TaCl_5$ (0.20 mmol, 72 mg) is placed in another 10-mL flask equipped with a three-way stopcock, the flask is flushed with dry nitrogen, toluene (5 mL) is added with a syringe, and the metal chloride is completely dissolved by keeping the flask at 80 °C for 15 min (the system is yellow and practically homogeneous). Following this, monomer solution (5 mL) is added with a syringe to the catalyst solution at 80 °C. Polymerization proceeds at 80 °C for 3 h, and the system gradually turns brown and solidifies over the period. The polymerization is quenched with a toluene/methanol mixture (4:1 v/v; 2 mL), and the system then becomes colorless. The monomer conversion, as determined by GC, (silicone DC550, 3 m, 90 °C) is 100%. The solution is then diluted with toluene (500 mL) and poured into methanol (5 L) under stirring to precipitate the polymer. After leaving the precipitate overnight, the polymer is separated by filtration, washed with methanol, and dried to constant weight. The polymer yield, as determined by gravimetry, is 100%. $M_w = 730\times10^3$, $M_n = 130\times10^3$ (GPC), and $[\eta]$ 5.43 dL/g (in toluene, 30 °C). The polymer appears as a white solid that is stable in air at room temperature. The most significant feature of this polymer is its gas permeability, which is higher than any other polymer; its oxygen permeability reaches approximately 10 000 barrers.[31]

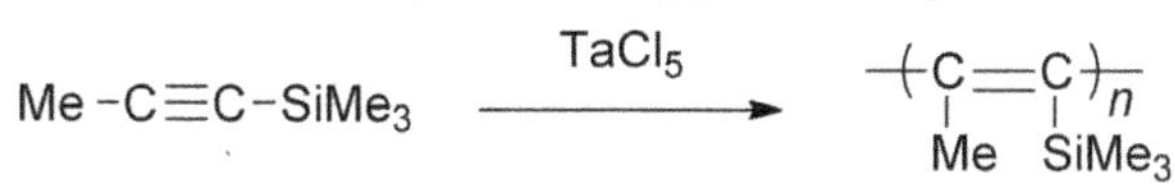

(3.23)

Example 23: Polymerization of 1-(Trimethylsilyl)-1-propyne with a $TaCl_5$–Ph_3Bi Catalyst

This example describes the synthesis of a particularly high molecular weight polymer. A catalyst solution is prepared by dissolving $TaCl_5$ (0.10 mmol, 36 mg) and triphenylbismuth (0.10 mmol, 44 mg) in toluene (5.0 mL), followed by aging at 80 °C for 15 min. Using this catalyst solution, polymerization is carried out in the same way as the case with the $TaCl_5$ catalyst alone. Polymer yield = 100% and $M_w = 4.0\times10^6$.[32]

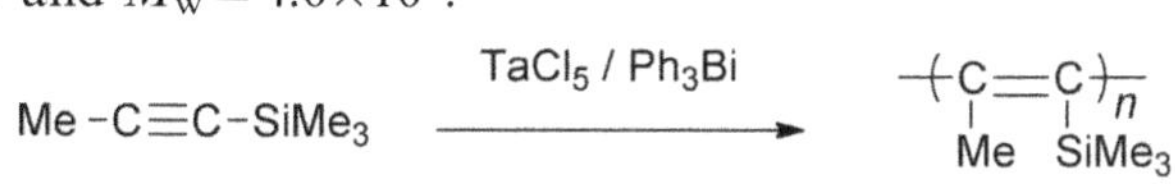

(3.24)

Example 24: Polymerization of 1-Chloro-1-octyne with a $MoCl_5$-Based Catalyst

The monomer solution is prepared by adding 1-chloro-1-octyne (6.0 mmol, 0.87 g, 0.95 mL), dodecane (0.35 mL; as internal standard of GC), and toluene

(4.7 mL) to a 10 mL flask. $MoCl_5$, (0.20 mmol, 55 mg) and n-Bu_4Sn (0.20 mmol, 69 mg, 66 µL) are dissolved in toluene (5.0 mL) in a 30-mL flask, and the solution is left at 30 °C for 15 min. Monomer solution (5 mL) is then added to the catalyst solution at 30 °C. After 24 h, polymerization is quenched by adding a mixture of toluene and methanol (4:1 v/v; 3 mL) and the monomer conversion is determined by GC (silicone DC 3 m, 150 °C). The reaction mixture is diluted with toluene (20 mL) and poured into methanol (500 mL) under stirring. The precipitated polymer is then separated by filtration and dried to constant weight. The polymer yield is determined by gravimetry to be 100%. $M_w = 870 \times 10^3$ and $M_n = 510 \times 10^3$ (GPC). The polymer appears as a white solid that is stable in air at room temperature.[33]

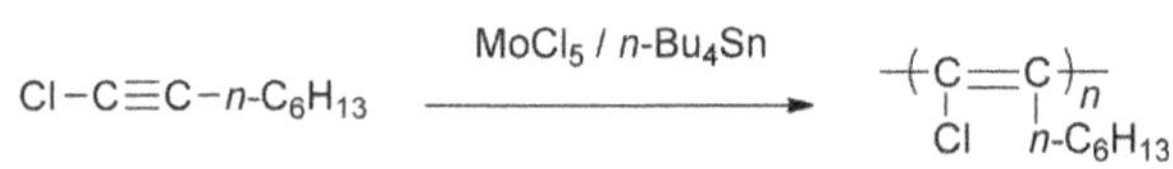

(3.25)

Example 25: Living Polymerization of 1-Chloro-1-octyne with a $MoOCl_4$-Based Catalyst

Living polymerization of 1-chloro-1-octyne can be achieved by using the $MoOCl_4$–n-Bu_4Sn–EtOH (1:1:1) catalyst. First, a glass tube is charged with 1-chloro-1-octyne (1.0 mmol, 145 mg, 0.159 mL), dodecane (0.050 mL, as internal standard for GC), and toluene (3.8 mL). The catalyst solution is prepared by adding $MoOCl_5$, (0.50 mmol, 127 mg), toluene (11.3 mL), and a toluene solution of n-Bu_4Sn (2.5 mL, 200 mM) to a separate glass tube, and the mixture is allowed to age at 30 °C for 15 min. A toluene solution of EtOH (1.25 mL, 200 mM) is then added to this solution, and the mixture is aged at 30 °C for a further 15 min. Polymerization is initiated by adding catalyst solution (6.0 mL) to the above monomer solution at 30 °C. The second, third, and fourth monomer feeds (1.0 mmol each) are injected in toluene (2 mL) at 5 min intervals. The polymerization is terminated with a mixture of methanol and toluene (1:4 v/v; 1 mL) 5 min after the final monomer addition. GC (silicone DC 3m, 150 °C) confirms complete reaction of the monomer. Further, it is confirmed by GC in control experiments that the monomer is completely consumed at each step. The polymerization mixture is diluted with toluene (30 mL) and poured into methanol (1 L) under stirring. The polymer precipitate is separated by filtration and dried to constant weight. The M_w/M_n remains at 1.15–1.25, and the polymer yield increases in proportion to the polymer yield. The initiator efficiency is fairly low at 0.025. The data of the first-stage polymerization are as follows: The monomer conversion is 100%, and the polymer yield is quantitative. $M_n = 14\,600$, $M_w/M_n = 1.16$ (GPC).[52]

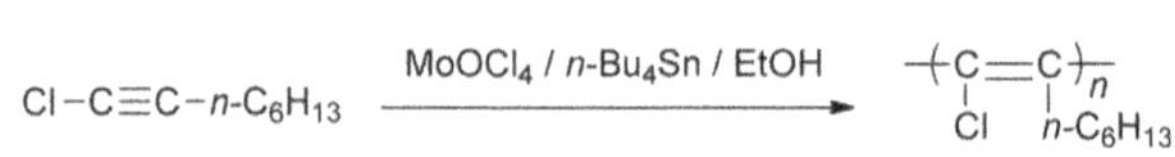

(3.26)

3.2.6 Characterization and Remarks

The absolute number-average molecular weight (M_n) of substituted polyacetylenes can be determined by vapor pressure and membrane osmometry

(VPO, MO), while the weight-average molecular weight (M_w) can be established using light scattering (LS). Each method has suitable molecular weight ranges: VPO $< 5\times10^4$, MO $= 5\times10^4$–5×10^5, LS $> 5\times10^4$. Since LS uses a He–Ne laser (633 nm), this technique is not suitable for the substituted polyacetylenes that absorb at this wavelength. GPC is a convenient method for determining M_n and M_w values simultaneously; however, the values are always relative as a polystyrene calibration is usually used. GPC is effective for checking whether a polymerization is living or not, because the polydispersity index (M_w/M_n) can be easily determined.

It has been demonstrated using NMR and IR that substituted polyacetylenes synthesized using transition-metal catalysts possess carbon–carbon alternating double and single bonds along the main chain, and do not contain triple bonds. The regioregular structure assumed for polymers derived from monosubstituted acetylenes like phenylacetylene and *t*-butylacetylene is reasonable from the viewpoint of steric hindrance. On the other hand, it is unexpected for poly[1-phenyl-2-[(*p*-trimethylsilyl)phenyl]acetylene] to have such a regioregular structure, although no information is available from NMR studies. The geometric structure of poly(phenylacetylene) can be revealed using IR and NMR spectroscopy; the Rh-based polymer has a selective *cis-transoidal* structure, while the W-based polymer has a *trans*-rich structure.[4,5] The geometric structure of poly(*t*-butylacetylene) can be clearly determined using NMR, with an all-*cis* polymer obtained if certain defined polymerization conditions are used.[53] On the other hand, the geometric structure of polymers from disubstituted acetylenes is generally difficult to determine, even with use of NMR. One exception is poly[1-(trimethylsilyl)-1-propyne], whose geometric structure can be deduced using ^{13}C NMR.[54]

The extent of conjugation of substituted polyacetylenes can be estimated using UV-visible spectroscopy. Aromatic monosubstituted acetylene polymers are colored from yellow to dark purple to varying extents, implying that their main chain is moderately conjugated. In contrast, poly(*t*-butylacetylene) is colorless, indicating a twisted structure due to the steric effects of the *t*-butyl group. Among the disubstituted acetylene polymers, poly(diphenylacetylene)s are colored yellow–orange owing to conjugation of the main chain with the aromatic rings, while many others are colorless, indicating that they are only slightly conjugated.

Unlike unsubstituted polyacetylene, many substituted polyacetylenes are soluble in common organic solvents such as toluene, $CHCl_3$, and THF, which is attributable to the interaction between solvent and side group, as well as their amorphous structure. The excellent solubility of substituted polyacetylenes enables measurement of polymer molecular weight and membrane preparation. Free-standing tough membranes are available by solution casting as long as the molecular weight of the substituted polymer is several hundred thousand or higher. Toluene, which evaporates at a moderate rate at ambient temperature, is useful as a casting solvent for such membrane preparation.

The stability of substituted polyacetylenes varies greatly depending on the type, bulkiness, and number of side groups; in general, disubstituted acetylene polymers tend to be more stable in air than the monosubstituted analogs.[55] For example, stability decreases in the order of poly(diphenylacetylene)s > poly(1chloro-2-phenylacetylene) ≈ poly(1choloro-1-octyne) > poly(1-phenyl-1-propyne) ≈ poly[1-(trimethylsilyl)-1-propyne] ≈ poly[*o*-(trimethylsilyl)phenylacetylene] ≈ poly(*t*-butylacetylene) > poly(phenylacetylene) > poly(1-hexyne). Specifically, poly(diphenylacetylene)s are stable up to approximately 400 °C, while poly(1-hexyne) is rapidly oxidized at room temperature.

With regards to the mechanical properties of substituted polyacetylenes, aromatic polymers like poly(diphenylacetylene) derivatives are generally hard and brittle, whereas aliphatic polymers with long alkyl chains like poly(2-octyne) are soft and ductile.[56,57] Considerations of mechanical properties are especially important when polymer membranes or fibers are required for the specific application.

Substituted polyacetylenes exhibit quite different properties from those of the unsubstituted polyacetylene. For instance, many substituted polyacetylenes are stable in air, soluble in various solvents, amorphous, and membrane forming. They are also less conjugated and practically electrically insulating unlike polyacetylene. Moreover, conjugation in substituted polyacetylenes plays an important role when their photoconductivity, photoluminescence, nonlinear optical properties, electrochromism, and certain other features are considered. It is further noted that the properties of substituted polyacetylenes vary greatly depending on the side group. When attempting the development of functional substituted polyacetylenes, all these properties should be taken into account.

References

1. F. Sanda, M. Shiotsuki and T. Masuda, in *Polymer Science: A Comprehensive Reference*, ed. K. Matyjaszewski and M. Moeller, Elsevier, Amsterdam, 2012, Vol. 3, Chapter 3.27, pp. 875–954.
2. F. Sanda, M. Shiotsuki and T. Masuda, *Polym. Chem. (RSC)*, 2011, **2**, 1044.
3. J. Liu, J. W. Y. Lam and B. Z. Tang, *Chem. Rev.*, 2009, **109**, 5799.
4. T. Masuda, *J. Polym. Sci., Part A: Polym. Chem.*, 2007, **45**, 165.
5. T. Masuda, F. Sanda and M. Shiotsuki, in *Comprehensive Organometallic Chemistry III*, ed. R. H. Crabtree and D. M. P. Mingos, Elsevier, Oxford, 2006, Chapter 11.18, p. 557.
6. T. Aoki, T. Kaneko and M. Teraguchi, *Polymer*, 2006, **47**, 4867.
7. J. W. Y. Lam and B. Z. Tang, *Acc. Chem. Res.*, 2005, **38**, 745.
8. M. G. Mayershofer and O. Nuyken, *J. Polym. Sci. Part A: Polym. Chem.*, 2005, **43**, 5723.
9. J. Sedlacek and J. Vohlidal, *Collect. Czech, Chem. Commun.*, 2003, **68**, 1745.
10. T. Masuda and F. Sanda, in *Handbook of Metathesis*, ed. R. H. Grubbs, Wiley-VCH, Weinheim, 2003, Vol. 3, 375–406.
11. A. Nakazato, I. Saeed, T. Katsumata, M. Shiotsuki, T. Masuda, J. Zednik and J. Vohlidal, *J. Polym. Sci., Part A: Polym. Chem.*, 2005, **43**, 4530.

12. M. Tabata, W. Yang and K. Yokota, *J. Polym. Sci., Part A, Polym. Chem.*, 1994, **32**, 1113.
13. Y. Kishimoto, M. Ito, T. Miyatake, T. Ikariya and R. Noyori, *Macromolecules*, 1995, **28**, 6662.
14. Y. Kishimoto, P. Eckerle, T. Miyatake, M. Kainosho, A. Ono, T. Ikariya and R. Noyori, *J. Am. Chem. Soc.*, 1999, **121**, 12035.
15. M. Miyake, Y. Misumi and T. Masuda, *Macromolecules*, 2000, **33**, 6636.
16. T. Masuda, K. Hasegawa and T. Higashimura, *Macromolecules*, 1974, **12**, 2065.
17. T. Masuda, N. Sasaki and T. Higashimura, *Macromolecules*, 1975, **8**, 717.
18. T. Masuda, K. Yamamoto and T. Higashimura, *Polymer*, 1982, **23**, 1663.
19. R. R. Schrock, S. Luo, J. C. Lee, N. C. Zanetti and W. M. Davis, *J. Am. Chem. Soc.*, 1996, **118**, 3883.
20. T. Masuda, Y. Okano and T. Higashimura, *Polym. J.*, 1980, **12**, 907.
21. J. Tabei, M. Shiotsuki, F. Sanda and T. Masuda, *Macromolecules*, 2005, **38**, 5860.
22. F. Sanda, S. Nishiura, M. Shiotsuki and T. Masuda, *Macromolecules*, 2005, **38**, 3075.
23. Y. Suzuki, M. Shiotsuki, F. Sanda and T. Masuda, *Macromolecules*, 2007, **40**, 1864.
24. A. Niki, T. Masuda and T. Higashimura, *J. Polym. Sci., Part A, Polym. Chem.*, 1987, **25**, 1553.
25. K. Tsuchihara, T. Masuda and T. Higashimura, *Macromolecules*, 1992, **25**, 5816.
26. Y. Shida, T. Sakaguchi, M. Shiotsuki, F. Sanda, B. D. Freeman and T. Masuda, *Macromolecules*, 2005, **38**, 4096.
27. T. Masuda, A. Niki, E. Isobe and T. Higashimura, *Macromolecules*, 1985, **18**, 2109.
28. T. Masuda, Y. Kuwane and T. Higashimura, *J. Polym. Sci., Polym. Chem. Ed.*, 1982, **20**, 1043.
29. T. Masuda, M. Yamagata and T. Higashimura, *Macromolecules*, 1984, **17**, 126.
30. T. Higashimura, Y.-X. Deng and T. Masuda, *Macromolecules*, 1982, **15**, 234.
31. T. Masuda, E. Isobe and T. Higashimura, *Macromolecules*, 1985, **18**, 841.
32. T. Masuda, E. Isobe, T. Hamano and T. Higashimura, *Macromolecules*, 1986, **19**, 2448.
33. T. Masuda, T. Yoshimura, K. Tamura and T. Higashimura, *Macromolecules*, 1987, **20**, 1734.
34. H.-P. Xu, B.-Y. Xie, W.-Z. Yuan, J.-Z. Sun, F. Yang, Y.-Q. Dong, A. Qin, S. Zhang, M. Wang and B. Z. Tang, *Chem. Commun.*, 2007, 1322.
35. W. Z. Yuan, J. Z. Sun, Y. Dong, M. Haussler, F. Yang, H. P. Xu, A. Qin, J. W. Y. Lam, Q. Zheng and B. Z. Tang, *Macromolecules*, 2006, **39**, 8011.
36. T. Fulghum, S. M. Abdul Karim, A. Baba, P. Taranekar, T. Nakai, T. Masuda and R. C. Advincula, *Macromolecules*, 2006, **39**, 1467.

37. J. Qu, T. Katsumata, M. Satoh, J. Wada, J. Igarashi, K. Mizoguchi and T. Masuda, *Chem. Eur. J.*, 2007, **13**, 7965.
38. G. Kwak, H.-Q. Wang, K.-H. Choi, K.-H. Song, S.-H. Kim, H.-S. Kim, S.-J. Lee, H.-Y. Cho, E.-J. Yu, H.-J. Lee, E.-J. Park and L.-S. Park, *Macromol. Rapid Commun.*, 2007, **28**, 1317.
39. S.-H. Yang, C.-H. Huang, C.-H. Chen and C.-S. Hsu, *Macromol. Chem. Phys.*, 2009, **210**, 37.
40. Q. Zeng, C. K. W. Jim, J. W. Y. Lam, Y. Dong, Z. Li, J. Qin and B. Z. Tang, *Macromol. Rapid Commun.*, 2009, **30**, 170.
41. H. Xu, J. Sun, A. Qin, J. Hua, Z. Li, Y. Dong, H. Xu, W. Yuan, Y. Ma, M. Wang and B. Z. Tang, *J. Phys. Chem. B*, 2006, **110**, 21701.
42. J. W. Y. Lam and B. Z. Tang, *Acc. Chem. Res.*, 2005, **38**, 745.
43. I. Saeed, M. Shiotsuki and T. Masuda, *Macromolecules*, 2006, **39**, 8567.
44. M. Shiotsuki, S. Kumazawa, N. Onishi and F. Sanda, *J. Polym. Sci., Part A, Polym. Chem.*, 2011, **49**, 4921.
45. M. Shiotsuki, A. Nakagawa, J. R. Castanon, N. Onishi, T. Kobayashi, F. Sanda and T. Masuda, *J. Polym. Sci., Part A, Polym. Chem.*, 2010, **48**, 5549.
46. T. Masuda, T. Hamano, K. Tsuchihara and T. Higashimura, *Macromolecules*, 1990, **23**, 1374.
47. T. Masuda, T. Hamano, T. Higashimura, T. Ueda and H. Muramatsu, *Macromolecules*, 1988, **21**, 281.
48. Y. Suzuki, M. Shiotsuki, F. Sanda and T. Masuda, *Chem. Asian J.*, 2008, **3**, 2075.
49. W. Zhang, M. Shiotsuki and T. Masuda, *Macromol. Commun.*, 2007, **28**, 1115.
50. J. Qu, T. Fujii, T. Katsumata, Y. Suzuki, M. Shiotsuki, F. Sanda, M. Satoh, J. Wada and T. Masuda, *J. Polym. Sci., Part A, Polym. Chem.*, 2007, **45**, 5431.
51. J. Qu, Y. Suzuki, M. Shiotsuki, F. Sanda and T. Masuda, *Macromol. Commun.*, 2007, **28**, 1115.
52. T. Masuda, T. Yoshimura and T. Higashimura, *Macromolecules*, 1989, **22**, 3804.
53. M. Nakano, T. Masuda and T. Higashimura, *Macromolecules*, 1994, **27**, 1344.
54. H. Izumikawa, T. Masuda and T. Higashimura, *Polym. Bull.*, 1991, **27**, 193.
55. T. Masuda, B. Z. Tang, T. Higashimura and H. Yamaoka, *Macromolecules*, 1985, **18**, 2369.
56. T. Masuda, B. Z. Tang, A. Tanaka and T. Higashimura, *Macromolecules*, 1986, **19**, 1459.
57. H. Seki, B. Z. Tang, A. Tanaka and T. Masuda, *Polymer*, 1994, **35**, 3456.

CHAPTER 4

Polyphenylenes

TAKAKAZU YAMAMOTO

Tokyo Institute of Technology, Chemical Resources Laboratory, Yokohama 226-8503, Japan
Email: tyamamot@res.titech.ac.jp

4.1 Introduction

Poly(*p*-phenylene), PPP, shown in Chart 4.1, and oligophenylene have long attracted the interest of organic, theoretical and polymer scientists because they are expected to have interesting chemical, electronic and optical functionalities owing to their having expanded π-conjugation systems along the polymer chain.[1–9]

Recently various derivatives of PPP and related polymers have been reported. For example, poly(9,9-dialkylfluorene-2,7-diyl)s, PFlu(9,9-R), and copolymers of dialkylfluorene[10–13] and poly(9,10-dialkyl-9,10-dihydrophenanthrene-2,7-diyl)s, $PH_2Ph(9,10\text{-}R)$,[14,15] shown in Chart 4.2, have been prepared and their chemical, electronic and optical functionalities have been investigated.

Polyfluorenes are described in Chapter 5. Various PPPs with heterocyclic side rings (*e.g.*, poly(quinoxaline-5,8-diyl)s[16] with a pyrazine side rings) have also been synthesized. These polymers will be described in Chapters 14 and 18. This chapter also deals with some PPPs with heterocyclic side rings.

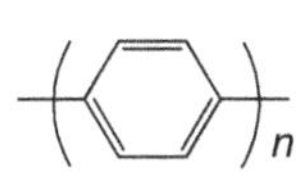

Chart 4.1 Poly(*p*-phenylene).

RSC Polymer Chemistry Series No. 9
Conjugated Polymers: A Practical Guide to Synthesis
Edited by Klaus Müllen, John R. Reynolds and Toshio Masuda

Published by the Royal Society of Chemistry, www.rsc.org

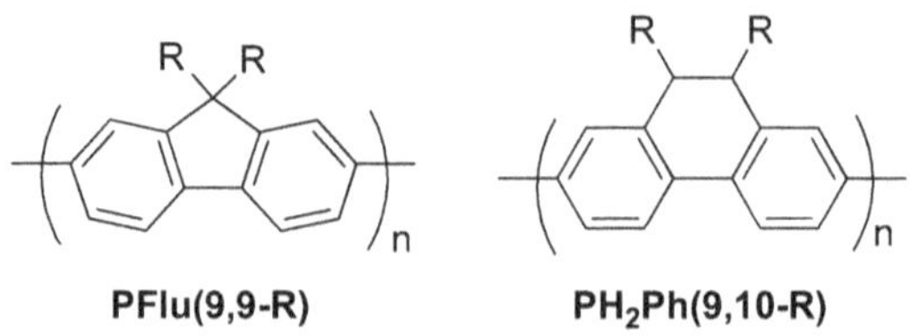

Chart 4.2 Poly(9,9-dialkylfluorene-2,7-diyl)s, PFlu(9,9-R), and poly(9,10-dialkyl-9,10-dihydrophenanthrene-2,7-diyl)s, PH_2Ph(9,10-R). R = H, alkyl, alkoxy, silyloxy, *etc.*

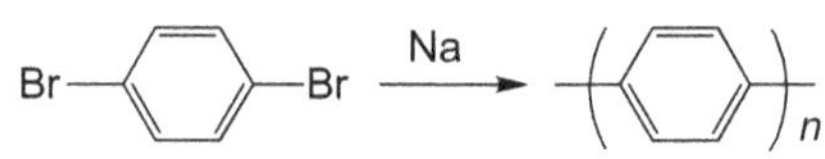

Scheme 4.1 Preparation of oligo-*p*-phenylene by reaction of *p*-dibromobenzene with Na.

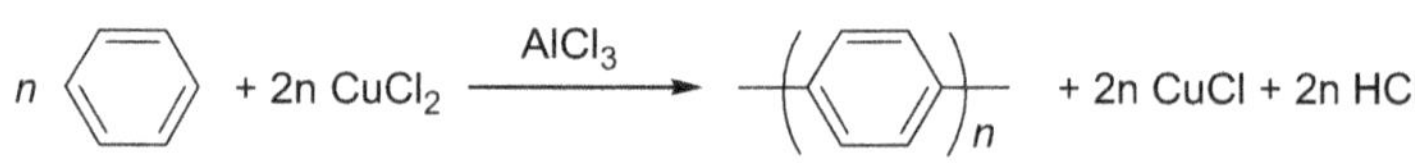

Scheme 4.2 Dehydro polymerization of benzene in the presence of $CuCl_2$ and $AlCl_3$.

4.1.1 Polyphenylenes without Side Chain

4.1.1.1 *Synthesis* via *Ullmann Reaction and Wurtz–Fittig Reaction*

Attempts have been made to prepare PPP and oligo-phenylenes *via* the Ullmann reaction and the Wurtz–Fittig reaction.[17–20] For example, treatment of *p*-dibromobenzene with Na gives an oligo-*p*-phenylene, as shown in Scheme 4.1.[17]

This oligo-*p*-phenylene is thought to have a degree of polymerization (DP) of 8 based on analytical data.[17]

4.1.1.2 *Synthesis by Oxidative Polymerization of Benzene*

Oxidative coupling of benzene proceeds using a Lewis-acid catalyst and an oxidant (the Kovacic method).[21–23] For example, benzene is converted into PPP by treatment with $CuCl_2$ and $AlCl_3$, as shown in Scheme 4.2.

The polymerization is thought to proceed *via* cationic processes. PPP prepared by this method essentially has a linear structure and DP of about 50;[8] this DP is comparable to that of PPP prepared *via* Grignard coupling.[8] Dehydro electrochemical oxidative polymerization of benzene has also been reported.[24]

4.1.1.3 *Synthesis* via *Intermediate Polymer*

Polymerization of 1,3-cyclohexadiene using a Ziegler catalyst gives a polymer consisting of cyclohexene units (*cf.* Scheme 4.3).

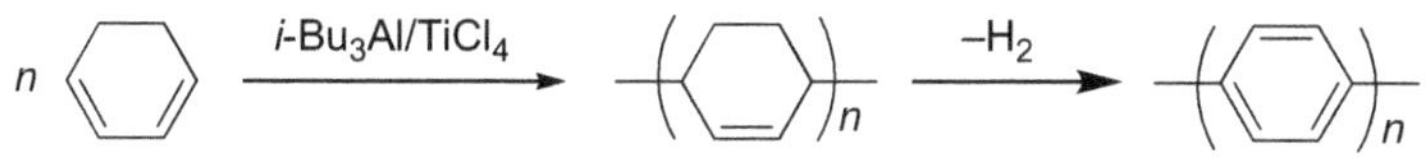

Scheme 4.3 Preparation and aromatization of poly(1,3-cyclohexadine).

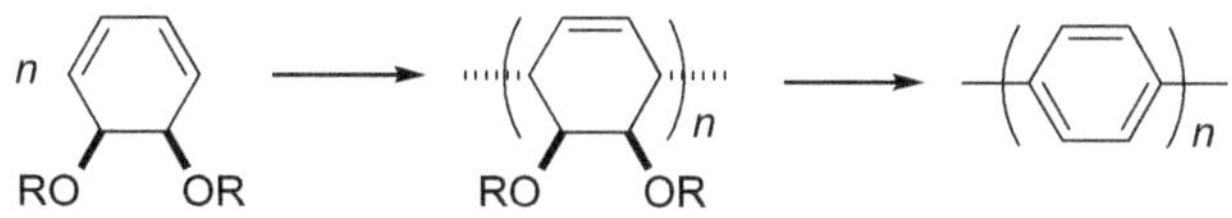

Scheme 4.4 Polymerization of 1,3-cyclohexadines and transformation of the obtained polymer to PPP. OR: OAc,[26,27] OTMS (trimethylsiloxy),[28,29] *etc*. In the case of OR = OTMS, the transformation is carried out *via* change of the OTMS group to OAc group.

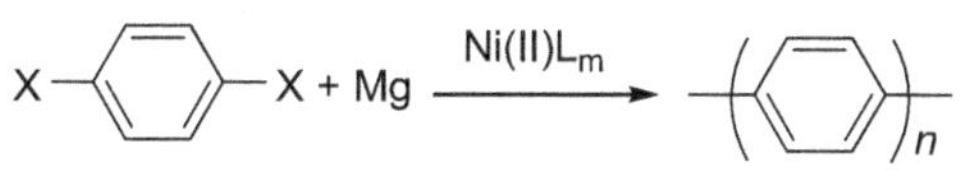

Scheme 4.5 Ni-catalyzed dehalogenative polycondensation of *p*-dihalobenzene with Mg. $Ni(II)L_m$: divalent Ni complexes such as $NiCl_2$(bpy) (bpy = 2,2′-bipyridyl) and $NiCl_2$(diphosphine). X = Br, Cl, *etc.*

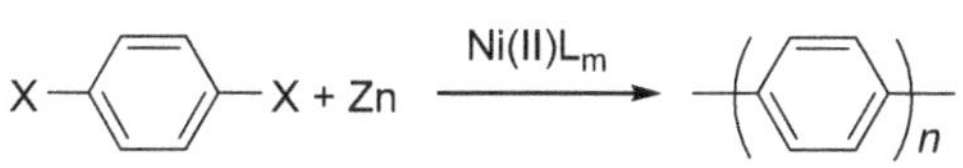

Scheme 4.6 Ni-catalyzed dehalogenative polycondensation of *p*-dihalobenzene with Zn.

This polymer is converted into PPP by reaction with dehydrogenation agents and thermal treatment.[25]

Polymerization of 5,6-disubstituted 1,3-cyclohexadine gives the corresponding polymer, which can also be converted into PPP, as shown in Scheme 4.4.

4.1.1.4 Synthesis by Organometallic Polycondensation

Organometallic C–C coupling reactions have been used for the preparation of PPP. For example, Ni-catalyzed dehalogenative polycondensation of *p*-dihalobenzene with Mg gives PPP as shown in Scheme 4.5.[30–32]

Various polymerization systems using Ni complexes as the catalyst and those using Ni complexes as the condensation agents have been developed for the preparation of PPP from *p*-dihalobenzene. Examples are:

(a) $Ni(II)L_m$-Zn catalytic system (cf. Scheme 4.6),[33]
(b) polymerization system using zero-valent Ni complex, $Ni(0)L_m$, as the condensation agent (cf. Scheme 4.7),[34]
(c) $Ni(II)L_m$–NaH catalytic system[35] and (d) $Ni(II)L_m$-e^- (electrochemical reduction with e^-) electrochemical polymerization system.[36]

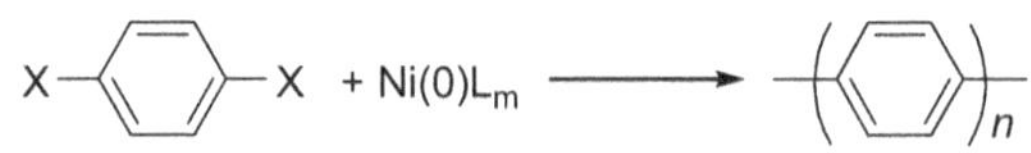

Scheme 4.7 Dehalogenative polycondensation of *p*-dihalobenzene.

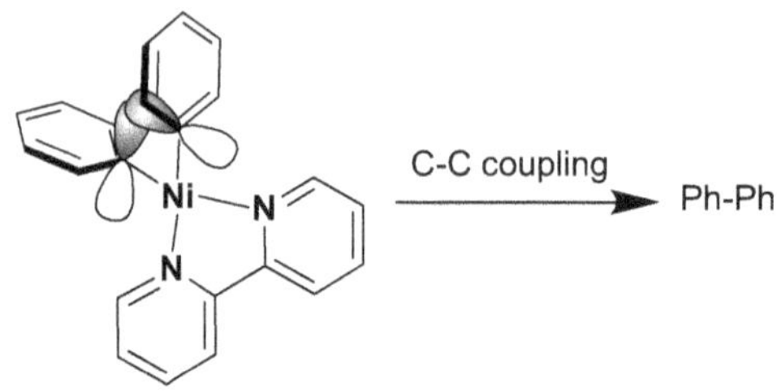

Scheme 4.8 C–C coupling on $Ni(C_6H_5)_2(bpy)$ to produce biphenyl Ph-Ph. This type of complex is isolable when the Ph group has electron-withdrawing substituents such as $-C_6F_5$, and its molecular structure has been revealed by X-ray crystallography.[37] bpy = 2,2′-bipyridyl.[7]

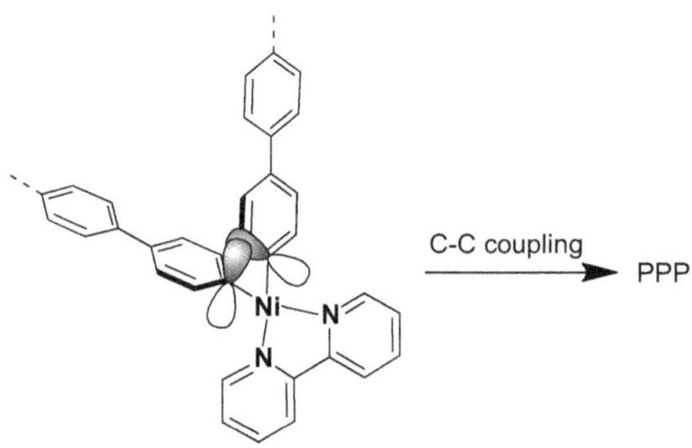

Scheme 4.9 C–C coupling on a Ni complex having two propagating oligo-*p*-phenylene groups to produce PPP.

Other transition-metal complexes such as Pd(II) complexes are also useful as the catalyst.

The Ni-catalyzed polycondensation is thought to proceed *via* C–C coupling (or reductive elimination) on an intermediate diorgano-nickel complex. Namely, the basic phenyl-phenyl coupling and the polycondensation are thought to proceed *via* the intermediates, as shown in Schemes 4.8 and 4.9.[7]

For the C–C coupling (or reductive elimination of R–R) on Ni, the following two factors have to be considered:

(a) The Ni–R (R = organic group) is thought to be polarized as $Ni^{\delta+}$–$R^{\delta-}$. On the other hand, the C–C coupling produces neutral R–R. Therefore, the C–C coupling on Ni complexes is thought to be accompanied by electron migration from R to Ni.

electron migration

$$Ni^{\delta+}\text{-}R^{\delta-}$$

For this reason, the Ni–Ph bond is thought to be more stable when Ph has electron-withdrawing substituent(s). Thus, $Ni(C_6F_5)_2(bpy)$ is isolated as a stable Ni complex.[37] However, attempts to isolate $Ni(C_6H_5)_2(bpy)$ have not been successful, and Ph–Ph is liberated in the attempted synthesis of $Ni(C_6H_5)_2(bpy)$.

These fundamental organometallic results are consistent with synthetic results of biphenyls using zero-valent nickel complexes ($Ni(0)L_m$: L = neutral ligand such as bpy and tertiary phosphine PR_3). The reaction of $Ni(0)L_m$ with X-phenyls (X–Ar; X = halogen) with various substituents gives the coupling product Ar–Ar or a nickel complex $Ni(Ar)_2L_m$ depending on the type of the reactant X–Ar, and the formation of Ar–Ar is difficult when Ar has electron-withdrawing substituent(s):[37–41]

$Ni(0)L_m + X-Ar \rightarrow Ar-Ar$ or Ni complex ($Ni(Ar)_2L_m$ is sometimes isolable[37])

X = halogen. Ar = aryl such as substituted phenyl.

When Br-phenyl has strongly electron-withdrawing substituent(s) such as $-NO_2$, the reaction does not give the C–C coupling products, substituted biphenyls, in good yields. On the other hand, Br-phenyls with other less electron-withdrawing substituents such as $-CH_3$ give the C–C coupling products (*e.g.*, $CH_3(C_6H_4)_2CH_3$) in high yields.

(b) The Ph–Ph coupling on Ni is thought to be assisted by π-electron interaction between the two Ph groups, as shown in Scheme 4.8; isolation of $Ni(Ph)_2(bpy)$ is not possible because of the presence of this electronic interaction to cause a facile Ph–Ph coupling. By contrast, in the case of dialkyl complexes such an electronic interaction is absent, and $Ni(alkyl)_2(bpy)$ (alkyl = Me, Et, *etc.*) is isolable.[42,43] $Ni(alkyl)_2(bpy)$ undergoes the C–C coupling to produce alkyl–alkyl (*e.g.*, butane from the Et complex) only at elevated temperatures (*e.g.*, above 100 °C).

Owing to the π-electron interaction, the C–C coupling of the Ph groups is thought to proceed readily, in spite of the higher electron-withdrawing ability of Ph than the alkyl group. Thus, isolation of $Ni(C_6H_5)_2(bpy)$ has not been successful. However, if the π-electron interaction is disturbed by choosing appropriate Ar group and neutral ligand L, $Ni(Ar)_2L_m$ might be isolated. Actually, such Ni complexes as $Ni(mesityl)_2L_m$ are isolabe.[44,45] In these complexes, the π-electron interaction seems to be weakened because of twisting of the Ni–phenyl bonds due to the presence of the *ortho*-substituent(s). This knowledge of the Ni complexes is consistent with synthetic results that aryl halides having two *ortho*-substituents (*e.g.*, 2-bromo-*m*-xylene[38]) do not couple in the reaction with $Ni(0)L_m$.[38,39]

4.1.2 Polyphenylenes with Side Chains or Side Rings

Polyphenylenes with side chains or side rings have usually been prepared *via* organometallic polycondensation. Examples are shown in Schemes 4.10 and 4.11.

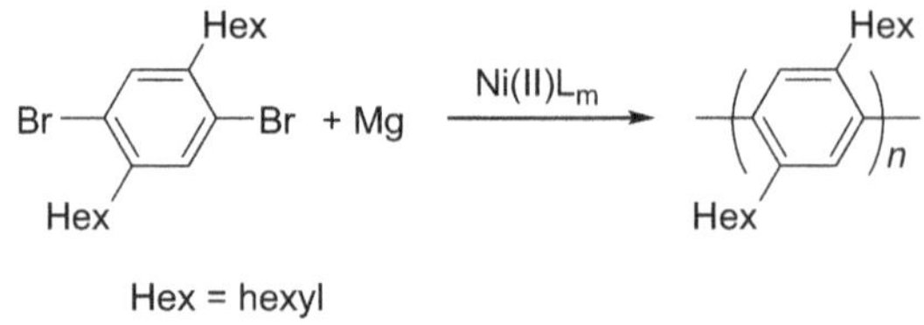

Scheme 4.10 Preparation of PPP with hexyl side chains.[46]

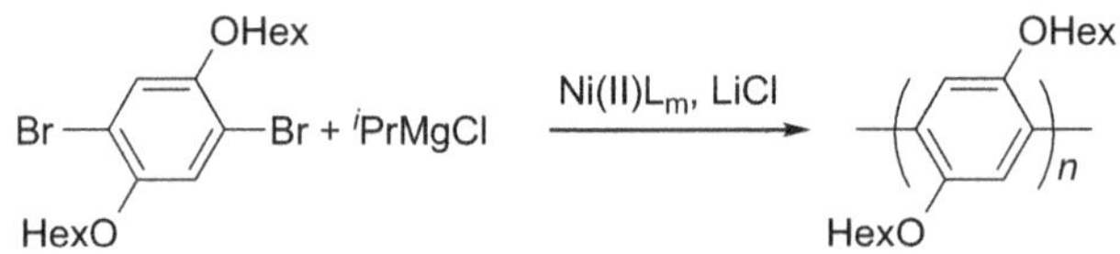

Scheme 4.11 Preparation of PPP with hexyloxy side chains.[47]

Various organometallic coupling reactions have been applied for the polycondensation. Examples are Kumada–Tamao coupling,[48] Suzuki–Miyaura coupling,[49] Migita–Kosugi–Stille coupling[50,51] and coupling using zero-valent nickel complexes.[37–41,52]

4.2 Experimental Procedures

4.2.1 Materials and General Polymerization Procedures

Many monomers for organometallic polycondensation are commercially available, which include *p*- and *m*-dihalobenzenes and diboronic compounds of benzene (*e.g.*, *p*-$C_6H_4(B(OH)_2)_2$ and its bis(pinacol) ester). Various dihalobenzenes with substituents (*e.g.*, 1,4-dibromo-2,5-dimethoxybenzene and 1,4-dibromo-2,5-dimethylbenzene) are also commercially available. However, some of the dihalobenzenes with substituents have to be synthesized, for example, using Br_2 and *N*-bromosuccinicimide (NBS).

Many Ni and Pd complexes are also commercially available. Examples are $NiCl_2$(dppe) (dppe = 1,2-bis(diphenylphosphino)ethane), $NiCl_2$(dppp) (dppp = 1,3-bis(diphenylphosphino)propane), $NiBr_2(PPh_3)_2$ (PPh_3 = triphenylphosphine), $Ni(PPh_3)_4$, $Ni(cod)_2$ (cod = 1,5-cyclooctadiene), $PdCl_2$(dppe), $PdCl_2(PPh_3)_2$, $Pd(OAc)_2(PPh_3)_2$ (Ac = acetyl) and $Pd(PPh_3)_4$. Divalent metal complexes can be stored under air. By contrast, zerovalent complexes such as $Ni(cod)_2$ are usually sensitive to air and have to be stored under inert gas such as N_2 and Ar.

List of abbreviations for ligands:

dppe = 1,2-bis(diphenylphosphino)ethane.
dppp = 1,3-bis(diphenylphosphino)propane.
PPh_3 = triphenylphosphine
dppf = 1,1′-bis(diphenylphosphino)ferrocene

bpy = 2,2′-bipyridyl
cod = 1,5-cyclooctadiene

Other abbreviations:

Ph = phenyl or *p*-phenylene
PPP and PMP = poly(*p*-phenylene) and poly(*m*-phenylene), respectively
DP = degree of polymerization
Ms = mesyl (-SO_2Me)
rt = room temperature
dL = 100 mL

Organometallic polycondensation is usually carried out under inert gas (*e.g.*, N_2 and Ar) using the Schlenk technique. When the Ni complex is employed as the condensation agent or catalyst, dry solvents such as THF, DMF and toluene are used.

When zero-valent Ni complexes (Ni(0)L_m) are used as the condensing agent for dihalo monomers, the Ni(0)L_m/monomer ratio strongly affects the molecular weight of the obtained polymers.[15] Zn used for the Ni-catalyzed polycondensation is usually activated by treating with acids and dried before use. Characterization of the obtained polymers by GPC is described in the Section 4.2.7.

4.2.2 Polyphenylenes without Side Chain

Example 1: PPP Synthesized with $CuCl_2$/$AlCl_3$ (Kovacic Method: cf. Scheme 4.2)[21,53]

The polymerization of benzene is carried out under N_2 in a 3-necked flask equipped with a paddle stirrer. After the dropwise addition of water (1 mL) to a mixture of benzene (1 mol, 78 g) and $AlCl_3$ (0.5 mol, 67 g), $CuCl_2$ (0.5 mol, 67 g) is introduced.

The temperature is increased to 35–37 °C and the reaction is allowed to proceed for 30 min at these temperatures. An acidic gas is evolved. The reaction mixture is then added to water and filtered. After treatment with 18% hydrochloric acid, the product is triturated repeatedly with boiling conc. hydrochloric acid, and washed with water until the washings are colorless. After two triturations with boiling 2 M NaOH solution, water washings are continued until the filtrate is colorless. Then the polymer is washed repeatedly with distilled water until the filtrate gives a negative test for Cl^- ion ($AgNO_3$ method). After being dried at 120 °C for 5 h, the product is in the form of finely divided, light brown solid. Yield is about 60%.[21] PPP is also prepared according to the procedure shown below.[53]

$AlCl_3$ (0.063 mol, 8.4 g) and $CuCl_2$ (0.031 mol, 4.2 g) are placed in a 100 mL round-bottomed flask equipped with a condenser and a $CaCl_2$ drying tube. Benzene (0.25 mol, 19.5 g) is added, and the mixture is stirred magnetically for 2 h at rt. The resulting black suspension is quickly filtered and washed with benzene (50 mL). After the crude polymer is dried under vacuum for 2 h, the PPP content is 10–15%, the remainder being inorganic salts. Purified PPP is obtained after workup; yield 0.75–0.85 g of light brown powder.[53] The polymerization reaction is expressed by Scheme 4.12. Example 1 is based on refs. 21 and 53.

Degree of polymerization (DP) of PPP prepared by the Kovacic method is estimated using nitrated PPP; DP = ca. 50 (by GPC; solvent = DMF; *vs.* polystyrene (PS) standards).[8]

Example 2: PPP Prepared *via* Poly(cyclohexadiene) (Ballard Method: *cf.* Scheme 4.4)[27]

Diacetate ester of 5,6-*cis*-dihydoxycyclohexa-1,3-diene (15 g, 76.5 mmol) and azobis(isobutyronitrile) AIBN (53 mg, 0.32 mmol) are placed in a 50-mL round-bottomed flask and degassed by pumping followed by flushing with N_2 three times. The reaction mixture is then heated to 70 °C and allowed to polymerize for 72 h (*cf.* Scheme 4.13). The resulting solidified reaction mass is dissolved in chloroform (100 mL) with stirring and the polymer (12 g, 89% yield) is recovered by precipitation into hexane (1 L). The radical polymerization can also be carried out using benzoyl peroxide, in solution systems and at other temperatures. A molecular weight of about 67 000 is estimated (by GPC; eluent = chloroform) for the polymer obtained with a radical initiator.

Polymerization of other derivatives of 5,6-*cis*-dihydoxycyclohexa-1,3-diene, such as benzoate and methoxycarbonyl derivatives, can be carried out analogously.

Aromatization of the obtained polymer to PPP is carried out both in the solid state and in solution, as shown in Scheme 4.14.

The aromatization in the solid is studied by thermogravimetric analysis (TGA) at a fixed heating rate, typically 10 °C min^{-1}, or isothermally at given

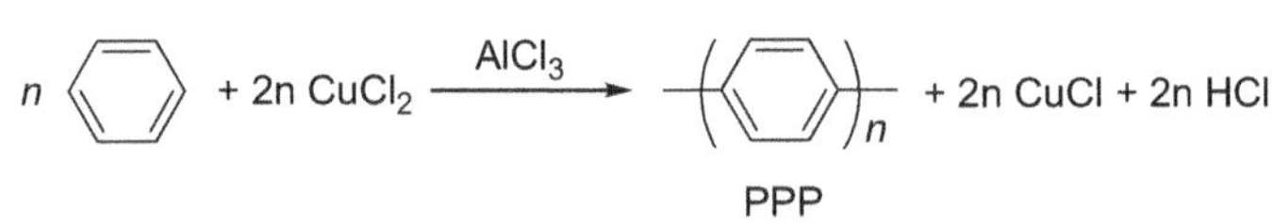

Scheme 4.12 Preparation of PPP using $CuCl_2$ and $AlCl_3$.

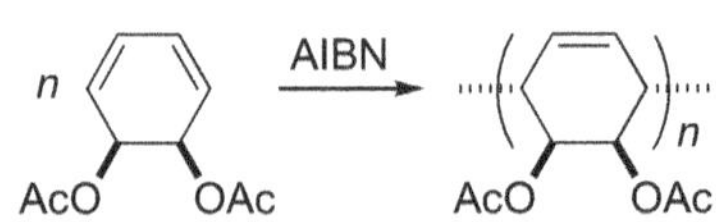

Scheme 4.13 Radical polymerization of diacetate ester of 1,3-hexadiene.

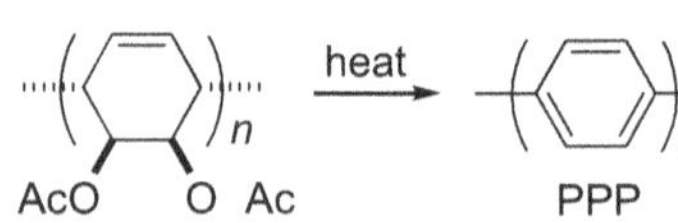

Scheme 4.14 Aromatization of the diacetate polymer to PPP.

temperature, typically at 300 °C. Aromatization in solution is performed by heating the polymer (5% w/v) in *N*-methylpyrrolidone (NMP) at 150–200 °C. The powder X-ray diffraction (XRD) pattern of the polymer show diffractions from PPP. Example 2 is based on ref. 27.

Example 3: PPP Prepared *via* Ni-Catalyzed C–C Coupling of Grignard Reagent (*cf.* Scheme 4.5)[31]

When 51 mg (0.18 mmol) of $NiCl_2$(bpy) is added to a mixture of the product obtained by the reaction of 11.8 g (50 mmol) of *p*-dibromobenzene with 1.22 g (50 mmol) of Mg in dry THF, a smooth polymerization starts. Polymerization is almost completed after refluxing the mixture for 1 h. After further refluxing for 4 h, the mixture is poured into 500 mL of ethanol. The polymer precipitated is collected over a glass filter, washed with diluted hydrochloric acid and ethanol, and dried in vacuum to yield 3.8 g (95%) of light yellow PPP (*cf.* Scheme 4.15)

Other Ni(II) complexes such as $NiBr_2(PPh_3)_2$, $NiCl_2$(dppf) (dppf = 1,1′-bis(diphenylphosphino)ferrocene[8]) and Ni(cod)$(PPh_3)_2$ also give PPP. Other ethereal solvents such as Bu_2O (Bu = butyl) and CPME (cyclopentyl methyl ether) are also useful. Addition of toluene to the ethereal solution sometimes gives PPP more smoothly. The DP of PPP thus prepared is estimated by GPC of nitrated PPP (eluent = DMF; *vs.* PS standards).[8] The DP of PPP obtained using $NiCl_2$(bpy) and a 5 : 1 mixture of Bu_2O and THF is thus estimated to be 45. The DP of PPP obtained using $NiCl_2$(dppf) and a 1 : 8 : 8 mixture of THF, CPME and toluene is estimated at 68. Poly(*m*-phenylene) PMP can be prepared analogously,[8,31] and its DP is estimated to be 45 from GPC (eluent = DMF; *vs.* PS standards) data of nitrated PMP. PMP assumes a helical structure in solid. Example 3 is based on ref. 31.

Example 4: Soluble Copolymer of *p*-Phenylene and *m*-Phenylene Prepared *via* Ni-Catalyzed C–C Coupling of Grignard Reagent[8,54]

The reaction of a 2 : 8 mixture of *p*-dibromobenzene (5.0 mmol, 1.18 g) and *m*-dibromobenzene (20 mmol, 4.72 g) with Mg (25 mmol, 0.61 g) in dry THF (35 mL) at rt gives a mixture of Grignard reagents. The addition of $NiCl_2$(bpy) (0.087 mmol, 25 mg) and stirring the reaction mixture for 40 h at 75 °C give the random copolymer **PP(a/b)** (a/b = 2/8 in this case), as shown in Scheme 4.16.

The reaction mixture is poured into 250 mL of ethanol, and the yellowish white precipitate is recovered by filtration, washed with diluted hydrochloric acid and ethanol, and dried under vacuum. Yield is 67%. This **PP(2/8)** is soluble in THF, DMF, DMSO, and chloroform. GPC (eluent = chloroform; *vs.* PS standards) data of this polymer show M_n (number-average molecular

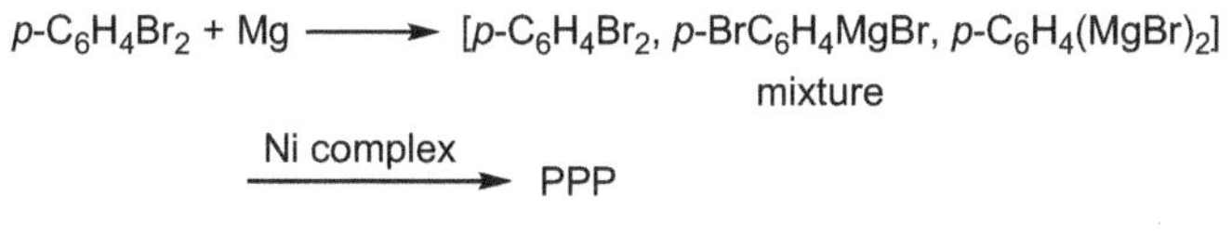

Scheme 4.15 Ni-catalyzed preparation of PPP *via* Grignard reagents.

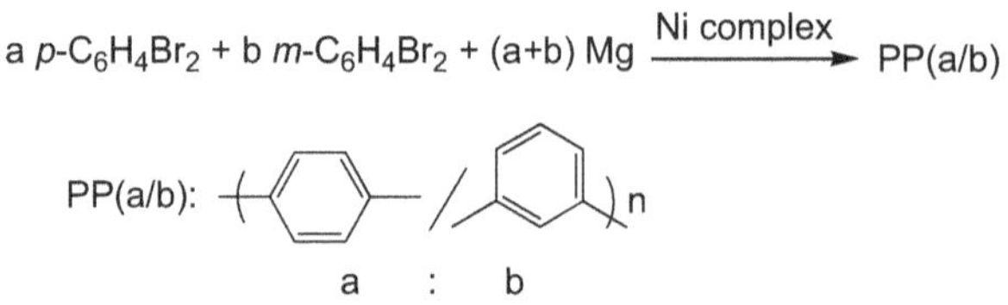

Scheme 4.16 Ni-catalyzed preparation of random copolymers of *p*-phenylene and *m*-phenylene. PP(2/8) is soluble in THF, chloroform, *etc.*

Table 4.1 Copolymerization of *p*-$C_6H_4X_2$ and *m*-$C_6H_4X_2$ (in *a* : *b* molar ratio) Using $NiCl_2L_m$ as the catalyst.[8]

Run	*a:b*[a]	*X*[b]	*Catalyst*[c]	*Solvent*[e]	*Yield (%)*	*DP*
1	1 : 9	Br	$NiCl_2$(dppp)	THF + TL (1 : 3)	75	39
2	2 : 8	Br	$NiCl_2$(dppp)	THF + TL (1 : 3)	80	29
3	2 : 8	Br	$NiCl_2$(dppf) + dppf[d]	THF + TL (1 : 3)	91	43
4	2 : 8	Br	$NiCl_2$(dppf) + dppf[d]	CPME	85	53
5	2 : 8	Br	$NiCl_2$(dppf) + dppf[d]	THF + CPME(1 : 16)	90	53
6	2 : 8	Br	$NiCl_2$(dppf) + dppf[d]	THF + CPME + TL (1 : 8 : 8)	For ball-like solid deposited[f] 53 (Fraction D 16)[g]	71 (M_n/M_w = 1.7) 159
7	2 : 8	Cl	$NiCl_2$(dppf) + dppf[d]	THF	89	54

[a]Mole ratio of *p*-$C_6H_4X_2$ to *m*-$C_6H_4X_2$ to prepare **PP(a/b)**.
[b]X in *p*-$C_6H_4X_2$ and *m*-$C_6H_4X_2$.
[c]dppp = 1,3-bis(diphenylphosphino)propane. dppf = 1,1′-bis(diphenylphosphino)ferrocene.
[d]1 mol of dppf per 1 mol of $NiCl_2$(dppf) is added.
[e]TL = toluene. CPME = cyclopentyl methyl ether.
[f]Ball-like solid deposited from the polymerization solvent.
[g]Higher molecular weight fraction is obtained using THF and methanol (cf. the text).

weight) of 2300, which corresponds to DP of 30, and the M_w (weight-average molecular weight) of 2780.

Ni-diphosphine complexes such as $NiCl_2$(dppf) also give **PP(a/b)** in various ethereal solvents, and the results are shown in Table 4.1

In run 6 in Table 4.1 a ball-like solid (curd) **PP(2/8)** is obtained. This solid shows a DP of 71, and fractionation of this product using mixtures of THF and methanol gives a higher molecular weight fraction (Fraction D) in 16% yield (based on the amount of the monomer). This Fraction D shows M_n of 12100 corresponding to DP of about 160. $M_w/M_n = 1.6$ in GPC (eluent = chloroform; *vs.* PS standards). This Fraction D contains 0.63% of Br, and treatment of Fraction D with $LiAlH_4$, according to the Müllen's method,[55] gives Br-free polymer; Anal. Calcd. for $H(C_6H_4)_{160}H$: C, 94.69; H, 5.31. Found: C, 94.82; H, 5.60; Br, 0.[8] Example 4 is based on refs. 8 and 54.

4.2.3 Polyphenylenes with Side Chains

Example 5: Preparation of PPP with Hexyloxy (OHex) Side Chains *via* Grignard Exchange Reaction[47]

Treatment of 1,4-dibromo-2,5-dihexyloxybenzene with 1 equiv of ^{i}PrMgCl in THF at rt for 24 h leads to a magnesium-bromine exchange reaction shown in Scheme 4.17.

$NiCl_2$(dppp) (1.8 mol%) is added, and the polymerization is carried out at rt to yield the polymer, PPP(2,5-OHex), as shown in Scheme 4.18.

GPC data (eluent = THF; *vs.* PS standards) show M_n of 3200 and M_w/M_n of 1.74. The molecular weight of PPP(2,5-OHex) increases (*e.g.*, to M_n of 14 600 (by GPC; eluent = THF; *vs.* PS standards)) when LiCl is added, and the molecular weight distribution becomes narrow under certain conditions, as shown in runs 3 and 4 in Table 4.2, which summarizes results of the polymerization.[47]

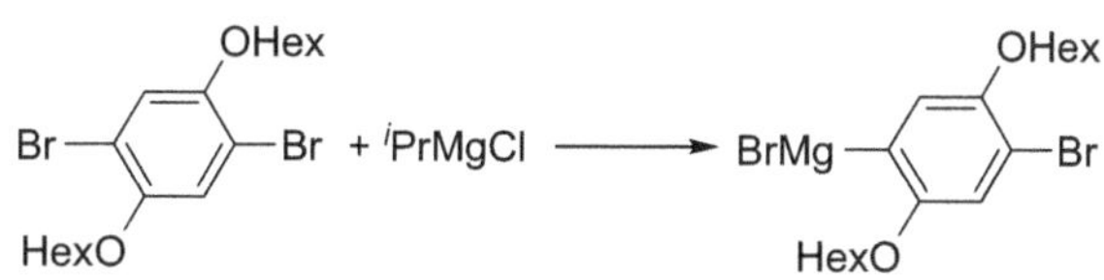

Scheme 4.17 Grignard exchange reaction.

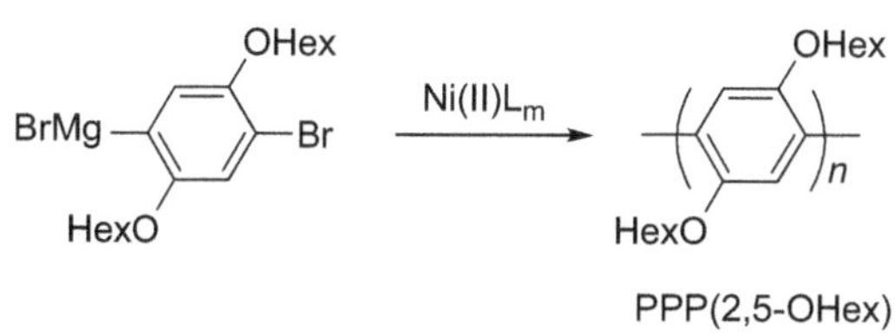

Scheme 4.18 Polymerization catalyzed by Ni(II) complex.

Table 4.2 Preparation of PPP(2,5-OHex) from Br-$C_6H_2(OHex)_2$-Br-*p* Using ^{i}PrMgBr and $NiCl_2L_m$.[a] Data for runs 1-4 are from ref. 47. Data for the preparation of poly(2,4-dibutoxy-*m*-phenylene) from Br-$C_6H_2(OBu)_2$-I-*m* using ^{i}PrMgBr and $NiCl_2L_m$ are also shown in run 5.[57]

Run	*$NiCl_2L_m$*[b]	*LiCl*[c]	*Time, h*	*Conversion, %*	*M_n*[d]	*M_w/M_n*
1	$NiCl_2$(dppp)	0	12	87	3200	1.74
2	$NiCl_2$(dppf)	0	46	93	4700	2.40
3	$NiCl_2$(dppp)	1.0	6	94	14600	1.33
4	$NiCl_2$(dppe)	1.0	1	92	12900	1.18
5[e,57]	$NiCl_2$(dppp)	4.7	24	62	25900	1.07

[a]Polymerization carried out using 1.0 mol of ^{i}PrMgBr per the dibromide monomer and 1.8% mol of $NiCl_2L_m$ at rt. for runs 1–4.
[b]dppp = 1,3-bis(diphenylphosphino)propane. dppe = 1,2-bis(diphenylphosphino)ethane.
[c]Amount of LiCl in equiv.
[d]Estimated by GPC (*vs.* polystyrene standard; eluent = THF).
[e]Data for the preparation of poly(2,4-dibutoxy-*m*-phenylene).[57]

Effects of LiCl in the polymerization[47] and in the Grignard formation reaction[56] have been discussed.

A similar $NiCl_2$(dppe)-catalyzed polymerization using Br-$C_6H_2(OC_4H_8C_6H_4SO_3CH_2CMe_3)_2$-Br as the monomer and LiCl as the additive gives the corresponding PPP-type polymer with M_n of 14 900 and M_w/M of 1.09 (by GPC; eluent = THF).[58] Block copolymerization is also possible.[58] Example 5 is based on refs. 47, 57 and 58.

Example 6: Preparation of PPP with Butoxy (OBu) Side Chains Using Pd-Catalyzed Suzuki-Type Condensation[59]

1,4-Dibromo-2,5-dihydroxybenzene Br-$C_6H_2(OH)_2$-Br-*p* is converted to 1,4-dibromo-2,5-dialkoxybenzene, Br-$C_6H_2(OR)_2$-Br-*p*, by the reaction with alkyl bromide RX (*e.g.*, butyl bromide) in the presence of KOH. Yield is typically 80%.

Br-$C_6H_2(OR)_2$-Br-*p* is converted to 4-bromo-2,5-dialkoxyphenylboronic acid *via* reactions with BuLi and boronic acid triisopropyl ester. Thus, **Monomer-1** (Br-$C_6H_2(OBu)_2$-$B(OH)_2$ (Bu = butyl)), shown in Chart 4.3 is obtained in 73% yield.

^{1}H NMR (acetone-d_6): δ: 7.42 and 7.25 (s, 2H), 7.15 (s, 2H), 4.13 and 4.05 (t, 4H), 1.89–1.70 (m, 4H), 1.61–1.41 (m, 4H), 0.99 and 0.98 (t, 6H).

Polymerization of 4-bromo-2,5-dialkoxyphenylboronic acids such as **Monomer-1** is carried out using $Pd(PPh_3)_4$ as the catalyst.

A mixture of 20 mL THF and 20 mL of 2 M aqueous Na_2CO_3 solution is degassed and 0.003 mol (1.04 g) of **Monomer-1** is dissolved in this mixture. $Pd(PPh_3)_4$ (0.5 mol%, 17 mg) is added, and the solution is vigorously stirred for 120 h under reflux to carry out the polymerization shown in Scheme 4.19. A second portion of $Pd(PPh_3)_4$ is added after 24 h. Since $Pd(PPh_3)_4$ is sensitive to light, the reaction vessel should be protected from direct sun light. After cooling under N_2, the mixture is poured into a five-fold volume of methanol. The polymer precipitates at once and is filtered off. It is redissolved in chloroform, the solution is filtered again and the polymer is isolated by

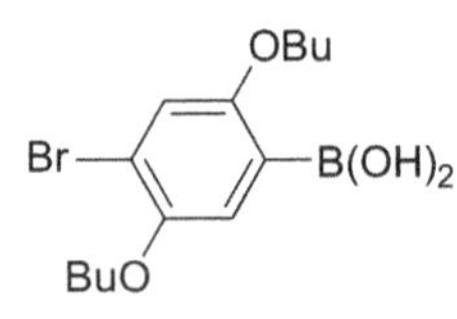

Chart 4.3 Structure of **Monomer-1**.

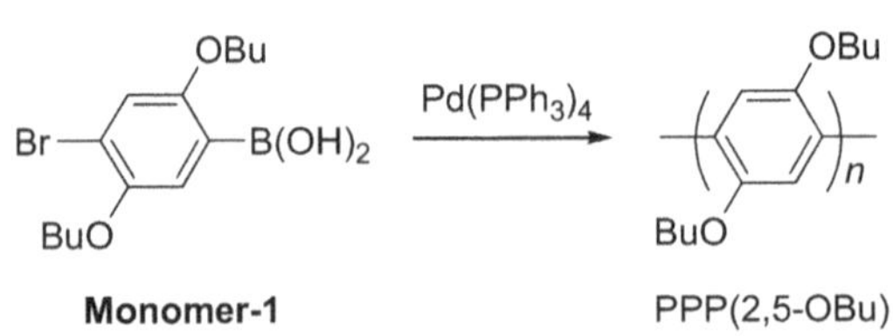

Scheme 4.19 Preparation of PPP with butoxy groups.

evaporation of the solvent. The polymer is obtained as a colorless powder by freeze-drying from benzene.

Yield > 90%. DP (VPO) = 30.[59] ^{1}H NMR ($CDCl_3$): δ: 7.10 (br, 2H), 7.03–6.94 (m)*, 4.02 (t)*, 3.93 (t (br), 4H), 1.88–1.77 (m)*, 1.76–1.59 (m, 4H), 1.59–1.47 (m)*, 1.47–1.27 (m, 4H), 1.04 (m)*, 0.97 (t, 6H); signals with * are caused by end-groups. Example 6 is based on ref. 59.

Example 7: Preparation of PPP with Dodecyl (Dod) Side Chains Using Pd-Catalyzed Suzuki-Type Condensation[60]

To a 500-mL three-neck round-bottom flask equipped with a reflux condenser and a 500-mL addition funnel, 12 g (23.89 mmol) of 2,5-didodecylbenzene-1,4-diboronic acid, 17.09 g (29.85 mmol) of 1,4-dibromo-2,5-dodecylbenzene, 25.43 g (240 mmol) of Na_2CO_3 and a magnetic stirring bar are added. The addition funnel is separated from the round-bottomed flask by an adapter containing a vacuum-tight stopcock. The apparatus is then evacuated (< 1 mbar) and filled with a N_2 atmosphere. This process is repeated four additional times, after which 0.138 g (0.12 mmol) of $Pd(PPh_3)_4$ is added. Solvents used for polymer preparation are treated and transferred so as to guarantee the strict exclusion of oxygen. Distilled water (200 mL) and toluene (120 mL) are separately refluxed for 1 h under a steady stream of N_2, which is bubbled into the solutions. The solvents are then cooled to rt and transferred to the addition funnel. In a separate apparatus, THF is maintained at reflux for at least 1 h under a N_2 atmosphere, after which 90 mL of distilled THF is withdrawn and combined with the other solvents in the addition funnel. N_2 is bubbled through the combined solvent mixture for an additional hour, after which the solvents are added to the reaction vessel. The addition funnel is then replaced by a rubber septum. The two-phase reaction is maintained overnight at reflux ($\approx$ 120 °C oil bath temperature) using the fastest possible stirring rate (*cf.* Scheme 4.20).

Additional quantities of diboronic acid are subsequently injected twice daily into the round-bottomed flask *via* the stopcock adapter equipped with a rubber septum. The diboronic acid is dissolved in the smallest quantity of distilled, deoxygenated THF necessary. The diboronic acid is added in the following increments: 1.5 g, 0.75 g, 0.45 g, 0.225 g, 0.075 g (1 : 1 monomer ratio, 4 days), followed twice daily by 10 additional injections of 0.045 g each over a 5-day period (1.03 : 1 monomer ratio, 9 days). The two-phase solution is then cooled and added to 500 mL of acetone. The precipitated polymer is filtered and washed twice in boiling water. The polymer is dried, dissolved by heating in 2 L

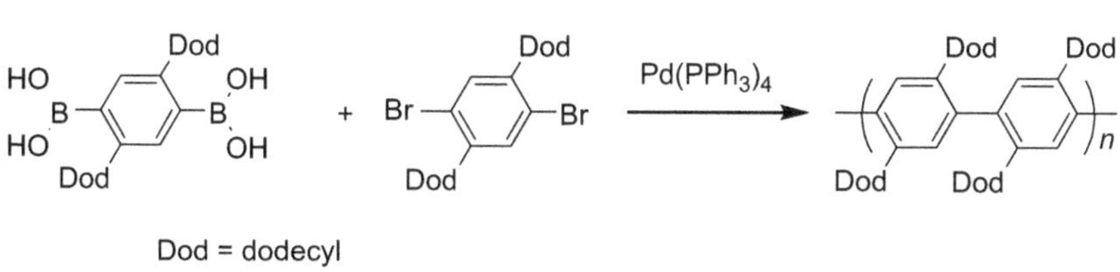

Scheme 4.20 Preparation of PPP with dodecyl side chains.

of toluene, filtered, and precipitated into a copious amount of acetone. The collected polymer is then extracted in a Soxhlet for 1 day with acetone. The polymer is additionally dissolved in toluene, precipitated into acetone, collected, and vacuum dried. Yield = 99%.[60] M_n (by GPC; eluent = 1,2-dichlorobenzene at 60 °C) = 39 400. M_w (by GPC under the same conditions) = 137 300. Inherent viscosity in 1,2-dichlorobenzene at 85 °C ($[\eta]_{inh}$) = 0.91 dL g^{-1}.

Similar Pd-catalyzed polycondensation has also been for the preparation of PPPs with ionic side chains or octyloxy side chains shown in Chart 4.4.[61–63]

For R = -$(CH_2)_3$-SO_3Na, yield = 64% and reduced viscosity = 0.1–0.2 dL g^{-1} at about c = 0.1 g dL^{-1}. λ_{max} = ca. 340 nm.

For R = -$(CH_2)_2NEt_2$ synthesized using $PdCl_2$(dppf) (1 mol%). Yield = 95%. M_n = 18 700 with M_w/M_n = 1.18 (by GPC; eluent = chloroform; *vs.* PS standards). λ_{max} = 350 nm. ^{1}H NMR ($CDCl_3$): δ: 7.71 (4H), 7.13 (2H), 4.11 (4H), 2.87 (4H), 2.62 (8H), 1.05 (m, 12H).

For R = n-C_8H_{17} synthesized using $Pd(PPh_3)_4$ (0.5 mol%); 1,4-dibromo-2,5-dioctyloxybenzene and a 1,4-di[1,3,2]dioxaborin-2-ylbenzene are used as the monomers. Yield = 66%. M_n = 8343 with M_w/M_n = 1.20 (by GPC; eluent = THF; *vs.* PS standards). Example 7 is based on refs. 60–64.

Example 8: Preparation of PPP with Tetrahydropyrene Repeating Units[55]

To a 100-mL Schlenk flask equipped with a stirring bar and containing $Ni(cod)_2$ (1.10 g, 3.77 mmol), bpy (0.65 g, 4.36 mmol) and cod (0.43 g, 4.01 mmol) under an Ar atmosphere are added DMF (7 mL) and toluene (10 mL) *via* a syringe. The ensuing deep blue/lilac solution is stirred at 70 °C for 30 min, and then a solution of **Monomer-2** (1.49 g, 2.54 mmol; *cf.* Scheme 4.21) in toluene (12 mL) is added, upon which the color of the reaction mixture turns to red/brown. The reaction mixture is then stirred at 75 °C under exclusion of light for 5 days. The polymerization is expressed by the equation in Scheme 4.21.

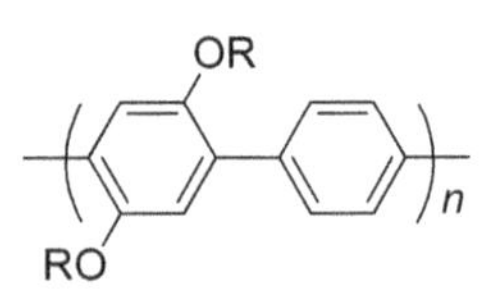

Chart 4.4 PPPs with ionic side chains or octyloxy side chains. R = -$(CH_2)_3$-SO_3Na,[61,62] -$(CH_2)_2NR'_2$,[63] or n-C_8H_{17}.[64]

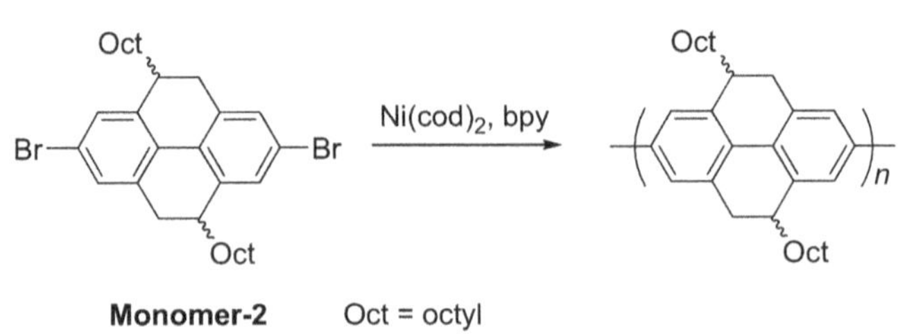

Scheme 4.21 Preparation of poly(dioctyltetrahydropyrene).

The mixture is cooled to rt, filtered through Celite®, and washed with 5 M aqueous HCl. The organic layer is separated and the aqueous portion extracted with chloroform (3×). The combined organic layers are successively washed with an aqueous solution of *N,N,N′,N′*-ethylenediaminetetraacetic acid (EDTA; pH = 3.8), an aqueous solution of EDTA (pH = 9) and water (3×). The organic layer is dried over $MgSO_4$ and the solvent evaporated. To remove Br end groups, the residue is then dissolved in toluene (70 mL) and added dropwise at rt to a suspension of $LiAlH_4$ (1.00 g, 26.3 mmol) in THF (20 mL) under an Ar atmosphere. The ensuing mixture is then heated under reflux for 24 h. After cooling to rt, the reaction mixture is quenched with 2 N aqueous H_2SO_4 (25 mL) and water (200 mL).

The organic layer is separated and the aqueous portion extracted with chloroform (3×). The combined organic layers are dried over $MgSO_4$. After evaporating the solvent, purification by column chromatography (alumina, chloroform) affords a solution of the poly(dioctyltetrahydropyrene) in chloroform, which is concentrated *in vacuo* to 10 mL. Then the polymer is precipitated by adding the ensuing solution dropwise to acetone. Filtration and removal of the solvent under reduced pressure affords 0.81 g (75%) of the polymer as a light yellow solid. $M_n = 17\,400$ (GPC performed after the first precipitation in acetone; *vs.* PS standards). $M_w = 35\,000$. ^{1}H NMR ($CDCl_3$, 500 MHz): δ: 7.43–7.47 (m, 4H), 2.86–3.16 (m, 6H), 1.30–1.88 (m, 28H), 0.90(bs, 6H). FTIR (KBr): ν: 2954.3, 2923.9, 2854.3, 1607.2, 1465.9, 1455.5, 1444.6, 871.4, 764.8, 742.7, 486.7, 477.4, 458.9. A second GPC, performed after the debromination and purification of the polymer shows identical values for M_n and M_w. Separation of a high molecular weight fraction (in an amount of 20%) with M_n and M_w of 33 300 and 55 000 (by GPC, *vs.* PS standards), respectively, is possible.[55] Example 8 is based on ref. 55.

Example 9: Preparation of PPP with 9,10-Dihydrophenathrene Repeating Units[15]

2,7-Dibromo-9,10-dioctyl-9,10-dihydrophenanthrene (415 mg, 0.738 mmol) and dry DMF (10 mL) are added to a mixture of $Ni(cod)_2$ (490 mg, 1.78 mmol), bpy (280 mg, 1.79 mmol), cod (220 μL, 1.77 mmol) and dry DMF (7 mL) in a Schlenk tube, and the mixture is stirred for 2 days at 60 °C. The polymer precipitate is washed with methanol, an aqueous solution of disodium ethylenediaminetetraacetate (Na_2EDTA) and water, and dried under vacuum to obtain a yellowish green powder of PPP with the dihydrophenanthrene repeating unit, **PH$_2$Ph(9,10-Oct)** (214 mg, 72% yield) shown in Scheme 4.22.

$M_n = 4600$. $M_w/M_n = 2.9$ (by GPC; eluent = 1,2,4-trichlorobenzene at 140 °C).[15] ^{1}H NMR ($CDCl_2CDCl_2$, 120 °C): δ: 7.93 (2H), 7.71 (2H), 7.58 (2H),

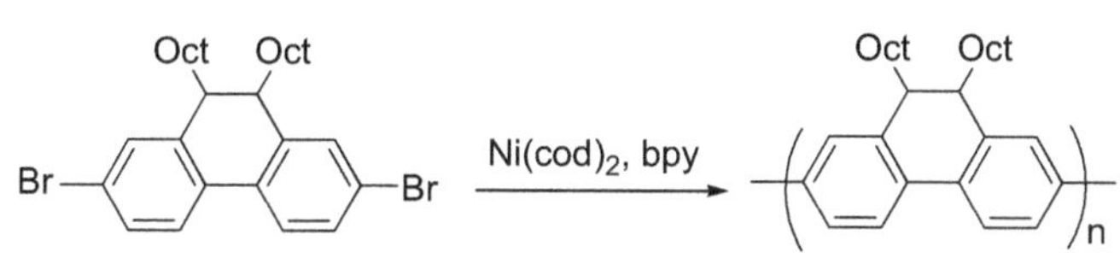

Scheme 4.22 Preparation of **PH$_2$Ph(9,10-Oct)**.

2.96 (2H), 1.8–1.1 (28H), 0.9 (6H). λ_{max} (*o*-dichlorobenzene (*o*-DCBZ)) = 387 and 416 nm. λ_{max} (film) = 385 and 412 nm.

A bis(2-ethylhexyl) analog is prepared analogously. $Ni(cod)_2$ (550 mg, 1.98 mmol) is added to dry DMF (5.5 mL) in a 50-mL Schlenk tube under Ar. To this mixture, bpy (310 mg, 1.98 mmol) and cod (242 μL, 1.98 mmol) are added, and the mixture is stirred. 2,7-Dibromo-9,10-bis(2-ethylhexyl)-9,10-dihydrophenanthrene (460 mg, 0.82 mmol) and dry DMF (13.3 mL) are added to the mixture, and this mixture is stirred for 2 days at 60 °C. The polymer precipitate is washed with methanol, an aqueous solution of Na_2EDTA and water (twice for each), and dried under vacuum to obtain a yellowish green powder of poly(9,10-dihydrophenanthrene) with 2-ethylhexyl side chains, **PH$_2$Ph(9,10-EtHex)**, shown in Chart 4.5.

Yield = 0.26 g (79%), M_n = 10 200. M_w/M_n = 1.6 (by GPC; eluent = chloroform; *vs.* PS standards).[15] ^{1}H NMR ($CDCl_2CDCl_2$, 120 °C): δ: 7.90–7.53 (6H), 3.05 (2H), 1.5–0.5 (34H). λ_{max} (*o*-DCBZ) = 391 and 414 nm. λ_{max} (film) = 391 and 410 nm.

Analogous polymers with dimethylalkylsilyloxy -OSiMe$_2$R side chains are also synthesized by this method; M_n = 69 000. M_w/M_n = 3.0 (by GPC; eluent = chloroform; *vs.* PS standards) for R = octadecyl; M_n = 16 000. M_w/M_n = 1.2 (by GPC; eluent = chloroform; *vs.* PS standards) for R = Me.

The polycondensation using Ni(0)L$_m$ usually gives a polymer with a higher molecular weight when a larger amount of the condensing agent Ni(0)L$_m$ is added.[15] For example, **PH$_2$Ph(9,10-OSiMe$_2$Ocd)** with the -OSiMe$_2$(octadecyl) side chains show M_ns of 9100, 69 000, and 6.0×10^5 in GPC when prepared at Ni(0)L$_m$/monomer ratios of 1.60, 1.94, and 2.36, respectively. Example 9 is based on ref. 15.

Example 10: Preparation of PPP and PMP with Side Chains Using a Ni/Zn System[65]

A 125-mL Schlenk tube is charged with $NiCl_2(PPh_3)_2$ (115 mg, 0.175 mmol), Zn (800 mg, 12.3 mmol), Et_4NI (675 mg, 2.63 mmol) and a magnetic stirring bar. The tube is sealed with a rubber septum and the contents are dried at 22 °C under vacuum (1×10^{-6} mmHg) for 24 h. After placing the contents under an Ar atmosphere, dry THF (1.0 mL) is added *via* a syringe through the rubber septum. The mixture is stirred at 22 °C for 5 min (the color of the mixture gradually changes to deep red-brown). Methyl 2,5-bis[(methylsulfonyl)oxy]-benzoate (568 mg, 1.75 mmol) in THF (0.5 mL) is added, and the mixture is heated at 67 °C for 10 h (*cf.* Scheme 4.23; R = Me).

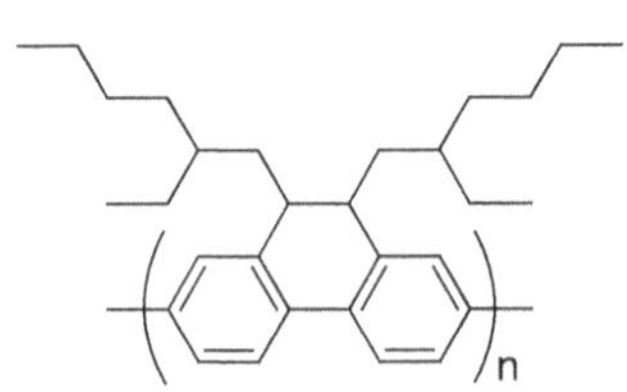

Chart 4.5 Structure of **PH$_2$Ph(9,10-EtHex)**.

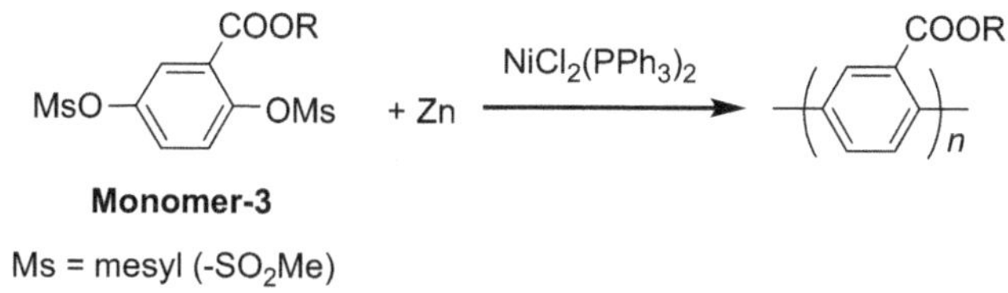

Scheme 4.23 Ni-catalyzed polymerization of 2,5-bismesylate monomer using Zn. R = Me.

Table 4.3 Preparation of substituted PPPs and PMPs Using $NiCl_2$-ligand(L)/Zn systems. Monomer = X-substituted phenylene-X-*p* or -*m*.

No.	Monomer[a]	L_m	Yield, %	M_n	M_w/M_n	Ref.
1	X = *p*-MsO R = 2-COOMe	PPh_3	75	3950	2.24	65
2	X = *p*-MsO R = 2-COOEtHex	PPh_3	96	7230	2.60	65
3	X = *p*-Cl R = 2-COOEtHex	PPh_3	87	12 905	2.88	65
4	X = *p*-Cl R = 2-COPh	PPh_3/bpy		26 700 $\eta_{inh} = 0.87$ dL g^{-1}		66
5	X = *p*-Cl R = 2-COPh-OPh	PPh_3	83	86 000	1.92	67
6	X = *m*-Cl R = $OC_{12}H_{25}$	PPh_3/bpy	87.3	9700		68
7	X = *m*-Cl R = OEtHex	PPh_3/bpy	70	11 902	2.0	69

[a]MsO = -OSO_2Me. R = substituent. EtHex = 2-ethylhexyl. Ph = phenyl or *p*-phenylene.

After cooling to 22 °C, the reaction mixture is poured into 100 mL of methanol acidified with 25 mL of conc. hydrochloric acid. The resulting precipitate is collected by filtration and dissolved in 2 mL of $CHCl_3$. The solution is filtered and poured into 100 mL of methanol. The precipitate is collected by filtration and dried *in vacuo* (75%). The polymer is purified by reprecipitation into methanol from a $CHCl_3$ solution before being analyzed by GPC ($M_n = 3950$, $M_w/M_n = 2.24$, DP = 29; eluent = THF; *vs.* PS standards). When a 2-ethylhexyl ester analog of **Monomer-3** is used (R = 2-methylhexyl), a polymer with M_n of 7230 and M_w/M_n of 2.6 is obtained.[65] Data of the preparation of substituted PPPs and PMPs (poly-*m*-phenylenes) from dihalogenated monomer using $NiCl_2L_m$ and Zn are summarized in Table 4.3. Example 10 is based on refs. 65–69.

4.2.4 Polyphenylenes with -N=N- or -NR- Binding Units

Example 11: Preparation of Diazobenzene Polymer Using Pd Catalyst[70,71]

A sealable 10-mL ampoule is charged with 1,4-bis(1,3,2-dioxaborinan-2-yl)-2,5-dihexylbenzene (166 mg, 0.40 mmol), 4,4′-diiodoazobenzene (174 mg,

0.40 mmol), $Pd(PPh_3)_4$ (13.9 mg, 0.012 mmol), NaOH (64 mg, 1.6 mmol), *N*,*N*-dimethylacetamide (5.0 mL), H_2O (0.80 mL) and a stirring bar. The mixture is degassed twice by the freeze–pump–thaw procedure, and then the ampoule is sealed off with a gas burner under vacuum, followed by stirring at 110 °C for 48 h to carry out the polymerization shown in Scheme 4.24.

Then the ampoule is opened, and the reaction mixture is poured into 100 mL of methanol. The precipitated solids (crude polymer product) are filtered and washed with H_2O/methanol (1 : 1) and then with methanol. After drying in a vacuum, the polymer is redissolved in toluene and reprecipitated from methanol to give the polymer as an orange powder in 100% yield. $M_n = 7700$. $M_w/M_n = 1.8$ (GPC; eluent = THF; *vs.* PS standards. ^{1}H NMR ($CDCl_3$): δ: 8.13–7.98 and 7.64–7.49 (4H and 4H), 7.32–7.20 (2H), 2.78–2.53 (4H), 1.65–1.47 (4H), 1.37–1.13 (12H), 0.96–0.80 (6H). λ_{max} (toluene) = 365 nm (+ 450 nm depending on conditions).

$Pd(PPh_3)_4$ also serves as a catalyst for the preparation of $[(C_6H_4\text{-}p)_mN(BOC)]_n$ (*e.g.*, $m = 4$, $M_n = 9200$ (by GPC; *vs.* PS standards) having the N(BOC) (BOC = *t*-butoxycarbonyl) group, instead of the -N=N- group, from Br-Ph-N(BOC)-Ph-Br and $Me_3Sn\text{-}Ph_{m-2}\text{-}SnMe_3$.[72] Thermal treatment of the BOC-protected polymer ($m = 4$) gives $[(C_6H_4\text{-}p)_4NH]_n$ which shows λ_{max} (in film) = 375 nm near that of powdery PPP.

Pd-catalyzed homopolymerization of 4,4′-diiodoazobenzene with hexyl side chains gives the homopolymer of azobenzene when bis(pinacolato)diboron is used as a condensation agent, as shown in Scheme 4.25.[71]

Yield = 96%. $M_n = 14\ 700$. $M_w/M_n = 2.4$ (by GPC; eluent = THF; *vs.* PS standards). Example 11 is based on refs. 70–72.

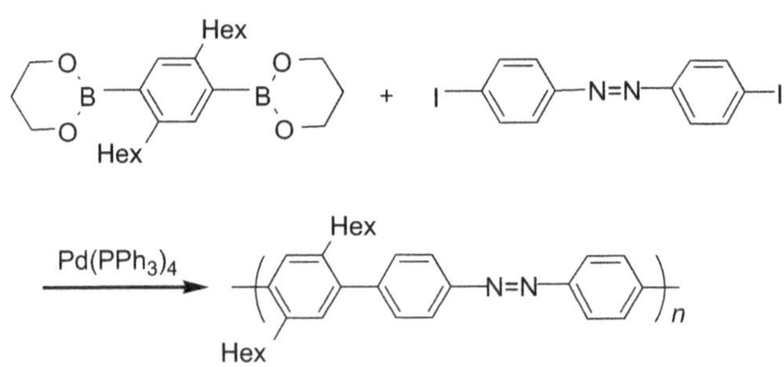

Scheme 4.24 Pd-catalyzed preparation of diazobenzene polymer.

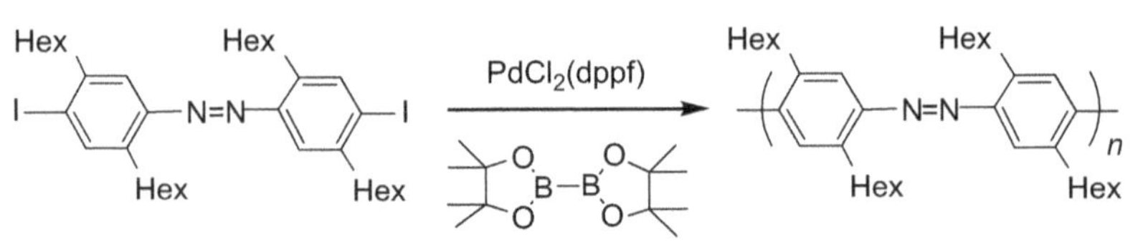

Scheme 4.25 Preparation of homopolymer of azobenzene.

Similar Pd-catalyzed preparation of π-conjugated polymers (*e.g.*, coplanar head-to-head type polythiophene with -C≡C-alkyl side chains, **HH-P(C≡CR)Th**[73]) is possible by using $R_3Sn\text{-}SnR_3$ (R = Me or Bu), instead of $(RO)_2B\text{-}B(OR)_2$, as the condensation agent.[73–75]

4.2.5 Polyphenylenes with Conjugated Side Rings

Example 12: Preparation of PPP with Triazole Side Rings Using Ni(0)L_m[76]

Monomer-4 (0.78 g, 1.3 mmol) shown in Scheme 4.26 is added to a toluene solution containing $Ni(cod)_2$ (1.32 g, 4.8 mmol), bpy (0.75 g, 4.8 mmol) and cod (0.58 g, 5.4 mmol) under N_2. The reaction mixture is stirred at 60 °C for 96 h to give a black precipitate (*cf.* Scheme 4.26).

The precipitate is washed with methanol, aqueous ammonia, an aqueous solution of Na_2EDTA, distilled water, a methanol solution of dimethylglyoxime, diluted hydrochloric acid and methanol, and it is dried under vacuum at 100 °C for 3 days to give **P[BTz(C12-CBz)]** shown in Scheme 4.26 as a red solid. Yield = 90%. $M_n = 4700$. $M_w/M_n = 1.51$ (by GPC; eluent = chloroform; *vs.* PS standards). λ_{max} ($CHCl_3$): 265, 296, 348, 452 nm. ^1H NMR ($CDCl_3$): δ: 8.90 (2H), 8.06 (6H), 7.4–7.1 (2H), 4.84 (2H), 4.18 (2H), 2.17 (2H), 1.78 (2H), 1.25 (16H). Example 12 is based on ref. 76.

Example 13: Preparation of PPP with Triazole Side Rings Using Pd Catalyst[76]

A methanol (10 mL) solution of 1,4-bis(1,3,2-dioxaborinan-2-yl)benzene (0.25 g, 1.0 mmol) is added to a mixture of toluene (20 mL) solution of **Monomer-4** (0.61 g, 1.0 mmol), catalytic amount of $Pd(PPh_3)_4$ (1.0 mg) and an aqueous solution of K_2CO_3 (2.0 M, 4.0 mL). After the reaction mixture is stirred at 80 °C for 4 days to carry out the polymerization shown in Scheme 4.27, the reaction mixture is poured into methanol to obtain a precipitate.

The crude product is separated by filtration and washed with water and methanol. The obtained polymer is dissolved in chloroform and reprecipitated in methanol to give a yellow powder of **P[BTz(C12-CBz)-Ph]** shown in Scheme 4.27. Yield = 62%. $M_n = 27\ 000$. $M_w/M_n = 1.77$ (by GPC; eluent = chloroform; *vs.* PS standards). λ_{max} ($CHCl_3$): 265, 296, 398 nm. ^1H NMR ($CDCl_3$): δ: 8.26

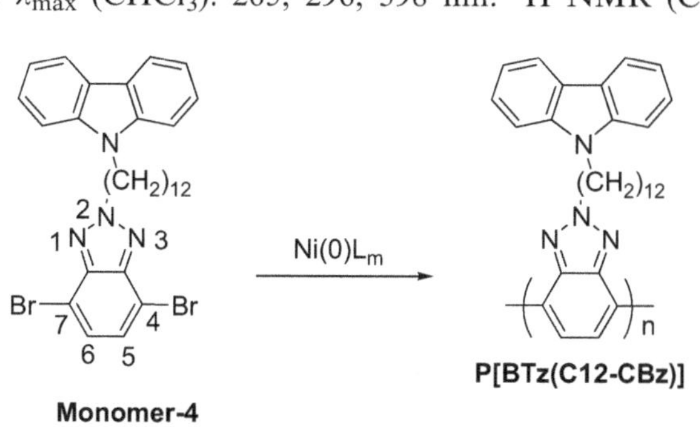

Scheme 4.26 Preparation of PPP with triazole side rings using Ni(0)L_m.

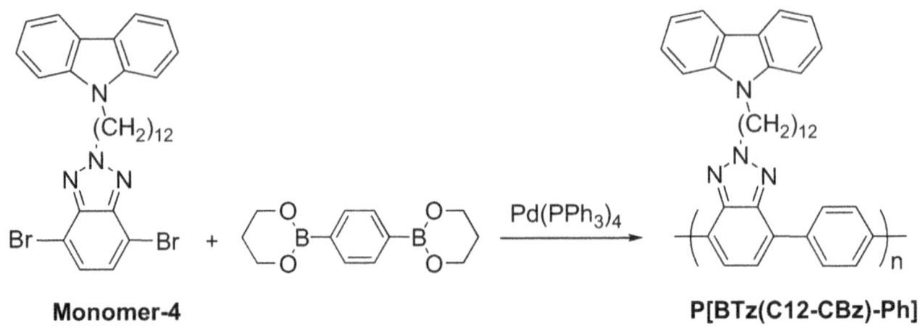

Scheme 4.27 Preparation of PPP with triazole side rings using Pd catalyst.

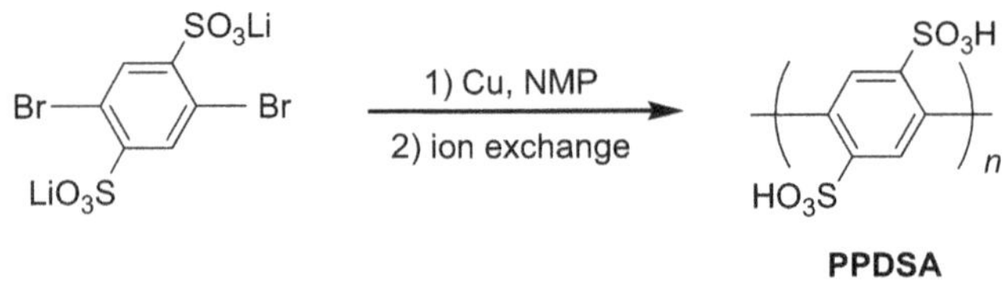

Scheme 4.28 Preparation of PPP with $-SO_3H$ side chains.

(2H), 8.05 (3H), 7.75 (1H), 7.5–7.3 (6H), 7.25 (2H), 4.81 (2H), 4.23 (2H), 2.17 (2H), 1.80 (2H), 1.25 (16H). Example 13 is based on ref. 76.

4.2.6 Polyphenylenes with $-SO_3M$ or $-NO_2$ Side Chains Prepared *via* Ullmann Coupling

Example 14: Preparation of PPP with $-SO_3Li$ Side Chains *via* Ullmann Coupling[77]

The Ullmann coupling can be applied for 1,4-dibromobenzene with $-SO_3Li$ groups. A dilithium salt of 1,4-dibromo-2,5-disulfobenzene ($Br-C_6H_2(SO_3Li)_2-Br$, 16.1 g, 0.04 mol) and freshly activated and predried Cu powder (25.4 g, 0.4 mol) are placed in a 1000 mL three-neck heavy-duty flask with a magnetic stirring bar, and dried at 135 °C under high vacuum (0.03 Torr) for 48 h. Freshly distilled NMP (750 mL) is added to the system using a double-tipped needle under Ar atmosphere. The mixture is degassed by bubbling Ar for 30 min and then polymerized at 140 °C for 36 h under vigorous stirring. After polymerization, the mixture is filtered. The solid is stirred at rt with 800 mL DI water for 24 h and the mixture is filtered to remove insoluble Cu powder and CuBr. The water solution is concentrated to ~50 mL and purified by ultra-filtration using 0.01 M HCl solution (2000 mL). The purified polymer solution is passed through an H^+ ion exchange column to remove any strongly bound cations. After H_2O is evaporated, the polymer is dried under vacuum (10 Torr) at 90 °C for 48 h to give **PPDSA** (7.8 g, yield = 84%) shown in Scheme 4.28.

PPDSA shows η_{sp}/c of approximately 0.6 dL g^{-1} at about $c = 0.4$ g dL^{-1} in DMF.

A similar dehalogenative polymerization using Cu is possible when dichloroanthraquinone with electron-accepting -NO_2 substituents, and poly-(anthraquinone-1,5-diyl) with two -NO_2 groups at the 4,8-positions is obtained ($M_n = 8600$, $M_w/M_n = 1.5$ (by GPC; eluent = DMF; *vs.* PS standards).[78,79] Example 14 is based on refs. 77–79.

4.2.7 Characterization and Remarks

PPP is insoluble, whereas copolymers of *p*-phenylene (PP) and *m*-phneylene (MP) in a certain range of PP/MP ratio are soluble (*cf.* Example 4).[8] Nitration of PPP makes this polymer soluble without crosslinking.[8] Introduction of side chains or side rings often makes PPP soluble. The number-average molecular weight (M_n) can be determined by vapor pressure osmometry, while weight-average molecular weight (M_w) can be determined by light scattering (LS).[80] According to LS, the degree of depolarization (ρ_v) of the polymer is obtained, and this gives information about the stiffness of the polymer in the solution.[8,81] LS uses a He-Ne laser (633 nm, commercially available apparatuses usually use this laser) or an Ar laser (488 nm), and is suited for polymers that do not absorb at these wavelengths. GPC is a convenient method to evaluate M_n and M_w. $CHCl_3$, DMF and THF are common eluents for GPC. Other solvents such as $(CF_3)_2CHOH$, aqueous solutions, *o*-dichlorobenzene (for high temperatures) and 1,2,4-trichlorobenzene (for high temperatures) are also used. π-Conjugated polymers often assume a stiff structure in solutions, and GPC using polystyrene standards may give a somewhat larger molecular weight than true molecular weight.

IR and NMR give information about the molecular structure of poly-phenylenes. For PPP, the out-of-plane vibration peak δ(C–H) appears at about 805 cm^{-1}. Because PPP is insoluble, measurement of NMR spectra in solutions is not possible. Solid ^{13}C NMR spectra of PPP can be normally obtained, and discussion on the IR, Raman and solid ^{13}C NMR spectra are made using ^{13}C-labbeled PPP.[82,83] Because the magnetic anisotropy at the phenylene-C is large, solid ^{13}C NMR spectra of PPP and its related polymers usually show strong side bands. Polymers with the PPP main chain show a main UV-vis absorption peak at about 380 nm.[7,53] When the polymer molecules aggregate, an additional peak (*e.g.*, at 416 nm[15]) appears presumably due to intermolecular electronic interaction. When the polymer has conjugated side rings (as in **P[BTz(C12-CBz)]** shown in Example 12), it sometimes shows a UV-vis peak at a longer wavelength. PPP and its derivatives usually show strong photo-luminescence (PL) and their PL peak appears at the position essentially agreeing with the onset position of the UV-vis absorption band. For example, poly(dioctyltetrahydropyrene) shown in Example 8 gives rise to a UV-vis peak at 385 nm and the onset position of the UV-vis absorption band is located at approximately 430 nm. The PL peak of this polymer appears at 425 nm, essentially agreeing with the onset position of the UV-vis absorption band.[55]

Dehalogenative organometallic polycondensation gives -C–H, -C–X, and -C–m (m = organometallic group such as -B(OR′)$_2$ and -SnR′$_3$) terminal units. The -C–H group is thought to be formed from a polymer–transition-metal terminus (*e.g.*, -polymer–Ni terminus), for example, during work-up using acids. The preparation of π-conjugated polymers using Ni(0)L$_m$ sometimes gives the polymers without the -C–X terminal (halogen free polymers), suggesting that most of the polymer ends are -polymer-Ni units in these cases. When the obtained polymer has the -C–X terminal, it can be converted to a halogen-free polymer by treating with $LiAlH_4$ according to the Müllen's method.[55] The Pd-catalyzed polycondensation between X–Ar–X and Me_3Sn–Ar′–SnMe_3 gives the -C–X and -C–SnMe_3 terminal groups, and data from elemental analysis and ^{1}H NMR spectra sometimes indicate a -C–X terminal-rich structure,[84] suggesting that the -C–SnMe_3 bond has a higher reactivity than the -C–X bond toward Pd. The C–C coupling (or reductive elimination) on the transition-metal complex (*cf.* Schemes 4.8 and 4.9) is thought to be a key step in the polycondensation.

Various Ni and Pd complexes can be used for the dehalogenative organometallic polycondensation. Ni complexes are reactive toward -C–Cl, -C–Br and -C–I bonds. However, Pd complexes are usually not reactive to the -C–Cl bond and the polycondensation is usually carried out with Br–Ar–Br monomers. In the case of the polycondensation with Ni complexes, pseudo halogen compounds such as MsO–Ar–OMs are also usable as the monomer (cf. Example 10).

In the dehalogenative organometallic polycondensation of X–Ar′–X (dihaloaromatic compound or dihaloheterocyle) using Ni(0)L$_m$, Ni(0)L$_m$ serves as a condensing agent. As described above, the molecular weight of the obtained polymer usually increases with increase in the Ni(0)L$_m$/X–Ar′–X ratio. In the Ni-promoted polymerization using Zn, addition of iodides such as NaI is important to proceed smooth polymerization.

PPP is a rigid polymer and assumes a herringbone-type packing in solid with $a = 7.8$ Å, $b = 5.6$ Å and $c = 4.2$ Å parameters; one *p*-phenylene unit has a 4.2 Å repeating height along the polymer chain ($a = 7.8$ Å, $b = 5.5$ Å and $c = 4.3$ Å parameters have also been reported). PPP with side chains sometimes form well-ordered structure in the solid state, judging by the observation of distinct X-ray diffraction peaks. Vacuum evaporation of PPP on substrates gives a layer where PPP molecules stand upright on the surface of the substrate.[34] Poly(*m*-phenylene) assumes a helical structure in the solid state.[85,86] Polyphenylenes are usually stable under air and show high thermal stability.

References

1. *Handbook of Conducting Polymers, 3rd edn.*, ed. T. A. Skotheim and J. R. Reynolds, CRC Press, Boca Raton, Florida, 2007.
2. *Organic Conductive Molecules and Polymers, Vol. 2*, ed. H. S. Nalwa, John Wiley, Chichester, UK, 1997.
3. *Conjugated Polymer Synthesis*, ed. Y. Chujo, Wiley-VCH, Weinheim, Germany, 2010.

4. B. Schmaltz, T. Weil and K. Müllen, *Adv. Mater.*, 2009, **21**, 1067.
5. K. Akagi, *J. Polym. Sci., Part A: Polym. Chem.*, 2009, **47**, 2463.
6. M. Abe and T. Yamamoto, In *Science of Synthesis, Houben-Weyl, Methods of Molecular Transformations, Vol. 45b*, ed. J. S. Slegel and Y. Tobe, Georg Thieme Verlag, Stuttgart, Germany, 2010; Chapter 45.15, p. 653.
7. T. Yamamoto, *Bull. Chem. Soc. Jpn.*, 2010, **83**, 431.
8. T. Yamamoto, M. Abe, B. Wu, B.-K. Choi, Y. Harada, Y. Takahashi, K. Kawata, S. Sasaki and K. Kubota, *Macromolecules*, 2007, **40**, 5504.
9. B. Schmaltz, T. Weil and K. Müllen, *Adv. Mater.*, 2009, **21**, 1067.
10. M. Fukuda, K. Sawada and K. Yoshino, *J. Polym. Sci., Part A: Polym. Chem.*, 1993, **31**, 2465.
11. U. Scherf and E. J. W. List, *Adv. Mater.*, 2002, **14**, 477.
12. O. Inganäs, F. Zhang and M. R. Anderson, *Acc. Chem. Res.*, 2003, **42**, 1731.
13. W. Lu, J. Kuwabara, T. Iijima, H. Higashimura, H. Hayashi and T. Kanbara, *Macromolecules*, 2012, **45**, 5128.
14. N. Saito, T. Kanbara, T. Sato and T. Yamamoto, *Polym. Bull.*, 1993, **30**, 285.
15. T. Yamamoto, R. Tokimitsu, T. Asao, T. Iijima, H. Fukumoto, T.-A. Koizumi, T. Fukuda and H. Ushijima, *Macromol. Chem. Phys.*, 2011, **212**, 2406.
16. T. Yamamoto, K. Sugiyama, T. Kushida, T. Inoue and T. Kanbara, *J. Am. Chem. Soc.*, 1996, **118**, 3930.
17. G. Goldschmiedt, *Monatsh. Chem.*, 1886, **7**, 40.
18. A. A. Berlin, *J. Polym. Sci.*, 1961, **55**, 621.
19. A. A. Berlin, Z. V. Popova and D. M. Yanovskii, *Vysokomol. Soyed.*, 1965, **7**, 569.
20. S. Ozasa, N. Hatada, Y. Fujioka and E. Ibuki, *Bull. Chem. Soc. Jpn.*, 1980, **53**, 2610.
21. P. Kovacic and A. Kyriakis, *J. Am. Chem. Soc.*, 1963, **85**, 454.
22. P. Kovacic and J. Oziomek, *J. Org. Chem.*, 1964, **29**, 100.
23. P. Kovacic and M. B. Jones, *Chem. Rev.*, 1987, **87**, 357.
24. T. Osawa and K. Yoshino, *Synth. Met.*, 1987, **17**, 601.
25. C. S. Marvel and G. E. Hartzell, *J. Am. Chem. Soc.*, 1959, **81**, 448.
26. D. G. H. Ballard, A. Courtis, I. M. Shirley and S. C. Taylor, *J. Chem. Soc., Chem. Commun.*, 1983, 954.
27. D. G. H. Ballard, A. C. Courtis, I. M. Shirley and S. C. Taylor, *Macromolecules*, 1988, **21**, 294.
28. D. L. Gin, V. P. Conticello and R. H. Grubbs, *J. Am. Chem. Soc.*, 1994, **116**, 10507.
29. D. L. Gin, V. P. Conticello and R. H. Grubbs, *J. Am. Chem. Soc.*, 1994, **116**, 10934.
30. T. Yamamoto and A. Yamamoto, *Chem. Lett.*, 1977, **6**, 353.
31. T. Yamamoto, Y. Hayashi and A. Yamamoto, *Bull. Chem. Soc. Jpn.*, 1978, **51**, 2091.
32. T. Yamamoto, *Prog. Polym. Sci.*, 1992, **17**, 1153.

33. T. Yamamoto, K. Osakada, T. Wakabayashi and A. Yamamoto, *Makromol. Chem. Rapid Commun.*, 1985, **6**, 671.
34. T. Yamamoto, A. Morita, Y. Miyazaki, T. Maruyama, H. Wakabayashi, Z.-H. Zhou, Y. Nakamura, T. Kanbara, S. Sasaki and K. Kubota, *Macromolecules*, 1992, **25**, 1214.
35. T. Yamamoto, N. Hayashida and T. Maruyama, *Macromol. Rapid Commun.*, 1997, **198**, 341.
36. J. F. Fauvarque, A. Digua, M. A. Petit and J. Savard, *Makromol. Chem.*, 1985, **186**, 2415.
37. T. Yamamoto, M. Abla and Y. Murakami, *Bull. Chem. Soc. Jpn.*, 2002, **75**, 1997.
38. M. F. Semmelhack, P. M. Helquist and L. D. Jones, *J. Am. Chem. Soc.*, 1971, **93**, 5908.
39. A. S. Kende, L. S. Liebeskind and D. M. Braitsch, *Tetrahedron Lett.*, 1975, **16**, 3375.
40. I. Colon and D. R. Kelsey, *J. Org. Chem.*, 1986, **51**, 2627.
41. T. Yamamoto, *Chem. Lett.*, 2012, **41**, 1422.
42. T. Saito, Y. Uchida, A. Misono, A. Yamamoto, K. Morifuji and S. Ikeda, *J. Am. Chem. Soc.*, 1966, **88**, 5198.
43. T. Yamamoto, A. Yamamoto and S. Ikeda, *J. Am. Chem. Soc.*, 1971, **93**, 3350.
44. J. Cámpora, M. del Mar Conejo, K. Mereiter, P. Palma, C. Pérez, M. L. Reyes and C. Ruiz, *J. Organomet. Chem.*, 2003, **683**, 220.
45. A. Klein, M. P. Feth, H. Bertagnolli and S. Zälis, *Eur. J. Inorg. Chem.*, 2004, 2784.
46. M. Rehahn, A.-F. Schlüter, G. Wegner and W. J. Feast, *Polymer*, 1989, **30**, 1054.
47. R. Miyakoshi, K. Shimono, A. Yokoyama and T. Yokozawa, *J. Am. Chem. Soc.*, 2006, **128**, 16012.
48. K. Tamao, K. Sumitani, Y. Kiso, M. Zembayashi, A. Fujioka, S. Kodama, I. Nakajima, A. Minato and M. Kumada, *Bull. Chem. Soc. Jpn.*, 1976, **49**, 1958.
49. N. Miyaura and A. Suzuki, *Chem. Rev.*, 1995, **95**, 2458.
50. M. Kosugi, M. Koshiba, H. Sano and T. Migita, *Bull. Chem. Soc. Jpn.*, 1985, **58**, 1075.
51. J. K. Stille, *Angew. Chem. Int. Ed. Engl.*, 1986, **25**, 508.
52. T. Yamamoto, S. Wakabayashi and K. Osakada, *J. Organomet. Chem.*, 1992, **428**, 223.
53. M. B. Jones, P. Kovacic and D. Lanska, *J. Polym. Sci.: Polym. Chem. Ed.*, 1981, **19**, 89.
54. T. Yamamoto, B. Wu, B.-K. Choi and K. Kubota, *Chem. Lett.*, 2000, **29**, 720.
55. M. Kreyenschmidt, F. Uckert and K. Müllen, *Macromolecules*, 1995, **28**, 4577.
56. A. Krasovskiy and P. Knochel, *Angew. Chem. Int. Ed.*, 2004, **43**, 3333.
57. K. Ohshimizu, A. Takahashi, T. Higashihara and M. Ueda, *J. Polym. Sci. Part A: Polym. Chem.*, 2011, **49**, 2709.

58. K. Umezawa, T. Oshima, M. Yoshizawa-Fujita, Y. Takeoka and M. Rikukawa, *ACS Macro Lett.*, 2012, **1**, 969. M_n and M_w of the polymer seem to be estimated by GPC (eluent = THF; vs. PS standards).
59. T. Vahlenkamp and G. Wegner, *Macromol. Chem. Phys.*, 1994, **195**, 1933.
60. T. F. McCarthy, H. Witteler, T. Pakula and G. Wegner, *Macromolecules*, 1995, **28**, 8350.
61. A. D. Child and J. R. Reynolds, *Macromolecules*, 1994, **27**, 1975.
62. S. Kim, J. Jackiw, E. Robinson, K. S. Schanze, J. R. Reynolds, J. Baur, M. F. Rubner and D. Boils, *Macromolecules*, 1998, **31**, 964.
63. P. B. Balanda, M. B. Ramey and J. R. Reynolds, *Macromolecules*, 1999, **32**, 3970.
64. S. Lightowler and M. Hird, *Chem. Mater.*, 2004, **16**, 3963.
65. V. Percec, J.-Y. Bae, M. Zhao and D. H. Hill, *Macromolecules*, 1995, **28**, 6726.
66. R. W. Phillips, V. V. Sheares, E. T. Samulski and J. M. DeSimone, *Macromolecules*, 1994, **27**, 2354.
67. I. Tonozuka, M. Yoshida, K. Kaneko, Y. Takeoka and M. Rikukawa, *Polymer*, 2011, **52**, 6020.
68. J. L. Reddinger and J. R. Reynolds, *Macromolecules*, 1997, **30**, 479.
69. D. G. D. Patel, Y.-y. Ohnishi, Y. Yang, S.-H. Eom, R. T. Farley, K. R. Graham, J. Xue, S. Hirota, K. S. Schanze and J. R. Reynolds, *J. Polym. Sci. Part B: Polym. Phys.*, 2011, **49**, 557.
70. A. Izumi, M. Teraguchi, R. Nomura and T. Masuda, *Macromolecules*, 2000, **33**, 5347.
71. A. Izumi, R. Nomura and T. Masuda, *Chem. Lett.*, 2000, **29**, 728.
72. M. Horie, I. Yamaguchi and T. Yamamoto, *Macromolecules*, 2006, **39**, 7493.
73. T. Yamamoto, T. Sato, T. Iijima, M. Abe, H. Fukumoto, T.-A. Koizumi, M. Usui, Y. Nakamura, T. Yagi, H. Tajima, T. Okada, S. Sasaki, H. Kishida, A. Nakamura, T. Fukuda, A. Emoto, H. Ushijima, C. Kurosaki and H. Hirota, *Bull. Chem. Soc. Jpn.*, 2009, **82**, 896.
74. J. Xu, S. C. Ng and H. S. O. Chan, *Tetrahedron Lett.*, 2001, **42**, 5327.
75. T. Yasuda, Y. Sakai, S. Aramaki and T. Yamamoto, *Chem. Mater.*, 2005, **17**, 6060.
76. A. Tanimoto and T. Yamamoto, *Macromolecules*, 2006, **39**, 3546.
77. K. Si, D. Dong, R. Wycisk and M. Litt, *J. Mater. Chem.*, 2012, **22**, 20907.
78. Y. Muramatsu and T. Yamamoto, *Chem. Lett.*, 1997, **26**, 581.
79. T. Yamamoto, Y. Muramatsu, B.-L. Lee, H. Kokubo, S. Sasaki, M. Hasegawa, T. Yagi and K. Kubota, *Chem. Mater.*, 2003, **15**, 4384.
80. P. J. Flory, *Principles of Polymer Chemistry*, Cornell University Press, Ithaca, 1953.
81. K. Kubota, H. Uraba, Y. Tominaga and S. Fujime, *Macromolecules*, 1984, **17**, 2096.
82. T. Yamamoto, I. Wataru, T. Kanbara, Y. Nakamura, M. Kikuchi and I. Ando, *Chem. Lett.*, 1992, **21**, 2001.

83. Y. Furukawa, H. Ohtsuka, M. Tasumi, I. Wataru, T. Kanbara and T. Yamamoto, *J. Raman Spectrosc.*, 1993, **24**, 551.
84. For example, T. Yamamoto, M. Yoshizawa, A. Mahmut, M. Abe, S.-I. Kuroda, T. Imase and S. Sasaki, *J. Polym. Sci. Part A: Polym. Chem.*, 2005, **43**, 6223
85. S. Kobayashi, S. Sasaki, M. Abe, S. Watanabe, H. Fukumoto and T. Yamamoto, *Macromolecules*, 2004, **37**, 7986.
86. D. J. Williams, H. M. Colquhoun and C. A. O'Mahoney, *J. Chem. Soc., Chem. Commun.*, 1994, 1643.

CHAPTER 5

Polyfluorenes

BYUNG JUN JUNG,[a] HONG-KU SHIM[b] AND DO-HOON HWANG*[c]

[a] Department of Materials Science and Engineering, The University of Seoul, Seoul 130-743, Korea; [b] Department of Chemistry, Korea Advanced Institute of Science and Technology, Daejeon 305-701, Korea; [c] Department of Chemistry, and Chemistry Institute for Functional Materials, Pusan National University, Busan 609-735, Korea
*Email: dohoonhwang@pusan.ac.kr

5.1 Introduction

A fluorene is an interesting aromatic system that is composed of a biphenyl connected by a methylene bridge, and does not contain any heteroatoms such as N, O, or S. The numbering for the carbon positions is shown in Chart 5.1. The five-membered ring from carbon 9 to carbon 13 in the center exhibits 6-electron aromaticity when the fluorene loses a proton at the 9-position. Therefore, the two protons at the 9-position are acidic, and the easily generated fluorenyl anion readily participates in nucleophilic substitution reactions with alkyl- or aryl-based electrophiles. The alkyl or aryl substituents are not located in the same plane as the fluorene rings and can function as good solubilizing groups in polymers derived from these compounds. In the synthesis of π-conjugated polymers, the solubility of the final polymer is almost always the primary concern of polymer chemists. Due to strong interactions in the π-conjugated domains, unsubstituted π-conjugated polymers usually have low solubility in common organic solvents, and their device applications as thin films are restricted. The introduction of the two side

RSC Polymer Chemistry Series No. 9
Conjugated Polymers: A Practical Guide to Synthesis
Edited by Klaus Müllen, John R. Reynolds and Toshio Masuda

Published by the Royal Society of Chemistry, www.rsc.org

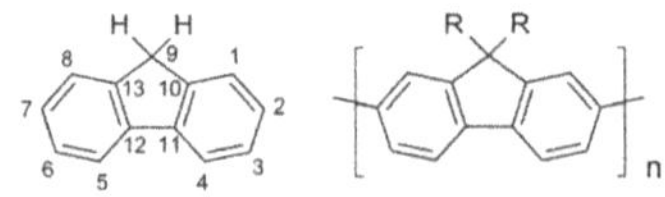

Chart 5.1 Fluorene and polyfluorene.

groups at the 9-position underpins the importance of the fluorene unit as a well-known monomer in π-conjugated polymers.

In the polymerization of fluorene, the polymer chain is propagated through connection at the 2- and 7-positions by aryl-aryl coupling; this configuration is similar to the *para*-position in biphenyls. Polyfluorenes (PFs) are analogous to polyphenylenes (PPs), and like PPs, they have wide bandgaps. Therefore, they are attractive and have been developed as blue light-emitting polymers. Numerous alternating or random copolymers based on fluorene have been designed and synthesized with other aromatic units for use in other applications. In addition, scientists are applying PFs with various functional side groups in new, emerging fields.

In this chapter, synthetic routes to fluorene monomers and fluorene polymerization methods are introduced, with a specific focus on homopolymers. However, the polymerization methods can be used for alternating or random copolymers. Recent developments in polymerization methods, including purification, are mentioned.

5.2 Experimental Procedures

5.2.1 Oxidative Polymerization

The oxidative polymerization of 9,9′-dialkylfluorenes[1] was initially introduced for soluble polyfluorenes. The procedure for the synthesis of the monomers and their polymerization to polyfluorenes is outlined in Scheme 5.1. The route was simple, but this polymerization provided only low molecular weight polyfluorenes with a degree of polymerization of ~10. Therefore, researchers investigated other polymerization methods to achieve high molecular weight polyfluorenes. This initial work on polyfluorenes provided the driving force for the development of blue light-emitting polymers. This synthetic route is still useful in other polymerizations with the same or similar starting materials.

5.2.1.1 Monomer Synthesis: 9,9-Dialkylfluorenes[1–3]

To a solution of fluorene (1.0 g, 6.0 mmol) in dry ether or THF (14 mL) at −78 °C is added dropwise *n*-butyllithium (12.6 mmol, 2.1 equiv, 2.5 M or 1.6 M in hexane). The color of the solution gradually changes to red. The mixture is stirred at −78 °C for 1 h, and an alkyl bromide (13.8 mmol, 2.3 equiv) in dry ether or THF (3 mL) is added dropwise. The solution is allowed to warm to room temperature and stirred for 3 h. The mixture is poured into water and extracted with ether. The organic extracts are washed with brine and dried over $MgSO_4$. The solvent is removed with a rotary

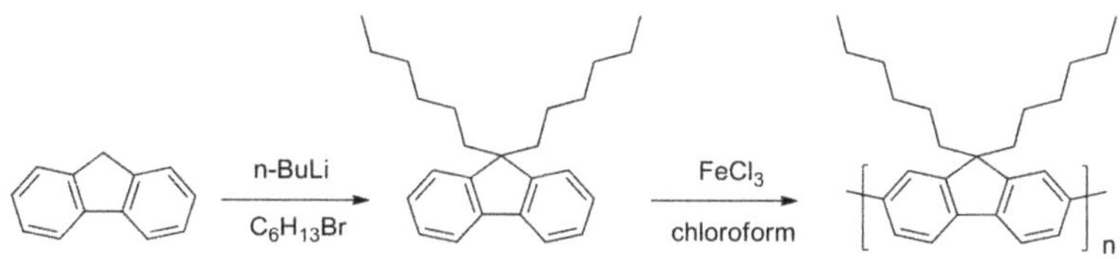

Scheme 5.1

evaporator. In the case of a long alkyl bromide or large-scale synthesis, the excess alkyl bromide is removed by vacuum distillation. Furthermore, the residue can be purified by column chromatography using *n*-hexane or petroleum ether, and/or recrystallized from ethanol to give the desired products. The products are white solids (9,9-dihexylfluorene, mp. 30–32 °C) or colorless oils (*e.g.*, 9,9-dioctylfluorene, mp. 23 °C) depending on the alkyl groups.

Now some 9,9-dialkylfluorenes are commercially available. This alkylation is often useful for the design and synthesis of new monomers with other alkyl chains or unsymmetrical alkyl groups. Monosubstituted 9-alkylfluorenes can be synthesized with 1 equiv *n*-butyllithium using the above procedure. Alternatively, these compounds can also be obtained by the base-mediated alkylation of fluorene and primary alcohols.[4] 9-Alkylfluorenes are possible side products in the foregoing procedure and affect both the optical and electrical properties of the final polymers.[5] Therefore, careful purification of the products to remove the 9-alkylfluorene impurities is often required. Recently, new multistep routes to 9,9-dialkylfluorenes were developed by Holmes's group.[4,6]

5.2.1.2 Polymer Synthesis: Poly(9,9-dialkylfluorene)[1]

Anhydrous iron(III) chloride (3–4 equiv, $FeCl_3$) is placed into a Schlenk flask under an inert atmosphere (glove box). Anhydrous chloroform is added into the flask, which is then connected to a Schlenk line under an inert atmosphere (Ar or N_2). A solution of 9,9-dialkylfluorene in anhydrous chloroform is slowly added to the stirred solution of $FeCl_3$ in chloroform. The mixture is stirred for 70 h at room temperature. The reaction mixture is then quenched by adding it to methanol. The precipitated solid material is separated by filtration and washed with methanol. The crude product is dissolved in chloroform and reprecipitated with methanol. The refined polymer is washed with methanol and dried at room temperature under vacuum overnight.

5.2.2 Yamamoto Coupling Polymerization

Monomer Synthesis

The Yamamoto polymerization is an AA type, which uses dibromide compounds as monomers. Therefore, a bromination step is required during monomer synthesis, as shown in Scheme 5.2. Initially, researchers extended the first method from the already known synthesis of 9,9-dialkylfluorenes. Lately, scientists have preferred the second route, which can be scaled more easily. When polymer chemists design new polymer structures, they often consider scalable synthetic

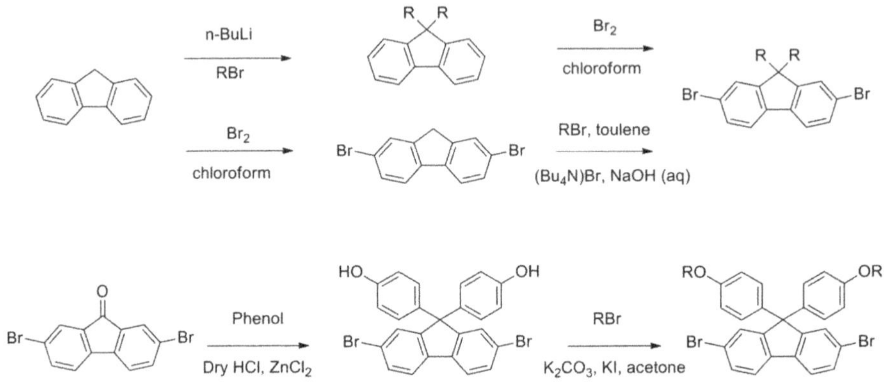

Scheme 5.2

routes for monomers, although the scale of the final polymerization may be less than 1 g or 1 mmol. Often, many trials are required for successful polymerization and the optimization of high molecular weights. In addition, adequate supply of pure monomers can provide source material for other copolymers.

5.2.2.1 2,7-Dibromo-9,9-dialkylfluorenes[2]

Bromine (4.13 mL, 80.6 mmol, 2.1 equiv) is slowly added to a solution of 9,9-dioctylfluorene (15.0 g, 38.4 mmol, 1 equiv) and iron(III) chloride (96 mg, 0.59 mmol) in $CHCl_3$ (58 mL) at 0 °C. A syringe is acceptable for adding small amounts of bromine. When adding a large amount of bromine, the use of a dropping funnel and a reflux condenser with a hydrogen bromide absorption trap is highly recommended. It is important that the reaction proceeds in the dark; therefore, the reaction flask needs to be wrapped with aluminum foil to avoid undesired aliphatic bromination elsewhere in the molecule. The solution is warmed to room temperature and stirred for 3 h. The resulting slurry is poured into water and washed with $Na_2S_2O_3$ (aq) until the red color disappears. The aqueous layer is extracted with $CHCl_3$, and the combined organic layers are dried over $MgSO_4$ to afford the desired product as a solid after concentration. If necessary, the product can be decolorized by charcoal and recrystallized from ethanol. The reported yields for 2,7-dibromo-9,9-dioctylfluorene[2] and 2,7-dibromo-9,9-dihexylfluorene[3] are over 99% and 75%, respectively. The bromination of 9,9-dialkylfluorenes can be also performed using excess bromine in DMF.[7,8]

5.2.2.2 2,7-Dibromofluorene

2,7-Dibromofluorene is now widely available commercially. 2,7-Dibromofluorene is prepared by the bromination of fluorene, and the method is similar to the bromination of 9,9-dialkylfluorenes.[9] Alternatively, *N*-bromosuccinimide (NBS)/dimethylformamide (DMF) can be used instead of $Br_2/CHCl_3$, and the reaction is conducted at reflux rather than r.t.[10] 2,7-Dibromofluorene is now commercially available.

5.2.2.3 2,7-Dibromo-9,9-dialkylfluorenes

2,7-Dibromo-9,9-dialkylfluorenes can be prepared by reacting 2,7-dibromofluorene with an alkyl bromide using a phase-transfer catalyst (PTC).[11] This method is easily scalable because *n*-BuLi is not employed. Therefore, a sufficient supply of 2,7-dibromo-9,9-dialkylfluorene monomers for Yamamoto polymerization, can be obtained.[10,12] 2,7-Dibromofluorene (10 g, 31 mmol), the phase-transfer catalyst tetra-*n*-butylammonium bromide (TBAB, 0.32 g, 1.0 mmol), and 1-bromooctane (10.7 g, 71 mmol, 2.3 equiv) are dissolved in toluene (90 mL). The solution is purged with N_2 or Ar to prevent formation of the 2,7-dibromofluorenone byproduct. Aqueous NaOH solution (90 mL, 50 wt%) is slowly added to the solution, and the reaction mixture is stirred at 60 °C for 4 h. After dilution with ethyl acetate, the organic layer is washed with water several times to remove excess NaOH. The layer can be additionally washed with saturated aqueous sodium bicarbonate solution. The organic layer is separated and dried with anhydrous $MgSO_4$. After the removal of the solvent, the residue is purified by column chromatography using *n*-hexane to remove 2,7-dibromofluorenone and/or recrystallized from ethanol to yield the desired products. The reported yield for 2,7-dibromo-9,9-dioctylfluorene is 83%.[10]

Alternatively, benzyl(triethyl)ammonium chloride/dimethyl sulfoxide (DMSO) can be used instead of TBAB/toluene.[13,14] DMSO is a polar aprotic solvent that accelerates S_N2 reactions with alkyl bromides. For 2,7-dibromo-9,9-dialkylfluorenes with different alkyl groups, the above alkylation is still possible under a controlled feeding ratio if the product can be separated by column chromatography as a result of polarity difference. For example, 2,7-dibromofluorene (1 equiv), 1-bromohexane (0.8 equiv) and 2-(2-bromoethoxy)tetrahydro-2H-pyran (1.3 equiv) generate 2-[2-(2,7-dibromo-9-hexylfluoren-9-yl)ethoxy]perhydro-2H-pyran as a major product. This can be converted to 2-(2,7-dibromo-9-hexylfluoren-9-yl)ethan-1-ol by deprotection under acidic conditions.[15] The alkylation of 2,7-dibromo-9-alkyl-9H-fluorene is another possible route for the preparation of 2,7-dibromo-9,9-dialkylfluorenes with different substituents.[4]

5.2.2.4 2,7-Dibromo-9,9-bis(4-n-octyloxyphenyl)fluorene[16]

A mixture of 2,7-dibromofluorenone (18 g, 53 mmol), phenol (20 g, 214 mmol), and zinc chloride (3g, 22 mmol) is charged in a 250 mL three-necked flask (Scheme 5.2, lower reaction). Dry HCl gas is supplied for 1 h at 60 °C with vigorous stirring. The resultant mixture is washed with toluene. The solid product, 2,7-dibromo-9,9-bis(4-hydroxyphenyl)fluorene, is obtained after filtration and vacuum drying (yield: 25 g, 92%). A mixture of 2,7-dibromo-9,9-bis(4-hydroxyphenyl)fluorene (25 g, 49 mmol), K_2CO_3 (20 g, 147 mmol), 1-bromooctane (28.4 g, 147 mmol), and KI (4 g, 24.5 mmol) is prepared in acetone (400 mL), and the reaction is refluxed for 2 days. The reaction mixture is extracted three times with chloroform and water. The organic layer is dried with anhydrous $MgSO_4$, and the solvent is removed by a rotary evaporator. The resultant crude product is purified by column chromatography (hexane/ethyl acetate = 30 : 1). The obtained solid is recrystallized from MeOH. (Yield: 28 g, 75%).

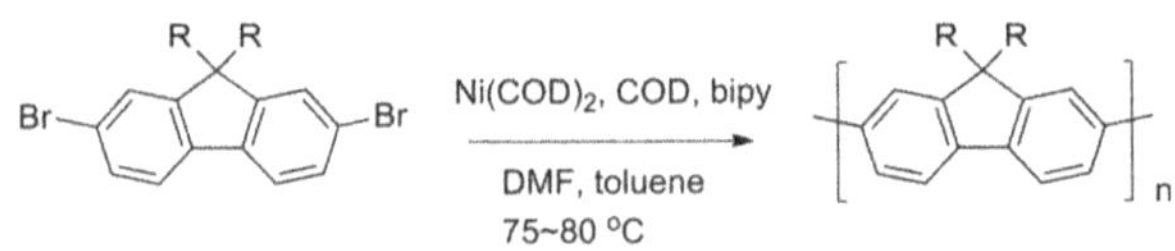

Scheme 5.3

5.2.2.5 *Polymer Synthesis*

A solution containing bis(1,5-cyclooctadiene)nickel(0) ($Ni(COD)_2$, which is handled in the glove box), 1,5-cyclooctadiene (COD), and 2,2′-dipyridyl in a 1 : 1 : 1 mol ratio (2.67 mmol each) in anhydrous DMF (5 mL) is heated at 80 °C for 30 min under argon. Monomers (1.35 mmol) dissolved in toluene (5 mL) are added dropwise to the solution. After 3 days of heating, excess end-capping reagent dissolved in anhydrous toluene (3 mL) is added to the flask. The reaction mixture is further heated at 80 °C for 24 h. After cooling to room temperature, the reaction mixture is poured into HCl (100 mL), acetone (100 mL), and methanol (100 mL), and stirred for 2 h. The filtered polymers are extracted with chloroform; K_2CO_3 solution is used instead of pure water until the aqueous layers are neutralized. The organic layer is separated, dried over $MgSO_4$, concentrated by solvent evaporation, and the residue is precipitated from methanol. The polymers can be further purified by Soxhlet extraction in methanol/acetone for 2 days. A final purification is performed by repeating the precipitation with methanol (Scheme 5.3). The amounts of solvent can be varied, because the different side groups in the monomers affect the solubilities of the final polymers in the reaction solvents. The chief concern is gelation during the polymerization due to insufficient solvent or the sudden emergence of high molecular weight polymer chains. Sometimes, an additional injection of toluene is advisable based on the viscosity of the reaction mixture.

Pei and Yang[7] reported the synthesis of polyfluorene using Zn/$NiCl_2$, and Miller's group[17] at IBM used $Ni(COD)_2$ as the reactive aryl-aryl coupling agent according to the Yamamoto polymerization[18] for other conjugated polymers. This polymerization has been widely used for the homopolymerization of polyfluorenes. The various monohaloaryl end-cappers, as shown Chart 5.2, are introduced to enhance electrical properties and stabilize blue emission. Table 5.1 shows a summary of reported polymerizations (Chart 5.3).

5.2.3 Suzuki Coupling Polymerization

The Suzuki polymerization is the AA–BB type, and the monomers are usually diarylbromide compounds and diarylboronate esters. It is possible to use diboronic acids as monomers, but the diboronate esters are generally used because they are more easily purified than diboronic acids.

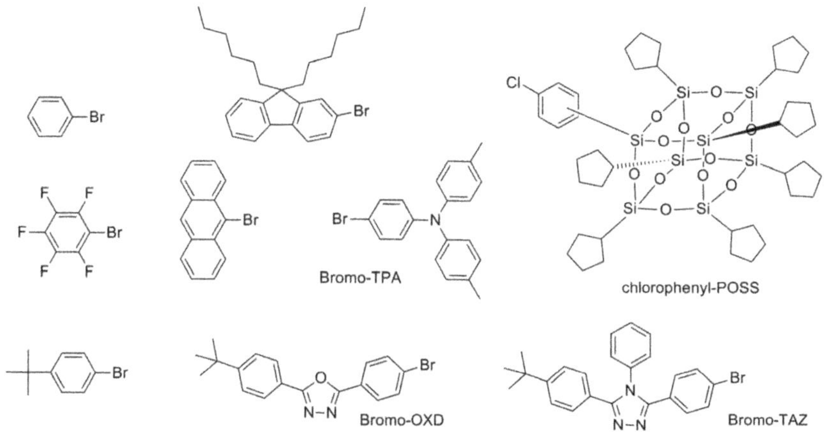

Chart 5.2 Chemical structures of various end-cappers.

5.2.3.1 2,7-Bis(4,4,5,5-tetramethyl-1,3,2-dioxaborolan-2-yl)-9,9-dioctylfluorene[2,28,29]

To a solution of 2,7-dibromo-9,9-dioctylfluorene (1.0 g, 1.8 mmol) in dry THF (14 mL) at −78 °C is added dropwise *n*-butyllithium (1.6 mL, 4.0 mmol, 2.5M in hexane, 2.2 equiv) (Scheme 5.4, upper). The pink-orange mixture is stirred at this temperature for 10 min, warmed at 0 °C for 7 min, and cooled again at −78 °C for 20 min. 2-Isopropoxy-4,4,5,5-tetramethyl-1,3,2-dioxaborolane (1.9 g, 2.1 ml, 10 mmol) is added rapidly by syringe, and the resulting solution is stirred at −78 °C for 6 h. At that time, the color changes to orange. The bath is allowed to reach room temperature very slowly while stirring for 24 h. The pale yellow mixture is poured into water and extracted with ether. The organic layer is washed with water and brine, dried over $MgSO_4$, and concentrated to give a pale yellow solid that is recrystallized from *n*-heptane (or *n*-hexane) to afford the desired product as white crystals (Yield: 0.94 g, 94%).[29]

For this reaction, the flask should be well dried before use. Fresh *n*-BuLi solution is recommended for the beginner, as the concentration of *n*-BuLi in hexane may change when it is not carefully sealed or maintained by other users. The correct *n*-BuLi concentration can be determined by titration. Lim *et al.*[30] used *t*-BuLi instead of *n*-BuLi and did not elevate the temperature to 0 °C after the injection of the reagent. Liu and Bazan used *t*-BuLi for the lithiation of 2,7-dibromo-9,9-bis(6′-bromohexyl)fluorene to minimize nucleophilic reaction of the aliphatic bromide on the side chain.[31] *t*-BuLi is a powerful but quite dangerous basic reagent. Therefore, *t*-BuLi should be used as a last resort, and if used, the amount should be small. Recently, a new method for the synthesis of boronic esters was developed using bis(pinacolato)diboron and a Pd catalyst.[32] This method is quite useful for precious dibromoaryl compounds that have been synthesized over several steps.[33] When the above procedure is carried out on a large scale (5–7 g of 2,7-dibromo-9,9-dioctylfluorene), the purification by column chromatography is often needed due to side-product formation.

Table 5.1 Polymerization of fluorenes with $Ni(COD)_2$.

Polymer	*Side chain, [dibromide] (in toluene)*	*Reaction temp/time*	*dibromide/nickel complex*	*End capper*	*yield, %*	M_w	*PDI*	*ref.*
PDHF	Hexyl,	80 °C/2–3 days	0.65	none	83	45 000	2.5	17
PDHF	Hexyl, 0.25 M	80 °C/1 day	0.57	2-bromo-9,9-di-n-hexylfluorene[a]	N/A	32 500	2.5	12
PDHF	Hexyl, 0.12 M	80 °C/3 days	0.67	bromopentafluorobenzene	62	82 000	2.3	15
PDHF	Hexyl, 0.12 M	80 °C/3 days	0.67	9-bromoantracene	67	105 300	3.4	19
PFO	Octyl, 0.16 M	80 °C/2 days	0.43	chlorophenyl-POSS	79	160 000[b]	N/A	20
PFO	Octyl, 0.17 M	80 °C/4 days	0.45	none	N/A	69 700	1.7	10
PFO	Octyl, 0.13 M	85 °C/2 days	0.64	1-bromo-4-tert-butylbenzene	60	194 000[c]	1.6	21
PFO	Octyl, 0.13 M	85 °C/2 days	0.64	Bromo-OXD	N/A	182 000[c]	2.0	21
PFO	Octyl, 0.13 M	85 °C/2 days	0.64	Bromo-TAZ	N/A	277 000[c]	1.9	21
PFO	Octyl, 0.15 M	75 °C/1 day	0.91	bromobenzene	79–90	25 000–39 800	1.7–2.6	6
PF2/6	Ethylhexyl	80 °C/5 days	N/A	none	N/A	211 000	1.7	14
PF2/6am4	Ethylhexyl	N/A	N/A	Bromo-TPA [a] (4%)	N/A	76 800	1.6	22
PF2/6	Ethylhexyl, 0.36 M	80 °C/3 days	N/A	9-bromoanthracene	50	79 000	3.2	23
PF2/6	Ethylhexyl, 0.078 M in THF	80 °C/3 days	0.42	N/A	50	222 000	1.7	24
PTPAF	TPA, 0.09 M	75 °C/1 day	0.92	bromobenzene	97	25 200	2.1	25
PBOPF	octyloxyphenyl, 0.36 M	80 °C/3 days	N/A	bromopentafluorobenzene	60	40 000	2.0	19
PF	Dendron, 0.02M	80 °C/3 days	0.34	bromobenzene	75	160 000	3.6	26
PF	Dendron, 0.16 M	80 °C/3 days	0.33	bromobenzene	52	103 000	2.0	14
PF7CY	-$(CH_2)_7$-C_6H_{12}, 0.27 M	80 °C/3 days	0.51	9-bromoanthracene	73	92 000	2.1	27

[a]The end cappers are added with the starting monomers at the beginning of the reaction.
[b]Molecular weight is estimated by elemental analysis, including the end groups.
[c]The reaction is performed in a glove box.

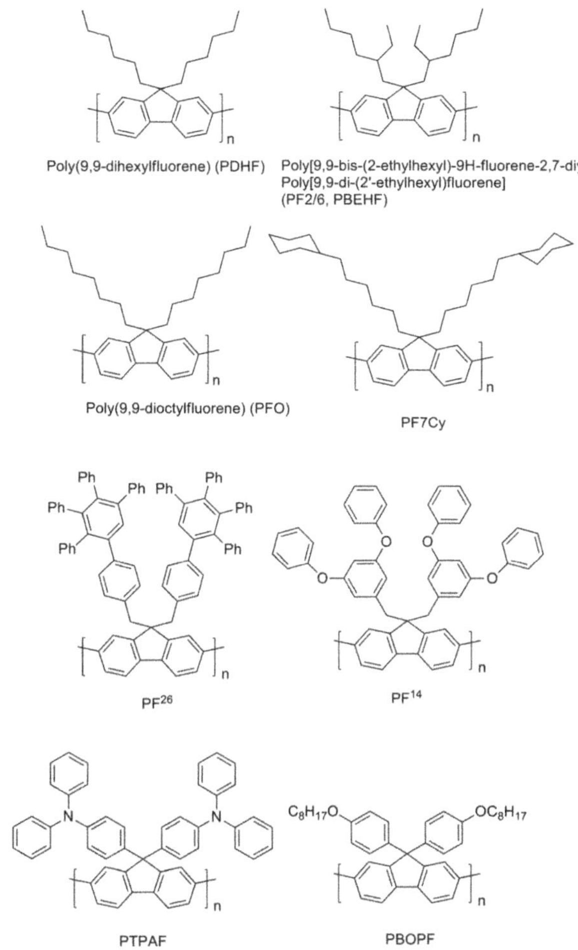

Chart 5.3 The structures of the polyfluorenes listed in Tables 5.1 and 5.3.

5.2.3.2 9,9-Dihexylfluorene-2,7-bis(trimethylene boronate)[34,35]

2,7-Dibromo-9,9-dihexylfuorene (16.27 g, 33 mmol) is treated with magnesium turnings (1.90 g, 80 mmol) in dry THF containing a catalytic amount of iodine to form the Grignard reagent (Scheme 5.4, lower). The Grignard reagent solution is slowly dropped into a stirred solution of trimethyl borate (38 mL, 330 mmol) in dry THF at −78 °C. The mixture is stirred at −78 °C for 2 h and then at room temperature for 2 days. The reaction mixture is then poured onto crushed ice containing sulfuric acid (5%) while stirring. The mixture is extracted with diethyl ether and the combined extracts are evaporated to give a white solid. Recrystallization of the crude product from hexane–acetone (1 : 2) affords pure 9,9-dihexylfluorene-2,7-diboronic acid (6.31 g, 44%) as a white solid. The diboronic acid is then reacted with propane-1,3-diol in toluene under

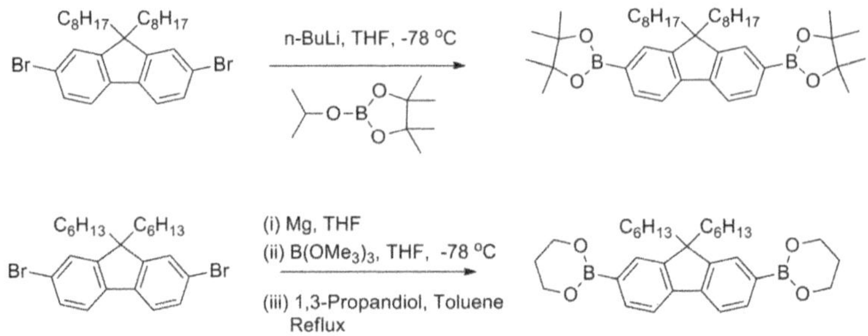

Scheme 5.4

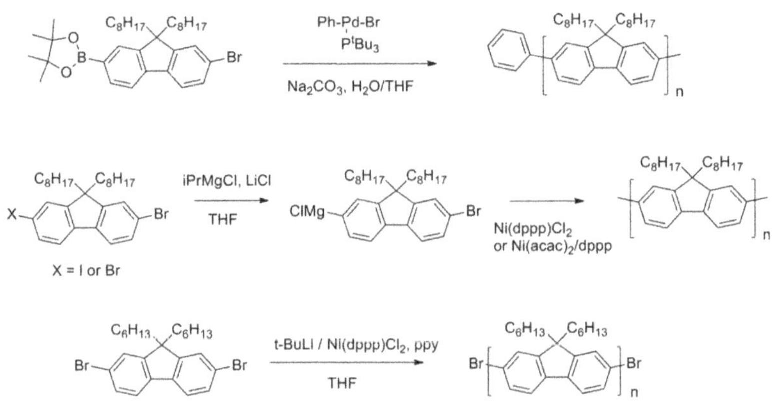

Scheme 5.5

reflux for 10 h. After work up, the crude product is recrystallized from hexane to afford white crystals (Yield: 5.50 g, 73%).

This monomer, also named 9,9-dihexylfluorene-2,7-diboronic acid bis(1,3-propanediol) ester, is commercially available. The propanediol or pinacol boronate esters of dialkylfluorenes with different alkyl chain lengths are also commercially available products, and these compounds can be used directly in the following Suzuki polymerization. However, the given synthetic procedures for boronate ester monomers will be useful for preparing new fluorene monomers, which are not commercially available.

5.2.3.3 Polymer Synthesis

Three principal types of Suzuki polymerizations[36] for polyfluorenes have been developed for homopolymers and copolymers by Dow, Cambridge Display Technology (CDT), and Covion, which was merged with Merck. Early detailed experimental procedures were reported in the patent literature (Scheme 5.5 and Table 5.2).[37–39] These and other groups have used these polymerizations for numerous polyfluorene derivatives (Table 5.3).

Table 5.2 Comparison of Suzuki coupling protocols for polyfluorenes.

	Method 1	*Method 2*	*Method 3*
Catalyst	$Pd(PPh_3)_4$	$Pd(PPh_3)_4$ or Pd(II) catalyst/Phosphine ligand	$Pd(PPh_3)_4$
PTC	Aliquat 336	Et_4NOH	–
Base	Na_2CO_3 or K_2CO_3	Et_4NOH	$K_3PO_4 \cdot H_2O$
Solvent	Toluene/Water	Toluene/Water	Toluene/Water/Ethanol

Table 5.3 Suzuki-type polymerization of fluorenes.

Polymer	*Method*	*Pd catalyst mol%*[a]	*[bromide monomer] in toluene (M)*	*Reaction condition*	*yield, %*	M_w	*PDI*	*ref.*
PFO	1[b]	0.5–1.5	0.50	Reflux, 2 days	N/A	40 000	1.7	2
PDHF	1[b]	3.0	0.10[c]	90 °C, 1–3 days	N/A	37 000	2.6	45
PFO	1	0.5	0.12	Reflux, 2.5 days	N/A	148 000	3.1	36
PFO	1	0.06	0.11[d]	Reflux, 2.5 days	95	–[e]	N/A	36
PFO	1	1.0	0.09[d]	Reflux, 2.5 days	81	100 000	2.8	30
PF[f]	1	1.0	0.04	Reflux, 2.5 days	60	35 000	1.6	46
PF[f]	1	3.7	0.06	Reflux, 3 days	95	160 000	1.5	47
PFO	2	6.3	0.04	90 °C, 1 day	86	21 000	3.5	6
PFO	2	0.3	0.10	Reflux, 2 days	N/A	204 000[g]	N/A	34
PFO	2	2.0	0.20	90 °C, 1 day	64	81 000	2.8	40
PFO	2[h]	1.5–2.0	0.10	Reflux, 2 days	85	101 000	2.2	43
PF[f]	3	1.5	0.33	87 °C, 4 days	84	67 000	2.3	39
PF[f]	3	1.5	0.33	87 °C, 2 days	77	115 000	2.4	44

[a]With respect to bromide monomer.
[b]method 1, but without phase-transfer catalyst.
[c]THF instead of toluene.
[d]bromide/boronate ester = 1 : 1.02.
[e]Greater than 148 000.
[f]Polymer structures are described in the reference.
[g]M_p (peak molecular weight).
[h]$Pd(OAc)_2$, tricyclohexylphosphine.

Method 1. Poly(9,9-dioctylfluorene) (PFO)[30,37]

2,7-Dibromo-9,9-dioctylfluorene (0.88 g, 1.61 mmol), 2,7-bis(4,4,5,5-tetramethyl-1,3,2-dioxaborolan-2-yl)-9,9-dioctylfluorene (1.06 g, 1.64 mmol or 1.03 g, 1.61 mmol) and tetrakis(triphenylphosphine)palladium (13.4 mg, 1 mol%) are dissolved in anhydrous toluene (16 mL). A phase-transfer catalyst, Aliquat 336 (65 mg, 0.16 mmol) is added to 2 M aqueous Na_2CO_3 (2.6 mL, 5.2 mmol). The solution is degassed by sparging with nitrogen for 15 min and transferred to the toluene solution *via* cannula. The reaction mixture is stirred and heated at reflux for 60 h until the reaction mixture becomes viscous. The end group is capped by heating the mixture for 12 additional hours with bromobenzene

(25 mg, 0.16 mmol) in anhydrous toluene (1 mL). When the same equivalents of monomers are used, an excess of phenylboronic acid or ester as another end capper is added. After further stirring and heating for 12 h, the reaction mixture is cooled to about 50 °C and added slowly to a vigorously stirred mixture consisting of methanol (160 mL) and 1 N aqueous HCl (7 mL). The polymer fibers are collected by filtration and precipitation from methanol and acetone. The polymer is purified by Soxhlet extraction in acetone for 2 days, and column chromatography using chloroform/toluene as an eluent to remove oligomers and catalyst residues. The precipitation procedure in toluene/methanol is then repeated several times. The final product is obtained after drying in vacuo at 60 °C, with a yield of 0.85 g (81%).[30]

Method 2. Poly(9,9-dioctylfluorene) (PFO)[6,38,40]

2,7-Dibromo-9,9-dioctylfluorene (0.548 g, 1.00 mmol), 2,7-bis(4,4,5,5-tetramethyl-1,3,2-dioxaborolan-2-yl)-9,9-dioctylfluorene (0.642 g, 1.00 mmol) and tetrakis(triphenylphosphine)palladium(0) (23 mg, 2 mol%) are dissolved in toluene (5 mL). The solution is degassed by sparging with nitrogen for 15 min. A solution of 1 M aqueous Et_4NOH (5 mL) is also degassed thoroughly. The monomer solution and aqueous base solution are added to a Schlenk tube and heated at 90 °C for 24–48 h. Bromobenzene is added to the mixture and stirred at 90 °C for 12 h. Phenylboronic acid is added to the mixture after which the mixture is stirred for 12 h. The mixture is cooled to room temperature and dropped by pipette into stirred methanol. The residue is redissolved in toluene and filtered through a plug of silica followed by reprecipitation in MeOH. A pale yellow amorphous solid (0.25 g, 64% yield) is collected by filtration and dried under vacuum.[40] CDT developed a Pd(II)/phosphine ligand catalystic system in place of the air-sensitive $Pd(PPh_3)_4$ catalyst.[41,42] This modified Suzuki method afforded PFO with molecular weights exceeding 100 000.[43]

Method 3. Poly(spirobifluorene)[39,44]

The boronic ester (10 mmol), the bromide monomers (10 mmol), and $K_3PO_4 \cdot H_2O$ (9.67 g, 42 mmol) are mixed in toluene (30 mL), water (15 mL) and ethanol (0.25 mL) and degassed for 30 min by passing N_2 through the mixture. $Pd(PPh_3)_4$ (175 mg, 0.15 mmol) is subsequently added under inert atmosphere. The suspension is stirred vigorously at 87 °C under N_2. After 2–4 days, a monoboronic ester is added, and the mixture is refluxed for another 6 h. Then, bromobenzene is added, and the mixture is refluxed for another 3 h. After cooling, the reaction mixture is diluted with toluene (200 mL) and stirred with NaCN (200 mL, 2% in water) for 3 h. The organic phase is washed with water and the polymer is precipitated by addition to ethanol (800 mL). The polymer is then dissolved in THF (200 ml) and reprecipitated from methanol (250 ml). The latter procedure is repeated. The polymer is collected by filtration and dried under vacuum.

5.3 Developments in Polymerization

5.3.1 Microwave-Assisted Polymerization

The conventional Yamamoto or Suzuki coupling conditions for polymerization involve heating around 80 °C or refluxing for several days to obtain high molecular weight polymers and high yields. Recently, microwave heating has enabled organic chemists to reduce reaction times for many synthetic methods. The equipment is not particularly expensive, and the method has spread throughout many synthetic groups. Carter[48] applied microwave heating in the Yamamoto coupling and obtained poly(9,9-dihexylfluorene) (PDHF) with M_n>100 000 within 10 minutes. The PDI was quite reasonable at 1.65–2.22. It is possible to insert a somewhat less-reactive monomer, such as an aryl dichloride, into polyfluorene using the microwave-assisted Yamamoto polymerization, even though the incorporation ratio of the aryl chloride unit in the random copolymer is less than the feeding ratio.[24] For Suzuki coupling polymerization, microwave heating is useful for the synthesis of an alternating polyfluorene copolymer.[49,50] However, it remains a challenge to obtain a high molecular weight polyfluorene homopolymer using $Pd(PPh_3)_4$ with microwave heating. The difficulty of polymerization may be due to the air-sensitive Pd[0] catalyst. In the case of alternating polyfluorene copolymers, $Pd(PPh_3)_2Cl_2$ is used instead of $Pd(PPh_3)_4$ Recently, Kudla *et al.*[51] reported polyfluorene homopolymers from a bifunctional monomer that had a bromide and a boronate ester as an AB type monomer. The catalyst was also $Pd(PPh_3)_2Cl_2$ and the reaction time was 12 min. The molecular weights (M_w) of the product poly(9-dodecyl-9-methylfluorene)s were between 22 400 and 25 000.

5.3.2 Other Synthetic Techniques

Recently, Seyler *et al.*[40] reported a continuous-flow methodology for Suzuki coupling polymerization. poly(9,9-dioctylfluorene) (PFO) using this method shows a quite high molecular weight of 62 000 within only 30 min. This reaction time is comparable to that of microwave-assisted polymerization.

5.3.3 New Polymerization Methods

Yamamoto polymerizations of AA types and Suzuki polymerizations of AA–BB types have provided moderate PDI values for polyfluorenes. Of course, PDIs can be improved after separating the low molecular weight polymer fraction through Soxhlet extraction and/or reprecipitation. From the standpoint of polymer synthesis, the chain-growth polymerization of polyfluorene is an interesting topic. The well-controlled chain-growth polymerization of AB type monomers provides narrower molecular weight distributions than other conventional polymerizations.[52–55] Yokoyama *et al.*[52] used Suzuki polymerization and prepared a special Pd catalyst, *t*-$Bu_3PPd(Ph)Br$, which contained

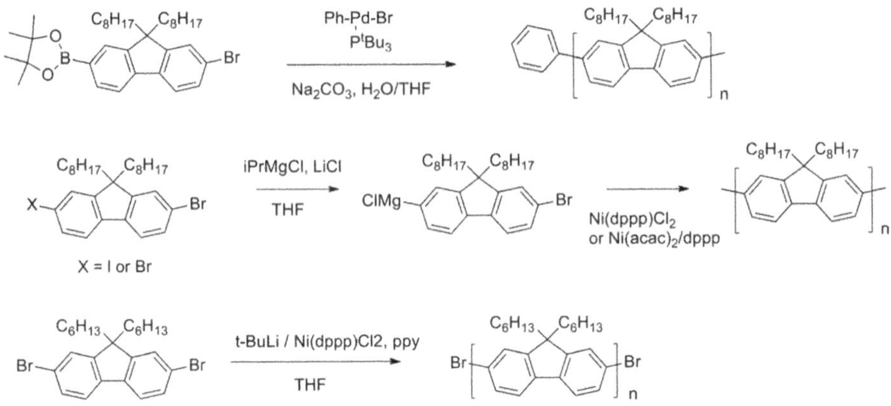

Scheme 5.6

its own initiator unit. In this case, they obtained PFO with $M_n = 17\,700$ and PDI = 1.33 after 30 min of reaction time. Without a termination step, the end group of the propagating polymer chain was mainly the bromide (Scheme 5.6).

Geng's group first reported a Grignard metathesis method (GRIM) for poly(9,9-dioctylfluorene), but the PDIs of PFOs ranged from 1.5 to 2.2.[56] McCullough's group, which is well known for its work on the GRIM of poly(thiophene), achieved a very small PDI (<1.2) and demonstrated the formation of block copolymers with other conjugated polymer chains through continuous chain extension to the end group of polyfluorene.[54] Recently, Geng's group developed nickel acetylacetonate/1,3-bis(diphenylphosphino)propane ($Ni(acac)_2$/dppp) as a new catalyst system instead of 1,3-Bis(diphenylphosphino)propane]dichloronickel(II) ($Ni(dppp)Cl_2$), and obtained PFO with $M_n = 62\,200$ and PDI = 1.2.[55] Moreover, they also successfully synthesized other polyfluorenes, [poly(9,9-bis(triphenylamine)fluorene) (PTPAF) and poly[9,9-bis(4′-*n*-octyloxyphenyl)fluorene] (PBOPF). These polymers also showed smaller PDIs than those of the polymers[16,25] from Yamamoto polymerization.

The Yamamoto polymerization using $Ni(COD)_2$ requires a nearly stoichiometric amount of the reagents. Therefore, the use of catalytic amounts of a Ni complex is of interest. Carter's group developed an organolithium-activated nickel-catalyzed polymerization using a Ni(II) complex, $Ni(dppp)Cl_2$.[57] Under the optimum conditions, the obtained PDHF showed $M_n = 33\ 400$ and PDI = 2.07. The above-mentioned methods are shown in Scheme 5.6.

5.3.4 Purification of Polymers

Polyfluorenes have been developed as promising candidates in organic electronics applications such as polymer light-emitting diodes (PLEDs) and polymer transistors. In the early stages of development, the relationships between polymer structure and device performance were investigated. This is still an important research area. As the organic materials have been put to practical use in the electronic industry area, their controlled qualities are required. Among them, the impurity content is a serious problem in polymeric materials. Small-molecule organic semiconductors can be purified by vacuum sublimation. However, it is difficult to apply this sublimation technique to polymers. Moreover, polyfluorenes have mostly been synthesized using Ni or Pd catalysis. These metallic impurities affect the electrical properties and have been identified as the main cause of poor operating stability. Therefore, many researchers have tried to remove the metallic impurities in the final polymers through washing with acidic water, repeated precipitation, Soxhlet extraction, or column chromatography.

Xiao *et al.*[58] successfully reduced the Ni content in the PFO from the Yamamoto polymerization from 350 ppm to 5–7 ppm. They used 5% ethylenediaminetetraacetic acid (EDTA) solution, granular KCN, or amino-functionalized silica as scavengers of metallic impurities. Tsuda *et al.*[59] accomplished the reduction of Pd content in the PFO from the Suzuki polymerization from 800 ppm to 20–30 ppm. They treated a toluene solution of PFO with 1% aqueous solutions of oxidizing agents such as hydrogen peroxide and potassium peroxymonosulfate (trade name, Oxone). The removal mechanism is related to the increased solubility of oxidized Pd(II) from Pd(0) in water. Liu *et al.*[60] introduced a water-soluble phosphine ligand instead of PPh_3 to reduce Pd impurities. After purification with amino-functionalized silica, they achieved 2–9 ppm Pd impurities. Patel *et al.*[61] used diethylammonium diethyldithiocarbamate (Et_2NH_2DDC) as a palladium scavenger in the final purification step. DCC derivatives have been used as chelating agents for transition-metal ions in environmental science and technology applications.[62]

5.4 Various Applications of Polyfluorenes

5.4.1 Polymer Light-Emitting Diodes (PLEDs)

Polyfluorenes have been attractive as blue light-emitting polymers. In contrast to their high fluorescence emissions, PLEDs of homopolyfluorenes have suffered from low device efficiencies, low maximum brightness, high operating voltages, and poor color purity. Many polymer chemists have tried to overcome poor device performance through the design of the end capper, side chain, and comonomer. In addition, performances have improved through collaboration with device physicists.[63,64] Beyond the blue emission, color tuning to red was possible through polyfluorene copolymers.[65,66] In 2003, Meerholz's group and their colleagues demonstrated RGB three-color PLEDs with high efficiency (over 1 cd/A for all colors).[44] For the purpose of lighting applications rather than display, polyfluorenes for white light-emitting diodes have been studied by the Shim and Wang groups.[67,68] The control of the feeding ratios for the green and red comonomers is important to produce white emissions. In order to incorporate very small contents of green and red monomers in the main polymer chain, diluted toluene solutions of the comonomers are often used in the polymerizations rather than the powders.[68,69]

Phosphorescent small-molecule organic light-emitting diodes (OLEDs) from Ir complexes were commercialized because of their high efficiency compared to fluorescent OLEDs. To achieve the high efficiency in PLEDs, there are trials of using fluorescent and phosphorescent emission by incorporating Ir complex to polyfluorenes.[4,70,71] In the blue-emitting polymers, there were other approaches using fluorene units, typically fluorenevinylene copolymers from Wittig and Heck polymerizations.[72–74] Later, Jin's group reported the successful Gilch polymerization of the fluorenevinylene homopolymer[75,76] (Chart 5.4–5.6).

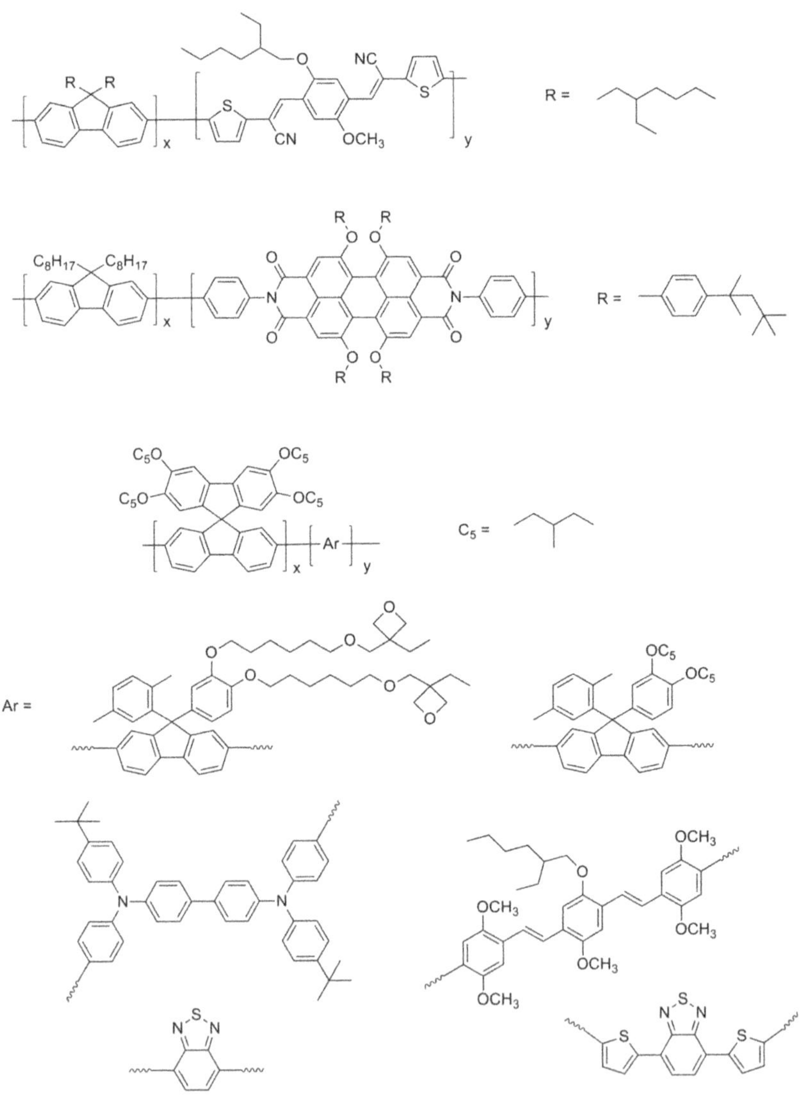

Chart 5.4 Color tuning of polyfluorenes.

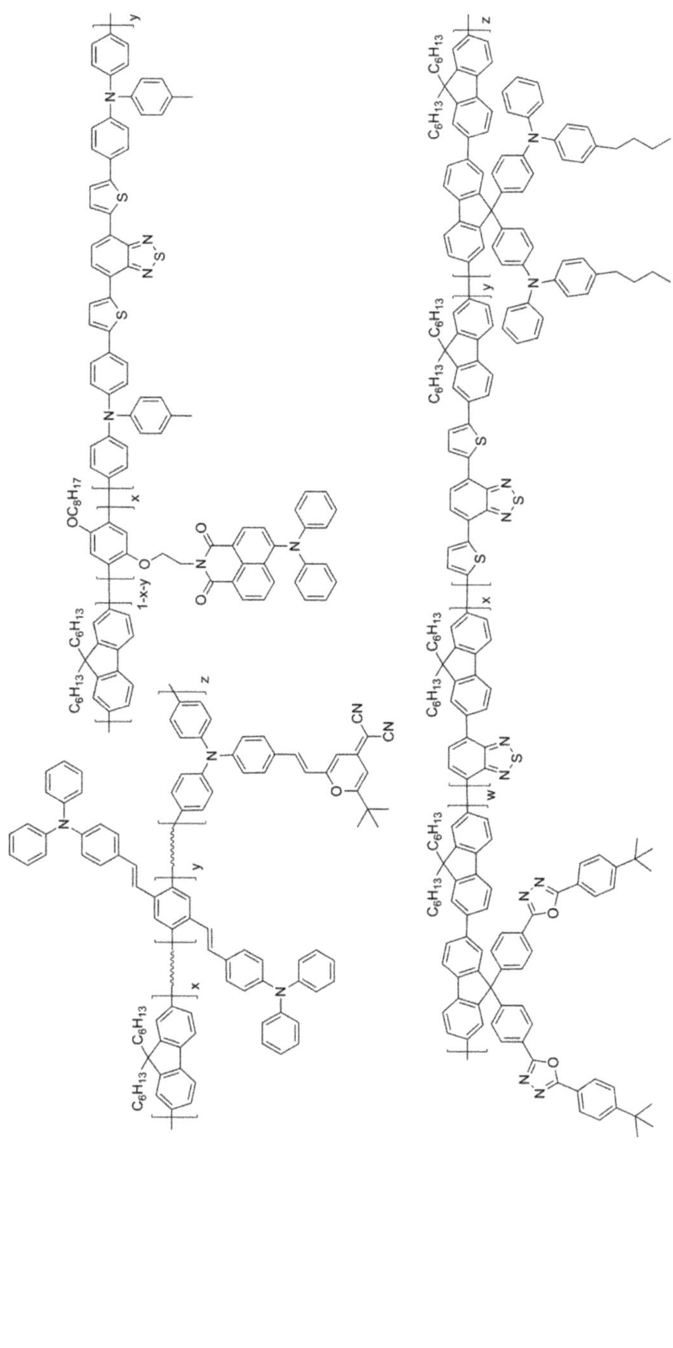

Chart 5.5 Polyfluorenes for white light-emitting diodes.

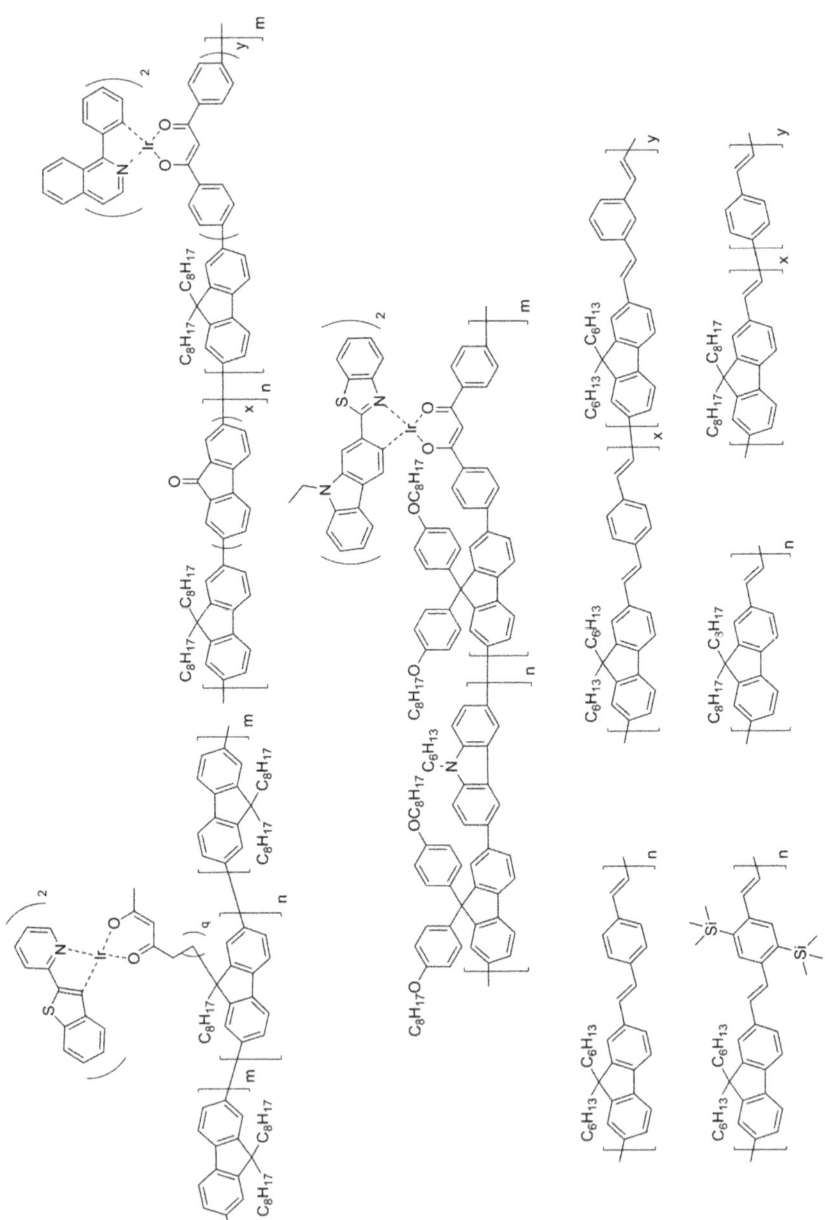

Chart 5.6 Phosphorescent polyfluorenes and blue polyfluorenevinylenes.

5.4.2 Sensing Applications

For biosensor applications, water-soluble poly(fluorene-co-phenylene)s were developed by Bazan's group,[31,77,78] and the ends of their side chains were quaternary ammonium salts. At first, monomers with side chains bearing terminal tertiary amines were used in the Suzuki polymerization, and the polymers were converted to the salt form.[77] Later, monomers bearing bromoalkyl side chains were used, and the halide functionality in the polymer was converted to the quaternary ammonium salt by trimethylamine treatment.[23] Using these polymers, Bazan's group successfully detected DNA labelled with dye from its fluorescence quenching through fluorescence resonance energy transfer (FRET).

Zhou *et al.* synthesized polyfluorenes with phosphonate groups in the side chain and demonstrated their use as metal ion sensors, especially for Fe^{3+}.[79] Kim *et al.* were able to detect 1.5 nM K^+ using polyfluorene copolymers and a molecular beacon[80] (Chart 5.7).

5.4.3 Polymer Transistors and Solar Cells

Among the many alternating copolymers developed by Dow,[81] poly(9,9-dioctylfluorene-co-bithiophene) (F8T2)[82] is as well known as poly(3-hexylthiophene) in polymer transistor materials. Shim's group synthesized poly(9,9-dioctylfluorene-alt-thieno[3,2-b]thiophene) (F8TT)[83] and poly(9,9-dioctylfluorene-alt-biselenophene) (F8Se2)[84] *via* the Suzuki coupling reaction and tested their transistors. The hole mobilities of F8TT and F8Se2 were slightly higher than that of F8T2. Today many polymers with other backbones show higher mobility than those of the polyfluorene copolymers. However, these are still attractive in some devices, which require air stability.

In polymer solar cells, poly(9,9-dialkylfluorene-alt-dithienylbenzothiadiazole)s (PFTBTs) have been developed to achieve power conversion efficiencies (PCEs) of 4.5%.[85–88] Upon changing the alkyl side chains, the performance of

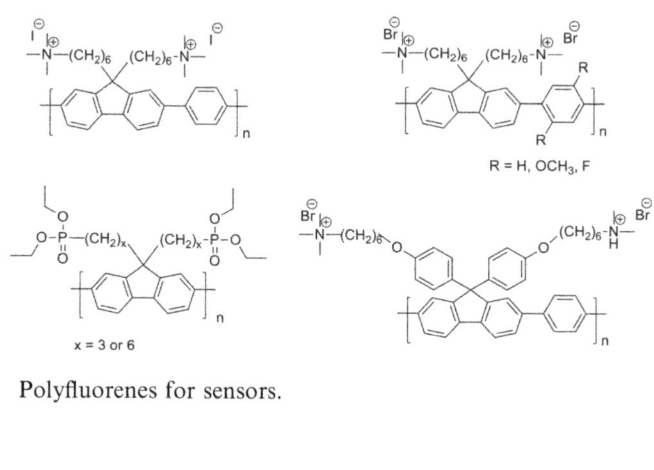

Chart 5.7 Polyfluorenes for sensors.

the PFTBTs was dramatically improved, with PCEs from 2.4% to 4.5%. Because a dithienylbenzothiadiazole (TBT) group does not have a solubilizing side chain, PFTBTs with hexyl and ethyl hexyl side chains have an M_n of 4800.[86] PF10TBT, with dodecyl side chains and BisMDO-PFDTBT, with 3,7-dimethyloctyl side chains, have M_n of 9700 and 20 000, respectively.[87,88] The electrical performance of a polymer device is usually dependent on the molecular weight. Therefore, the design of monomers to achieve high molecular weight polymers is important for polymeric electronic materials. Zheng *et al.*[33] reported a ladder-type tetra-*p*-phenylene-containing copolymer (P3FTBT6) that showed a higher open-circuit voltage in polymer solar cells than those of PFTBTs. Its PCE was comparable to that of BisMDO-PFDTBT at 4.5%. In this case, M_n of P3FTBT6 could be as high as 14 200 due to the three fused fluorene unit with six hexyl side chains (Chart 5.8).

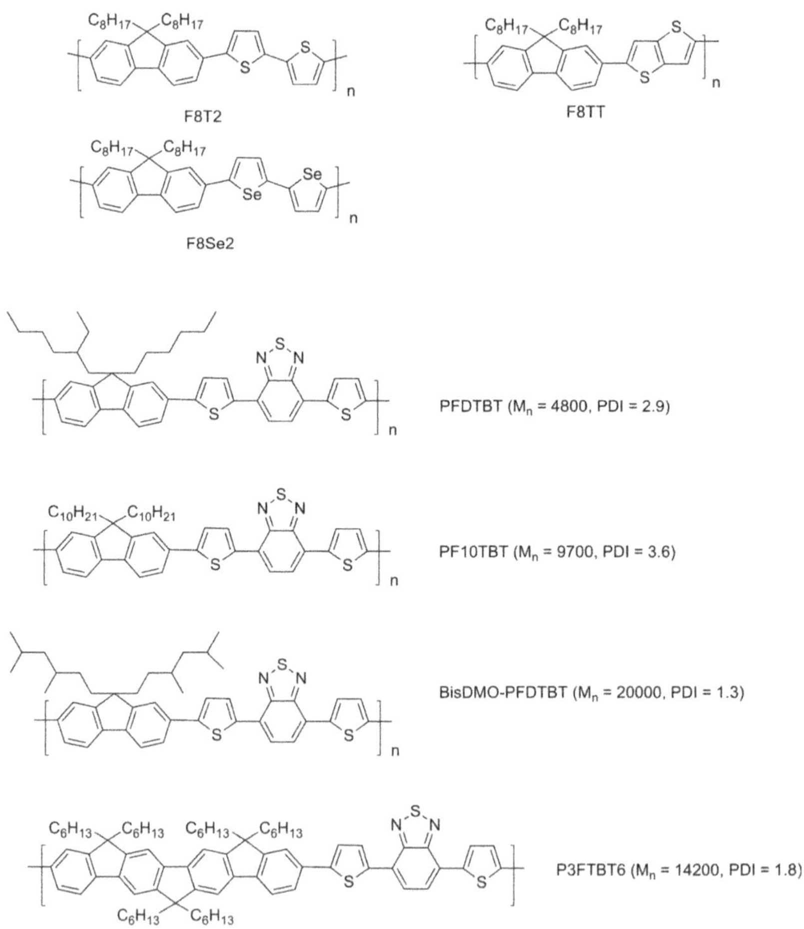

Chart 5.8 Polyfluorenes for transistors and solar cells.

5.4.4 Electron Injection Layers in Organic Electronic Devices

Polyfluorene derivatives for sensor applications are soluble in water or polar solvents such as alcohols. These solubility characteristics provide the chance to fabricate a bilayer device in polymer electronics, in which the active polymers are soluble in common organic solvents and not soluble in polar solvents. Cao's group demonstrated enhanced PLEDs using alcohol-/water-soluble polyfluorene derivatives.[89,90] These polyfluorene derivatives acted as an effective electron injection layer (EIL). Zhou *et al.* reported the use of polyfluorenes with phosphonate groups (PF-EPs) as the EIL layer in PLEDs.[91] Jen's group synthesized poly[9,9-bis(6′-(diethanolamino)hexyl)-fluorene] (PFN-OH) and demonstrated that the simply processed PFN-OH could replace the commonly used CsF layer in phosphorescent PLEDs and show comparable or better device performance.[92,93] Ye *et al.* reported that the phosphorescent PLED with a Li_2CO_3-doped PF-EP showed enhanced performance compared to PLEDs with a neutral PF-EP.[94] In n-channel organic transistors, Bazan's group introduced their poly(fluorene-co-phenylene)s with quaternary ammonium salts as carrier injection layers.[95] They determined the reduced energy barriers from these polyelectrolyte interlayers by ultraviolet photoelectron spectroscopy (UPS). When polyelectrolyte interlayers were used in organic solar cells, an improvement in the V_{oc} was observed, but the overall performance (PCE) was not significantly improved due to the trade-off between other performance parameters.[96] Finally, Bazan's group and their collaborators successfully demonstrated enhanced performance in organic solar cells[97] using poly(9,9-bis(2-ethylhexyl)fluorene)-b-poly[3-(6-trimethylammoniumhexyl)thiophene] (PF2/6-b-P3TMAHT), a polyelectrolyte block copolymer reported by Gutacker *et al.*[98] (Chart 5.9).

5.5 Remarks

Polyfluorenes have been developed over the last two decades, based on well-established polymerization methods. Today, the monomers for poly(9,9-dioctylfluorene), a representative polyfluorene, are commercially available. The adaptation of catalysts and the use of anhydrous solvents may be needed for successful polymerizations. New comonomers and functional side chains still offer many choices with respect to the chemical structure. Controlling the polydispersity index (PDI), preparing block copolymers, and investigating the formation of impurities have only recently become active research area. The successful pathway to useful devices is continuing through the combined efforts of many polymer chemists, device physicists, and engineers. From polyfluorenes, investigation of new applications and technological innovations are expected in the very near future.

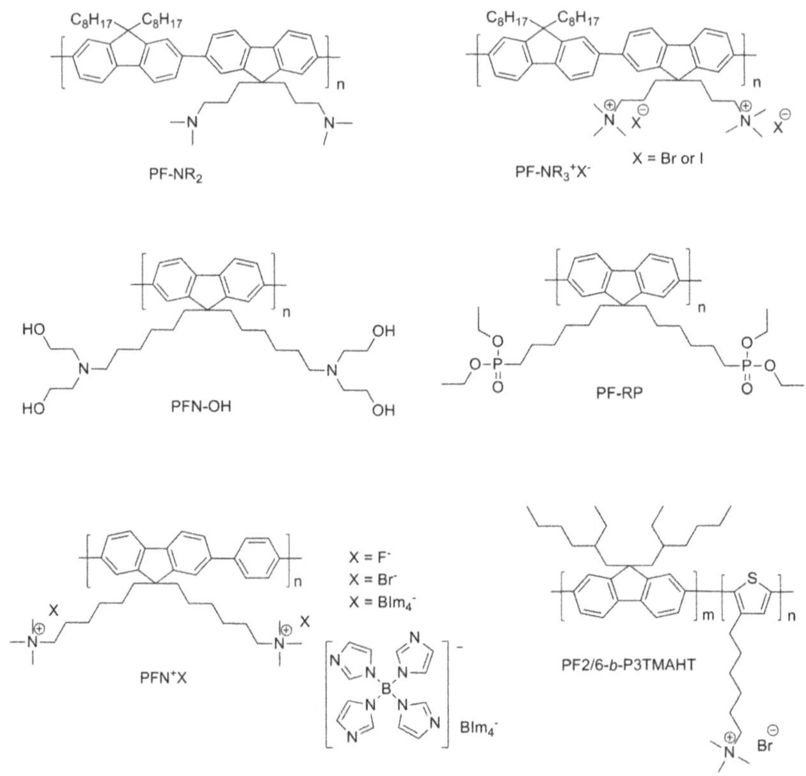

Chart 5.9 Polyfluorenes for electron-injection layers.

References

1. M. Fukuda, K. Sawada and K. Yoshino, *J. Polym. Sci., Part A: Polym. Chem.*, 1993, **101**, 2465.
2. Ranger, D. Rondeau and M. Leclerc, *Macromolecules*, 1997, **30**, 7686.
3. S. Destri, M. Pasini, C. Botta, W. Porzio, F. Bertini and L. Marchiò, *J. Mater. Chem.*, 2002, **12**, 924.
4. N. R. Evans, L. S. Devi, S. S. K. Mak, S. E. Watkins, S. I. Pascu, A. Köhler, R. H. Friend, C. K. Williams and A. B. Holmes, *J. Am. Chem. Soc.*, 2006, **128**, 6647.
5. U. Scherf and E. J. W. List, *Adv. Mater.*, 2002, **14**, 477.
6. S. Y. Cho, A. C. Grimsdale, D. J. Jones, S. E. Watkins and A. B. Holmes, *J. Am. Chem. Soc.*, 2007, **129**, 11910.
7. Q. Pei and Y. Yang, *J. Am. Chem. Soc.*, 1996, **118**, 7416.
8. M. Kreyenschmidt, G. Klaerner, T. Fuhrer, J. Ashenhurst, S. Karg, W. D. Chen, V. Y. Lee, J. C. Scott and R. D. Miller, *Macromolecules*, 1998, **31**, 1099.
9. D. W. Price and J. M. Tour, *Tetrahedron*, 2003, **59**, 3131.

10. C. Xia and R. C. Advincula, *Macromolecules*, 2001, **34**, 5854.
11. E. P. Woo, M. Inbasekaran, W. Shiang and G. R. Roof, WO 9705184, 1997.
12. J. I. Lee, G. Klaerner and R. D. Miller, *Chem. Mater.*, 1999, **11**, 1083.
13. M. Grell, W. Knoll, D. Lupo, A. Meisel, T. Miteva, D. Neher, H.-G. Nothofer, U. Scherf and A. Yasuda, *Adv. Mater.*, 1999, **11**, 671.
14. D. Marsitzky, R. Vestberg, P. Blainey, B. T. Tang, C. J. Hawker and K. R. Carter, *J. Am. Chem. Soc.*, 2001, **123**, 6965.
15. H. J. Cho, B. J. Jung, N. S. Cho, J. Lee and H. K. Shim, *Macromolecules*, 2003, **36**, 6704.
16. J. H. Lee and D. H. Hwang, *Chem. Commun.*, 2003, 2836.
17. M. Kreyenschmidt, G. Klaerner, T. Fuhrer, J. Ashenhurst, S. Karg, W. D. Chen, V. Y. Lee, J. C. Scott and R. D. Miller, *Macromolecules*, 1998, **31**, 1099.
18. T. Yamamoto, A. Morita, Y. Miyazaki, T. Maruyama, H. Wakayama, Z. H. Zhou, Y. Nakamura, T. Kanbara, S. Sasaki and K. Kubota, *Macromolecules*, 1992, **25**, 1214.
19. J. H. Lee, H. J. Cho, B. J. Jung, N. S. Cho and H. K. Shim, *Macromolecules*, 2004, **37**, 8523.
20. S. Xiao, M. Nguyen, X. Gong, Y. Cao, H. B. Wu, D. Moses and A. J. Heeger, *Adv. Funct. Mater.*, 2003, **13**, 25.
21. M. C. Hung, J. L. Liao, S. A. Chen, S. H. Chen and A. C. Su, *J. Am. Chem. Soc.*, 2005, **127**, 14576.
22. T. Miteva, A. Meisel, W. Knoll, H. G. Nothofer, U. Scherf, D. C. Muller, K. Meerholz, A. Yasuda and D. Neher, *Adv. Mater.*, 2001, **13**, 565.
23. D. H. Hwang, S. K. Kim, M. J. Park, B. W. Koo, I. N. Kang, S. H. Kim and T. Zyung, *Chem. Mater.*, 2004, **16**, 1298.
24. F. Galbrecht, X. H. Yang, B. S. Nehls, D. Neher, T. Farrell and U. Scherf, *Chem. Commun.*, 2005, 2378.
25. C. Ego, A. C. Grimsdale, F. Uckert, G. Yu, G. Srdanov and K. Müllen, *Adv. Mater.*, 2002, **14**, 809.
26. S. Setayesh, A. C. Grimsdale, T. Weil, V. Enkelmann, K. Müellen, F. Meghdadi, E. J. W. List and G. Leising, *J. Am. Chem. Soc.*, 2001, **123**, 946.
27. Y. S. Suh, S. W. Ko, B. J. Jung and H. K. Shim, *Opt. Mater.*, 2003, **21**, 109.
28. M. Ranger and M. Leclerc, *Chem. Commun.*, 1997, 1597.
29. M. Pasini, S. Destri, W. Porzio, C. Botta and U. Giovanella, *J. Mater. Chem.*, 2003, **13**, 807.
30. E. Lim, B. J. Jung and H. K. Shim, *Macromolecules*, 2003, **36**, 4288.
31. B. Liu and G. C. Bazan, *J. Am. Chem. Soc.*, 2006, **128**, 1188.
32. L. Zhu, J. Duquette and M. Zhang, *J. Org. Chem.*, 2003, **68**, 3729.
33. Q. Zheng, B. J. Jung, J. Sun and H. E. Katz, *J. Am. Chem. Soc.*, 2010, **132**, 5394.
34. W. L. Yu, J. Pei and W. Huang, *Chem. Commun.*, 1999, 1837.
35. B. Liu, W. L. Yu, Y. H. Lai and W. Huang, *Macromolecules*, 2000, **33**, 8945.

36. N. Miyaura and A. Suzuki, *Chem. Rev.*, 1995, **95**, 2457.
37. M. Inbasekaran, W. Wu and E. P. Woo, US 5,777,070, 1998.
38. C. Towns and R. O'Dell, WO 0053656, 2000.
39. K. Treacher, H. Becker, P. Stoessel, H. Spreitzer, A. Falcou, A. Parham, and A. Buesing WO 02077060A1, 2002; US 7288617B2, 2007.
40. H. Seyler, D. J. Jones, A. B. Holmes and W. H. Wong, *Chem. Commun.*, 2012, **48**, 1598.
41. C. Towns, P. Wallace, I. Allen, T. Pounds and L. Murtagh, US 7173103 B2, 2007.
42. K. L. Chan, M. J. McKiernan, C. R. Towns and A. B. Holmes, *J. Am. Chem. Soc.*, 2005, **127**, 7662.
43. E. G. Wang, C. Li, Y. Q. Mo, Y. Zhang, G. Ma, W. Shi, J. B. Peng, W. Yang and Y. Cao, *J. Mater. Chem.*, 2006, **16**, 4133.
44. C. D. Müller, A. Falcou, N. Reckefuss, M. Rojahn, V. Wiederhirn, P. Rudati, H. Frohne, O. Nuyken, H. Becker and K. Meerholz, *Nature*, 2003, **421**, 829.
45. G. Vamvounis, G. L. Schulz and S. Holdcroft, *Macromolecules*, 2004, **37**, 8897.
46. M. J. Park, J. I. Lee, H. Y. Chu, S. H. Kim, T. Zyung, J. H. Eom, H. K. Shim and D. H. Hwang., *Synth. Met.*, 2009, **159**, 1393.
47. Y. H. Tseng, P. I. Shih, C. H. Chien, A. K. Dixit, C. F. Shu, Y. H. Liu and G. H. Lee, *Macromolecules*, 2005, **38**, 10055.
48. K. R. Carter, *Macromolecules*, 2002, **35**, 6757.
49. B. S. Nehls, U. Asawapirom, S. Füldner, E. Preis, T. Farrell and U. Scherf, *Adv. Funct. Mater.*, 2004, **14**, 352.
50. A. Tsami, X.-H. Yang, T. Farrell, D. Neher and E. Holder, *J. Polym. Sci., Part A: Polym. Chem.*, 2008, **46**, 7794.
51. C. J. Kudla, N. Koenen, W. Pisula and U. Scherf, *Macromolecules*, 2009, **42**, 3483.
52. A. Yokoyama, H. Suzuki, Y. Kubota, K. Ohuchi, H. Higashimura and T. Yokozawa, *J. Am. Chem. Soc.*, 2007, **129**, 7236.
53. M. C. Stefan, A. E. Javier, I. Osaka and R. D. McCullough, *Macromolecules*, 2009, **42**, 30.
54. A. E. Javier, S. R. Varshney and R. D. McCullough, *Macromolecules*, 2010, **43**, 3233.
55. A. Sui, X. Shi, S. Wu, H. Tian, Y. Geng and F. Wang, *Macromolecules*, 2012, **45**, 5436.
56. L. Huang, S. Wu, Y. Qu, Y. Geng and F. Wang, *Macromolecules*, 2008, **41**, 8944.
57. S. Jhaveri, J. J. Peterson and K. R. Carter, *Macromolecules*, 2008, **41**, 8977.
58. S. S. Xiao, C. Qiu and C. X. Qiu, US 6894145 B2, 2005.
59. Y. Tsuda, Y. Morishita, S. Nomura, S. Tai, Y. Hoshi, S. Funyuu, M. L. Marroco, F. J. Motamedi, and L. S. Wang, US 8242235 B2, 2012.
60. C. Liu, A. Repoley and B. Zhou, *J. Polym. Sci., Part A: Polym. Chem*, 2008, **46**, 7268.

61. D. G. Patel, K. R. Graham and J. R. Reynolds, *J. Mater. Chem.*, 2012, **22**, 3004.
62. C. M. Wai, S. F. Wang, Y. Liu, V. Ldpez-Avila and W. F. Beckert, *Talanta*, 1996, **43**, 2083.
63. A. W. Grice, D. D. C. Bradley, M. T. Bernius, M. Inbasekaran, W. W. Wu and E. P. Woo, *Appl. Phys. Lett.*, 1998, **73**, 629.
64. M. Gross, D. C. Muller, H. G. Nothofer, U. Scherf, D. Neher, C. Brauchle and K. Meerholz, *Nature*, 2000, **405**, 661.
65. N. S. Cho, D. H. Hwang, J. I. Lee, B. J. Jung and H. K. Shim, *Macromolecules*, 2002, **35**, 1224.
66. C. Ego, D. Marsitzky, S. Becker, J. Zhang, A. C. Grimsdale, K. Müllen, J. D. MacKenzie, C. Silva and R. H. Friend, *J. Am. Chem. Soc.*, 2003, **125**, 437.
67. S. K. Lee, D. H. Hwang, B. J. Jung, N. S. Cho, J. Lee, J. D. Lee and H. K. Shim, *Adv. Funct. Mater.*, 2005, **15**, 1647.
68. J. Liu, Q. Zhou, Y. Cheng, Y. Geng, L. Wang, D. Ma, X. Jing and F. Wang, *Adv. Mater.*, 2005, **17**, 2974.
69. C. Y. Chuang, P. I. Shih, C. H. Chien, F. I. Wu and C. F. Shu, *Macromolecules*, 2007, **40**, 247.
70. K. Zhang, Z. Chen, C. Yang, Y. Tao, Y. Zou, J. Qin and Y. Cao, *J. Mater. Chem.*, 2008, **18**, 291.
71. M. J. Park, J. Kwak, J. Lee, I. H. Jung, H. Kong, C. Lee, D. H. Hwang and H. K. Shim, *Macromolecules*, 2010, **43**, 1379.
72. H. N. Cho, D. Y. Kim, Y. C. Kim, J. Y. Lee and C. Y. Kim, *Adv. Mater.*, 1997, **9**, 326.
73. H. N. Cho, J. K. Kim, C. Y. Kim, N. W. Song and D. Kim, *Macromolecules*, 1999, **32**, 1476.
74. T. Ahn, S. Y. Song and H. K. Shim, *Macromolecules*, 2000, **33**, 6764.
75. S. H. Jin, H. J. Park, J. Y. Kim, K. Lee, S. P. Lee, D. K. Moon, H. J. Lee and Y. S. Gal, *Macromolecules*, 2002, **35**, 7532.
76. D. H. Hwang, J. M. Lee, S. Lee, C. H. Lee and S. H. Jin, *J. Mater. Chem.*, 2003, **13**, 1540.
77. M. Stork, B. S. Gaylord, A. J. Heeger and G. C. Bazan, *Adv. Mater.*, 2002, **14**, 361.
78. B. S. Gaylord, A. J. Heeger and G. C. Bazan, *J. Am. Chem. Soc.*, 2003, **125**, 896.
79. G. Zhou, G. Qian, L. Ma, Y. Cheng, Z. Xie, L. Wang, X. Jing and F. Wang, *Macromolecules*, 2005, **38**, 5416.
80. B. Kim, I. H. Jung, M. Kang, H. K. Shim and H. Y. Woo, *J. Am. Chem. Soc.*, 2012, **134**, 3133.
81. M. T. Bernius, M. Inbasekaran, J. O'Brien and W. Wu, *Adv. Mater.*, 2000, **12**, 1737.
82. H. Sirringhaus, T. Kawase, R. H. Friend, T. Shimoda, M. Inbasekaran, W. Wu and E. P. Woo, *Science*, 2000, **290**, 2123.
83. E. Lim, B. J. Jung, J. Lee, H. K. Shim, J. I. Lee, Y. S. Yang and L. M. Do, *Macromolecules*, 2005, **38**, 4531.

84. Y. M. Kim, E. Lim, I. N. Kang, B. J. Jung, J. Lee, B. W. Koo, L. M. Do and H. K. Shim, *Macromolecules*, 2006, **39**, 4081.
85. M. Svensson, F. Zhang, O. Inganäs and M. R. Andersson, *Synth. Met.*, 2003, **135-136**, 137.
86. M. Svensson, F. L. Zhang, S. C. Veenstra, W. J. H. Verhees, J. C. Hummelen, J. M. Kroon, O. Inganäs and M. R. Andersson, *Adv. Mater.*, 2003, **15**, 988.
87. L. H. Slooff, S. C. Veenstra, J. M. Kroon, D. J. D. Moet, J. Sweelssen and M. M. Koetse, *Appl. Phys. Lett.*, 2007, **90**, 143506.
88. M. H. Chen, J. Hou, Z. Hong, G. Yang, S. Sista, L. M. Chen and Y. Yang, *Adv. Mater.*, 2009, **21**, 4238.
89. F. Huang, H. B. Wu, D. L. Wang, W. Yang and Y. Cao, *Chem. Mater.*, 2004, **16**, 708.
90. H. B. Wu, F. Huang, Y. Q. Mo, W. Yang, D. L. Wang, J. B. Peng and Y. Cao, *Adv. Mater.*, 2004, **16**, 1826.
91. G. Zhou, Y. H. Geng, Y. X. Cheng, Z. Y. Xie, L. X. Wang, X. B. Jing and F. S. Wang, *Appl. Phys. Lett.*, 2006, **89**, 233501.
92. F. Huang, Y. H. Niu, Y. Zhang, J. W. Ka, M. S. Liu and A. K. Y. Jen, *Adv. Mater.*, 2007, **19**, 2010.
93. Y. Zhang, F. Huang, A. K. Y. Jen and Y. Chi, *Appl. Phys. Lett.*, 2008, **92**, 063303.
94. T. L. Ye, M. R. Zhu, J. S. Chen, D. G. Ma, C. L. Yang, W. F. Xie and S. Y. Liu, *Org. Electron.*, 2011, **12**, 154.
95. J. H. Seo, A. Gutacker, B. Walker, S. Cho, A. Garcia, R. Yang, T.-Q. Nguyen, A. J. Heeger and G. C. Bazan, *J. Am. Chem. Soc.*, 2009, **131**, 18220.
96. C. He, C. Zhong, H. Wu, R. Yang, W. Yang, F. Huang, G. C. Bazan and Y. Cao, *J. Mater. Chem.*, 2010, **20**, 2617.
97. J. H. Seo, A. Gutacker, Y. M. Sun, H. B. Wu, F. Huang, Y. Cao, U. Scherf, A. J. Heeger and G. C. Bazan, *J. Am. Chem. Soc.*, 2011, **133**, 8416.
98. A. Gutacker, S. Adamczyk, A. Helfer, L. E. Garner, R. C. Evans, S. M. Fonseca, M. Knaapila, G. C. Bazan, H. D. Burrows and U. Scherf, *J. Mater. Chem.*, 2010, **20**, 1423.

CHAPTER 6

Poly(carbazolylene)s

SUNG JU CHO AND ANDREW C. GRIMSDALE*

School of Materials Science and Engineering, Nanyang Technological University, 50 Nanyang Avenue, Singapore 639798
*Email: acgrimsdale@ntu.edu.sg

6.1 Introduction

Carbazole (**1**) is a tricyclic fused aromatic hydrocarbon consisting of two benzene rings fused to a pyrrole ring. It may also be thought of as a bridged biphenylene system with a nitrogen bridge as compared to the carbon bridge in fluorene (Chart 6.1). The standard numbering of the positions is as shown with the nitrogen being sometimes referred to as the 9-position. While carbazole is isoelectronic with fluorene the presence of the electron-rich nitrogen makes it more readily oxidizable, or to use an alternative terminology it is a better hole acceptor, than fluorene, due to having a higher HOMO energy. Carbazole derivatives including poly(carbazolylene)s have similar potential applications to their fluorene analogs (see Chapter 5) in electronic devices such as light-emitting diodes (LEDs) and solar cells.[1–5] The main differences in their properties arise from the higher HOMO energies of the carbazoles, making them better hole acceptors, and in the inability of the nitrogen to be oxidized, whereas the 9-position of fluorene is readily oxidized to form ketones, which affect the emission color of fluorene-based materials in LEDs.

Besides the difference in orbital energies the presence of the nitrogen also makes the chemistry of carbazole vary significantly from that of fluorene. The nitrogen acts as an *ortho-*, *para-*directing activating unit, so that whereas fluorene undergoes electrophilic substitution at the 2,7-positions, carbazole reacts at the 3- and 6-positions first, and if those are blocked then at the 1- and

RSC Polymer Chemistry Series No. 9
Conjugated Polymers: A Practical Guide to Synthesis
Edited by Klaus Müllen, John R. Reynolds and Toshio Masuda

Published by the Royal Society of Chemistry, www.rsc.org

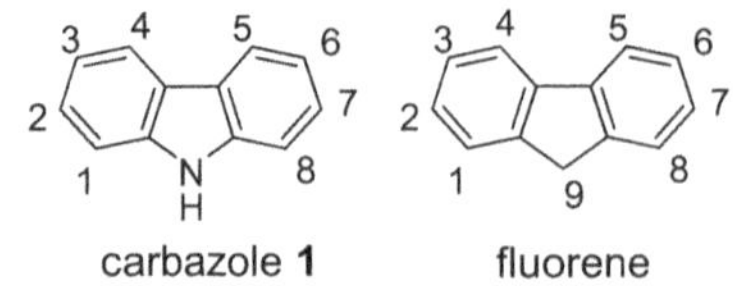

Chart 6.1

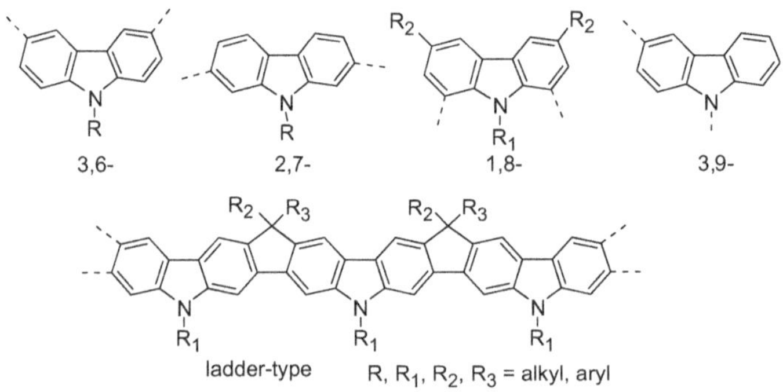

Chart 6.2

8-positions. As a result 2,7-substituted carbazoles are inaccessible from carbazole and must be made by total synthesis, as will be shown below. In addition to substitution of the benzene rings, attachment of aryl groups at the nitrogen is generally straightforward using aryl amination chemistry. As a result, 4 types of conjugated polycarbazolylenes are possible (Chart 6.2) with, respectively, 3,6- 1,8-, 2,7- and 3,9-substituted carbazoles. Of these only the poly(2,7-carbazole)s retain all-carbon conjugation, with the others having conjugation go through the nitrogen bridgehead atom. Because of the electron-rich nature of the 3- and 6-positions in 2,7-carbazoles, their polymers are susceptible to degradation processes arising from oxidation at these positions during passage of charge through them as occurs in devices such as LEDs. This can be overcome using ladder-type polymers in which the carbazoles are doubly linked through both the 3,6- and the 2,7-positions. These materials retain the through-carbon conjugation seen in the 2,7-carbazole-based polymers but the substitution of the 3- and 6-positions precludes oxidation there. These structures are attractive candidates for optoelectronic device materials since the rigid structure of ladder-type conjugated polymers usually results in small Stokes shifts and sharply resolved vibronic features, as well as high luminescence efficiencies.

6.1.1 Poly(3,6-carbazolylene)s

Poly(3,6-carbazolylene)s are the most readily accessible polymers of carbazole and are readily synthesized starting from carbazole. In order to obtain solubility, an alkyl group must be attached at the nitrogen. The synthesis of poly(*N*-alkyl-3,6-carbazolylene)s (**2**) appears possible in two steps starting from carbazole (Scheme 6.1). Alkylation of carbazole (**1**) is readily accomplished using an alkyl halide and base (potassium carbonate, sodium hydride and sodium hydroxide have all been used for this reaction). Oxidative polymerization of the *N*-alkylcarbazole (**3**) should then give the desired polymers **2**. However, oxidation of **3** with iron(III) chloride is reported to produce only a mixture consisting mainly of the dimer with small amounts of higher oligomers.[6] Whether better results could be obtained from use of other oxidants remains unexplored.

To obtain polymers it is necessary to start from the commercially available 3,6-dibromocarbazole (**4**) (Scheme 6.2). Alkylation of this by same methods as for carbazole gives the monomers **5**,[7–11] which can then be reductively polymerized to the desired polymers **2**. The monomers **5** can also be reacted with suitable bisboronates to make either the homopolymers **2** or a wide range of alternating copolymers using Suzuki or other crosscoupling reactions.[3] The polymerization of the monomers **5** has been carried by electrochemical reduction using catalytic nickel(II), but the resulting polymers are of low molecular weight (*ca.* 3 kDa) corresponding to only 15–20 units.[12] More usually a Yamamoto-type polycondensation using stoichiometric amounts of a commercially available bis(cycloocta-1,5-dienyl)nickel(0) has been employed, despite the cost of the reagent. Using standard reaction conditions in which the monomer is added to the nickel(0) complex produces relatively low molecular weights (<10 kDa), but by reversing the order of addition of reagents polymers number-average molecular weights of >50 kDa can be obtained.[7] Using branched alkyl chains increases the solubility of the polymers and thus leads to higher molecular weight polymers. The polymers **2** have also been obtained by Suzuki condensation of **5** with a carbazole-3,6-bisboronate but are of only moderate molecular weight (<10 kDa),[13] so the Yamamoto procedure is to be preferred as it involves fewer steps and can produce higher molecular weight polymers.

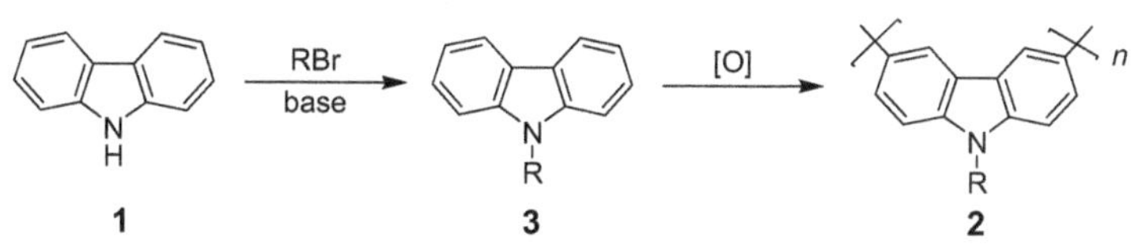

Scheme 6.1

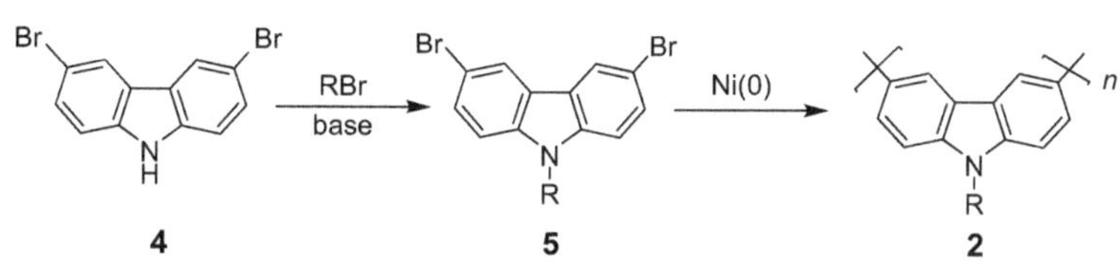

Scheme 6.2

These polymers because of their limited conjugation have large bandgaps and produce blue emission.[14] Because of their large bandgaps and high-lying triplet energy levels the homopolymers **2** (and some copolymers) have been investigated as host materials for blue-emitting phosphors in LEDs.[3,15] Despite the apparently limited conjugation a recent study found that the intra- and intermolecular charge-carrier mobility in poly(3,6-carbazolylene)s **2** was higher than in their 2,7-carbazolylene analogs.[13]

6.1.2 Poly(1,8-carbazolylene)s

In order to prepare 1,8-substituted carbazoles it is necessary to first substitute the more sterically accessible 3- and 6-positions. As shown in Scheme 6.3 the 1,8-diiodide **6** can be obtained in good yield from *N*-alkyl carbazole **3** by Friedel–Crafts alkylation with *tert*-butyl chloride followed by iodination.[16] Reductive coupling of this monomer under standard Yamamoto conditions then produces the homopolymer **7**.[17] The molecular weight, however, is low ($M_w = 1.4$ kDa) indicating that the material is an oligomer. This is presumably because of the steric congestion at the 1,8-positions due to the presence of the nearby alkyl chain. Higher molecular weight copolymers are obtainable by coupling **6** with other units by Suzuki, Sonogashira or other crosscoupling reactions, but here also the molecular weights are only moderate (< 10 kDa). The steric repulsion at the 1- and 8-positions also leads to torsion along the polymer chain so these materials show absorption edges in the blue region of the spectrum while their emission is blue-green as they emit in both the blue and green regions of the spectra.

6.1.3 Poly(2,7-carbazolylene)s

Whereas the previous two classes of polymers can be synthesized starting from carbazole, the monomers for poly(2,7-carbazolylene)s 2,7-dihalocarbazoles such as the dibromide **8**, have to be made by total synthesis. The most efficient

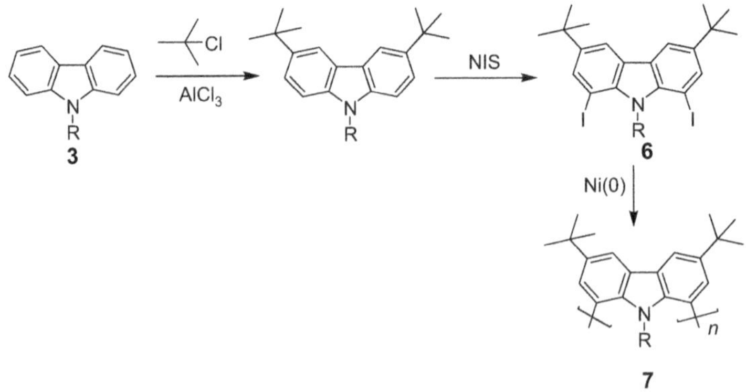

Scheme 6.3

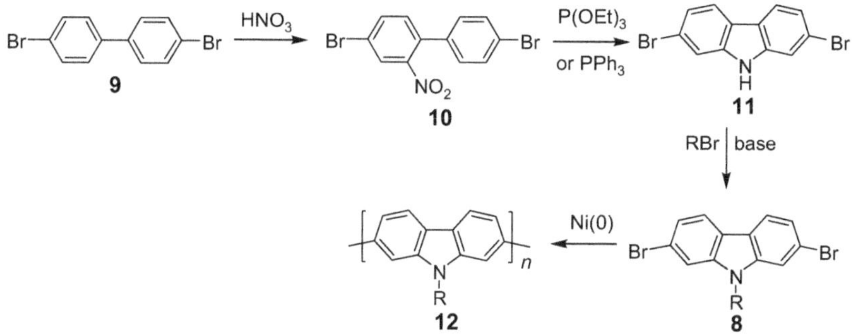

Scheme 6.4

synthesis of **8** is shown in Scheme 6.4.[18] It requires only three high yielding steps from a commercially available dibromobiphenyl **9**. The key step is the reductive ring closure of the nitrobiphenyl **10** to form the dihalocarbazole **11**. This was initially performed using triethylphosphite – the Cadogan method[19] – but more recently a new procedure using triphenylphosphine has been introduced. This method is more convenient than the Cadogan method as the phosphine reagent is less toxic, excess phosphine and the phosphine oxide byproduct are more easily removed from the reaction mixture, and the yields are frequently higher.[20] Alkylation of **11** using an alkyl halide and base (sodium hydride or potassium carbonate have both been used) then gives the desired monomer **8**.[18,21–29] The alkylation can also be performed using an alkyl tosylate with potassium hydroxide as base.[30] It has recently been reported that performing the ring closure by reduction of the nitrobiphenyl with a trialkyl phosphite in the presence of an aromatic nitro compound leads to partial alkylation of the product.[31] These reagents were found to alkylate carbazole in 55% yield, indicating that the reaction proceeds in two steps in the one pot, and suggesting that with further optimization it may be possible to produce the *N*-alkylcarbazoles **8** efficiently in a one-pot process. The polymers **12** are then readily accessible by Yamamoto polycondensation.[29,32] To obtain high molecular weights it has been found necessary to use long, preferably branched alkyl chains, *e.g.* 2-decyltetradecyl, as these make the polymers readily soluble. Thus, the polymer **12** with decyltetradecyl side chain combines weight-averaged molecular weights (M_w) of 75.5 kDa with good solubility,[29] whereas other materials with shorter alkyl chains show molecular weights of below 20 kDa with only marginal solubility.[32]

Poly(2,7-carbazolylene)s are large-bandgap blue-emitting materials whose emission spectra closely match those of polyfluorenes but that do not display the long-wavelength emission seen from the latter.[3] As was mentioned above, their charge-carrier mobilities have been reported to be lower than their 3,6-carbazolylene analogs,[13] which was somewhat unexpected, and was attributed to their having too crystalline a nature, leading to formation of grain boundaries in thin films; further studies will be needed to further elucidate this matter. They have also been used as electron-donating materials in bulk-heterojunction

(BHJ) solar cells.[5] Due to their absorption lying only in the higher energy part of the spectrum the resulting devices display modest efficiencies (<1%); curiously the efficiencies are significantly higher with nonfullerene acceptor molecules such as a perylene dye,[29] rather than with the fullerene derivatives usually used as acceptors in such devices, probably due to the much stronger absorption by the former. By copolymerizing 2,7-carbazoles with other units it is possible to tune their bandgap so as to produce materials with emission colors other than blue,[3] or which act as much more efficient donor materials in solar cells,[5] due to their having absorption over a much wider range of the visible spectrum. The copolymerization of a carbazole as an electron-rich donor unit with an electron-accepting unit in particular has been used to make medium- or low-bandgap materials for such applications. The polymer that has to date demonstrated the highest efficiency in devices is the benzodithiazole copolymer **13** whose synthesis is shown in Scheme 6.5 as an illustration of the synthesis of such copolymers in general.[30] The *N*-alkylcarbazole **8** is converted into the bisboronate **14** by lithiation followed by addition of a boron-pinacol ester. This is then reacted with one equivalent of the dihalide **15** to produce the alternating copolymer **13**. Note that as with all such polycondensations it is vital to maintain strict control of the relative stoichiometry of the two reagents in order to achieve high molecular weights. In this case the number (M_n) and weight-averaged (M_w) molecular masses were determined by GPC to be 37 kDa and 73 kDa, respectively. With further optimization the molecular weights might be improved and/or the polydispersity lowered. The latter is important as low polydispersity seems generally to be favorable for controlling polymer properties and optimizing device performance.

Some doubts have been raised about the electrochemical stability of poly(2,7-carbazolylene)s due to the high electron density at the free 3- and 6-positions, making oxidation at these sites facile and leading to possible crosslinking or

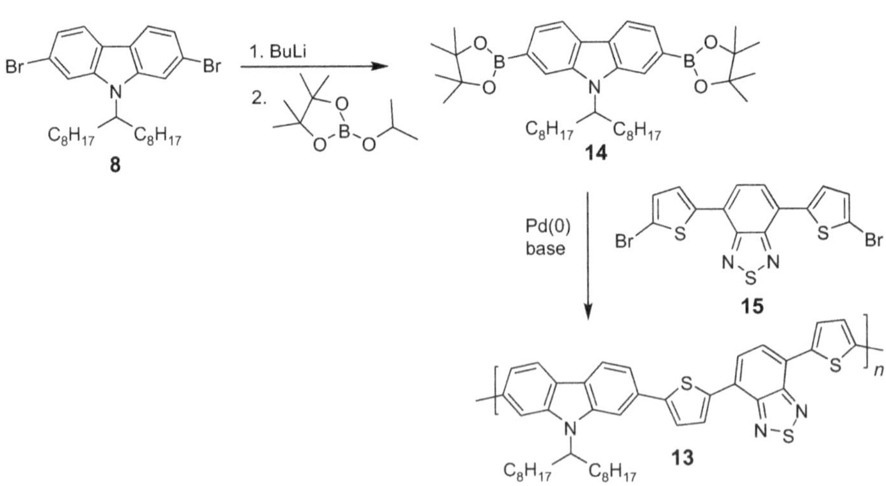

Scheme 6.5

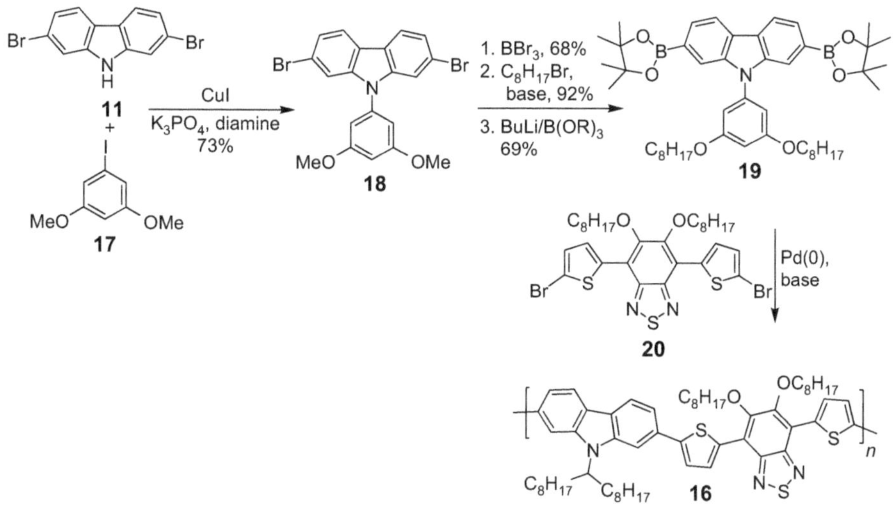

Scheme 6.6

other reactions through the cations formed at these sites. The stability of the *N*-alkyl groups is also under question and it is thought that *N*-aryl carbazoles might be more stable. Arylation of the nitrogen is possible through either copper-mediated Ullmann coupling or palladium-catalyzed Buchwald–Hartwig aryl amination methods. An aryl iodide is needed to arylate 2,7-dibromocarbazole as the more reactive iodide can be reacted under mild conditions without risk of self-condensation of the carbazole. An example of the synthesis of an arylcarbazole-based copolymer **16** is shown in Scheme 6.6 as illustrative of the methods used. Coupling of dibromocarbazole **11** with 3,5-dimethoxyiodobenzene **17** catalyzed by copper(I) produces the *N*-arylcarbazole in good yield (73%).[33] Demethylation, followed by alkylation was performed to introduce longer alkyl chains to improve the solubility of the final product. The synthesis of the diboronate **19** and the condensation with the acceptor unit **20** follow the same procedure as for the *N*-alkylcarbazole polymer **13** in Scheme 6.5. The alkoxy side chains on **20** further improve the solubility of the polymer, which helps explain why the M_n of 65 kDa obtained was higher than for **13** above. This polymer when used as a donor in a BHJ solar cell with a fullerene has produced an efficiency of 3.96%, which matches the values seen for **13** in similar devices.

6.1.4 Poly(3,9-carbazolylene)s

The problem of self-condensation in synthesis of *N*-aryl-dihalocarbazoles alluded to above can be exploited to produce 3,9-carbazole-based polymers, as shown in Scheme 6.7. The polymer **21** is obtained directly by treatment of 3-iodocarbazole **22** with copper(I).[34] Due to the lack of solubilizing groups the material has low solubility and thus low molecular weight ($M_n = 2.36$ kDa,

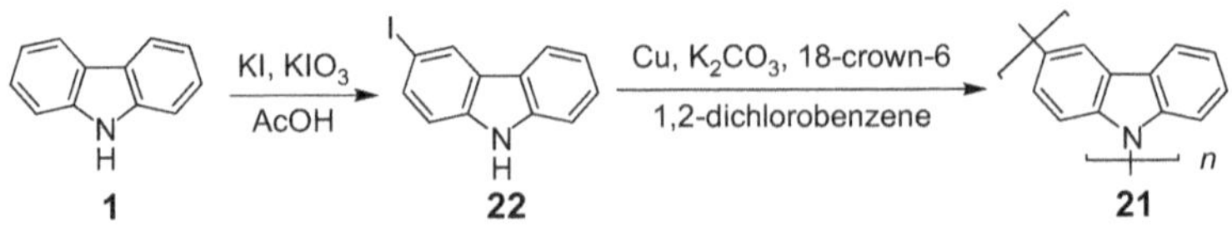

Scheme 6.7

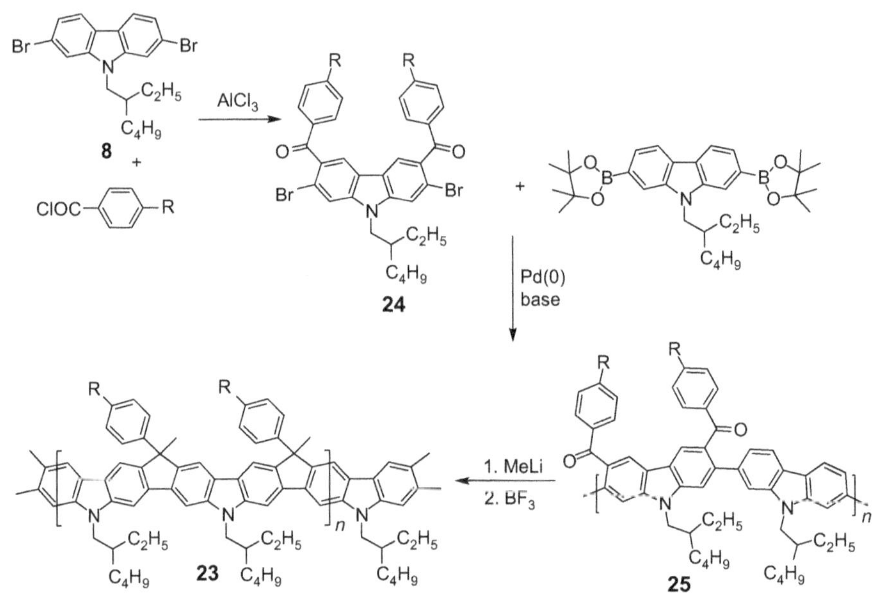

Scheme 6.8

$M_w = 5.07$ kDa). This polymer has an ionization potential of 5.45 eV suggesting it may be suitable for use as a hole-transport material, but poor film quality prevented measurement of its charge carrier mobility. It should be possible to make soluble polymers by applying this methodology to 6-alkyl-3-iodocarbazole. Also using 3,6-diiodocarbazole this type of reaction could be used to make hyperbranched or dendritic macromolecules with possible application as hole transporters or matrices for emitters.

6.1.5 Ladder-Type Polycarbazolylenes

One obvious way to get around the problem of oxidation at the 3- and 6-positions in poly(2,7-carbazolylene)s is to substitute these sites. If this is done in a single-stranded polymer, steric interaction between the substituents on adjacent units would cause torsion of the backbone with loss of conjugation, producing blue shifts in absorption and emission and also reduction in chain packing and charge-carrier mobilities. This can be overcome by making doubly stranded ladder-type polymers **23**, as shown in Scheme 6.8.[35] Acylation of the

2,7-dihalocarbazole **8** using a acid halide or anhydride occurs at the 3,6-positions to give a diacyl adduct **24**. This process only occurs efficiently with aryl acid halides or anhydrides and use of a substituted benzoic acid derivative enhances the solubility of the resulting monomer and polymers derived from it. Suzuki coupling of **24** with a diboronate then produces the alternating copolymer **25**. This had a molecular weight (M_n) by GPC of 17 kDa corresponding to a degree of polymerization of *ca.* 33 units, whereas the corresponding polymer made using a dibenzoylcarbazole had a low molecular weight of only 4.3 kDa (*ca.* 11 units). Treatment of **25** with methyl lithium converted all the ketone units to carbinols as shown by the disappearance of the ketone stretch in the infrared spectrum. Treatment with boron trifluoride then induced ring closure to form the desired ladder-type polymers **23**. These were found to undergo aggregation in solution producing broad peaks in the ^1H NMR spectra, which complicated structural analysis. However, by diluting the solutions sufficiently, it was possible to obtain absorption and emission spectra with small Stokes' shifts, and the sharp, well-resolved features characteristic of ladder-type materials, confirming the structures. These polymers displayed blue-green emission with emission maxima between 470 and 480 nm and ionization potentials of 5.2 to 5.3 eV making them suitable for use as hole-transporting emissive materials in OLEDs. They have been tested as donor materials for BHJ solar cells, but the device results were very poor, probably due to poor charge transport.[5]

6.2 Experimental Procedures

6.2.1 Materials and General Procedures

All reagents and solvents are obtained from commercial sources and used without further purification. Column chromatography is performed on silica. GPC data are reported against polystyrene standards unless otherwise stated, and uncorrected. It is known that absolute molecular weights of conjugated rigid-rod-like polymers obtained by comparison with such standards are inaccurate due to differences between the hydrodynamic volumes of rigid-rod and coil-like polymers, so the molecular weight values should be used for comparative purposes only. Molecular weights are given in units of kDa (1 kDa = 1000 g/mol).

6.2.2 Poly(3,6-carbazolylene)s

Example 1: Synthesis of *N*-Alkylated 3,6-Dibromocarbazoles

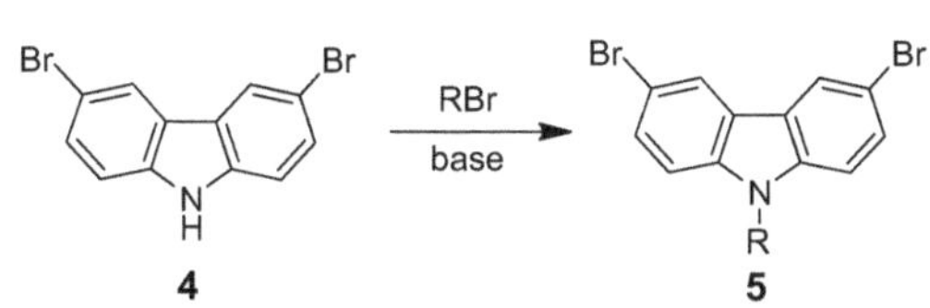

(6.9)

Table 6.1 Alkylation of 3,6-dibromocarbazole.

Alkyl bromide	*Base*	*Solvent*	*Temp, Time*	*Yield, %*	*Ref.*
1-bromodecane	K_2CO_3	DMF	50 °C, 24 h	85	7
3,7-dimethyloctyl bromide	K_2CO_3	DMF	50 °C, 24 h	91	7
(R) 3,7-dimethyloctyl bromide	K_2CO_3	DMF	50 °C, 24 h	92	8
(S) 3,7-dimethyloctyl bromide	K_2CO_3	DMF	50 °C, 24 h	81	8
1-bromotetradecane	NaH	THF/DMF (3:1)	0 °C to r.t., 12 h	86	9
1-bromohexane	KOH, TBAB	H_2O	r.t., 30 min	95	10
1-bromopentane	NaH	DMF	r.t., 6 h	95	11

The mixture of 3,6-dibromocarbazole (10 mmol), alkyl bromide (11 mmol) and potassium carbonate (30 mmol) in anhydrous dimethylformamide (DMF, 20 mL) was heated at 50 °C for 24 h under argon, and then water (200 mL) was added. Dichloromethane (3×50 mL) was used to extract the product. The organic layer was washed with water (2×100 mL) and dried over anhydrous $MgSO_4$. The solvent was removed to give the raw product, which was purified by column chromatography using hexane as the eluent to give a colorless viscous liquid.[7] Yield: R = n-decyl, 85%; R = 3,7-dimethyloctyl, 91% (Table 6.1).

Example 2: Synthesis of Low Molecular Weight Poly(*N*-alkylcarbazole) *via* Yamamoto Coupling

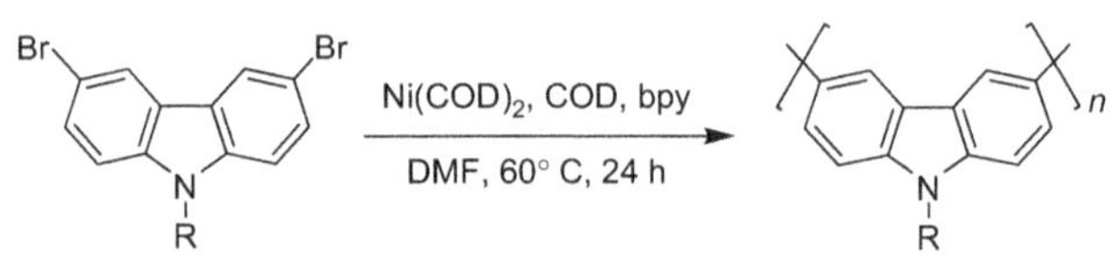

(6.10)

Bis(cycloocta-1,5-dienyl)nickel(0) ($Ni(COD)_2$, 12 mmol), cycloocta-1,5-diene (COD, 10 mmol), and 2,2′-bipyridine (bpy, 12 mmol) were mixed in 100 mL of anhydrous DMF under argon. The solution was heated at 60 °C for 0.5 h, and then a solution of the carbazole monomer (R = decyl, 10 mmol) in anhydrous DMF (20 mL) was added. The reaction was maintained at 60 °C for 24 h in the dark. The resulting mixture was poured into methanol (1 L) with magnetic stirring. A gray solid was collected by reduced-pressure filtration, and then dried under vacuum overnight at room temperature. The raw product was dissolved in tetrahydrofuran (THF, 200 mL). Insoluble particles were removed by filtration through a membrane filter with the pore size of 0.5 μm to give a transparent yellow solution. After the addition of methanol (200 mL), a white solid precipitated, which was separated by filtration and dried under vacuum overnight at room temperature to give the pure product.[7] (GPC M_n 1.7 kDa, M_w 2.6 kDa).

Example 3: Synthesis of High Molecular Weight Poly(*N*-alkylcarbazole) *via* Reverse Yamamoto Coupling

$Ni(COD)_2$ (12 mmol), COD (10 mmol), and bpy (12 mmol) were dissolved in anhydrous DMF (100 mL) under argon. The solution was heated at 60 °C for 30 min, and then slowly added to the solution of carbazole monomer (10 mmol) in anhydrous DMF (20 mL). The reaction was maintained at 60 °C for 24 h in the dark. The resulting mixture was poured into 1 L of methanol with magnetic stirring. A gray solid was collected by reduced pressure filtration, and then dried under vacuum overnight at room temperature. The raw product was dissolved in THF (200 mL). Insoluble particles were removed by filtration through a membrane filter with the pore size of 0.5 μm to give a transparent yellow solution. After the addition of isopropanol (100 mL), a white solid precipitated, which was separated by centrifugation and dried under vacuum overnight at room temperature to give the pure product.[7] Yield = 61% (R = decyl), 67% (R = 3,7-dimethyoctyl). GPC (R = decyl: M_n 53.4 kDa, M_w 91.0 kDa; R = 3,7-dimethyloctyl: M_n 84.2 kDa, M_w 120.0 kDa).

6.2.3 Poly(1,8-carbazolylene)s

Example 4: 3,6-Di-*tert*-butyl-*N*-hexadecylcarbazole

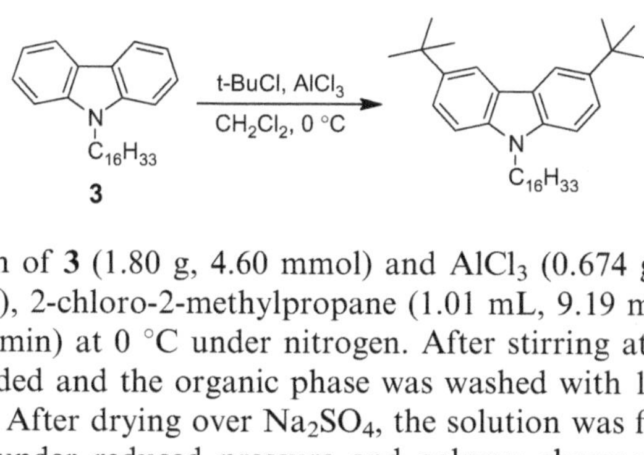

3

(6.11)

To a suspension of **3** (1.80 g, 4.60 mmol) and $AlCl_3$ (0.674 g, 5.06 mmol) in CH_2Cl_2 (14 mL), 2-chloro-2-methylpropane (1.01 mL, 9.19 mmol) was added dropwise (>20 min) at 0 °C under nitrogen. After stirring at 20 °C for 16 h, CH_2Cl_2 was added and the organic phase was washed with 1 M aq. HCl followed by brine. After drying over Na_2SO_4, the solution was filtered. Removal of the solvent under reduced pressure and column chromatography of the residue (hexane/CH_2Cl_2 2:1) afforded 3,6-di-*tert*-butyl-*N*-hexadecylcarbazole (2.11 g, 91%) as a yellow oil.[16]

Example 5: 3,6-Di-*tert*-butyl-1,8-diiodo-*N*-hexadecylcarbazole

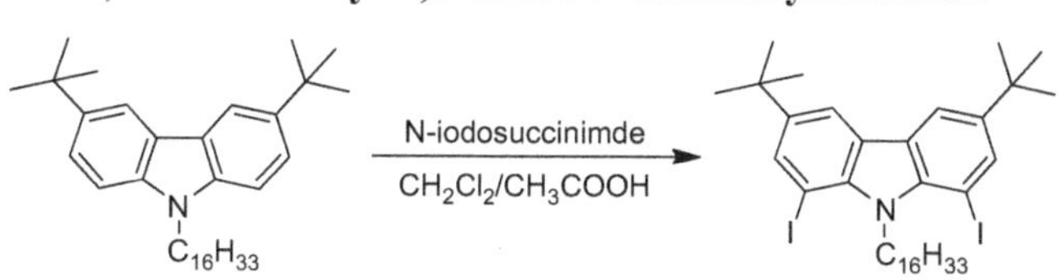

6

(6.12)

To a solution of 3,6-di-*tert*-butyl-*N*-hexadecylcarbazole (0.743 g, 1.47 mmol) in CH_2Cl_2 (5 mL) and CH_3COOH (5 mL), *N*-iodosuccinimide (0.682 g, 3.03 mmol) was added and the mixture was stirred at 20 °C for 16 h. CH_2Cl_2

was added and the organic phase was washed with aq. $NaHCO_3$ and water. After drying over Na_2SO_4, the solution was filtered. Removal of the solvent under reduced pressure and column chromatography of the residue (hexane) afforded **6** (0.802 g, 72%) as a yellow oil.[16]

Example 6: Synthesis of Poly(3,6-di-*tert*-butyl-*N*-hexadecyl-1,8-carbazolylene)

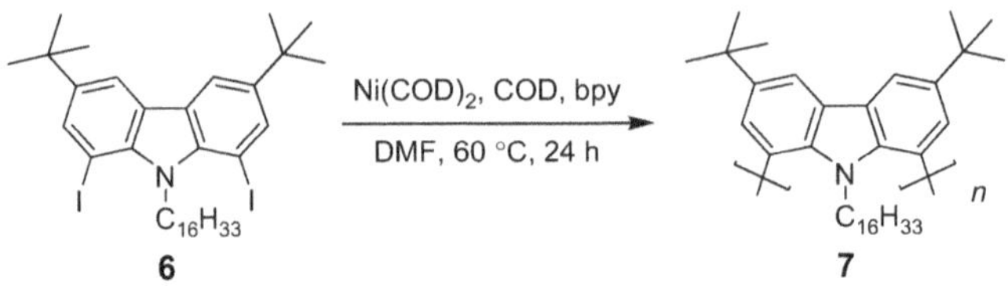

(6.13)

To a solution of **6** (303 mg, 0.397 mmol) in dry DMF (1.4 mL), a solution of COD, (48.5 μL, 0.395mmol), $Ni(COD)_2$ (130 mg, 0.476 mmol), and bpy (75 mg, 0.47 mmol) in dry DMF (4.1 mL) was added dropwise at 60 °C. After stirring at 60 °C for 24 h, the solvent was reduced *in vacuo* and poured into MeOH. The precipitate was collected, and the polymer was purified by reprecipitation in MeOH and then dried *in vacuo* (69.6 mg, 35%).[17] GPC (M_n 1.1 kDa, M_w 1.3 kDa).

6.2.4 Poly(2,7-carbazolylene)s

Example 7: 4,4′-Dibromo-2-nitrobiphenyl

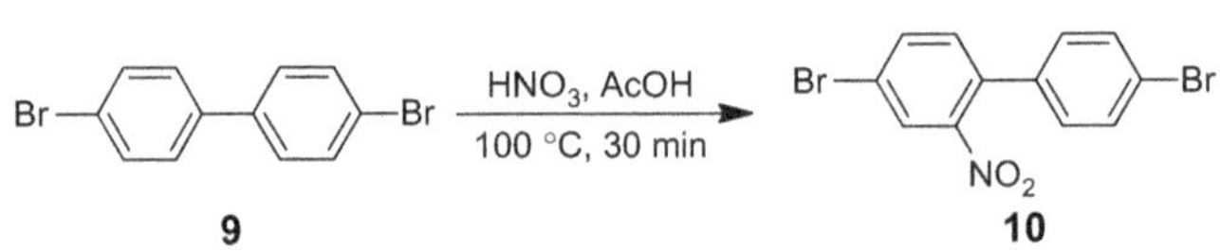

(6.14)

To a solution of 4,4′-dibromobiphenyl (**9**; 20 g, 0.064 mol) in glacial AcOH (300 mL) at 100 °C was added slowly a mixture of fuming HNO_3 (100%, 92.5 mL) and H_2O (7.5 mL). On heating the reaction mixture for 30 min at 100 °C, the initially formed precipitate redissolved. The solution was cooled down and the resulting yellow paste was collected by filtration. Recrystallization from EtOH afforded the title compound as a yellow solid (21.52 g, 91%).[18]

Example 8: Synthesis of 2,7-Dibromocarbazole with Triethyl Phosphite (Cadogan Synthesis)

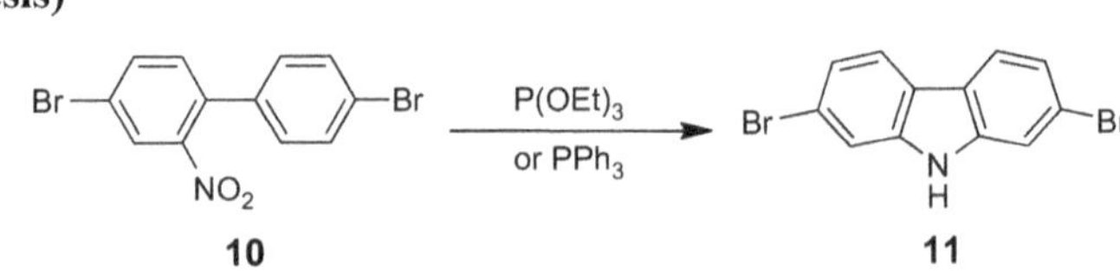

(6.15)

A mixture of compound **10** (20.77 g, 58.2 mmol) and triethyl phosphite (75 mL) was heated under reflux for 18 h in an inert atmosphere. The excess of triethyl phosphite was distilled off and the product was purified by column chromatography (5–20% EtOAc in hexane) to provide 10.6 g of the title compound **11** as a white solid (56%).[18]

Example 9: Synthesis of 2,7-Dibromocarbazole with Triphenylphosphine

A mixture of 4,4′-dibromo-2-nitrobiphenyl (**10**, 1.02 g, 2.86 mmol) and PPh_3 (2.5 equiv.) was taken up in 1,2-dichlorobenzene (*o*-DCB) (2 mL per mmol of **10**) under N_2 and heated to reflux, with vigorous stirring, using a heating mantle equipped with a sand bath. The reaction was stopped upon complete consumption of the nitrobiphenyl starting material, at which point the reaction was cooled and the solvent stripped under high vacuum. The residue was either chromatographed directly, or slurried with P950 ligroin (~2.5 mL per mL of *o*-DCB used) to precipitate PPh_3O. Upon removal of the PPh_3O by filtration, the filtrate was concentrated. Chromatography of the residue (75:25 P950 ligroin:CH_2Cl_2) gave the product **11** as a flaky, lustrous, off-white solid (0.70 g, 75% yield).[20]

Example 10: Synthesis of *N*-Alkylated 2,7-Dibromocarbazoles

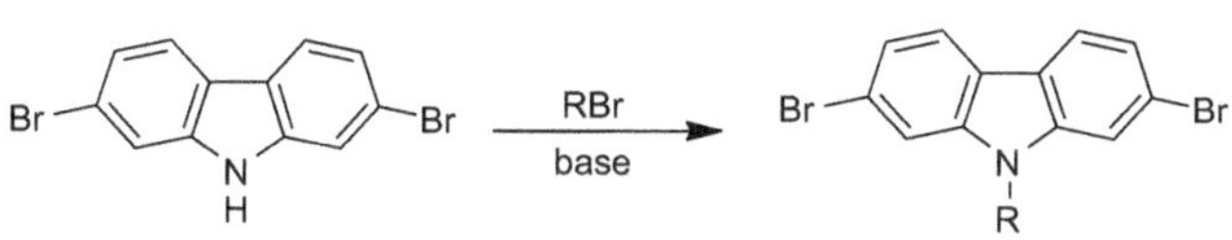

11 → **8** (6.16)

To a mixture of compound **11** (10 g, 31 mmol) and anhydrous DMF (100 mL) was added slowly NaH (1.73 g, 60% w/w dispersion in mineral oil, 43 mmol). After 30 min, 2-ethylhexyl bromide (7.13 mL, 40 mmol) was added and the solution was stirred for 18 h under argon. The reaction was quenched with H_2O and extracted with CH_2Cl_2. The organic fractions were dried ($MgSO_4$) and the solvent was removed under reduced pressure. The residue was purified by column chromatography (10% EtOAc in hexane) to give the title compound **8**. Yield: R = 2-ethylhexyl, 12.9 g, 96%[18] (Table 6.2).

Table 6.2 Alkylation of 2,7-dibromocarbazole.

Alkyl bromide	*Base*	*Solvent*	*Temp, Time*	*Yield, %*	*Ref.*
2-ethylhexyl bromide	NaH	DMF	r.t., 18 h	96	18
2-ethylhexyl bromide	K_2CO_3	DMF	80 °C, 24 h	80	21
2-ethylhexyl bromide	K_2CO_3	DMF	80 °C, 24 h	78	22
1-bromooctane	K_2CO_3	DMF	80 °C, 17 h	84	23
1-bromooctane	K_2CO_3	DMF	80 °C, 12 h	82	24
1-bromooctane	NaH	DMF	r.t.	n.a	25
1-bromohexane	NaH	DMF	r.t.	n.a	25
1-bromohexane	NaH	DMF	r.t., 13 h	89	26
1-bromodecane	NaH	DMF	0 °C to r.t., 24 h	71	27
1-bromododecane	NaH	DMF	r.t., overnight	99	28
11-(bromomethyl)-tricosane	NaH	DMF	70 °C, 18 h	88	29
9-heptadecane *p*-toluenesulfonate	KOH	DMSO	r.t., 8h	87	30

n.a = not available.

Example 11: Synthesis of Poly[*N*-(2′-decyltetradecyl)carbazole]-2,7-diyl *via* Yamamoto Coupling

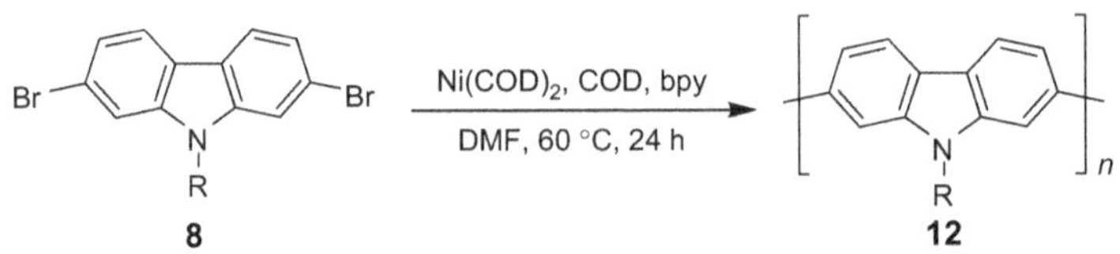

(6.17)

$Ni(COD)_2$ (515 mg, 1.87 mmol), COD (0.23 mL, 1.87 mmol), and bpy (293 mg, 1.87 mmol) were dissolved in dry DMF (3 mL) and dry toluene (3 mL) under argon. This mixture was heated for 30 min at 80 °C and then 2,7-dibromo-*N*-(2-decyltetradecyl)carbazole (1 g, 1.08 mmol) and bromobenzene (2.6 mL, 0.025 mmol) dissolved in toluene (6 mL) were added. The reaction was stirred for 72 h under argon at 80 °C. The reaction mixture was poured into methanol/conc. HCl (1 : 1) and the precipitated product was filtered off. The polymer was dissolved in chloroform, washed with a saturated EDTA solution, and the oligomers were extracted in a Soxhlet apparatus using acetone as solvent to give the title polymer as a light-yellow solid (318 mg, 84%).[29] GPC (M_n 39.1 kDa, M_w 75.5 kDa).

Example 12: Synthesis of 2,7-Carbazolebisboronate

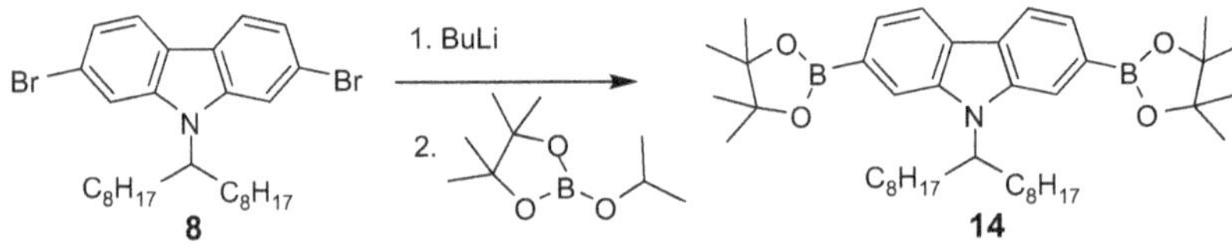

(6.18)

To a solution of compound **8** (6.0 g, 10.7 mmol) in THF (107 mL) in a flame-dried 250 mL flask at −78 °C was added dropwise *n*-butyllithium (2.5 M in hexane, 8.7 mL, 21.8 mmol). The mixture was stirred at −78 °C for 1 h and 2-isopropoxy-4,4,5,5-tetramethyl-1,3,2-dioxaborolane (4.8 mL, 4.4 g, 23.4 mmol) was added rapidly to the solution. After one additional hour at −78 °C, the resulting mixture was warmed to room temperature and stirred overnight. The mixture was poured into water, extracted with diethyl ether four times and dried over magnesium sulfate. The solvent was removed under reduced pressure, and the residue was purified by recrystallization in methanol/acetone (ca. 10 : 1) to obtain the bisboronate product as white crystals (6.1 g, Yield: 87%). m.p. 128–130 °C.

Example 13: Synthesis of Alternating Copolymer of 2,7-Carbazole with Acceptor Unit

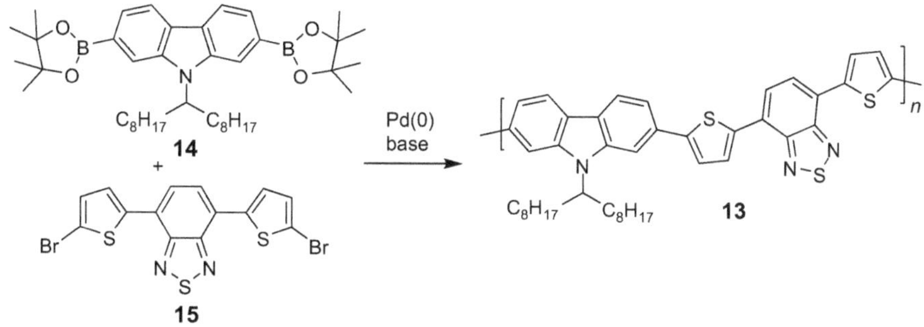

(6.19)

In a 50-mL flame dried flask, compound **14** (723.3 mg, 1.1 mmol), 4,7-di(2′-bromothien-5′-yl)-2,1,3-benzothiadiazole (**15**) (504.0 mg, 1.1 mmol), tris(dibenzylideneacetone)dipalladium(0) (5.0 mg, 0.006 mmol) and tri(*o*-tolyl)phosphine (6.7 mg, 0.02 mmol) were dissolved in a mixture of degassed toluene (11.0 mL) and degassed 20% aqueous tetraethylammonium hydroxide (3.7 mL). The reaction mixture was vigorously stirred at 90–95 °C. After a few hours, the mixture became very viscous and toluene (5.0 mL) was added. After 72 h, bromobenzene (12 μL, 0.1 mmol) was added to the reaction then one hour later, phenylboronic acid (13.4 mg, 0.1 mmol) was added and the reaction refluxed overnight to complete the end-capping reaction. The polymer was purified by precipitation in methanol/water (10:1), filtered through 0.45 μm nylon filter and successively washed in a Soxhlet apparatus with acetone, hexanes, toluene, dichloromethane, and then chloroform. The chloroform fraction (550–600 mL) was reduced to 40–50 mL under reduced pressure, then the product was precipitated from methanol/water (10:1, 500 mL), collected by filtration through a 0.45-μm nylon filter and finally air dried overnight. Yield = 178.1 mg (23%). GPC ($M_n = 37$ kDa, $M_w = 73$ kDa).

Example 14: *N*-9′-(3,5-Dimethoxyphenyl)-2,7-dibromocarbazole (18a)

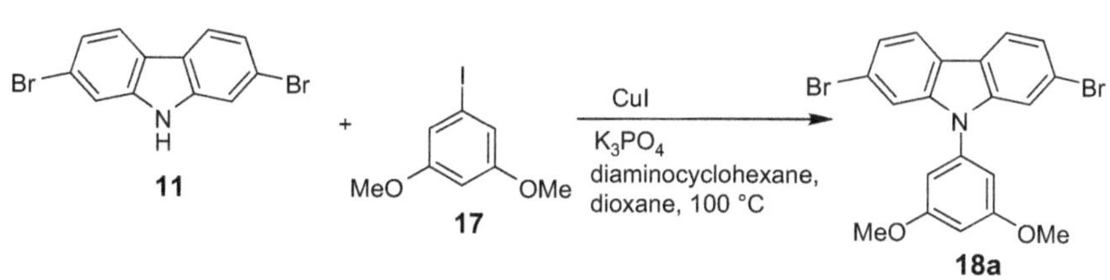

(6.20)

To a 100-mL round-bottomed flask was charged 2,7-dibromocarbazole (**11**, 4.0 g, 12.3 mmol, 1.0 eq), 1-iodo-3,5-dimethoxybenzene (**17**, 3.9 g, 14.8 mmol, 1.2 eq), potassium phosphate (7.82 g, 36.9 mmol, 3.0 eq), copper iodide (0.59 g, 3.1 mmol, 0.25 eq). The solid mixture was purged with argon before adding anhydrous dioxane (30 mL). The mixture was stirred at room temperature for 15 min, then (±)-*trans*-1,2-diaminocyclohexane (0.72 mL, 0.5 eq) was added rapidly and the reaction was heated at reflux overnight. The mixture was then filtered and washed with dichloromethane. After removing the solvent, the crude product was first purified by column chromatography (15% DCM in hexane as eluent), then the as-obtained solid was dissolved in a small amount of boiling chloroform. The product **18a** (4.2 g, 73%) was obtained as an off-white solid by precipitation from chloroform solution in methanol (200 mL). The product contains a small amount (<4%) of a monoiodo impurity formed by halide exchange at C2 or C7, which could not be separated.

Example 15: *N*-9′-(3,5-Dioctyloxyphenyl)-2,7-dibromocarbazole (18b)

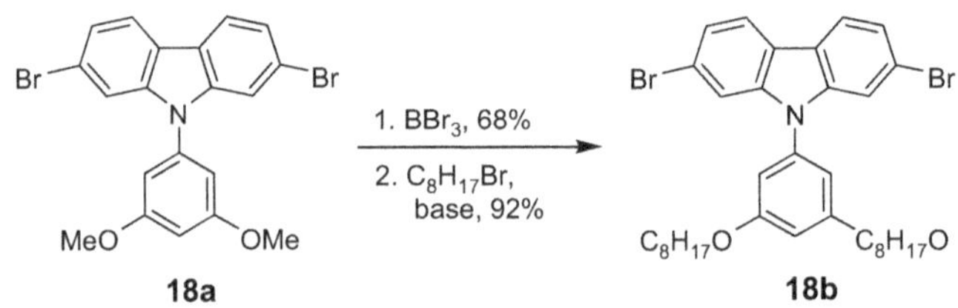

(6.21)

To a 100-mL flask was added compound **18a** (4.0 g, 8.68 mmol, 1.0 eq) and DCM (50 mL). The clear solution was purged with argon for 30 min before being cooled to −78 °C using an acetone/dry ice bath. BBr_3 (6.52 g, 26.0 mmol, 3.0 eq) was dissolved in DCM (20 mL) under argon atmosphere and the solution was added dropwise to the mixture over 10 min. The reaction was allowed to warm to room temperature and stirred for 48 h. The reaction mixture was then quenched with water carefully, extracted with DCM three times. The organic layer was washed with water and dried over magnesium sulfate. The solvent was removed under reduced pressure. The residue was purified by column chromatography (silica gel, 5% ethyl acetate in DCM as eluent). *N*-9′-(3,5-dihydroxyphenyl)-2,7-dibromocarbazole (2.56 g, 68%) was obtained as a white solid.

A suspension of *N*-9′-(3,5-dihydroxyphenyl)-2,7-dibromocarbazole (1.5 g, 3.5 mmol, 1.0 eq), 1-bromooctane (3.6 mL, 21.0 mmol, 6.0 eq), and potassium carbonate (5.8 g, 42 mmol, 12.0 eq) in dry DMF (15 mL) was degassed three times. The mixture was stirred at 70 °C for 5 h monitored by thin-layer chromatography (TLC). The suspension was then cooled to room temperature and filtered off. The filtrate was directly loaded on a silica gel column, first using pure hexane as eluent to remove DMF, then using 10% DCM in hexane as eluent. Compound **18b** (2.1 g, 92%) was collected as a white powder after removal of solvents.

6.2.5 Poly(3,9-carbazolylene)s

Example 16: Synthesis of 3-Iodocarbazole

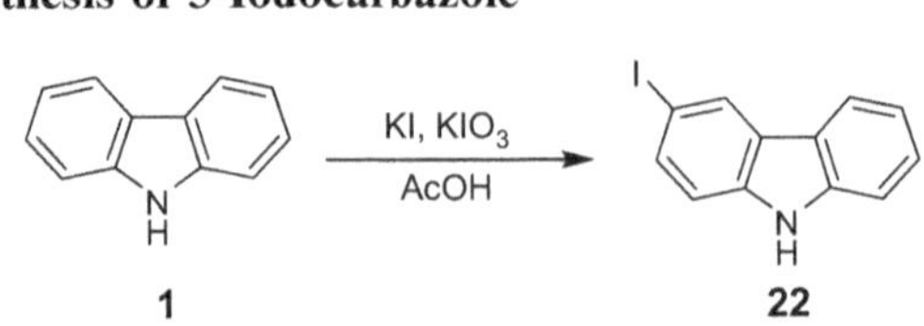

(6.22)

Carbazole (**1**) (16.7 g, 95.7 mmol) was dissolved in boiling glacial acetic acid (260 mL), and potassium iodide (11 g, 66.4 mmol) added. The solution was cooled somewhat, finely powdered potassium iodate (16 g, 66.4 mmol) added, and the mixture then boiled until it acquired a clear straw-colored tint (10 min). The hot solution was decanted from the insoluble potassium iodate and allowed to cool slowly to 45 °C. The faintly brown, nacreous plates that separated were rapidly filtered off and recrystallized from glacial acetic acid (or alcohol), the solution being allowed to cool to 45 °C and filtered as before from the white, nacreous plates (10–12 g) of 3-iodocarbazole.[36]

Example 17: Synthesis of Poly(3,9-carbazole)

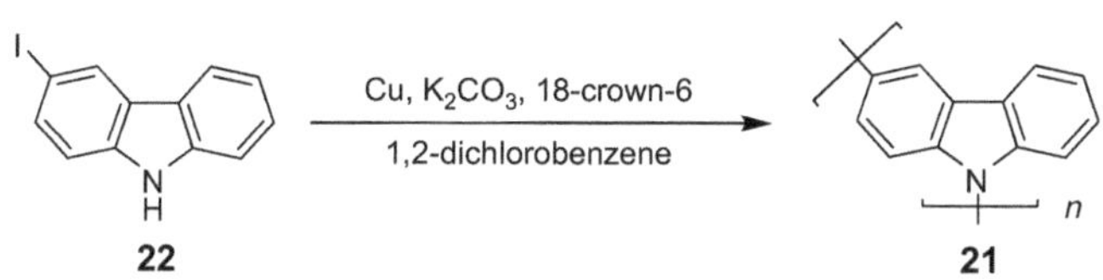

(6.23)

1.64 g (5.58 mmol) of 3-iodo-9*H*-carbazole, 0.71 g (11.16 mmol) of copper powder, 3.08 g (22.32 mmol) of potassium carbonate and 0.15 g (0.56 mmol) 18-crown-6 in 1,2-dichlorobenzene (20 mL) were refluxed for 36 h in a nitrogen atmosphere. The inorganic components were then filtered off while hot and product was precipitated from hexane several times. The solid residue was extracted with hot ethanol to remove 18-crown-6. Yield: 0.8 g (87%) of brown powder. GPC (M_n 2.36 kDa, M_w 5.07 kDa).[34]

6.2.6 Ladder-Type Poly(carbazolylene)s

Example 18: Synthesis of 3,6-Diaroyl-2,7-dibromo-*N*-alkylcarbazole

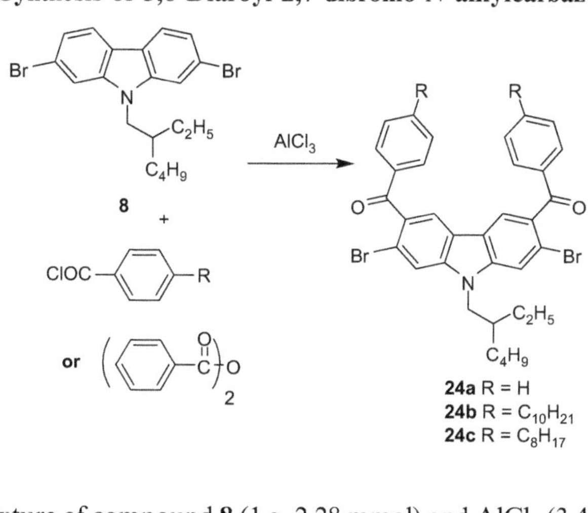

(6.24)

a) To a mixture of compound **8** (1 g, 2.28 mmol) and $AlCl_3$ (3.4 g, 25.5 mmol) in nitrobenzene (14 mL) was added benzoic anhydride (1.54 g, 6.8 mmol) at 0 °C. The reaction mixture was stirred for 8 h and then quenched with ice. The inorganic precipitate was dissolved in 2 M HCl and the product was extracted with Et_2O (3 × 50 mL). The combined organic fractions were dried ($MgSO_4$) and the solvent was removed under reduced pressure. The product was first purified by column chromatography (20% EtOAc in hexane) and then recrystallization from EtOAc to afford **24a** as a white solid (1.1 g, 76%); mp 201–203 °C.

b) To a mixture of compound **8** (1 g, 2.28 mmol) and $AlCl_3$ (0.73 g, 5.5 mmol) in 1,2-dichloroethane (3 mL) was added slowly, 4-octylbenzoyl chloride (1.35 g, 4.8 mmol) at r.t. The mixture was stirred for 4 h at 50 °C and then quenched with ice. The inorganic precipitate was dissolved in 2 M HCl and the product was extracted with CH_2Cl_2. The organic fractions were dried ($MgSO_4$) and the solvent was removed under reduced pressure. The product was purified by column chromatography (10% EtOAc in hexane) to provide **24c** as a yellowish solid (1.6 g, 75%); mp 90–93 °C.

Example 19: Synthesis of Precursor Alternating Copolymer

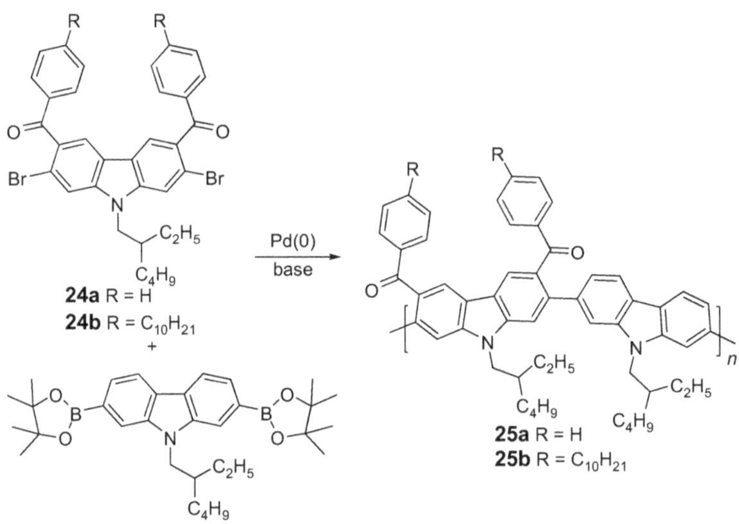

(6.25)

a) *N*-(2-ethylhexyl)-2,7-dibromo-3,6-dibenzoylcarbazole **24a** (607 mg, 0.94 mmol), *N*-(2-ethylhexyl)-2,7-bis(4,4,5,5-tetramethyl-1,3,2-dioxaborolane-2-yl)carbazole (500 mg, 0.94 mmol) and $Pd(PPh_3)_4$ (43 mg, 0.038 mmol) were dissolved in a degassed mixture of aqueous K_2CO_3 (2 M, 3.5 mL) and THF (5 mL). The reaction was stirred under reflux for 5 d and then poured into a mixture of methanol/HCl (250 mL/50 mL [2 M]). The precipitate was filtered off and washed with methanol to give 660 mg (91%) of **25a** as a yellow solid. FT-IR (KBr): 2955, 2924, 2854, 1658, 1597, 1461, 1377, 1268, 1066 cm^{-1}. GPC ($M_n = 4.30$ kDa, $M_w = 9.2$ kDa).

b) Monomer **24b** (1 g, 1.08 mmol), *N*-(2-ethylhexyl)-2,7-bis(4,4,5,5-tetramethyl-1,3,2-dioxaborolane-2-yl)carbazole (604 mg, 1.08 mmol) and $Pd(PPh_3)_4$ (62 mg, 0.038 mmol) were dissolved in a degassed mixture of aqueous K_2CO_3 (2 M, 3.5 mL), toluene (5 mL) and 1-butanol (1 mL). The reaction was stirred for 3 d at 80 °C and then poured into a mixture of methanol/HCl (250 mL/50 mL [2 M]). The precipitate was filtered off and washed with methanol followed by Soxhlet extraction with acetone to give 1.02 g (90%) of **25b** as a gray polymer. FT-IR (KBr): 2957, 2925, 2853, 1659, 1600, 1458, 1325, 1262, 1200, 1176 cm^{-1}. GPC ($M_n = 17.0$ kDa, $M_w = 27.0$ kDa).

Example 20: Synthesis of Ladder-Type Polycarbazole

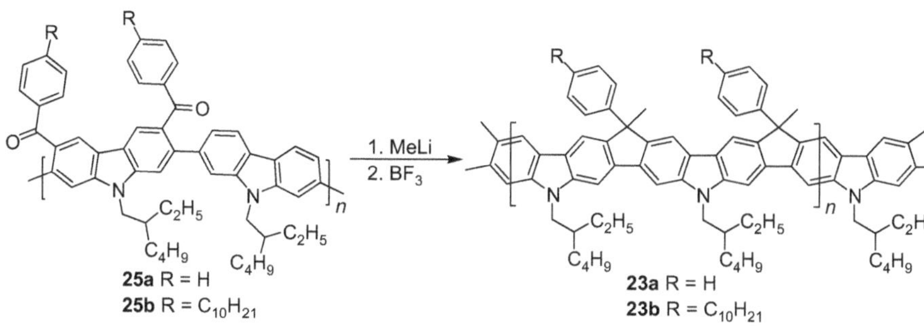

(6.26)

a) A solution of polymer **25a** (100 mg) in THF (20 mL) was cooled to −78 °C. MeLi (2.65 mL, 4.25 mmol, 1.6 M) was added quickly and then the reaction was slowly warmed up (18 h). The reaction was quenched by adding ethanol, water and hydrochloric acid and then extracted with dichloromethane. The organic fractions were dried with $MgSO_4$ and the solvent was removed under reduced pressure to give a yellow polymer (97 mg, 94%). FT-IR (KBr): 3547, 2955, 2924, 2856, 1601, 1456, 1375, 1321, 1249, 1066, 698, 455 cm^{-1}. GPC ($M_n = 4.9$ kDa, $M_w = 10.0$ kDa).

The product was dissolved in dichloromethane (10 mL) and $BF_3 \cdot Et_2O$ (1 mL, 8.1 mmol) was added. After 5 min at r.t. the reaction was quenched with ethanol (20 mL) and extracted with dichloromethane. The organic fractions were dried with $MgSO_4$ and the solvent was removed under reduced pressure to give **23a** as a brown powder (84 mg, 88%). GPC ($M_n = 5.4$ kDa, $M_w = 10.5$ kDa).

b) Treatment of **25b** (200 mg) with methyl lithium using the same procedure as for **25a** gave a yellow solid (200 mg, 98%). FT-IR (KBr): 3550, 2957, 2924, 2852, 1601, 1457, 1321, 1248, 1193, 1075 cm^{-1}. GPC ($M_n = 17.0$ kDa, $M_w = 32.0$ kDa). This polymer was ring-closed by the same method as for **25a** to give **23b** as a brown solid (164 mg, 83%). GPC ($M_n = 13.5$ kDa, $M_w = 23.0$ kDa).

References

1. J. V. Grazulevicius, P. Strohriegl, J. Pielichowski and K. Pielichowski, *Prog. Polym. Sci.*, 2003, **28**, 1297.
2. J.-F. Morin, M. Leclerc, D. Ades and A. Siove, *Macromol. Rapid Commun.*, 2005, **26**, 761.
3. A. C. Grimsdale, K. L. Chan, R. E. Martin, P. G. Jokisz and A. B. Holmes, *Chem. Rev.*, 2009, **109**, 987.
4. Y.-J. Cheng, S.-H. Yang and C.-S. Hsu, *Chem. Rev.*, 2009, **109**, 5868.
5. J. Li and A. C. Grimsdale, *Chem. Soc. Rev.*, 2010, **39**, 2399.
6. D. B. Romero, M. Schaer, M. Leclerc, D. Adès, A. Siove and L. Zuppiroli, *Synth. Met.*, 1996, **80**, 271.
7. Z.-B. Zhang, M. Fujiki, H.-Z. Tang, M. Motonaga and K. Torimitsu, *Macromolecules*, 2002, **35**, 1988.
8. Z.-B. Zhang, M. Motonaga, M. Fujiki and C. E. McKenna, *Macromolecules*, 2003, **36**, 6956.
9. A. M. Geyer, E. S. Wiedner, J. B. Gary, R. Gdula, N. C. Kuhlmann, M. J. A. Johnson, B. D. Dunietz and J. W. Kampf, *J. Am. Chem. Soc.*, 2008, **130**, 8984.
10. C. He, Q. G. He, Q. Chen, L. Shi, H. M. Cao, J. G. Cheng, C. M. Deng and T. Lin, *Tetrahedron Lett.*, 2010, **51**, 1317.
11. J. F. Xing, W. Q. Chen, J. Gu, X. Z. Dong, N. Takeyasu, T. Tanaka, X. M. Duan and S. Kawata, *J. Mater. Chem.*, 2007, **17**, 1433.
12. E. Ngblio, D. Ades, C. Chevrot and A. Siove, *Polym. Bull.*, 1990, **24**, 17.
13. Y. Yasutani, Y. Honsho, A. Saeki and S. Seki, *Synth. Met.*, 2012, **162**, 1713.
14. K. Lmimouni, C. Legrand and A. Chapoton, *Synth. Met.*, 1998, **97**, 151.
15. A. van Dijken, J. J. A. M. Bastiaansen, N. M. M. Kiggen, B. M. W. Langeveld, C. Rothe, A. Monkman, I. Bach, P. Stössel and K. Brunner, *J. Am. Chem. Soc.*, 2004, **126**, 7718.
16. T. Michinobu, H. Osako and K. Shigehara, *Macromol. Rapid Commun.*, 2008, **29**, 111.
17. T. Michinobu, H. Osako and K. Shigehara, *Macromolecules*, 2009, **42**, 8172.
18. F. Dierschke, A. C. Grimsdale and K. Müllen, *Synthesis*, 2003, 2470.
19. J. I. G. Cadogan, M. Cameron-Wood, R. K. Mackie and R. J. G. Searle, *J. Chem. Soc.*, 1965, 4831.
20. A. W. Freeman, M. Urvoy and M. E. Criswell, *J. Org. Chem.*, 2005, **70**, 5014.
21. Q. D. Zheng, S. C. Chen, B. Zhang, L. X. Wang, C. Q. Tang and H. E. Katz, *Org. Lett.*, 2011, **13**, 324.
22. S. Qiu, L. L. Liu, B. L. Wang, F. Z. Shen, W. Zhang, M. Li and Y. G. Ma, *Macromolecules*, 2005, **38**, 6782.
23. K. Brunner, A. van Dijken, H. Borner, J. J. A. M. Bastiaansen, N. M. M. Kiggen and B. M. W. Langeveld, *J. Am. Chem. Soc.*, 2004, **126**, 6035.
24. J. Bouchard, S. Wakim and M. Leclerc, *J. Org. Chem.*, 2004, **69**, 5705.

25. P. Gao, X. L. Feng, X. Y. Yang, V. Enkelmann, M. Baumgarten and K. Müllen, *J. Org. Chem.*, 2008, **73**, 9207.
26. L. L. Rong, Q. C. Liu, J. Tang and Z. G. Chi, *Heterocycles*, 2010, **81**, 977.
27. F. A. Lemasson, T. Strunk, P. Gertsel, F. Hennrich, S. Lebedkin, C. Barner-Kowollik, W. Wenzel, M. M. Kappes and M. Mayor, *J. Am. Chem. Soc.*, 2011, **133**, 652.
28. Y. L. Chen, W. G. Huang, C. H. Li and Z. S. Bo, *Macromolecules*, 2010, **43**, 10216.
29. J. Li, F. Dierschke, J. Wu, A. C. Grimsdale and K. Müllen, *J. Mater. Chem.*, 2006, **16**, 96.
30. N. Blouin, A. Michaud and M. Leclerc, *Adv. Mater.*, 2007, **19**, 2295.
31. D. Vyprachticky, I. Kminek, V. Pokorna and V. Cimrova, *Tetrahedron*, 2012, **68**, 5075.
32. A. Iraqi and I. Wataru, *Chem. Mater.*, 2004, **16**, 442.
33. X. Liu, W. Wen and G. C. Bazan, *Adv. Mater.*, 2012, **24**, 4505.
34. S. Grigalevicius, J. V. Grazulevicius, V. Gaidelis and V. Jankauskas, *Polymer*, 2002, **43**, 2603.
35. F. Dierschke, A. C. Grimsdale and K. Müllen, *Macromol. Chem. Phys.*, 2004, **205**, 1147.
36. S. H. Tucker, *J. Chem. Soc.*, 1926, 546.

CHAPTER 7

Poly(phenylenevinylene)s

WALLACE W. H. WONG,* HELGA SEYLER AND ANDREW B. HOLMES

School of Chemistry, Bio21 Institute, University of Melbourne, 30 Flemington Road, Parkville, Victoria 3010, Australia
*Email: wwhwong@unimelb.edu.au

7.1 Introduction

Poly(phenylenevinylene), PPV **1**, is one of the most investigated conjugated polymer material with over 2000 publications on this topic to-date (Figure 7.1). One of the earliest reports of PPV synthesis was published in 1960 where an oligomer of PPV was isolated *via* a Wittig reaction.[1] However, research into PPV materials did not take off until the first report of electroluminescence from PPV **1** in 1990.[2] The field of PPV materials was further boosted by the introduction of solution-processable derivatives in the early 1990s.[3] A range of photophysical and electronic properties have been achieved by varying the functional groups along the PPV backbone. The absorption and emission maxima of PPV materials can vary through much of the visible spectrum depending on substituents (Figure 7.1). The effective conjugation length for solid state fluorescence in PPV is between 8 and 10 repeat units.[4] Thin films of unsubstituted PPV **1** exhibit bright yellow-green fluorescence with emission maximum at 551 nm. Both electron-donating and -withdrawing groups will change the photophysical properties of PPV derivatives. Frequently reported examples include solution-processable MEH-PPV **2** and CN-PPV **3** (Figure 7.1). PPV materials without appropriate solubilising substituents are difficult to process and thin films are typically prepared *via* precursor routes (*vide infra*). PPV materials have been prepared in a variety of molecular weight

RSC Polymer Chemistry Series No. 9
Conjugated Polymers: A Practical Guide to Synthesis
Edited by Klaus Müllen, John R. Reynolds and Toshio Masuda

Published by the Royal Society of Chemistry, www.rsc.org

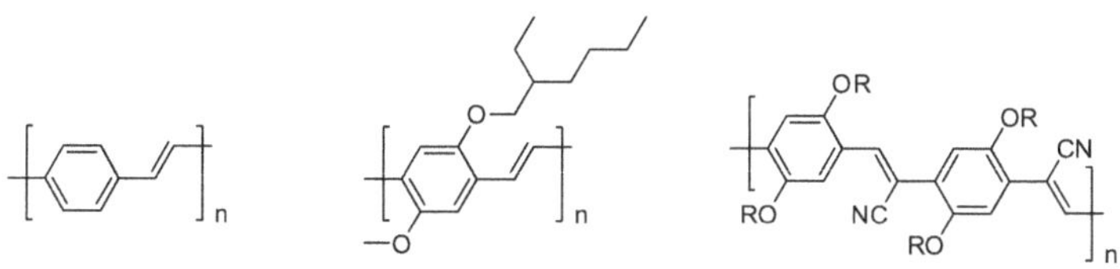

Figure 7.1 Chemical structure of unsubstituted PPV **1** and two frequently reported PPV derivatives, MEH-PPV **2** and CN-PPV **3**.

distributions from 10 to more than 10 000 repeat units. The primary function of PPV is in semiconducting applications, such as organic light-emitting diodes (OLED)[5] and organic solar cells (OSC).[6] For details and discussions on structure–property relationships and performance in devices, readers are directed to a number of reviews[5–7] and books.[8–11]

7.2 Methods of Synthesis

There are a substantial number of methods for the preparation of PPV materials. There is a popular division between so-called precursor routes and nonprecursor routes in the literature.[12] In precursor methods, polymerization of monomeric compounds does not lead directly to the target conjugated PPV product. Typically, a second treatment of the precursor polymer is performed to create the ethylene unit or to aromatise the ring, resulting in the conjugated polymer structure of PPV. One of the key advantages of precursor methods is that target PPV materials with low processability can be handled at the precursor polymer stage. While most precursor polymers are obtained *via* radical/anionic polymerization, there are also examples of palladium-catalyzed and ring-opening metathesis polymerization methods (*vide infra*). In this chapter, the preparation methods are classified according to the polymerization mechanism.

7.2.1 Precursor Routes *via* Radical/Anionic Polymerization Mechanism

Early syntheses of PPV were achieved *via* either radical or anionic polymerization mechanisms. There are several common elements in the radical/anionic polymerization of precursor PPV materials. An intermediate active monomer is typically generated by the treatment of the starting monomer compound **4** with a base (Scheme 7.1). With the elimination of the leaving group (L), the *p*-quinodimethane intermediate **5** is produced that can then be polymerized *via* the radical or anionic mechanism giving the precursor polymer. The substituent (P) on the precursor material imparts solubility and processability to the polymer. The mechanism of the polymerization is still poorly understood in many systems. In the radical mechanism, two reactive quinodimethane intermediates dimerise forming a diradical species **6** (Scheme 7.1). This diradical will

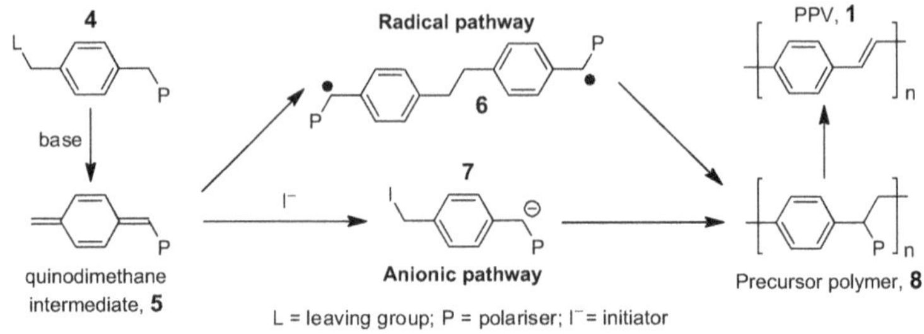

Scheme 7.1 Radical/anionic polymerization pathways for PPV preparation.

then undergo a free-radical polymerization giving relatively high molecular weight polymers. In the anionic pathway, the initiator (I) is usually the base in the reaction and the resulting polymer molecular weight is usually low as a consequence of the large amounts of base present (Scheme 7.1). It has been found that the mechanism, either radical or anionic, is not only dependent on the monomer starting materials but also solvent, concentration, temperature, and additives. In fact, both polymerization mechanisms can be in operation at the same time in many cases leading to bimodal polymer molecular weight distributions. More detailed discussion and studies on the polymerization mechanisms can be found in the literature.[13–15] The following section is divided into the different precursor monomers and polymers. Comparisons in their method of preparation, stability and reactivity are presented (Scheme 7.2).

7.2.1.1 Sulfonium Precursor

The first synthesis of a sulfonium precursor polymer was described by Wessling and Zimmermann in a patent disclosed in 1968.[16] Since then, this synthesis method has been extensively studied by several research groups leading to significant modifications and optimization.[17–23] Although a variety of dialkyl sulfonium precursors **9** can be used for polymerization (Scheme 7.2a), cyclic sulfonium moieties, tetrahydrothiophene in particular, are preferred as undesired side reactions are suppressed in the subsequent conversion step for these systems.[18] Sulfonium monomers can be readily prepared by the reacting bis(halomethyl)benzene with thiol compounds. Typically, the polymerization is induced with the addition of slightly less than one equivalent of hydroxide base to a solution of monomer in a protic solvent (*e.g.* methanol/water) at around 0 °C. It is important to note that the addition of one equivalent of base leads to base-induced elimination of some sulfonium groups on the precursor polymer. As a result of the radical/anionic polymerization mechanism, these reactions must be performed under an inert atmosphere. For substituted phenylene monomers with low solubility or miscibility with protic solvents, the polymerization can be achieved in a biphasic system (*e.g.* water/mineral oil) using a high-speed stirrer or homogenizer.[24]

Purification of precursor sulfonium polymer **10** is achieved by dialysis to remove small oligomeric material (typically <5000 g/mol depending on dialysis

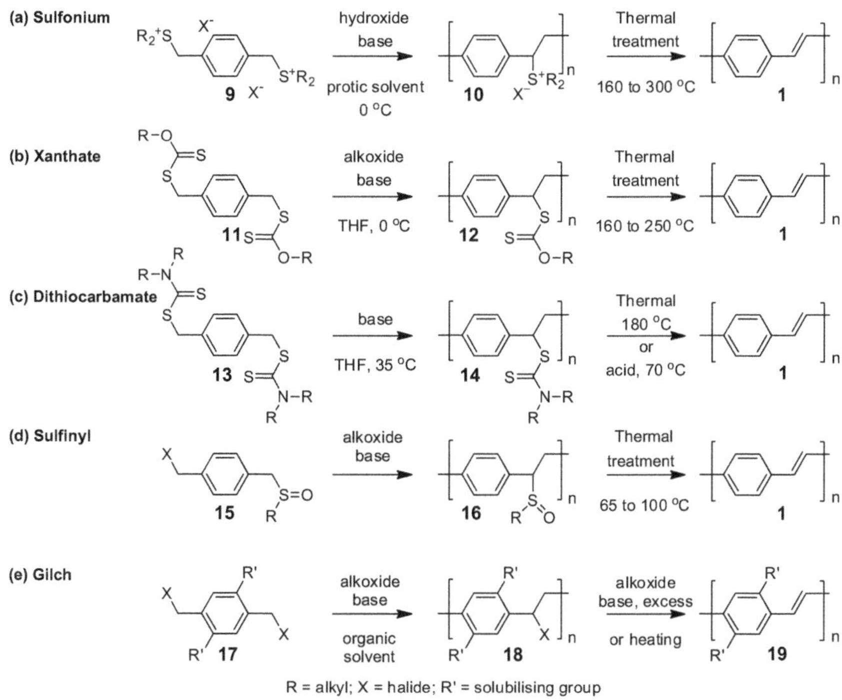

Scheme 7.2 General representation of five precursor routes *via* radical/anionic polymerization.

membrane used). The polyelectrolyte nature of the sulfonium polymers prevents direct molecular weight analysis using size exclusion chromatography (SEC). Conversion of the polymer in refluxing methanol to the neutral methoxy-substituted species enabled molecular weight determination with SEC.[25] The purified mainly aqueous sulfonium polymer solution is often processed directly into thin films but the shelf life of these precursor polymers can be improved by the removal of water and redispersion in methanol.[22] Solutions of sulfonium polymers (with number average molecular weight, M_n, of ~100 000 g/mol) are extremely viscous even at concentrations of 1% w/w. This viscosity facilitates thin-film deposition processes (*e.g.* spin-coating, doctor blade, *etc.*) with methanolic polymer solutions giving films of higher uniformity.[22]

Conversion of the sulfonium precursor polymer **10** to PPV **1** is typically achieved by heating thin films to as high as 300 °C in high vacuum (10^{-6} mbar) for a period of several hours.[2,22] The elimination conditions have a substantial influence on the properties of the thin film material particularly in device applications. Studies on the elimination process have shown a complex multistep mechanism[26] but it is instructive to remember that the elimination products from a sulfonium chloride precursor polymer are HCl and thioethers. The size of the sulfonium group has been shown to influence both the intrachain and interchain order of the resulting PPV material.[27,28] The temperature for the elimination reaction can be reduced if alternative counteranion (*e.g.* 100 °C for bromide) for the sulfonium polymer is used.[23,29] The properties of the polymer films are also affected by the atmosphere in which the elimination step is performed. The presence of oxygen is thought to promote the formation of carbonyl defects[30,31] while an argon atmosphere can reduce elimination temperature to 160 °C allowing fabrication of PPV films on flexible polymeric substrate.[32–34] Room-temperature elimination is also possible with the addition of protic acids but this leads to doped films that must be dedoped to recover the semiconducting properties.[35,36] Elimination of the sulfonium group has also been achieved using UV irradiation,[37–39] laser[40,41] and microwave heating.[42]

7.2.1.2 Xanthate and Dithiocarbamate Precursors

Development of xanthate[43] and dithiocarbamate[44] derivatives overcomes several drawbacks of the sulfonium monomer (Scheme 7.2b and c). Xanthates and dithiocarbamates are easily prepared by the reaction of bis(halomethyl)benzene with alkylxanthate and dialkyldithiocarbamate salts respectively. Both precursors are stable at room temperature and soluble in organic solvents. This means the polymerization of these monomers can be performed in organic solvents (*e.g.* THF) with the addition of alkoxide base (*e.g.* potassium *tert*-butoxide). For the dithiocarbamate precursor, lithium bis(trimethylsilyl)amide can be used as the base and the polymerization proceeds at 35 °C.[45] The elimination temperature of these precursor polymers is typically lower than that of the sulfonium polymers with xanthate elimination at 160–250 °C and dithiocarbamate at 180 °C. It has been found that elimination of dithiocarbamate gave materials with reduced structural defects. Both xanthate and dithiocarbamate routes avoid the corrosive acid byproducts (HCl) present in the sulfonium elimination. This is particularly advantageous in device fabrication as acids have a negative impact on indium tin oxide electrodes and interfaces.[46]

7.2.1.3 Sulfinyl Precursor

The sulfinyl precursor route is distinct from other radical/anionic polymerizations as the monomer is nonsymmetric (Scheme 7.2d).[47,48] This means the polymer synthesis is more controlled and it is possible to achieve regioregular PPV materials.[49] However, the monomers are more difficult to prepare (Scheme 7.3).[50] Starting from the bis-sulfonium monomer **20**, reaction with one

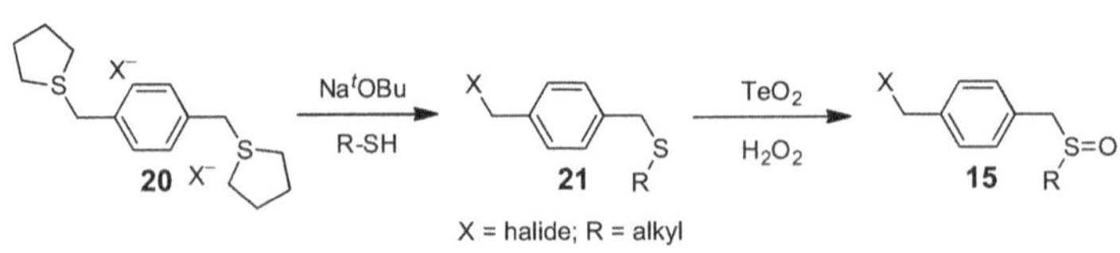

Scheme 7.3 Synthesis of nonsymmetric sulfinyl monomer **15**.

equivalent of sodium *tert*-butoxide and one equivalent of an alkanethiol gives the thiol ether intermediate **21** that can then be partially oxidized to the sulfinyl monomer **15** with TeO_2 catalyst and hydrogen peroxide as oxidant.[51] Polymerization of the sulfinyl monomer can occur in a variety of solvents including alcohols in the presence of an alkoxide base. The sulfinyl precursor polymers are stable at room temperature and can be processed in a range of organic solvents. A further advantage of the sulfinyl precursor is the low temperature (65–100 °C) required in converting the precursor polymer to PPV.[52]

7.2.1.4 Gilch Route

Perhaps the most convenient method for the synthesis of PPV is one first described by Gilch and Wheelwright in 1966.[53] Symmetric 1,4-bis(chloromethyl)benzene was treated with excess potassium *tert*-butoxide in organic solvents to give PPV. The mechanism of the polymerization has been subject to some debate but there is now a general consensus on a radical chain-growth mechanism.[13] Although most literature reports of the Gilch method performed the polymerization and elimination reactions in one step, it is possible to treat the monomer **17** with slightly less than one equivalent of alkoxide base to give the precursor polymer **18** (Scheme 7.2e). The precursor polymer **18** can then be converted to PPV **19** by the addition of more alkoxide base or by heating (300 °C, 1 h). As unsubstituted PPV is insoluble, the Gilch method is most often used to prepare solution processable derivatives of PPV in one step from the functionalized monomer **17**. There are some disadvantages to the Gilch method. Strong physical gelation is commonly observed during polymerization and it is difficult to disentangle the polymer chains.[54] Structural defects on the resulting PPV material are also more prevalent especially compared to the sulfinyl precursor route.[55–57] Head-to-head and tail-to-tail couplings can occur during polymerization along with the desired head-to-tail reaction (Scheme 7.4). Despite these disadvantages, the Gilch route is commonly used for PPV synthesis in industry and has been demonstrated in continuous processing.[58]

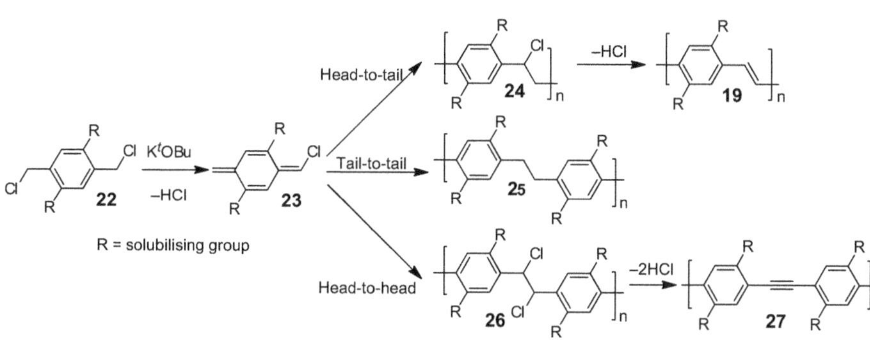

Scheme 7.4 Possible structural defects in PPV materials synthesized using the Gilch method.

7.2.2 Direct Routes to PPVs: Step-growth Polycondensation

Step-growth polycondensations are particularly attractive for the preparation of PPVs. The great advantages of the polymerization method include their mild reaction conditions, high yields and thus access to materials with less defects and predictable properties. In addition, polycondensations are highly compatible with functional groups and can provide control over molecular design, allowing the fine tuning of the final macromolecular optoelectronic properties. The broad scope of step-growth polycondensation reactions for the preparation of PPVs is illustrated by the numerous examples of homo- and copolymers accessible *via* this route.[5] In contrast to the precursor approach, direct routes to PPVs allow the preparation of strictly alternating copolymers and do not require postpolymerization modifications, which can result in undesirable side reactions.[59]

The preparation of the PPV skeleton can be generally classified into two main strategies: (i) olefinations methods to prepare the vinylene double bond and (ii) transition-metal-catalyzed single bond formation between the arene and vinylene components.[60] The former includes Wittig-, Wittig–Horner- and Knoevenagel-type condensations, whereas the latter approach was demonstrated *via* Heck, Stille, Suzuki–Miyaura and McMurry coupling reactions (Scheme 7.5).

The monomers often require multistep synthesis, particularly for the case of asymmetric lateral substitution and AB-type monomers[8] and the resulting polymers are generally characterized by low to moderate number average molecular weights. However, direct routes to PPVs allow the access to soluble, fusible and thus processable functional material, which can be cast from solution into thin films. Furthermore, regio- and stereoregular polymers can be prepared *via* some of these approaches. The following section summarizes the most relevant synthetic approaches to PPV-skeleton. A comprehensive review of reported functionalized homo- and copolymers was reported earlier by Grimsdale *et al.*[5]

7.2.2.1 Wittig and Wittig–Horner Condensations

Polycondensations based on Witting and its related Horner reactions are widely applied in the synthesis of PPVs for the preparation of homopolymers and alternating copolymers (Scheme 7.5a and b). The Wittig method was first reported in the 1960s by McDonald and Campbell for the preparation of unsubstituted PPVs from the bisphosphonium salt **29** and bisaldehyde **28**.[1] The scope of the reaction was further broadened with the introduction of solubilizing groups to the monomers, allowing the preparation of copolymers, asymmetric lateral functionalization, as well as the incorporation of ortho- and metalinkages.[5] The reaction requires the presence of a strong base, is fast and tolerant to a broad range of functional groups, however, it delivers a mixture of *cis*- and *trans*-isomers,[61] which can potentially influence the fluorescence quantum efficiency of the polymer. The degree of stereoselectivity varies with

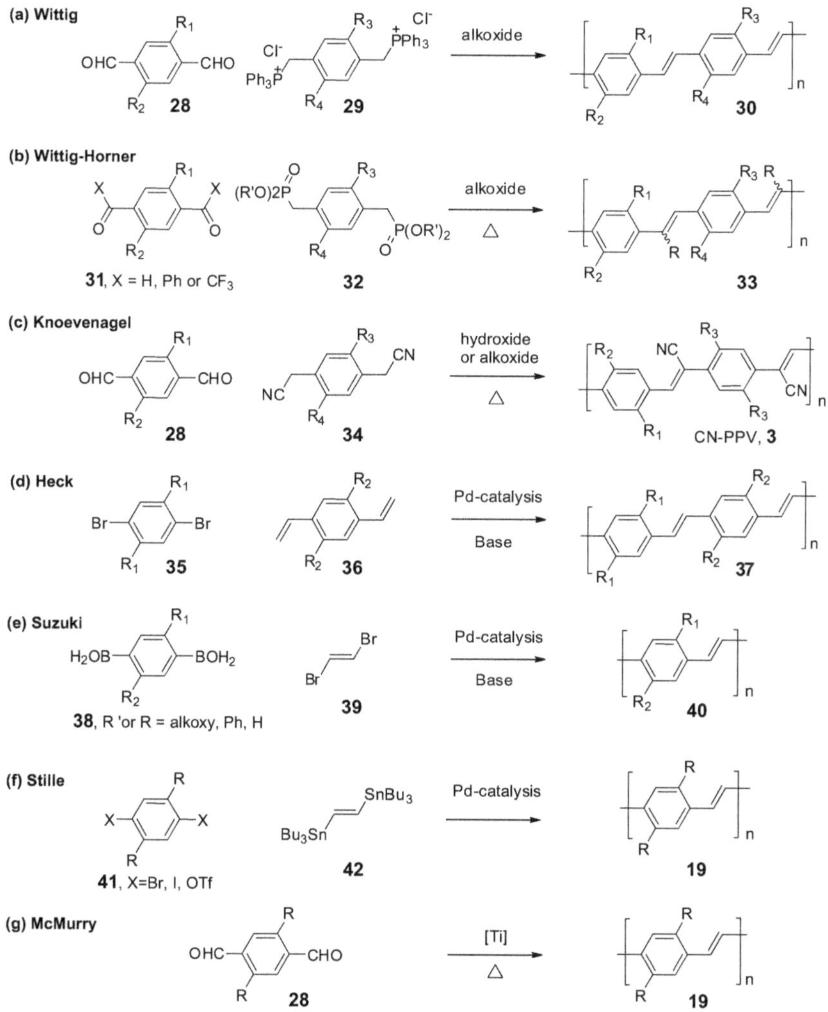

Scheme 7.5 General representation of the direct routes to PPVs *via* olefination and transition metal-catalyzed coupling reactions.

the reaction conditions, the nature of the ylide and carbonyl components as well as substituents and careful control of these parameters has been shown to deliver over 80% of bonds in *cis*-configuration.[62] Additional thermal, iodine treatment or UV-vis irradiation will result in isomerization of the bonds.

The Horner approach involving the use of phosphonates[61] is generally preferred over the phosphonium salts due to its higher reactivity of the ylide intermediate and facile removal of water-soluble phosphonate byproduct.[63] The resulting polymers exhibit higher degrees of polymerization and predominately *trans* configuration of double bonds.[8,64] This approach was successfully applied to the synthesis of regioregular MDMO-PPV, MEH-PPV and other PPVs with different lateral functionalization from asymmetrically substituted AB-[65] or AA/BB-[66] type monomers. The superior properties of regioregular MDMO-PPV were demonstrated by the significant improvement of the photovoltaic device performance.[67] When the aromatic dialdehyde is replaced by a diketone, PPVs with functionalized vinylene linkages can be prepared (Scheme 7.5b), as demonstrated by Hörhold and Helbig[68] and Lux *et al.*[69]

7.2.2.2 Knoevenagel Condensation

Terephthaldehydes are also starting materials for the synthesis of PPVs *via* Knoevenagel condensation with benzylic nitriles.[70,71] This approach delivers poly(cyanoterephthalilydene)s CN-PPVs from a wide range of functionalized monomers and affords solution processable homo- and copolymers with high electron affinities (Scheme 7.5c).[72–75] This synthetic strategy is not only particularly tolerant to functional groups, but the monomers are also readily available or easily accessible from the corresponding dihalomethylene derivatives. The polymerization takes place typically in the presence of an excess of a strong base (tetrabutylammonium hydroxide or potassium *tert*-butoxide) in a mixture of aprotic and protic solvents (typically poor solvents for the polymers as reaction medium). Molecular weights obtained by those methods are generally low to moderate, ranging between 5 to 20 kg/mol. These conditions can limit the molecular weight of the product and can favor side reaction as Michael addition *via* the reaction of nucleophiles to the vinylene bond. In this regard, a ruthenium-catalyzed Knoevenagel offers an alternative approach to CN-PPVs under mild reaction conditions and was successfully applied in a broad range of solvents.[76] A high molecular weight polymer (70 kg/mol) was obtained *via* standard reaction conditions for a water-soluble PPV-derivative, containing pendant oligoethylene glycol and sulfonic side chains.[77]

7.2.2.3 Heck Coupling

One of the most important routes for the synthesis of PPVs is the palladium-catalyzed arylation of olefins by aryl halides (Scheme 7.5d). In general, homogeneous transition-metal catalysis offers advantages over other methods due to its mild reaction conditions and hence is highly tolerant to functional groups, produces a low level of defects and offers high control over stereoselectivity (*cis*/*trans* ratio).[78,79] A drawback of this approach is the poor control of the regioselectivity, thus leading to the formation of 1,1-substituted vinylene defects in the polymer backbone and eventually crosslinking.[79,80] However, the side reaction can be minimized by the careful selection of reaction conditions and substitution pattern of the monomers. Heck coupling was first applied to the synthesis of PPVs in the 1980s from *p*-bromostyrene as well as *via* copolymerization of *p*-dibromobenzene with *p*-divinylbenzene or gaseous ethylene.[81–83] However, a limited degree of polymerization was achieved with poor solubilizing groups as methyl, trifluoremethyl, methoxy and phenyl moieties.

With the introduction of alkoxy side chains of different lengths (butyl to hexadecyl), soluble PPVs of (up to 21 kg/mol) were successfully synthesized, leading to materials with attractive liquid-crystalline properties.[84] The applicability of this methodology for the preparation of soluble regioregular polymers was later demonstrated by Lui *et al.* from a 2-alkoxy-4-bromostyrene monomer.[85] Branched and hyperbranched phenylene vinylenes from AB and AB2 and AA/BB monomer systems are also accessible *via* this route.[86–88] The broad scope of the Heck reaction makes the method a convenient strategy for the preparation of alternating copolymers, allowing a precise introduction of functional groups and hence the fine tuning of the materials' final optoelectronic properties. Numerous publications report the access to copolymers containing for example cyano-, oxadiazole-, tetrazole-, amine-, nitro-, and glucose-substituted PPVs,[89] as well as cationic-, phosphonates[90] or mesogenic side chain at the benzene or vinylene moieties and the incorporation of kinks *via* ortho linkages.[5]

Standard reaction conditions involve the use of catalytic palladium acetate and tri-2-tolylphosphine in DMF or DMA in the presence of a tertiary amine under an inert atmosphere and elevated temperatures. Under these conditions, number-average molecular weights of approximately 10–20 kg/mol are obtained.

7.2.2.4 Suzuki–Miyaura, Stille and McMurry Reactions

Suzuki–Miyaura, Stille and McMurry reactions were reported as suitable methods for the preparation of PPVs, however, these strategies were explored to a lesser extent than the Heck coupling. The first Suzuki-type polycondensation was reported by Koch and Heitz by the reaction of symmetrically and asymmetrically substituted aryldiboronic acids with *trans*-dibromoethylene to afford soluble PPVs (Scheme 7.5e) with number-average molecular weights of up to 12 kg/mol. Silver (I) oxide, bis(acetonitrile)palladium (II) chloride and triphenylarsine were used as base and catalyst system, respectively.[91] Even though the polymerizations were performed under mild conditions, biaryl defects were identified in the PPV polymers resulting from homocoupling of arylboronic acid monomers. More recently, alkoxy-substituted *cis*-PPVs containing >95% stereoregular vinyl linkages was prepared by Katayama *et al.* *via* Suzuki–Miyaura polycondensation of (Z,Z)-bis(2-bromoethenyl) arenes and arenediboronic acids. Standard biphasic conditions employing tetrakis(triphenylphosphine)-palladium(0), potassium hydroxide and a phase-transfer agent afforded number average molecular weights of up to 9 kg/mol.[92]

Soluble low molecular weight poly(2,5-dialkoxy-l,4-phenylenevinylene) has been prepared *via* a Stille coupling reaction between 2,5-dialkoxy-1,4-diiodobenzene (or triflates) and *E*-1,2-bis(tributylstannyl)ethene in the presence of palladium complexes (Scheme 7.5f).[93] The unsatisfactory degree of polymerization was attributed to the poor reactivity of the ethylene monomer.[94] Later, Chiavarone *et al.* prepared a cyclophane substituted PPV prepared *via* this method.[95] The deoxygenative (McMurry) coupling of 2,5-dihexylterephthaldehydes yields PPVs (Scheme 7.5g) of moderate number average molecular weight (*ca.* 8 kg/mol) with estimated 0.4 *cis*/*trans* ratio when treated with low-valent titanium.[96]

7.2.3 Ring-Opening Metathesis Polymerization

Ring-opening metathesis polymerization (ROMP) has been used to prepare PPV materials. The first report was described by Grubbs and coworkers where a bicycle[2.2.2]octadiene monomer **43** was treated with a ruthenium olefin metathesis catalyst to give a precursor polymer **44** (Scheme 7.6a).[97] A number-average molecular weight (M_n) of up to 63 000 g/mol and polydispersity of 1.3 were obtained for the precursor polymer. From the ^1H NMR spectrum of the precursor polymer **44**, it is possible to estimate a *cis*/*trans* double-bond ratio of 1 : 1.[97] Conversion of the precursor polymer **44** to PPV **1** was achieved by heating at 280 °C. The infrared (IR) spectrum of PPV indicated that the mixture of *cis* and *trans* double bonds in the precursor polymer **44** was mostly isomerized to *trans* double bonds during thermal elimination treatment. Subsequently, the same group reported the polymerization of another monomer, 2,3-dicarboxybarrelene **45**, *via* ROMP (Scheme 7.6b).[98] Controlled polymerization was demonstrated whereby the molecular weight of the resulting polymer was directly related to the monomer to catalyst ratio. A high M_n of 50 000 g/mol was obtained for these polymers with narrow polydispersity of

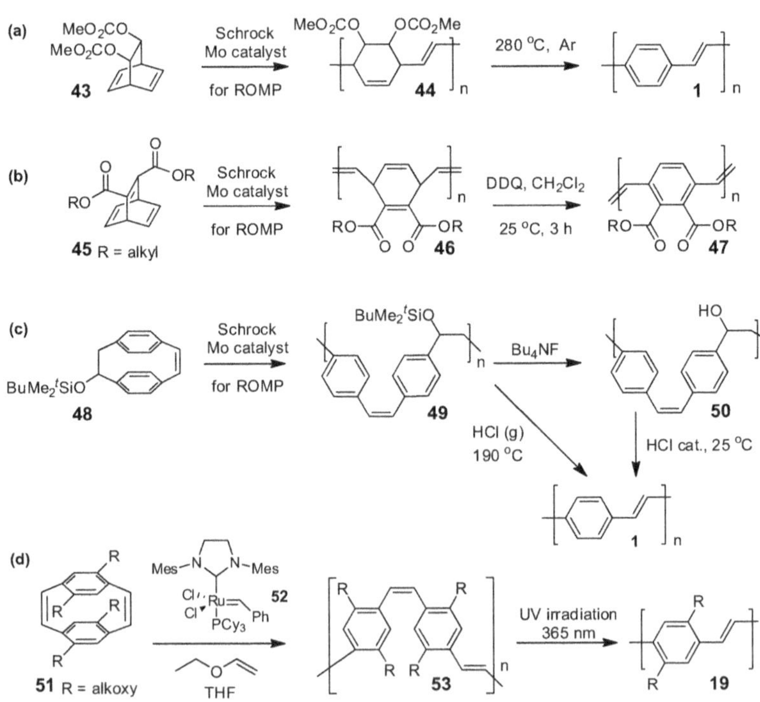

Scheme 7.6 Ring-opening metathesis polymerization (ROMP) for the preparation of various PPV derivatives.

around 1.2. Precursor polymers **46** generated from ROMP of 2,3-dicarboxy-barrelene **45** were converted to PPV derivatives **47** by oxidation with 2,3-dichloro-5,6-dicyano-1,4-benzoquinone (DDQ) at room temperature. Interestingly, the mixture of *cis* and *trans* double bonds in the precursor polymer **46** isomerized to only *trans* double bonds during oxidation as observed in both ^{1}H NMR and IR analyses.[98]

It has also been reported that siloxy paracyclophene **48** can undergo ROMP to obtain PPV precursor polymers (Scheme 7.6c).[99] Living polymerization of monomer **48** was initiated by Schrock's molybdenum catalyst giving precursor polymer **49** with M_n of 59 000 g/mol and PDI of 1.3 at a monomer to catalyst ratio of 200 : 1. Precursor polymer **49** is solution processable and is stable up to 400 °C. Conversion of precursor polymer **49** to *trans*-PPV **1** was achieved at 190 °C under a stream of HCl (g) and N_2 (g). The conversion temperature can be decreased to 25 °C if siloxy precursor polymer **49** was first deprotected with tetrabutylammonium fluoride (Scheme 7.6c).[99] A more recent example of ROMP synthesis of PPV was achieved using a paracyclophane-diene monomer **51** (Scheme 7.6d).[100] Unlike the other ROMP routes, soluble monomer **51** was subjected to a second-generation Grubbs catalyst **52** leading directly to PPV polymer **53**. Molecular weight control was also demonstrated with this method by varying the monomer to catalyst ratio from 10 : 1 to 30 : 1. This is indicative of a living polymerization mechanism.[100] Interestingly, ROMP of monomer **51** led to an alternating *cis*/*trans* polymer microstructure. The same group has subsequently reported the synthesis of MEH-PPV by microwave-assisted heating ROMP.[101] High molecular weight ranges were obtained using 3rd-generation Grubbs catalyst with microwave heating at 80 °C for 1 h. The alternating *cis*/*trans* polymer structure from the ROMP reaction was readily converted to all-*trans* MEH-PPV by UV irradiation at 365 nm (Scheme 7.6d). The photophysical properties of the MEH-PPV polymers indicated that the ROMP-synthesized MEH-PPV had less structural defects than polymers obtained *via* the Gilch route. In a further study, the living nature of ROMP was used to prepare well-defined block copolymers of PPV.[102] While ROMP synthesis of PPV derivatives typically provides the best control, the preparation of the starting monomers is difficult with moderate to low yields. This is especially a problem for the strained cyclophane monomers **48** and **51**.

7.2.4 Alternative Routes

Two other methods are available for the preparation of PPV films (Scheme 7.7). There are a number of reports on PPV film formation by chemical vapor deposition (CVD).[103–105] By heating bis(halomethyl)benzene **54** under vacuum, the monomer was carried into the gas phase and subsequently pyrolyzed to form films of halo-substituted precursor polymer **57** (Scheme 7.7). Precursor polymer films were converted to PPV by further heating under high vacuum. It is also possible to perform the conversion at a lower temperature by using UV irradiation. It has been reported that less xylylene defects on PPV films was obtained if α,α-dihaloxylene monomer **55** was used. Cyclophane **56** has also

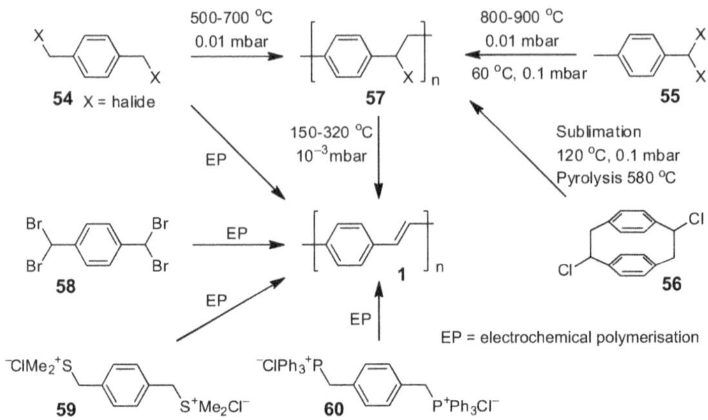

Scheme 7.7 Chemical vapor deposition (CVD) and electrochemical polymerization (EP) methods for the preparation of PPV films.

been subjected to CVD to give PPV films.[106] Electrochemical polymerization (EP) is another method to prepare PPV films.[107] A number of monomer precursors can be used in electrochemical polymerizations (Scheme 7.7). In most cases, the monomer has a leaving group substituent, allowing facile conversion to the quinodimethane intermediate.

7.2.5 Comparison of Syntheses and Summary

As one can observe from the above sections, no other family of conjugated polymers boasts such a broad range of synthetic methods as PPV materials. Consequently, the structural diversity is huge.[5] Detailed comparisons of the various methods, such as conversion efficiency, molecular weight distribution and photophysical and device properties of the materials, will not be attempted here. Rather, a summary of the advantages, disadvantages and features of the preparation methods is provided. This section serves as a guide to choosing a preparation method depending on requirements of the resulting materials.

The first decision in choosing a synthetic method for a PPV material is the way in which the material will be processed (Scheme 7.8). The precursor routes will enable the preparation of solvent-resistant and more durable thin films of PPV. This is particularly desirable if a multilayer device structure is required for the application. When choosing different precursor methods, it is important to assess the criteria of the application. Most precursor methods involve a thermal elimination step to convert the precursor polymer to the PPV material. Sulfonium precursors require higher-temperature elimination compared to sulfinyl precursors. This makes the sulfinyl route compatible with deposition on plastic substrates. Another factor to consider in precursor methods is the nature of the elimination byproducts. Sulfonium precursors convert to PPV with elimination of acids, such as HCl or HBr, which has been shown to be detrimental to device performance. Xanthate and dithiocarbamate routes involve the elimination of amine and CO_2 and CS_2, respectively.

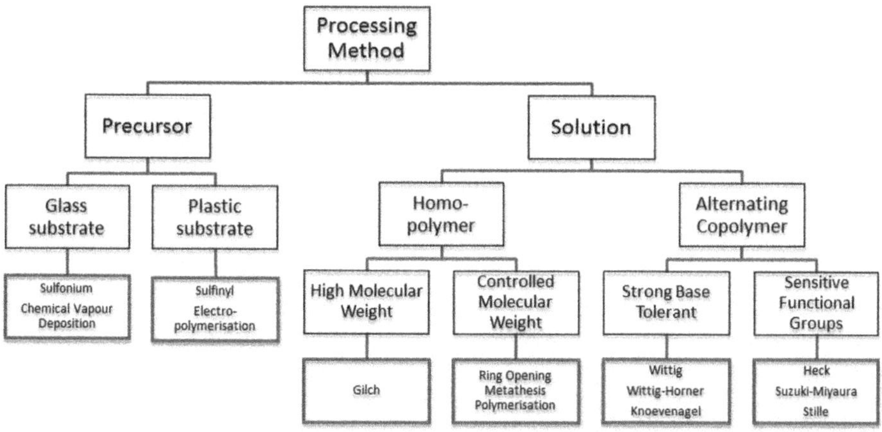

Scheme 7.8 A guide to choosing synthetic methods for the preparation of PPV materials.

For solution-processable PPV materials, the composition of the polymer and the sensitivity of the functional groups will determine the synthetic method (Scheme 7.8). Using easy-to-prepare α,α′-dihaloxylene monomers, the Gilch route is the most convenient method for the synthesis of high molecular weight homopolymers of PPV. If controlled molecular weight materials are desired, ring-opening metathesis polymerization (ROMP) offers narrow polydispersity PPVs *via* living polymerization. However, the structural complexity of ROMP monomers makes them difficult to prepare. Alternating copolymers are best synthesized using AA/BB monomers *via* step-growth polymerization. Wittig, Wittig–Horner and Knoevenagel reactions require the use of strong bases, such as KO^tBu, while transition-metal-catalyzed coupling reactions are typically more suitable for materials with sensitive functional groups.

7.3 Experimental Procedures

7.3.1 General Experimental Requirements

For most of the polymerization reactions, an inert atmosphere is required to achieve high molecular weight distributions. In some methods, anhydrous reaction conditions are also desirable. All solution-processable PPV derivatives can be examined using a variety of analytical techniques. Molecular weight distributions are typically obtained by gel permeation chromatography (GPC) analysis using refractive-index detection calibrated against polystyrene standards. In cases where the PPV derivative is not solution processable, the molecular weight of the precursor polymer is determined prior to conversion. Polymers are also commonly characterized by NMR, UV-vis and IR spectroscopy and elemental analysis. ^{1}H NMR and IR spectroscopy methods are particularly useful in determining the *cis*/*trans* double-bond configuration of PPV materials.

7.3.2 Example Procedures

7.3.2.1 *Sulfonium Precursor Route*[22]

Precursor sulfonium polymer **10** (Scheme 7.2a, $R_2=(CH_2)_4$): To a solution of the sulfonium monomer **9** ($R_2=(CH_2)_4$) (5.0 g, 14 mmol) in methanol (38 mL) cooled to 0 °C is added dropwise aqueous sodium hydroxide (0.4 M; 36 mL, 14 mmol) over 20 min. The reaction mixture is stirred for a further 60 min at 0 °C and then neutralized with hydrochloric acid (0.4 M). The solution is then dialyzed against water (3×2000 mL) over 3 days, after which the solvent is completely removed. The residue (~1.3 g) is redissolved in methanol (75 mL). NOTE: The molecular weight of the polyelectrolyte product cannot be determined by GPC. Converting the sulfonium polymer to the methoxy derivative enables molecular weight measurement giving M_n 97 500, M_w 560 000 and PDI 5.7.

Thermal conversion of **10** to PPV **1**: Thin films of precursor polymer **10** are heated at 220 °C for 12 h *in vacuo* forming PPV **1**. Thin-film samples of **1** are analyzed using infrared and UV-vis spectroscopy.

7.3.2.2 *Gilch Route*[58,108]

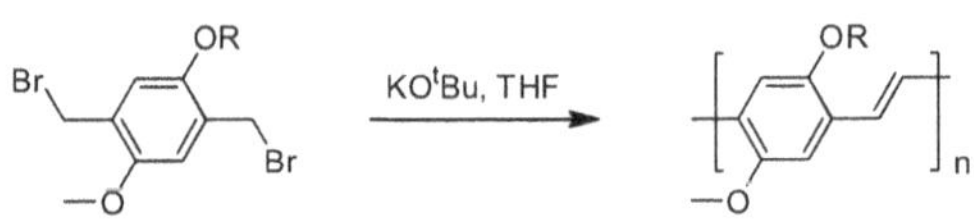

61, R = 2 -ethylhexyl MEH-PPV, **2**

Poly(1-methoxy-4-(2-ethylhexyloxy)-*p*-phenylenevinylene) MEH-PPV **2**: Potassium *tert*-butoxide (5 mL, 1 M) is added to anhydrous THF (20 mL). α,α′-Dibromo-2-methoxy-5-(2-ethylhexyloxy)xylene **61** (0.5 g, 1.2 mmol) in anhydrous THF (5 mL) is added dropwise using a syringe pump at a rate of 20 mL/h. The reaction is allowed to stir for 5 h at 25 °C and the product is precipitated in MeOH. A red amorphous solid (0.25 g, 77% yield) is collected by filtration and dried under vacuum. GPC data: M_n 70 000; M_w 200 000; PDI 2.8. ^{1}H NMR ($CDCl_3$, 500 MHz) δ ppm: 0.92 (br m, -CH_3), 1.01 (br m, -CH_3), 1.37 (br m, -CH_2-); 1.54 (br m, -CH_2-), 1.83 (br m, -CH-), 3.9–4.1 (br m, -OCH_2- and -OCH_3), 7.20 (br m, vinyl-H), 7.4–7.5 (br m, ArH).

7.3.2.3 *Wittig–Horner Olefination*[66]

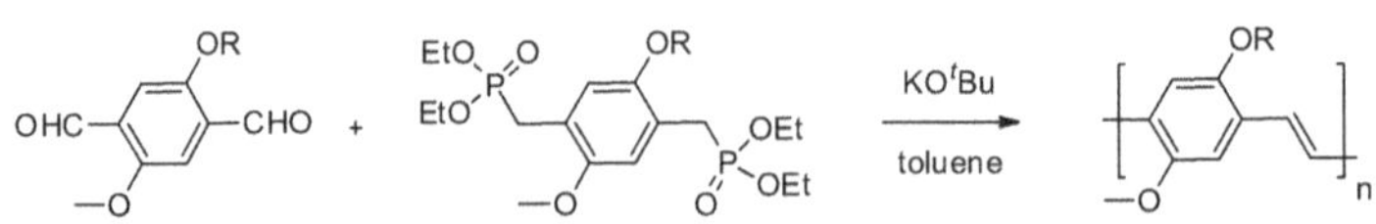

62, R = 2-ethylhexyl **63**, R = 2-ethylhexyl MEH-PPV, **2**

Aldehyde **62** (2.34 g; 8 mmol) and phosphanate **63** (4.29 g; 8 mmol) are dissolved in toluene (100 mL) whilst stirring and heating at reflux under argon.

To this solution is added solid potassium *tert*-butoxide (2.64 g; 24 mmol) in one portion. After 5 h at reflux the reaction mixture is diluted with additional toluene and quenched with aqueous HCl (50 mL). The organic layer is then separated, washed several times with distilled water, and dried in a Dean–Stark apparatus. The resulting toluene solution is filtered, reduced to a minimum, and precipitated into methanol. The polymer is extracted with methanol for 5 h, and precipitated from toluene/methanol. Yield: 2.7 g (66%), orange-red fibers. (C17H24O2)*n* (260.4)*n* Calc. C, 78.42; H, 9.29; Found C, 77.82; H, 9.41. GPC (g/mol): $M_n = 9100$, $M_w = 25\ 000$; VPO (g/mol) $M_n = 12\ 300$. FTIR (cm^{-1}, KBr): 3056 (Ph), 2956, 2926, and 2862 (aliphatic CH), 1598 and 1500 (Ph breathing), 1204, 968 (*trans*-vinylene). ^{1}H-NMR ($CDCl_3$): δ: 7.5–6.8 (m; H4), 3.9–3.5 (m; H5), 1.8–0.8 (m; H15). ^{13}C-NMR ($CDCl_3$): δ 151.1, 126.7, 123.3, 122.5, 110.3, 108.9, 71.6, 56.4, 39.8, 30.7, 29.0, 24.0, 23.1, 14.1, 11.4.

7.3.2.4 Knoevenagel Reaction[72]

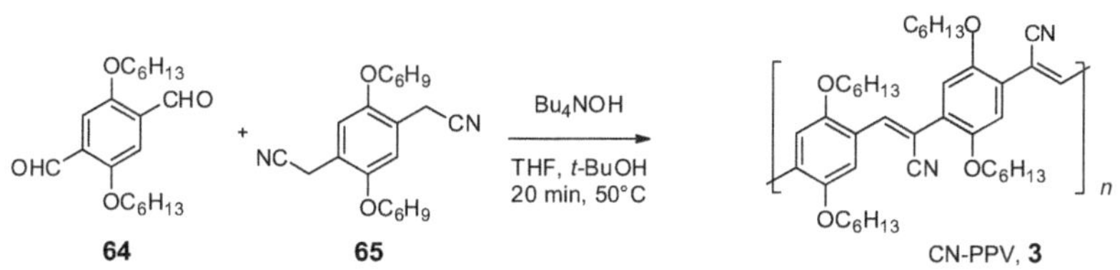

Equimolar quantities of the aldehyde **64** and nitrile **65** are treated with 5 mol% Bu_4NOH in tetrahydrofuran and *tert*-butyl alcohol (*t*-BuOH) (1 : 1) at 50 °C for 20 min, and the resulting red precipitate is filtered and washed with methanol. ^{1}H-NMR: ($CDCl_3$, 250 MHz) δ: 0.7–2.0 (22H, m), 4.09 (4H, m), 7.16 (1H, s), 7.93 (1H, s), 8.10 (1H, s). GPC (g/mol): 4000.

7.3.2.5 Heck Coupling[84]

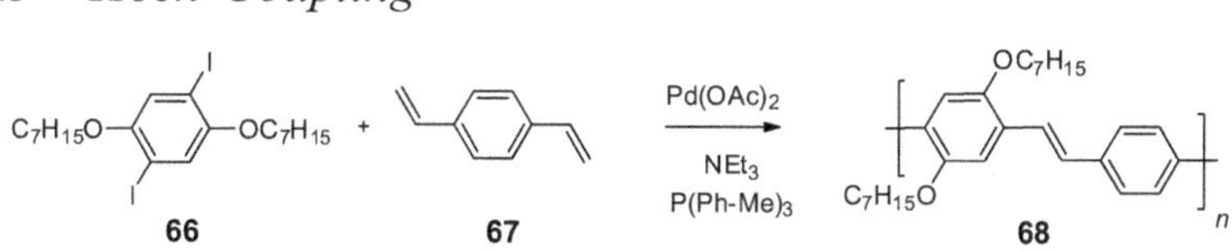

Triethylamine (0.35 mL, 2.5 mmol) is added to a solution of *p*-divinylbenzene **67** (130 mg, 1 mmol), iodo monomer **66** (1 mmol), $Pd(OAc)_2$ (9 mg, 0.04 mmol), and tri-*o*-tolylphosphine (69 mg, 0.2 mmol) in 5 mL of DMF. The reaction mixture is heated at 100 °C for 5 h under nitrogen atmosphere and poured into 20 mL of methanol. The precipitated polymer is collected by filtration and further purified by redissolving in a minimum amount of hot chloroform and precipitating into acetone. The resulting polymer is extracted with methanol for 24 h and dried under vacuum at 40 °C for 2 days (yield 87%).

^{1}H-NMR ($CDCl_3$) δ: 0.95 (broad, -CH_3, 6H), 1.35 (brad, -$(CH_2)_2$-, 8H), 1.42 (broad, -CH_2-, 4H), 1.85 (broad, -CH_2-, 4H), 4.05 (broad, O-CH_2-, 4H), 7.10 (broad, vinyl proton, 4H), 7.5 (broad aromatic protons, 6H). Anal. Calcd for C30H40O2: C, 82.28; H, 9.32. Found C, 82.55; H, 9.67. GPC (g/mol): $M_n = 18\,500$, $M_w = 49\,300$.

7.3.2.6 *Suzuki–Miyaura Coupling*[91]

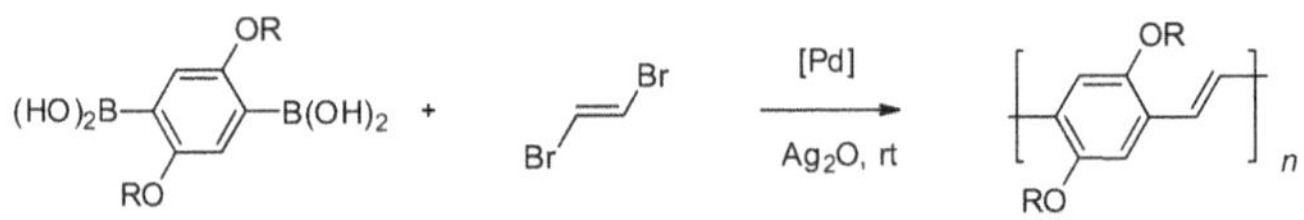

38, R = 2-ethylhexyl **39** **40**, R = 2-ethylhexyl

A 100-mL three-necked vessel with a magnetic stirrer is charged with 60 mL of dry THF and 15 mL of water. The solution is saturated with argon. Arylboronic acid **38** (5 mmol), 1,2-dibromoethene **39** (5 mmol), triphenylarsine (0.4 mmol), bis(acetonitrile)palladium (II) chloride (0.1 mmol) and silver(I) oxide (15 mmol) are added. The reaction mixture is stirred for 16 h at room temperature. The polymers are poured into 800 mL of methanol. The precipitate is filtered off and extracted with toluene in a Soxhlet apparatus for 16 h. The resulting solution is concentrated to about 50 mL and precipitated into methanol (500 mL). The polymer is filtered off, dissolved in toluene and filtered through aluminum oxide. After concentrating the toluene solution to about 50 mL, the polymer is precipitated into 500 mL of methanol and dried under reduced pressure at 50 °C for 15 h. Yield: 1.59 g (82%), $\eta_{inh} = 0.93$ dL/g. GPC (g/mol): $M_n = 12\,300$. UV-vis (λ max, THF) = 499 nm.

7.3.2.7 *Ring-Opening Metathesis Polymerization*[100]

Synthesis of *cis*/*trans* alternating PPV **53** (Scheme 7.6d): In a nitrogen-filled Radleys GreenHouse, paracyclophanediene monomer **51** (50.1 mg, 0.07 mmol) is dissolved in dry THF (0.4 mL). A second-generation Grubbs catalyst **52** (2 mg) is added and the reaction is heated at 68 °C for 36 h. The reaction mixture is cooled to room temperature and excess ethyl vinyl ether (2 mL) is added to quench the reaction. After stirring for a further 4 h at room temperature, the reaction mixture is concentrated *in vacuo*. The product is redissolved in chloroform and filtered through a short plug of silica to remove the catalyst. The solvent is removed under reduced pressure and the resulting solid exhaustively washed with methanol. GPC data: $M_n = 21\ 600$; $M_w = 23\,600$; PDI = 1.1. ^{1}H-NMR ($CDCl_3$): δ(ppm) 0.70–0.88 (br t, 12H), 1.06–1.87 (br m, 48H), 3.41–3.60 (br m, 4H), 3.86–4.06 (br m, 4H), 5.12–5.17 (d, 0.02H), 5.60–5.67 (d, 0.02H), 6.57–6.72 (br s, 2H), 6.72–6.92 (br s, 2H), 6.92–7.18 (br s, 2H), 7.20–7.50 (br s, 2H).

References

1. R. N. McDonald and T. W. Campbell, *J. Am. Chem. Soc.*, 1960, **82**, 4669–4671.
2. J. H. Burroughes, D. D. C. Bradley, A. R. Brown, R. N. Marks, K. Mackay, R. H. Friend, P. L. Burn and A. B. Holmes, *Nature*, 1990, **347**, 539–541.
3. F. Wudl and G. Srdanov, US Patent 5189136, 19901212; *Chem. Abstr.* 1993, **118**, 255575.
4. R. Schenk, H. Gregorius, K. Meerholz, J. Heinze and K. Müllen, *J. Am. Chem. Soc.*, 1991, **113**, 2634–2647.
5. A. C. Grimsdale, K. Leok Chan, R. E. Martin, P. G. Jokisz and A. B. Holmes, *Chem. Rev.*, 2009, **109**, 897–1091.
6. S. Guenes, H. Neugebauer and N. S. Sariciftci, *Chem. Rev.*, 2007, **107**, 1324–1338.
7. H.-K. Shim and J.-I. Jin, *Adv. Polym. Sci.*, 2002, **158**, 193–243.
8. Conjugated Polymers: Theory, Synthesis, Properties, and Characterization, *Handbook of Conducting Polymers*, ed. T. A. Skotheim and J. R. Reynolds, 3rd edn, CRC Press, Taylor and Francis, Boca Raton FL 2007, pp. 4.1–4.22.
9. K. Müllen and U. Scherf, ed., *Organic Light-Emitting Devices: Synthesis, Properties and Applications*, Wiley-VCH, Weinheim, 2006.
10. D. F. Perepichka, I. F. Perepichka, H. Meng and F. Wudl, in *Organic Light Emitting Materials and Devices*, ed. Z. Li and H. F. Meng, CRC Press, Boca Raton, 2007, pp. 45–293.
11. T. P. Nguyen and P. Destruel, in *Handbook of Luminescence, Display Materials and Devices*, ed. H. S. Nalwa and L. S. Rohwer, American Scientific Publishers, Stevenson Ranch, California, 2003, vol. 1, pp. 1–129.
12. T. Junkers, J. Vandenbergh, P. Adriaensens, L. Lutsen and D. Vanderzande, *Polym. Chem.*, 2012, **3**, 275–285.
13. T. Schwalm, J. Wiesecke, S. Immel and M. Rehahn, *Macromol. Rapid Commun.*, 2009, **30**, 1295–1322.
14. I. Cosemans, L. Hontis, D. Van Den Berghe, A. Palmaerts, J. Wouters, T. J. Cleij, L. Lutsen, W. Maes, T. Junkers and D. J. M. Vanderzande, *Macromolecules*, 2011, **44**, 7610–7616.
15. L. Hontis, M. Van Der Borght, D. Vanderzande and J. Gelan, *Polymer*, 1999, **40**, 6615–6617.
16. R. A. Wessling and R. G. Zimmerman, US Patent 3401152, 19661103.
17. R. A. Wessling, *J. Polym. Sci. Polym. Symp.*, 1985, **72**, 55–66.
18. R. W. Lenz, C.-C. Han, J. Stenger-Smith and F. E. Karasz, *J. Polym. Sci., Part A: Polym. Chem.*, 1988, **26**, 3241–3249.
19. R. Garay and R. W. Lenz, *Makromol. Chem., Suppl.*, 1989, **15**, 1–7.
20. S. Tokito, T. Momii, H. Murata, T. Tsutsui and S. Saito, *Polymer*, 1990, **31**, 1137–1141.
21. F. R. Denton, A. Sarker, P. M. Lahti, R. O. Garay and F. E. Karasz, *J. Polym. Sci., Part A: Polym. Chem.*, 1992, **30**, 2233–2240.

22. P. L. Burn, D. D. C. Bradley, R. H. Friend, D. A. Halliday, A. B. Holmes, R. W. Jackson and A. Kraft, *J. Chem. Soc., Perkin Trans*, 1992, **1**, 3225–3231.
23. R. O. Garay, U. Baier, C. Bubeck and K. Müllen, *Adv. Mater.*, 1993, **5**, 561–564.
24. M. J. Cherry, S. C. Moratti, A. B. Holmes, P. L. Taylor, J. Grüner and R. H. Friend, *Synth. Met.*, 1995, **69**, 493–494.
25. D. A. Halliday, P. L. Burn, R. H. Friend, D. D. C. Bradley and A. B. Holmes, *Synth. Met.*, 1993, **55**, 902–907.
26. H. V. Shah and G. A. Arbuckle, *Macromolecules*, 1999, **32**, 1413–1423.
27. J. H. F. Martens, D. A. Halliday, E. A. Marseglia, D. D. C. Bradley, R. H. Friend, P. L. Burn and A. B. Holmes, *Synth. Met.*, 1993, **55**, 434–439.
28. P. L. Burn, D. D. C. Bradley, A. R. Brown, R. H. Friend, D. A. Halliday, A. B. Holmes, A. Kraft and J. H. F. Martens, in *Electronic Properties of Polymers (Springer Ser. Solid-State Sci. 107)*, ed. H. Kuzmany, M. Mehring and S. Roth, Springer, Heidelberg, 1992, pp. 293–297.
29. A. Beerden, D. Vanderzande and J. Gelan, *Synth. Met.*, 1992, **52**, 387–394.
30. F. Papadimitrakopoulos, K. Konstadinidis, T. M. Miller, R. Opila, E. A. Chandross and M. E. Galvin, *Chem. Mater.*, 1994, **6**, 1563–1568.
31. V. H. Tran, V. Massardier, T. P. Nguyen and J. Davenas, *Polymer*, 1996, **37**, 2061–2065.
32. M. Herold, J. Gmeiner, W. Riess and M. Schwoerer, *Synth. Met.*, 1996, **76**, 109–112.
33. M. Herold, J. Gmeiner and M. Schwoerer, *Acta Polym.*, 1996, **47**, 436–440.
34. M. Herold, J. Gmeiner and M. Schwoerer, *Acta Polym.*, 1994, **45**, 392–395.
35. Y. Xia, A. G. MacDiarmid and A. J. Epstein, *Adv. Mater.*, 1994, **6**, 293–295.
36. V. Massardier, A. Guyot and V. H. Tran, *Polymer*, 1994, **35**, 1561–1563.
37. S. Taguchi and T. Tanaka, EP Patent 261991, 19870928; *Chem. Abstr.* **1988**, *109*, 160610.
38. W. Schmid, R. Dankesreiter, J. Gmeiner, T. Vogtmann and M. Schwoerer, *Acta Polym.*, 1993, **44**, 208–210.
39. J. Bullot, B. Dulieu and S. Lefrant, *Synth. Met.*, 1993, **61**, 211–215.
40. A. Torres-Filho and R. W. Lenz, *J. Polym. Sci., Part B: Polym. Phys.*, 1993, **31**, 959–970.
41. S. Y. Paik, S. H. Kwon, O. J. Kwon, J. S. Yoo and M. K. Han, *Synth. Met.*, 2002, **129**, 101–105.
42. A. Torres-Filho and R. W. Lenz, *J. Appl. Polym. Sci.*, 1994, **52**, 377–386.
43. S. Son, A. J. Lovinger and M. E. Galvin, *Polym. Mater. Sci. Eng.*, 1995, **72**, 567–568.
44. A. Henckens, I. Duyssens, L. Lutsen, D. Vanderzande and T. J. Cleij, *Polymer*, 2006, **47**, 123–131.

45. J. Vandenbergh, J. Wouters, P. J. Adriaensens, R. Mens, T. J. Cleij, L. Lutsen and D. J. M. Vanderzande, *Macromolecules*, 2009, **42**, 3661–3668.
46. S.-C. Lo, A. K. Sheridan, I. D. W. Samuel and P. L. Burn, *J. Mater. Chem.*, 1999, **9**, 2165–2170.
47. F. Louwet, D. Vanderzande and J. Gelan, *Synth. Met.*, 1992, **52**, 125–130.
48. F. Louwet, D. Vanderzande and J. Gelan, *Synth. Met.*, 1995, **69**, 509–510.
49. M. M. de Kok, T. P. Nguyen, P. Molinie, A. J. J. M. van Breemen, D. J. Vanderzande and J. M. Gelan, *Synth. Met.*, 1999, **102**, 949–950.
50. A. J. J. M. van Breemen, D. J. M. Vanderzande, P. J. Adriaensens and J. M. J. V. Gelan, *J. Org. Chem.*, 1999, **64**, 3106–3112.
51. E. Kesters, M. M. de Kok, R. A. A. Carleer, J. H. P. B. Czech, P. J. Adriaensens, J. M. Gelan and D. J. Vanderzande, *Polymer*, 2002, **43**, 5749–5755.
52. E. Kesters, L. Lutsen, D. Vanderzande, J. Gelan, T. P. Nguyen and P. Molinié, *Thin Solid Films*, 2002, **403–404**, 120–125.
53. H. G. Gilch and W. L. Wheelwright, *J. Polym. Sci. Part A: Polym. Chem.*, 1966, **4**, 1337–1349.
54. T. Schwalm and M. Rehahn, *Macromol. Rapid Commun.*, 2008, **29**, 33–38.
55. H. Becker, H. Spreitzer, W. Kreuder, E. Kluge, H. Vestweber, H. Schenk and K. Treacher, *Synth. Met.*, 2001, **122**, 105–110.
56. H. Becker, O. Gelsen, E. Kluge, W. Kreuder, H. Schenk and H. Spreitzer, *Synth. Met.*, 2000, **111–112**, 145–149.
57. H. Becker, H. Spreitzer, K. Ibrom and W. Kreuder, *Macromolecules*, 1999, **32**, 4925–4932.
58. H. Seyler, D. J. Jones, A. B. Holmes and W. W. H. Wong, *Chem. Commun.*, 2012, **48**, 1598–1600.
59. B. R. Hsieh, Y. Yu, A. C. VanLaeken and H. Lee, *Macromolecules*, 1997, **30**, 8094–8095.
60. U. Scherf, in *Carbon Rich Compounds II, Macrocyclic Oligoacetylenes and Other Linearly Conjugated Systems*, ed. A. de Meijere, Springer, Berlin/Heidelberg, 1999, vol. 201, pp. 163–222.
61. V. H.-H. Hörhold and J. Opfermann, *Macromol. Chem.*, 1970, **131**, 105–132.
62. F. Wang, F. He, Z. Q. Xie, Y. P. Li, M. Hanif, M. Li and Y. Ma, *Macromol. Chem. Phys.*, 2008, **209**, 1381–1388.
63. R. Kiebooms, R. Menon and K. Lee, in *Hanbook of Advanced Electronic and Photonic Materials and Devices*, ed. H. S. Nalwa, Academic Press, USA, 2001, p. 26.
64. S. Pfeiffer and H.-H. Hörhold, *Macromol. Chem. Phys.*, 1999, **200**, 1870–1878.
65. Y. Suzuki, K. Hashimoto and K. Tajima, *Macromolecules*, 2007, **40**, 6521–6528.
66. S. Pfeiffer and H.-H. Hörhold, *Synth. Met.*, 1999, **101**, 109–110.
67. K. Tajima, Y. Suzuki and K. Hashimoto, *J. Phys. Chem. C*, 2008, **112**, 8507–8510.

68. H. H. Hörhold and M. Helbig, *Macromol. Chem. Macromol. Symp.*, 1987, **12**, 229–258.
69. A. Lux, A. B. Holmes, R. Cervini, J. E. Davies, S. C. Moratti, J. Grüner, F. Cacialli and R. H. Friend, *Synth. Met.*, 1997, **84**, 293–294.
70. R. W. Lenz and C. E. Handlovits, *J. Org. Chem.*, 1960, **25**, 813–817.
71. V. W. Funke and E. C. Schütze, *Macromol. Chem.*, 1964, **74**, 71–91.
72. N. C. Greenham, S. C. Moratti, D. D. C. Bradley, R. H. Friend and A. B. Holmes, *Nature*, 1993, **365**, 628–630.
73. S. C. Moratti, R. Cervini, A. B. Holmes, D. R. Baigent, R. H. Friend, N. C. Greenham, J. Grüner and P. J. Hamer, *Synth. Met.*, 1995, **71**, 2117–2120.
74. S.-A. Chen and E.-C. Chang, *Macromolecules*, 1998, **31**, 4899–4907.
75. E. G. J. Staring, R. C. J. E. Demandt, D. Braun, G. L. J. Rikken, Y. A. R. R. Kessener, A. H. J. Venhuizen, M. M. F. van Knippenberg and M. Bouwmans, *Synth. Met.*, 1995, **71**, 2179–21780.
76. J. Liao and Q. Wang, *Macromolecules*, 2004, **37**, 7061–7063.
77. W. Zhang, L. Zhu, J. Qin and C. Yang, *J. Phys. Chem. B*, 2011, **115**, 12059–12064.
78. A. Hilberer, H.-J. Brouwer, B.-J. van der Scheer, J. Wildeman and G. Hadziioannou, *Macromolecules*, 1995, **28**, 4525–4529.
79. M. Brenda, A. Greiner and W. Heitz, *Macromol. Chem.*, 1990, **191**, 1083–1100.
80. S. Klingelhöfer, C. Schellenberg, J. Pommerehne, H. Bässler, A. Greiner and W. Heitz, *Macromol. Chem. Phys.*, 1997, **198**, 1511–1530.
81. Asahi Glass Co., Japan Patent 57,207,618, 1981; Chem. Abstr., 99, 23139k (1983).
82. W. Heitz, W. Brügging, L. Freund, M. Gailberger, A. Greiner, H. Jung, U. Kampschulte, N. Nießner, F. Osan, H.-W. Schmidt and M. Wicker, *Macromol. Chem.*, 1988, **189**, 119–127.
83. A. Greiner and W. Heitz, *Macromol. Chem. Rapid Commun.*, 1988, **9**, 581–588.
84. Z. Bao, Y. Chen, R. Cai and L. Yu, *Macromolecules*, 1993, **26**, 5281–5286.
85. Y. Liu, P. M. Lahti and F. La, *Polymer*, 1998, **39**, 5241–5244.
86. L. Huo, J. Hou, Y. Zhou, M. Han and Y. Li, *J. Appl. Polym. Sci.*, 2008, **110**, 1002–1008.
87. S.-J. Lim, D. Y. Seok, B.-K. An, S. D. Jung and S. Y. Park, *Macromolecules*, 2005, **39**, 9–11.
88. J. A. Mikroyannidis, *Macromolecules*, 2002, **35**, 9289–9295.
89. A. Takasu, K. Iso, T. Dohmae and T. Hirabayashi, *Biomacromolecules*, 2006, **7**, 411–414.
90. S.-J. Yoon, S. Samal and T.-H. Yoon, *Eur. Polym. J.*, 2010, **46**, 2282–2289.
91. F. Koch and W. Heitz, *Macromol. Chem. Phys.*, 1997, **198**, 1531–1544.
92. H. Katayama, M. Nagao, T. Nishimura, Y. Matsui, Y. Fukuse, M. Wakioka and F. Ozawa, *Macromolecules*, 2006, **39**, 2039–2048.
93. F. Babudri, S. R. Cicco, G. M. Farinola, F. Naso, A. Bolognesi and W. Porzio, *Macromol. Rapid Commun.*, 1996, **17**, 905–911.

94. Z. Bao and L. Yu, *Proc. SPIE*, 1995, **2528**, 210–218.
95. L. Chiavarone, G. M. Farinola, A. Losacco, M. Striccoli, L. Torsi, G. Scamarcio, M. Sibilano, S. R. Cicco, F. Babudri and F. Naso, *Adv. Sci. Tech.*, 1999, **27**, 271–278.
96. M. Rehahn and A. D. Schlüter, *Macromol. Chem. Rapid Commun.*, 1990, **11**, 375–379.
97. V. P. Conticello, D. L. Gin and R. H. Grubbs, *J. Am. Chem. Soc.*, 1992, **114**, 9708–9710.
98. M. W. Wagaman and R. H. Grubbs, *Macromolecules*, 1997, **30**, 3978–3985.
99. Y.-J. Miao and G. C. Bazan, *J. Am. Chem. Soc.*, 1994, **116**, 9379–9380.
100. C.-Y. Yu and M. Turner, *Angew. Chem. Int. Ed.*, 2006, **45**, 7797–7800.
101. A. M. Spring, C.-Y. Yu, M. Horie and M. L. Turner, *Chem. Commun.*, 2009, 2676–2678.
102. C.-Y. Yu, J. W. Kingsley, D. G. Lidzey and M. L. Turner, *Macromol. Rapid Commun.*, 2009, **30**, 1889–1892.
103. E. G. J. Staring, D. Braun, G. L. J. A. Rikken, R. J. C. E. Demandt, Y. A. R. R. Kessener, M. Bouwmans and D. Broer, *Synth. Met.*, 1994, **67**, 71–75.
104. O. Schäfer, A. Greiner, J. Pommerehne, W. Guss, H. Vestweber, H. Y. Tak, H. Bässler, C. Schmidt, G. Lüssem, B. Schartel, V. Stümpflen, J. H. Wendorff, S. Spiegel, C. Möller and H. W. Spiess, *Synth. Met.*, 1996, **82**, 1–9.
105. K. M. Vaeth and K. F. Jensen, *Adv. Mater.*, 1997, **9**, 490–493.
106. S. Iwatsuki, M. Kubo and T. Kumeuchi, *Chem. Lett.*, 1991, 1071–1074.
107. J. H. P. Utley and J. Gruber, *J. Mater. Chem.*, 2002, **12**, 1613–1624.
108. T. Schwalm, J. Wiesecke, S. Immel and M. Rehahn, *Macromolecules*, 2007, **40**, 8842–8854.

CHAPTER 8

Poly(p-phenyleneethynylene)s and Poly(aryleneethynylene)s

UWE H. F. BUNZ

Organisch-Chemisches Institut, Ruprecht-Karls Universität Heidelberg, Im Neuenheimer Feld 270, 69120, Heidelberg, FRG
Email: Uwe.bunz@oci.uni-heidelberg.de

8.1 Introduction

Poly(paraphenyleneethynylene)s (PPE) and poly(aryleneethynylene)s (PAE) are attractive classes of polymeric materials with unusual structures and surprising optical properties. The PPEs were first synthesized by Giesa and Schulz in 1990[1] by the Pd-catalyzed coupling reaction (Sonogahsira reaction)[2] of aromatic diynes with aromatic dibromides. Later Moroni and LeMoigne[3] discovered that PPEs of better quality could be obtained by adding THF to the reaction mixture of the Sonogashira coupling. The authors claimed that the degree of polymerization (P_n) of their PPEs was large and in the hundreds, however, analysis of their published ^{13}C NMR spectra suggests that PPEs with a P_n of around 20 repeat units had formed. About the same time Bunz and coworkers[4] reported the synthesis of novel arylene ethynylenes displaying high emissive quantum yields. Also, the group of Wrighton at MIT was active in the early exploration of the PPEs.[5,6] and published a series of attractive papers, in which the synthesis and characterization of dialkoxy-PPEs was described.

An important development in this area was the synthesis of PPEs with endo receptors.[7] Coupling of **1a** to **2** (Chart 8.1), gives PPEs **3** with an M_n of up to 2.5×10^5. The polydispersity (PDI) for different samples of **3** ranged from 1.8 to 2.9, not unusual for this type of polycondensation reactions, An ideal

RSC Polymer Chemistry Series No. 9
Conjugated Polymers: A Practical Guide to Synthesis
Edited by Klaus Müllen, John R. Reynolds and Toshio Masuda

Published by the Royal Society of Chemistry, www.rsc.org

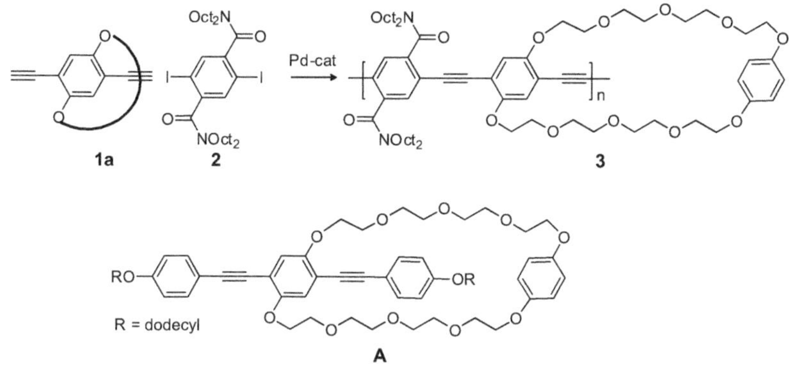

Chart 8.1

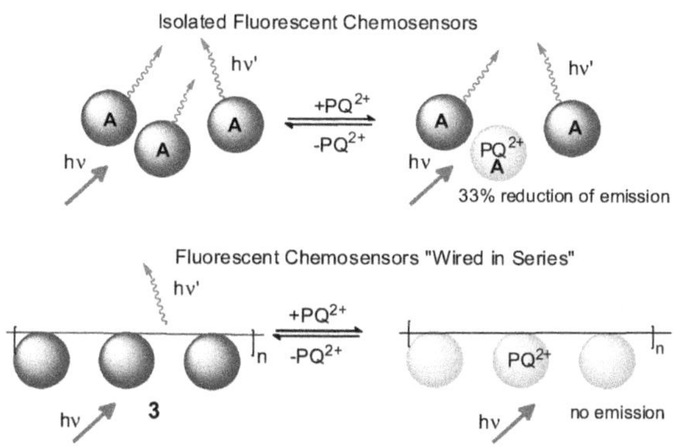

Chart 8.2

polycondensation polymer, however, would show a Flory–Schultz distribution with a PDI = 2.

PPE **3** was used to investigate the "molecular wire" effect. A quenching experiment of **3** and a model compound **A** was performed with paraquat (PQ^{2+}). Chart 8.2 shows the principle. Both fluorophores **A** and **3** were deployed in solutions in which the concentration of phenyleneethynylene units was equal. However, the observed Stern–Volmer constants for **A** and for **3** differed by a factor of 63, with the polymer being more sensitive towards the quencher at the same concentration on a per monomer base. Swager explained this effect (Chart 8.2) and contended that an exciton is delocalized over a large segment of the PPE chain, and therefore can be quenched by a paraquat molecule that is rather far away on the chain. The enhancement of around 60 suggests that one exciton "services" around 60 monomer units, *i.e.* increases the sensitivity by this factor. This discovery was an important contribution to the understanding and the engineering of useful properties in PPEs and their applications, particularly in sensory schemes.

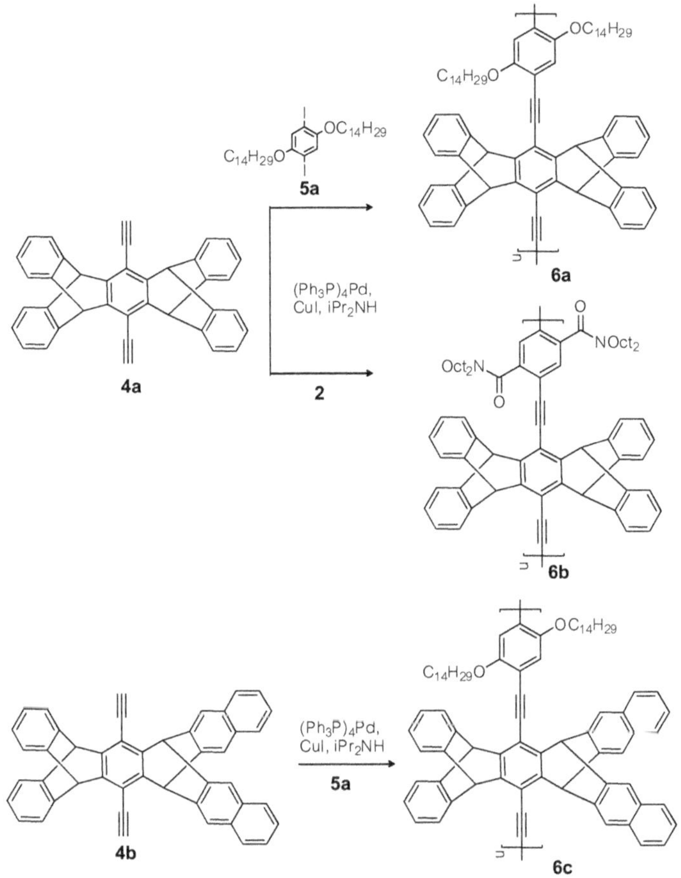

Chart 8.3

An extension of this work led to[8] the pentiptycene-based PPEs **6a–c**, suitable for the detection of nitroaromatic compounds. These PPEs **6** form in the Pd-catalyzed coupling of **4** to **2** or **5a** (Chart 8.3). Dinitrotoluene is an electron-poor arene and its vapors efficiently quench the fluorescence of the PPE **6a** efficiently. DNT is also the major decomposition product of TNT and therefore **6a** senses explosives in a chemical-nose-type gadget. The quenching of PPEs′ fluorescence is superbly suited to sense and detect a number of other analytes.

A practical application for quenching was shown for the detection of heavy-metal ions such as mercury and lead, using a sugar-substituted PPE **11**, the synthesis of which is shown in Chart 8.4.[9] Despite the oligoethylene glycol units and the sugar, the polymers were soluble in DMF but not in water. Nevertheless, the mercury and the lead ions strongly coordinated to the sugar moieties to quench the fluorescence of **11**. Later, it was demonstrated[10] that carboxylate substituted PPEs are also efficient as aqueous sensors for transition metals. When used in combination with proteins, the simple

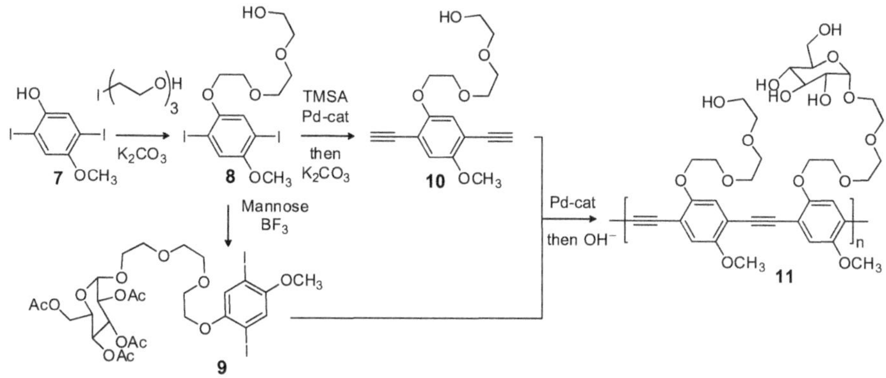

Chart 8.4

carboxylate-PPE (such as **23**, *vide infra* Chart 8.10) leads to an exquisitely sensitive and selective, yet easily obtained mercury sensor.[11]

PPEs have been around now for more than 20 years, and a number of reviews have appeared dealing with this topic (Table 8.1). PAEs are successful in the solubilization of carbon nanotubes,[12] and in the detection of lectins[13] and bacteria.[14] Ionic PPEs are used in the presence of cationic gold nanoparticles. The formed PPE-particle constructs are nonfluorescent, as the cationic gold nanoparticles strongly quench the emission of PPEs. These constructs detect proteins, cell states, bacteria and other analytes through the disruption of the complexes under turn on of the fluorescence and its quantification. A review about this topic has appeared.[15] However, even a collection of chemically different, water-soluble PPEs in an array format is able to discern different proteins quite well even without the presence of gold nanoparticles.[16] PPEs are now applied in a wide array of different electronic and sensory applications. In the next parts of this review we will cover how to make PPEs and look into significant experimental details of how to produce them.

8.1.1 Alkyl-PPEs and PAEs by Alkyne Metathesis

The first approach to make PPEs by alkyne metathesis (ADIMET, acyclic diyne metathesis) was disclosed by Weiss *et al.*[17] Monomer **12a** (Chart 8.5) was treated with a preformed Schrock carbyne complex $(t\text{BuO})_3\text{WC-tBu}$ to give PPE **13a** of moderate molecular weights. The relatively poor solubility of dihexyl-PPE **13a** in organic solvents is the probable reason for the modest molecular weight. Bunz and coworkers performed productive metathesis of dipropynyldialkylbenezenes **12a–d**[18] and naphthalenes[19] into the corresponding PPEs **13** using a simple "shake and bake" approach, in which mixtures of molybdenum hexacarbonyl with different phenols give catalytically active mixtures at temperatures around 120 °C in 1,2-dichlorobenzene. Addition of the dipropynyldialkylbenzene **12** at this temperature and reaction for 12–28 h gives the PPEs always in almost quantitative yields and, if the solubilizing

Table 8.1 Reviews dealing with PPEs and PAEs.

Entry	*Author(s)*	*Year*	*Journal citation*	*Title*
1	U. H. F. Bunz	2000	*Chem. Rev.* **100**, 1605	Poly(aryleneethynylene)s: Syntheses, properties, structures, and applications
2	U. H. F. Bunz	2001	*Acc. Chem. Res.* **34**, 998	Poly(p-phenyleneethynylene)s by alkyne metathesis
3	U. H. F. Bunz	2005	*Adv. Polym. Sci.* **177**, 1	Synthesis and structure of PAEs
4	U. H. F. Bunz	2009	*Macromol. Rapid Commun.* **30**, 772	Poly(aryleneethynylene)s
5	J. Z. Liu, J. W. Y. Lam, and B. Z. Tang	2009	*Chem. Rev.* **109**, 5799	Acetylenic polymers: syntheses, structures, and functions
6	F. Babudri, G. M. Farinola, and F. Naso	2004	*J. Mater. Sci.* **14**, 11	Synthesis of conjugated oligomers and polymers: the organometallic way
7	G. Voskerician and C. Weder	2005	*Adv. Polym. Sci.* **177**, 209	Electronic properties of PAEs
8	W. Zhang and J. S. Moore	2007	*Adv. Synth. Catal.* 349, 93	Alkyne metathesis: Catalysts and synthetic applications
9	Z. Juan and T. M. Swager	2005	*Adv. Polym. Sci.* **177**, 151	Poly(arylene ethynylene)s in chemosensing and biosensing
10	U. H. F. Bunz and V. M. Rotello	2010	*Angew. Chem.* **49**, 3268	Gold nanoparticle-fluorophore complexes: sensitive and discerning "noses" for biosystems sensing
11	U. H. F. Bunz, J. N. Wilson, and G. C. Bangcuyo	2005	*ACS Symp. Series.* **888**, 147	Chromicity in poly(aryleneethynylene)s

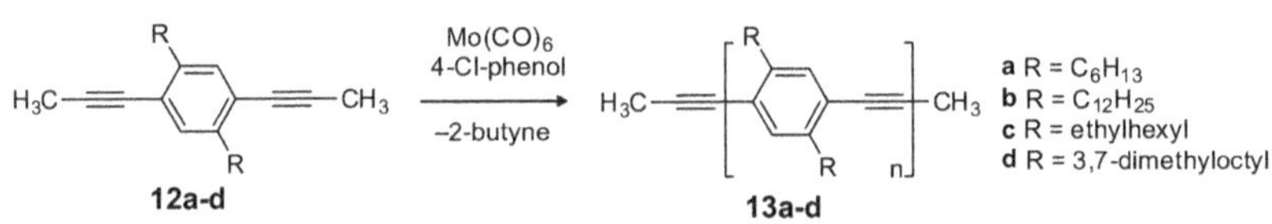

Chart 8.5

dodecyl groups are used, also with high molecular weight. The *in situ* formed catalyst is probably a metal carbyne, but it is of unknown structure. The isolated PPEs **13a–d** are brilliantly yellow, microcrystalline, meltable powders that display lc phases[20] and their absorption and emission spectra were studied in great detail.[21] Dialkyl and dialkoxy-PPEs have been examined as successful emitters in light-emitting diodes in a variety of different configurations.[22]

Alkyne metathesis is the "silver bullet" to obtain high molecular weight, pure dialkyl-PPEs **13**. If dialkoxy-dipropynylarenes are used, under the same conditions, dialkoxy-PPEs **18** form but with a much lower degree of polymerization, perhaps due to the preferred coordination of the alkoxy-groups to the *in situ* formed metathesis catalyst. A way[23] to increase the molecular weight involves pressurized preactivation of the catalytic "witches brew" of molybdenum hexacarbonyl and chlorophenol using 3-hexyne at enhanced temperatures. The much more active catalyst metathesizes dipropynyldialkoxy-benzenes easily into dialkoxy-PPEs **18** with a P_n of up to 140 units according to gel permeation chromatography (GPC). The activated catalyst allows also to metathesize vinyl-containing model compounds efficiently into their dimers. However, dipropynylated thiophenes are not metathesized by these catalysts, only more active, defined molybdenum precatalysts give poly(3-alkyl-thienyleneethynylene)s.[24]

8.1.2 Alkyl-PPEs by Pd Catalysis[25]

While high molecular weight dialkyl-PPEs **13** are obtained by ADIMET, Pd-catalysis can also be used to make dialkyl-PPEs (Charts 8.6–8.8). However, in most cases the formed PPEs **16** display defect structures that are due to the presence of diyne units. These defect structures are difficult to control, but seem to arise from oxidative dimerization and Pd-catalyzed dehalogenation with or without concomitant phosphonium salt formation.

Starting either from **14a** or from **17** (Chart 8.7), Pd-catalyzed condensation furnished **16a** in good yields but only relatively low degree of polymerization.[3]

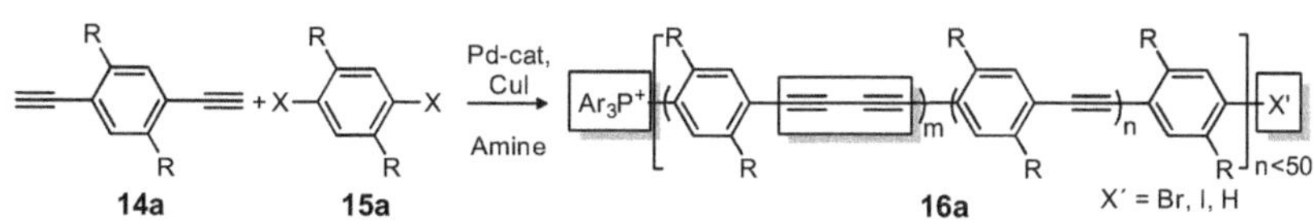

Chart 8.6

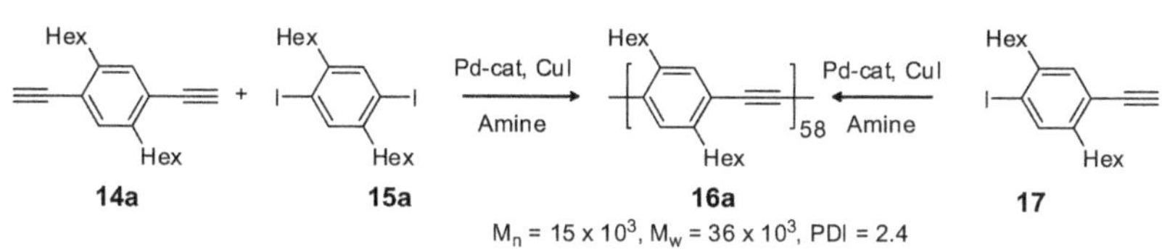

Chart 8.7

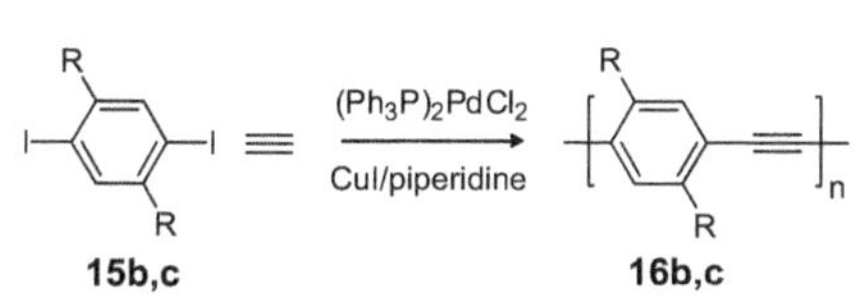

substituent key	Pd-cat (%)	yield (%)	P_n	M_w/M_n
b R = ethylhexyl	0.2	87	259	2.7
ethylhexyl	0.1	72	181	3.6
c octyl	0.2	56	91	1.8
octyl	0.1	85	127	4.1

Chart 8.8

The use of triethylamine and of 5 mol% of the Pd catalyst were not ideal, as we later found out, and therefore the P_n values of **16a** hovered around 20–60 PE units.

An important discovery was that the decrease of Pd loading leads to considerably better results and PPEs **16** form then with fewer defect structures. A piperidine–toluene mixture that works both as base and also as solvent is advantageous. The most efficient way to prepare PPEs **16** uses a small excess of acetylene gas in a premeasured fashion in a reaction vessel of known volume. PPEs with a P_n of up to 250 and polydispersities (PDI) ranging from 1.8–3.6 (Chart 8.8) are produced.

8.1.3 Alkoxy-PPEs by Pd Catalysis

Synthesis of dialkoxy-PPEs **18** employs different reaction conditions. Wrighton and coworkers,[26] coupled **5** to **16** in the presence of diisopropylamine and toluene with $(Ph_3P)_4Pd$ as catalyst. The authors reported PPEs **18** with an M_n of around 10^2 kD and noted that higher molecular weight polymers can be formed but are insoluble in common organic solvents. 1,4-Dibromo-2,5-dialkoxybenzenes are less effective coupling partners even in the presence of THF as cosolvent, where PPEs with a high concentration of defects and relatively low molecular weights are isolated.[27]

The acetylene gas method that we used for the synthesis of dialkyl-PPEs **16** is also powerful in the synthesis of dialkoxy-PPEs **18b,c** (Chart 8.9), which are formed in high yields and with a P_n of up to 400 repeat units according to GPC; 0.1–0.2 mol% Pd-catalyst suffice to give PPEs of excellent quality and large size.

As **18** are easily prepared it is also possible to attach ester groups to such a polymer by treating **19** with ethylbromoacate (Chart 8.10). Alkynylation into **21** and coupling to diiodide **20** then furnishes the organo-soluble PPE **22**, which

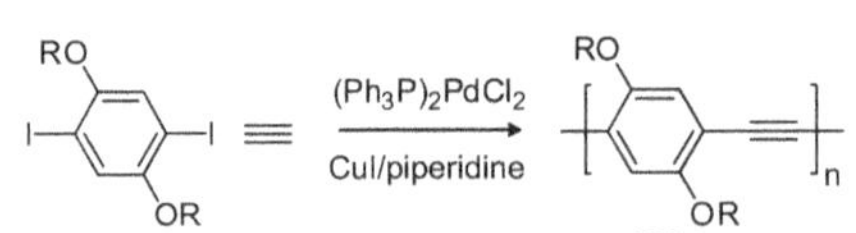

5b,c **18b,c**

substituent key		Pd-cat (%)	yield (%)	P_n	M_w/M_n
b	R = hexyl	0.2	74	55	1.4
	hexyl	0.1	63	33	3.2
c	ethylhexyl	0.2	92	316	2.4
	ethylhexyl	0.1	88	104	3.1

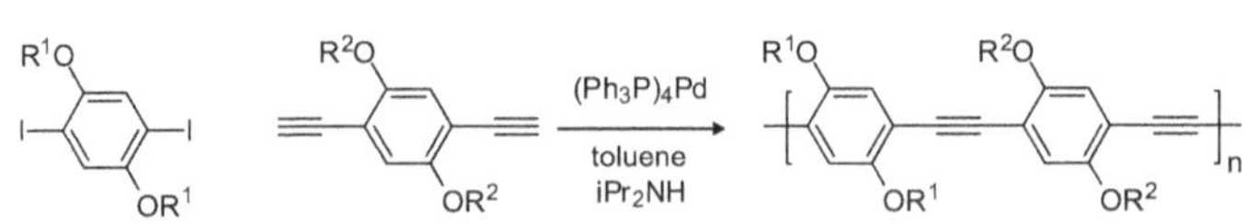

5 **1b** **18d-g**

substituent key: **5,18 d** $R^1 = C_{16}H_{33}$, **e** $R^1 = C_8H_{17}$, **f** $R^1 = C_4H_9$, **g** $R^1 = CH_3$ **h** R^1, $R^2 = C_{12}H_{25}$

$R^2 = C_{18}H_{33}$

Chart 8.9 Synthesis of dialkoxy-PPEs **18b,c** using the acetylene gas method.

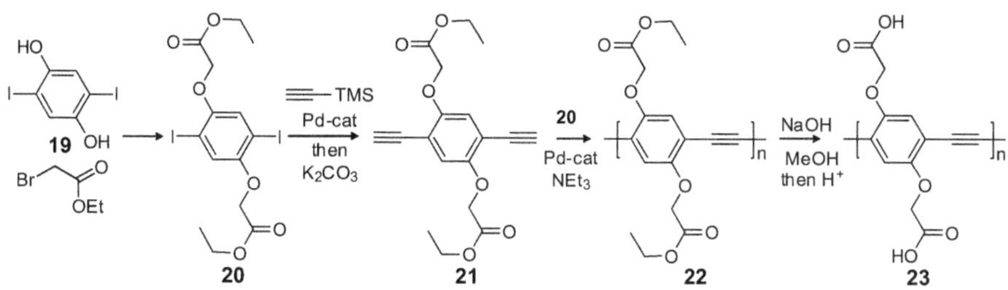

Chart 8.10

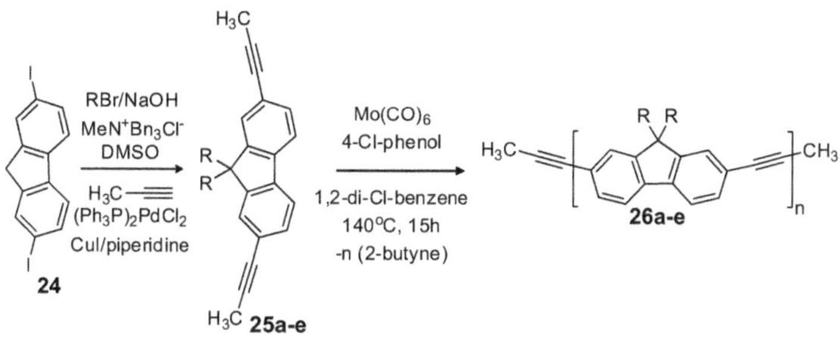

25/26	substituent	yield (%)	M_n (10^3)	M_w (10^3)	P_n	M_w/M_n
a R =	hexyl,	23	14	60	40	4.2
b	octyl,	74	12	67	28	5.8
c	dodecyl,	69	10	43	19	4.4
d	2-ethylhexyl,	34	17	61	41	3.6
e	(*S*)-3.7-dimethyloctyl	65	10	51	20	5.4

Chart 8.11

saponifies with sodium hydroxide solution into the anionic carboxylate PPE **23** in 93% yield and an (M_n of 1.2×10^4, PDI of 4.3, for **22**) after dialysis. PPE **23** is useful and can be easily generated in gram quantities and features a fluorescence quantum yield $\phi = 0.08$ in water.

8.1.4 Poly(fluorenyleneethynylene)s by Alkyne Metathesis

A variation to the theme of alkyne methathesis is the production of poly-(dialkylfluorenyleneethynylene)s **20**[28] (Chart 8.11) using $Mo(CO)_6$ and 4-chlorophenol mixtures in 1,2-dichlorobenzene. The low yield of **26a** is due to its scant solubility. If octyl or dodecyl groups are used, the yields are higher but **26b–e** are still relatively short with typical molecular weights M_n around $1–2 \times 10^4$. The polydispersities are quite high. However, **26b–e** are solid-state fluorescent and display fascinating optical properties upon aggregation.

8.1.5 Other PAEs by Pd Catalysis

There have been a large number of PAEs prepared by Pd catalysis, all showing attractive optical or electronic properties. Quinoline-containing[29] PAEs **28** sense metal cations and are significantly acidochromic. PAEs **28** were obtained by either coupling the diiodide **27a** to the dialkynyl **15b** to give the dialky-substituted **28a** or by coupling **5b** to the diyne **27b** to obtain the dialkoxy-PAE **28b** (Chart 8.12).

It is also possible to insert either quinoxaline[30] or benzothiadiazole into PAEs[31] It is advantageous to produce the dialkynylated heterocycles and couple them to **5c** (Charts 8.13 and 8.14). The somewhat decreased reactivity of the dihalogenated quinoxaline and benzothiadiazoles are possibly due to peri interactions that decrease the reactivity of the larger dihalides. Therefore, it is better to use the dialkynylated species **30**, **33** and **36** in the formation of heterocyclic PAEs such as **31**, **34** or **37**.

PAEs **31** and **34** form in yields around 90–95% with a P_n of 40–110 (GPC). In the case of the benzothiadiazole containing PAE **37** (Chart 8.14) the molecular weight and the yields are a bit lower. We also tried to repeat the

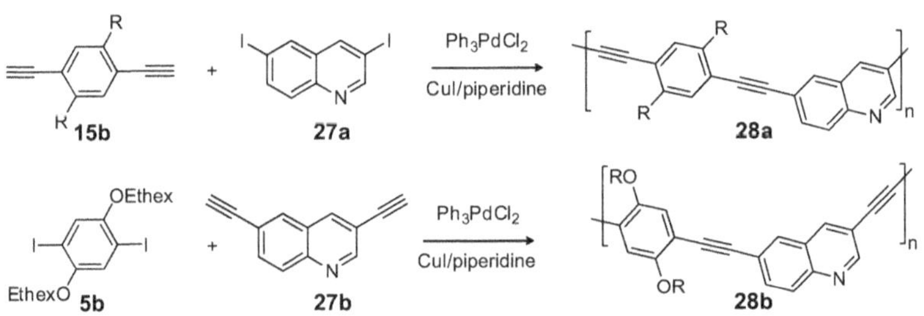

Chart 8.12

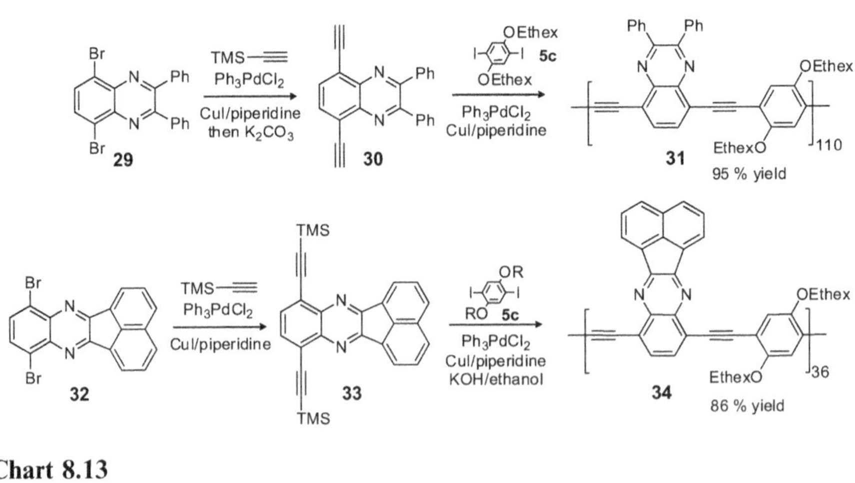

Chart 8.13

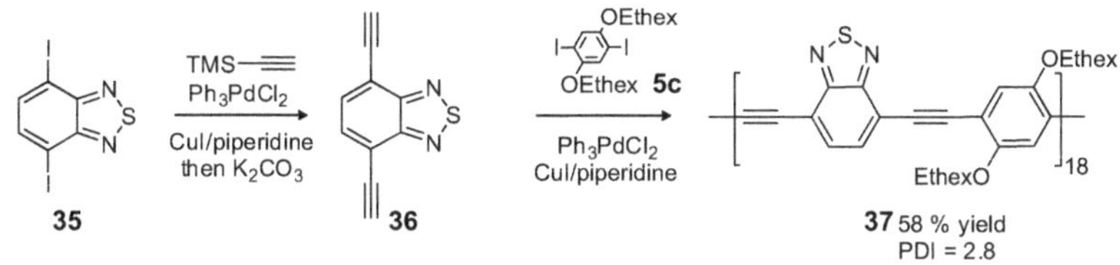

Chart 8.14

experimental procedure from Yamamoto, where dibromobenzothiadiazole is coupled to a dialkynylated dialkoxybenzene, but failed even after additional optimization efforts. However, the authors did produce PAEs of the type **37** as their spectroscopic (^{13}C NMR, UV-vis) data were identical to those obtained by us for **37**.[32]

8.1.6 Side-Chain Functionalized PPEs by Pd-Catalyzed Coupling

An important class of PPEs carry further polymers on their side chains. There are two different strategies to make such polymers (Charts 8.15 and 8.16). One can first prepare a prepolymer such as **40** by treating **38** with caprolactone (**39**) in the presence of a tin catalyst and then polymerize macromonomer **40** by Pd catalysis in the presence of acetylene into **41**[33] obtained as fluorescent material with a high degree of polymerization.

Alternatively (Chart 8.16) one can prepare the TEMPO-substituted monomer **44** from **42** and perform a Sonogashira coupling using 1,4-bis(ethylhexyl)-2,5-diethynylbenzene **14b** to obtain the nitroxide-functionalized polymer **45**.[34] Surprisingly, the presence of the nitroxide units does not at all inhibit the polymerization; PPE **45** forms in almost quantitative yields with a P_n of 29. Addition of an excess of styrene to **45** shows that around 12 styrene units can be incorporated per repeat, before the nitroxide group in **46** becomes inactive, even though a large excess of styrene was added to **45**.

8.2 Experimental Procedures

8.2.1 Materials and General Polymerization Procedures

1,4-Didodecylbenzene is obtained by Kumada coupling of 1,4-dichlorobenzene with dodecylMgBr in diethylether and $dpppNiCl_2$ as catalyst. In larger preparations it is well advised to decrease the catalyst loading to 0.1 mol% or less. Addition of the catalyst in batches is recommended to avoid runaway coupling. The product is purified by high-vacuum distillation followed by crystallization from methanol or ethanol as colorless crystalline solid. The reaction of other Grignard reagents gives any desired dialkylbenzene. Primary and secondary Grignard reagents work particularly well.

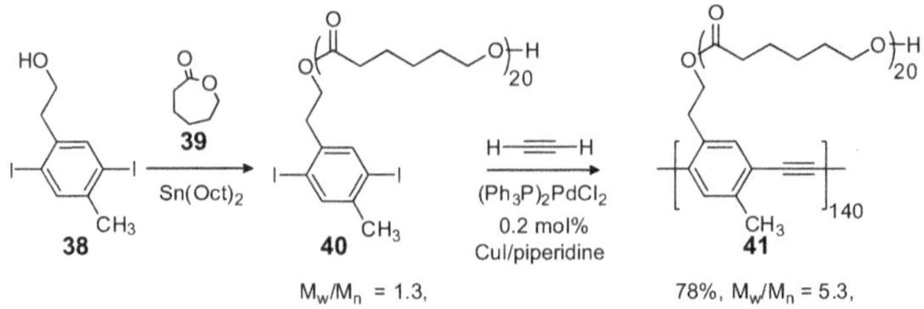

Chart 8.15

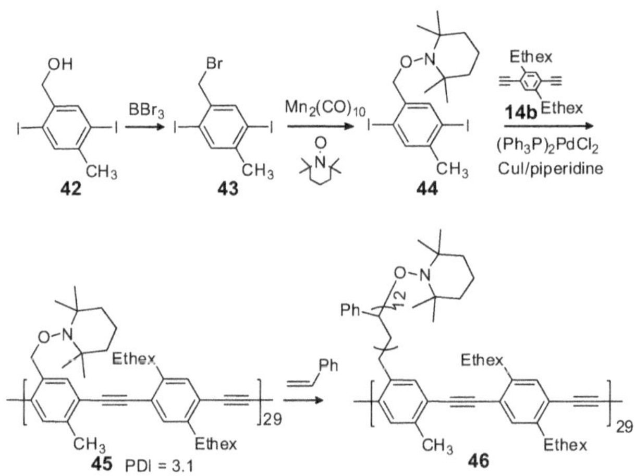

Chart 8.16

1,4-Didodecyloxybenzene: Hydroquinone is heated to reflux for 18 h with an excess of 1-bromododecane (2.5 equivalents) in acetone or butanone and potassium carbonate as base. The reaction mixture is cooled and filtered to remove the potassium carbonate and potassium bromide, the solids are washed with hexanes. Acetone and hexanes are removed using a rotary evaporator, bromododecane is distilled off under high vacuum. The remaining product is purified by crystallization from ethanol. Alternatively, one can prepare the dialkoxybenzenes by reaction of hydroquinone with KOH and an excess of bromoalkane in DMSO at room temperature for 16 h. Here, aqueous workup is followed by extraction with hexanes. Both preps work well for any primary halides to give dialkoxybenzenes.

1,4-Didodecyl-2,5-diiodobenzene (15d): 1,4-Didodecylbenzene (57.3 g, 138 mmol), KIO_4 (18.0 g, 78.3 mmol), I_2 (40.6 g, 0.160 mol),120 mL of acetic acid, 100 mL of $CHCl_3$, 40 mL of concentrated H_2SO_4, and 40 mL of H_2O are stirred for 14 h at 90 °C. The reaction may not be fully finished after this time. A small sample workup and a ^1H NMR will show if insufficiently iodinated material is left. If that is the case one should add more KIO_4 and I_2 and let stir at 90 °C for another 16–24 h to test again. After sufficient time, all of the didodecylbenzene will be transformed. After the reaction has finished, it is cooled to 4 °C, the mixture is neutralized with sodium bicarbonate solution, washed with aqueous sodium sulfite solution, and extracted 3 times with 100 mL of hexanes. Recrystallization from ethanol provides colorless **15d** (35.5 g, 39%).

Other dialkyldiiodobenzenes **15** are produced using the same procedure, which is also useful for making dialkoxydiiodobenzenes. The latter are more easily obtained by this route as the dialkoxybenzenes are more reactive towards electrophilic iodination.

General Procedure for the Preparation of the Dipropynylbenzenes 12

A 250-mL flame-dried Schlenk flask containing 1,4-diiodo-2,5-dialkylbenzene **15** (10.0 mmol), $Pd(PPh_3)_2Cl_2$ (35 mg, 50 μmol), and CuI (19 mg, 0.1 mmol) is degassed and evacuated. After the addition of 1 atm of propyne, piperidine (25 mL) is introduced into the reaction vessel *via* syringe through a septum. A colorless precipitate forms, and the clear green solution changes its color to yellow. One atmosphere of propyne is added after 2 and 4 h, respectively. The reaction mixture is stirred for a total of 7 h. For workup, 50 mL of hexanes and 50 mL of H_2O are added to the solution and the aqueous layer is extracted 3 times with 50 mL of hexanes. The combined organic layers are washed 3 times with 50 mL of H_2O and dried over $MgSO_4$. Evaporation of solvent yields a crude product, which is purified *via* filtration through a silica gel column with hexanes or hexanes/CH_2Cl_2 as eluent followed by recrystallization from ethanol.

2,5-Dihexyl-1,4-dipropynylbenzene 12a: Application of the general procedure to **15a** (4.98 g, 10.0 mmol) furnishes colorless needles (2.83 g, 88%; eluent hexanes; recrystallization); mp 44 °C.

2,5-Dinonyl-1,4-dipropynylbenzene 12e: Application of the general procedure furnishes colorless (83%; eluent hexanes; recrystallization) needles; mp 62 °C.

2,5-Didodecyl-1,4-dipropynylbenzene12b: Application of the general procedure furnishes colorless (96%; eluent hexanes) needles; mp 66–68 °C.

The corresponding dipropynyldialkoxybenzenes are prepared accordingly.

8.2.1.1 Alkyne Metathesis, General Remarks

In a typical reaction, the dipropynyldialkylbenzene **12** (0.500 g) and the catalyst system consisting of $Mo(CO)_6$ (5 mol%)/substituted phenol (1 equiv of either *para*-chlorophenol or *para*-trifluorocresol) are dissolved in 30 mL of 1,2-dichlorobenzene and stirred at temperatures between 120–170 °C overnight, removing butyne by a slow stream of nitrogen. The formed fluorescent solution is cooled and any precipitated polymer dissolved by the addition of CH_2Cl_2. The organic layer is washed with 100 mL of each H_2O, 10% NaOH, and 25% HCl. Addition of methanol precipitates yellow PPE **13**, which is filtered and vacuum dried. This reaction works very well for all dialkyldipropynylbenzenes **12**.

Activation of the metathesis catalytic mixture *via* hexyne. A cylindrical glass high-pressure tube (Ace Glass, Inc., dried overnight) was charged under N_2 with 0.1 equiv. $Mo(CO)_6$, 1 equiv. of freshly distilled 4-chlorophenol, and 1,2-dichlorobenzene (dried over molecular sieves). For polymers, the solution was kept concentrated. Four to five equivalents of 3-hexyne was added and the tube was sealed with a teflon cap. After heating to 130 °C for 2 h, the cap was removed and replaced with an upside-down rubber septum, which was secured with copper wire. A 20-gauge syringe needle was placed in the septum to serve as a gas outlet and a slow stream of N_2 was introduced *via* cannula. This mixture was heated at 130 °C for an additional hour. The starting dipropynylated arene, dissolved in 1,2-dichlorobenzene, was added *via* syringe. Overnight heating at 130 °C with continued N_2 purge gave a dark brown solution after 24 h. The mixture was cooled and diluted with CH_2Cl_2. Washing with 10% NaOH, 10% HCl, and H_2O followed by removal of solvent *in vacuo* gave the products that were purified by precipitation. When using 1,4-dihexoxy-2,5-dipropynylbenzene, a high yield of the corresponding dihexoxy-PPE **18b** ($P_n = 140$) was obtained as brilliantly orange material.

8.2.2 Didodecyl-PPE 13b by Alkyne Metathesis of 12b

Several experiments performed with 1,4-didodecyl-2,5-dipropynylbenzene at 150 °C with trifluorocresol as cocatalyst furnished didodecyl-PPE **13b** in excellent isolated yields (95–98%) with very high molecular weights (P_n 200–640, GPC *vs.* polystyrene) and polydispersities ranging from 3–7. All samples of **13b** display melting points around 130–160 °C and feature homeotropically aligned smectic liquid crystalline phases.[35]

8.2.3 Sonogashira Reactions, General Remarks

The success of Sonogashira reactions for the synthesis of PPEs is dependent upon the choice of halide, base, monomer concentration and catalyst. The combinations of aromatic iodides with piperidine as solvent and either toluene, THF or occasionally dichloromethane as cosolvents are preferable. If aromatic bromides are used, piperidine underperforms. Here, either triethylamine or Hünig's base in combination with THF as cosolvent seem to be the best option. Generally, aromatic iodides are considerably better for the synthesis of PPEs and PAEs than bromides. Another important issue is the concentration of the reactants, which should be as high as possible to ensure high molecular weights of the formed PAEs. During the polycondensation reaction the concentration of the active end groups sinks dramatically. If one produces a 20-mer, the concentration of the reactive species (end groups(!)) has decreased to 5% of the original monomer concentration. As Pd catalysts for this reaction, $(Ph_3P)_2PdCl_2$ or $(Ph_3P)_4Pd$ are the exclusive contenders. Our group predominantly employs the Pd(II) species, as it is more stable for storing and not air sensitive at all. Pd catalysts are deployed at around 0.1–0.5 mol%

concentration. Lower Pd loadings often give PAEs of better quality if one works at high monomer concentrations.

8.2.3.1 1,4-Bisethylhexyl-PPE 14b by Sonogashira Methods Using Acetylene Gas

A nitrogen-flushed tubular 25 mL Schlenk flask (total volume 37 mL) was charged with 0.772 g (1.39 mmol) of **15b**, 2 mg of $(Ph_3P)_2PdCl_2$ (0.2 mol%), and 1 mg of CuI (0.4 mol%). Toluene and piperidine, 1.5 mL each, were introduced with a syringe to wash down the sides of the flask. The flask was capped with a previously unused septum and degassed by three freeze–thaw cycles. After the final freezing and evacuation, the reaction was thawed with cold water (heating will introduce piperidine and toluene vapor into the evacuated Schlenk flask). The sidearm was purged with nitrogen and then fitted with a balloon slightly inflated with acetylene. The stopcock was briefly opened, allowing the acetylene gas to replace the vacuum but not to be absorbed into the thawing reaction mixture (approximate volume of acetylene gas is 34 mL or approximately 1.4 mmol). The reaction was allowed to stir at 50 °C for 36 h. Aqueous workup and precipitation into methanol furnishes the desired Bisethylhexyl-PPE after careful drying under high vacuum with spectroscopic properties described in ref.17.

8.2.3.2 Ester-PPE 22 and Carboxylate PPE 23

1,4-Diiodo-2,5-dihydroquinone 19 (3.62 g, 10.0 mmol),1 2-butanone (50 mL), potassium carbonate (13.8 g, 100 mmol), and ethyl bromoacetate (6.68 g, 40.0 mmol, commercially available, Aldrich) were placed in a 250 mL round-bottomed flask and heated to reflux for 24 h. The mixture was allowed to cool to room temperature and the solvent was removed. The crude solid was crystallized from a mixture of ethyl acetate and hexane to yield pure **20** as a colorless solid (3.61 g, 68%).

Synthesis of 21: Diiodide **20** (2.14 g, 4.00 mmol) and TMS-acetylene (1.00 g, 10.0 mmol) were dissolved in THF (20 mL) and triethylamine (5 mL) in a 50-mL round-bottomed flask. The solution was purged with nitrogen for 15 min and $(Ph_3P)_2PdCl_2$ (28 mg, 0.04 mmol), and CuI (7.6 mg, 0.04 mmol) were added. The mixture was allowed to stir at room temperature for 24 h. The mixture was filtered to remove the ammonium salts and the solvent mixture was removed in vacuo. The solid residue was dissolved in THF and a solution of tetrabutylammonium fluoride (1 M) was added. The reaction mixture was stirred for 10 min at room temperature and the solvent was then removed *in vacuo*. The product **21** was isolated on a silica gel column (ethyl acetate/hexane 1:3) as a pale yellow solid in 55% yield (727 mg).

Polymer 22: Monomer **21** (0.214 g, 0.400 mmol) and monomer **20** (0.145 g, 0.440 mmol) were dissolved in chloroform (4 mL) and triethylamine (1 mL) in an oven-dried Schlenk flask. $(Ph_3P)_2PdCl_2$ (1.4 mg, 0.002 mmol) and CuI (0.4 mg, 0.002 mmol) were added to the flask. The reaction mixture was stirred

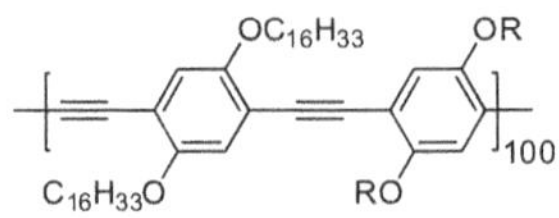

R = 50% octyl and 50% methyl
randomly distributed

18h

Chart 8.17

under nitrogen at 50 °C for 72 h. The solution was slowly added to ether (100 mL). The precipitate was washed with hot ether. An orange solid (**22**) was obtained in 93% yield (220 mg). The M_n of **22** was 12×10^3 (GPC) with an $M_w/M_n = 4.3$.

Polymer 23: A solution of **22** (150 mg) and sodium hydroxide (0.45 g) in methanol (25 mL) was stirred at reflux temperature for 24 h and cooled in a refrigerator overnight. The precipitate was filtered and washed with methanol and ether yield **23** (130 mg) as a yellow powder. The polymer was purified by dialysis using 1 kD MWCO cellulose membrane and recovered by freeze drying. The quantum yield of **23** in water is $\phi = 8.0\%$. ^{1}H NMR (D_2O): δ 7.08 (s, 2H), 4.53 (s, 4H).

8.2.3.3 Dialkoxy-PPEs[36]

The preparation of **18h** (Chart 8.17) is directly taken from: T. M. Swager, C. J. Gil, M. S. Wrighton, *J. Phys. Chem.*, 1995, **99**, 4891. **1,4-Diethynyl-2,5-dihexadecyloxybenze (1b)**: 0.0123 mol of 1,4-diiodo-2,5-di(hexadecyloxy)benzene, 0.6 mmol of CuI, and 0.6 mmol of $(Ph_3P)_2PdCl_2$ were dissolved in 150 mL of diisopropylamine at room temperature; 0.0271 mol of trimethylsilylacetylene was added to the mixture in a time frame of 15 min. Ammonium iodide salts crash out. At the end of the addition the solution turns dark brown. This mixture was heated (2 h), diluted with toluene, and filtered. The solvent was removed on a rotary evaporator. The remaining brown solid was dissolved in toluene and passed through a 4-cm plug of silica gel. Removal of the solvent yielded a brown solid that was recrystallized twice from EtOH/$CHCl_3$ to give 92% of pure 1,4-bistrimethylsilylethynyl-2,5-dihexadecyloxybenze as a colorless crystalline solid, mp 61–63 °C; 100 mL of methanol and 2 mL of 20% KOH were added to a rapidly stirred solution of 0.0114 mol of this material in 200 mL of THF at room temperature. The mixture initially gelled and then after a couple of minutes became homogeneous and a precipitate formed. The mixture was stirred for 5 h and then 200 mL of methanol was added. The mixture was filtered and the solvent removed by rotary evaporation. The remaining solid was recrystallized twice from hexane with charcoal to yield colorless crystalline **1b**.

Polymer **18h**: 0.3295 mmol of **1b**, 0.1647 mmol of **5g**, 0.1647 mmol of **5e**, 0.016 mmol of $(Ph_3P)_4Pd$, and 0.016 mmol of CuI were combined in 6 mL of diisopropylamine and 14 mL of toluene under Ar. The reactants were heated at 70 °C for 14 h to yield a noticeably viscous, fluorescent mixture. After cooling

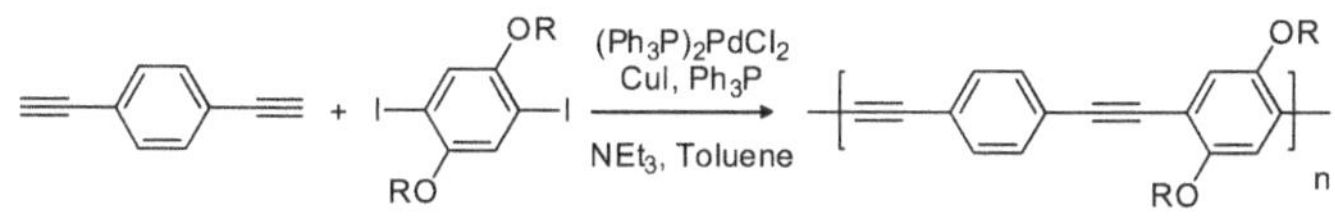

5h, R = $C_{12}H_{25}$

Chart 8.18

the material was added to 200 mL of rapidly stirred acetone. The resulting precipitate was collected, rinsed with acetone, hot CH_3CN, hot EtOH, and hexane, and dried at 40 °C overnight to give a light yellow green solid **18h** in 96% yield. The ^1H NMR gave no signals that could be attributed to end groups. $M_n = 9.8 \times 10^4$ by GPC (PDI = 1.5).

Synthesis of a Mixed Alkoxy-PPE

This preparation (Chart 8.18) was directly taken from H. Li, D. R. Powell, R. K. Hayashi, R. West, *Macromolecules* 1998, **31**, 52.

1,4-Diethynylbenzene (0.0875 g, 0.694 mmol), 2,5-diiodo-1,4-bis(dodecyloxy)benzene (0.485 g, 0.694 mmol), Et_3N (5 mL), toluene (50 mL), a catalytic amount of CuI, $Pd(PPh_3)_2Cl_2$, and PPh_3 were added into a 100-mL three-necked flask equipped with condenser and magnetic stir bar. The mixture was stirred for 24 h at room temperature, warmed to 40 °C for 24 h, and refluxed for 24 h. After the reaction mixture was cooled to room temperature, the salts were filtered, and the solvent was removed under vacuum. The resulting solid was column chromatographed on silica gel, eluting with THF to remove catalysts. The first component was collected and precipitated by dropping the solution into 200 mL of 2-propanol and dried at 120 °C under vacuum, giving 0.34 g (85%) of the desired PPE: $M_n = 15 \times 10^3$, $P_n = 24$ by ^1H NMR.

8.2.4 Polyfluorenyleneethynylene by ADIMET

Propynylation of 2,7-diiodo-9,9-didodecyl-9H-fluorene was performed using a controlled amount of propyne gas resulting in **25c** (89%) as a clear oil.

Polymerization: 25c (0.500 g, 0.860 mmol), $Mo(CO)_6$ (0.021 g, 0.086 mmol), 4-chlorophenol (0.111 g, 0.860 mmol), and 1,2-dichlorobenzene (20 mL) were held at 140 °C for 25 h under a steady stream of N_2. The cooled reaction mixture was dissolved in 75 mL of CH_2Cl_2, washed with diluted base, acid, and H_2O, and then poured into MeOH for precipitation. The resulting yellow solid **26c** (69%) was obtained by filtering a #4 frit; The PAE **25c** was dried under high vacuum at 40–90 °C bath temperature. The other poly-(fluorenyleneethynylene)s were prepared analogously.

8.2.5 Quinoxaline-Containing PAEs

Preparation of 5,8-bis(trimethylsilylethynyl)-2,3-diphenylquinoxaline. To a flame-dried 50-mL Schlenk flask, **29** (0.586 g, 1.33 mmol), $(Ph_3P)_2PdCl_2$ (46.7 mg, 0.066 mmol), CuI (52.3 mg, 0.275 mmol), THF (5 mL) and triethylamine (10 mL) were added under a slow stream of nitrogen. Trimethylsilylacetylene (0.270 g, 2.75 mmol) was added *via* syringe, and the reaction was stirred overnight at 80 °C. The reaction was poured into a mixture of dichloromethane (100 mL) and water (150 mL). The organic layer was washed three times with water (100 mL) followed by 10% NH_4OH (150 mL). The aqueous layer was washed three times with dichloromethane (30 mL). The combined organic layers were dried over anhydrous sodium sulfate, and the solvent was evaporated to give a brown powder. Purification by column chromatography with eluent hexanes/dichloromethane (80/20) furnished colorless crystals (0.601 g, 95%). mp: 230 °C (decomp.).

Preparation of 5,8-diethynyl-2,3-diphenylquinoxaline 30. To a 100-mL round-bottomed flask, 5,8-bis(trimethylsilylethynyl)-2,3-diphenylquinoxaline (0.670 g, 1.41 mmol) was added to a solution of methanol (50 mL) and potassium carbonate (5.00 g, 36.2 mmol) and stirred overnight at room temperature. The solution was poured into a mixture of dichloromethane (50 mL) and water (100 mL). The organic layer was washed three times with water (75 mL). The aqueous layer was washed three times with dichloromethane (30 mL) and dried over sodium sulfate. The solvent was evaporated to give a yellow powder **30** (0.396 g, 90%). mp: 183 °C (decomp.).

PAE 31. To a flame-dried 50-mL Schlenk flask, **30** (0.210 g, 0.636 mmol), **5c** (0.339 g, 0.578 mmol), $(Ph_3P)_2PdCl_2$ (100 mg, 0.142 mmol), CuI (50.0 mg, 0.260 mmol), and toluene (6 mL) were added under a slow stream of nitrogen. Piperidine (6 mL) was added to the reaction *via* syringe. The reaction mixture was stirred and kept at 30 °C for 24 h. The solution was poured into a mixture of chloroform (75 mL) and water (100 mL). The organic layer was washed three times with water (100 mL) followed by 50% NH_4OH (150 mL). The aqueous layer was washed twice with chloroform (20 mL), and the organic layers were combined. To the resultant organic layer, methanol (400 mL) was added to precipitate the polymer. Suction filtration followed by drying under high vacuum produced a dark red polymer **31** with a green metallic luster (0.400 g, 96%). GPC (polystyrene standards): P_n 110, $M_w/M_n = 3.0$.

Preparation of 33. To a flame-dried 50-mL Schlenk flask, **32** (0.990 g, 2.40 mmol), $(Ph_3P)_2PdCl_2$ (100 mg, 0.142 mmol), CuI (100 mg, 0.521 mmol), Ph_3P (150 mg, 5.71 mmol), and triethylamine (20 mL) were added under a slow stream of nitrogen. Trimethylsilylacetylene (0.610 g, 6.21 mmol) was added *via* syringe, and the reaction was stirred overnight at 80 °C. Then, the reaction was poured into a mixture of dichloromethane (100 mL) and water (150 mL). The organic layer was washed three times with water (100 mL) followed by 10% NH_4OH (150 mL). The aqueous layer was washed three times with dichloromethane (30 mL), and the organic layers were combined, dried over anhydrous sodium sulfate, and the solvent was evaporated to give a light brown powder. Purification by column chromatography with eluent hexanes/dichloromethane (80/20) furnished colorless crystals of **33** (0.882 g, 83%). MP: 243 °C (decomp.).

PAE 34. To a flame-dried 50-mL Schlenk flask, **33** (0.250 g, 0.559 mmol), **5c** (0.330 g, 0.562 mmol), $(Ph_3P)_2PdCl_2$ (50.0 mg, 0.071 mmol), CuI (50.0 mg,

0.260 mmol), and THF (5 mL) were added under a slow stream of nitrogen. A saturated ethanol/KOH solution (1 mL) was also added. Piperidine (7 mL) was introduced to the reaction *via* syringe. The reaction was stirred and held at 30 °C for 48 h. The solution was poured into a mixture of chloroform (75 mL) and water (100 mL). The organic layer was washed three times with water (100 mL) followed by 50% NH_4OH (150 mL). The aqueous layer was washed twice with chloroform (20 mL), and the organic layers were combined. To the resultant organic layer, methanol (400 mL) was added to precipitate the polymer **34**. Suction filtration followed by drying under high vacuum produced an orange polymer **34** (0.305 g, 86%); GPC (polystyrene): $P_n = 37$, $M_w/M_n = 1.44$.

8.2.6 Benzothiadiazole-Containing PAE

4,7-Bis(trimethylsilyl)ethynyl-2,1,3-benzothiadiazole. Under inert conditions, **35** (4.46 g, 11.5 mmol), $(Ph_3P)_2PdCl_2$ (554 mg, 0.575 mmol), CuI (100 mg, 0.529 mmol), toluene (5 mL), and piperidine (20 mL) were combined. Trimethylsilylacetylene (2.45 g, 24.9 mmol) was added dropwise to the stirred solution. After 16 h at ambient temperature the reaction mixture was poured into dichloromethane (50 mL) and water. The organic layer was washed twice with water (75 mL), 25% NH_4OH (100 mL), and water (100 mL), dried over magnesium sulfate. The solvent is stripped off. Purification by crystallization from methanol at –78 °C gives green-yellow crystals of the product (2.12 g, 55%).

4,7-Diethynyl-2,1,3-benzothiadiazole 36. To a 250-mL round-bottom flask, 4,7-bis(trimethylsilyl)-ethynyl-2,1,3-benzothiadiazole (2.00 g, 6.08 mmol), potassium carbonate (approximately 2.5 g), methanol (125 mL), and water (1 mL) were added and stirred overnight at room temperature. The solution was poured into dichloromethane (50 mL) and water (100 mL). Aqueous workup and crystallization from methanol at –78 °C furnished the green-yellow crystalline **36** (0.971 g, 86%).

Preparation of PAE 37. Under inert conditions, **36** (0.310 g, 1.68 mmol), **5c** (0.908 g, 1.54 mmol), $(Ph_3P)_2PdCl_2$ (74.0 mg, 0.077 mmol), CuI (15.0 mg, 0.077 mmol), dichloromethane (5 mL), and piperidine (15 mL) were added and heated to 85 °C for 48 h. Dilution with chloroform (100 mL), washing with water (2 × 100 mL), 25% NH_4OH (100 mL), and water (100 mL), and drying with $MgSO_4$ are followed by removal of the solvent under vacuum. The resulting solid was thoroughly washed with hexanes (until clear) followed by extensive washings with methanol (until clear) to furnish a red-flaky solid **37** with green-metallic luster followed by drying under high vacuum; PAE **37** is well soluble in chloroform and dichloromethane (0.412 g, 58%); GPC (polystyrene): $P_n = 19$, $M_w/M_n = 2.78$.

8.2.7 Quinoline-Containing PAEs

3,6-Diiodoquinoline 27a. In a 1-L round-bottomed flask, quinoline (50.0 g, 387 mmol), iodine (100 g, 394 mmol) KIO_4 (50.0 g, 0.217 mol), acetic acid (200 mL),

CH_2Cl_2 (150 mL), concentrated H_2SO_4 (60 mL), and H_2O (60 mL) are stirred for 96 h at 90 °C. The solution is neutralized with sodium bicarbonate solution, washed with aqueous sodium sulfite, and extracted with 3 × 100 mL of hexanes. Crystallization from ethanol produces yellow crystals of diiodoquinoline (2.61 g, 18%). Mp: 155 °C.

3,6-Bis(trimethylsilylethynyl)quinoline. A flame-dried 1000-mL Schlenk flask containing **27a** (11.0 g, 34.7 mmol), $(PPh_3)_2PdCl_2$ (0.150 g, 2.14 mmol) CuI (0.100 g, 0.525 mmol), and piperidine (10 mL) is degassed and evacuated. After dropwise addition of TMS-acetylene (5.50 g, 55.0 mmol) the resulting mixture is stirred for 16 h at ambient temperature. Addition of hexanes (150 mL) is followed by aqueous workup and purification by column chromatography to furnish a clear oil (84% yield; eluent hexanes/CH_2Cl_2 80 : 20). The adduct was subsequently deprotected, and only checked for purity by ^{13}C NMR. ^{13}C NMR ($CDCl_3$): δ 153.08, 146.43, 138.47, 133.20, 131.40, 129.60, 126.91, 122.42, 118.20, 104.42, 101.99, 99.04, 96.56, 0.141.

3,6-Diethynylquinoline 27b. To a 500-mL round-bottomed flask, 8-bis-(trimethylsilylethynyl)quinoline (7.00 g, 21.8 mmol) and an excess amount of K_2CO_3 (15.0 g, 108 mmol) was added to methanol (200 mL) and stirred overnight. Addition of hexanes (150 mL) is followed by aqueous workup and purification by crystallization from ethanol to yield brownish crystals of **27b** (4.67 g, 94%). Mp 116–118 °C.

General Procedure for the Synthesis of the Quinoline Polymers. In a flame-dried 500-mL Schlenk flask under nitrogen are placed **27b** and **5b**, 1 mol% of $(Ph_3P)_2PdCl_2$, 1 mol% of CuI, and piperidine (15 mL). The reaction mixture was stirred for 48 h. For workup, chloroform (3 × 50 mL) was added. The solution was washed with water and NH_4OH to remove the transition-metal-containing catalyst residues. Removal of solvent and addition of methanol (400 mL) precipitates the polymers, which are isolated by filtration and vacuum dried.

Polymer 28a. 1,4-Diethynyl-2,5-bis(ethylhexyl)benzene **15b** (2.00 g, 5.70 mmol) and **27a** (2.00 g, 5.25 mmol) were reacted according to the general procedure to furnish a green-yellow powder **28a** (2.33g, 93%), which was well soluble in chloroform. GPC (polystyrene standards): $P_n = 37$, PDI = 2.75.

Polymer 28b. Application of the general procedure to **27b** (0.500 g, 2.82 mmol) and 1,4-diiodo-2,5-bis(2-ethylhexyloxy)benzene **5b** (1,55 g, 2.64 mmol) furnished an orange powder **28b** (1.18 g, 88%), which was not very soluble in chloroform or any other organic solvent. GPC (polystyrene standards): $P_n = 95$, PDI = 3.13.

8.2.8 Grafted PPEs

Polymer 41

Polyester grafts: An oven-dried Schlenk flask cooled under nitrogen was charged with 2,5-diiodo-4-methylphenethylalcohol **38** (2.00 g, 5.16 mmol), ε-caprolactone **39** (11.8 g, 103 mmol), and tin(II)-2-ethylhexanoate (127 mg,

313 μmol). The flask was heated to 110 °C with stirring for 12 h. The viscous product was diluted with chloroform (10 mL), then precipitated into 300 mL of methanol. The diiodo-polyester **40** was obtained as a colorless solid. Yield 6.56 g, 48%; GPC (*vs.* polystyrene standards in chloroform): $M_n = 2400$, $M_w/M_n = 1.3$.

Polymer 41: 40 (3.29 g, 1.39 mmol) was combined with piperidine (1.5 mL), THF (1.5 mL), $(Ph_3P)_2PdCl_2$ (2.0 mg, 2.8 μmol, 0.2 mol%), and CuI (1.0 mg, 5.3 μmol, 0.4 mol%) in a Schlenk flask (37 mL). Acetylene gas (34 mL, 1.4 mmol) was added through the purged side arm with a balloon. The reaction solidified after 36 h. The resultant polymer was filtered over a cotton plug using dichloromethane as a solvent before precipitating into methanol. Polymer **41** was collected over a fritted funnel, redissolved in dichloromethane and precipitated again. A bright yellow polymer **41** was obtained followed by drying under high vacuum (2.31 g, 77% yield); GPC (*vs.* polystyrene standard in chloroform): $M_n = 3.4 \times 10^5$. Repeat = 2400 g/mol, $P_n = 140$, $M_w/M_n = 5.3$.

Polystyrene grafts: Polymer 46

Synthesis of 43. To a nitrogen-purged 250-mL round-bottomed flask were placed **42** (6.50 g, 17.4 mmol) and toluene (50 mL); BBr_3 (5.30 g, 21.2 mmol) was added *via* syringe. The reaction was stirred at room temperature for 3 h under an inert atmosphere. The reaction mixture was poured into ice-cold water, and the organic layer was washed with water (3×times 100 mL) and then dried over $MgSO_4$. The solvent was subsequently removed in vacuo. The crude product was purified by crystallization from hexanes to yield a colorless solid **43** (6.12 g, 81%), Mp = 157–158 °C.

Synthesis of 44. To a 100-mL Schlenk tube were added **43** (2.18 g, 5.00 mmol), tetramethylpiperidine-*N*-oxide (1.24 g, 7.94 mmol), $Mn_2(CO)_{10}$ (0.390 g, 1.00 mmol), and toluene (20 mL). The tube was degassed and then filled with nitrogen. The flask was irradiated by a broadband sunlight lamp for 24 h. The reaction mixture was filtered and purified by column chromatography (hexanes) to obtain **44** as a colorless solid (1.03 g, 40%). Mp = 126–128 °C

Synthesis of PAE 45. A 50-mL Schlenk flask was charged with **44** (513 mg, 1.00 mmol), **14b** (351 mg, 1.00 mol), THF (5 mL), piperidine (1.50 mL), $(Ph_3P)_2PdCl_2$ (2 mg, 2.8 μmol), and CuI (1 mg, 5.3 μmol). The flask was degassed and filled with nitrogen. The reaction was allowed to stir at room temperature for 24 h. The reaction mixture was filtered over silica on a fritted funnel with hexane. After the hexanes were evaporated off, the crude product was dissolved in chloroform (10 mL) and was slowly added into methanol (300 mL). A bright yellow solid, **45** (603 mg, 94%) was obtained followed by drying under high vacuum; GPC (*vs.* polystyrene standards in chloroform): $M_n = 18 \times 10^3$, $P_n = 31$, $M_w/M_n = 3.1$. Quantum yield (in chloroform): $\phi = 0.78$.

Synthesis of PAE 46. An oven-dried 50-mL Schlenk flask was charged with **45** (200 mg, 0.329 mmol counted by the TEMPO functional group), styrene (5 mL), and acetic anhydride (20 mg, 0.196 mmol). The flask was degassed and filled with nitrogen. The reaction was allowed to stir at 135 °C for 3 h during which the reaction mixture solidified. Chloroform (20 mL) was added into the flask to dissolve the solid, and the solution was precipitated into methanol (200 mL). A very lightly yellow polymer was obtained (631 mg, 13% with respect to used styrene); GPC (*vs.* polystyrene standards in chloroform): $M_n = 56 \times 10^3$, $M_w/M_n = 3.2$. Quantum yield (in chloroform): $\phi = 0.32$.

8.2.9 Characterization and Concluding Remarks

The molecular weight of PPEs is in most cases determined by gel permeation chromatography (GPC), and even though it is not an absolute method and the obtained molecular weights are obtained only in comparison to polystyrene, GPC is a most convenient method to determine the molecular weight and also polydispersity of the synthesized PPEs and PAEs. Several groups have dealt with the question of errors introduced by GPC.[37] Bunz and coworkers could show that the apparent molecular weight of samples that were reduced by catalytic hydrogenation[38] decreased by a factor of around 1.4. Later Höger and coworkers[39] demonstrated on a series of dialkylpoly(*para*-phenylenebutadiynylene)s (Chart 8.19) that the overestimation of the molecular weight using GPC is significant. According to their measurements and can reach a factor of 3.8 when approaching infinity. However, these numbers have to be taken with some caution, as in the case of a polymer there is M_n and M_w; as in PPEs, made by polycondensation methods the PDI is relative large, it is not clear if the factor of 3.8 has to be used for M_n or M_w; different factors need to be taken into account for each value. Höger's results would explain the often high polydispersities (PDI) that one can obtain when investigating dialkyl-PPEs made by alkyne metathesis. However, as everybody uses GPC for the determination of molecular weights of PPEs and PAEs it is now standard. One must be aware that the numbers that one obtains from GPC should be taken with some caution.

The PAEs are always characterized by their optical signatures, *i.e.* their UV-vis and emission spectra in solution. For oligomers that are longer than ten repeat units, the optical properties converge, *i.e.* for "real polymers" *i.e.* $P_n > 15$ the absorption and emission spectra are generally independent of their molecular weight. PPEs are generally blue or turquoise emitting species in solution.[40] To obtain PAEs with red-shifted emission colors one has to incorporate other aromatic building blocks such as thiophene, quinoxaline, or benzothiadiazole, *etc*. These PAEs can show emission maxima in the yellow or orange range. An important qualitative measure to evaluate if a specific sample of a PPE is defect free, is their simple visual appearance in the solid state. Dialkyl-PPEs should be brilliantly yellow powders without a brownish or grayish tinge,

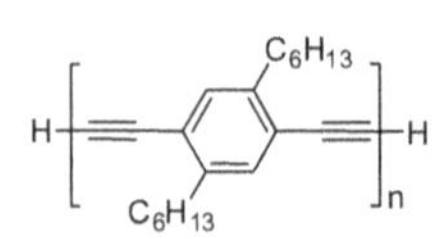

Chart 8.19

while dialkoxy-PPEs should be brilliantly orange in appearance. Here too, the presence of discolorations suggests significant concentration of defect sites. ^{1}H NMR and ^{13}C NMR spectra of PAEs give significant information on the purity of the materials. In particular, ^{13}C NMR spectra are important to discover if end groups are still visible or if large amount of diyne defects are present. In poorly soluble PAEs the signals for the aromatic backbone can be weak and therefore then are perhaps suited for structure determination. In such cases ^{13}NMR cannot to detect structural defects, unless they affect more than 10–20% of the polymer.

Dialkyl-PPEs are stable and melt at temperatures that range between 130-240°C without oxidative degradation. Dialkoxy-PPEs do not melt as they normally oxidize or decompose before reaching melting temperatures. Overall, PPEs and PAEs are easily produced, attractive materials that are supremely useful for sensory and other applications and display fascinating structural and optical properties.

References

1. R. Giesa and R. C. Schulz, *Makromol. Chem.*, 1990, **191**, 857.
2. For an overview of the Sonogashira reaction see: (a) R. Chinchilla and C. Najera, *Chem. Rev.*, 2007, **107**, 874; (b) E. Negishi and L. Anastasia, *Chem. Rev.*, 2003, **103**, 1979; (c) K. Sonogashira, *J. Organomet. Chem.*, 2002, **653**, 46; (d) S. Thorand and N. Krause, *J. Org. Chem.*, 1998, **63**, 8551.
3. M. Moroni, J. Le Moigne and S. Luzzati, *Macromolecules*, 1994, **27**, 562.
4. T. Mangel, A. Eberhardt, U. Scherf, U. H. F. Bunz and K. Müllen, *Macromol. Rapid Commun.*, 1995, **16**, 571.
5. D. Ofer, T. M. Swager and M. S. Wrighton, *Chem. Mater.*, 1995, **7**, 418.
6. C. Weder and M. S. Wrighton, *Macromolecules*, 1996, **29**, 5157.
7. Q. Zhou and T. M. Swager, *J. Am. Chem. Soc.*, 1995, **117**, 12593.
8. J. S. Yang and T. M. Swager, *J. Am. Chem. Soc.*, 1998, **120**, 11864.
9. I. B. Kim, B. Erdogan, J. N. Wilson and U. H. F. Bunz, *Chem. Eur. J*, 2004, **10**, 6247.
10. I. B. Kim, A. Dunkhorst, J. Gilbert and U. H. F. Bunz, *Macromolecules*, 2005, **38**, 4560.
11. (a) I. B. Kim and U. H. F. Bunz, *J. Am. Chem. Soc.*, 2006, **128**, 2818; (b) Y. Zhang, C. B. Murphy and W. E. Jones, *Macromolecules*, 2002, **35**, 630.
12. J. Chen, H. Y. Liu, W. A. Weimer, M. D. Halls, D. H. Waldeck and G. C. Walker, *J. Am. Chem. Soc.*, 2002, **124**, 9034.
13. I. B. Kim, J. N. Wilson and U. H. F. Bunz, *Chem. Commun.*, 2005, 1273.
14. M. D. Disney, J. Zheng, T. M. Swager and P. H. Seeberger, *J. Am. Chem. Soc.*, 2004, **126**, 13343.
15. U. H. F. Bunz and V. M. Rotello, *Angew. Chem. Int. Ed.*, 2010, **49**, 3268.
16. O. R. Miranda, C. C. You, R. Phillips, I. B. Kim, P. S. Ghosh, U. H. F. Bunz and V. M. Rotello, *J. Am. Chem. Soc.*, 2007, **129**, 9856.

17. K. Weiss, A. Michel, E. M. Auth, U. H. F. Bunz, T. Mangel and K. Müllen, *Angew. Chem.*, 1997, **36**, 506.
18. L. Kloppenburg, D. Jones and U. H. F. Bunz, *Macromolecules*, 1999, **32**, 4194.
19. N. G. Pschirer, M. E. Vaughn, Y. B. Dong, H. C. zur Loye and U. H. F. Bunz, *Chem. Commun.*, 2000, 85.
20. L. Kloppenburg, D. Jones, J. B. Claridge, H. C. zur Loye and U. H. F. Bunz, *Macromolecules*, 1999, **32**, 4460.
21. T. Miteva, L. Palmer, L. Kloppenburg, D. Neher and U. H. F. Bunz, *Macromolecules*, 2000, **33**, 652.
22. (a) N. G. Pschirer, T. Miteva, U. Evans, R. S. Roberts, A. R. Marshall, D. Neher, M. L. Myrick and U. H. F. Bunz, *Chem. Mater.*, 2001, **13**, 2691; (b) C. A. Breen, J. R. Tischler, V. Bulovic and T. M. Swager, *Adv. Mater.*, 2005, **17**, 1981; (c) C. Schmitz, P. Posch, M. Thelakkat, H. W. Schmidt, A. Montali, K. Feldman, P. Smith and C. Weder, *Adv. Funct. Mater.*, 2001, **11**, 41; (d) K. T. Nielsen, K. Bechgaard and F. C. Krebs, *Macromolecules*, 2005, **38**, 658.
23. G. Brizius and U. H. F. Bunz, *Org. Lett.*, 2002, **4**, 2829.
24. W. Zhang and J. S. Moore, *Macromolecules*, 2004, **37**, 3973.
25. J. N. Wilson, S. M. Waybright, K. McAlpine and U. H. F. Bunz, *Macromolecules*, 2002, **35**, 3799.
26. T. M. Swager, C. J. Gil and M. S. Wrighton, *J. Phys. Chem.*, 1995, **99**, 4886.
27. See ref. 3 and for a discussion of the used amine in the Sonogashira couplings see: M. Alami, F. Ferri and G. Linstrumelle, *Tetrahedron Lett* 1993, **34**, 4603.
28. N. G. Pschirer and U. H. F. Bunz, *Macromolecules*, 2000, **33**, 3961.
29. C. G. Bangcuyo, M. E. Rampey-Vaughn, L. T. Quan, S. M. Angel, M. D. Smith and U. H. F. Bunz, *Macromolecules*, 2002, **35**, 1563.
30. C. G. Bangcuyo, J. M. Ellsworth, U. Evans, M. L. Myrick and U. H. F. Bunz, *Macromolecules*, 2003, **36**, 546.
31. C. G. Bangcuyo, U. Evans, M. L. Myrick and U. H. F. Bunz, *Macromolecules*, 2001, **34**, 7592.
32. T. Morikita, I. Yamaguchi and T. Yamamoto, *Adv. Mater.*, 2001, **13**, 1862.
33. Y. Q. Wang, B. Erdogan, J. N. Wilson and U. H. F. Bunz, *Chem. Commun.*, 2003, 1624.
34. Y. Q. Wang, J. N. Wilson, M. D. Smith and U. H. F. Bunz, *Macromolecules*, 2004, **37**, 9701.
35. L. Kloppenburg, D. Jones, J. B. Claridge, H. C. zur Loye and U. H. F. Bunz, *Macromolecules*, 1999, **32**, 4460.
36. H. Li, D. R. Powell, R. K. Hayashi and R. West, *Macromolecules*, 1998, **31**, 52.
37. P. M. Cotts, T. M. Swager and Q. Zhou, *Macromolecules*, 1996, **29**, 7323.

38. H. L. Ricks, U. H. Choudry, A. R. Marshall and U. H. F. Bunz, *Macromolecules*, 2003, **36**, 1424.
39. D. Mossinger, S. S. Jester, E. Sigmund, U. Müller and S. Höger, *Macromolecules*, 2009, **42**, 7974.
40. C. E. Halkyard, M. E. Rampey, L. Kloppenburg, S. L. Studer-Martinez and U. H. F. Bunz, *Macromolecules*, 1998, **31**, 8655.

CHAPTER 9

Polythiophenes

DAHLIA HAYNES[a] AND RICHARD McCULLOUGH*[b]

[a] Carnegie Mellon University, Department of Chemistry, Mellon Institute, Pittsburgh PA 15213, USA; [b] Harvard University, 1350 Massachusetts Avenue, Holyoke Center, Suite 836, Cambridge, MA 02138, USA
*Email: Richard_McCullough@harvard.edu

9.1 Introduction

Since the evolution of conjugated polymers and their use as semiconductors, Polythiophenes have remained one of the most intensely researched materials in the field of '*organic electronics*'. Due to the relatively facile and well-established synthetic modifications of the corresponding monomers, oligothiophenes (oligo-PTs) and polythiophenes (PTs) have been vastly explored in a variety of applications such as organic field-effect transistors (OFETs),[1–8] photo- and electroluminescent diodes (OLEDs),[9–12] organic photovoltaics (OPVs),[13–18] and sensing devices in medical and biological fields.[19–21] Various synthetic methods have been used to achieve a wide variety of structural variations that showcase the attractive potential of oligo- and polythiophenes.

Over the years, many synthetic strategies have been employed in the development of well-defined oligothiophenes and excellent reviews on the evolution of synthetic scopes have been published.[22–29] Oligothiophenes have typically been synthesized by oxidative homocoupling procedures,[22–24] ring closure from acyclic precursors[25,26] and metal-catalyzed C–C coupling reactions.[27–29] As thiophene chemistry progressed, success in new synthetic methods begat more enthusiasm for polythiophene research, with the promise of enhanced semiconductor performance coupled to the processability of polymer-based materials. In this chapter, we take a different approach to some of the excellent

RSC Polymer Chemistry Series No. 9
Conjugated Polymers: A Practical Guide to Synthesis
Edited by Klaus Müllen, John R. Reynolds and Toshio Masuda

Published by the Royal Society of Chemistry, www.rsc.org

reviews on polythiophene research. Our intention is to provide highlights of the synthetic and polymerization details throughout the history of polythiophenes with detailed experimental procedures.

9.1.1 Synthesis of Nonsubstituted Polythiophenes (PT)s

Synthesis of nonsubstituted polythiophene was first attempted in 1967, prepared through the use of protic acid.[30] These electrophilic reactions were believed to result in mostly 2,5-couplings of the thiophene monomer. However, the structure of the poly(2,5-thienylene) polymer (**1** in Chart 9.1) was not clearly elucidated due to lack of solubility. In the early 1980s, Yamamoto *et al.* and Lin and Dudek reported another synthetic strategy involving Ni catalysts employing metal-catalyzed polycondensation polymerization of the 2,5-dibromothiophene Grignard **2** generated *in situ*.[31,32] Both groups reported the use of Ni catalysts as being the most effective in the polymerization of **1**. High electric conductivity was achieved ranging from 10^{-11} (ohm cm)$^{-1}$ of the predoped polymer to 10^{-4} (ohm cm)$^{-1}$ upon doping with iodine, a magnitude of 7 increase. Only solid-state characterization such as FTIR and elemental analysis were reported. Limited characterization details of these compounds were due to the insolubility of the material resulting in at least 3% impurities.

Wudl and workers did further work on the metal-catalyzed polycondensation polymerization using di-iodo derivatives **3**.[33] They proved that purified 2,5-dihalides thiophenes resulted in partially soluble PT polymers with an approximate molecular weight of 4000. Upon doping with AsF_6, the electrical conductivity of the polymer was increased by a factor of 10 to values of 10 or greater (ohm cm)$^{-1}$. The improved polymerization was determined to be due to the increased purity of the di-iodo thiophene starting material and milder polymerization conditions. IR and UV-Vis spectra and ESR suggested formation of 2,5-coupled polythiophene. An additional synthetic method of PTs involved bis-lithium derivatives **4**, which underwent oxidative coupling with copper chloride.[34] This synthetic route was believed to eliminate terminal functional groups at the end of the polymer chain resulting in higher conductivities. Table 9.1 shows a summary of the synthetic methods of nonsubstituted PTs and selected properties.

Electrochemical synthetic methods were also attempted around this time, yet this approach was limited due to solubility issues of the polymer.[35–38] Electrochemical synthesis of polythiophene was typically performed on metal substrates, the most common being platinum, silver, iron and copper electrodes

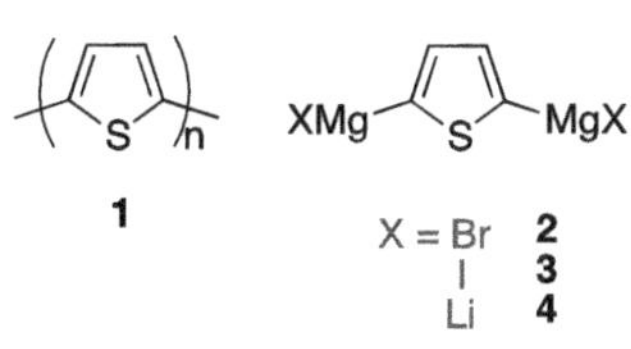

Chart 9.1

Table 9.1 Selected polymerization methods for polythiophenes.

Authors	*Catalyst*	*Synthetic Method*	*Conductivity*	*Yield*
Upadhyay[30]	F_3CCOOH	Acidic	n/a	60%
Yamamato[31]	$NiCl_2(bpy)$	Cat. C–C coupling	3.4×10^{-4}	61%
Dudek[32]	$Ni(AcAc)_2$	Cat. C–C coupling	0.1 ohm^{-1} cm^{-1}	70%
Wudl[33]	$NiCl_2(dppp)_2$	condensation polymerization	10 ohm^{-1} cm^{-1}	94.0
Sannicoló[34]	$CuCl_2$	Li-oxidative coupling	4.8 S cm^{-1}	25–50%

in the presence of ionic salts. IR, CV and electron microscopy data showed similar characteristics to chemical synthetic methods. High conductivities were achieved, yet insolubility remained an issue as with other synthetic methods for nonsubstituted polythiophenes.

9.1.2 Synthesis of Regioirregular Substituted Polythiophenes (PST)

The insolubility of nonsubstituted polythiophenes prompted researchers to expand the PT family by adding solubilizing groups to the parent monomer. It was first reported that adding a flexible side chain to the 3-position in the thiophene monomer significantly improved the solubility of the resultant polymer (see Figure 9.1).[39,40] Similar to the synthetic method of Wudl and workers for nonsubstituted PTs, Elsenbaumer *et al.* used Ni-catalyzed cross coupling polymerization for a variety of 3- and 4-alkyl substituted thiophene monomers. Solution-processable substituted PTs with various alkyl chain substituents were fully characterized, and a block copolymer consisting of poly(3-methylthiophene)-*co*-poly(3-butylthiophene) had the most promising conductivities (~ 50 ohm^{-1} cm^{-1}) as a freestanding film.[39]

This success brought on an onset of various synthetic methodologies for the synthesis of substituted PTs, particularly alkyl substitution at the 3-positions. Synthetic strategies similar to that for PT polymerizations were attempted with varied influence on the resultant polymers' properties.[41–47] For example, Kaeriyama and workers undertook electrochemical polymerization of 3-methylthiophene and 3-ethylthiophene to produce a series of poly(3-alkyl substituted thiophenes) (P3ATs) with variable conductivities.[41] Further improvement was achieved using $FeCl_3$-catalyzed reactions of alkyl-substituted thiophene monomers that gave processable P3ATs with high molecular weight compared to electrochemical methods and chemical coupling of halogenated thiophene derivatives.[44] It was shown that poly(3-decylthiophene) (PDT) made from this $FeCl_3$-catalyzed method gave processable free standing films with conductivities as high as 5 ohm^{-1} cm^{-1}. Despite the achievable high molecular weights and increased solubility obtained through the various synthetic procedures, it was noted that there was no regiocontrol over the polymerization and modeling studies prompted an effort to target regiochemistry in substituted PTs.

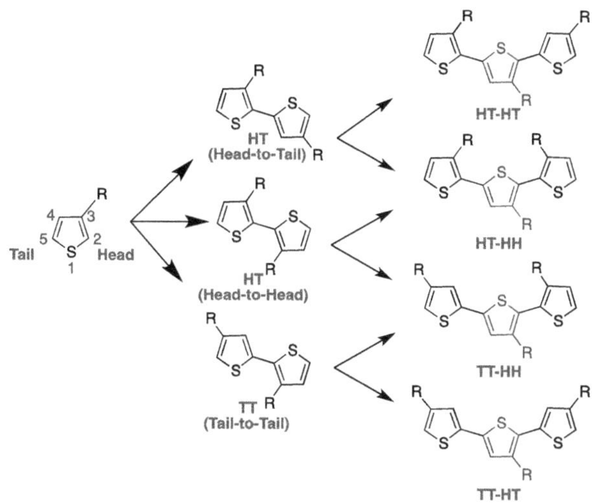

Figure 9.1 Description of regioisomers orientations and notations. Adapted from references 49 and 50.

Polythiophenes are typically coupled at the 2,5-position, with substitution normally at the 3- or 4-position (Figure 9.1). Due to the asymmetrical coupling of 3-substituted PTs, the coupling of two monomers can lead to a number of regiochemical isomers. Figure 9.1 shows two and three monomer combinations coupled together along with their current accepted notations. Typically, the sterically crowded 2-position is noted as the "head" and the 5-position noted as the "tail". These monomers can then be joined in different configurations known as dyads, such as head-to-head (HH), tail-to-tail (TT), and head-to tail (HT). The orientations can become more intricate when discussing tryads, wherein the notation between 3-substituted thiophene monomers can be noted as HT-HT, HT-HH, TT-HH and TT-HT configurations.[48–50]

The previous polymerization methods mentioned afforded poor regiochemical control, thus covalent linkages between subsequent monomers gave a high percentage of regioirregularity. It was observed that this nonspecific control of isomer couplings resulted in detrimental effects in the electronic properties of the polymer.[51,52] In particular, due to steric hindrance occurring in either HH or TT orientations, twisting and reduced planarity could alter the π-stacking ability, resulting in a loss of energy-related properties. Elsenbaumer and coworkers determined the regioirregularity of P3ATs using NMR experiments.[48,51] The regiochemistry was investigated by comparing the chemical environment of the ring substituents on neighboring monomers. These techniques observe the protons on the 4-position and α-methylene groups of the substituents with H^1 NMR and the carbons on the ring with C^{13} NMR were used to determine the regiochemical configuration. The HT linkage enables 3D macromolecular organization that promotes delocalization throughout the polymer backbone and enhanced interchain physics due to the organized self-assembly that results in efficient charge transport and ambipolar properties. This concept prompted an emphasis on research designed to develop synthetic procedures towards regioregular polythiophenes.

9.1.3 Synthesis of Regioregular Poly(Substituted Thiophenes) (*rr*-P3STs)

The synthesis of regioregular (*rr*-P3STs) was initially investigated using previously established Ni- and Pd- crosscoupling methods. One of the first successful attempts involved Kumada-style coupling that afforded regiospecific PATs with high molecular weights. This synthesis scope, known as the McCullough method, affords > 90% HT coupling. It involves the generation of 2-bromo-5-bromomagnesio-3-alkylthiophenes using $Ni(dppp)_2$ catalysts (Figure 9.2).[53,54] Subsequently, Poly(3-dodecylthiophene) (PDDT) synthesized using this synthetic method showed conductivities of up to 1000 S cm^{-1} compared to 20 S cm^{-1} reported for $FeCl_3$-catalyzed PDDT.[55] A modification of this reaction involved using $ZnCl_2$ as a replacement for the Grignard reagent $MgBr_2•Et_2O$, and examples showed control over the end-group functionalization of the polymer.[56] A summary of selected *rr*-P3ATs and their properties synthesized by this method is shown in Table 9.2.

Shortly after, the Rieke group reported upon the synthesis of *rr*-P3ATs using a Negishi-style coupling with organozinc reagents.[57–60] The novelty of this approach involved using a highly reactive zinc compound ("Rieke Zinc")[61,62] to treat 2,5-dihalothiophenes in order to generate the organometallic isomers whose yield ratio is dependent upon reaction conditions (Figure 9.2). The selective nature of both the catalysts and C–C coupling reactions gives highly favorable structurally homogenous P3ATs with > 98% HT orientations. Ni catalysts offered better regiochemical control over the polymerization compared to Pd catalysts. This was attributed to the smaller ionic radius and high steric bulkiness of the surrounding ligands.[57] Thus, dependent upon the desired regiochemistry, the use of either Ni or Pd could yield P3ATs with variable

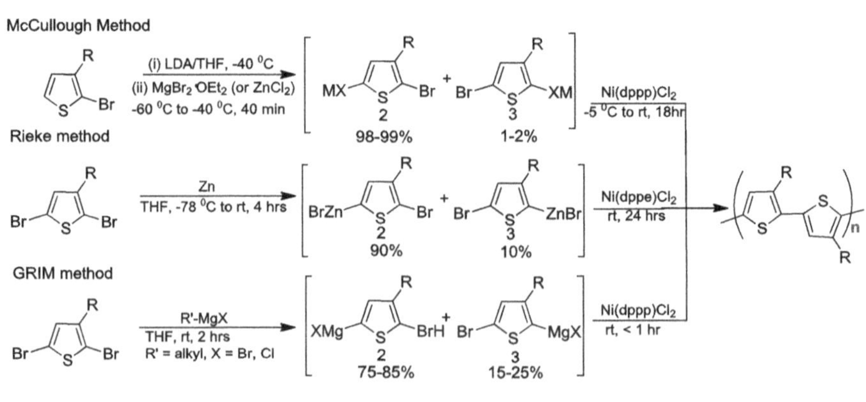

Figure 9.2 Common synthetic procedures for regioregular poly (3-alkyl thiophenes).

Table 9.2 Selected *rr*-side-chain-functionalized P3ATs and properties.

Polymer configuration 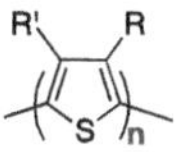	*Method*[a]	*Catalyst*	*Conductivity*	λ_{max}[e]	*Yield* %	M_n[b]	% *rr*[c] *HT*
R′=H, R=C_4H_9	A	Ni(dppp)Cl_2	100	438	69	– 24.4 K	93[54]
	B	Ni(dppe)Cl_2	–[d]	449	73		97[59]
R′=H, R=C_6H_{13}	A	Ni(dppp)Cl_2	200	442	36	10 K	98[54]
	B	Ni(dppe)Cl_2	–[d]	456	81	25.5 K	98.5[59]
	D	Pd(PPh_3)$_4$	–[d]	449	50	16.1 K	96[67]
R′=H, R=C_8H_{17}	A	Ni(dppp)Cl_2	200	446	65	24.6 K	97[54]
	B	Ni(dppe)Cl_2	–[d]	451	–[d]	34.6 K	98.5[59]
	E	Pd(OAc)$_2$	–[d]	448	55	27 K	96[68]
R′=H, R=$C_{10}H_{21}$	B	Ni(dppe)Cl_2	–[d]	447	–[d]	30.5 K	98.5[59]
R′=H, R=$C_{12}H_{25}$	A	Ni(dppp)Cl_2	1000 S cm^{-1}	450	33	11.6 K	91[53]
	B	Ni(dppe)Cl_2	–[d]	453	–[d]	30.2 K	98.5[59]
	C	Ni(dppp)Cl_2	–[d]	565	65	35 K	≥99[63]
R′=H, R=SC_8H_{17}	B	Ni(dppe)Cl_2	600 S cm^{-1}	510	–[d]	4.4 K	> 90[60]
R′=H, R= 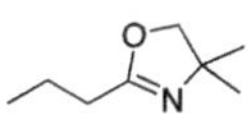	D	Pd(PPh_3)$_4$/ CuO	–[d]	–[d]	80	–[d]	90[66]
R′=H, R= 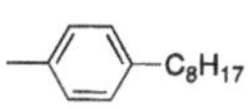	F	$FeCl_3$	4 S cm^{-1}	493	80	23 K	94[70]
R′=CH_3, R= (CH_2CH_2O)$_m$$CH_3$	F	$FeCl_3$	–[d]	550	33	22.5 K	95[73]
R′=H, R = 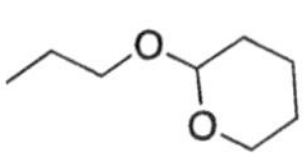	A	Ni(dppp)Cl_2	–[d]	439	38	10.7 K	–[74]
R′=H, R= 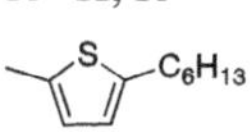	A	Ni(dppp)Cl_2	–[d]	465	36	55 K	90[75]
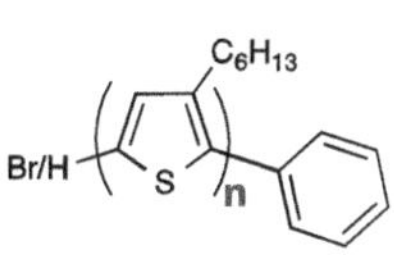	G	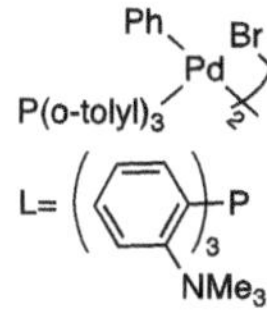	–[d]	–[d]	99	9 K	99[76]
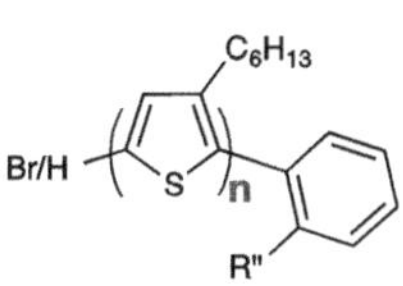	H	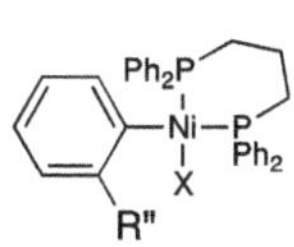	–[d]	–[d]	90	9.8 K	99[77]

[a]Method (A) McCullough Method, (B) Rieke Method, (C) Grim Method, (D) Stille Coupling, (E) Suzuki Coupling, (F) $FeCl_3$-catalyzed coupling, (G) Direct arylation, (H) Ni-catalyzed coupling.
[b]Determined by GPC in THF of $CHCl_3$.
[c]Determined by ^{1}H NMR.
[d]Not reported.
[e]Determined by UV-Vis.

degrees of regioregularity. High conductivities (10^{-6} units for undoped P3ATs) and small bandgaps (1.7 eV) were reported, compared to that of regioregular P3ATs (Table 9.2).

In 1999, a much more economically feasible approach to *rr*-P3ATs was developed. The innovation involved the use of commercially available and inexpensive Grignard reagents that could selectively couple to the dibromothiophene monomers while eliminating the need for harsh highly reactive metals and cryogenic temperatures.[63,64] This Grignard metathesis method (GRIM) occurs through a metal–halogen exchange to form a mixture of reactive monomer isomers in 85 : 15 ratios (Figure 9.2). The use of the Ni(dppp)Cl_2 catalysts selectively reacts with the 2-bromo-5-bromomagnesio-3-alkylthiophene isomer yielding *rr*-P3ATs. Polymerization and subsequent purification resulted in *rr*-P3ATs with 96–100% HT couplings, molecular weights in the range of 20–35 K and polydispersities of 1.2–1.4.

In addition to the McCullough, Rieke and GRIM methods, other methods have been used towards the synthesis of structurally homogenous *rr*-P3ATs. Stille Pd-catalyzed crosscoupling reactions were used with a variety of heterocyclic halides and halobenzenes.[65–67] The addition of a coreagent such as CuO was found to improve reaction times and yields. It was noted that the Ni catalyst could coordinate with the oxaline protective group during the synthesis of poly(3-propionic acid thiophene) resulting in low yields of < 5% and molecular weights under 6 K.[66] The use of Stille conditions with CuO reagents as adapted by Gronowitz *et al.*[65] provided desired results. In addition, the well-known Suzuki crosscoupling reaction was employed for the synthesis of *rr*-P3ATs.[68,69] Similar to the Stille methods, this reaction tends to be of greater advantage when additional functional side groups are present on the polymer. In the presence of palladium (II) acetate, greater than 96% HT couplings were achieved for poly(3-octylthiophene). Anderson *et al.*[70,71] and Bongini and coworkers[72] also reported the regioselective polymerization of P3ATs using $FeCl_3$. The reaction conditions involved slow addition of the $FeCl_3$ that limited the Fe^{3+}/Fe^{2+} ratio allowing for a more selective polymerization. A range of phenyl-substituted polythiophenes was studied and a few resulted in high regioregularity (>90%) and were attributed to the interaction of the π-overlap of the substituted group to the polythiophene chain.

More recently, regioregular 3-substituted polythiophenes with added functionalities including liquid-crystalline and/or chiral side chains have been synthesized with modified traditional procedures.[78–82] Kijima and coworkers synthesized *rr*-poly(2-phenylnaphthalene polythiophenes) (Chart 9.2) **5** using two different polymerization methods that resulted in a change in the polymer's properties.[78] The two methods, dehalogenative polycondensation using Ni(cod)$_2$ and Ni-catalyzed chain-growth polymerization gave significant differences in solubility, UV-Vis absorption and self-assembly. A modified McCullough method was used to synthesize **6** with an M_n of 5.8 K and a 24.5% yield.[81] This achiral, regioregular (poly[3-{4-((R)-4-ethyl-2-oxazolin-2-yl)phenyl}thiophene]) polymer formed an optically active macromolecular complex in the presence of chiral amines or chiral PTs.[81]

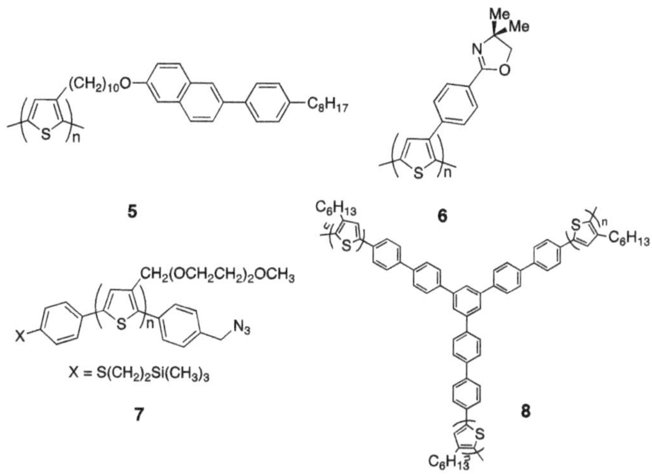

Chart 9.2

In addition, regioregular substituted polythiophenes bearing end-group functionalities have been synthesized using *in situ* methods *via* catalyst modifications.[83–87] Suzuki–Miyaura type polymerizations of 2-bromo-5-(4,4,5,5-tetramethyl-1,3-di-(2-dioxaborolane-2-yl))-3-hexylthiophene with end-capped 4-(azidomethyl) phenylboronic acid pinacol ester groups gave benzyl azide-functionalized poly(3-hexylthiophenes) **7** that could serve as a building block for polymer or hybrid macromolecular structures.[86] In another report, Kumada-style catalyst transfer polymerization was used to externally initiate Y-shaped macromolecular architectures of *rr*-P3HTs.[87] Trifunctional complex-based initiators enabled the synthesis of C_6H_3-$(C_6H_4$-$C_7H_6)_3$-tethered *rr*-P3HT **8** that can pave the way for multidimensional polymer architectures that could be of interest in a variety of '*organic electronic*' applications.

9.2 Experimental Procedures

This section will describe some of the synthetic methods of selected PTs and PSTs. Examples of the different synthetic routes that are used to achieve thiophene-based polymers with variable properties will be noted. The latter examples will also include monomer synthesis as this leads into the synthetic path chosen in many of the cases. The experimental procedures and characterization details will be included.

9.2.1 Nonsubsitiuted Polythiophenes

Example 9.1: Metal-Catalyzed Polycondensation Polymerization of Poly(2,5-thienylene)

A mixture of a 1/1 molar ratio of 2,5-dibromothiophene and magnesium in THF was prepared and a solution of metal acetylacetonate dissolved in a small

Table 9.3 Polymerization of 2,5-dibromothiophene using different metal catalysts.

Exp	*2,5-dibromothiophene (mmole)*	M_g *(mmole)*	*Catalyst (mmole)*	*Reaction time (h)*	*Yield (g)*
1	24.8	24.7	$Ni(AcAc)_2$ (0.098)	4.1	.919
2	28.8	25	$Fe(AcAc)_2$ (0.187)	2.8	.374
3	24.8	24.9	$Co(AcAc)_2$ (0.0886)	3.3	.151
4	21	21	$NiCl_2(bpy)$ 0.07	3	61

amount of tetrahydrofuran (THF) was added. The addition of catalyst caused an immediate formation of a black product. The reaction mixture was then heated for several hours and precipitated into 500 ml of a 0.2 M HCl/MeOH solution. The polymer then underwent repetitive washings with cold methanol and water until free of halide ions resulting in a clear solution. The resultant black product was dried at 40 °C under vacuum. Table 9.3 lists details of the different metal catalysts and reaction equivalents.

Example 9.2: Oxidative Coupling Polymerization of Poly(2,5-thienylene)

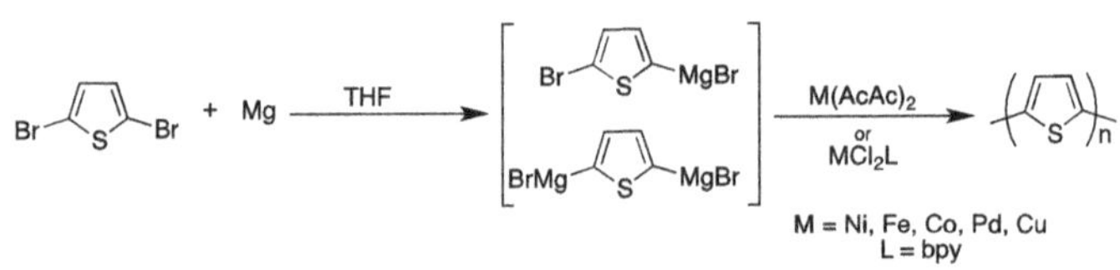

(9.1)

Two equivalents of BuLi (1.6 M solution in hexane) was added to a dry ether solution under N_2 at −10 °C; slight precipitation of the bis-lithiated bithiophene species occurred during the reaction. A slurry of $CuCl_2$ (2 equivalents) in dry anisole was added slowly into the reaction and the reaction temperature gradually raised to 100 °C, while the Et_2O was distilled off. After overnight reaction, the mixture was quenched with water to yield a brownish precipitate, which was collected by precipitation. The precipitate was then dissolved in concentrated HCl, washed with aqueous NH_3 solution and then rinsed with water. The polymer then underwent Kumagawa extraction with MeOH and $CHCl_3$ for 48 h, respectively.

Example 9.3: Electrochemical Polymerization of Poly(2,5-thienylene)

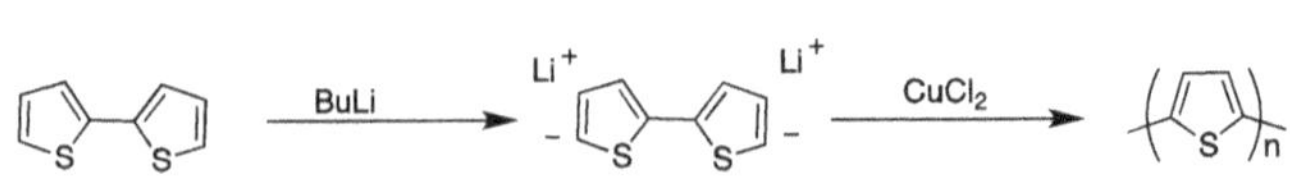

(9.2)

A solution of 0.01 M thiophene and 0.1 M $TEABF_4$ in acetonitrile was degassed using argon. In a 3-electrode, single-compartment cell, polythiophene films were grown on a platinum or other metal working electrode with a gold counterelectrode and sodium chloride saturated calomel reference electrode. Electrolyte solutions using different salts were also used.

9.2.2 Syntheses of Regioirregular Substituted Polythiophenes

Example 9.4: Ni-Catalyzed Crosscoupling Reactions of Regioirregular P3ATs using 2,5-Diiodo-alkylthiophenes

Alkylthiophenes were made according to Tamao *et al.* and distilled under N_2.[88] 3-hexylthiophene was synthesized by coupling the Grignard reagent of 1-bromohexane (70 mmol) to 3-bromothiophene (50 mmol) using Ni(dppp)Cl_2 (0.3 mmol) as catalyst in dry ether. The reaction was refluxed for 6 h, quenched with MeOH and then distilled to give the final product in 67% yield. The distilled product was then added to a solution of iodine (I_2) in 18 mL of $CHCl_3$, 3 mL of conc. HNO_3 and 3 mL of H_2O and refluxed for 16 h. The diodo-3-hexyl monomer was purified through a silica gel column with hexanes with a yield of 71%. Magnesium metal (24 mmol) was added to a solution of 2,5-diiodo-3-hexylthiophene (10 mmol) in dry ether. After adding 10 mmol of 2,5-diiodo-3-hexylthiophene, the reaction was allowed to proceed with the addition of Ni(dppp)Cl_2. The polymerization was quenched after 24 h with MeOH/H_2O and then poured into 5% HCl/MeOH to precipitate. Purification occurred through multiple washings with 5% HCl/MeOH solutions (50% yield).

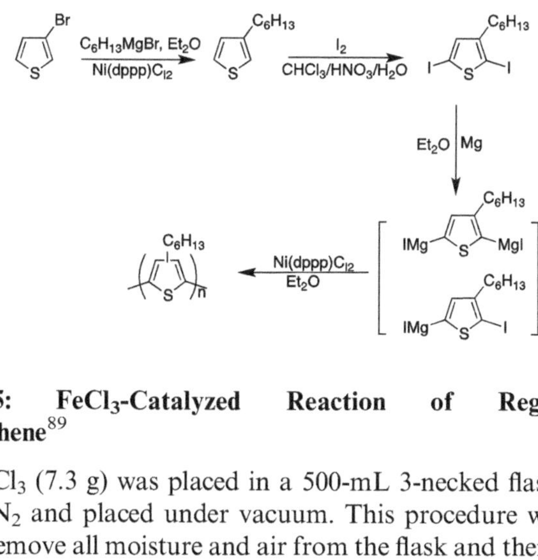

(9.3)

Example 9.5: FeCl$_3$-Catalyzed Reaction of Regioirregular 3-Octadecylthiophene[89]

Anhydrous $FeCl_3$ (7.3 g) was placed in a 500-mL 3-necked flask and heated to 100 °C under N_2 and placed under vacuum. This procedure was repeated two more times to remove all moisture and air from the flask and then allowed to cool under N_2 pressure. $CHCl_3$ (75 mL) was added to the flask and 0.015 moles of 3-octadecylthiophene was added dropwise and stirred for 2 days at room temperature. The viscous reaction mixture was then precipitated into MeOH, filtered and washed repeatedly with MeOH and H_2O. Soxhlet extractions with MeOH and acetone then drying under vacuum at 60 °C yielded a brown polymer (93%).

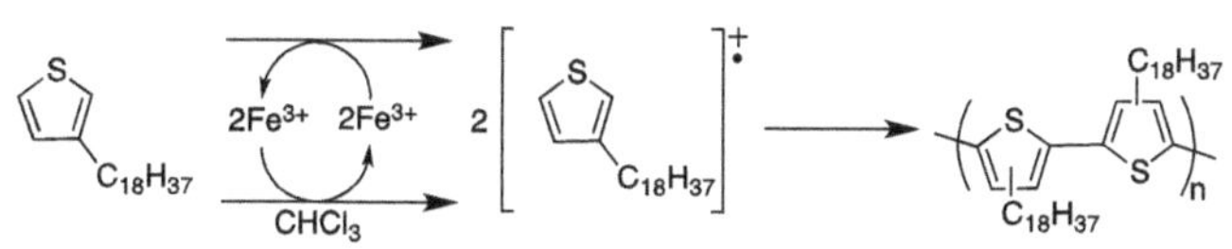

(9.4)

Example 9.6: Heck Crosscoupling of Regioirregular Poly(3-dodecylthienylene vinylenes)[90]

To a 500-mL flask, 93.41 mmol of 2-bromo-3-dodecylthiophene was added to 250 mL of $CHCl_3$. Iodine (51.4 mmol) and iodobenzene diacetate (56 mmol) were added and the reaction allowed to stir at room temperature for 1.5 h. The solution was poured over 150 mL of aqueous sodium thiosulfate, the aqueous layer washed with 150 ml of $CHCl_3$ and the organic layer dried over sodium sulfate. Solvent was removed *in vacuo* and the iodobenzene byproduct removed by Kugelrohr distillation at 60 °C and 0.1 Torr. The final product was filtered through a silica plug using hexanes as an eluent and removed *in vacuo* yielding 41.25 g of 2-bromo-3-dodecylthiophene. The 2-bromo-3-dodecyl-5-iodothiophene product (13.7 mmol) was dissolved in 60 mL of toluene and the solution sparged with N_2 for 20 min. $Pd_2(dba)_3$ (2.5 mol%) and Ph_3P (10 mol%) was added and the system placed in a 60 °C oil bath. Tributyl(vinyl)tin was then added (14.2 mmol) and the solution stirred for 6 h. The reaction was cooled and filtered through a silica plug using toluene. The filtrate was washed with 100 mL of water and the organic layer dried over sodium sulfate, filtered and rotovapped to yield 2-Bromo-3-dodecyl-5-vinylthiophene (88%). The 2-Bromo-3-dodecyl-5-vinylthiophene monomer (7.56 mmol) was added to a solution of tributylamine (1.3mL) in 100 mL of NMP and sparged with N_2 for 20 min. The catalysts $Pd(OAc)_2$ and $P(o\text{-tol})_3$ were added and the reaction placed in a 100 °C oil bath and stirred for 12 h. The reaction was quenched by pouring over 300 mL of methanol, collected by filtration through a Soxhlet thimble and extracted with methanol, hexanes, and chloroform using the Soxhlet procedures to obtain poly(dodecylthiophene vinylenes) in 24% yield.

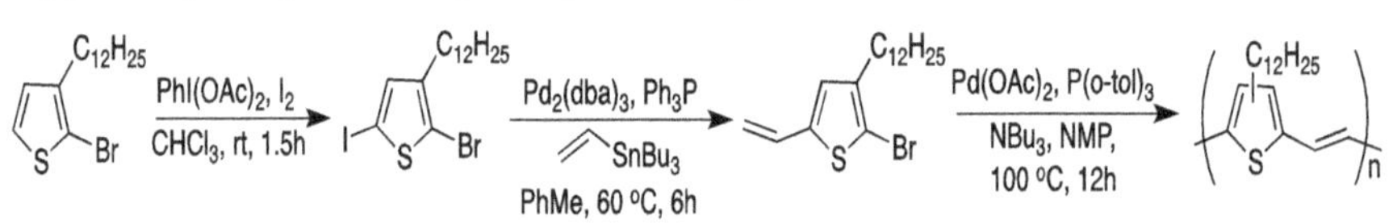

(9.5)

Example 9.7: $ZnCl_2$-Catalyzed Coupling of 3, 3′-(2,2′-Dimethoxyethoxy)-bithiophene[47]

Seven equivalents of 3, 3′-(2,2′-Methoxyethoxy) bithiophene monomer and 10 equivalents of dry LiCl were added to THF in a Schlenk flask (60 mM) and the Schlenk flask was submerged in a sonicator bath at 0 °C. Two equivalents of BuLi was added dropwise *via* syringe and the reaction sonicated for 30 min while the temperature was raised to 20 °C. 2.2 equivalents of $ZnCl_2$ was dissolved in a minimal amount of THF and added to the reaction. The Schlenk flask was removed from the sonicator bath and stirred magnetically. One equivalent of 2,5-dibromo-3-decylthiophene and 3 mol% $Pd(PPh_3)_4$ were dissolved in a small amount of THF and the solution added to the Schlenk flask, and the reaction mixtures heated to reflux. Removal of THF during the reaction was accomplished *via* a dry argon purge until a final concentration of

80 mM was obtained. After the concentrated solution refluxed overnight, it was precipitated in MeOH, filtered, rinsed repeatedly with MeOH, and dried overnight under reduced pressure (98% yield).

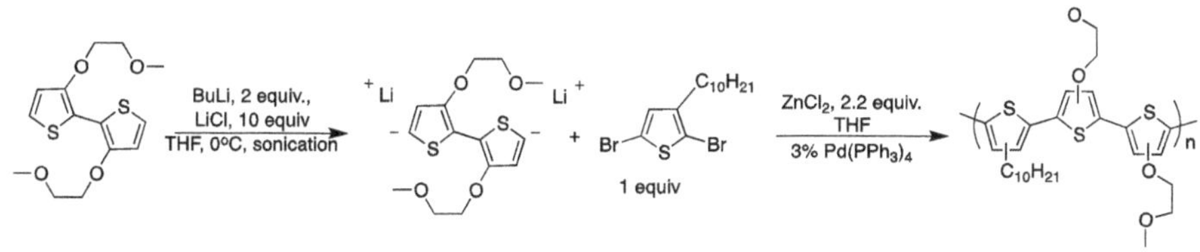

(9.6)

9.2.3 Syntheses of Regioregular Substituted Polythiophenes

Example 9.8: Synthesis of Poly(3-alkylthiophenes) (PAT) Using the McCullough Method[54]

Diisopropylamine (15 mmol) was added to 75 ml of freshly distilled THF (0.2 M) in a dry round-bottomed flask. 2.5 M *n*-BuLi (15 mmol) was then added to the flask at RT and the reaction cooled to –40 °C and stirred for 40 min. The reaction was then cooled to –78 °C and 2-bromo-3-alkylthiophene (15 mmol) added. The temperature was gradually increased to –40 °C and stirred for 40 min. Before addition of $MgBr_2{\bullet}Et_2O$ (15 mmol), the reaction was cooled to –60 °C and stirred for 20 min. The temperature was brought up to –40 °C and stirred for an additional 15 min. The reaction was then slowly allowed to warm up to –5 °C and Ni(dppp)Cl_2 (0.072 mmol) was added. The reaction was allowed to warm up to RT, and stirred overnight (12–18 h). The reaction mixture was quenched by pouring into MeOH (~300 mL), the solid collected by filtration was washed with MeOH, H_2O and MeOH again. The product then underwent Soxhlet extractions using MeOH and hexanes for purification. The polymer was extracted with chloroform; subsequent evaporation of $CHCl_3$ yielded 36% of the corresponding poly(alkylthiophene) (Table 9.4).

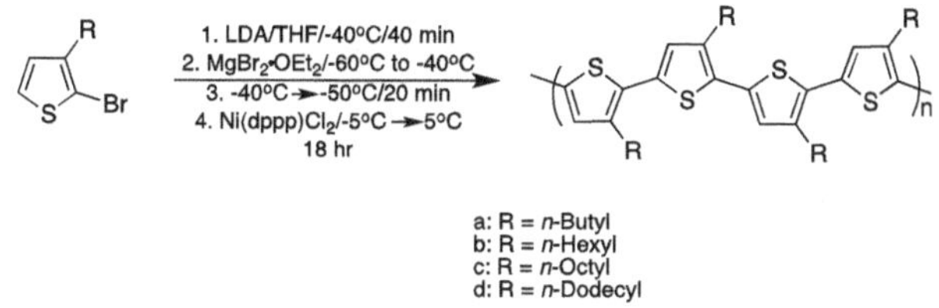

(9.7)

Example 9.9: Synthesis of Poly(3-alkylthiophene Substituted Thiophene) using Modified $ZnCl_2$ McCullough Method[75]

5-Bromo-5″-hexadecyl-[2,2′;3′,2″]terthiophene (**3TC16Br**) was synthesized according to reference 75. To a round-bottomed flask, **3TC16Br** (.906 mmol, 1 equiv.) was added to 10 mL of THF and cooled to –55 °C. LDA (0.2 M in THF, 0.861 mmol, 0.95 equiv.) was added slowly and the reaction stirred for

Table 9.4 Polymerization of P3ATs using the McCullough method.

R Group	$M_w{}^a$	*PDI*[a]	*% RR*[b]	$\lambda_{max}{}^c$	*Yield %*
n-Butyl	–[d]	–[d]	93	438	69
n-hexyl	10 K	1.6	98	442	36
n-octyl	24 K	1.98	97	446	65
n-dodecyl	–[d]	–[d]	95	450	33

[a]Determined by GPC (eluent $CHCl_3$ or THF, polystyrene standards).
[b]Determined by ^1H NMR.
[c]Determined by UV-Vis.
[d]Not reported.

1 h. $ZnCl_2$ (1.05 equiv.) was then added to the reaction and stirred for 15 min before being brought to RT. The reaction was stirred for 30 min at RT, followed by addition of Ni(dppp)Cl_2 (0.045 mmol, 5 mol%). The reaction was allowed to proceed for 10 min followed by quenching with MeOH. The polymer was washed repeatedly with MeOH and dried under vacuum to yield 36%.

Example 9.10: Rieke Synthesis of Poly(3-alkylthiophenes)[57]

10.0 mmol of the appropriate 2,5-dibromo-3-alkylthiophene was dissolved in 20 mL of THF and added to a freshly prepared active Rieke Zinc (Zn*) catalyst solution (11 mmol in 40 mL of THF) *via* cannula (see reference 59 for preparation of Zn*) at –78 °C. The reaction was stirred at –78 °C for an hour and then gradually raised to 0 °C, after which 0.2 mol% of Ni(dppe)Cl_2 (0.02 mmol) in 20 mL of THF was added by cannula while maintaining a 0 °C temperature. The reaction was allowed to stir for 24 h; a dark purple precipitate formed over the course of the reaction. The reaction was quenched by pouring into a beaker of 100 mL of MeOH and 50 mL 2N HCl, the solid isolated and the filtered product washed repeatedly with MeOH/HCL and dried. Purification of the polymer by Soxhlet extractions with MeOH, then hexanes afforded <67% yield (Table 9.5).

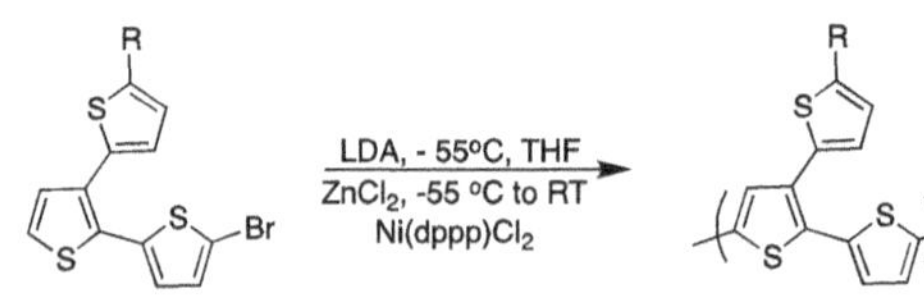

(9.8)

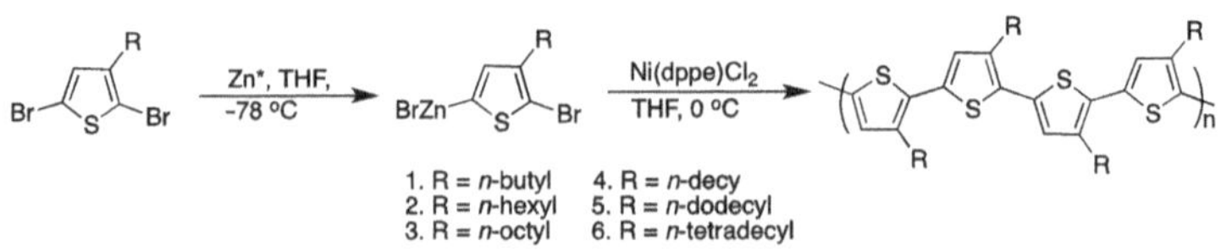

(9.9)

Example 9.11: Grignard Metathesis of Regioregular Poly(3-alkylthiophenes)[64]

4.33 mmol of 2,5-Dibromo-3-hexylthiophene was added to 25 mL of THF (0.17 M solution) in a round-bottomed flask. Isopropylmagnesium bromide (4.13 mmol, 1.0 M solution in diethyl ether) was then added to the reaction and heated to reflux for an hour. Ni(dppp)Cl_2 (0.5 mol %) was added and the

Table 9.5 Polymerization of P3ATs using the Rieke method.

R Group	M_w[a]	*PDI*[a]	*% RR*[b]	λ_{max}[c]	*Yield %*
n-Butyl	insoluble	insoluble	97	449	80
n-hexyl	25.5 K	1.48	≥98.5	456	82
n-octyl	34.6 K	1.13	≥98.5	451	79
n-decyl	30.5 K	1.39	≥98.5	447	77
n-dodecyl	34.7 K	1.42	≥98.5	453	71
n-tetradecyl	30.2 K	1.46	≥98.5	450	67

[a]Determined by GPC (eluent $CHCl_3$ or THF, polystyrene standards).
[b]Determined by ^1H NMR.
[c]Determined by UV-Vis.

reaction proceeded for 45 min. The reaction was quenched by pouring into 120 mL of MeOH and then filtered through a Soxhlet thimble, which was then subjected to Soxhlet extractions with methanol, hexanes, and chloroform. The polymer was recovered from the chloroform fraction by evaporation and dried under vacuum to yield 71% of polymer.

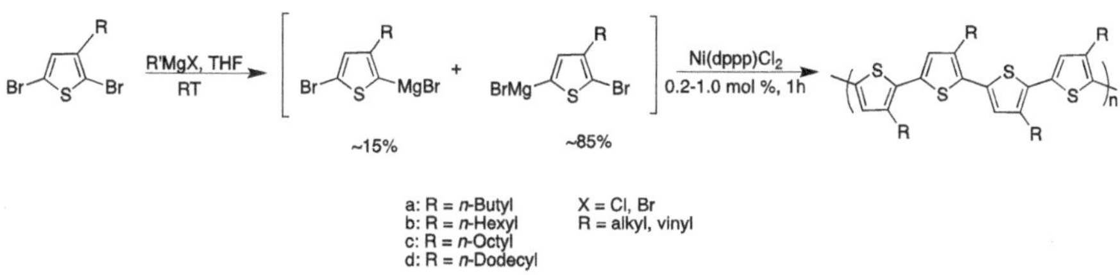

(9.10)

Example 9.12: CuO-Modified Stille Pd^0-Catalyzed Reaction of 3-(2(4,4-Dimethyloxazolin-2-yl)ethyl)thiophene[66]

2-Bromo-3-(2(4,4-dimethyloxazolin-2-yl)ethyl)thiophene (40 mmol), TMEDA (44 mmol) and anhydrous Et_2O (100 mL) were added to a round-bottomed flask under Ar and cooled in a salt water/ice bath. LDA (40 mmol) was added and the reaction allowed to warm to RT, followed by heating to reflux for an hour. The reaction was then placed in an ice/salt water bath and 1.0 M Me_3SnCl (45 mmol) in THF was added slowly over 10 min. The reaction mixture was allowed to stir for 45 min, brought to RT and stirred overnight under Ar. The reaction was quenched with 50 mL of aqueous NH_4Cl, the organic and aqueous solutions separated, and the organic phase washed with water and brine. The organic phase was then dried over $MgSO_4$, filtered and solvent removed *in vacuo* to yield a clear oil (88% yield). 5-trimethylstannyl-2-Bromo-3-(2(4,4-dimethyloxazolin-2-yl) ethyl)thiophene (3 mmol) was added to a slurry of $Pd(PPH_3)_4$ (5 mol%) and CuO (3 mmol) in 100 mL of DMF. The reaction was stirred under Ar at 110 °C for 16 h, cooled to RT, and quenched by pouring into MeOH (~300 mL). The polymer was then filtered and washed repeatedly with MeOH and H_2O. Purification was done by Soxhlet extractions of methanol and hexanes, dissolved in chloroform and evaporated to give a yield of 50%.

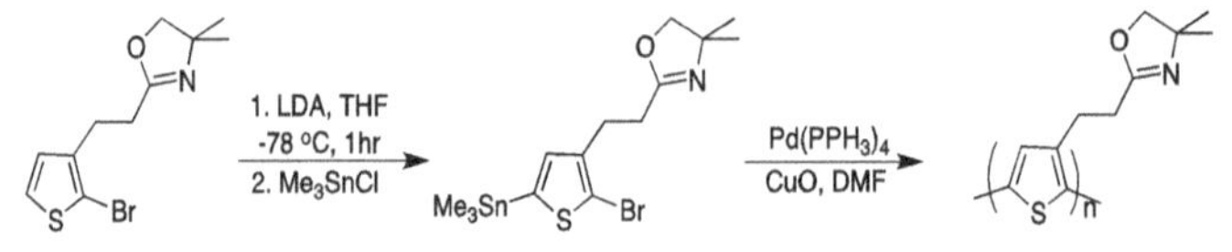

(9.11)

Example 9.13: Regioregular Poly(3-Octylthiophene) *via* Suzuki Coupling Reaction[68]

2-Iodo-3-octylthiophene was added dropwise to a cooled solution of LDA (12 mmol) in anhydrous THF (–78 °C) and stirred for 30 min. The reaction was then brought to –40 °C and stirred for an additional 40 min, followed by the addition of $B(OMe)_3$ (45 mmol). The reaction was then warmed to RT and poured onto 200 mL of iced HCL (1M) and extracted with ether 3 times. The combined ether extract was washed with water and brine and dried of Na_2SO_4 in the presence of 2,2-dimethyl-1,3-propanediol and stirred at RT for 30 min. The mixture was filtered, the filtrate evaporated and the product purified by gravity chromatography with hexanes and then eluted with $CHCl_3$ (70% yield). [1′,3′-(2,22′-dimethylpropylene)]-2-iodo-3-octyl-5-thienylboronate (0.58 mmol) and 1 mol% of $Pd(OAc)_2$ were added to a degassed (60/2 vol./vol.) mixture of THF and ethanol in a round-bottomed flask under Ar. The reaction was stirred for 10 min, followed by addition of K_2CO_3 (1.5 equiv) and immediately followed by 1 mL degassed water. The reaction was sparged with Ar and refluxed overnight. The flask was cooled to RT, quenched by pouring into 250 mL MeOH and filtered through a Soxhlet thimble. The product was then Soxhlet extracted with methanol, hexanes, and chloroform. The polymer was recovered from the $CHCl_3$ fraction *via* evaporation to yield 55% polymer.

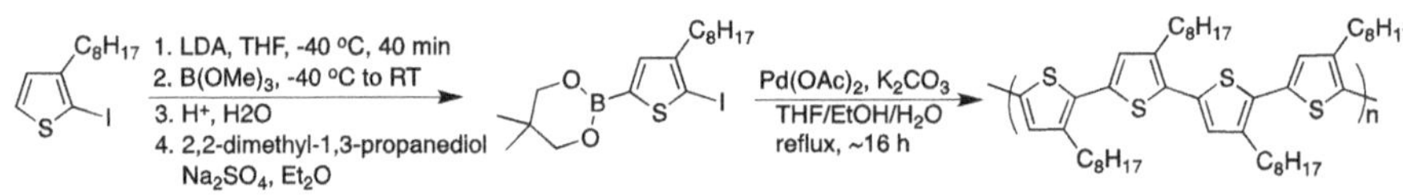

(9.12)

Example 9.14: Synthesis of *In Situ* Heterotrimethyl Silyl/Azide End-Capped Poly(alkylthiophenes) using Modified Suzuki–Miyaura Coupling

The modified Pd catalyst **1** was synthesized according to the procedure reported in reference 86. 2-bromo-5-(4,4,5,5-tetramethyl-1,3-2-dioxaborane-2-yl)-3-hexylthiophene[86] **2** (0.43 mmol) and the modified Pd catalyst **1** (2 mol%) was dissolved in 5 mL anhydrous THF in a rubber septum-capped vial in a glovebox. The vial was removed from the glovebox and 2 mL 2M K_2CO_3 was added to the reaction and stirred for 1 h at RT. 4-(azidomethyl) phenylboronic acid pinacol ester[86] **3** was added to the reaction and stirred overnight. The polymer was quenched in MeOH, washed with acetone (100 mL) and hexanes (100 mL). The polymer was washed with THF until the filtrate was clear and then dried under vacuum. The polymer was purified using column chromatography using toluene to give a yield of 83%.

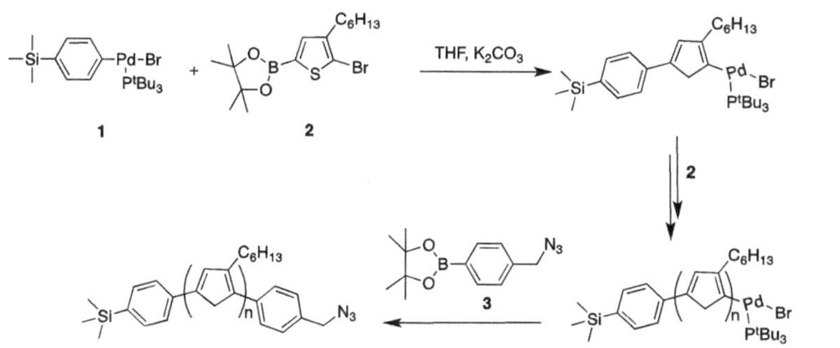

(9.13)

Example 9.15: Synthesis of Regioregular Phenyl-Substituted Polythiophenes with $FeCl_3$[71]

The phenyl-substituted thiophene monomers were synthesized according to Gronowitz[91] The desired monomer (7.68 mmol) was dissolved in 40 mL freshly distilled $CHCl_3$ and stirred. A suspension of $FeCl_3$ in 100 mL freshly distilled $CHCl_3$ was stirred for 30 min and slowly added to the monomer solution over a period of 2 to 6 h at RT, followed by stirring for another hour. The ratio of monomer : $FeCl_3$ should be maintained at 1:6 in the final solution. The reaction was quenched by precipitation into MeOH, filtered and washed with MeOH. Dedoping of the polymer occurred by dissolving in a solution of $CHCl_3$, concentrated NH_3 and 0.05 M EDTA. The solution was then filtered, concentrated and then precipitated into MeOH. Purification by Soxhlet extraction with diethyl ether yields 80% of polymer (Table 9.6).

Table 9.6 Selected polymerization characteristics of $FeCl_3$-catalyzed *rr*-phenyl-substituted PTs.

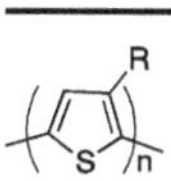 R =	$M_n{}^a \times 10^{-3}$ *g/mol*	$M_w{}^a \times 10^{-3}$ *g/mol*	%*RR*[b]	$\lambda_{max}{}^c$
3-4-octylbenzene	13	35	94	482
3-4-(heptyloxy)benzene	1.8	47	low	408
3-4-methoxyethoxy)ethoxy)benzene	7.8	25	88	476
3-4-1-phenyloctan-1-one	16	36	85	454
3-4 -(4-ethylhexyl)oxybenzene	1.5	2.4	low	424
3-3-(4-ethylhexyl)oxybenzene	2.6	5.4	low	–
3-2-(2-ethylhexyl)oxybenzene	19	150	92	539
3-2,5 dioctylbenzene	28	63	high	530
3-2-methoxy-4-octylbenzene	18	40	90	470
3-2,5-methoxyethoxy)ethoxy)benzene	21	135	high	519
3-2,5-(heptyloxy)benzene	2.4	7.1	–	–

[a]Determined by GPC (eluent THF, polystyrene standards).
[b]Determined by ^1H NMR.
[c]Determined by UV-Vis.
[d]Not determined.

9.2.4 General Procedures and Characterization

Many of the simple monomers can be purchased commercially. Facile Grignard coupling affords the 3-substitued thiophene derivatives, which can be purified by distillation procedures. The more complex monomers or halide derivatives are synthesized *via* various techniques; dependent upon the type of synthetic method and are usually noted in the polymerization reaction schemes. The most common polymerization techniques involve anhydrous solvents, with stable catalysts. THF is the most commonly used solvent for both substituted and nonsubstituted polythiophenes.

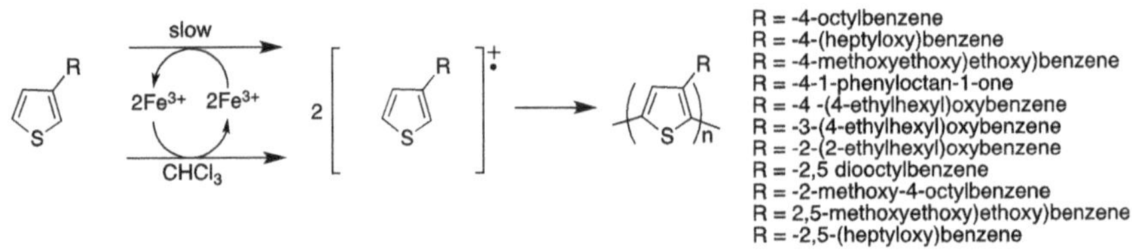

(9.14)

Quenching of polymers is traditionally done by pouring into MeOH, followed by Soxhlet extractions with MeOH and hexanes. Sensitive substituted groups capable of cleaving in hot solvents forgo this procedure for repeated cold washings with MeOH until the filtrate is clear. The MeOH is used to deactivate the catalysts and remove impurities, while the hexanes extractions are normally used to remove low molecular weight product. Typical contaminants can include inorganic salts, phosphines, or other ligands, unreacted monomer and oligomers that may be detrimental to the polymers electronic properties. The polymer is typically dissolved with $CHCl_3$, followed by evaporation to yield the polymer.

Gel permeation chromatography (GPC) is used to determine the degree of distribution, number and weight-average molecular weight (M_n and M_w), while ^1H NMR determines the more accurate degree of polymerization (DP_n). The dyads and tryads (see Figure 9.1) enable quantitative analysis to the upfield shift of the HH configurations and is used to determine the regioregularity of the polymer. Other characterization techniques such as UV-Vis, AFM and thermal analyses are used to probe structure property relationships and morphology that are typical in polymer studies.

9.3 Conclusion and Outlook

This chapter showcases the synthetic strategies for both nonsubstituted and substituted polythiophenes. To date, many of these synthetic procedures are still being used for traditional polymerizations of PTs and PSTs, while variations in the synthetic methodology are dependent upon the monomer configuration or end-group functionality of the polymer. For instance, modifications to the catalyst, particularly ligand exchanges in Pd and Ni complexes can alter the regiochemistry of the resultant polymer or provide polymers bearing specific end-capped functionality; while other synthetic routes employed, offer new generations of PT-based materials that can advance the '*organic electronics*' era. The examples mentioned in the chapter should inspire the creation of new or modification of catalysts and/or procedures to ensure the increased feasibility of polythiophenes as one of the most promising semiconductor materials for electronic devices.

References

1. H. Sirringhaus, M. Bird, T. Richards and N. Zhao, *Adv. Mater.*, 2010, **22**, 3893–3898.
2. Pang, Yi. (ed) (2010) *Design and Synthesis of Conjugated Polymers*, Wiley-VCH Verlag GmbH & Co. KGaA, Weinheim.
3. M. J. Malachoski and J. Żmija, *Opto-Electron. Rev.*, 2010, **18**(2), 121–136.
4. B. de Boer and A. Facchetti, *Polym. Rev.*, 2008, **48**(3), 423–431.
5. G. Horowitz, *Adv. Mater.*, 1998, **10**, 365–377.
6. Y. Guo, G. Yu and Y. Liu, *Adv. Mater.*, 2010, **22**, 4427–4447.
7. A. Fachetti, *Chem. Mater.*, 2011, **23**, 733–758.
8. A. Pron and P. Rannou, *Prog. Polym. Sci.*, 2002, **27**, 135–190.
9. Y. Park and R. C. Advincula, *Chem. Mater.*, 2011, **23**, 4273–4294.
10. C. Li and Z. Bo, *Polymer*, 2010, **51**, 4273–4294.
11. C. R. McNeill and N. C. Greenham, *Adv. Mater.*, 2009, **21**, 3840–3850.
12. F. Laquai, Y-S. Park, J-J. Kim and T. Basche, *Macromol. Rapid Commun.*, 2009, **30**, 1203–1231.
13. J. Roncali, *Acc. Chem. Res.*, 2009, **42**, 1719–1730.
14. Brabec, C., Dyakonov, V., and Scherf, U. (ed.) *Organic Photovoltaics*, Wiley-VCH GmbH & Co. KGaA, Weinheim, 2008.
15. J. Weickert, R. B. Dunbar, H. C. Hesse, W. Wiedemann and L. Schmidt-Mende, *Adv. Mater.*, 2011, **23**, 1810–1828.
16. C. J. Brabec, M. Heeney, I. McCulloch and J. Nelson, *Chem. Soc. Rev.*, 2011, **40**, 1185–1199.
17. M. O'Neill and S. M. Kelly, *Adv. Mater.*, 2011, **23**, 566–584.
18. X. Zhan and D. Zhu, *Polym. Chem.*, 2010, **1**, 409–419.
19. S. W. Thomas III, G. D. Joly and T, M. Swager, *Chem. Rev.*, 2007, **107**, 1339–1386.
20. D. T. McQuade, A. E. Pullen and T. M. Swager, *Chem. Rev.*, 2000, **100**, 2357–2574.
21. T. M. Swager, *Acc. Chem. Res.*, 1998, **31**, 201–207.
22. K. Johno, Y. Tanaka, T. Koike and M. Akita, *Dalton Trans.*, 2011, **40**, 8089–8091.
23. J. T. Henssler, X. Zhang and A. J. Matzger, *J. Org. Chem.*, 2009, **74**, 9112–9119.
24. S. Tanaka, S. Tamba, D. Tanaka, A. Sugie and A. Mori, *J. Am. Chem. Soc.*, 2011, **133**(42), 16734–16737.

25. H. Wynberg, A. Logothetis and D. VerPloeg, *J. Am. Chem. Soc.*, 1957, **79**, 1972–1975.
26. J. Kagan and S. K. Arora, *Heterocycles*, 1983, **20**, 1941–1943.
27. P. E. Fanat, *Synthesis*, 1974, 9–21.
28. S. Gronowitz and D. Peters, *Heterocycles*, 1990, **30**, 645–658.
29. G. T. Crisp, *Synth. Commun.*, 1989, **19**, 307–316.
30. M. Armour, A. G. Davies, J. Upadhyay and A. Wasserman, *J. Polym. Sci. Pt. A*, 1967, **5**, 1527–1538.
31. T. Yamamato, K. Sanechika and A. Yamamato, *J. Polym. Sci. Polym. Lett. Ed.*, 1980, **18**, 9–12.
32. J. W. P. Lin and L. P. Dudek, *J. Polym. Sci. Polym. Chem. Ed.*, 1980, **18**, 2869–2873.
33. M. Kobayashi, J. Chen, T-C. Chung, F. Moraes, A. J. Heeger and F. Wudl, *Synth. Met.*, 1984, **9**, 77–86.
34. A. Berlin, G. A. Pagani and F. Sannicolò, *J. Chem. Soc., Chem. Commun.*, 1986, **22**, 1663–1664.
35. G. Tourillon and F. Garnier, *J. Electroanal. Chem.*, 1982, **135**, 173–178.
36. R. J. Waltman, J. Bargon and A. F. Diaz, *J. Phys. Chem.*, 1983, **87**, 1459–1463.
37. G. Tourillon, E. Dartyge, A. Fontaine and A. Jucha, *Phys. Rev. Lett.*, 1986, **57**(5), 603–606.
38. S. Aeiyach, A. Kone, M. Dieng, J-J. Aaron and P-C. Lacaze, *J. Chem. Soc., Chem. Commun.*, 1991, 822–824.
39. R. L. Elsenbaumer, K. Y. Jen and R. Oboodi, *Synth. Met.*, 1986, **15**, 169–174.
40. K. Y. Jen, G. G. Miller and R. L. Elsenbaumer, *J. Chem. Soc. Chem. Commun.*, 1986, **17**, 1346–1347.
41. M. Sato, S. Tanaka and K. Kaeriyama, *J. Chem. Soc. Chem. Commun.*, 1986, **11**, 873–874.
42. R. Sugimoto, S. Takeda, H. B. Gu and K. Yoshino, *Chem. Exp.*, 1986, **1**(11), 635–638.
43. S. Hotta, S. D. D. V. Rughooputh, A. J. Heeger and F. Wudl, *Macromolecules*, 1987, **20**(1), 212–215.
44. M. Pomerantz, J. J. Tseng, H. Zhu, S. J. Sproull, J. R. Reynolds, R. Uitz, H. J. Arnott and M. I. Haider, *Synth. Met.*, 1991, **41**(3), 825–830.
45. T. Yamamoto, A. Morita, Y. Miyakazi, T. Maruyama, H. Wakayuma, Z. H. Zhou, Y. Nakamura, T. Kanbara, S. Sasaki and K. Kubota, *Macromolecules*, 1992, **25**, 1214–1223.
46. H. Mao, B. Xu and S. Holdcroft, *Macromolecules*, 1993, **26**, 1163–1169.
47. M. J. Marsella, P. J. Carroll and T. M. Swager, *J. Am. Chem. Soc.*, 1995, **117**(39), 9832–9841.
48. M.-A. Stao and H. Morii, *Macromolecules*, 1991, **24**, 1196–1200.
49. R. D. McCullough, The Chemistry of conducting polythiophenes: From synthesis to self-assembly to intelligent materials, in *Handbook of Oligo- and Polythiophenes*, ed. Fichou, D. John Wiley and Sons Ltd., West Sussex, UK, 1999, pp. 1–44.

50. M. Jeffries-El, and R. D. McCullough Regioregular Polythiophenes, in *Handbook of Conducting Polymers: Conjugated Polymers, Theory, Synthesis, Properties and Characterization*, ed. Skotheim, T. A, and Reynolds, J. R., Taylor and Francis Group, Boca Raton, Fl, 2007, pp. 9.1–9.49.
51. J. P. Aime, F. Bargain, M. Schott, H. Eckhardt, G. G. Miller and R. L. Elsenbaumer, *Phys. Rev. Lett.*, 1989, **62**, 55–58.
52. H. Sirringhaus, P. J. Brown, R. H. Friend, M. M. Nielsen, K. Bechgaard, Langeveld-Voss, A. J. H. Spiering, R. A. J. Janssen, E. W. Meijer, P. Herwig and D. M. de Leeuw, *Nature*, 1999, **401**, 685–688.
53. R. D. McCullough and R. D. Lowe, *J Chem. Soc. Chem. Commun.*, 1992, **1**, 70–72.
54. R. D. McCullough, R. D. Lowe, M. Jayaraman and D. L. Anderson, *J. Org. Chem.*, 1993, **58**, 904–912.
55. A. O. Patil, A. J. Heeger and F. Wudl, *Chem. Rev.*, 1988, **88**, 183–200.
56. J. Liu and R. D. McCullough, *Macromolecules*, 2002, **35**, 9882–9889.
57. T. Chen and R. D. Rieke, *J. Am. Chem. Soc.*, 1992, **114**, 10087–10088.
58. T. Chen and R. D. Rieke, *Synth. Met.*, 1993, **60**, 157–177.
59. T. Chen, X. Wu and R. D. Rieke, *J. Am. Chem. Soc.*, 1995, **117**, 233–244.
60. X. Wu, T. Chen and R. D. Rieke, *Macromolecules*, 1996, **29**, 7671–7677.
61. R. D. Rieke, *Acc. Chem. Res.*, 1977, **10**, 301–305.
62. R. D. Rieke, P. T. Li, T. P. Burns and S. T. Uhm, *J. Org. Chem.*, 1981, **46**, 4324–4326.
63. R. S. Loewe, S. M. Kheronsky and R. D. McCullough, *Adv. Mater.*, 1999, **11**(3), 250–258.
64. R. S. Loewe, P. C. Ewbank, J. Liu, L. Zhai and R. D. McCullough, *Macromolecules*, 2001, **34**(13), 4324–4333.
65. S. Gronowitz, P. Bjork, J. Malm and A-B. Hörnfeldt, *J. Organomet. Chem.*, 1993, **460**, 127–129.
66. R. D. McCullough and P. C. Ewbank, *Synth. Met.*, 1997, **84**, 311–312.
67. A. Iraqi and G. W. Barker, *J. Mater. Chem.*, 1998, **8**(1), 25–29.
68. S. Guillerez and G. Bidan, *Synth. Met.*, 1998, **93**, 123–126.
69. I. A. Liversedge, S. J. Higgins, M. Giles, M. Heeney and I. McCulloch, *Tetrahedron Lett.*, 2006, **47**, 5143–5146.
70. M. R. Andersson, D. Selse, M. Berggren, H. Järvinen, T. Hjertberg, O. Inganäs, O. Wennerström and J-E. Österholm, *Macromolecules*, 1994, **27**, 6503–6506.
71. M. R. Andersson, W. Mammo, T. Olinga, M. Svensson, M. Theander and O. Inganäs, *Synth. Met.*, 1999, **101**, 11–12.
72. G. Barbarella, M. Zambianchi, R. Di Toro, M. Colonna Jr., D. Iarossi, F. Goldoni and A. Bongini, *J. Org. Chem.*, 1996, **61**, 8285–8292.
73. I. Lévesque and M. LeClerc, *Chem. Mater.*, 1996, **8**, 2843–2849.
74. J. Yu and S. Holdcroft, *Macromolecules*, 2000, **33**, 5073–5079.
75. T. V. Richter, C. H. Braun, S. Link, M. Scheuble, E. J. W. Crossland, F. Stelzl, U. Würfel and S. Ludwigs, *Macromolecules*, 2012, **45**, 5782–5788.
76. Q. Wang, M. Wakioka and F. Ozawa, *Macromol. Rapid. Commun.*, 2012, **33**, 1203–1207.

77. H. A. Bronstein and C. Luscombe, *J. Am. Chem. Soc.*, 2009, **131**, 12894–12895.
78. M. Watanabe, K. Tsuchiya, T. Shinnai and M. Kijima, *Macromolecules*, 2012, **45**, 1825–1831.
79. I. Tahr-Djebbar, F. Nekelson, B. Heinrich, B. Donnio, D. Guillon, D. Kreher, F. Mathevet and A. Attias, *Chem. Mater.*, 2011, **23**, 4653–4656.
80. L. Chien, Y. Chien, D. Zhou, F. Zha, D. Li and K. Yao, *Macromol. Chem. Phys.*, 2011, **212**, 24–41.
81. H. Goto, Y. Yokochi and E. Yashima, *Chem. Commun.*, 2012, **48**, 3291–3293.
82. L. A. P. Kane-Macguire and G. G Wallace, *Chem. Soc. Rev.*, 2010, **39**, 2545–2576.
83. M. Jeffries-El, G. Sauve and R. D. McCullough, *Adv. Mater.*, 2004, **16**, 1017–1019.
84. M. Jeffries-El, G. Sauve and R. D. McCullough, *Macromolecules*, 2005, **38**, 10346–10352.
85. V. Senkovsky, N. Khanduyeva, H. Komber, U. Oertel, M. Stamm, D. Kuckling and A. Kiriy, *J. Am. Chem. Soc.*, 2007, **129**, 6626–6632.
86. J. K. Lee, S. Ko and Z. Bao, *Macromol. Rapid Commun.*, 2012, **33**, 938–942.
87. M. Yuan, K. Okamoto, H. A. Bronstein and C. Luscombe, *ACS Macro Lett.*, 2012, **1**, 392–395.
88. K. Tamao, S. Kodama, I. Nakajima, M. Kumada, A. Minato and K. Suzuki, *Tetrahedron*, 1982, **38**, 3347–3352.
89. Y. Liu, Y. Xu and D. Zhu, *Macromol. Chem. Phys.*, 2001, **202**, 1010–1015.
90. R. D. Loewe and R. D. McCullough, *Chem. Mater.*, 2000, **12**, 3214–3221.
91. S. Gronowitz, *Heterocycles*, 1990, **30**, 645–658.

CHAPTER 10

Poly(oxythiophene)s

ANIL KUMAR,*[a,b,c,d] SREELEKHA P. GOPINATHAN[d] AND REKHA SINGH[a]

[a] Department of Chemistry; [b] Centre for Excellence in Nanoelectronics (CEN); [c] National Centre for Photovoltaic Research and Education (NCPRE); [d] Centre for Research in Nanotechnology and Science (CRNTS), Indian Institute of Technology Bombay, Mumbai 400076, India
*Email: anilkumar@iitb.ac.in

10.1 Introduction

The intrinsically conducting polymers (ICPs) based on oxythiophene constitutes a large area in the domain of polythiophenes. The oxythiophenes differ from alkylthiophenes since the oxygen atom is directly attached to the thiophene ring at 3- or 4-positions. This alters the optical, electrical and electrochemical properties.[1] The polymerization of oxythiophene-based monomers is achieved either by the oxidative or by the transition-metal-assisted polymerization. During the oxidative polymerization process, the obtained polymer is in the oxidized (doped) form, whereas it is obtained in the reduced (undoped) form in the transition-metal-assisted polymerization. The method of polymerization has significant impact on the optical and electronic properties of the resulting polymers.[2] This chapter outlines several structural variants of oxythiophenes as classified in various categories. These poly(oxythiophene)s could be synthesized using various polymerization processes that are given below and are also listed in the Charts.

RSC Polymer Chemistry Series No. 9
Conjugated Polymers: A Practical Guide to Synthesis
Edited by Klaus Müllen, John R. Reynolds and Toshio Masuda

Published by the Royal Society of Chemistry, www.rsc.org

10.1.1 Polymerization Processes

Mentioned below are broadly eleven ways of synthesizing these polymers;

Method A: Electrochemical Polymerization
Method B: Solution Polymerization
Method C: *In situ* Chemical Oxidative Polymerization
Method D: Dispersion Polymerization
Method E: Vapor-Phase Polymerization
Method F: Solid-State Polymerization
Method G: Solvent-Free Chemical Polymerization
Method H: Grignard Metathesis (GRIM) Polymerization
Method I: Direct C–H Arylation Polymerization
Method J: Stille Polymerization
Method K: Yamamoto Polymerization

10.1.2 Poly(3-oxythiophene)s

In this category, the monomers are based on 3-alkoxy-substituted thiophene with optional nonalkoxy substituent at 4-position as given in Chart 10.1.

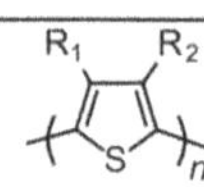

Substituent		Method of polymerization	References
$R_1 = H$, $R_2 =$			
$-OCH_2OCH_2OCH_3$, $-O(CH_2)_2OC_4H_9$		B	[5-6]
$-OCH_2CH_2CH_2SO_3^-$, $-OCH_2CH_2CH_2N^+(C_2H_5)_3$		B	[7]
$-OC_nH_{2n+1}$	$n = 1, 2, 4, 6, 7, 8, 10$	B, H, K	[1, 5-6, 8-10]
$-OCH(CH_3)CH_2CH_2CH_2CH_2CH_2CH_3$		K	[10]
$-OCH_2CH_2CH(CH_3)CH_2CH_2CH_2CH(CH_3)_2$		H	[12]
$-(OCH_2CH_2)_nOCH_3$	$n = 2, 3$	B, H	[2, 6]
$-(OCH_2CH_2)_3OCH_2CH_3$		H	[2]
$R_1 = CH_3$, $R_2 =$			
$-OCH_2CH_2OC_4H_9$		B	[5-6]
$-OC_nH_{2n+1}$	$n = 4, 7, 8$	B	[5-6, 13]
$R_1 = R_2 =$			
$-OC_nH_{2n+1}$	$n = 1, 4, 5, 8, 12$	A, B	[13-15]
$-OCH_2CH(C_2H_5)CH_2CH_2CH_2CH_3$		B	[15]

Chart 10.1 Structure variations in poly(3-alkoxythiophene)s and poly(3,4-dialkoxythiophene)s.[1,2,5–10,12,13–15]

Various monoalkoxy substituted polythiophenes are the earliest examples reported in this category.[3] These are followed by the 3-alkoxy-4-alkyl substituted thiophenes, introduced by Hoechst AG group, that result in conductivity enhancement as compared to 3-alkoxythiophenes.[4]

10.1.3 Poly(3,4-dioxythiophene)s

Poly(3,4-dioxythiophene)s are divided into two subgroups: a) poly(3,4-dialkoxythiophene)s having acyclic-dialkoxy substituent as outlined in Chart 10.1, and b) poly(3,4-alkylenedioxythiophene)s having a cyclic-dioxy substituent. Poly(3,4-alkylenedioxythiophene)s are further subdivided into two broad categories: a) Poly(3,4-ethylenedioxythiophene)s (PEDOT)s (Chart 10.2) including the bis-3,4-ethylenedioxythiophene-arylene series described in Chart 10.3; b) Poly(3,4-propylenedioxythiophene)s (PProDOT)s (Chart 10.4). Other derivatives of poly(3,4-alkylenedioxythiophene)s that cannot be classified in one of the above category are listed in Chart 10.5.

10.1.3.1 Poly(3,4-alkylenedioxythiophene)s

Polymers based on 3,4-alkylenedioxythiophenes were first reported in the late 1980s and were found to exhibit much improved properties as compared to acyclic dioxythiophenes. The first example reported in this category is poly(3,4-methylenedioxythiophene).[16] PEDOT, that came later, was found to exhibit low oxidation potential, enhanced stability (chemical and thermal) in both reduced as well as oxidized form and good transparency in its conducting form.[17] These desirable properties make it one of the most successful commercial conducting polymer to date.

10.1.3.1.1 Poly(3,4-ethylenedioxythiophene)s. PEDOT is one of the most commercially successful ICP as it is used for various applications.[19–21] Due to its insolubility, it can be processed through the aqueous dispersion in the presence of anionic polyelectrolyte, polystyrene sulfonic acid sodium salt (PSS-Na). Various substituents on EDOT (Chart 10.2) are introduced in order to enhance the solubility and to tune the bandgap of resulting polymer.

10.1.3.1.2 Poly(3,4-propylenedioxythiophene)s. The next alkylenedioxythiophene in the series after EDOT is 3,4-propylenedioxythiophene (ProDOT).[22] Polymers based on ProDOT and its derivatives are studied mainly for electrochromic applications. Various substituents (Chart 10.4) are introduced in order to fine tune the final properties, *e.g.* alkyl substitution for increasing solubility in organic solvents,[61] alkylsulfonate for aqueous solubility,[62] and propargyl/azide for inserting clickable linkage.[64]

Several other 3,4-alkylenedioxythiophenes based polymers have also been reported, by different groups, and are listed separately in Chart 10.5.

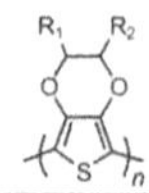		
Substituent	Method of polymerization	References
$R_1 = H$, $R_2 =$		
H	A, D, C, E, F	[18, 22, 89, 93, 94, 100]
$-C_nH_{2n+1}$ $n = 1, 4, 6, 8, 10, 12, 14, 16$	A, B	[24-29]
$-CH_2OC_nH_{2n+1}$ $n = 6, 8, 10, 12, 14, 16$	B	[30]
$-C_6H_5$	A	[25-26]
$-CH_2SH$, $-CH_2SCH_2CH_2CN$	A	[21]
$-CH_2OH$	A	[32]
-COOH	A	[33]
$-CH_2O(CH_2)_nSO_3Na$ $n = 3, 4$	A, B	[34, 35]
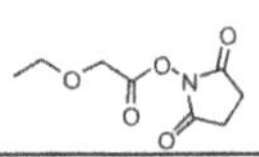	A	[36]
$-CH_2OCH_2(CH_2OCH_2)_3CH_2N_3$	A	[36]
$-CH_2(OCH_2CH_2)_nNH_2$ $n = 3, 4$	A	[37]
$-CH_2OCH_2COOH$	A	[32, 38]
$-CH_2(OCH_2CH_2)_3OH$	A	[38]
$-CH_2(OCH_2CH_2)_nOCH_2CH_3$ $n = 2$	A	[38]
$-CH_2OCOCH_2CH_2COOH$	A	[32]
$-CH_2OCOC_7H_{15}$	A	[38]
$-CH_2OBz$	A	[39]
$-CH_2S(CH_2)_5CH_3$, $-CH_2S(CH_2)_2(CF_2)_5CF_3$	A	[31]
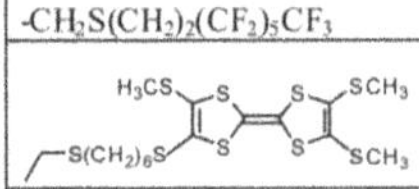	A	[31]
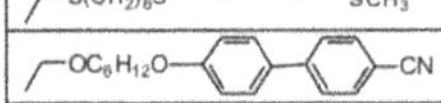	A	[41]
$-CH_2N_3$	A	[40]
$R_1 = R_2 =$		
$-C_6H_{13}$	A	[40]

Chart 10.2 Structure variations in poly(3,4-ethylenedioxythiophene)s.[18,21,22,24–41,89,93,94,100]

10.1.4 Miscellaneous Poly(oxythiophene)s

Some of the poly(oxythiophene)s cannot be classified either as poly(3-oxythiophene)s or as poly(3,4-dioxythiophene)s and are listed in Chart 10.6.

10.2 Experimental Procedures

Monomers based on oxythiophenes are polymerized by either oxidative or transition-metal-assisted polymerization. The oxidative polymerization is

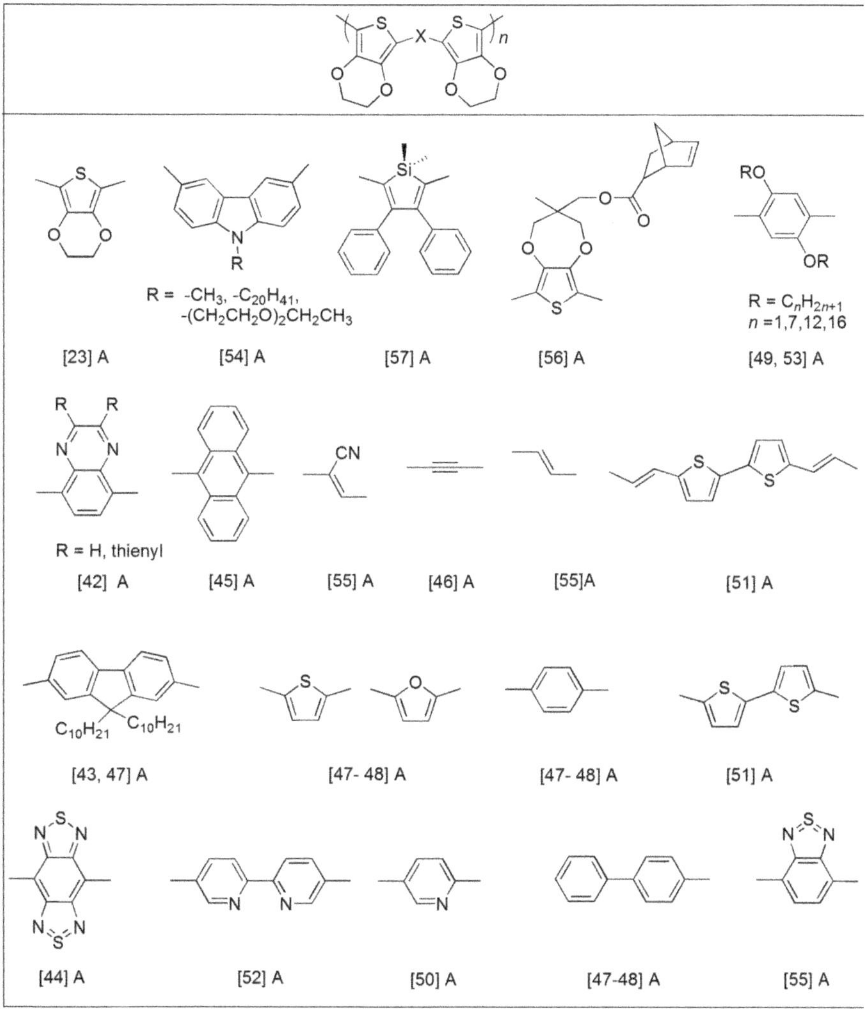

Chart 10.3 Structure variations in poly(Bis-EDOT-arylene)s.[23,42–57]

carried out by means of either electrochemical or chemical oxidation whereas a transition-metal-assisted polymerization is carried out by means of metal-catalyzed C–C bond formation. Electrochemical polymerization results in the formation of thin films directly on the substrate and hence is suitable for insoluble polymers. However, it is essential here to have a conducting substrate in order to get thin films. On the other hand, chemical oxidation is generally preferred for polymers that are soluble. The chemical oxidative polymerization process is generally simple and tolerant to pendant functional groups and does not demand highly pure monomers and solvents as required for transition-metal-assisted polymerization. Polymers based on oxythiophenes are generally characterized by various spectroscopic methods (for molecular structure and

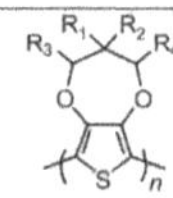

Substituent	Method of polymerization	References
$R_1 = R_3 = R_4 = H, R_2 =$		
H	A	[22-23 , 58]
$-C_nH_{2n+1}$ $n = 1, 6$	A, B	[25, 59, 60]
$-CH_2OH$	A	[38-39]
$-CH_2OBz$	A	[39]
$-CH_2OCH_2CH_2CH_2SO_3Na$	B	[62]
$R_1 = CH_3, R_3 = R_4 = H, R_2 =$		
$-CH_2OH$	A	[39]
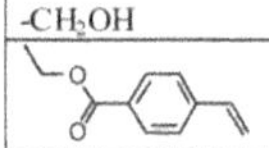	A	[63]
$-CH_2OCH_2CCH$ 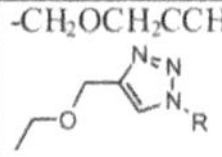 R = $-CH_2COONa$, $-C_6H_{13}$, $-(CH_2)_3N(CH_3)_3I$, $-(CH_2)_2O(CH_2)_2OH$	A, B	[64]
$-CH_2OBz$	A	[65]
$-CH_2OCH_2BiPh$	A	[65]
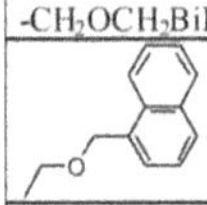	A	[65]
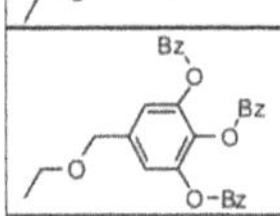	A	[65]
$R_3 = R_4 = H, R_1 = R_2 =$		
$-C_nH_{2n+1}$ $n = 1, 2, 4, 6, 10$	A, G, H, I	[59-61, 67-68]
-Bz	A	[69]
$-CH_2COOH$	B	[70]
$R_1 = R_2 = H, R_3 = R_4 =$		
$-C(CH_3)_3$, $-C_6H_{13}$, $-CH(CH_3)_2$, $-CH_3$	A, B	[71]

Chart 10.4 Structure variations in poly(3,4-propylenedioxythiophene)s.[22,23,25,38,39,58–65,67–71]

optical properties), electrochemical methods (for redox properties), gel permeation chromatography (GPC) (for number-average (M_n) and weight-average molecular weight (M_w) and polydispersity (PD). These polymers are generally soluble in solvents like tetrahydrofuran (THF), toluene, halogenated hydrocarbons, dimethylformamide (DMF), *etc*. The polymers with pendant anionic or cationic or oligoethyleneoxy groups are soluble in water. In general, these polymers are not stable in dedoped form in the presence of oxygen due to doping. Therefore, most of the applications of poly(oxythiophene)s result from

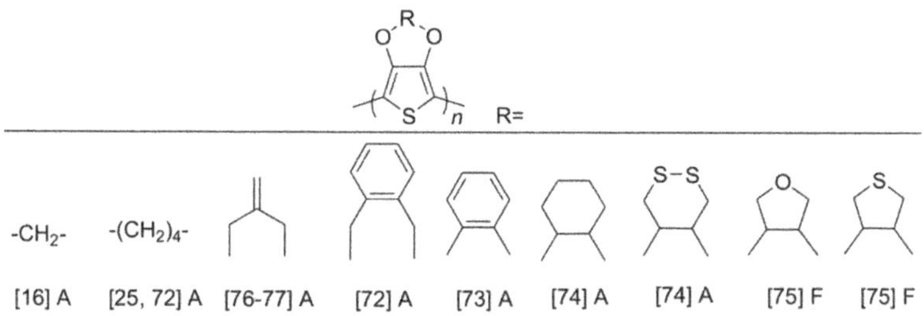

Chart 10.5 Miscellaneous poly(3,4-alkylenedioxythiophene)s.

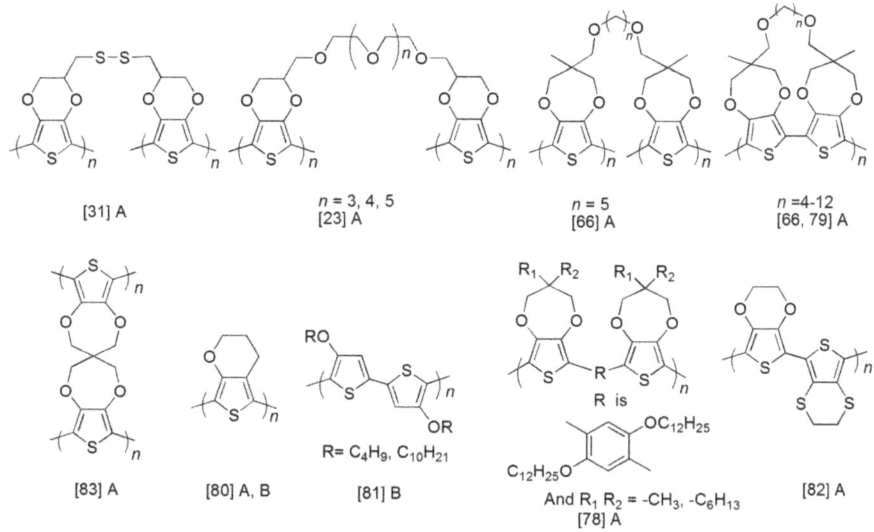

Chart 10.6 Miscellaneous poly(oxythiophene)s.

their doped state, which is highly stable, transparent and conducting. This air doping results in underestimation of molecular weights using GPC as the partially doped (due to exposure to oxygen) polymer adsorbs on to the GPC column. Therefore, polymers are freshly dedoped before GPC analysis. Homopolymers based on oxythiophene exhibit a dark blue color in the reduced state and light transparent green to transparent in the oxidized state, making them ideal for electrochromic displays. The optical bandgap of these polymers is calculated from the low-energy side onset of the absorption in UV-Vis spectrum. Cyclic voltammetry (CV) gives the redox behavior (oxidation and reduction potentials) of the polymer. The onset of the oxidation peak corresponds to the HOMO energy level $\{-(E_{(\text{ox. } vs. \text{ NHE})}+4.6)\}$ in eV. The onset of the reduction peak corresponds to the LUMO energy level $\{-(E_{(\text{red. } vs. \text{ NHE})}+4.6)\}$ in eV. The difference of the onset of oxidation and reduction peaks in CV studies gives the electrochemical bandgap. The LUMO energy level can also be calculated using the optical bandgap and the HOMO energy level (LUMO = HOMO + Optical Bandgap).

10.2.1 Oxidative Polymerization

10.2.1.1 Electrochemical Polymerization [Method A]

Electrochemical polymerization results in the synthesis of a thin film on the electrode that is used for further electrochemical and optical characterization. Since this process is very simple and all the standard electrochemical studies can be carried out using minimum amounts of monomer, this is a method of choice when the final polymer is insoluble or the monomer synthesis is tedious or expensive. The solvent for electropolymerization is chosen in such a way that the monomer is soluble, whereas the polymer is insoluble. The electrode surface, nature of the electrolyte and other polymerization conditions play a significant role in the morphology of the polymeric film. It is possible to synthesize various nanostructures including nanofibers, nanodots, nanonetwork, and nano- to microsize tubes of functionalized PEDOTs and PProDOTs.[38] Oxythiophenes can be electropolymerized more easily due to their low oxidation potential as compared to alkylthiophenes.[23]

Commonly used solvents for electropolymerization of organic soluble monomers are acetonitrile, propylene carbonate, chloroform and dichloromethane. Several aqueous combinations have also been explored either in the presence of various additives such as sodium dodecylsulfate (SDS), polystyrene sulfonate (PSS) or dissolving the monomer by acoustic emulsification.[84,85,88] Tetrabutylammonium perchlorate (TBAP), tetrabutylammonium tetrafluoroborate (TBAF) and lithium perchlorate are the most common electrolytes. Monomers are purified either by fractional distillation in the case of liquid monomers or by recrystallization in the case of solid monomers and solvents used are freshly distilled. Electrochemical polymerization is carried out either by potentiodynamic mode (cyclic voltammetry) or by potentiostatic mode, wherein a constant potential (higher than the monomer oxidation potential) is applied for a fixed duration. Acetonitrile is the most commonly used solvent along with TBAP electrolyte, wherein the concentration of electrolyte (100 mM) is generally kept an order of magnitude higher than the monomer concentration (10 mM).

Example 1: Electropolymerization of EDOT in Acetonitrile

A solution of 0.0142 g (0.1 mmol) of freshly distilled EDOT and 0.342 g (1 m mol) of TBAP in 10 ml of acetonitrile is added to a three-electrode glass cell fitted with Ag/AgCl, platinum (Pt) button (2.0 mm radius) and Pt wire as the reference, working and counterelectrode, respectively. The potential of the working electrode is increased from 0.0 to +0.8 V, at a scan rate of 100 mV/s, for a period of 10 s followed by a further increase to 1.3 V. The initial step to 0.8 V allows the double-layer charging of the electrode/electrolyte solution interface that minimizes the distortion of the polymerization current transient by double-layer charging. The thickness of the resulting film on the electrode surface is controlled by the number of potential cycles during the electropolymerization. This method is also performed in potentiostatic mode wherein the potential is held constant at 1.4 V for a fixed duration.[86]

Example 2: Electropolymerization of EDOT in Sodium Dodecylsulfate/Water

0.071 g (0.5 mmol) EDOT, 0.20 g (0.7 mmol) SDS (Critical micellar concentration (CMC) = 0.002 M) and 0.106 g (1 mmol) lithium perchlorate are taken in an electrochemical cell with 10 mL water. PEDOT films are electrochemically deposited on the Pt working electrode from the above solution using cyclic voltammetry method. Cyclic voltammograms are recorded between −0.5 and +1.2 V vs. saturated calomel electrode (SCE) in micellar solution. The first rapid growth of the anodic current at 0.76 V *vs.* SCE corresponds to the oxidation potential of EDOT, which is 280 mV lower relative to that in acetonitrile (E_{ox} = 1.04 V). The electroactivity of the resulting polymer (9% loss of electroactivity after 50 redox cycles) was found to be similar to the one polymerized using acetonitrile.[87]

Example 3: Electropolymerization of EDOT in Aqueous Electrolytes by Tandem Acoustic Emulsification

1.775 mg (0.0125 mmol) EDOT is added to 25 mL of aqueous solution containing 2.66 g (25 mmol) lithium perchlorate in a glass beaker. The ultrasonication of the above mixture is conducted with an ultrasonic stepped horn (13 mm diameter, titanium alloy) connected to a 20-kHz oscillator for 5 min. This gives rise to a milky solution that is further subjected to ultrasonication with 1.6 MHz using an ultrasonic transducer (16 W cm^{-2}) connected to a cylindrical Pyrex glass tube (diameter, 24 mm; length, 75 mm) for 5 min. This results in a clear solution. Further ultrasonication with 2.4 MHz by an ultrasonic transducer (7 W cm^{-2}) connected to a cylindrical Pyrex glass tube (diameter, 24 mm; length, 75 mm) for 5 min gives a very clear transparent and stable dispersion of EDOT with a droplet size of 82 nm in water. The electropolymerization of EDOT emulsion is carried out by applying a constant potential of +1.4 V *vs.* SCE. ITO (1×3 cm^2) is used as the working electrode, a platinum plate (2×2 cm^2) as the counterelectrode and SCE as the reference electrode. The conductivity of electrodeposited films from this aqueous dispersion is 298 S cm^{-1}.[88]

10.2.1.2 Chemical Polymerization

10.2.1.2.1 Solution Polymerization [Method B]. In general, the chemical oxidative polymerizations are carried out using iron (III) compounds as oxidizing agent in various aqueous and organic solvents followed by dedoping of the polymer using hydrazine hydrate and final purification by reprecipitation in methanol or other nonsolvents. Soxhlet extraction is performed in order to remove oliogomeric impurities. The ratio of monomer to oxidant plays an important role in the properties of the final polymers obtained.

Example 4: Chemical Oxidative Polymerization of 3-Heptoxythiophene

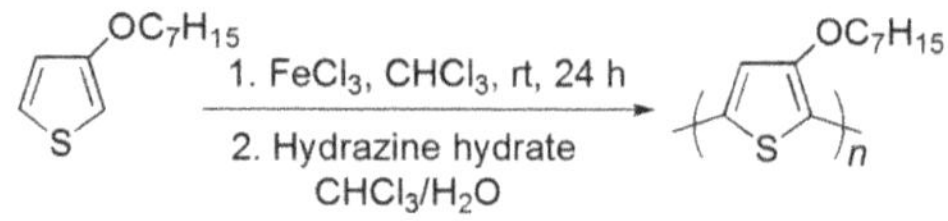

(10.1)

0.594 g (3 mmol) of 3-heptoxythiophene is taken in 30 mL of chloroform in a round-bottomed flask and 1.944 g (12 mmol) of ferric chloride is added to this solution at room temperature. The reaction mixture is then stirred under nitrogen for 24 h followed by precipitation of the resulting polymer in excess of methanol. The polymer is filtered and washed several times with methanol to remove unreacted monomer and oxidant followed by stirring with hydrazine hydrate in the presence of water and chloroform for dedoping. The chloroform layer is separated and washed several times with water. The polymer is obtained by removing chloroform under reduced pressure, which is further purified by Soxhlet extraction initially with methanol and finally with chloroform. The chloroform fraction is concentrated to get the final pure polymer in reduced form. The weight-average molecular weight (M_w) and polydispersity (PD), as determined by GPC, of this polymer is 2100 Da and 1.6, respectively.[6]

Example 5: Chemical Oxidative Polymerization of 1,3-di-*t*-butyl-ProDOT (1,3-ProDOT-TB$_2$)

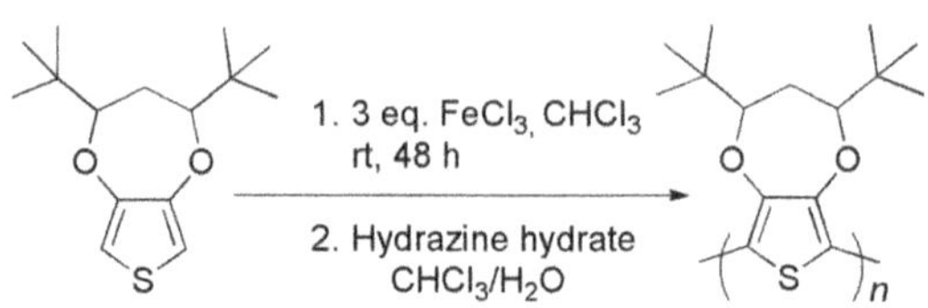

(10.2)

0.804 g (0.3 mol) of 1,3-ProDOT-TB$_2$ is taken in a round-bottomed flask in 30 mL dry chloroform and 1.46 g (0.9 mol) of ferric chloride is added to this solution. The reaction mixture is stirred for 48 h at room temperature followed by precipitation of the polymer in an excess of methanol. The polymer is filtered and then washed several times with methanol to remove unreacted monomer and oxidant followed by stirring with hydrazine hydrate in the presence of water and chloroform for dedoping. The chloroform layer is washed several times with water before chloroform is removed under reduced pressure. The resulting dedoped polymer is further purified by Soxhlet extraction initially with methanol and finally with chloroform. The chloroform fraction is concentrated to get the final pure polymer in reduced form. The weight-average molecular weight of this polymer is 10.6 kDa as determined using GPC analysis.[71]

10.2.1.2.2 *In Situ* Chemical Oxidative Polymerization [Method C]. In order to obtain good-quality thin films of the insoluble polymers directly on the nonconducting substrates, *in situ* chemical oxidative polymerization method is generally utilized. The *in situ* method employs the spreading of monomer and oxidant mixture on the substrate by spin- coating followed by heating to get polymer in the form of thin film. A commonly used oxidizing agent for the *in situ* oxidative polymerization is ferric tosylate. However, various other oxidants such as ferric chloride, phosphomolybdic acid hydrate, *etc.*, have also been explored.[90] The rate of polymerization and composition of the reaction mixture are the critical parameters that determine the quality of the thin polymeric film formed. Various weak bases such as imidazole, pyridine, *etc.*, are used to reduce the polymerization rate, which results in more ordered and compact thin films with higher conductivity.[89] Generally solvents with high boiling temperature result in the formation of good-quality films.

Example 6: *In Situ* Polymerization of EDOT using Ferric Tosylate as Oxidant

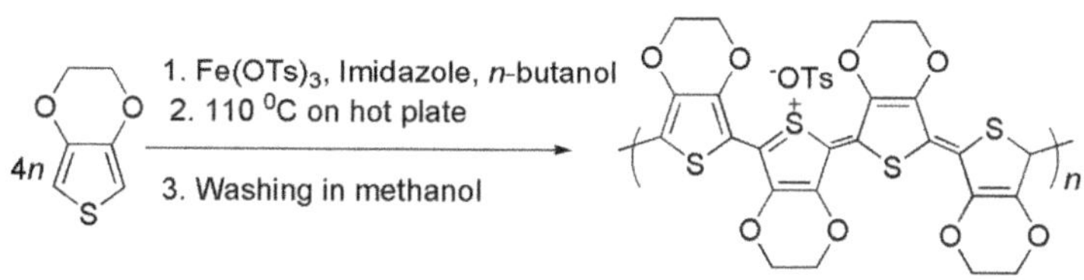

(10.3)

4.2 g (6.2 mmol) of ferric tosylate and 0.48 g (7 mmol) imidazole are dissolved by heating at 100 °C, separately in 3 mL and 1.2 mL *n*-pentanol, respectively. 0.5 g (3.5 mmol) of freshly distilled EDOT monomer is added to the imidazole solution followed by the addition of ferric tosylate solution to the reaction mixture. This reaction mixture is spin coated onto a clean glass substrate and then heated the substrate at 110 °C for 2 min under atmospheric conditions. This results in the formation of sky blue films that are further washed with methanol and air dried. The conductivity of the resulting thin film is 750 S cm^{-1}.[89]

Example 7: *In Situ* Metastable Chemical Polymerization of EDOT

Two separate solutions, one containing 0.142 g (1 mmol) of EDOT monomer in 5 mL acetonitrile and the other containing 1.825 g (1 mmol) of phosphomolybdic acid hydrate (PMA) oxidant in 5 mL acetonitrile are mixed and spin coated onto a glass substrate at 2000 rpm for 10 s. The resulting brown/black films are allowed to dry at room temperature for 30 min followed by rinsing in 3% v/v isopropylalcohol/dichloromethane for 30 min resulting in the formation of conductive, transmissive blue films. The conductivity of these PEDOT films is in the range of 0.02–0.47 S cm^{-1}.[90]

10.2.1.2.3 Dispersion Polymerization [Method D]. Dispersion polymerization of EDOT monomer in the presence of anionic polyelectrolyte, sodium salt of polystyrene sulfonic acid, results in the formation of a stable aqueous dispersion of PEDOT that is used to prepare thin films for various applications.[19] This method is simple, economically viable and environmental friendly. In fact, aqueous dispersion of PEDOT in the presence of PSS (available under the trade name of CLEVIOS-P) is one of the most successful commercial products based on any conducting polymer reported to date. PEDOT nanotubes are synthesized by dispersion polymerization using dioctyl sodium sulfosuccinate (AOT) as a template in reverse microemulsion polymerization.[91] PEDOT latex particles are synthesized by polymerizing EDOT in the presence of water-soluble polymers, such as polyethyleneoxide (PEO) containing EDOT as the chain ends resulting in the syntheses of stable latex particles consisting of random block copolymers of PEDOT and PEO.[92]

Example 8: Synthesis of PEDOT-PSS Dispersion

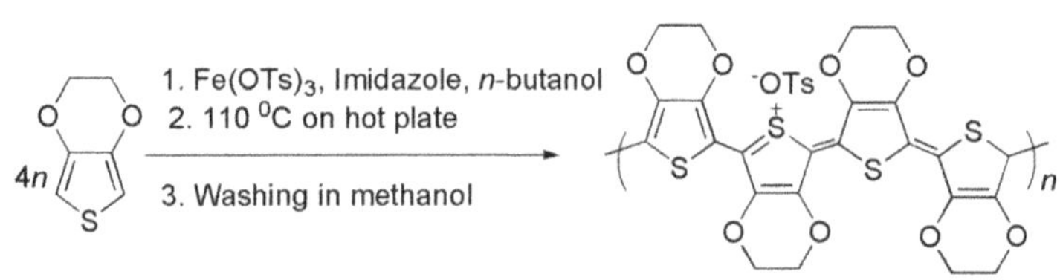

(10.4)

10.65 g (75 mmol) of EDOT, 439 g of 6 wt% PSS solution and 21.4 g (104 mmol) of sodium perxoydisulfate are dissolved in 2 L water. After initial stirring at room temperature for 10 min, 0.187 g (0.46 mmol) of ferric sulfate is added and the mixture is stirred vigorously for 24 h. This results in the formation of a stable blue-colored dispersion that is purified by dialysis to remove low molecular weight species and ion-exchange resins for inorganic ions.[93]

Example 9: Synthesis of PEDOT Nanotubes by Reverse Emulsion Polymerization

8.5 g (19.12 mmol) of dioctyl sodium sulfosuccinate is dissolved in 70 mL of *n*-hexane followed by addition of a solution of 1.6 g (10.0 mmol) ferric chloride in 1.0 mL of water. The resulting orange-colored reaction mixture is stirred for 5 min followed by addition of 0.5 g (3.52 mmol) EDOT monomer and the reaction is further continued for 3 h. The resulting blue/black precipitate of PEDOT nanotubes is filtered and washed with methanol and acetonitrile followed by drying under dynamic vacuum for 12 h at 80 °C to get a navy-blue powder (80% yield). These PEDOT nanotubes exhibit a conductivity of 3–6 S cm^{-1} (compressed pellet). These, as-synthesized, PEDOT nanotubes are readily dispersed in common organic solvents and used to prepare films on variety of substrates.[91]

Example 10: Synthesis of PEDOT Latex Particles

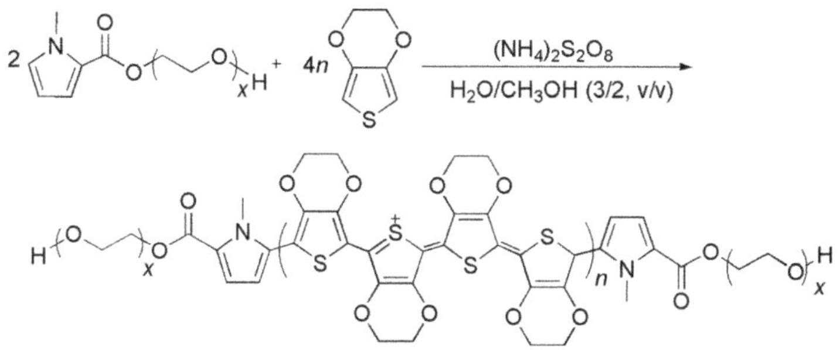

(10.5)

500 mg (3.5 mmol) of EDOT is added to a solution of 125 mg of reactive stabilizer (α-N-methyl pyrrole-PEO, $M_n = 10\,000$ g•mol^{-1}) in 40 mL of 3 : 2 v/v water/methanol mixture. A solution of ammonium persulfate (4.36 g dissolved in 10 mL of the water/methanol mixture) is introduced in this reaction mixture in one shot followed by stirring for 72 h at room temperature. The resulting blue dispersion is centrifuged at 10 000 rpm at 5 °C for 30 min. The supernatant is carefully decanted, and the dark blue sediment is redispersed in the water/methanol mixture (3 : 2 v/v). This redispersion and centrifugation cycle is repeated three times in order to ensure the complete removal of inorganic salts as well as nonincorporated reactive stabilizer. The particle size of the resulting PEDOT latex particles vary from 200 to 700 nm by changing the oxidant equivalents and molecular weight of the reactive stabilizer. [92]

10.2.1.2.4 Vapor-Phase Polymerization [Method E]. Vapor-phase polymerization (VPP) helps in preparing thin conducting film directly on different substrates, such as fibers and plastic sheets, by polymerizing the volatile monomer in the vapor phase.[94,95,97] The typical VPP is carried out in a closed chamber wherein the volatile monomer is brought into this chamber using air or nitrogen/argon during the polymerization. The substrate on which the film is to be deposited is coated with the oxidant solution (ferric chloride or ferric tosylate in *n*-butanol or ethanol). After drying, the oxidant-coated substrates are transferred to the polymerization chamber followed by introduction of the monomer vapor. Films of various thicknesses can be prepared by either changing the oxidant layer thickness or varying the monomer-vapor expose time. Various weak bases (*e.g.*, pyridine, imidazole) are added to the oxidant to inhibit the unwanted acid-initiated polymerization that leads to nonconjugated polymer.[96] The oxidant can also be sublimed on the substrate which eliminates the use of any solvent processing step.[99] Use of bromine vapors as oxidants eliminates the need of post-polymerization washing to remove unreacted oxidant and also results in polymer with better conductivity.[98]

Example 11: Vapor-Phase Polymerization of EDOT Using Ferric Tosylate in the Presence of Poly(ethylene glycol-propylene glycol-ethylene glycol) (PEG-PPG-PEG) Block Copolymer

PEDOT films are synthesized on 5×5 cm^2 glass substrates that are washed using a mild detergent, followed by an ethanol and high-purity water rinse and finally cleaned using air plasma. The oxidant solution (1 : 1 molar ratio of ferric tosylate and block copolymer in *n*-butanol) is spin coated on these substrates at a speed of 1900 rpm for 15 s followed by heating on a hotplate at 70 °C for 2 min to help evaporate the solvent. This results in the formation of a thin oxidant layer on substrate with thickness in the range 6–10 μm. Samples are removed from the hotplate and immediately placed into a 115 L vacuum oven set to 35 °C. The chamber is pumped down to 45 mbar and the internally mounted knife valve and externally mounted bleed valve are opened allowing ingress of EDOT monomer and water vapor, respectively. Samples are removed after 25 min and placed on a 70 °C hotplate for 2 min to anneal the polymer. The sample is then rinsed in an ethanol bath to remove unreacted oxidant, unbound surfactant, and residual unreacted monomer. The samples are finally dried in an oven at 70 °C for 30 min to remove any residual solvent. These PEDOT films show very high conductivity ~3400 S cm^{-1} with > 80% visible transmittance.[99]

Example 12: Vapor-Phase Polymerization of EDOT using Bromine

The glass substrate is heated to 80 °C in a polymerization chamber maintained at 150 mTorr. Two source jars with bromine and EDOT are connected, independently, to the chamber through transfer lines. The monomer source jar and transfer line are heated to 150 °C and simultaneously allow the vapors of bromine and EDOT to enter the chamber for 40 min. The resulting thin (40 nm) films of PEDOT are used for further characterization and show conductivity of 380 ± 8 S cm^{-1}. The thickness of thin films of PEDOT can be controlled by varying the deposition time.[98]

10.2.1.2.5 Solid-State Polymerization [Method F]. Solid-state polymerization (SSP) is a spontaneous topochemical polymerization that occurs only in crystalline form of the monomer without involvement of any external catalyst.[100] Various examples are known for PEDOT and the substituted PEDOTs synthesized by SSP.[75] Generally, this polymerization process is slow at room temperature but the rate can be enhanced at slightly higher temperature (up to the melting point of the monomer) leading to halide-doped conducting polymer while retaining the crystalline arrangement of the monomer. This type of polymerization occurs only in the crystalline form of the monomer as it is restricted by the specific relative arrangement of neighboring monomers.

Example 13: Solid-State Polymerization of 2,5-Dibromo-3,4-ethylenedioxythiophene

Crystalline 2,5-dibromo-3,4-ethylenedioxythiophene, DBEDOT, is incubated at 60 °C for 24 h followed by drying under vacuum to remove excess of Br_2.

This results in the formation of dark color polymer and is used for further studies. The conductivity of the PEDOT polymer is 33 S cm^{-1}.[100]

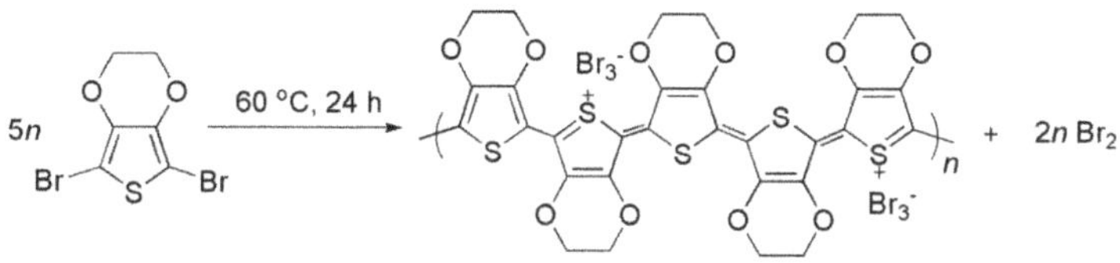

(10.6)

10.2.1.2.6 Solvent-Free Chemical Polymerization [Method G]. Typically this method of polymerization involves mixing the monomer with an excess of oxidant in a mortar and pestle followed by heating it at an elevated temperature. Ultrahigh molecular weight is obtained for poly(dihexyl-ProDOT)[68] which is the highest reported to date for any oxidative polymerization of oxythiophenes. Various oxidants such as ferric chloride, ferric tosylate, and bromine can be used.

Example 14: Solvent-Free Chemical Polymerization of Dihexyl-Substituted ProDOT (ProDOT-hex$_2$)

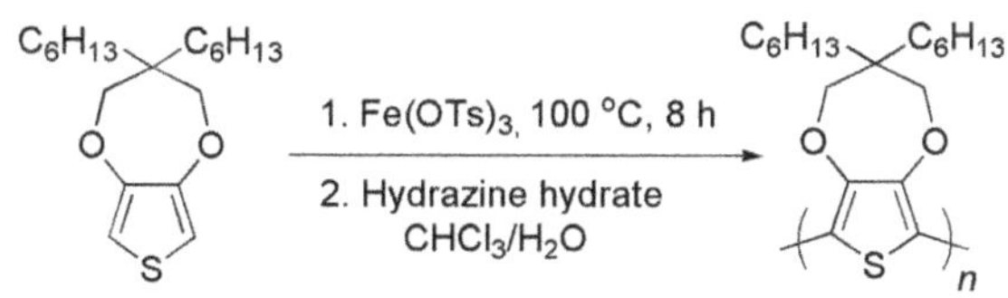

(10.7)

4.2 g (60 mmol) of ferric tosylate and 0.5 g (15 mmol) of the ProDOT-hex$_2$ monomer are mixed in a mortar. This reaction mixture is heated in an oven at 100 °C for 8 h resulting in the formation of dark colored powder indicative of the polymerization. The reaction mixture is washed with methanol several times to remove unreacted monomer and oxidant. The washed polymer is dedoped by stirring with hydrazine hydrate in chloroform and water. The chloroform layer is separated and washed twice with water followed by precipitating the dedoped polymer in methanol. The precipitated polymer is centrifuged and purified further by Soxhlet extraction using methanol to remove oligomers followed by chloroform. Chloroform is evaporated under reduced pressure to get the pure polymer as a dark purple solid. A polymer with 132 kDa molecular weight (M_w) and 1.7 polydispersity (PD) is obtained in 90% yield.[68]

10.2.2 Transition-Metal-Assisted Polymerization

Transition-metal-assisted polymerization of oxythiophenes is carried out using various transition-metal catalysts known for C–C bond formation reactions. Transition-metal-assisted polymerization methods are more stringent in terms of reaction conditions but give better quality materials (high molecular weight and regioregularity) compared to oxidative polymerization methods.[2,67] The common drawbacks in many transition-metal-assisted polymerization routes are the use of organometallic catalysts that are expensive, formation of toxic byproducts, require highly pure monomers, moisture-free conditions and non-compatibility of various functional side chains. It also involves additional synthetic steps for the monomer, and yields are comparatively low (<70%).[101–104]

10.2.2.1 Grignard Metathesis (GRIM) Polymerization [Method H]

Grignard metathesis polymerization (GRIM) is a very versatile method and was originally developed to synthesize regioregular poly(3-alkylthiophene)s. However, this method requires dry solvent and is not compatible with many functional groups. Polymerization proceeds upon addition of a nickel catalyst and the feed ratio of monomer to catalyst decides the final molecular weight of the polymer. GRIM polymerization produces high molecular weight regioregular polymers and is considered to be of living nature and hence allows the synthesis of end-functionalized polythiophenes.[103,105] Miozzo *et al.* and Sheina and McCullough have reported the synthesis of copolymer of EDOT using the GRIM method.[106,107] ProDOT and its derivatives are also polymerized by Nielsen and Bjørnholm using the GRIM method.[60]

Example 15: Synthesis of Poly(3-hexyloxythiophene) by GRIM Polymerization

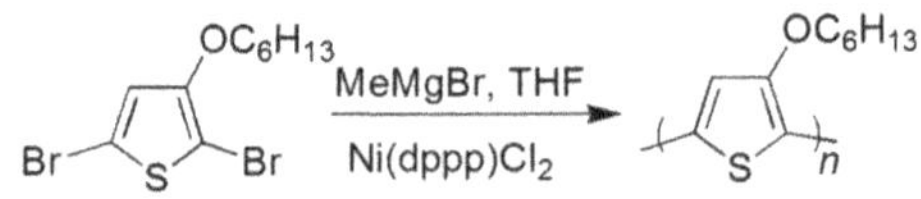

(10.8)

3.75 g (10.9 mmol) of 2,5-dibromo-3-hexyloxythiophene is added *via* syringe to a dry 3-neck 250 mL round-bottomed flask fitted with reflux condenser containing 55 mL of freshly distilled THF. 3.6 mL (10.9 mmol) of methylmagnesium bromide solution in hexane is added slowly *via* syringe followed by heating the reaction mixture to reflux for 1 h. After this period, the flask is allowed to cool to room temperature for 30 min followed by the addition of 0.059 g (0.11 mmol) of [1,3-bis(diphenylphosphino)propane]dichloronickel(II) (Ni(dppp)Cl_2). The reaction mixture is refluxed for a further 1 h. The solution is cooled down to room temperature and poured into methanol (200 mL) to precipitate the polymer. The polymer is filtered using a cellulose extraction thimble and the sample is air dried overnight. This polymer is further purified by Soxhlet extraction using hexane and methanol. Methanol is removed by distillation to get the final polymer as a dark blue material in 62% yield. The solid obtained is mostly intractable in common solvents (THF, chloroform) indicating the doped nature. The MALDI-MS studies indicate a degree of polymerization of 18 for the soluble fraction. The polymer exhibits a higher absorption maximum and lower solubility as compared to the corresponding alkyl analogs.[108]

10.2.2.2 Direct C–H Arylation Polymerization [Method I]

This method is based on the direct C–H arylation of thiophenes catalyzed by palladium.[67] Unlike GRIM polymerization, this method does not require organometallic intermediates and hence it is compatible with various pendant functional groups and is not sensitive to moisture.[67,102] Various side reactions and the presence of brominated and debrominated polymers are the limiting factors in this polymerization.[101] Schipper and Fagnou used C-H arylation for the synthesis of poly(alkylthiophenes).[109]

Example 16: Copolymerization of 2,5-Dibromo-EDOT and ProDOT-hex$_2$ using C–H Arylation

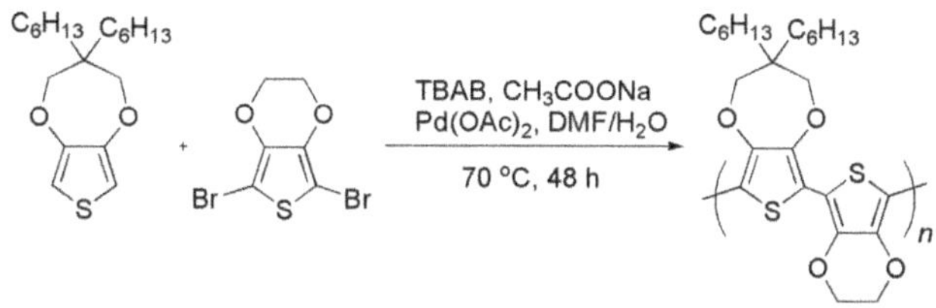

(10.9)

0.324 g (1 mmol) of ProDOT-hex$_2$, 0.322 g (1 mmol) of tetrabutylammonium bromide (TBAB) and 0.328 g (4 mmol) of sodium acetate are taken in a three-neck round-bottomed flask containing 80 mL of dry DMF, equipped with a reflux condenser and addition funnel. The reaction mixture is stirred at room temperature for 15 min followed by the addition of 0.299 g (1 mmol) of 2,5-dibromo-3,4-ethylenedioxythiophene and 22.4 mg (0.1 mmol, 10 mol%) of palladium acetate. The reaction mixture is heated to 70 °C under nitrogen atmosphere for 48 h. The color of the reaction mixture turns to dark red within 20 min and finally to purple, indicating the formation of polymer. The polymer is precipitated by pouring the reaction mixture into methanol. The resulting polymer is centrifuged and washed thrice with fresh methanol and subjected to Soxhlet extraction using methanol, acetone and finally with chloroform. The final polymer is obtained as a dark purple solid after removal of chloroform at reduced pressure. The molecular weight of the polymer is found to be 11.4 kDa with a PD of 2.5.[67]

10.2.2.3 Stille Polymerization [Method J]

This method is used to make HH-TT poly (3-alkoxythiophene) and involves the reaction of distannyl derivative with dibromoderivative of the monomer in presence of palladium (II) catalysts under mild reaction conditions.[11]

Example 17: Polymerization of [3-(3,7-Dimethyloctyloxy)-2-(3-(3,7-dimethyloctyloxy)thio-phenyl)thiophene] using Stille Coupling

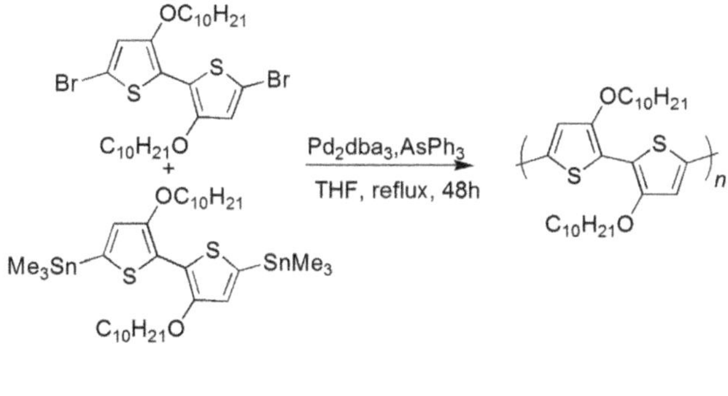

(10.10)

191 mg (300 μ mol) of dibrominated and 241 mg (300 μ mol) of distannylated derivatives of [3-(3, 7-dimethyloctyloxy)-2-(3-(3, 7-dimethyloctyloxy)thiophen-2-yl)thiophene] are taken in a round-bottomed flask in 8 mL of dry THF. Add 6.9 mg (7.5 μ mol) of $Pd_2(dba)_3$ and 18.4 mg (60.0 μ mol) of $AsPh_3$ are added to this solution and refluxed the resulting reaction mixture for 48 h. The reaction mixture is allowed to cool to room temperature and poured into an excess of methanol to precipitate the polymer. The precipitated polymer is filtered and further purified by Soxhlet extraction by acetone, hexane and chloroform. The chloroform-soluble fraction is reprecipitated in methanol to get the final polymer. The final polymer is found to be regioregular and semicrystalline in nature with a number-average molecular weight of 47 kDa and a PD of 3.0.[11]

10.2.2.4 Yamamoto Polymerization [Method K]

Yamamoto polymerization involves the homopolymerization of a dihalogenated alkoxythiophene monomer catalyzed by nickel (0) using DMF as solvent in the presence of suitable coligands.[10]

Example 18: Polymerization of 3-Octoxy-2,5-dibromothiophene

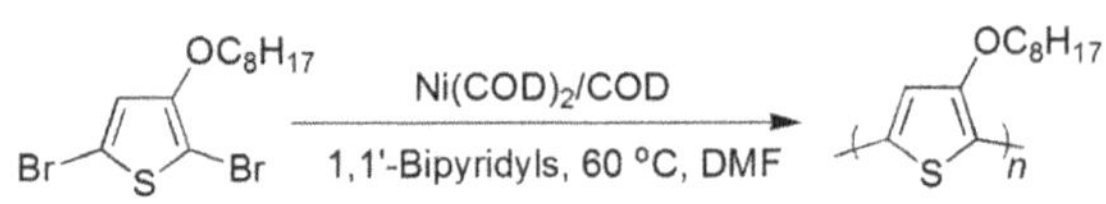

(10.11)

A mixture of 1,1′-bipyridyls (0.85 g, 5.4 mmol), bis(cyclooctadiene)nickel, $(Ni(COD)_2)$, (1.50 g, 5.4 mmol), cyclooctadiene (COD) (0.48 g, 4.5 mmol) in 15 mL of DMF is added to a solution of 1.68 g (4.5 mmol) of 3-octoxy-2,5-dibromothiophene in 18 mL DMF under an argon atmosphere. The reaction mixture is stirred at 60 °C for 18 h and 200 mL of acidic methanol is added to quench the reaction. The solid formed is filtered and washed with acidic methanol followed by aqueous solution of hot ethylenediaminetetraacetic acid (EDTA) and finally with water. Crude product is dissolved in chloroform and precipitated in methanol to get the final pure polymer as a blue violet solid. The polymer is obtained in good yields (70%) and is readily soluble in organic solvents (THF, chloroform, toluene). The weight-average molecular weight of the polymer was found to be 4.6 kDa with a PD of 1.35.[10]

10.3 Concluding Remarks

The low oxidation (*p*-doping) potential of poly(oxythiophene)s along with high conductivity, transparency and stability of doped state make them ideal as a transparent conductor and as a hole-transport layer in various optoelectronic devices. On the other hand, alkyl-substituted EDOT and ProDOT derivatives are suitable for electrochromic applications. However, PEDOT remains one of the most extensively studied polymers in this class due to the easy commercial access of EDOT monomer and processable PEDOT polymers (Clevios-P).

Currently, PSS is used as a polyelectrolyte to stabilize the PEDOT dispersion. However, use of PSS in these dispersions, has certain drawbacks such as its nonconducting nature and also it limits the shelf life as it releases sulfuric acid (desulfonation of styrene sulfonic acid side chain) with time. Various efforts need to be directed towards possible replacements of PSS or to synthesize PEDOT powders that can be dispersed easily in water for final processing. Another direction could be to develop new additives or formulations to further improve the properties of the existing PEDOT dispersion such as conductivity, transparency and electrochromic contrast. Recent efforts in improving the conductivity of Clevios-P ($\sim$3000 S cm^{-1}) while retaining the transparency (>85%) in the visible region are very significant.[110] These results open up a whole new area of applications of PEDOT as a potential replacement of ITO in the field of transparent conductors.

We acknowledge Department of Information Technology, Department of Science and Technology and Ministry of New and Renewable Energy, India for support. Rekha Singh thanks Council of Scientific and Industrial Research, India for Senior Research Fellowship.

References

1. K. Elschner, S. Kirchmeyer, W. Lovenich, U. Merker and K. Reuter, *PEDOT: Principles and Applications of an Intrinsically Conductive Polymer*, CRC Press, Taylor and Francis Group, LLC, 2012.
2. E. E. Sheina, S. M. Khersonsky, E. G. Jones and R. D. McCullough, *Chem. Mater.*, 2005, **17**, 3317.
3. M. Feldhues, T. Mecklenburg, P. Wegener and G. Kaempf, EP257573A1, 1987.
4. M. Feldhues, G. Kämpf, H. Litterer, T. Mecklenburg and P. Wegener, *Synth. Met.*, 1989, **28**, 487.
5. S. A. Chen, H. P. Kung, C. C. Tsai, C. N. Hou and C. S. Liao, *Synth. Met.*, 1993, **55**, 582.
6. S. A. Chen and C. C. Tsai, *Macromolecules*, 1993, **26**, 2234.
7. J. Lukkari, M. Salomäki, A. Viinikanoja, T. Ääritalo, J. Paukkunen, N. Kocharova and J. Kankare, *J. Am. Chem. Soc.*, 2001, **123**, 6083.
8. G. Koeckelberghs, M. Vangheluwe, K. V. Doorsselaere, E. Robijns, A. Persoons and T. Verbiest, *Macromol. Rapid Comm.*, 2006, **27**, 1920.
9. C. Shi, Y. Yao, Yang and Q. Pei, *J. Am. Chem. Soc.*, 2006, **128**, 8980.
10. Z. B. Zhang and M. Fujiki, *Polym. J. (Tokyo, Jpn.)*, 2001, **33**, 597.
11. M. Vangheluwe, T. Verbiest and G. Koeckelberghs, *Macromolecules*, 2008, **41**, 1041.
12. K. Van den Bergh, J. De Winter, P. Gerbaux, T. Verbiest and G. Koeckelberghs, *Macromol. Chem. Phys.*, 2011, **212**, 328.
13. G. Daoust and M. Leclerc, *Macromolecules*, 1991, **24**, 455.
14. M. Fall, L. Assogba, J. J. Aaron and M. M. Dieng, *Synth. Met.*, 2001, **123**, 365.

15. Z. J. Qi, B. Wei, Y. M. Sun, X. M. Wang, F. Kang, M. X. Hong and L. L. Tang, *Polym. Bull.*, 2011, **66**, 905.
16. H. J. Ahonen, J. Kankare, J. Lukkari and P. Pasanen, *Synth. Met.*, 1997, **84**, 215.
17. F. Jonas, G. Heywang and W. Schmidtberg, DE3813589A1, 1989.
18. G. Heywang and F. Jonas, *Adv. Mater.*, 1992, **4**, 116.
19. F. Jonas and W. Krafft, EP440957A2, 1991.
20. F. Jonas, W. Krafft and B. Muys, *Macromol. Symp.*, 1995, **100**, 169.
21. W. Krafft, F. Jonas, B. Muys and D. Qintens, DE4211461A1, 1993.
22. M. Dietrich, J. Heinze, G. Heywang and F. Jonas, *J. Electroanal. Chem.*, 1994, **369**, 87.
23. L. Groenendaal, G. Zotti, P. H. Aubert, S. M. Waybright and J. R. Reynolds, *Adv. Mater.*, 2003, **15**, 855.
24. A. Kumar and J. R. Reynolds, *Macromolecules*, 1996, **29**, 7629.
25. A. Kumar, D. M. Welsh, M. C. Morvant, F. Piroux, K. A. Abboud and J. R. Reynolds, *Chem. Mater.*, 1998, **10**, 896.
26. M. Wolfs, T. Darmanin and F. Guittard, *Macromolecules*, 2011, **44**, 9286.
27. S. Garreau, G. Louarn, G. Froyer, M. Lapkowski and O. Chauvet, *Electrochim. Acta*, 2001, **46**, 1207.
28. B. Sankaran and J. R. Reynolds, *Macromolecules*, 1997, **30**, 2582.
29. L. Groenendaal, G. Zotti and F. Jonas, *Synth. Met.*, 2001, **118**, 105.
30. P. Schottland, O. Stéphan, P. Y. Le Gall and C. Chevrot, *J. Chim. Phys.*, 1998, **95**, 1258.
31. M. Balog, H. Rayah, F. Le Derf and M. Salle, *New J. Chem.*, 2008, **32**, 1183.
32. E. M. Ali, E. A. Kantchev, H. h. Yu and J. Y. Ying, *Macromolecules*, 2007, **40**, 6025.
33. J. Kim, J. C. Cho, L. Povlich and D. C. Martin, WO2008106194A1, **2008**.
34. C. A. Cutler, M. Bouguettaya, T. S. Kang and J. R. Reynolds, *Macromolecules*, 2005, **38**, 3068.
35. K. Krishnamoorthy, M. Kanungo, A. V. Ambade, A. Q. Contractor and A. Kumar, *Synth. Met.*, 2001, **125**, 441.
36. S. C. Luo, E. Mohamed Ali, N. C. Tansil, H. h. Yu, S. Gao, E. A. B. Kantchev and J. Y. Ying, *Langmuir*, 2008, **24**, 8071.
37. N. C. Tansil, E. A. Kantchev, Z. Gao and H. h. Yu, *Chem. Commun.*, 2011, **47**, 1533.
38. S. Luo, J. Sekine, B. Zhu, H. Zhao, A. Nakao and H. Yu, *ACS Nano*, 2012, **6**, 3018.
39. A. Kros, R. J. M. Nolte and N. A. J. M. Sommerdijk, *J. Polym. Sci., Part A: Polym. Chem.*, 2002, **40**, 738.
40. H. B. Bu, G. Goetz, E. Reinold, A. Vogt, R. Azumi, J. L. Segura and P. Baeuerle, *Chem. Commun.*, 2012, **48**, 2677.
41. K. Krishnamoorthy, M. Kanungo, A. Q. Contractor and A. Kumar, *Synth. Met.*, 2001, **124**, 471.
42. A. Durmus, G. E. Gunbas and L. Toppare, *Chem. Mater.*, 2007, **19**, 6247.

43. F. Larmat, J. R. Reynolds, B. A. Reinhardt, L. L. Brott and S. J. Clarson, *J. Polym. Sci. A Polym. Chem.*, 1997, **35**, 3627.
44. T. T. Steckler, K. A. Abboud, M. Craps, A. G. Rinzler and J. R. Reynolds, *Chem. Commun.*, 2007, 4904.
45. A. Yildirim, S. Tarkuc, M. Ak and L. Toppare, *Electrochim. Acta*, 2008, **53**, 4875.
46. G. Zotti, G. Schiavon, S. Zecchin and A. Berlin, *Synth. Met.*, 1998, **97**, 245.
47. R. J. Mortimer, A. L. Dyer and J. R. Reynolds, *Displays*, 2006, **27**, 2.
48. J. R. Reynolds, A. Kumar, J. L. Reddinger, B. Sankaran, S. A. Sapp and G. A. Sotzing, *Synth. Met.*, 1997, **85**, 1295.
49. J. A. Irvin, I. Schwendeman, Y. Lee, K. A. Abboud and J. R. Reynolds, *J. Polym. Sci. A Polym. Chem.*, 2001, **39**, 2164.
50. A. A. Argun, P. H. Aubert, B. C. Thompson, I. Schwendeman, C. L. Gaupp, J. Hwang, N. J. Pinto, D. B. Tanner, A. G. MacDiarmid and J. R. Reynolds, *Chem. Mater.*, 2004, **16**, 4401.
51. M. Turbiez, P. Frere, P. Blanchard and J. Roncali, *Tetrahedron Lett.*, 2000, **41**, 5521.
52. G. Zotti, S. Zecchin, G. Schiavon and A. Berlin, *J. Electroanal. Chem.*, 2001, **506**, 106.
53. J. A. Irvin, F. Piroux, M. C. Morvant, V. L. Robertshaw, A. Angerhofer and J. R. Reynolds, *Synth. Met.*, 1999, **102**, 965.
54. G. A. Sotzing, J. L. Reddinger, A. R. Katritzky, J. Soloducho, R. Musgrave, J. R. Reynolds and P. J. Steel, *Chem. Mater.*, 1997, **9**, 1578.
55. J. Roncali, P. Blanchard and P. Frere, *J. Mater. Chem.*, 2005, **15**, 1589.
56. A. Nantalaksakul, K. Krishnamoorthy and S. Thayumanavan, *Macromolecules*, 2009, **43**, 37.
57. Y. Lee, S. Sadki, B. Tsuie and J. R. Reynolds, *Chem. Mater.*, 2001, **13**, 2234.
58. C. L. Gaupp, D. M. Welsh, R. D. Rauh and J. R. Reynolds, *Chem. Mater.*, 2002, **14**, 3964.
59. D. M. Welsh, A. Kumar, E. W. Meijer and J. R. Reynolds, *Adv. Mater.*, 1999, **11**, 1379.
60. C. B. Nielsen and T. Bjørnholm, *Macromolecules*, 2005, **38**, 10379.
61. D. M. Welsh, L. J. Kloeppner, L. Madrigal, M. R. Pinto, B. C. Thompson, K. S. Schanze, K. A. Abboud, D. Powell and J. R. Reynolds, *Macromolecules*, 2002, **35**, 6517.
62. V. Jain, R. Sahoo, S. P. Mishra, J. Sinha, R. Montazami, H. M. Yochum, J. R. Heflin and A. Kumar, *Macromolecules*, 2008, **42**, 135.
63. P. T. Dirlam, W. J. Chung, P. J. Costanzo and J. Pyun, *Polym. Prepr. (Am. Chem. Soc., Div. Polym. Chem.)*, 2011, **52**.
64. J. Sinha, R. Sahoo and A. Kumar, *Macromolecules*, 2009, **42**, 2015.
65. S. P. Mishra, R. Sahoo, A. V. Ambade, A. Q. Contractor and A. Kumar, *J. Mater. Chem.*, 2004, **14**, 1896.
66. R. M. Walczak, J. Cowart, K. A. Abboud and J. R. Reynolds, *Chem. Commun.*, 2006, 1604.

67. A. Kumar and A. Kumar, *Polym. Chem.*, 2010, **1**, 286.
68. A. Kumar, R. Singh, S. P. Gopinathan and A. Kumar, *Chem. Commun.*, 2012, **48**, 4905.
69. K. Krishnamoorthy, A. V. Ambade, M. Kanungo, A. Q. Contractor and A. Kumar, *J. Mater. Chem.*, 2001, **11**, 2909.
70. P. M. Beaujuge, C. M. Amb and J. R. Reynolds, *Adv. Mater.*, 2010, **22**, 5383.
71. T. Dey, M. A. Invernale, Y. Ding, Z. Buyukmumcu and G. A. Sotzing, *Macromolecules*, 2011, **44**, 2415.
72. D. M. Welsh, A. Kumar, M. C. Morvant and J. R. Reynolds, *Synth. Met.*, 1999, **102**, 967.
73. S. Roquet, P. Leriche, I. Perepichka, B. Jousselme, E. Levillain, P. Frere and J. Roncali, *J. Mater. Chem.*, 2004, **14**, 1396.
74. D. Caras-Quintero and P. Bauerle, *Chem. Commun.*, 2004, 926.
75. M. Lepeltier, J. Hiltz, T. Lockwood, F. Belanger-Gariepy and D. F. Perepichka, *J. Mater. Chem.*, 2009, **19**, 5167.
76. D. Yoon, S. Ko, Y. Lee, S. Lee, M. Pyo and K. Zong, *J. Electrochem. Soc.*, 2008, **155**, 209.
77. T. M. Nguyen, S. Cho, N. Varongchayakul, D. Yoon, J. Seog, K. Zong and S. B. Lee, *Chem. Commun.*, 2012, **48**, 2725.
78. E. M. Galand, J. K. Mwaura, A. A. Argun, K. A. Abboud, T. D. McCarley and J. R. Reynolds, *Macromolecules*, 2006, **39**, 7286.
79. R. M. Walczak, J. S. Cowart and J. R. Reynolds, *J. Mater. Chem.*, 2007, **17**, 254.
80. S. Inagi and T. Fuchigami, *Synth. Met.*, 2008, **158**, 782.
81. K. Faid, R. Cloutier and M. Leclerc, *Macromolecules*, 1993, **26**, 2501.
82. I. F. Perepichka and D. F. Perepichka, *Handbook of Thiophene-Based Materials: Application in Organic Electronics and Photonics*, John Wiley and Sons Ltd., 2009.
83. B. D. Reeves, B. C. Thompson, K. A. Abboud, B. E. Smart and J. R. Reynolds, *Adv. Mater.*, 2002, **14**, 717.
84. R. Asami, M. Atobe and T. Fuchigami, *J. Am. Chem. Soc.*, 2005, **127**, 13160.
85. R. Asami, T. Fuchigami and M. Atobe, *Langmuir*, 2006, **22**, 10258.
86. H. Randriamahazaka, V. Noel and C. Chevrot, *J. Electroanal. Chem.*, 1999, **472**, 103.
87. N. Sakmeche, J. J. Aaron, M. Fall, S. Aeiyach, M. Jouini, J. C. Lacroix and P. C. Lacaze, *Chem. Commun.*, 1996, 2723.
88. K. Nakabayashi, F. Amemiya, T. Fuchigami, K. Machida, S. Takeda, K. Tamamitsu and M. Atobe, *Chem. Commun.*, 2011, **47**, 5765.
89. Y. H. Ha, N. Nikolov, S. K. Pollack, J. Mastrangelo, B. D. Martin and R. Shashidhar, *Adv. Funct. Mater.*, 2004, **14**, 615.
90. S. L. McFarlane, B. A. Deore, N. Svenda and M. S. Freund, *Macromolecules*, 2010, **43**, 10241.

91. X. Zhang, J. S. Lee, G. S. Lee, D. K. Cha, M. J. Kim, D. J. Yang and S. K. Manohar, *Macromolecules*, 2005, **39**, 470.
92. M. Mumtaz, S. ü. Lecommandoux, E. Cloutet and H. Cramail, *Langmuir*, 2008, **24**, 11911.
93. F. Louwet, L. Groenendaal, J. Dhaen, J. Manca, J. Van Luppen, E. Verdonck and L. Leenders, *Synth. Met.*, 2003, **135–136**, 115.
94. J. Kim, E. Kim, Y. Won, H. Lee and K. Suh, *Synth. Met.*, 2003, **139**, 485.
95. P. A. Levermore, L. Chen, X. Wang, R. Das and D. C. Bradley, *Adv. Mater.*, 2007, **19**, 2379.
96. B. Winther-Jensen and K. West, *Macromolecules*, 2004, **37**, 4538.
97. J. P. Lock, S. G. Im and K. K. Gleason, *Macromolecules*, 2006, **39**, 5326.
98. H. Chelawat, S. Vaddiraju and K. Gleason, *Chem. Mater.*, 2010, **22**, 2864.
99. M. V. Fabretto, D. R. Evans, M. Mueller, K. Zuber, P. Hojati-Talemi, R. D. Short, G. G. Wallace and P. J. Murphy, *Chem. Mater.*, 2012, **24**, 3998.
100. H. Meng, D. F. Perepichka and F. Wudl, *Angew. Chem. Int. Ed.*, 2003, **42**, 658.
101. J. Hassan, E. Schulz, C. Gozzi and M. Lemaire, *J. Mol. Catal. A: Chem.*, 2003, **195**, 125.
102. H. Zhao, C. Y. Liu, S. C. Luo, B. Zhu, T. H. Wang, H. F. Hsu and H. h. Yu, *Macromolecules*, 2012, **45**, 7783.
103. R. S. Loewe, S. M. Khersonsky and R. D. McCullough, *Adv. Mater.*, 1999, **11**, 250.
104. I. Osaka and R. D. McCullough, *Acc. Chem. Res.*, 2008, **41**, 1202.
105. M. C. Stefan, M. P. Bhatt, P. Sista and H. D. Magurudeniya, *Polym. Chem*, 2012, **3**, 1693.
106. L. Miozzo, N. Battaglini, D. Braga, L. Kergoat, C. Suspène and A. Yassar, *J. Polym. Sci. A Polym. Chem.*, 2012, **50**, 534.
107. E. E. Sheina and R. D. McCullough, 228th ACS National Meeting, Philadelphia, PA, United States, August 22–26, 2004.
108. E. E. Sheina, S. M. Khersonsky, E. G. Jones and R. D. McCullough, *Polym. Prepr. (Am. Chem. Soc., Div. Polym. Chem.)*, 2003, **44**, 843.
109. D. J. Schipper and K. Fagnou, *Chem. Mater.*, 2011, **23**, 1594.
110. Y. Xia, K. Sun and J. Ouyang, *Adv. Mater.*, 2012, **24**, 2436.

CHAPTER 11

Polypyrroles

PIERRE AUDEBERT* AND FABIEN MIOMANDRE*

PPSM – CNRS UMR 8531 – Ecole Normale Supérieure de Cachan, 61 Avenue Président Wilson, 94235 CACHAN, France
*Email: audebert@ppsm.ens-cachan.fr; fabien.miomandre@ppsm.ens-cachan.fr

11.1 Introduction

Polypyrrole (PPy) is probably the most often electrochemically synthesized polymer within the huge family of the conducting polymers, while for most other monomers (*e.g.* thiophene, aniline, *etc.*) the chemical syntheses are dominant. In addition, both electrochemical and chemical syntheses are quite symmetric in their approach, since with one exception, the chemical syntheses of PPy all use an oxidative route, thus mimicking the electrochemical process.

We have therefore chosen to introduce the synthesis of PPy in two major sections, describing, respectively, the electrochemical and chemical syntheses. Beyond their common features, the electrochemical section plays a large role in deposition on non-noble metals, a specifically 'electrochemical' part, and to nanostructuration, another hot target for electrochemical methods. On the other hand, the chemical section describes a lot of composites, an activity that has fully emerged in the last two decades. This review is not extensive, but gives the reader a couple of concrete examples of almost every type of synthesis of this outstanding polymer, which, more than any other, still triggers today a lot of active research in many application fields.

RSC Polymer Chemistry Series No. 9
Conjugated Polymers: A Practical Guide to Synthesis
Edited by Klaus Müllen, John R. Reynolds and Toshio Masuda

Published by the Royal Society of Chemistry, www.rsc.org

11.2 Polypyrrole Electrosynthesis

11.2.1 Overview

Electrochemical methods have been widely used for the synthesis of conjugated polymers since they represent an easy, clean and versatile way of obtaining the targeted materials as coatings on conductive substrates. The subject was recently reviewed by Heinze *et al.* both from the theoretical and experimental points of view.[1] Theoretical aspects focusing on the electrochemistry of PPy are also discussed in a recent book chapter.[2] In the following sections will be briefly discussed the choice of experimental conditions suited for PPy deposition under various forms and for various applications.

11.2.1.1 General Features: Methods, Conditions, Electrolytic Media

From a general point of view, it can be asserted that PPy is, among the 'historical' conducting polymers, one of the best candidates for electrosynthesis. Indeed the moderate oxidation potential of pyrrole, as well as its solubility in polar solvents including water, constitutes great advantages for this kind of synthesis. The experimental procedure relies on the choice of:

- adequate solvent and electrolyte salts;
- appropriate concentration of monomer and monomer *vs.* electrolyte ratio;
- potential or current signal and electrolysis time.

Concerning solvent and electrolyte, typical conditions associate a polar solvent (water, acetonitrile, DMF, propylene carbonate, *etc.*) with a non-nucleophilic and nonoxidizable anion like a perchlorate, hexafluorophosphate or tetrafluoroborate salt. The electrolyte cation is usually chosen to fit solubility requirements (*e.g.* lithium or sodium in water, alkylammonium in organic solvents). After several years of investigation, it was admitted that the best properties were obtained using an organic solvent like acetonitrile or propylene carbonate containing a small amount (typically 1%) of water.[3] Indeed the deprotonation step of the coupled cation radicals (*e.g.* the oxidized dimer and subsequent oligomers) involved in the electrosynthesis mechanism requires a weak base in the medium.[1] The anion also plays a critical role in the final properties of the electrosynthesized PPy and, *e.g.*, arylsulfonate salts in organic solvent have been demonstrated to generate well-organized structures leading to high conductivity of the final material.[4] In the case of PPy synthesized in an aqueous environment, pH is an influencing parameter. It was shown that moderately acidic medium tends to increase the polymerization rate and lead to fewer defect-containing macromolecular structures. This is because the competing water oxidation with oxygen gas evolution is slowed down.

More recently, new electrolytic media appeared like BF_3-etherate[5] (a Lewis acid medium used to initiate cationic polymerization), supercritical carbon dioxide[6] (a medium with a very low viscosity) or ionic liquids. The latter are particularly suitable for electropolymerization due to their wide potential range of stability and chemical inertness, albeit their conductivity is usually lower than classical electrolytes due to their higher viscosity. A lot of studies dealing with PPy electrosynthesis in ionic liquids appeared in the literature during the last decade, after the pioneering work of Wallace and coworkers.[7] In some cases, a striking increase of the PPy electroactivity was found compared to usual organic solvents and mainly related to the very smooth morphology of the film.[7]

The choice of potential or current signals applied during electrosynthesis is of course crucial to adjust the morphology and the thickness of PPy, and thus its resulting properties. Classical signals can be sorted into potentiodynamic, potentiostatic and galvanostatic methods, corresponding, respectively, to scanning potential, constant potential and constant current applied for certain duration. It is generally acknowledged that potentiodynamic methods lead to more adhesive, homogeneous shiny films than the other methods,[8] while the latter give a better control of the thickness. More sophisticated signals have also been used recently, especially those using pulse techniques, since the 'on' and 'off' pulse durations are able to control the conjugation length, chain orientation and incorporation rate of the compensating charges during the polymer backbone formation and oxidation (see eqn (11.1)). The use of very short 'on' current pulses lead to PPy with a much higher capacitance compared to the classical galvanostatic synthesis.[9] Similar results were also obtained for PPy electrosynthesized in water with sulfonate dopant anions.[10,11]

11.2.1.2 Electrosynthesis on Oxidizable Metals

Electrodeposition of PPy on non-noble metals is a real challenge with a high impact in the field of anticorrosion. The advantage of this technique over other deposition methods (*e.g.* spin coating) is the accurate control of the polymer film thickness through the injected charge. However, the difficulty comes from the oxidation potential of the monomer being more positive than that of metals apart from platinum or gold, so that classical electrosynthesis conditions would lead to metal dissolution. The trick to achieving electropolymerization on oxidizable metals consists in finding a ligand able to complex the metal salt into a low-solubility compound, leading to a strong decrease in the metal oxidation kinetics. In other words, a passivation layer keeping the accessibility to the electrode open for the oxidation of the organic electroactive species (pyrrole and its oligomers) must be electrogenerated (see eqn (11.1)).

$$M^{2+} + \frac{2}{q}Y^{q-} + nH_2O \rightarrow MY_{2/q} \cdot nH_2O \tag{11.1}$$

where M is the divalent metal (Cu, Zn, Fe, *etc.*) and Y is the ligand bearing the charge q.

Oxalate, salicylate, and tartrate salts appeared to be good candidates for this purpose and have been successfully used for PPy deposition on copper,[12–15] zinc[16] or iron.[17] The electrodeposited PPy films show good adherence properties at low pH and low current density, thus demonstrating efficient protection against metal corrosion.

11.2.1.3 Electrosynthesis of Functionalized PPy

Functionalized PPy can be produced using either polymerization of functionalized monomers or postfunctionalization of PPy. Usually, chemical functions are introduced *via* covalent linking on the monomer because this approach leads to a much higher amount in the final material. An alternative way is to introduce this function *via* the dopant ions (mainly the anion). Concerning the monomer functionalization, the substitution can be performed either on the nitrogen position (after deprotonation) or on the 3,4 positions of the pyrrole ring. In both cases, the conditions for the modified pyrrole electropolymerization need to be adjusted, depending on various factors like steric hindrance, planarity distortion and possible electronic effects due to the substituents. The monomer reactivity is often lowered so that higher potentials or even copolymerization with non-substituted pyrrole are required. An exception to that rule comes from substitution at the 3,4 positions by electron-donating groups like alkoxy or alkyleneoxy,[18,19] which leads to polymers electrosynthesized at lower potentials and containing less conjugation defects (no α–β or β–β coupling in that case).

Several examples of electrosynthesized PPy functionalized by redox-active groups like quinone derivatives,[20–22] metallocene[23,24] or various metal complexes[25] can be found in the literature. Other important functional PPy for bioanalytical applications can be cited, like biotinylated PPy for amperometric biosensors,[26] oligonucleotide derivatized PPy for DNA biochips[27] or PPy-glucose oxidase for glucose sensors.[28]

11.2.1.4 Nanostructuration, Patterning, Nanocomposites

The electrosynthesis of PPy under other morphologies than classical films has become a research field for almost two decades. Most of the interest of the scientific community focused on nanostructuration, as it opens the way to reach new dimensionality (nanotubes, nanowires, *etc.*) as well as combining conjugated polymers with other nanomaterials like carbon nanotubes or nanoparticles. Synthetic methods in that field can be divided into template-based and template-free methods.[29] The former methods, pioneered by Martin's group,[30] generally use inorganic porous membranes (silica,[31] alumina,[32,33] *etc.*) or insulating polymers[34] deposited on a conducting substrate. The solution containing the monomer and electrolyte is then electropolymerized in the template pores or voids, while in a second step, the template is dissolved (generally by basic chemical treatment in the case of inorganic membranes or organic solvent for polymers) leaving the place to the nanostructured polypyrrole. Using this kind of method PPy nanowires were also obtained using an electrogenerated insulating polymer template and subsequent electropolymerization in the voids of this insulating layer.[35] Template-free methods appeared more recently as they generally require tricky conditions to generate nanostructuration. These conditions are generally based on the use of well-chosen electrolyte salts, *e.g.* phosphate,[36] which lead to PPy nanofibrillar structures through hydrogen bonding, or coupling electropolymerization with hydrogen or oxygen gas evolution to generate nanowires.[37–39]

Electrochemical methods can also be used to design PPy with specific micrometer-size patterns on a surface, allowing designing all-organic electronic microcircuits.[40] A general method to do so uses SAMs deposited on a conducting surface by microcontact printing techniques. PPy is electrodeposited either on the uncovered electrode areas (positive pattern) or on the SAM-covered areas (negative pattern) depending on the medium (aqueous or non-aqueous) and the terminal functions on the SAM.[41]

Finally, nanocomposites incorporating electrogenerated PPy have known a growing interest in the last decade, especially those based on carbon nanotubes (CNT). Indeed CNT-PPy hybrids exhibit very high capacitance that makes them promising materials for supercapacitors.[42] Several methods can be envisaged for the obtention of CNT-PPy hybrids: electropolymerization on a CNT-array-modified electrode,[43,44] coelectrodeposition from mixtures of surfactant stabilized aqueous CNT dispersion with the monomer[45,46] or from mixtures in an ionic liquid layer on the electrode,[47] and alternatively direct electrodeposition of PPy on CNT in a cavity microelectrode.[48]

Apart from carbonaceous materials, other examples of nanocomposites can be found: gold nanoparticles can be incorporated into PPy in a full electrochemical process using oxidation–reduction cycles.[49] Alternatively a synthetic route consists in electroplating gold on PPy nanowires synthesized by a template electrochemical method.[50]

11.2.2 Experimental Procedures

11.2.2.1 General Procedure and Materials

All the electrosyntheses described below use a three-electrode electrochemical cell, with a working electrode (usually platinum, gold or vitreous carbon), a counterelectrode and a reference electrode. The working electrode needs to be polished before use on specific clothes using micrometer-size alumina powders, then wiped and rinsed in dichloromethane before use. The reference electrode can be either a real $Ag^+(CH_3CN)/Ag$ reference connected to the solution by a frit, or an AgCl/Ag wire directly dipped in the solution. In any case, it is recommended to check the potential *vs.* ferrocene prior and after the experiments. The counterelectrode is usually an inert metal coil or high area plate.

The electrosyntheses are performed in the presence of a supporting electrolyte, which is a highly soluble electrolyte salt in the solvent used for electropolymerization. It is preferred to use either high-purity (Fluka Puriss quality) salts or recrystallized ones.

The general procedure consists in applying an electrochemical signal to the working electrode dipped in the solution containing the electrolyte and the monomer for an appropriate time. Then, the electrode is washed and rinsed, and the electrolysis medium is replaced by a monomer-free electrolyte solution in order to check the polymer electroactivity.

11.2.2.2 Electrosynthesis of PPy Films on Inert Electrodes

Example 1: PPy in 'Wet' Acetonitrile (Adapted from ref. 3)

A solution containing 0.05 M of the appropriate pyrrole monomer and 0.1 M of electrolyte salt (*e.g.* tetrabutylammonium hexafluorophosphate) in acetonitrile (with 1% water) is prepared. Oxygen traces are removed by argon bubbling and the solution injected into the electrochemical cell containing a Pt surface as the working electrode, a Pt wire as the counterelectrode and a saturated calomel electrode as the reference electrode. PPy is grown by applying a constant potential near the beginning of the pyrrole oxidation wave (typically +800 mV for PPy) and the thickness of the film is controlled by the amount of charge passed (8 mC cm^{-2} for 20 nm in the case of a 0.5 cm^2 Pt surface). Thicker films (20 μm) can be obtained using larger Pt surfaces (7 cm^2) and a constant current density of 500 μA/cm^2. After the films are formed, the cell is rinsed several times with electrolyte solutions and refilled again to check the electroactivity of the film in the monomer-free electrolyte solution. It is recommended to keep oxygen out of the cell during all the procedure including the washings.

Example 2: PPy in Ionic Liquids (Adapted from ref. 7)

A 0.1 M solution of freshly distilled pyrrole in dried and degassed 1-ethyl-3-methylimidazolium trifluoromethanesulfonylimide (TFSI) is prepared. The electrochemical polymerization is performed by cycling the working electrode (Pt) from −1 V to +1 V at 100 mV/s (reference: Ag wire; counter: Pt coil). Typical cyclic voltammograms are obtained as shown in Figure 11.1.

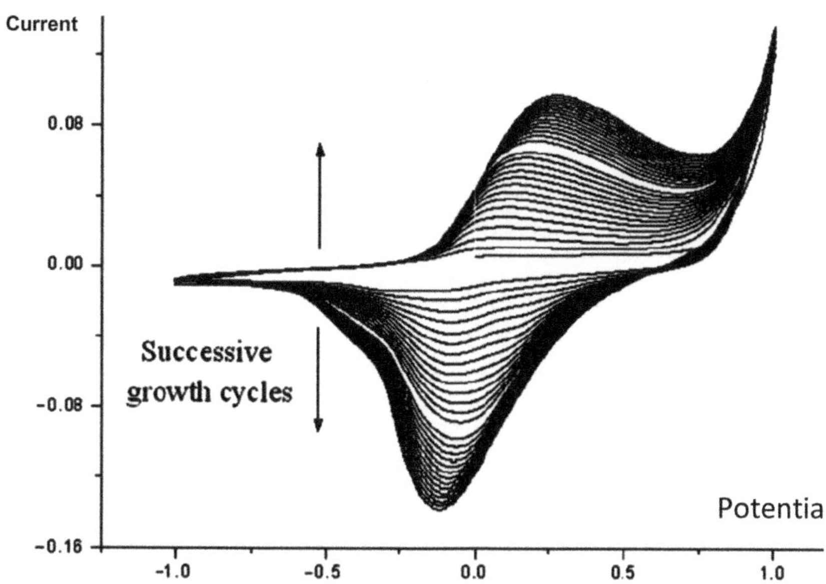

Figure 11.1 Successive cyclic voltammograms for the synthesis of PPy in a TFSI ionic liquid (adapted from ref. 7).

After the synthesis, films are washed with acetonitrile and dried at 30 °C for 48 h before further characterizations.

Example 3: High-Capacitance PPy using Pulse Potential Signal[9]

Pyrrole (0.1 M) and sulfuric acid (0.5 M) are mixed together in a three-electrode electrochemical cell. A polished graphite plate is used as the working electrode, Pt and AgCl/Ag, respectively, as counter and reference electrodes. Electro-polymerization is performed by applying current density pulses between 0 and 4 mA cm^{-2}. The total polymerization time is kept constant (200 s), as well as the rest time ('off' pulse) equal to 100 ms. The 'on' pulse time is varied between 5 ms and 200 s. When this pulse time is equal to 10 ms, nice PPy flakes (see Figure 11.2) are obtained with sizes around 10 μm and the specific capacitance reaches values as high as 400 F/g, twice the value of the corresponding non-pulsed galvanostatic method during the same time.

11.2.2.3 Electrosynthesis of PPy Film on Oxidizable Electrodes

Example 4: PPy on Copper Electrode[12]

A working electrode made of a copper rod (5 mm diameter) and platinum disk (3 mm diameter) sealed in a Teflon holder using epoxy resin is realized. This electrode is polished with aqueous suspensions of alumina (1 μm) and then polarized at −1 V *vs.* SCE. The polymerization medium is an aqueous solution containing the pyrrole monomer (0.5 M) and sodium salicylate (1 M). PPy electrodeposition on copper is performed either potentiostatically (0.6 V) or galvanostatically (0.5 to 5 mA cm^{-2}) for 1 min. After the synthesis, the PPy film is rinsed with water and its electroactivity checked by cycling from −1 V to +0.5 V in a monomer-free solution, with a pH adjusted to 6 by addition of salicylic acid or soda.

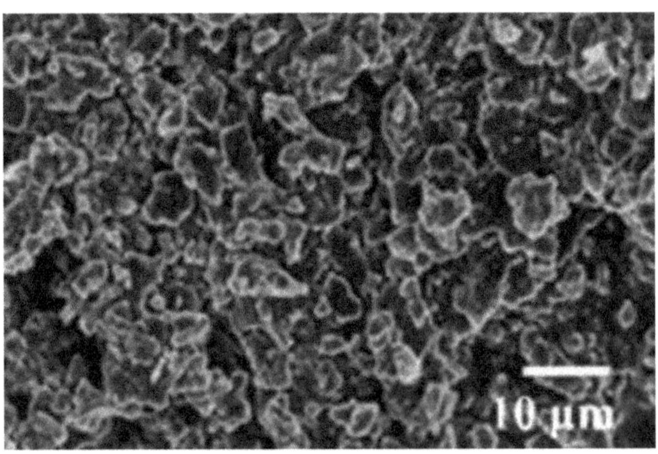

Figure 11.2 PPy flakes obtained by pulse potential signal (adapted from ref. 9).

Example 5: PPy on Zinc Electrode[16]

A solution of 0.3 M pyrrole in acetonitrile containing 0.1 M tetra-ethylammonium paratoluene sulfonate is poured in the electrochemical cell containing a SCE reference electrode and a Pt wire counterelectrode. The working electrode is a rectangular zinc sheet ($10\,cm^2$), polished with abrasive paper, rinsed with acetone that has undergone a chemical treatment with sodium sulfide (the zinc electrode is soaked into a 0.2 M aqueous solution of sodium sulfide for 12 h). Electropolymerization is achieved by repeated cycles between −1 V and +1.3 V. Alternatively, galvanostatic electrodeposition can be implemented but the current density needs to be higher than $1\,mA\,cm^{-2}$. Potentiostatic electrodeposition works for potentials in the range +0.7 to 1 V. Smooth and adherent PPy films are obtained for current densities in the range $1–2.5\,mA\,cm^{-2}$.

11.2.2.4 Electrosynthesis of Functionalized PPy

Example 6: PPy Functionalized by a Ruthenium Complex

First, a ruthenium complex with a pyrrole functionalized ligand, like 4-(2-pyrrol-1-yl-ethyl)pyridine (**1**) (Chart 11.1) is synthesized by mixing a ruthenium salt ($RuCl_3$) with the appropriate ligands (bipyridine and the pyrrole modified pyridine **1**) in an organic solvent. A 0.5 mM solution of this complex in acetonitrile containing 0.1 M tetrabutylammonium perchlorate is prepared in a three-electrode electrochemical cell. The working electrode is a Pt disk electrode polished using an alumina suspension. The reference electrode is Ag^+/Ag and the counterelectrode is a Pt wire. Electropolymerization is achieved by successive cycles from −0.4 to +1.3 V. After rinsing the electrode and dipping into pure electrolyte, a symmetrical signal corresponding to a Ru^{II}/Ru^{III} redox couple is observed with all the features of an immobilized system (*i.e.* peak current proportional to the scan rate), showing that the redox properties of the metallic center are kept in the modified electrode. The number of deposited layers can be inferred from the integration of the redox peak of this Ru^{II}/Ru^{III} redox couple.

Example 7: Electropolymerization of Alkylenedioxypyrrole[51]

Ethylenedioxypyrrole (EDOP, **2**) and Propylenedioxypyrrole (PRODOP, **3**) are synthesized from dimethyl-3,4-dihydroxypyrrole-2,5-dicarboxylate in five

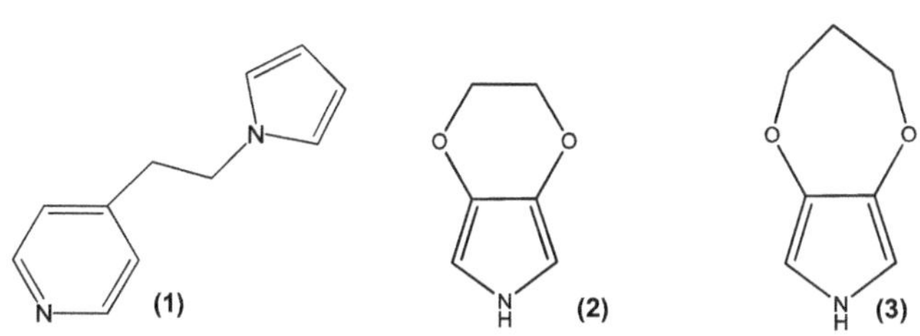

Chart 11.1 4-(pyrrol-1-yl-ethyl)pyridine (**1**), EDOP (**2**) and ProDOP (**3**).

steps.[51] Then, a 10^{-2} M solution of the monomer in propylene carbonate (PC) + 0.1 M lithium perchlorate is prepared. The electrochemical cell is filled with this solution, and a Pt disk electrode (working electrode), a Pt wire (counterelectrode), an Ag^+/Ag reference electrode are inserted into the cell. Electropolymerization is performed by cycling the potential from 0 to +0.8 V at 20 mV/s. The upper limit is typically 100 mV lower than for PPy. After the synthesis, the working electrode is rinsed and cycled in a monomer free electrolyte solution (PC + lithium perchlorate), displaying a reversible redox system near −0.4 V (CV should be recorded from −1.5 to +0.5 V). Alternatively, the electrosynthesis can be performed potentiostatically at +0.87 V (the potential is nearly the same for both monomers) from the same solution. A current *vs.* time trace shows a classical peak shape followed by a plateau, corresponding to the nucleation-growth mechanism, with a short induction period and a lower current density in the case of EDOP.

11.2.2.5 Electrosynthesis of Nanostructured PPy and Nanocomposites

Example 8: PPy Nanowires by Electrogenerated Template-Assisted Method[35]

An electrochemical cell containing 0.05 M tetraethylammonium perchlorate (TEAP) and 2.5 M ethyl acrylate in DMF is prepared. The working electrode (platinum or vitreous carbon) is polarized at −1.8 V (*vs.* a Pt pseudoreference) until the current drops to zero. Then, a second potential scan from 0 to −1.8 V is performed to complete the electrode passivation. After careful washing with pure DMF to remove any soluble unreacted species, the modified electrode is dipped into a second electrochemical cell containing 0.05 M TEAP and 0.1 M pyrrole in DMF. Electropolymerization is realized at constant current (0.5 mA, 400 s) and a chronopotentiogram $E(t)$ is recorded. A sharp increase of potential is observed followed by a peak and a weak decrease until stabilization.

Example 9: PPy Nanowires without Template[38]

An aqueous solution containing 0.15 M pyrrole, 0.2 M sodium hydrogen carbonate ($NaHCO_3$) and 10^{-3} M sodium perchlorate is prepared. A Pt (or vitreous carbon) working electrode, a SCE reference electrode and a Pt coil counterelectrode are dipped in the cell containing the previous solution. Electropolymerization is performed by applying a constant potential of 0.8 V for at least 600 s and the chronoamperogram $i(t)$ is recorded. This curve usually shows a continuous decrease followed by a plateau. An array of PPy nanowires growing from a thin overoxidized PPy film is obtained at the end (see Figure 11.3). The length of the nanowire can be monitored by the electrosynthesis time, while their density is related to the applied potential.

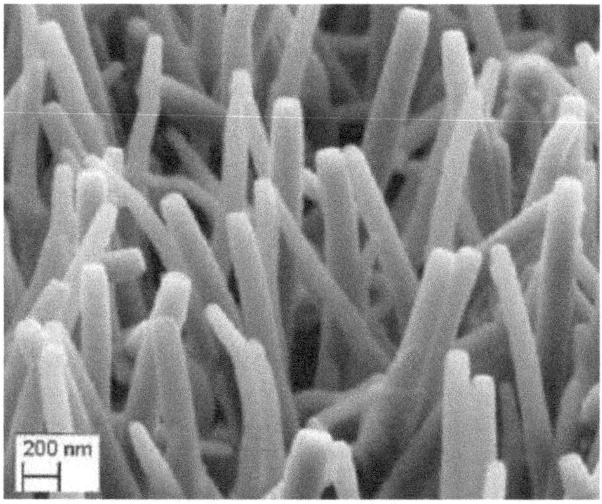

Figure 11.3 PPy nanowires obtained by a template-free electrochemical method (adapted from ref. 38).

Example 10: PPy–CNT Nanocomposite[46]

CNTs, synthesized by CVD in the presence of alumina supported iron catalyst, are treated by concentrated HCl to remove the remaining catalyst, heated to eliminate the amorphous carbon, and treated again with concentrated HCl. The electrolytic medium for the electropolymerization is made of 0.1 M SDS, 0.025 M pyrrole and 0.17 g/L CNT. The working electrode is a gold plate, the counterelectrode is a platinum wire, and the reference electrode is a SCE. Electropolymerization is achieved by scanning the potential between −0.2 and +0.8 V at 50 mV/s. SDS acts both as a surfactant for CNT dispersion and a supporting electrolyte. CNT wrapped by PPy appears on the gold electrode surface, and the PPy thickness can be adjusted by the number of cycles up to several hundreds of nanometers.

Example 11: PPy–Au Nanocomposite[49]

A solution containing pyrrole (10^{-2} M), $KAu(CN)_2$ (5 mM) as a source of gold, $NaClO_4$ (0.1 M) as a supporting electrolyte and a small amount of KCN ($5\ 10^{-5}$ M) added to achieve stability of the gold cyanide complex, is prepared. A 50-mL glass cell and a three-electrode system are used with a glassy carbon (GC) disc as the working electrode. The GC electrode surface is polished with 1 and 0.3 μm alumina powder and finally rinsed with deionized water. The counter and reference electrodes are a Pt wire or a Pt foil and a saturated calomel electrode (SCE), respectively. Care is taken to separate the Pt counterelectrode to avoid any platinum deposition at the working electrode. The solution is desaerated by argon bubbling before electrodeposition and pH is adjusted to 8 using soda. A sequence of short pulses (1 s, 100 mV) between +0.75 V and −1.6 V is applied to electrodeposit successively PPy and

Au. The total duration at both potentials is respectively 30 s (PPy) and 600 s (Au). A typical waveform is the following:

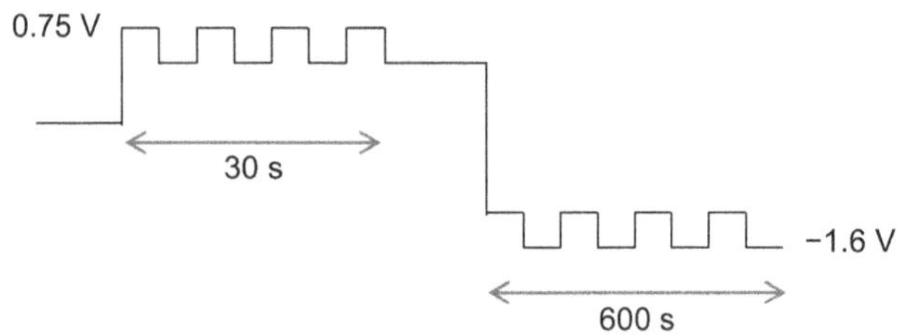

With this signal, Au platelets of approximately 50 nm diameter are obtained on the PPy film.

11.3 Polypyrroles – Chemical Syntheses

11.3.1 Overview

Contrariwise to other polyheterocycles, PPy cannot be easily formed by classical organometallic heterocyclic coupling reactions (*e.g.* Stille, Suzuki, Kumada, *etc.*), which works perfectly with thiophenes and several related heterocycles. Actually, only one report of a PPy prepared through metal-catalyzed coupling (in this case Kumada type) has been published by Yokozawa and coworkers.[52] They polymerized 2,5-dibromo-*N*-hexylpyrrole and obtained the polymer as a brown oil with a chain length of about 45 monomer units. However, nothing is said about the electrochemical response, or on the conductivity of the polymer in the doped state. The monomer was purified by low-pressure distillation, and the authors did not comment on its stability (which might be low given the low oxidative resistance of alkyl pyrroles). The same paper also describes copolymers based on phenyl and alkylpyrrole moieties.[53]

There are many reasons for this special situation concerning pyrrole monomers; first, the high electron-rich character of the pyrrole ring is a drawback to metal-catalyzed coupling and lowers the yields, which is very detrimental in polymer chemistry. But even more importantly, the acid character of the hydrogen borne by the pyrrole nitrogen just forbids the preparation of most of the organometallic derivatives indispensable to the reaction; still, this could be overcome by the use of *N*-alkyl derivatives of pyrroles (although these generally lead to less-conducting polymers). However, a last drawback is that bromopyrroles are unstable and cannot be handled in pure state.[54] (Interestingly, in several cases their decomposition leads to polypyrroles, which is a marginal but interesting chemical route to polypyrrole and partially brominated polypyrroles.[54]) For all these reasons, the (almost) only path employed for the synthesis of chemical PPy is the traditional oxidation route, which mimics the electrochemical oxidation, only substituting chemical oxidants to the electrode.

Therefore, the preferred oxidants are 'one-electron' oxidants, the best examples of the kind being oxidizing metal salts (and the most common among them, Fe^{III} salts[55]), which likely form initially the pyrrole cation-radical, which can then couple according to a previously demonstrated mechanism,[56] leading to the polymer after several successive steps. The global outcome of the reaction is illustrated by eqn (11.2). However, other oxidants like ammonium persulfate $(NH_4)_2S_2O_8$ have also been used,[57] despite the fact that the oxidation mechanism with this class of reagents is certainly much more complicated and definitely different from pure outer sphere one-electron exchange.

$$n(\mathrm{HP}y\mathrm{H}) + n\delta\mathrm{X}^- \rightarrow \left(\mathrm{P}y^{\delta+}, \delta\mathrm{X}^-\right)_n + (2+\delta)ne^- + 2n\mathrm{H}^+ \qquad (11.2)$$

In this section, we will detail the classical syntheses of 'chemical' PPy, first in organic media, then in water, since many efforts have been made to use this environment-friendly solvent. After some insights on substituted polypyrroles, the case of composites will be reported, since numerous works now describe the synthesis of PPy blends and composites with a lot of matrices.

11.3.2 Classical Polypyrrole Synthesis through Pyrrole Oxidation

11.3.2.1 Syntheses in Organic Media

The usual processes consist in adding a solution of oxidant to a solution of pyrrole under vigorous mixing.[55,58–62] The reactional medium most of the time turns quickly black and a precipitate forms, after which the polymer is filtered and washed. If nothing more is added, the polypyrrole obtained is doped with the ions coming from the oxidant, *e.g.* $FeCl_4^-$. A quickly discovered, very useful variant consists in introducing another ion, innocent towards the polymerization reaction, but that can insert as a dopant in the polymer.[57,60] This is particularly effective with sulfonate ions,[60] and has allowed the preparation of many polypyrroles with improved stability and conductivity, since these sulfonate ions are inserted as dopants inside the polymer, preferentially to the ions (*e.g.* chloride) coming from the oxidant (although the reason for this is still basically not really understood!).

Various oxidants are effective to polymerize pyrrole and pyrrole derivatives in organic media, but, as previously stated, the preferred ones are ferric salts, and especially ferric chloride, which is a cheap and readily available reagent. Ferric perchlorate is an even more effective reagent,[58] since its redox potential in acetonitrile ($E^\circ = 1.3$ V *vs.* SCE) is much higher than ferric chloride (the reason being the stabilization of Fe^{III} by chloride ions), but its high price has limited its use. Many other metal-based oxidants have been used, including Cu^{II} salts,[63] Au^{III} and Ag^{I} [64] species (in that case the aim was to make composites between the polymer and the noble metal). However, no other oxidant seems to be more effective than ferric chloride in the presence of sulfonate ions, which is a technique almost universally used in recent chemical PPy syntheses, and is known to provide polymers with a high stability allied with a reasonable conductivity (1–200 S/cm according to the published works[65,66]).

11.3.2.2 Syntheses in Aqueous Media

Since water is the cheaper and cleaner available solvent, and since in addition the relatively low oxidation potential of pyrrole is compatible with polymerization in aqueous medium, many attempts have been made in this direction, with appreciable success. The main way is the use of surfactants,[67,68] which permits the solubilization of pyrrole in aqueous medium,[69] and therefore its oxidation. In addition, it allows the use of purely water-soluble oxidants, like permanganate.[70] Emulsion polymerization seems to be the preferred approach, and this approach has continuously drawn researchers' attention until now.[71,72] However, recently some groups have started to use pure water solutions, adding for example progressively the pyrrole, to minimize the difficulties linked to the relatively low solubility of the monomer[70,73] (however, a cosolubilizing agent is always necessary). This solution can lead to the formation of composites with the second component being the byproduct issued from the oxidant, like a PPy/MnO_2 nanocomposite in the case where $KMnO_4$ is chosen as the oxidant.[70] However, it should be noted that, in comparison with PPy produced in organic medium, the one produced in aqueous solutions or emulsion is always noticeably less conducting, rarely exceeding a few S/cm. This is in clear contrast with other conducting polymers like polyaniline, and is likely to be due to the formation of hydroxylated products that disrupt conjugation in the polymer.

A clever amelioration of the oxidative polymerization method occurred when Miyata and coworkers, mimicking the electrochemical synthesis conditions, had the idea to control the oxidizing power of the oxidant solution by adjusting its redox potential.[66] This was made by adding controlled amounts of the associated reductant (the first example described was adding $FeCl_2$ to a $FeCl_3$ oxidant solution).

11.3.2.3 Syntheses Involving Biphasic Systems

Since most oxidants are soluble in aqueous media, and pyrrole most often in organic media, a natural way to prepare the polymer was at an organic/aqueous interface. Systems involving successive impregnation have also been examined with success.[66] A few studies, for example, describe PPy synthesis at the interface of an aqueous solution containing an oxidant (*e.g.* a ferric salt or ammonium persulfate), and an organic solution containing the pyrrole monomer.[74] However, the polymer prepared this way does not seem to display outstanding properties compared to the standard ones.

Emulsion polymerization of pyrrole in a water/hexane system, in the presence of a surfactant has also been described and shown to form well-shaped nanotubes, stemming from the previous organization of the hexane-surfactant phase.[75]

11.3.2.4 Synthesis of Polypyrrole Composites

11.3.2.4.1 Polypyrrole–Polymer Composites. While electropolymerization through a polymer membrane had been tried very quickly after standard PPy synthesis (see Section 11.2.1.4) chemical syntheses were published somewhat later. The first purely chemical approach in this direction was initiated by Bidan,[55] quickly followed by Miyata and coworkers,[66] and consists in the impregnation of a solid matrix (*e.g.* a polymer) with pyrrole, followed by dipping into an aqueous oxidant solution. A highly conductive chemical composite PMMA/PPy sample has been produced by Miyata and coworkers, applying to the composite preparation their previously developed method of controlling the redox potential of the oxidant solution. Mandal *et al.* later demonstrated that, when working only with aqueous solutions of both pyrrole and iron chloride, and especially using poly(vinylacetate) as the host matrix, a real interpenetrated network of PPy and the host polymer could be formed.[76]

A noticeable improvement was further brought by De Paoli and coworkers,[77–79] who reversed the impregnation order by first incorporating the oxidant inside the host matrix, and further treating it by pyrrole vapor instead of a pyrrole-containing solution. This is a quite interesting approach, since no backdiffusion of the oxidant in the monomer solution is possible, thus leaving a much cleaner polymerization reaction. However, despite the mechanical properties of the composite being occasionally quite remarkable, the conductivity of the samples were never higher than a few S/cm.

11.3.2.4.2 Anchoring Polypyrrole On Glass. This approach has been used by Attias and coworkers[80] to prepare glass plates and fibers coated with a thin uniform PPy film. However, these authors started by depositing a monolayer of a pyrrole monomer having a siloxane functional group able to anchor to the glass substrate, followed by oxidant impregnation, and finally treatment by pyrrole vapor. The PPy layer obtained in this way was quite uniform and highly conductive; however, due to its very small thickness, it had to be protected from atmosphere to retain its original after-synthesis properties. The same authors also prepared a sandwich-type composite in a simpler way, alternating PPy and xerogel layers deposition.[81]

11.3.2.5 Miscellaneous

A marginal, although interesting chemical route to PPy is the decomposition of 3-bromopyrrole and 3,4-dibromopyrrole.[54] These chemicals can be easily obtained, either from the decarboxylation of the corresponding 2-pyrrole carboxylic acids, or by deprotection of the N-silylated precursors. They are quite stable in solutions, but do polymerize very quickly in the pure state, with evolvement of gaseous hydrobromic acid. The resulting polymers have been produced and characterized. They were found to correspond to doped PPy, very likely by bromide ions.

Chemically prepared PPy can also be processed in various shapes, and especially a wide variety of micro- and nanostructured composites have been prepared. An important amount of work is due to Armes and coworkers, who systematically studied the formation of PPy composites formed by chemical oxidation of pyrrole, essentially by iron salts, in various colloidal solutions.[67] This results into the formation of colloidal suspensions of the host substrate (silica,[82] polystyrene, silica-coated magnetite, *etc.*) impregnated with PPy. However, the conductivity of pressed pellets of these composites is low, mostly below 1 S/cm. More recently, a wood saw dust/PPy composite has been reported,[61] and applied to the selective removal of chromium (VI) from waste water. PPy/MnO_2 composite (see above) has been shown to be effective for use as an active material in supercapacitors.

There are also some scattered reports of chemically prepared micro- and nanostructured pure PPys,[75] despite electrochemical syntheses are much more effective regarding this goal. Demoustier-Champagne *et al.* reported the formation of 15-μm diameter monodisperse PPy tubules by simultaneous diffusion of pyrrole and oxidant into a porous polycarbonate membrane.[73] Finally, PPy microfibers were prepared from chloroform solutions by electrospinning.[62]

11.3.3 Experimental Procedures

11.3.3.1 General Procedures

A very general procedure for standard oxidative polypyrroles preparation can be given as follows. First, it is preferable to distillate the monomer under reduced pressure,† or recrystallize if it is a solid (note that even a low melting point solid like 3,4-dimethylpyrrole can be recrystallized in pentane in a dry ice bath). Then, the monomer is dissolved into a polar solvent (acetonitrile, ethanol, water or their mixtures, *etc.*) and a solution of oxidant is quickly added under vigorous stirring (to avoid local overoxidation). The black polymer is filtered off, and can be washed, either by successive addition/stirring steps with the same solvent, or, preferably, with the help of a Soxhlet apparatus.

11.3.3.2 Organometallic Polypyrrole Synthesis

Example 12: Synthesis of Poly(N-hexylpyrrole) by Organometallic Pathway

a) Preparation of 2,5-dibromo-1-hexylpyrrole (adapted from ref. 52).

A solution of *N*-hexylpyrrole (6.00 g, 40 mmol) in THF (150 mL) is stirred at –78 °C under an argon atmosphere, and *N*-bromosuccinimide (NBS) (15.5 g, 87 mmol) is added. The mixture is stirred at –78 °C for 2 h and once diluted with brine, dried over anhydrous Na_2SO_4, and concentrated under reduced pressure. Diethyl ether is added to the residue,

†Note from the authors: instead of distillation, it is also possible to purify pyrroles by simply flowing them through dry alumina column.

and the mixture is filtered. The filtrate is washed with 10% aqueous NaOH (twice) and water.

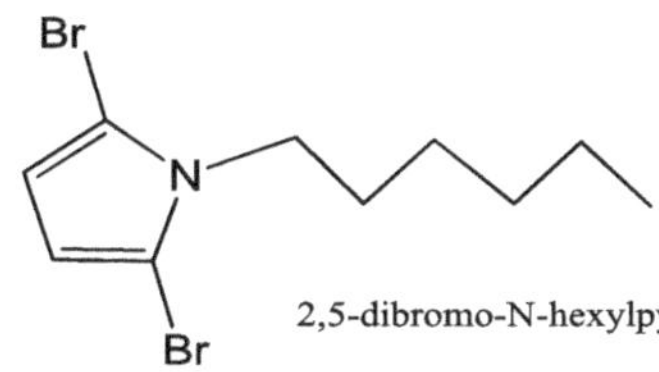

2,5-dibromo-N-hexylpyrrole

Purification of the crude product by distillation (71.5 °C/0.16 mmHg) gives the title compound (4.56 g, 37%) as a colorless oil. ^{1}H NMR (500 MHz, CDCl3) δ 6.16 (s, 2 H), 3.95 (t, $J=7.7$ Hz, 2 H), 1.67 (quint, $J=7.7$ Hz, 2 H), 1.38–1.28 (m, 6 H), 0.89 (t, $J=6.9$ Hz, 3 H); ^{13}C NMR (126 MHz, CDCl3) δ 111.6, 100.6, 47.3, 31.3, 30.2, 26.1, 22.5, 14.0; IR (neat) 3132, 2930, 2858, 1637, 1516, 1458, 1421, 1377, 1281, 1107, 904, 741 cm^{-1}; EIMS *m*/*z* (relative intensity) 311 (26, M+ containing two ^{81}Br), 309 (50, M+ containing one ^{81}Br and one ^{79}Br), 307 (27, M+ containing two ^{79} Br), 120 (100).

b) Preparation of poly(*N*-hexylpyrrole) (adapted from ref. 53).

General procedure: All glass apparatus are dried prior to use. Addition of reagents into the reaction flask and withdrawing a small aliquot of the reaction mixture for analysis are carried out *via* a syringe from a three-way stopcock with a stream of nitrogen. A round-bottomed flask equipped with a three-way stopcock is heated under reduced pressure, and then cooled to room temperature under a nitrogen atmosphere. Monomer precursor 2,5-dibromo-*N*-hexylpyrrole (312 mg, 1.01 mmol), diphenylphenanthroline (dppe) (11.7 mg, 0.0294 mmol, 2.9 mol%) and naphthalene (used as an internal standard for GC analysis, 42 mg, 0.33 mmol) are placed in the flask, and the atmosphere in the flask is replaced by argon. Into the flask is added dry THF (5.0 mL) *via* a syringe, and the mixture is stirred at 0 °C. Isopropyl magnesium chloride i-PrMgCl (2.0 M solution in THF, 0.48 mL, 0.96 mmol) is added *via* a syringe, and the mixture is stirred at room temperature for 24 h (conversion of monomer = 80%). Then, the mixture is cooled to 0 °C, and a suspension of Ni(dppe)Cl_2 (13.5 mg, 0.0256 mmol, 2.5 mol%) in THF (5.0 mL) is added *via* a syringe. After the reaction mixture is stirred at 0 °C for 9 h (conversion of 2-bromopyrrol-5-ylmagnesium bromide = 82%), 5 M hydrochloric acid is added and the mixture is extracted with CH_2Cl_2. The combined organic layers are washed with saturated aqueous $NaHCO_3$ and water, dried over anhydrous Na_2SO_4, and concentrated under reduced pressure to give a crude product (241 mg, $M_n = 6800$, $M_w/M_n = 1.09$). The crude product is dissolved in a small amount of CH_2Cl_2, and the solution is poured into CH_3OH ($CH_2Cl_2/CH_3OH = 1/100$) under vigorous stirring to remove the low molecular-weight compounds. The precipitated polymer is collected and dried under reduced pressure to give the final product as a dark red-brown oil (91 mg, 60%, $M_n = 7110$, $M_w/M_n = 1.09$). ^{1}H NMR (500 MHz, $CDCl_3$) δ 6.29 (br s, 2 H), 3.80–3.62 (br m, 2 H), 1.44–1.37 (br m, 2 H), 1.18 (quint, $J = 7.2$ Hz, 2 H), 1.12–1.00 (m, 4 H), 0.81 (t, $J = 7.2$ Hz, 3 H); ^{13}C NMR (126 MHz, $CDCl_3$) δ 125.6, 110.8, 45.1, 31.3, 26.4, 22.5, 13.9.

11.3.3.3 Pure Oxidative Polypyrrole Synthesis

Example 13: Polymerization of Pyrrole Derivatives by Iron^{III} Perchlorate in Organic Solvent (Adapted from ref. 58)

The polymerization is achieved by rapidly mixing a solution of a given pyrrole monomer, 0.1–0.2 M in acetonitrile, and a solution of ferric perchlorate, in the proportion of 1 mole of monomer for 2.4 moles of oxidant. The black polymer precipitates, then is collected by filtration, washed once in the filter and then in a Soxhlet apparatus. This procedure has been described to work not only with pyrrole, but especially also with several high potential halogenated pyrrole derivatives.

Example 14: Polymerization of Pyrrole Derivatives by Iron^{III} Chloride in Organic Solvent (Adapted from refs. 60–62 and 66)

A solution of pyrrole (10% in chloroform) is added to a methanol solution (30%) of $FeCl_3$ under vigorous stirring. The reaction is left for one hour and the polymer is collected by filtration and washed thoroughly with methanol before drying. In a variant of this protocol, a given amount of $FeCl_2$ is added to the oxidant solution so as to define precisely its oxidizing potential.

Example 15: Polymerization of Pyrrole Derivatives by Persulfate in Water (Adapted from refs. 67–69)

Pyrrole is dried with CaH_2 for 24 h, followed by distillation under reduced pressure. Ammonium persulfate is used as an oxidant and dodecylbenzene sulfonate as the codopant. 0.15 mol of dodecylbenzene sulfonic acid and 0.3 mol of pyrrole are dissolved in 500 mL of distilled water with vigorous stirring. 0.06 mol of ammonium persulfate in 100 mL of distilled water are slowly added to the above solution whose temperature is maintained at 0 °C. The reaction is carried out for 40 h and terminated by pouring methanol. The resultant PPy powder is filtered and washed sequentially with distilled water, methanol and acetone several times, followed by filtering and drying in a vacuum oven at 25 °C for 12 h. The apparent polymerization yield is 42%. This PPy can be dissolved by sonication into *m*-cresol in reasonable amounts. Another report describes the use of sodium di(2-ethylhexyl)sulfosuccinate in analogous conditions.

Example 16: Polymerization of Pyrrole Derivatives by Iron Sulfate in Water (Adapted from refs. 71, 72)

PPy is prepared by chemical polymerization in a deionized water solution. The polymerization begins when adding 100 mL of a 0.1 or 0.05 M $Fe_2(SO_4)_3$

solution in water into 100 mL of a stirred water solution containing pyrrole (0.365 M) and a surfactant (examples of reported surfactants: 40 wt% aqueous solution of sodium alkylnaphthalenesufonate (mean molecular weight: 338), sodium dodecylbenzenesulfonate (molecular weight: 349) and 40 wt% aqueous solution of sodium alkylsufonate (mean molecular weight: 328)). The surfactant concentration is 2.2–2.5$\times 10^{-2}$ M. After the prescribed polymerization time, PPy is filtered from the solution with a funnel and a filter paper, and thoroughly washed with deionized water until the filtrate indicates neutral. After being further washed with ethanol several times, PPy is dried in vacuum at about 40 °C.

Example 17: Polymerization of Pyrrole at a Water/Organic Solvent Interface (Adapted from ref. 74)

Polymerization of pyrrole is performed by putting a 20-mL chloroform solution of pyrrole into 100-mL beakers. The concentrations of pyrrole can vary in the range 2.0 to 0.1 M. Then, 20 mL of a 0.2 M aqueous solution of ammonium persulfate is poured slowly on the chloroform solution within 10 s. The interface between chloroform and water starts to be colored and becomes black due to the formation of PPy at the interface. After 10 min, a film with a thickness of 3–4 μm is formed and removed with a glass or a Teflon plate. Then, a fresh interface appears and a new PPy film starts to be produced. A sufficient amount of PPy is obtained by repeating the film-removing process described above. As-formed polymer films are rinsed with distilled methanol and dried in vacuum at 80 °C for 12 h.

Example 18: Nanotubes from Pyrrole Polymerization in Reverse Water–Organic Emulsion (Adapted from ref. 75)

To fabricate PPy nanotubes, 9 g (20.3 mmol) of sodium bis(2-ethylhexyl) sulfosuccinate, commercially known as AOT, is added to 40 mL of hexane at room temperature. AOT forms reversed micelles due to its double-tailed structure with a relatively small hydrophilic head group. Then, 1 mL of 9.0 M aqueous $FeCl_3$ solution is added to the AOT:hexane mixture. $FeCl_3$ helps the formation of rod-shaped micelles by decreasing the 'second critical micelle concentration (CMC II)' value and increasing the solvent ionic strength. The anionic polar head group of AOT extracts metal cations from the aqueous solution to the reverse micellar phase. In other words, Fe^{3+} ions are concentrated in the reversed micelles and capped by the anionic polar head groups of AOT, therefore maintaining the rod shape of AOT. Then, 0.5 g (7.5 mmol) of pyrrole monomers are added into the solvent, and pyrrole monomers are polymerized by Fe^{3+} ions outside the rod-shaped micelles. The polymerization is allowed to proceed for 3 h. The resultant product is moved to a separation funnel and an excess of ethanol is added to remove the surfactants. PPy nanotubules precipitate after 2 h, and the upper solution containing surfactants is discarded.

11.3.3.4 Syntheses of PPy Composites

a) Syntheses from solutions, suspensions or micellar dispersions.

Example 19: Solution Synthesis of a Polycarbonate/PPy Composite in Organic Medium (Adapted from ref. 55)

A solution containing 10 w/w% of polycarbonate in chloroform is prepared, and pyrrole (5 to 30 w/w%) is then added. A saturated solution of $FeCl_3$ in methanol is prepared separately and added to the previous solution under vigorous stirring; the quantity of ferric chloride is always in a 2.2 to 1 ratio with pyrrole. The mixture is stirred for one day, and then the polymer is collected by filtration and thoroughly washed with methanol and water before being dried at 50 °C in oven.

Example 20: Synthesis of Polypropylene/PPy Composite Microparticles in Water/Alcohol (Adapted from ref. 77)

Polypropylene particles (5 g) of about 35 μm in diameter are dispersed in 160 mL of a water:methanol mixture (volume ratio 5:2) and then $FeCl_3$ is added. Pyrrole is separately dissolved in 20 mL methanol and injected dropwise under vigorous stirring. The pyrrole concentration range may vary from 1 to 10% w/w. The mole ratio of $FeCl_3$ to pyrrole is 2.3. The reactional mixture is then stirred for a few hours. The reaction product is filtered off and washed successively with distilled water and methanol. The composite particles are dried in a vacuum oven at 50 °C for 6 h.

Example 21: Solution Synthesis of a Micellar Water-Soluble Polymer/PPy Composite in Water (Adapted from ref. 67)

In a typical experiment 1.0 mL pyrrole (vacuum-distilled from KOH and stored in the dark at −30 °C under argon prior to use) is added to a stirred solution of 5.47 g $FeCl_3$ and 0.3–1.0 g stabilizer (hydroxypropyl cellulose, poly(vinylalcohol), poly(vinylpyridine), poly(ethyleneoxide) or their copolymers) in 100 mL of deionized water at 25 °C. The reaction mixture is stirred for at least 24 h and then centrifuged until a clear yellow-green supernatant is obtained (the centrifugation time and speed vary according to the particle size of individual samples). The resulting sediments are washed several times with 2–3 mL water and allowed to dry under ambient conditions for several days.

b) Syntheses by impregnation or vapor-phase diffusion.

Example 22: Synthesis of a PVC/PPy Composite by Vapor-Phase Polymerization (Adapted from ref. 79)

Polyvinyl chloride (PVC) films impregnated with 20 wt% of $FeCl_3$ are cast on a platinum foil (1 cm^2) from a tetrahydrofuran (THF) solution containing PVC (M_n: 70000) and $FeCl_3$ in appropriate proportions. PVC/PPy blends with thicknesses of 20 and 100 μm are obtained by exposure of the corresponding PVC/$FeCl_3$ films to pyrrole vapors for 64 h in a closed vessel under ambient temperature and pressure. After the polymerization, the blends are immersed for 24 h in acetonitrile (dry and distilled) in order to eliminate iron chloride and residues.

Example 23: Synthesis of an Interpenetrated PVA/PPy Network by Impregnation (Adapted from ref. 76)

$FeCl_3$-impregnated films of poly(vinylacetate) (PVA) are cast in polyethylene Petri dishes from solutions containing PVA (750 mg) and anhydrous $FeCl_3$ (250 mg, *i.e.* 1.54 mmol) in 20 mL THF. The films are first dried under vacuum in a desiccator. The cast films are then dried at 60 °C in vacuum for 48 h. The $FeCl_3$-impregnated films are then dipped into aqueous solutions of pyrrole (5.8 mmol, 20 mL) and kept immersed in the solution for times ranging from 0.5 to 24 h. The films are then withdrawn, washed and soaked in pure water (renewed several times), for a period of 72 h to remove unreacted $FeCl_3$. The composite films thus obtained are then dried under vacuum at 60 °C for 72 h. Complete polymerization yields a composite with about 6.5% PPy by weight.

Example 24: Synthesis of Covalently Anchored PPy Coating on a Pyrrole-Silane-Treated Glass (Adapted from ref. 80)

i. Glass substrate preparation.

The glass substrates are cleaned carefully through the following procedure: they are successively sonicated in methanol and acetone, then refluxed in dichloromethane (in a Soxhlet apparatus) and finally dipped in concentrated (130 v/v) hydrogen peroxide solution before a last sonication in acetone and drying in air. Then, the [7-(pyrrol-1-yl)-4-aza-6-hydroxy]-1-heptyltriethoxysilane (SP1) coating solution is prepared by mixing for 35 min 0.375 g of SP1 and 0.25 mL of acetic acid into 25 mL of deionized water. The solution is stirred 35 min before use. The clean glass foils are then dipped in the pyrrole-silane coating solution for 35 min under weak stirring and nitrogen bubbling. They are then sonicated in deionized water and acetone successively and dried under argon.

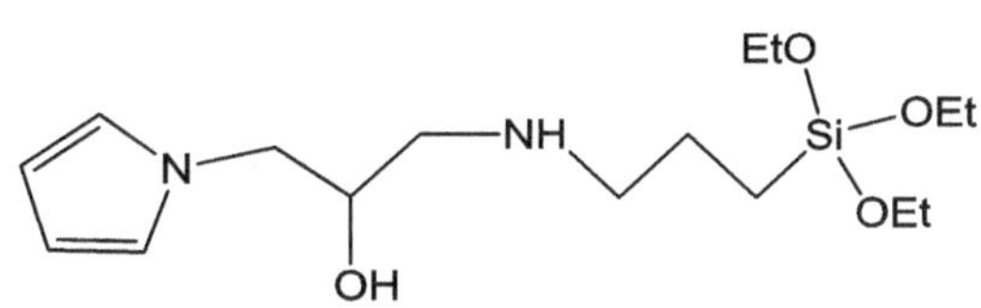

[7-(pyrrol-1-yl)-4-aza-6-hydroxy]-1-heptyltriethoxysilane (SP1)

ii. PPy deposition.

On the coated glass foils is carefully deposited an oxidizing solution made of 1.22 g (7.52 mmol) of ferric chloride and 4.28 g (22.5 mmol) of *p*-toluene sulfonic acid in 20 mL of 80% aqueous propanol. Excess solution is quickly poured out of the foil that is then transferred into a vessel saturated with pyrrole vapor for 5 min (this process has been recognized to lead to very homogeneous pyrrole layers of controlled thickness). The foil is then sonicated in water and acetone successively and dried under vacuum overnight. This process is also effective with glass fibers.

Example 25: Synthesis of a Composite PC/PPy Membrane by Mutual Diffusion (Adapted from ref. 73)

A porous polycarbonate membrane (50 to 100 μm thickness, nanopores ranging from 15 to 100 nm) is used as a dividing wall cell. In the first compartment, an aqueous pyrrole solution (0.5 M) is introduced and allowed to stand for 20 min. After that time, a solution of the oxidant (0.5 M $FeCl_3$ in water) is introduced in the second compartment, and the two solutions are allowed to diffuse through the membrane for a given time to produce the polymer. Once the desired time is reached (from several minutes to one hour), the membrane is removed and thoroughly rinsed with distilled water.

Example 26: One-Step Synthesis of a PPy/MnO_2 Nanocomposite (from ref. 70)

Toluene sulfonic acid (TSA), (2.408 g, 0.014 mol) and pyrrole (1 mL, 0.014 mol) are dispersed ultrasonically in 50 mL deionized water and stirred vigorously to form a homogeneous solution. Separately, $KMnO_4$ (2.212 g, 0.014 mol) is dissolved in 30 mL deionized water, and then slowly poured into the pyrrole:TSA mixture at 0 °C. The redox reaction is conducted without stirring in an ice bath for 24 h. Afterwards, the composite is collected by filtration and washed three times with methanol and deionized water thoroughly, to remove excess monomer and water-soluble impurities. The dark MnO_2/PPy precipitate is finally dried under vacuum at 65 °C for 12 h and weighed. A TSA-doped PPy is prepared by a procedure similar to that of the MnO_2/PPy composite synthesis except that $FeCl_3 \cdot 6\,H_2O$ is employed instead of $KMnO_4$ as the oxidant.

References

1. J. Heinze, B. A. Frontana-Uribe and S. Ludwig, *Chem, Rev.*, 2010, **110**, 4724.
2. P. Audebert and F. Miomandre, In *Handbook of Conducting Polymers 3rd edn*, eds, T. Skotheim and J. R. Reynolds, CRC Press, Boca Raton, 2007.
3. A. J. Downard and D. Pletcher, *J. Electroanalyt. Chem.*, 1986, **206**, 139.
4. H. An, Y. Haga, T. Yuguchi and R. Yosomiya, *Angew. Makromolek. Chem.*, 1994, **218**, 137.

5. X. B. Wan, X. R. Liu, G. Xue, L. X. Jiang and J. J. Hao, *Polymer*, 1999, **40**, 4907.
6. H. Yan, T. Sato, D. Komago, A. Yamaguchi, K. Oyaizu, M. Yuasa and K. Otake, *Langmuir*, 2005, **21**, 12303.
7. J. M. Pringle, J. Efthimiadis, P. C. Howlett, D. R. MacFarlane, A. B. Chaplin, S. B. Hall, D. L. Officer, G. G. Wallace and M. Forsyth, *Polymer*, 2004, **45**, 1447.
8. T. F. Otero and E. DeLaretta, *Synth. Met.*, 1988, **26**, 79.
9. R. K. Sharma, A. C. Rastogi and S. B. Desu, *Electrochem. Commun.*, 2008, **10**, 268.
10. J. P. Wang, Y. L. Xu, J. Wang, X. F. Du, F. Xiao and J. B. Li, *Synth. Met.*, 2010, **160**, 1826.
11. J. Zhang, L. B. Kong, H. Li, Y. C. Luo and L. Kang, *J. Mater. Sci.*, 2010, **45**, 1947.
12. A. M. Fenelon and C. B. Breslin, *Electrochim. Acta*, 2002, **47**, 4467.
13. A. C. Cascalheira, S. Aeiyach, P. C. Lacaze and L. M. Abrantes, *Electrochim. Acta*, 2003, **48**, 2523.
14. T. Tuken, B. Yazici and M. Erbil, *Progr. Organ. Coat.*, 2004, **51**, 152.
15. L. M. M. dos Santos, J. C. Lacroix, K. I. Chane-Ching, A. Adenier, L. M. Abrantes and P. C. Lacaze, *J. Electroanalyt. Chem.*, 2006, **587**, 67.
16. M. Bazzaoui, E. A. Bazzaoui, L. Martins and J. I. Martins, *Synth. Met.*, 2002, **128**, 103.
17. M. Bazzaoui, J. I. Martins, T. C. Reis, E. A. Bazzaoui, M. C. Nunes and L. Martins, *Thin Solid Films*, 2005, **485**, 155.
18. C. L. Gaupp, K. Zong, P. Schottland, B. C. Thompson, C. A. Thomas and J. R. Reynolds, *Macromolecules*, 2000, **33**, 7051.
19. G. Zotti, S. Zecchin and G. Schiavon, *Chem. Mater.*, 2000, **12**, 2996.
20. P. Audebert, G. Bidan and M. Lapkowski, *J. Chem. Soc.-Chem. Commun.*, 1986, 887.
21. P. Audebert, G. Bidan and M. Lapkowski, *J. Electroanalyt. Chem.*, 1987, **219**, 165.
22. J. Grimshaw and S. D. Perara, *J. Electroanalyt. Chem.*, 1990, **281**, 125.
23. A. Haimerl and A. Merz, *Angew. Chem.-Int. Ed.Engl.*, 1986, **25**, 180.
24. M. A. Vorotyntsev, M. Graczyk, A. Lisowska-Oleksiak, J. Goux and C. Moise, *J. Solid State Electrochem.*, 2004, **8**, 818.
25. A. Deronzier and J. C. Moutet, *Coord. Chem. Rev.*, 1996, **147**, 339.
26. S. Cosnier, M. Stoytcheva, A. Senillou, H. Perrot, R. P. M. Furriel and F. A. Leone, *Analyt. Chem.*, 1999, **71**, 3692.
27. T. Livache, B. Fouque, A. Roget, J. Marchand, G. Bidan, R. Teoule and G. Mathis, *Analyt. Biochem.*, 1998, **255**, 188.
28. D. Belanger, J. Nadreau and G. Fortier, *J. Electroanalyt. Chem.*, 1989, **274**, 143.
29. Y. Z. Long, M. M. Li, C. Z. Gu, M. X. Wan, J. L. Duvail, Z. W. Liu and Z. Y. Fan, *Prog. Polym. Sci.*, 2011, **36**, 1415.
30. C. R. Martin, *Acc. Chem. Res.*, 1995, **28**, 61.
31. K. Fukami, F. A. Harraz, T. Yamauchi, T. Sakka and Y. H. Ogata, *Electrochem. Commun.*, 2008, **10**, 56.

32. X. Li, M. Lu and H. Li, *J. Appl. Polym. Sci.*, 2002, **86**, 2403.
33. R. Xiao, S. I. Cho, R. Liu and S. B. Lee, *J. Am. Chem. Soc.*, 2007, **129**, 4483.
34. S. Demoustier-Champagne, J. Duchet and R. Legras, *Synth. Met.*, 1999, **101**, 20.
35. C. Jerome and R. Jerome, *Angew. Chem.-Int. Ed.*, 1998, **37**, 2488.
36. J. F. Zang, C. M. Li, S. J. Bao, X. Q. Cui, Q. L. Bao and C. Q. Sun, *Macromolecules*, 2008, **41**, 7053.
37. Y. S. Yang and M. X. Wan, *J. Mater. Chem.*, 2001, **11**, 2022.
38. C. Debiemme-Chouvy, *Electrochem. Commun.*, 2009, **11**, 298.
39. S. Gupta, *App. Phys. Lett.*, 2006, **88**.
40. C. B. Gorman, H. A. Biebuyck and G. M. Whitesides, *Chem. Mater.*, 1995, **7**, 526.
41. S. H. Cho, K. T. Song and J. Y. Lee, in *Handbook of Conducting Polymers 3rd edn*, eds, T. Skotheim and J. R. Reynolds, CRC Press, Boca Raton, 2007.
42. C. Peng, S. W. Zhang, D. Jewell and G. Z. Chen, *Prog. Natural Sci.-Mater. Int.*, 2008, **18**, 777.
43. M. Gao, S. Huang, L. Dai, G. Wallace, R. Gao and Z. Wang, *Angew. Chem. Int. Ed.*, 2000, **39**, 3664.
44. J. H. Chen, Z. P. Huang, D. Z. Wang, S. X. Yang, W. Z. Li, J. G. Wen and Z. F. Ren, *Synth. Met.*, 2001, **125**, 289.
45. G. Z. Chen, M. S. P. Shaffer, D. Coleby, G. Dixon, W. Z. Zhou, D. J. Fray and A. H. Windle, *Adv. Mater.*, 2000, **12**, 522.
46. X. T. Zhang, J. Zhang and Z. F. Liu, *Carbon*, 2005, **43**, 2186.
47. X. Q. Lin and Y. H. Xu, *Electrochim. Acta*, 2008, **53**, 4990.
48. M. Bozlar, F. Miomandre and J. B. Bai, *Carbon*, 2009, **47**, 80.
49. T. Rapecki, M. Donten and Z. Stojek, *Electrochem. Commun.*, 2010, **12**, 624.
50. O. Reynes and S. Demoustier-Champagne, *J. Electrochem. Soc.*, 2005, **152**, D130.
51. P. Schottland, K. Zong, C. L. Gaupp, B. C. Thompson, C. A. Thomas, I. Giurgiu, R. Hickman, K. A. Abboud and J. R. Reynolds, *Macromolecules*, 2000, **33**, 7051.
52. A. Yokoyama, A. Kato, R. Miyakoshi and T. Yokozawa, *T.* Macromolecules, 2008, **41**, 7271.
53. T. Yokozawa, Y. Nanashima and Y. Ohta, *ACS Macro Lett.*, 2012, **1**, 862.
54. P. Audebert and G. Bidan, *Synth. Met.*, 1986, **15**, 9.
55. G. Bidan, *J. Chim. Phys.*, 1989, **1**, 86.
56. C. P. Andrieux, P. Audebert, P. Hapiot and J. M. Savéant, *J. Phys. Chem.*, 1991, **95**, 10158.
57. J. Y. Lee, D. Y. Kim and C. Y. Kim, *Synth. Met.*, 1995, **74**, 103.
58. P. Audebert and G. Bidan, *Synth. Met.*, 1986, **14**, 71.
59. Y. Kudoh, *Synth. Met.*, 1996, **79**, 17.
60. E. J. Oh and K. S. Yang, *Synth. Met.*, 2001, **119**, 109.
61. R. Ansari and N. K. Fahim, *React. Funct. Polym.*, 2007, **67**, 367.

62. T. S. Kang, S. W. Lee, J. Joo and J. Y. Lee, *Synth. Met.*, 2005, **153**, 61.
63. C.-F. Liu, D.-K. Moon, T. Maruyama and T. Yamamoto, *Polym. J.*, 1992, **25**, 775.
64. S. Jing, S. Xing, L. Yu and C. Zhao, *Mater. Lett.*, 2007, **61**, 4528.
65. S. Machida, S. Miyata and A. Techagumpuch, *Synth. Met.*, 1989, **31**, 311.
66. T. Yoshikawa, S. Machida, T. Ikegami, A. Techagumpuch and S. Miyata, *Polym. J.*, 1990, **22**, 1.
67. S. P. Armes, M. Aldissi and S. F. Agnew, *Synth. Met.*, 1989, **28**, 837.
68. Y. Kudoh, K. Akami and Y. Matsuya, *Synth. Met.*, 1998, **95**, 191.
69. S. P. Armes, M. Aldissi, G. C. Idzorek, P. W. Keaton, L. J. Rowton, G. L. Stradling, M. T. Collopy and D. B. McColl, *J. Colloid Interface Sci.*, 1991, **141**, 119.
70. Z. H. Dong, Y. L. Wei, W. Shi and G. A. Zhang, *Mater. Chem. Phys.*, 2011, **131**, 529.
71. A. Reung-U-Rai, A. Prom-Jun, W. Prissanaroon-Ouajai and S. Ouajai, *J. Met., Mater. Miner.*, 2008, **18**, 27.
72. S. J. Peighambardoust and B. Pourrabas, *Macromol. Symp.*, 2007, **247**, 99.
73. S. D. Demoustier-Champagne, J. Duchet and R. Legras, *Synth. Met.*, 1999, **101**, 20.
74. Y. S. Lu, G. Shi, C. Li and Y. Liang, *J. Appl. Polym. Sci.*, 1998, **70**, 2169.
75. J. Jang and H. Yoon, *Chem. Commun.*, 2003, 720.
76. M. Chakraborty, D. C. Mukherjee and B. M. Mandal, *Synth. Met.*, 1999, **98**, 193.
77. R. A. Zoppi and M.-A. De Paoli, *Polymer*, 1996, **37**, 1999.
78. R. A. Zoppi and M.-A. De Paoli, *J. Electroanalyt. Chem.*, 1997, **437**, 175.
79. T. Matencio, V. Mano, M. I. Felisserti and M.-A. De Paoli, *Electrochim. Acta*, 1994, **39**, 1393.
80. F. Faverolle, A. J. Attias, B. Bloch, P. Audebert and C. P. Andrieux, *Chem. Mater.*, 1998, **10**, 740.
81. H. Cattey, P. Audebert and A. J. Attias, *Synth. Met.*, 1998, **93**, 127.
82. S. P. Armes, S. Gottesfeld, J. G. Beery, F. Garzon and S. F. Agnew, *Polymer*, 1991, **32**, 2325.

CHAPTER 12

Polyanilines

JACOB TARVER* AND YUEH-LIN LOO*

Princeton University, Department of Chemical and Biological Engineering, Princeton, NJ 08540, USA
*Email: jtarver@princeton.edu; lloo@princeton.edu

12.1 Introduction

The first synthesis of what is now known to be polyaniline (PANI) was reported in 1862 as the product of the electrochemical oxidation of aniline in sulfuric acid.[1] Studies carried out in 1910–1911 correctly identified a molecular sequence involving alternating amine/imine and benzenoid/quinoid moieties, and revealed that oscillations can occur between PANI's distinct oxidation states.[2,3] Due to the scientific community's poor comprehension of macromolecules at that time, however, the polymeric nature of PANI was not identified. Progress remained dormant until a publication in the 1960 s reporting that this material exhibited long-lasting electrical conductivity following exposure of its stable half-oxidized/half-reduced form, known as emeraldine base (EB) of polyaniline, to a protonic acid.[4] Further interest was generated following Heeger, MacDiarmid, and Shirakawa's Nobel Prize winning work that identified the physical mechanisms of hole transport in conductive polymer systems.[5–7] The realization that PANI's stable proton-doping mechanism had many advantages over redox doping exhibited by alternative conducting polymers propelled PANI towards the forefront of conductive polymer research. Current research has established PANI as a rival to commercially available poly(3,4-ethylenedioxythiophene) (PEDOT) systems in many electronic applications.

A new wave of interest in PANI occurred in the mid-1990 s when MacDiarmid and Epstein identified a process termed secondary doping in

RSC Polymer Chemistry Series No. 9
Conjugated Polymers: A Practical Guide to Synthesis
Edited by Klaus Müllen, John R. Reynolds and Toshio Masuda

Published by the Royal Society of Chemistry, www.rsc.org

which a solvent is introduced to induce restructuring of PANI, thereby dramatically enhancing its electrical conductivity.[8–11] X-ray diffraction (XRD) analysis of secondary-doped PANI films reveal a large increase in crystallinity persisting even after the complete evaporation of the solvent.[11] Rheological studies on dilute solutions of this form of PANI show a three-fold increase in reduced viscosity following secondary doping, indicative of a transition from a compact-coil structure to an expanded coil conformation.[8] The increase in crystallinity in the solid state and polymer-coil expansion in solution both suggest enhanced conjugation of PANI chains. The significant increase in conductivity associated with this structural extension arises from increased electron delocalization along PANI backbones, and has consequently been exploited in subsequent PANI research.[12]

In its neutral base form, PANI is characterized by three distinct oxidation states. Of the three neutral forms of PANI, the blue intermediate EB oxidation state (chemical structure shown in Chart 12.1a) is the most stable under ambient conditions. The leucoemeraldine base (LB, chemical structure shown in Chart 12.1b) and pernigraniline base (PB, chemical structure shown in Chart 12.1c) oxidation states result from the reduction and oxidation, respectively, of EB. The former is colorless or yellow in aerobic environments and develops a light-green tint as it slowly oxidizes, while the latter is violet and undergoes degradation in strongly oxidizing environments to form benzoquinone-terminated fragments. The unique aspects of PANI emerge upon exposure of the EB oxidation state to strong acids, inducing the formation of the green emeraldine salt form of PANI (ES, chemical structure shown in Chart 12.1d) that exhibits a conductivity that is up to ten orders of magnitude higher than its unprotonated counterpart.[4]

The structure of ES is difficult to elucidate directly. Spin-relaxation studies, however, strongly suggest that the structure of ES depicted in Chart 12.1d, in which semiquinone radical cations are separated by individual mers, is favored over the structure in which cations are present on adjacent mer units.[13–15] The latter case would lead to bipolaronic conduction bands and would render the

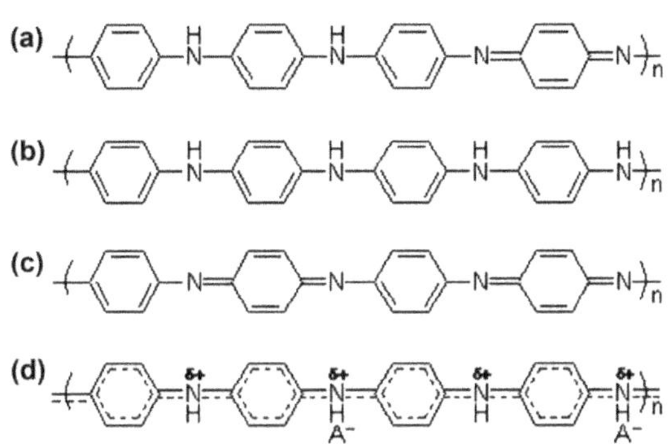

Chart 12.1 The chemical structure of (a) emeraldine base (EB), (b) leucoemeraldine base (LB), (c) pernigraniline base (PB), and (d) emeraldine salt (ES), where A represents the dopant's counterion.

material diamagnetic – a property not observed in the ES form of PANI. A combination of electron paramagnetic resonance, photoinduced adsorption, and Raman spectroscopy measurements has firmly established that ES's conductivity is the result of delocalization of these segregated semiquinone radical cations along the polymer backbone.[13] These distinct radical cations are in turn stabilized by acidic environments and allow charge transport through a uniformly distributed polaronic conduction band.[13]

Early studies on PANI employed small-molecule acids, such as hydrochloric acid (HCl) or sulfuric acid, as dopants for inducing the formation of the ES state. Although capable of inducing electrical conductivities on the order of 10^1 S/cm, small-molecule dopants typically render PANI insoluble in conventional solvents. Polymer acids have thus been used to overcome many of the processing shortcomings associated with their small-molecule counterparts. By incorporating acidic and hydrophilic functional groups, polymer acids can serve as dopants, while simultaneously conferring partial solubility to the PANI-polymer acid complex. The inclusion of large volume fractions of electrically insulating polymer acids, however, degrades the electrical properties of PANI. The conductivity of polymer-acid doped PANI systems typically fails to exceed 1 S/cm.

A number of reviews exist describing the synthesis of PANI in many of its different oxidation states, as well as the direct synthesis of its doped ES form, and the various advantages and disadvantages of differing procedures.[16–20] This chapter will focus on the experimental aspects of the polymerization of various forms of PANI and its derivatives, and will detail secondary doping techniques for enhancing the properties of PANI postsynthesis.

12.1.1 Synthesis Mechanism

The polymerization of aniline is initiated by the oxidation of an aniline monomer to form a resonantly stabilized anilinium radical. Polymerization proceeds in a head-to-tail fashion in acidic media through preferential coupling with other anilinium radical cations at the *para* position. Synthesis at elevated pH yields coupling at the *ortho* and *meta* positions; such structural defects result in poor electrical properties.[17] Through further oxidation, this newly formed dimer ejects two electrons and two protons to form a new radical that can accept another anilinium radical cation. Propagation subsequently proceeds through the repeated ejection of two electrons and two protons followed by the *para* coupling of more anilinium radicals. This polymerization mechanism is depicted in Scheme 12.1.[19]

12.2 Experimental Procedures

Synthesis of polyaniline is most often achieved through the direct polymerization of commercially available aniline monomers. Synthesis is usually performed in acidic media to yield the doped ES form of PANI that are coupled in the *para* – as opposed to *ortho* or *meta* – positions. The direct syntheses of

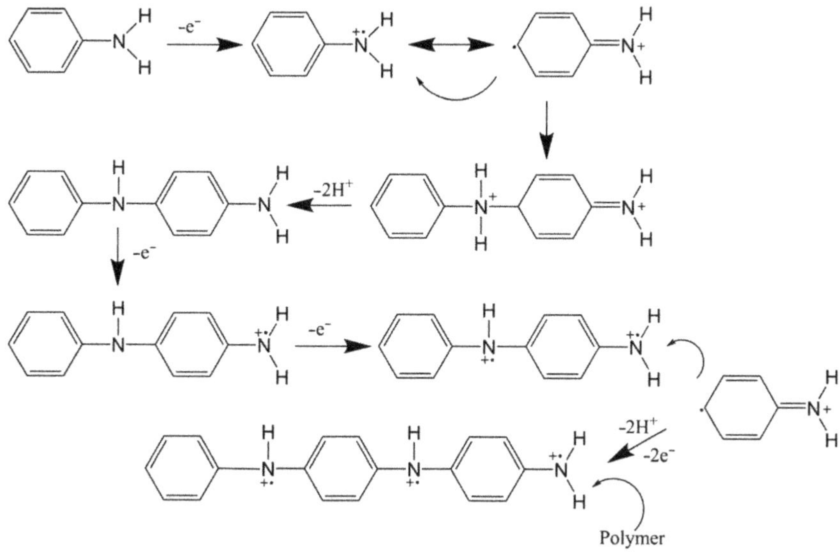

Scheme 12.1 Mechanism for polymerization of aniline in its doped ES form. Adapted from ref. 19.

electrostatically neutral forms of PANI in neutral and basic media have been reported,[21] but these procedures have not been extensively utilized.[16]

Prior to polymerization, aniline is often purified through distillation. Purified aniline should subsequently be deaerated under nitrogen or argon to prevent oxygen-induced autoinitiation. For chemical polymerization, oxidizing agents must also often be purified, dried, and deaerated prior to their addition to the acidic aniline solution; a syringe is typically used to transfer deaerated solutions containing the oxidizing agent into deaerated solutions containing aniline. Both chemical and electrochemical syntheses should take place under nitrogen or argon.

Deaerated water is the most common solvent for PANI synthesis due to the aqueous solubility of aniline, common acids (both organic and inorganic), and many oxidizing agents. In addition, the poor solubility of PANI's ES form leads to facile separation following synthesis. The aqueous polymerization of aniline is an exothermic reaction. As a consequence, care should usually be taken to regulate the temperature of the reaction vessel during PANI synthesis. The synthesis of PANI is generally quenched by exposing the contents of the reaction vessel to oxygen. Due to the poor aqueous solubility of PANI's as-synthesized ES form, the isolated product is typically deprotonated into the more-soluble EB oxidation state *via* exposure to bases for subsequent molecular characterization.

12.2.1 Chemical Polymerization of Anilines

The three most important experimental factors identified for controlling the chemical synthesis of PANI are reaction temperature, pH, and dopant identity.[19,22] The influence of these three parameters on PANI's molecular characteristics and overall yield are emphasized the following examples.

Example 1: Influence of Temperature (–35 °C to 0 °C) on the Synthesis of HCl-Doped PANI with Sodium Dichromate in Aqueous Solution

A 150 mL aqueous solution containing 0.1 mol (9.31 g) of aniline and 0.294 mol (29.0 g) of HCl (dopant) is prepared and cooled to 0 °C. For reactions at subzero temperatures, minimal amounts of lithium chloride (LiCl) are added to this solution to suppress the freezing temperature of the solution below the desired reaction temperature. A 27.6 mL solution containing 0.0417 mol (12.42 g) of sodium dichromate dihydrate (oxidizing agent) is prepared and slowly injected into the reaction mixture at 5 g/h, yielding a final molar ratio of 1 : 3.94 : 0.417 aniline : HCl : oxidant. The reaction is allowed to proceed for 45 h at the pre-determined reaction temperature. The reaction contents (green) are then filtered, washed with water, and deprotonated to the EB state by stirring in 200 mL of a 35% ammonium hydroxide solution for 1 h. The deprotonated product is filtered, washed with water followed by isopropanol, and dried *in vacuo* at 60 °C for 48 h. The resulting PANI has number-average molecular weights (M_n) spanning 8.3–20.9 kilodaltons and 78.5–97.6% yields, as measured by gel permeation chromatography (GPC) using *N*-methylpyrrolidone (NMP) as an eluent. Molecular weights are reported as polyvinylpyridine (PVP) equivalence. The molecular weight of PANI synthesized using this recipe generally increases as the reaction temperature is lowered; the maximum molecular weight occurs when PANI synthesis is carried out at –22.0 ± 3 °C (detailed in Table 12.1).[23]

Example 2: Influence of Temperature (–35 °C to +18 °C) on the Synthesis of HCl-Doped PANI with Ammonium Peroxydisulfate in Aqueous Solution

An aqueous solution is prepared by adding 0.1 mol (9.31 g) of aniline to 100 g of 1.0 M HCl in a jacketed reaction vessel. The pH of this solution is adjusted to 1.0 with additional HCl. For reactions proceeding above 0 °C, a solution containing 51.5 g of distilled water and 0.125 mol (28.5 g) of ammonium peroxydisulfate (APS; oxidant) is stirred directly into the reaction mixture and the

Table 12.1 Chemical synthesis conditions and resulting PANI characteristics using sodium dichromate as an oxidizing agent. Adapted from ref. 23.

Temp (°C)	*Wt. LiCl (g)*	*Wt. PANI (g)*	*Yield (%)*	M_w *(Dalton)*	M_n *(Dalton)*	*PDI* (M_w/M_n)
0.0	0.0	8.90	97.6	104000	20900	5.0
– 15.0	20.6	8.80	96.2	159000	20700	7.7
– 20.0	25.5	8.83	96.3	181000	17770	10.2
– 23.0	28.4	9.10	99.2	203000	23900	8.5
– 25.0	30.3	8.92	97.1	169000	19700	8.6
– 27.0	32.3	8.79	95.7	99000	13100	7.6
– 29.0	34.3	8.73	94.2	129000	15600	8.3
– 31.0	36.5	8.25	88.0	125000	14000	8.9
– 33.0	38.7	7.77	82.0	116000	9900	11.7
– 35.0	41.2	7.46	78.5	101000	8300	12.1

Table 12.2 Chemical synthesis conditions and resulting PANI characteristics using APS as oxidizing agent. Adapted from ref. 24.

Temp. (°C)	*Wt. LiCl (g)*	*Wt. PANI (g)*	*Yield (%)*	M_p *(Dalton)*	M_w	M_n	*PDI* (M_w/M_n)
18	0.0	8.46	93.0	19400	29700	4180	7.1
0	0.0	8.12	89.4	86000	122000	20400	6.0
– 10.0	10.35	8.35	91.8	107000	148000	28400	5.2
– 15.0	13.91	8.42	92.4	137000	159000	26800	5.9
– 20.0	17.12	8.54	93.4	133000	154000	24600	6.2
– 22.5	18.83	8.50	92.7	142000	178000	25400	7.0
– 25.0	20.76	8.84	96.7	153000	209000	43500	4.8
– 27.5	23.03	9.11	99.2	145000	190000	31700	6.0
– 35.0	25.79	8.60	93.6	138000	176000	27500	6.4
– 35.0	33.53	8.15	87.9	135000	166000	19100	8.6

contents are allowed to react for 30 min. For reactions proceeding below 0 °C, minimal amounts of LiCl are added to the reaction contents to suppress the freezing temperature of the solution below the desired reaction temperature. After cooling to the desired temperature and under continuous stirring, an APS solution identical to that described above is added dropwise to the reaction vessel over the course of 14.25 h. Afterwards, the reaction vessel is left to stir for a total of 46 h. The resulting products of both syntheses are filtered and washed with water, and then deprotonated to the EB oxidation state by stirring in a 100 mL aqueous solution of 33 w/v% ammonia for 24 h. The products are again filtered, washed with water and isopropanol, and dried *in vacuo* at 60 °C for 24 h. The yield for PANI that is synthesized above 0 °C is 93.0%; a purple powder with M_n of 4.18 kilodaltons and a polydispersity index (PDI) of 7.1, as determined by GPC (NMP eluent, PVP standards) results. The yield of PANI that is synthesized at-and-below 0 °C ranges between 89.4–99.2% and the M_n and PDI span 19.1–43.5 kilodaltons and 4.8–8.6, respectively. Synthesis following this recipe yields the highest molecular weight and narrowest molecular weight distribution when the reaction temperature is kept at –25.0 °C. Additional molecular characteristics of PANI from this synthesis are described in Table 12.2.[24]

Example 3: Influence of Temperature (–50 °C to 50 °C) and HCl Concentration on the Synthesis of HCl-Doped PANI with Ammonium Peroxydisulfate in the Solid State and in Aqueous Solutions

A 7.8-mL aqueous solution containing 0.2 M (0.259 g) of aniline hydrochloride and a 2.0 mL solution containing 1 M (0.456 g) of APS are prepared separately, yielding a mole ratio of 1 : 1 : 1 aniline : HCl : APS upon mixing. A separate reaction containing reactants in a 1 : 6 : 1 mole ratio of aniline : HCl : APS is prepared for comparison. For reactions above –10 °C, in which the polymerization proceeds in the solution state, the aniline/HCL and APS solutions are equilibrated at the desired temperature prior to mixing. For reactions below – 10 °C, in which the polymerization proceeds in the solid state, reactants are mixed at 0 °C, frozen at –78 °C, and brought back up to and equilibrated at the desired temperature for solid-state polymerization. Polymerization is allowed to proceed until the liquid or solid-state mixture becomes green, indicating the presence of PANI in the ES state. The PANI is subsequently collected on a filter. For PANI synthesized above 10 °C, the collected material is washed with water and acetone, while PANI that is synthesized below −10 °C is washed with 1.0 M HCl and acetone. All products are then dried *in vacuo* at 60 °C. The PANI is dissolved in NMP and deprotonated to the EB oxidation state for GPC measurements (NMP eluent, polystyrene standards). PANI synthesized at temperatures below 50 °C and above −50 °C yield 70–80% conversion with weight-average molecular weight (M_w) spanning 26.6–75.3 kilodaltons and PDI spanning 2.2–3.0, regardless of the acidity of the synthesis medium. The molecular weight of PANI is generally observed to increase as the reaction temperature decreases; crystallinity is observed to grow with increased molecular weight and increasing synthesis medium acidity. Additional experimental results are described in Table 12.3.[25]

Example 4: Polymerization of Alkyl- and N-alkyl-Substituted Anilines

Aqueous, 133 mL solutions of 1.0 M HCl and 0.133 mol of either *N*-ethylaniline (16.1 g), *N*-methylaniline (14.3 g), 2-ethylaniline (16.1 g), or 3-ethylaniline (16.1 g) are prepared. Each is mixed with a solution prepared by dissolving 0.133 mol of APS into 80 mL of water and the combined solution is allowed to react for 3, 12, 24, and 24 h, respectively, at temperatures between 20–30 °C. The filtered contents are then washed with water, dedoped to the EB oxidation state with an excess of a 37% ammonia solution, and washed again with water and ethanol (with the exception of *N*-methylaniline, for which only water is used in washing) after the completion of the reaction. The resulting blue powders are dried to provide yields between 57.0–85.1%, as summarized in Table 12.4.[26]

Table 12.3 Chemical synthesis conditions and resulting pH-dependent PANI characteristics. Adapted from ref. 25.

Temp. (°C)	*0.2 MHCI* $10^{-3}M_w$	M_w/M_n	*1.2 MHCI* $10^{-3}M_w$	M_w/M_n
− 50	182.6	3.1	71.5	2.7
− 30	70.7	3.0	75.3	2.6
− 10	47.5	2.9	51.0	2.6
0	42.2	2.5	37.0	2.3
10	30.3	2.4	27.8	2.3
30	26.6	2.2	32.0	2.2
50	18.2	2.5	25.0	2.1

Table 12.4 Summary of reaction conditions and results for polymerization of alkyl- and *N*-alkyl-substituted anilines. Adapted from ref. 26.

Monomer	*Reaction time (h)*	*Product*	*Yield (%)*
N-ethylaniline	3	N-ethyl-PANI-EB	71.3
N-methylaniline	12	N-methyl-PANI-EB	57.0
2-Ethylaniline	24	2-Ethyl-PANI-EB	85.1
3-Ethylaniline	24	3-Ethyl-PANI-EB	67.0

Example 5: Polymerization of o-Methoxyaniline

A homogenous solution of *o*-methoxylaniline is formed by thoroughly stirring 0.0548 mol (6.75 g) of the monomer in 75 mL of 1.0 M HCl at 0 °C. Synthesis is quickly initiated by adding 0.0126 mol (2.88 g) of APS dissolved in 50 mL of 1.0 M HCl over the course of 1 min. The solution is constantly stirred and maintained between 0–3 °C while monitoring the open-circuit potential using a platinum electrode and a saturated calomel electrode for reference. After 2 h, the open-circuit potential plateaus at 0.34 V *vs.* SCE, at which point the reaction is terminated. Following synthesis, the precipitate is filtered and washed with acetone until the filtrate is colorless. The resulting powder is dedoped to the EB oxidation state by stirring in excess 0.1 M ammonium hydroxide for 48 h. The material is again filtered and washed with water before drying *in vacuo* for 24 h to yield poly(*o*-methoxyaniline) with a M_w of 30.8 kilodaltons and a PDI of 1.71 as determined by GPC (NMP eluent, polystyrene standards).[27]

Example 6: Polymerization of Self-Doped 2-Methoxyaniline-5-sulfonic Acid

A reaction mixture is prepared by dissolving 0.2 mol (40.6 g) of 2-methoxyaniline-5-sulfonic acid into a 100 mL aqueous solution containing 0.25 mol (19.8 g) of pyridine at 4 °C. A separate 180 mL aqueous solution containing 0.2 mol (45.6 g) of APS is prepared and added to the reaction mixture, and the resulting contents are allowed to stir for 15 h at room temperature. The precipitate is then collected, washed with methanol, and dried *in vacuo* at 30 °C for 12 h. The resulting self-doped sulfonated PANI derivative, whose chemical structure is depicted in Chart 12.2, exhibits a PEO-equivalent molecular weight of 10^4 Daltons, as measured by GPC. The yield from the reaction is 96%.[28]

Example 7: Synthesis of Ring-Sulfonated PANI in Protonated and Non-protonated Forms

EB (0.5 g), synthesized inhouse or purchased commercially, is dissolved in 40 mL of 30% fuming sulfuric acid by stirring at 5 °C. After 2 h of stirring, the solution is precipitated by slowly pouring the reaction contents into 200 mL of methanol, followed by 100 mL of acetone, at 10–20 °C. The green precipitate is washed repeatedly with methanol until the filtrate has a pH of 7 and dried *in vacuo* for 24 h to yield ring-sulfonated, protonated PANI. Elemental analysis and Fourier-transform infrared (FTIR) spectroscopy confirm that the synthesized polymer is consistent with the chemical structures shown in Chart 12.3. To generate ring-sulfonated, nonprotonated PANI, 0.5 g of the previously synthesized protonated form is dissolved in 20 mL of 0.1 M sodium hydroxide

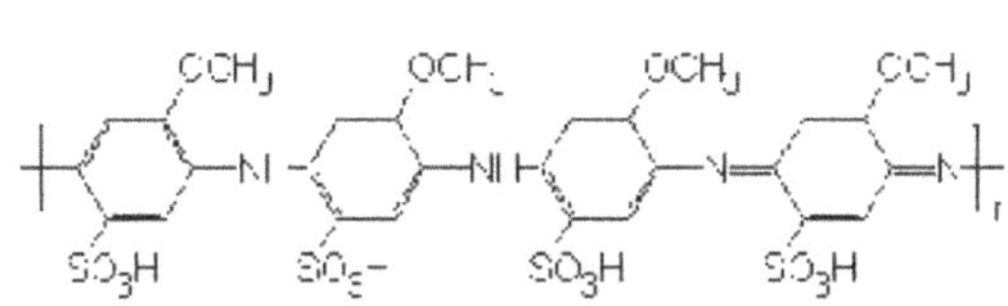

Chart 12.2 Chemical structure of poly(2-methoxyaniline-5-sulfonic acid).

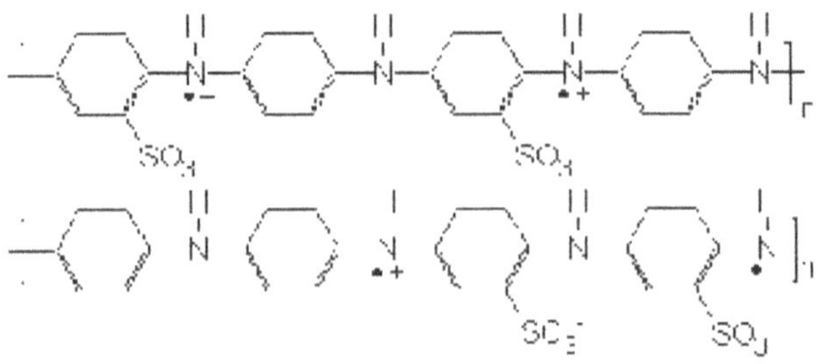

Chart 12.3 Chemical structures of ring-sulfonated, protonated PANI.

by stirring for 2 h at room temperature. The vessel containing the contents is slowly evacuated, and subject to 24 h *in vacuo* at room temperature to yield purple ring-sulfonated, nonprotonated PANI.[29,30]

Example 8: Synthesis of the PB Form of PANI

EB (1.0 g), synthesized inhouse or purchased commercially, is dissolved in a solution comprising 100 mL of NMP and 5 mL of glacial acetic acid. A separate solution comprising 0.72 g of purified *m*-chloroperbenzoic acid in 10 mL of glacial acetic acid is added in 1 mL increments every 10 min. The solution is dark purple after 15 min of stirring at room temperature. The purple powder is precipitated by adding 10 mL of triethylamine. The precipitate is filtered and stirred in 60 mL of acetone for 30 min prior to filtering again. After repeating this procedure, the precipitate is washed with 200 mL of acetone followed by 200 mL of diethyl ether to yield PB in the form of a partially crystalline brownish-purple powder. No evidence of chain scission is observed during the synthesis, and as a consequence, the molecular characteristics of the PB powder remain unchanged from the starting EB.[31]

12.2.2 Electrochemical Polymerization of Anilines

The electrochemical polymerization of anilines has several distinct advantages over the chemical synthesis of PANI. Rigid control of the oxidizing potential applied during synthesis provides the unique ability to moderate the polymerization's initiation and termination processes in a manner that chemical reagents are poorly suited for.[32] PANI produced through electrochemical polymerization is, consequently, not contaminated by unreacted oxidizing agents or other adulterants that can decrease the purity of chemically synthesized PANI.[17,19] In addition, electrochemical synthesis obviates the need for postsynthesis processing in several applications; techniques such as scanning electrochemical microscopy and electrochemical dip-pen lithography can be used to pattern PANI directly onto substrates.[33,34] As with other electrochemical syntheses, care must be taken to ensure strict cleaning of the electrodes and experimental apparatus prior to use. Working electrodes, for instance, are often mechanically polished in order to remove surface contamination and to reduce surface roughness by using a series of diamond or alumina particle suspensions in which the particle size is successively decreased.

Example 9: Potentiostatic Polymerization of Aniline in HCl

Aniline (0.15 mL, 0.153 g) is thoroughly mixed with HCl (1 M) in a 223 mL single-compartment, three-electrode aqueous cell. A 0.7 cm diameter, 15.0 cm length graphite rod is used as a working electrode; a 13.0 cm length, 2.0 cm width, and 0.03 cm thick stainless steel plate is used as a counterelectrode; a saturated calomel electrode (SCE) is used as a reference electrode. Aniline is polymerized potentiostatically by applying a 0.7 V (vs. SCE) potential difference between the working and counterelectrodes for 2 h. The PANI formed on the graphite electrode is collected, washed with distilled water, filtered, and dried *in vacuo* at room temperature. Excess HCl is removed by dedoping the PANI to the EB oxidation state with excess ammonia. The dedoped PANI is redoped with 1 : 1 weight ratio of *p*-toluenesulfonic acid, filtered, washed with distilled water, and again dried *in vacuo* at room temperature. Integration of FTIR peaks measured on a pressed pellet consisting of potassium bromide and the collected powder in a 10 : 1 ratio indicate a quinoid:benzenoid ratio of 0.8 : 1.0; this observation is consistent with the powder being PANI in its ES state. Scanning electron micrographs (SEM) reveal particulate aggregates having sizes of 20–40 μm.[32]

Example 10: Potentiodynamic Polymerization of Aniline in Sulfuric Acid

Aniline is potentiodynamically polymerized in a one-compartment, three-electrode cell using a platinum working electrode (0.5 cm^2) and an SCE reference electrode. Aniline is dissolved in 0.1 M sulfuric acid solution. The potential of the working electrode is swept continuously between –0.2 V and 0.8 V *vs.* SCE at 100 mV/s until 140 mC is consumed, yielding PANI in the ES oxidation state that strongly adheres to the platinum electrode. The coated electrodes are thoroughly rinsed with water and stored in 0.1 M sulfuric acid. SEM micrographs reveal a uniform and smooth film.[35]

Example 11: Galvanostatic Polymerization of Aniline in HCl

Aniline (1 M) is mixed thoroughly in a 2 M solution of HCl. A platinum plate and a platinum-coated quartz plate are used as working and counterelectrodes, respectively, in a single-compartment, two-electrode cell under a N_2 atmosphere. Aniline is polymerized galvanostatically on the surface of the electrodes by applying a constant current of 0.1 mA/cm^2 until 90 mC/cm^2 of charge is consumed. Multiple-beam interferometry measurements show that the film generated at 90 mC/cm^2 is approximately 100-nm thick, while IR spectroscopy performed on the film reveal that the galvanostatically synthesized PANI exists in a mixture of ES and PB oxidation states.[36]

12.2.3 Template Polymerization of Anilines

Template polymerization of aniline onto macromolecular acids is a strategy that has achieved a significant level of success with regards to overcoming PANI's limited solubility. Template polymerization of aniline and aniline derivatives proceeds through the same mechanism described for the traditional synthesis of PANI and can be performed *via* chemical or electrochemical routes, with modifications to previously established chemical and electrochemical procedures to incorporate a desired template. The template typically serves the role of PANI's protonic dopant, and as a consequence of its coordination of aniline monomers during polymerization, can be used to guide the synthesis of PANI along or within the template's structure. Importantly, the template introduces dispersability of PANI in aqueous solutions or common organic solvents, though often at the expense of diminished electrical conductivity. Due to intrinsic mass-transfer limitations, it is usually impractical to isolate and characterize PANI from the PANI/template complexes. Consequently, these materials are often characterized in terms of the composite's properties.

Example 12: Enzymatic Polymerization of Poly(styrene sulfonic acid)-Doped PANI

A 30 mL, 0.1 M sodium phosphate buffer solution (pH 4.3) containing poly-(styrene sulfonate) sodium salt, (NaPSS; 6 mM, based on monomer units) and aniline (6 mM, 0.0168 g) is prepared. A 0.2 mL, 10 mg/mL horseradish peroxidase (HRP) solution is added to the reaction mixture to serve as a renewable oxidizing agent. To activate HRP, 9 mL of a 0.02 M hydrogen peroxide solution is added dropwise to the reaction vessel over the course of 1.5 h under constant stirring. Hydrogen peroxide oxidizes compounds in a two-electron reaction, and as a consequence, is ordinarily unable to provide the single electron required for radicalization of aniline monomers or oligomers. In enzymatic polymerization, however, the hydrogen peroxide serves to oxidize HRP in a two-electron reaction; the HRP in turn sequentially oxidizes two aniline monomers or growing oligomers *via* one-electron reactions and returns to its native state. The reaction mixture is allowed to stir for 1 h prior to overnight dialysis against deionized water that has been adjusted to pH 4.3, yielding a stable solution of PSS-doped PANI in the ES oxidation state. Mass-transfer limitations preclude the isolation of PANI; however, GPC measurements (with pH 7 phosphate buffer as eluent) obtained at different stages of PANI's synthesis indicate that the retention time of PANI:PSS complex consistently shifts towards lower values throughout the reaction and thus suggest continued PANI growth.[37]

Example 13: Enzymatic Polymerization of Micellar-Templated PANI

Aqueous micelle solutions are prepared by adding sodium dodecylbenzenesulfonic acid (10 mM, 0.104 g) at concentrations above the critical micelle concentration (~1.6 mM) to 30 mL of a 0.1 M phosphate buffer solution under continuous stirring. Aniline (10 mM, 0.028 g) is added and stirred until dissolved. The solution is adjusted to a pH of 4.3 by the addition of 1.0 M HCl, followed by the addition of 0.2 mL of HRP (10 mg/mL). The polymerization of aniline is initiated by the dropwise addition of an amount of dilute hydrogen peroxide (20 mM) equivalent in moles to aniline over the course of 1.5 h under continuous stirring. HRP gains two electrons upon being oxidized by hydrogen peroxide, and subsequently oxidizes two aniline monomers to return to its native state. The solution is allowed to stir for an additional hour to yield a stable solution of micellar-templated PANI in the ES oxidation state.[38]

Example 14: Template Synthesis of PANI with Poly(2-acrylamido-2-methyl-1-propanesulfonic Acid)

A series of 50 mL, 100 mL, and 250 mL aqueous solutions containing 0.028 mol (2.61 g) of aniline, 0.028 mol (5.80 g) of poly(2-acrylamido-2-methyl-1-propanesulfonic acid) (PAAMPSA; based on monomer units), and 0.0252 mol (5.75 g) of APS are separately prepared, immersed in ice baths, and deaerated under nitrogen. The aniline solution is added to the vessel containing the PAAMPSA and thoroughly mixed prior to slowly adding the APS dropwise, bringing the reactants to a 1:1:0.9 molar ratio of aniline:PAAMPSA mer:APS. The resulting contents are stirred constantly while immersed in an ice bath for 24 h. During the course of polymerization, the solution undergoes sequential color changes from transparent, to blue, and finally to green. The PANI-PAAMPSA complex is subsequently precipitated in the form of particles by slowly adding 3.0 L of acetone to the reaction contents; adding the reaction contents to the acetone, alternatively, renders PANI-PAAMSPA intractable. The precipitate is filtered, washed repeatedly with excess acetone, and dried *in vacuo* for 24 h at room temperature to yield PANI-PAAMPSA particles in the ES oxidation state. These particles can subsequently be dispersed aqueously at 5 wt% concentrations due to PAAMPSA's hydrophilicity.[39] The size and size distribution of the resulting PANI-PAAMPSA particles, as measured by dynamic light scattering and transmission electron microscopy, scale with the molecular weight and molecular weight distribution of the PAAMPSA template. The details of the size and size distributions of the PANI-PAAMPSA particles are summarized in Table 12.5 along with the molecular weight and molecular-weight distributions of the PAAMPSA template.[40] The

Table 12.5 Molecular properties of PAAMPSA used to template synthesis PANI and the average hydrodynamic diameter of the corresponding PANI-PAAMPSA particles. Adapted from ref. 40.

PAAMPSA $M_n(kg/mol)$	*PDI*	*PANI-PAAMPSA* $D_h(nm)$
30	1.16	560 ± 11
45	1.43	543 ± 9
67	1.14	626 ± 13
83	1.53	699 ± 17
106	1.53	774 ± 16
117	1.20	739 ± 15
150	1.25	782 ± 18
255	1.64	908 ± 18
287	1.28	820 ± 16
724	1.64	1232 ± 15

characteristics of the particle have profound consequence on PANI-PAAMPSA's conductivity, with films comprising the smallest particles with the narrowest size distribution exhibiting the highest conductivities.[40]

12.2.4 Secondary Doping of Polyanilines

The ability to induce large and persistent improvements in the electrical properties of PANI postsynthesis through temporary exposure to an additional reagent is known as secondary doping. Although originally identified in PANI, it has since been applied to a number of other conductive polymer systems, in particular poly(3,4-ethylenedioxythiophene) complexes. Secondary dopants act on PANI by catalyzing structural relaxation. This relaxation is marked by orders-of-magnitude increases in PANI's electrical conductivity and significant changes in the optical properties of the ES state. In addition, secondary doping has been reported to enhance electrochemical kinetics upon switching between PANI's multiple oxidation states.[41]

Example 15: Secondary Doping of Camphorsulfonic Acid-Doped PANI with *m*-Cresol

Emeraldine base oxidation state is doped at room temperature with camphorsulfonic acid (CSA) at a 1:0.5 molar ratio of PANI's mer units to CSA by mixing using a mortar and pestle to induce the formation of PANI's fully doped ES form. The resulting powder is stirred into *m*-cresol at 0.3–3 wt%, placed in a double-helix mixing apparatus, and mixed at room temperature for 24 h, yielding a dark-green, viscous solution. *m*-Cresol has been shown to relax PANI's structure from its compact coil configuration to an expanded coil configuration. Films cast from these solutions exhibit electrical conductivity as high as 200 S/cm, compared to 0.1 S/cm for PANI prior to secondary doping with *m*-cresol.[11,42,43]

Example 16: Secondary Doping of PANI-PAAMPSA with Dichloroacetic Acid

PANI-PAAMPSA powder in the ES state is aqueously dispersed and stirred for 7–10 days. Uniform films are cast from this dispersion by spin coating and subsequently dried for 24 h. These films are subsequently immersed in dichloroacetic acid (DCA) at 100 °C and aggressively agitated with the film parallel to the surface of the DCA for three minutes. Excess DCA is removed from the films by blotting the side of the substrate. The film is quickly transferred onto a hot plate at 170 °C for 30 min, and any residual DCA that emerges to the surface is immediately blotted away. Atomic force microscope (AFM) micrographs reveal significant smoothening of the film surface (RMS decrease from 60.3 to 4.3 nm), while X-ray photoelectron spectroscopy measurements indicate a large increase in the concentration of PANI's nitrogen environments that are associated with polaronic character rather than ionic associations with PAAMPSA. The conductivity of PANI-PAAMPSA increases from 0.4 S/cm to 40 S/cm following secondary doping with DCA, thereby providing a route to realize the conductivities on par with small-molecule-doped PANI while maintaining the aqueous dispersability and processability of polymer-acid-doped systems.[12]

12.2.5 Characterization and Remarks

As a consequence of the extensive conjugation of PANI in the ES form in which it is most often synthesized, characterization of its molecular characteristics is usually indirect, commonly on the dedoped but more soluble EB oxidation state. Determination of molecular weight and molecular weight distribution of the EB form of PANI and PANI derivatives is most often accomplished by GPC using NMP as the eluent and calibrated against polystyrene or PVP standards. Characterization of PANIs synthesized through electrochemical methods typically focuses on their redox properties, and relies heavily upon cyclic voltammetry and spectroscopic tools. Determination of the molecular characteristics of PANI that is template synthesized is complicated by the difficulty encountered in separating PANI from its template; accordingly, template-synthesized PANIs are typically described in terms of the characteristics of the PANI/template complex rather than PANI by itself; tools used to characterize template-synthesized PANI include dynamic light scattering and electron microscopy.

Much interest in PANI research stems from its rich redox chemistry. The three neutral oxidation states (LB, EB, and PB) and doped ES form of PANI can be quickly differentiated by probing PANIs electronic structure through ultraviolet-visible/near-infrared (UV-Vis/NIR) and mid-infrared (IR) spectroscopy. Representative UV-Vis/NIR spectra of LB, EB, PB, and ES (before and after secondary doping) are shown in Figures 12.1a–e, respectively. The spectrum of PANI's colorless LB state exhibits an overwhelming peak around 330 nm associated with PANI's benzenoid moiety.[44] The spectrum of the blue EB oxidation state maintains similar peak positions, but with decreased absorbance near 330 nm and increased absorbance near 630 nm as a consequence of its half-oxidized and half-reduced nature.[13,45] Upon oxidation to the purple PB oxidation state of PANI, the peak near 630 nm in the accompanying UV-vis/NIR spectrum blue-shifts to 550 nm.[31] The electronic environment of the green ES form of PANI departs significantly from its neutral forms. Its spectrum exhibits a large peak near 800 nm associated with its polaronic character, as well as the emergence of a distinct shoulder at 420 nm.[39] A unique characteristic of PANI that has undergone secondary doping is the presence of a strongly absorbing free-carrier tail that extends well into the infrared range, consistent with its highly delocalized electronic character in highly conductive PANI samples.[8,43]

As a consequence of the strong interrelationship between PANI's optical properties and its redox chemistry, PANI possesses polyelectrochromic abilities. In strongly acidic electrolytes, PANI that is doped with small-molecule acids can stably and reversibly switch between its green ES and transparent LB forms;[46] at higher pH, PANI's ES state is dedoped to EB and the material's electroactivity is lost.[47] Transitions from ES to the purple PB form of PANI involve the loss of protons and thus their electrochemistry is pH dependent. In the acidic media required to stabilize switching between ES and LB, however, the oxidation potential required to access PB is beyond the PANI's degradation

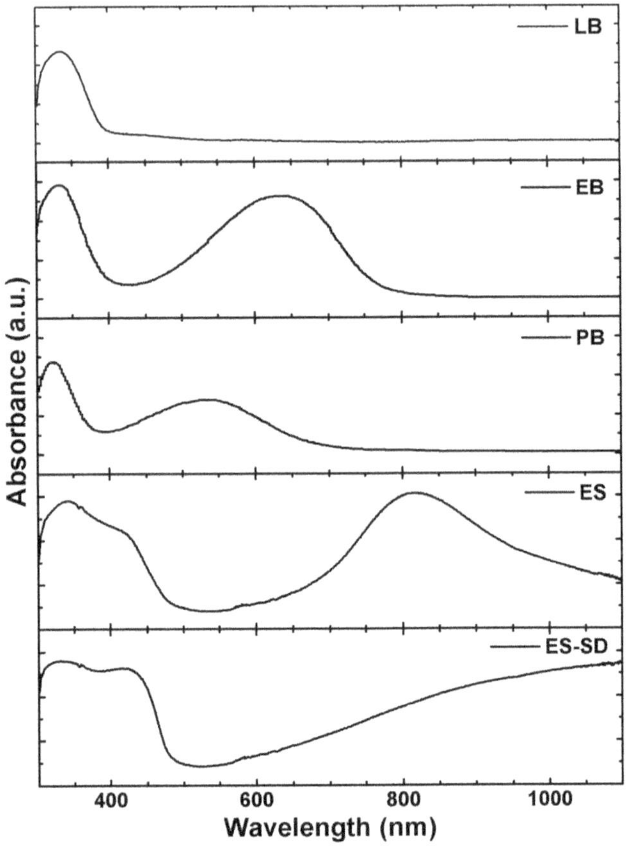

Figure 12.1 Representative UV-Vis/NIR spectra of PANI's different forms: (top to bottom) LB, EB, PB, as-synthesized ES, and secondary doped form of ES. Adapted from refs. 12, 31, 39, 44, and 45.

potential.[48] Accordingly, small-molecule-doped PANI is poorly suited to harness PANI's full polyelectrochromic range. Polymer-acid dopants have been found to stabilize PANI's ES form at elevated pH.[49] Mildly acidic electrolytes can consequently be utilized, thereby lowering the oxidation potential necessary to access PB below that of PANI's degradation potential while maintaining access to LB.[50] Polymer-acid doped PANI is thus able to fully exploit the material's polyelectrochromic ability.[41,51]

References

1. H. Letheby, *J. Chem. Soc.*, 1862, **15**, 161–163.
2. A. G. Green and A. E. Woodhead, *J. Chem. Soc.*, 1910, **98**, 2388–2403.
3. A. G. Green and A. E. Woodhead, *J. Chem. Soc.*, 1912, **101**, 1117–1123.
4. R. De Surville, M. Jozefowicz, L. T. Yu, J. Pepichon and R. Buvet, *Electrochim. Acta*, 1968, **13**, 1451–1458.

5. A. J. Heeger, *Angew. Chem. Int. Ed.*, 2001, **40**, 2591–2611.
6. A. G. MacDiarmid, *Angew. Chem. Int. Ed.*, 2001, **40**, 2581–2590.
7. H. Shirakawa, *Angew. Chem. Int. Ed.*, 2001, **40**, 2575–2580.
8. J. K. Avlyanov, Y. Min, A. G. MacDiarmid and A. J. Epstein, *Synth. Met.*, 1995, **72**, 65–71.
9. A. G. MacDiarmid, J. C. Chiang, M. Halpern, W. S. Huang, S. L. Mu, N. L. D. Somasiri, W. Q. Wu and S. I. Yaniger, *Mol. Cryst. Liq. Cryst.*, 1985, **121**, 173–180.
10. A. G. MacDiarmid, J. C. Chiang, W. S. Huang, B. D. Humphrey and N. L. D. Somasiri, *Mol. Cryst. Liq. Cryst.*, 1985, **125**, 309–318.
11. A. G. MacDiarmid and A. J. Epstein, *Synth. Met.*, 1995, **69**, 85–92.
12. J. E. Yoo, K. S. Lee, A. Garcia, J. Tarver, E. D. Gomez, K. Baldwin, Y. M. Sun, H. Meng, T.-Q. Nguyen and Y.-L. Loo, *Proc. Natl. Acad. Sci. USA*, 2010, **107**, 5712–5717.
13. A. G. MacDiarmid and A. J. Epstein, *Faraday Discuss. Chem. Soc.*, 1989, **88**, 317–332.
14. P. M. McManus, S. C. Yang and R. J. Cushman, *J. Chem. Soc., Chem. Commun.*, 1985, 1556–1557.
15. G. E. Wnek, *Synth. Met.*, 1986, **15**, 213–218.
16. S. Ameen, M. Shaheer Akhtar and M. Husain, *Sci. Adv. Mater.*, 2010, **2**, 441–462.
17. S. Bhadra, D. Khastgir, N. K. Singha and J. H. Lee, *Prog. Polym. Sci.*, 2009, **34**, 783–810.
18. A. A. Syed and M. K. Dinesan, *Talanta*, 1991, **38**, 815–837.
19. E. Genies, A. Boyle, M. Lapkowski and C. Tsintavis, *Synth. Met.*, 1990, **36**, 139–182.
20. E. Kang, K. Neoh and K. Tan, *Prog. Polym. Sci.*, 1998, **23**, 277–324.
21. E. Rebourt, J. A. Joule and A. P. Monkman, *Synth. Met.*, 1997, **84**, 65–66.
22. Y. Cao, P. Smith and A. J. Heeger, *Synth. Met.*, 1992, **48**, 91–97.
23. P. M. Adams, L. Abell, A. Middleton and A. P. Monkman, *Synth. Met.*, 1997, **84**, 61–62.
24. P. N. Adams, P. J. Laughlin, A. P. Monkman and A. M. Kenwright, *Polymer*, 1996, **37**, 3411–3417.
25. J. Stejskal, A. Riede, D. Hlavatá, J. Prokeš, M. Helmstedt and P. Holler, *Synth. Met.*, 1998, **96**, 55–61.
26. F. Cataldo and P. Maltese, *Eur. Polym. J.*, 2002, **38**, 1791–1803.
27. L. H. C. Mattoso, A. G. MacDiarmid and A. J. Epstein, *Synth. Met.*, 1994, **68**, 1–11.
28. S. Shimizu, T. Saitoh, M. Uzawa, M. Yuasa, K. Yano, T. Maruyama and K. Watanabe, *Synth. Met.*, 1997, **85**, 1337–1338.
29. J. Yue, Z. H. Wang, K. R. Cromack, A. J. Epstein and A. G. MacDiarmid, *J. Am. Chem. Soc.*, 1991, **113**, 2665–2671.
30. J. Yue and A. J. Epstein, *J. Am. Chem. Soc.*, 1990, **112**, 2800–2801.
31. Y. Sun, A. G. MacDiarmid and A. J. Epstein, *J. Chem. Soc., Chem. Commun.*, 1990, 529–531.

32. S. Bhadra, N. K. Singha and D. Khastgir, *J. Appl. Polym. Sci.*, 2007, **104**, 1900–1904.
33. J. Zhou and D. O. Wipf, *J. Electrochem. Soc.*, 1997, **144**, 1202–1207.
34. S. Jegadesan, R. C. Advincula and S. Valiyaveettil, *Adv. Mater.*, 2005, **17**, 1282–1285.
35. A. F. Diaz and J. A. Logan, *J. Electroanal. Chem. Interf.*, 1980, **111**, 111–114.
36. T. Kobayashi, H. Yoneyama and H. Tamura, *J. Electroanal. Chem. Interf.*, 1984, **161**, 419–423.
37. W. Liu, J. Kumar, S. Tripathy, K. J. Senecal and L. Samuelson, *J. Am. Chem. Soc.*, 1998, **121**, 71–78.
38. W. Liu, A. L. Cholli, R. Nagarajan, J. Kumar, S. Tripathy, F. F. Bruno and L. Samuelson, *J. Am. Chem. Soc.*, 1999, **121**, 11345–11355.
39. J. E. Yoo, J. L. Cross, T. L. Bucholz, K. S. Lee, M. P. Espe and Y.-L. Loo, *J. Mater. Chem.*, 2007, **17**, 1268–1275.
40. J. E. Yoo, W. P. Krekelberg, Y. Sun, J. D. Tarver, T. M. Truskett and Y.-L. Loo, *Chem. Mater.*, 2009, **21**, 1948–1954.
41. J. Tarver and Y.-L. Loo, *Chem. Mater.*, 2011, **23**, 4402–4409.
42. Y. Cao and P. Smith, *Polymer*, 1993, **34**, 3139–3143.
43. Y. Min, Y. Xia, A. G. MacDiarmid and A. J. Epstein, *Synth. Met.*, 1995, **69**, 159–160.
44. K. G. Neoh, E. T. Kang and K. L. Tan, *J. Polym. Sci., Part B: Polym. Phys.*, 1993, **31**, 395–401.
45. J. Y. Shimano and A. G. MacDiarmid, *Synth. Met.*, 2001, **123**, 251–262.
46. T. Kobayashi, H. Yoneyama and H. Tamura, *J. Electroanal. Chem. Interf.*, 1984, **177**, 281–291.
47. S.-A. Chen and H.-T. Lee, *Macromolecules*, 1995, **28**, 2858–2866.
48. A. Watanabe, K. Mori, Y. Iwasaki, Y. Nakamura and S. Niizuma, *Macromolecules*, 1987, **20**, 1793–1796.
49. N. Mano, J. E. Yoo, J. Tarver, Y.-L. Loo and A. Heller, *J. Am. Chem. Soc.*, 2007, **129**, 7006–7007.
50. J. Tarver, J. E. Yoo, T. J. Dennes, J. Schwartz and Y.-L. Loo, *Chem. Mater.*, 2008, **21**, 280–286.
51. J. Tarver, J. E. Yoo and Y.-L. Loo, *Chem. Mater.*, 2010, **22**, 2333–2340.

CHAPTER 13

Si–Si Bond Polymers, Oligomers, Molecules, Surface, and Materials

MICHIYA FUJIKI

Nara Institute of Science and Technology, Graduate School of Materials Science, 8916-5 Takayama, Ikoma, Nara 630-0101, Japan
Email: fujikim@ms.naist.jp

13.1 Hierarchy of the Si–Si Bond Family: From Gaseous SiH_4 to Crystal Silicon

Silicon (Si) is the second most abundant element (Clarke number approximately 26%) on the earth and exists in an oxidized form (SiO_2). Although the silanes (SiH_4 and Si_2H_6) and chlorosilanes ($HSiCl_3$, $SiCl_4$ and Me_2SiCl_2) are used widely in many technical applications, crystalline silicon (*c*-Si), polycrystalline silicon (*poly*-Si), amorphous silicon (*a*-Si:H) and silicon dioxide (SiO_2) are the most fundamental materials for use in microelectronics, optoelectronics and photonics. However, the first isolable silicon derivatives are most likely the organopolysilanes that are prepared by Kipping and coworkers in the early 1920s by Wurtz-type polycondensation of diphenyldichlorosilane with sodium in inert aromatic solvents.[1–3] Thus far, there are various types of Si–Si bond materials ranging from zero-dimensional (0D) materials and molecular compounds, one-dimensional (1D) polysilane and oligosilanes, and two-dimensional (2D) and sheet-like layered polymers, such as network-like

RSC Polymer Chemistry Series No. 9
Conjugated Polymers: A Practical Guide to Synthesis
Edited by Klaus Müllen, John R. Reynolds and Toshio Masuda

Published by the Royal Society of Chemistry, www.rsc.org

Table 13.1 Dimensions of Si–Si bond skeletons and optical bandgaps (E_g in eV).

3D	*2D*	*1D*	*0D*
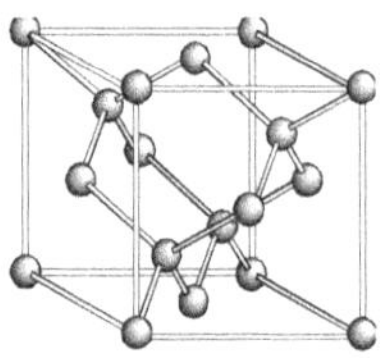	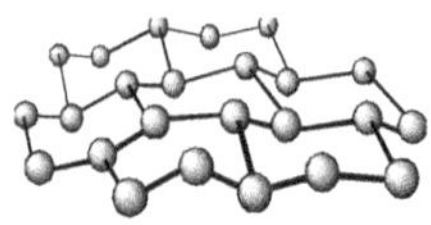	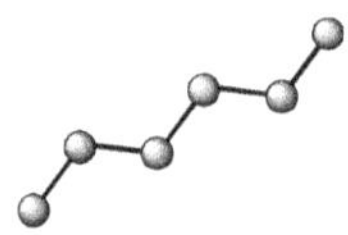	SiH_4, $SiMe_4$, $SiCl_4$
inorganic solids $E_g \sim 1.1$ eV	sheet-like polymers $E_g \sim 2$ eV	chain-like polymers $E_g \sim 4$ eV	molecules $E_g \sim 7$ eV

polysilyne and siloxene, to three-dimensional (3D) and tailored materials, including *c*-Si, *poly*-Si, *a*-Si:H, and porous silicon (*por*-Si) (Table 13.1).

It is possible for a carbon atom to form various covalent bonds with elements such as C, H, O, N, and S, including single bonds (C–C, C–O, C–H, C–N, C–S), double bonds (C=C, C=O, C=N, C=S), triple bonds (C≡C, C≡N), and aromatics. The nature of the chemical bonds leads to a rich variety of compounds that are thermally and kinetically stable and stable in the presence of air, organic polymers and graphite/diamond, which are classified as inorganics. By referring to the nature of organic chemistry and the uniqueness of organosilane compounds, organosilicon chemists and materials scientists have designed and produced the corresponding organosilane analogs.[9–12] The present chapter outlines a family hierarchy of organosilicon compounds and materials that have been produced with successful experimental procedures.

13.2 Polymerization Techniques

13.2.1 Wurtz-Type Condensation – The Most Versatile Method

Although Kipping attempted to produce tetraphenyldisilene by Na reduction of diphenydichlorosilane while refluxing in benzene and toluene, he isolated soluble crystalline perphenylcyclosilanes $(Ph_2Si)_n$ ($n = 4$ and 6) and insoluble $(Ph_2Si)_n$ (perhaps a chain-like polymer).[1–3] In 1949, Burkhard reported a similar insoluble, infusible white solid that was prepared by Na reduction of dimethyldichlorosilane in refluxing benzene and toluene.[4] This solid might be characterized as poly(dimethylsilane). When these reactions are carried out with Li in THF in place of Na in aromatic hydrocarbons, respectively, perphenylcyclosilanes $(Ph_2Si)_n$ ($n = 4$–6) and permethylcyclosilanes $(Me_2Si)_n$ ($n = 6$) are successfully isolated with excellent yields.[5–8] These pioneering studies led to the so-called Wurtz-type coupling method that provides the means to prepare a family of organosilane molecules and polymers with a choice of substituents, solvent, some additives, and alkaline metals.[9–12] These coupling methods, with modern modifications, are applied commonly to generate Si–Si single bonds, Si=Si double bonds, and Si≡Si triple bonds from

Table 13.2 Wurtz-type reduction of organochlorosilanes in inert solvents.

Metals	*Conditions*	*Solvents, additives, and remarks*
Li	no additives	THF
Na	no additives	hydrocarbons with higher bp (toluene, *n*-heptane)
Na	additives	diglyme, crown ether
Na–K alloy	ultrasound	*n*-pentane
K	graphite-K (KC_8)	THF

halogenated organosilane starting materials in inert solvents under controlled conditions. The original Wurtz-type condensation method needed a long reaction time but gave a poor yield. To improve the yield and to reduce the reaction time, several modifications are employed. The advantages and disadvantages of the original and modified Wurtz condensation methods are listed in Table 13.2.

The disadvantages of the Wurtz method are limited to the use of organochlorosilanes, solvents, and additives that do not contain any functional groups that can react with alkaline metals. The organochlorosilane should not contain O–H, N–H, S–H, NO_2, C=O, C(=O)O, Br, and I, except the C–O–C group. Regardless of this limitation, the Wurtz-type reduction and its modifications are the most versatile methods for producing the desired organosilicon compounds and materials, with well-designed substituents and proper protecting groups, in excellent yields.

13.2.2 Electrochemical Reduction

Compared to the Wurtz method, electrochemical reduction with alkaline-earth metals is a milder alternative coupling reaction with the corresponding organochlorosilane. The silanes that can react with Mg are not preferred for this approach. Regardless of this limitation, the electrochemical method is applicable to phenyl-containing chlorosilane because Mg and Mg–Cu are not reactive when in contact with water and air. The water- and/or air-sensitive reactions are not welcome for technical reasons.

13.2.3 Ring-Opening Reactions with Precursors

Highly strained monocyclic and/or bicyclic compounds containing Si–Si bonds can undergo ring-opening reactions with the help of organolithium and form chain-like polysilane and block copolymers. Highly strained masked disilene, initiated by anionic catalysts, can undergo ring-opening polymerization. Although six-membered cyclosilanes such as $(Ph_2Si)_6$ and $(Me_2Si)_6$ are less sensitive to ring-opening reactions, smaller cyclosilanes such as $(R_2Si)_n$ ($R = Ph$, Me, H, $n = 4$ and 5) undergo ring-opening reactions that are initiated by chemical reagents and thermolysis/photolysis due to their inherent strain energy. Well-defined cyclosilane derivatives are suitable precursors for various polysilanes and organosilane compounds. Highly strained three-membered cyclosilane $(Ar_2Si)_3$, which has the ability to produce silylene Ar_2Si by photolysis, led to the synthesis of the first Si=Si double bond compound, $Ar_2Si{=}SiAr_2$ (Ar = 2,4,6-trimethylphenyl or mesityl) three decades ago.[13]

13.2.4 Dehydrogenative Coupling with Organometallic Catalysts

Many organochlorosilanes are utilized for reductive coupling with alkaline metals and for electrochemical reductions with alkaline-earth metals. However, certain organohydrosilanes are used as silicon sources for oxidative coupling reactions with organometallic catalysts such as Zr, Ti, Hf, Rh, and Ir. Although primary aryltrihydrosilanes such a $PhSiH_3$ give moderately higher molecular weight polymers, secondary diaryldihydrosilanes such a Ph_2SiH_2 do not. Alkyltrihydrosilanes and dialkyldihydrosilanes, including *n*-octylSiH_3 and di-*n*-octylSiH_2, are used to produce oligosilanes using Wilkinson catalyst, $(Ph_3P)_3RhCl$.

13.2.5 Postpolymerization toward Functionalization

Due to the inherent limitation of those functional groups that react with alkali and alkaline-earth metals, postpolymerization is applicable to a broad range of functionalization. The most important issue in these reactions is the possibility of side reactions that break down the Si–Si bonds in organosilicon molecules and polymers.[9–12]

13.2.6 Chemical Modification of Si–H Bonds at the Surface of Crystal Silicon

The use of Si–H coatings to chemically modify the surface of *c*-Si is an alternative postpolymerization approach for the introduction of functional groups and a means to provide stability with respect to the contact of water and air at the *c*-Si surface. The *c*-Si surface offers ideal 2D Si-skeleton lattices or Si–Si bond nanosheets with puckered hexagonal lattices. Various functional groups are successfully employed. A coating of Si–H on *c*-Si is formed using aqueous HF to treat *c*-Si and remove SiO_2 layers.

13.2.7 Deintercalation from the Zintl Phase – An Ideal Two-Dimensional Si Skeleton

As an alternative to chemical reduction, deintercalation from Zintl phases under controlled hydrolysis at lower temperatures produced the corresponding Si–Si bond solids. A typical Zintl phase is calcium silicide, $CaSi_2$, which is produced by mixing Ca and Si with a small amount of Fe at elevated temperatures in a vacuum. The $CaSi_2$ has high electrical conductivity and metallic behavior and is similar to graphite intercalated with potassium (C_8K). $CaSi_2$ has ideal 2D structure, but negatively charged Si–Si bond polymer nanosheets are intercalated with positively charged Ca. The formal charges are those of an ionic solid composed of $[(Si_6)^{2-}]_{1/3}//Ca^{2+}$. $CaSi_2$ is similar to sila-graphene, which comprises Si–Si single bonds. By removing Ca^{2+} from $CaSi_2$, the Si layer opens the optical bandgap, forming an insulator that is able to emit in the visible region. Recently, in order to produce layered Si with H, controlled hydrolysis of $(Ca_xMg_{1-x})Si_2$ followed by substitution with organic groups is used to produce single Si–Si bond sheets bearing organic groups that are almost soluble in organic solvent.

13.2.8 Thermolysis

Crude *c*-Si is produced commercially by the thermal decomposition of $HSiCl_3$. Hydrogenated *a*-Si:H is produced by the chemical vapor deposition (CVD) of SiH_4 with and without plasma assisted external electromagnetic forces. Thus, *a*-Si is regarded as an insoluble highly crosslinked Si–Si bond polymer with SiH and SiH_2 termini. By applying a Kr-F excimer laser to an *a*-Si:H film, polycrystalline silicon (*poly*-Si) films are formed by eliminating H atoms and thermal annealing. Because the elimination of H atoms as H_2 gas occurred above 500 °C, higher temperatures are needed at the surface of *a*-Si:H for a short period. Because the wavelength of the Kr-F laser (248 nm) matches the Siσ-Siσ* transitions of *a*-Si:H and *poly*-Si, the elimination of H atoms and thermal annealing are efficient.

As an alternative to thermal decomposition, CVD, and laser-assisted H elimination, the flash pyrolysis of soluble alkylpolysilyne in a vacuum is reported recently. Unless they are under vacuum conditions, polysilanes and polysilyne may convert to silicon carbide (SiC). The flash process, which is a rapid removal of volatile organic substances and H_2 gas, enables the control of the dimension of Si structure from 2D to 3D. Flash pyrolysis of alkylpolysilyne in a vacuum above 500 °C leads to the formation of *poly*-Si with a minimal amount of SiH termini, although the size of the crystals is limited to the range between several μm and several nm.

In 2006, Shimoda *et al.* demonstrated the preparation of the first solution-processable *poly*-Si thin film with a high carrier mobility by means of XeCl excimer laser pyrolysis of an *a*-Si film (Chart 13.1).[14] The film is obtained by a

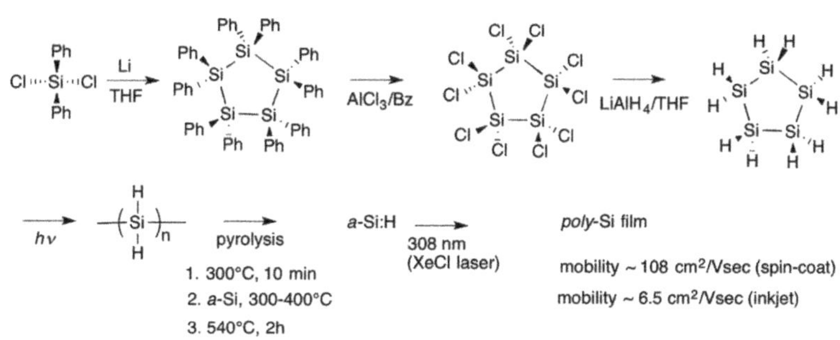

Chart 13.1

three-step pyrolysis of a Si–Si bond linear polymer, $(SiH_2)_n$, at 300–550 °C in an oxygen-free glove box. The hydrogenated polysilane is produced by the four-step synthesis, including a photoinduced ring-opening process based on diphenyldichlorosilane as a starting material.

In the following section, the preparation of a rich variety of compounds in the Si–Si bond family by applying the above preparation methods is demonstrated briefly.

13.3 Features of Si–Si Bond Family Members

13.3.1 Chain-Like Polysilanes and Oligosilanes

13.3.1.1 Poly(diarylsilane)s

Poly(diphenylsilane) is insoluble and infusible due to the absence of flexible substituents on the aromatic rings.[1,2] By simply introducing medium length alkyl and alkoxy moieties at *p*- and *m*-positions on the aromatic rings, the corresponding poly(diarylsilane)s become soluble in common organic solvents, although the isolated yield of the desired polymers is at most 5% (Chart 13.2).[15,16] By carefully adding a few drops (approximately 50 μL) of diglyme to several g of diaryldichlorosilane during the polycondensation reaction, the polymer yield of polymer is much improved and reached 20–30% due to the activation of the sodium surface. By introducing chiral (*S*)-2-methylbutyl moieties at *p*- and *m*-positions on the phenyl rings, the corresponding polysilane become optically active due to the adoption of a helical conformation in organic solvents.[17–19] The number of chiral groups per Si repeat unit, solution temperature and achiral diarylsilane comonomer determines the preferential screw sense regardless of the (*S*)-side chain

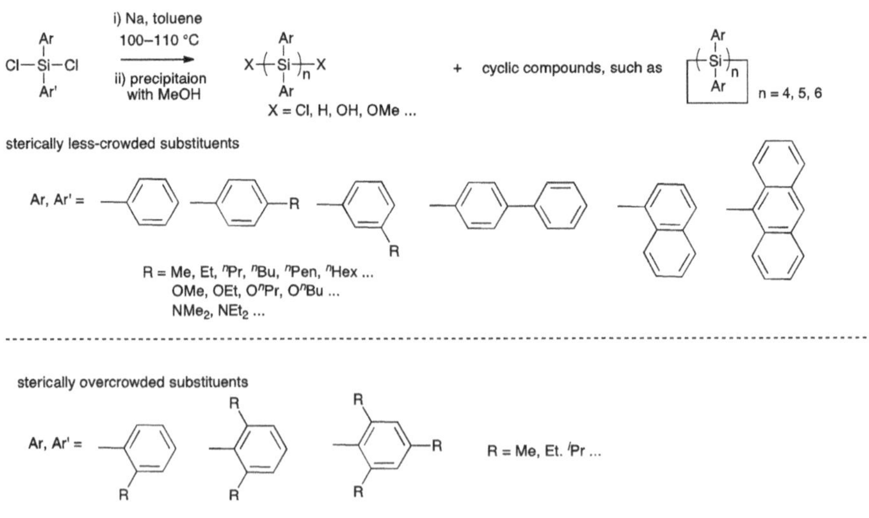

Chart 13.2

chirality. Mono- and disubstitutions at *o*-positions on the aromatic rings, however, prevent the production of the corresponding polymers due to crowding and steric repulsions between the rings. This fundamental knowledge and understanding can support the design of exotic and highly strained organosilane molecules, which include Si=Si double and Si≡Si triple bonds, as shown below.

13.3.1.2 Poly(dialkylsilane)s

Poly(dimethylsilane) is an insoluble infusible polymer due to the absence of flexible alkyl substituents on the Si atoms.[4] By simply replacing with alkyl side group, such as ethyl and longer *n*-alkyl groups including *n*-propyl, *n*-butyl, ... and *n*-octadecyl, the corresponding poly(dialkylsilane)s become highly soluble in common organic solvents. However, the yields of the desired polymers are poor and polymodal molecular weight distributions (MWD) are observed (Chart 13.3).[9,10] By carefully adding a few drops of diglyme and 1–2% of crown ethers to several grams of dialkyldichlorosilane during polycondensation reactions, the polymer yields improves by approximately 70% due to the activation of the sodium surface and monomodal MWDs are obtained.[20,21]

By introducing β- and γ-branched chiral alkyl moieties on Si atoms, the corresponding poly(dialkylsilane)s become optically active due to the adoption of a thermally stable and thermoresponsive 7/3 helical conformation in organic solvents.[22–24] When β- and γ-branched achiral dialkyldichlorosilanes are used, the resulting polysilanes adopt optically inactive 7/3 helical conformation due

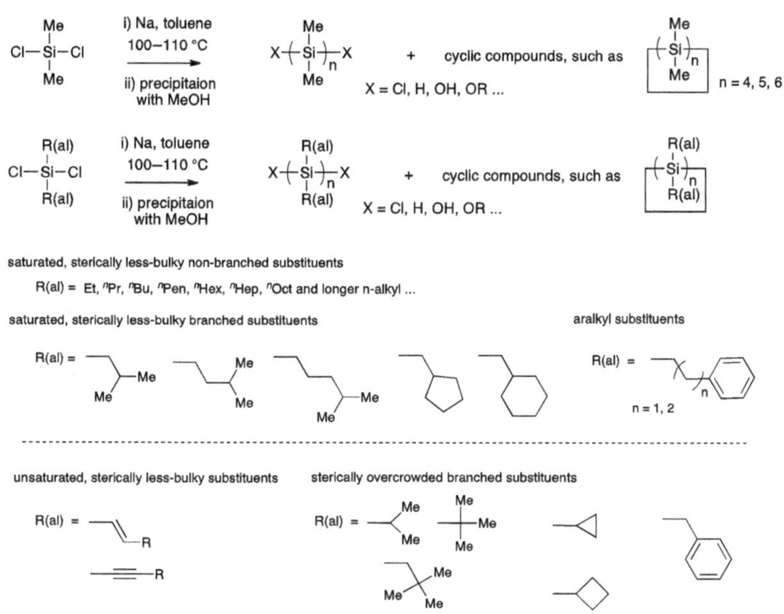

Chart 13.3

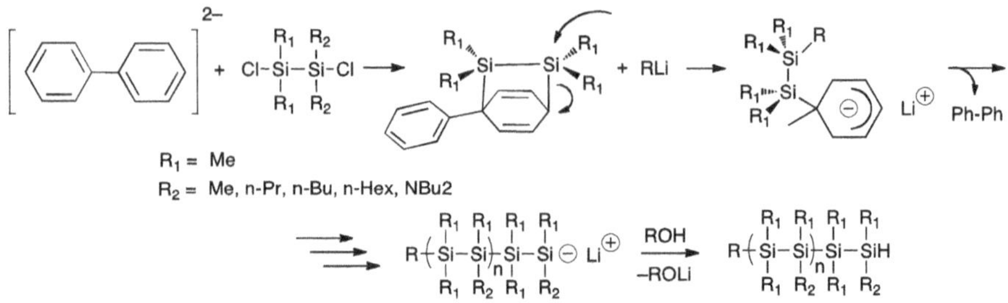

Chart 13.4

to the adoption of thermally labile, right-handed (*P*-) and left-handed (*M*-) 7/3 helical conformations in organic solvents. By designing γ-branched chiral, γ-branched achiral and β-branched achiral alkyl moieties per Si repeat units, the preferential screw sense is altered and the helical conformation is amplified in solution due to strong cooperative interactions.

However, α-branched alkyl side groups such as isopropyl group prevent the production of the corresponding high polymers due to crowding and steric repulsions between the substituents. The isopropyl group is a very useful substituent for generating ladder-like double-strand oligosilanes.

To control the molecular weight (MW) and MWD, Sakurai and coworkers developed an anionic polymerization method with masked disilene, a highly strained precursor with biphenyl (Chart 13.4).[25–28] Although the substituents are limited to alkyl and certain dialkylamino groups, this method made it possible to design well-defined Si–Si sequences, leading to diblock polymers with organic polymer blocks such as poly(triphenylmethylmethacrylate) by means of chiral anionic catalysts.

13.3.1.3 Poly(alkylarylsilane)s

Due to the insoluble and infusible nature of poly(diphenyl)silane and poly-(dimethylsilane), organopolysilanes were long believed to be insoluble in ordinary organic solvents and poorly characterized. By introducing certain irregularities such as asymmetric substituents (Me and Ph) and block-type sequences in the Si–Si main chain, alkylarylpolysilanes with high MW became highly soluble in most organic solvents.[29–32] The use of solvent-compatible side chains leads to the production and functionalization of various soluble poly-(alkylarylsilane)s (Chart 13.5).

13.3.1.4 Poly(hydroarylsilane)s and their Derivatives

The Wurtz reduction is useful for producing organopolysilane without hydrogen substituents. However, Wurtz coupling does not tolerate many functional groups that are present in some monomer starting materials and intended polymer products. In contrast, transition-metal-catalyzed oxidation may be an alternative route for producing polysilanes containing the Si–H

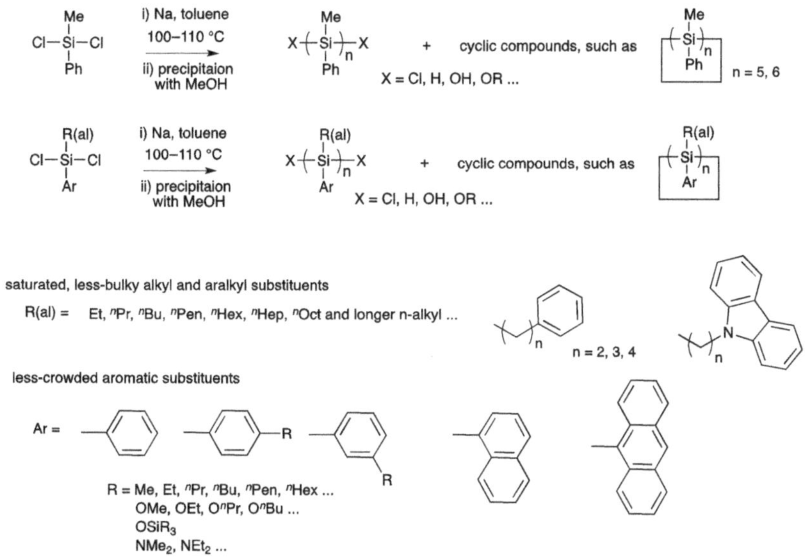

Chart 13.5

moiety.[33–46] The Si–H group may be a useful and versatile functional moiety that can facilitate the introduction of several functional groups (NO_2, NH_2, OH, CN, Cl, *etc*.). The Si–H containing polymers are soluble polymeric models of insoluble inorganics with Si–H groups, including silicide, *c*-Si etched with HF and *a*-Si:H, as discussed below. To overcome this difficulty, Harrod and coworkers,[33,34] Corey and coworkers,[35–38] Tilly and coworkers,[39–40] and Waymouth and coworkers[41–43] developed the dehydrogenative coupling reaction of trihydrosilanes and dihydrosilanes by means of organometallic catalysts to yield polysilanes and oligosilanes (Chart 13.6).

In 1985 and 1986, Harrod and coworkers reported the first dehydrogenative coupling of primary organosilanes by Cp_2TiMe_2 and Cp_2ZrMe_2 to yield the corresponding polymer $(PhSiH)_n$ ($n = 1$ to 10).[33,34] At the same time, Tilly and coworkers developed that several $Cp'_2M(SiR_3)R'$ compounds (M = Ti, Zr; Cp′ = Cp, Cp*; R = Me, Ph, $SiMe_3$; R′ = Cl, alkyl, silyl) are efficient polymerization (DP = 10–20) catalysts of $PhSiH_3$.[35] Corey and coworkers reported that the condensation reactions of secondary silane (Ph_2SiH_2) by $(Ph_3P)_3RhCl$ (Wilkinson's catalyst) and Cp_2MMe_2 (M = Ti, Zr) to yield the corresponding dimer and oligomers.[36] When metallocene dichlorides $(Cp^R)_2MCl_2$ (M = Ti, Zr; Cp^R = Cp carrying a 1,1,2,2-tetramethylmethyl group) are applied to $PhSiH_3$, more high MW (up to 4000) product is obtained.[36] Analysis with $^{29}Si\{H\}$-DEPT-NMR reveals that the poly(phenylsilane)s adopt an almost atactic microsequence in the main chain.

Waymouth and coworkers reported the stereoregular dehydrogenative polymerization of $PhSiH_3$ by *rac*-(EBI)$ZrMe_2$ (EBI = ethylenebis(indenyl)) and $Cp_2Zr(H)Cl$ derivatives.[41] They showed that, in the presence of CCl_4 and CBr_4 by free-radical reaction, the introduction of Cl and Br is possible.[42] These

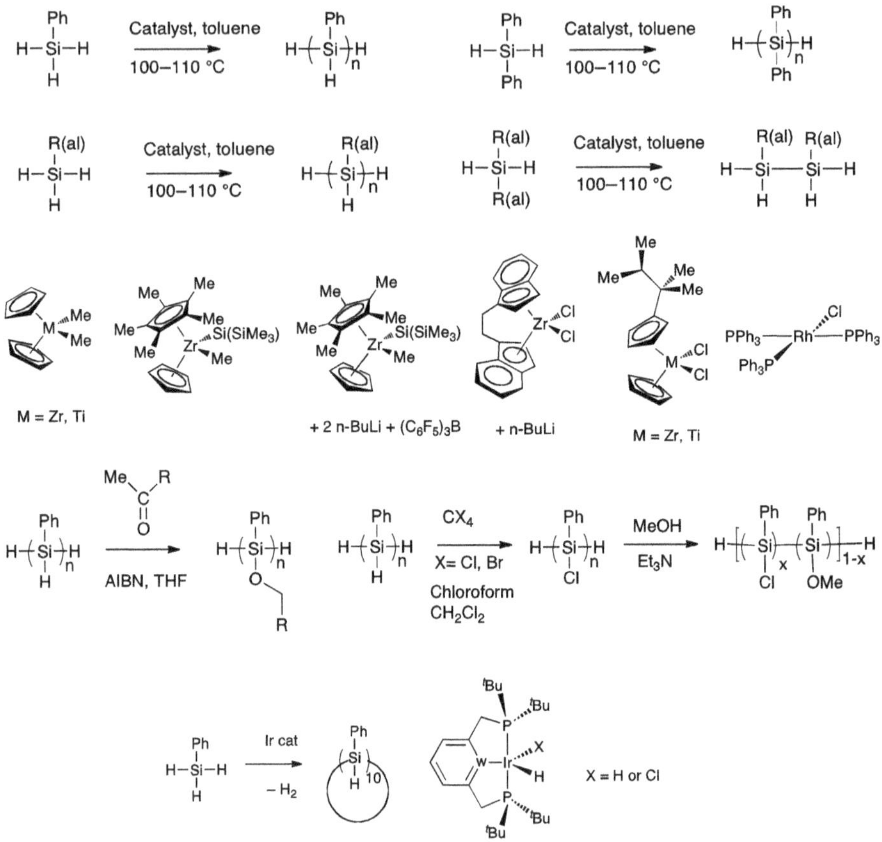

Chart 13.6

halogenated polysilanes are used to introduce methoxy and other alkoxy groups by substituting alcohol in the presence of triethylamine, by substituting alkylketone with AIBN and by substituting methyl groups by a Grignard coupling reaction.[43]

Although this alkoxylation reaction requires a three-step synthesis from $PhSiH_3$, Woo and coworkers achieves straightforward one-pot Si–Si/Si-O dehydrocoupling and synthesis of aryltrihydrosilane using titanocenes and zirconocenes with red-aluminum, resulting in the one-pot synthesis of poly-(alkoxyarylsilane) under milder conditions with excellent total yields.[44]

Although those titanocene and zirconocene catalysts provide $(PhHSi)_n$ with relatively high MW (up to 10 000), their catalytic activity is not sufficient to produce higher MW polymers when alkylltrihydrosilanes are used. The Wilkinson catalyst is an effective means to selectively produce dimers and 6–8mers from *n*-octyltrihydrosilane (*n*-$OctSiH_3$). Rosenberg *et al.* reported that Wilkinson's catalyst efficiently produces dimers and trimers of Ph_2SiH_2.[45,46] More recently, Waterman and coworkers selectively produce *cyclic*-10mers *of* PhHSi units from $PhSiH_3$ by using a designed Ir complex.[47]

13.3.1.5 Poly(chlorosilane)s and Poly(alkoxysilane)s

In another approach, Koe *et al.* reported a new synthetic route to polysilanes bearing two chlorines, $(SiCl_2)_n$, by means of a photolytic ring-opening reaction of *cyclo*-$(SiCl_2)_4$,[48] followed by a substitution reaction with the corresponding achiral and chiral alcohol in the presence of amine (Chart 13.7).[49] The resulting dialkoxypolysilanes show λ_{max} at 336–347 nm, which is red-shifted by 10–20 nm relative to the corresponding Poly(dialkylsilane)s. This shift may arise from electronic perturbations between the silicon (σ) and oxygen lone pair (n) orbitals.

13.3.1.6 Poly(fluoroalkylsilane)s

In the 1990s, Fujino *et al.* successfully obtained the first fluoroalkylpolysilane, poly(methyl-3,3,3-trifluoropropylsilane) by Wurtz coupling of methyl-3,3, 3-trifluoro- propyldichlorosilane in hot *n*-alkanes (Chart 13.8).[50,51] Although the yields are less than 5%, the results are surprising because molten Na exhibited few effects on the fluoroalkylsilane and/or the resulting polymer even under the reaction conditions. The use of crown ether has no effect on the production of this polymer. This result leads to several trifluoropropyl containing copolymers with dialkyldichlorosilane comonomers, yielding multiblock-like sequences in a one-pot polymerization.[52–57] These fluoroalkylpolysilanes had several unique functionalities, *e.g.*, DC electric field chromism, sensitivity toward fluorides and

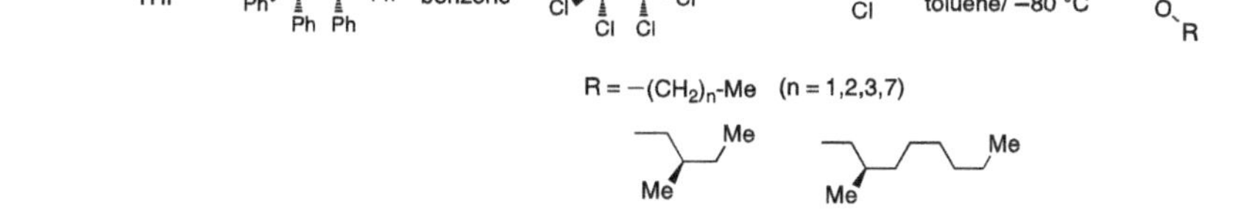

Chart 13.7

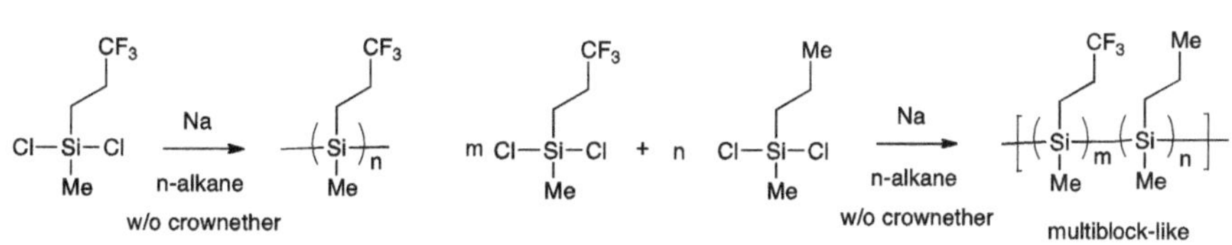

Chart 13.8

nitroaromatics, rod-coil transitions, stabilization of helical structures and organogelation owing to intra-/intermolecular CF/Si interactions.

13.3.1.7 Polysilole

Chain-like polymers containing five-member π-conjugated molecules such as polythiophene and polypyrrole are the most typical π-conjugated polymers. Chain-like Si–Si bond polymers containing five-member Si-containing molecules such as polysilole are prepared successfully in recent years by several research groups (Chart 13.9).[58–61] In 1999, West and coworkers reported the first synthesis of polysilole oligomers ($n = 15$) by 2-equimolar amounts of Li-based Wurtz reaction with tetraphenyl-1,1-dichlorosilole as starting monomer.[58] Trögler and coworkers developed this preparation by the dihydrocoupling reaction of tetraphenyl-1,1-dihydrosilole with 1 mol% of Wilkinson's catalysts, $(PPh_3)_3RhCl$ and $(PPh_3)_4Pd$.[59,60] Tang and coworkers prepared several substituted siloles by using Wilkinson's catalysts.[61]

13.3.1.8 Electrochemical Methods

It is possible to prepare three typical π-conjugated polymers, polythiophene, polypyrrole and polyaniline, by means of electrochemical oxidation of the corresponding monomers in the presence of an electrolyte at ambient temperature. In 1990, Canham discovered that photoluminescent *por*-Si materials could be formed by an electrochemical method that used a *p*-type *c*-Si wafer.[62,63] Electrochemical methods are among the safest, most environmentally friendly and technically favorable processes compared to Na/Li-based Wurtz coupling reactions (Chart 13.10).[64–68]

In the 1990s, Shono and coworkers first demonstrated the electrochemical reduction polymerization with a Mg electrode for several dialkyldichlororsilanes and tetraalky1,2-dichloroldisilanes in THF.[64–66] Okano *et al.* reported the usefulness of dimethoxyethane in place of THF.[67] Watanabe demonstrated that the electrochemical method with ultrasound allows the preparation of nanosize crystalline Si particles (*nc*-Si) capped with organic groups when tetrachlorosilane is used as the silicon source.[69]

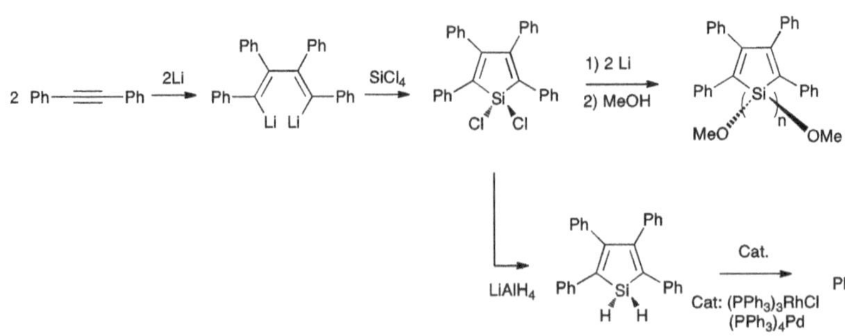

Chart 13.9

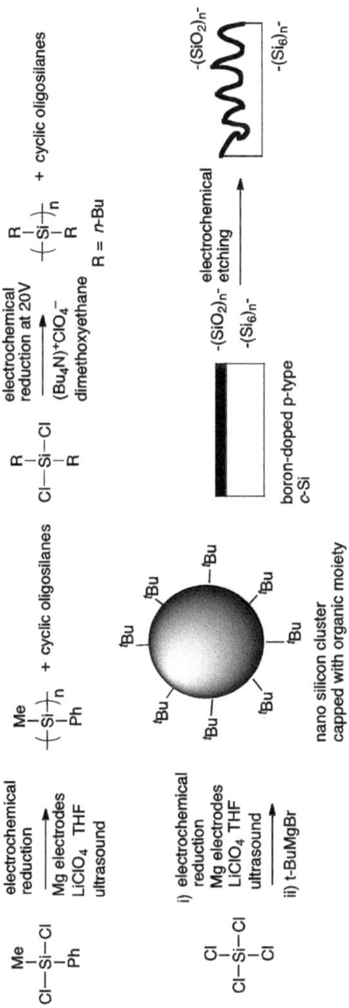
electrochemical reduction
Mg electrodes
LiClO4 THF
ultrasound
+ cyclic oligosilanes
electrochemical reduction at 20V
(Bu4N)+ClO4−
dimethoxyethane
+ cyclic oligosilanes
R = n-Bu
i) electrochemical reduction
Mg electrodes
LiClO4 THF
ultrasound
ii) t-BuMgBr
nano silicon cluster capped with organic moiety
boron-doped p-type c-Si
-(SiO2)n-
-(Si6)n-
electrochemical etching

Chart 13.10

13.3.2 Cyclic Four-Membered Oligosilanes and Ladder Oligosilanes

In the 1980s and 1990s, West,[70,71] Masamune and coworkers,[72] and Hengge and Janoschek[73] completed a comprehensive review of cyclosilanes. In recent years, the chemistry of cyclic oligosilane compounds has received much attention in experimental and theoretical studies. Matsumoto *et al.* developed the ladder-like oligosilanes consisting of fused *cyclo*-tetrasilanes due to the uniqueness of the isopropyl groups (Chart 13.11).[74]

Coupling di-isoproyltetrachlorodisilane and tetra-isoproyldichlorodisilane or hexa-isoproyl-tetrachlorotetrasilane with Li-based Wurtz coupling in THF yields the corresponding ladder oligosilanes, including two isolable stereochemical isomers, the *anti*- and *syn*-forms. In a manner similar to the thermal transformation from metastable *cis*- to stable *trans*-polyacetylenes, the *syn*-ladder oligosilanes are irreversibly transformed to the *anti*-ladder form.

13.3.3 Network-Like Organopolysilanes (Organopolysilyne)

Chain-like organopolysilanes are regarded as soluble polymeric models of 1D silicon semiconductors. Little progress has been made toward the synthesis of 2D silicon semiconductors and coagulated sila-graphene consisting of fused six-membered rings (Chart 13.12).

Bianconi and coworkers reported the first successful synthesis of several unique *n*-hexylpolysilyne, which is a Si–Si bond polymer network and a soluble polymeric model of 2D silicon and *a*-Si polymers, by means of Wurtz condensation of *n*-hexyltrichlorosilane with a liquid Na–K alloy with the application of high-power ultrasonic waves (USW).[75,76] Although this Na-K experiment with USW has disadvantages regarding safety, the resulting polysilyne is assumed to be a collection of various microstructures and the polysilyne reveals novel physical properties. Furukawa *et al.* developed a milder, safer synthesis of several polysilyenes with alkyl and aralkyl groups using Na and a catalytic amount of crown ether in hot toluene.[77] Jenneskens and

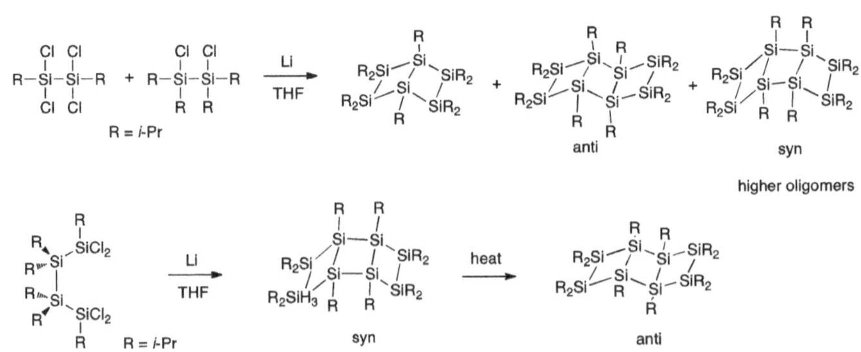

Chart 13.11

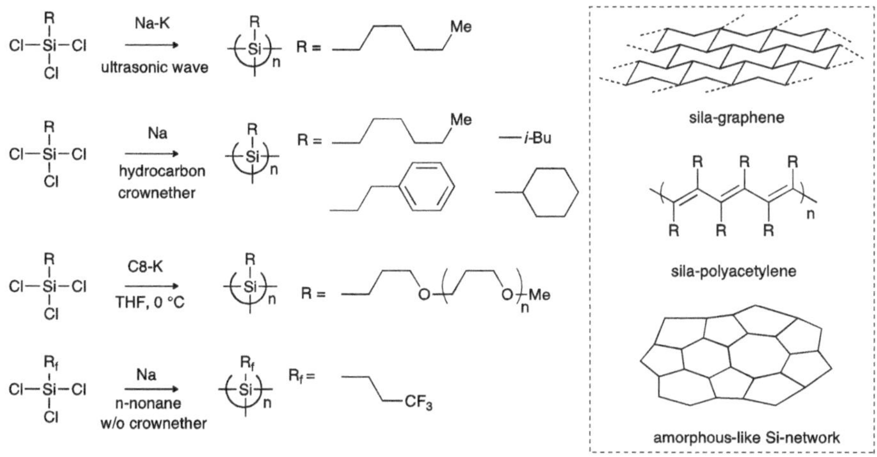

Chart 13.12

coworkers used graphite-K in place of Na–K and Na-crown ether.[78] Crown ether is not needed to produce fluoroalkyltrichlorosilane when fluoroalkylated polysilyne is reduced with Na in inert solvents.[79]

13.3.4 Chemical Modification of Ideal Two-Dimensional Si–Si Polymers with $CaSi_2$

Although the Wurtz condensation of alkyltrichlorosilanes cannot produce an ideal 2D Si sheet, the 2D layered $(Si_6H_6)_n$ is regarded as the best starting material to prepare nonoxidative Si nanosheets by organic modification and exfoliation.[80–82] Recent reports showed that $(Si_6H_6)_n$ can be prepared from calcium disilicide ($CaSi_2$),[83] where each Si-layer consists of 2D corrugated Si(111) planes (Chart 13.13). Although $(Si_6H_6)_n$ is a yellow-colored powder and exists as a suspension in organic solvents, its color changes to orange after it is stirred with *n*-decylamine. In place of $CaSi_2$, an alloy of $Ca_{0.85}Mg_{0.15}Si_2$ is a more efficient starting crystal due to its superior chemical reactivity with HCl.

13.3.5 Highly Strained Persila-Polyhedra

Extremely highly strained C–C framework compounds such as cyclopropane, tetrahedrane, cubane and prismane have been synthetic targets among organic chemists for a long time. Cyclopropane, tetrahedrane and cubane are highly symmetric compounds with three-fold axes. In 1964, Eaton and Cole performed the first synthesis of cubane.[84] In 1973, Kats and Acton reported the first production of prismane.[85] In 1981, Eaton *et al.* reported the first production of pentaprismane.[86] In 1978, Maier *et al.* successfully isolated tetrahedrane with three-fold *tert*-butyl substituents by photolysis of the corresponding 2-cyclopentadienone derivative.[87] Tetrahedrane is an isomer of cyclobutadiene.

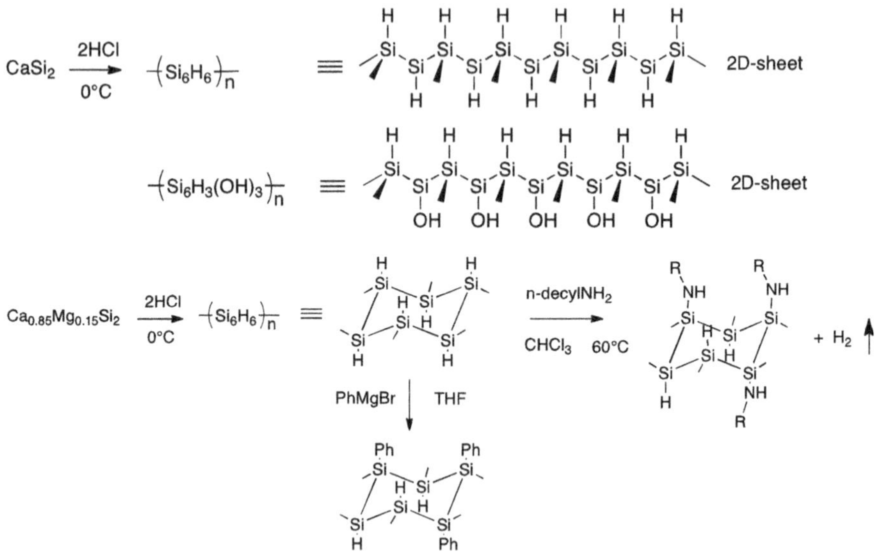

Chart 13.13

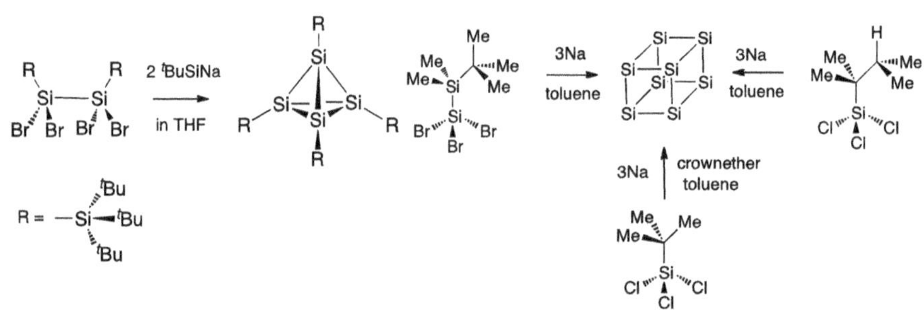

Chart 13.14

These pioneering studies stimulate theoretical and synthetic work on the corresponding silicon analogs (Chart 13.14). Nagase used *ab initio* calculations to predict that tetrahedrane, cubane and prismanes with Si–Si frameworks have lower strain energies than their carbon analogs.[88] By applying the chemistry and theory of highly strained organic compounds, the choice of bulky alkyl substituents (shape, symmetry, degrees of steric hindrance and electronic stabilization) with thermolysis/photolysis reactions plays a key role in the successful isolation of highly strained and/or kinetically stable strained organosilicon compounds.

Wiberg *et al.* successfully isolated the first tetrahedrane compound made of four Si elements and stabilized with three-fold electron-donating tri-*tert*-butylsilyl (named supersilyl) by reacting tetrabromodisilane with *tert*-butylsilyl Na in THF.[89] Matsumoto *et al.* chose other bulky and electron donating *tert*-butyldimethylsilyl groups to produce soluble, air-sensitive octasilacubane by

Li-based Wurtz reduction with the corresponding trichlorosilane in THF.[90] In place of the *tert*-butyldimethylsilyl group, the 1,1,2-trimethylpropyl group is possible to produce the corresponding cubane.[91] Furukawa *et al.* used a three-fold bulky *tert*-butyl group to produce soluble octasilacubane by Na-based Wurtz reduction in hot toluene with *tert*-butyltrichlorosilane (available commercially!) with the help of a small amount of 12-crown-4.[92] Recently, Maier *et al.* reported the successive synthesis of thermally stable tetrasilahedron by introducing four trimethylsilyl side groups.[93]

13.3.6 Exotic Unsaturated Multiple Si–Si Bond Compounds

Most importantly, the careful choice of bulky and electron-donating substituents definitively plays a key role in the successful isolation of kinetically stable Si–Si compounds containing Si=Si, Si≡Si, and Si=Si=Si bonds. The first kinetically stable Si=Si compound is isolated by photolysis of trisilane carrying two 2,4,6-trymethylphenyl (mesityl) groups, which are two-fold symmetry bulky aromatics, by West, *et al.* three decades ago[13] (Chart 13.15).

Since that first success, the synthetic approach for Si=Si compounds was developed greatly with the help of the rational design of substituents and reactions. As summarized in Chart 13.16, the synthetic methods are classified into four categories;[94] (i) photolysis of diorganotetramethyltrisilane (precursor 1), (ii) photolysis of *cyclo*-hexaorganotrisilane (precursor 2), (iii) alkaline metal reduction of dihalo-diorganosilane (precursor 3) and (iv) alkaline-metal

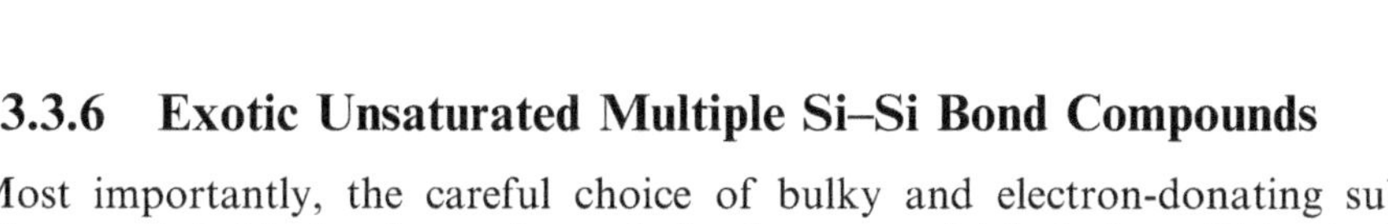

Chart 13.15

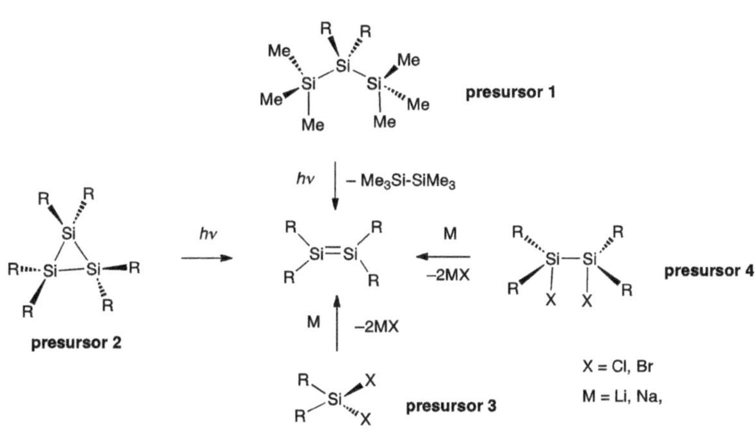

Chart 13.16

reduction of dihalo-tetraorganodisilane (precursor 4). The Si=Si compounds became a large family of organodisilenes that includes more than 100 isolated Si=Si compounds.[94,95]

On the basis of the knowledge of disilene chemistry, Kira and coworkers reported the first production of a four-membered Si ring, including one Si=Si and three Si–Si bonds, from tetrasilacyclobutene with a dimethyl-*tert*-butylsilyl group.[96,97] Weidenbruch *et al.* isolated tetrasila-1,3-butadine bearing 2,4,6-triisoproylphenyl (Tip) groups.[98] Kira and coworkers further obtained trisila-allene stabilized with well-designed substituents containing trimethylsilyl moieties.[99,100] The three compounds are depicted in Chart 13.17.

However, the ultimate challenge is to isolate a kinetically stable compound with a Si≡Si triple bond. Sekiguchi *et al.* designed bis-trimethylsilyl-isopropylsilyl groups, which are analogs of supersilyl groups, by reacting the corresponding tetrabromotetrasilane with milder reducing agents of potassium graphite (KC_8) in THF.[101] The disilyne with silyl groups is an isomer of tertahedrane. Wiberg *et al.* designed similar bulky electron donating $Si(Dis)_2(i\text{-}Pr)$ [Dis = $CH(SiMe_3)_2$] groups to stabilize the disilyne.[102] Tokitoh and coworkers reported that an extremely bulky aromatic group (2,6-bis[bis-(trimethylsilyl)methyl]-4-[tris(trimethylsilyl)methyl]phenyl) (called BbT) is used to produce disilyne with silylated substituents at remote positions.[103,104] The three syntheses are illustrated in Chart 13.18.

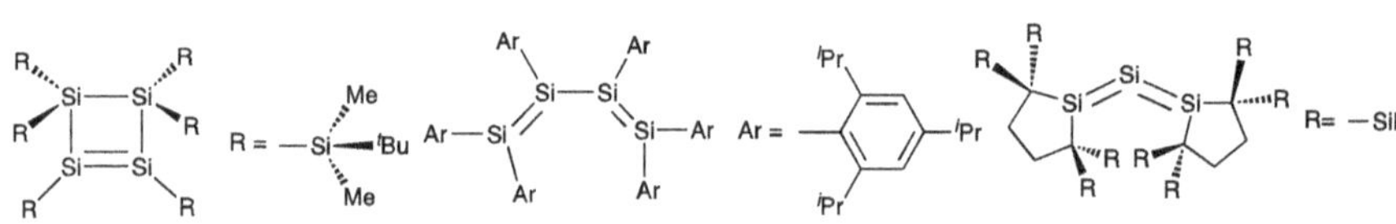

Chart 13.17

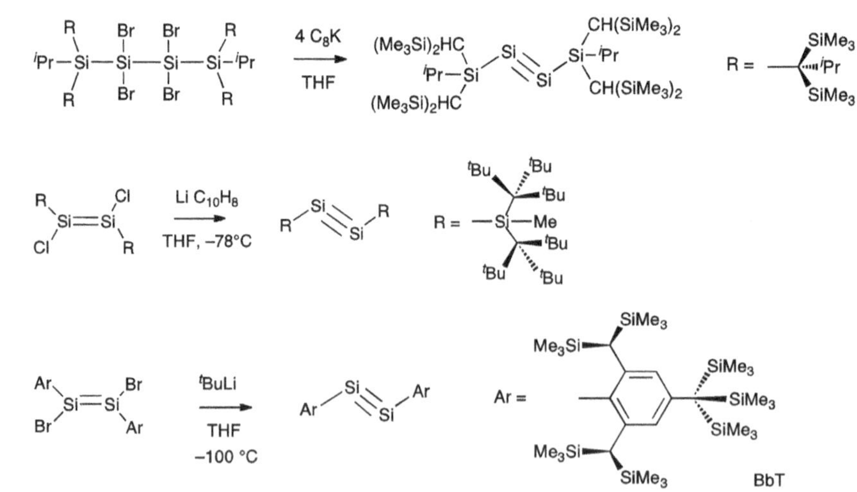

Chart 13.18

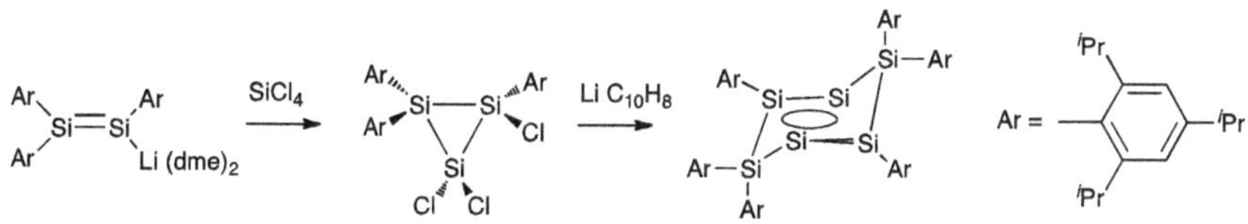

Chart 13.19

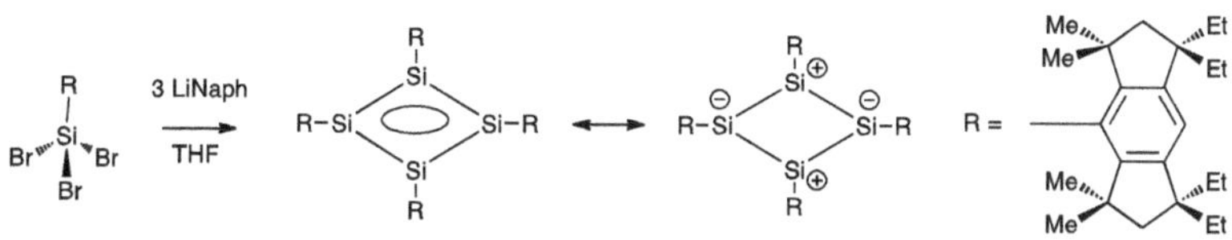

Chart 13.20

More recently, Scheschkewitz and coworkers reported the synthesis of an exotic tricyclic isomer stabilized with six 2,4,6-triisopropylphenyl (named Tip) substituents,[105] which is an attempt toward production of the first hexasilabenzene (Chart 13.19). In the crystalline form, the tricyclic ring adopted a cyclohexane-like puckered geometry, which is significantly distorted from a planar benzene-like geometry. The isomer showcased a unique aromaticity, called dismutational aromaticity, supported by three ^{29}Si resonances and high-level density functional theory calculations. This result can lead to the first isolation of an ideal hexasilabenzene with higher molecular symmetry (six- and three-fold) by designing more elaborate substituents and reaction routes.

Matsuo and coworkers reported that another exotic aromatic system consisting of a four-membered framework, stabilized by bulky 1,1,7,7-tetraethyl-3,3,5,5-tetramethyl-s-hydrindacen-4-yl (named EMind) groups, is prepared from the corresponding (EMind)$SiBr_3$ treated with 3eq lithium naphthalenide in THF.[106,107] In the crystal phase, the four-membered ring adopted a two-fold, rhombic polar planar geometry by distorting from a four-fold square, less-polar planar geometry due to a second-order Jahn–Teller distortion in a spontaneous symmetry-breaking manner (Chart 13.20).

These recent exciting outcomes inspire the hope of isolating dream compounds with more extended Siσ-Siπ conjugating polymers, oligomers and molecules that contain more plural Si–Si, Si=Si, and Si≡Si skeletons stabilized with substituents, which are carefully designed kinetically and electronically. As displayed in Chart 13.21, such silicon analogs in the distant future might be polyacetylene, polydiacetylene, nanotube, fullerene, and graphene.

13.3.7 Surface Modification of Si–C Linkages from *c*-Si with Si–H Bond

In microelectronic and photonic applications, *c*-Si wafers are used widely. The most common surfaces of *c*-Si wafers have Si(100) and Si(111) orientations. When exposed to air, surface layers of fresh *c*-Si undergo rapid auto-oxidation,

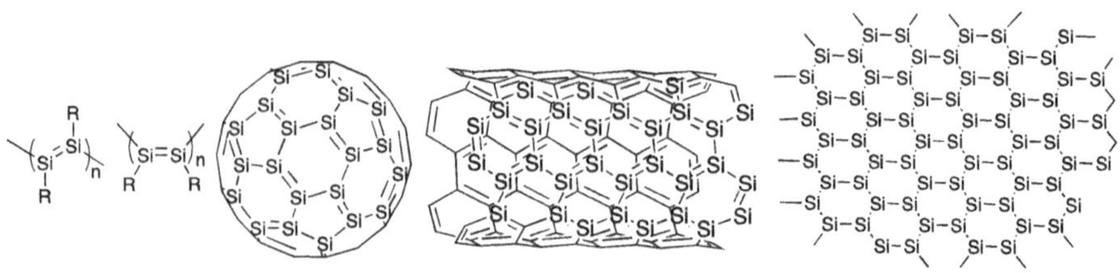

Chart 13.21

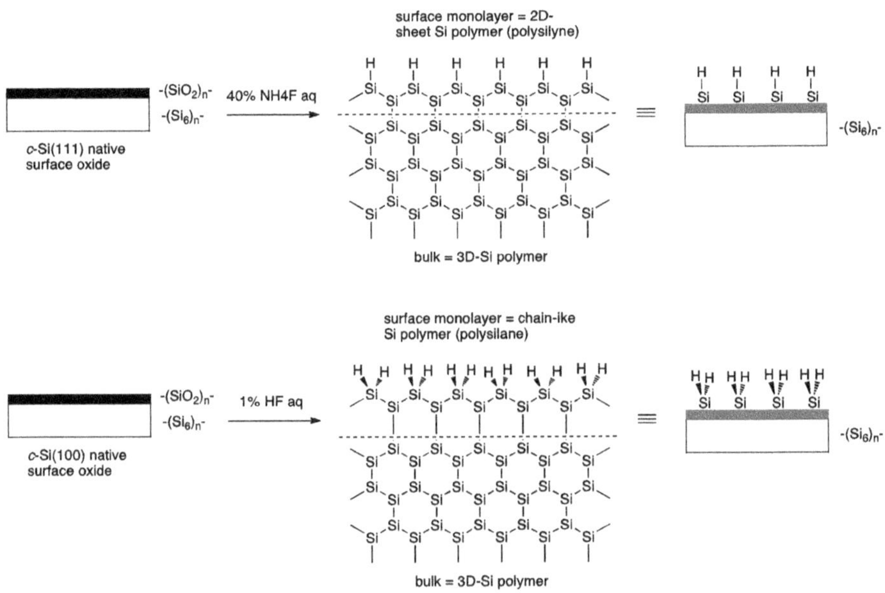

Chart 13.22

leading to the formation of thin SiO_2 layers. The oxide, however, can be removed chemically with fluoride ions. The surface of Si(111) is regarded as an ideal 2D Si nanosheet with coagulated cyclohexasilane (polysilyne), $(Si_6H_6)_n$. The Si(100) surface would be 1D Si with trans-zigzag chain-like polysilane with H side groups, $(Si_2H_4)_n$, as shown in Chart 13.22. These hydride surfaces on *c*-Si wafers become alternative sources of material suitable for surface functionalized Si-based semiconductors, although hydride surfaces are insoluble and difficult to exfoliate in common organic solvents.[108–111]

Hydrosilylation of unsaturated organic compounds, including C=C double and C≡C triple bonds, to form Si–H-terminated *c*-Si surfaces is possible to provide alkenyl and alkyl termination (Chart 13.23). The first hydrosilylation to hydride-passivated *c*-Si(111) surfaces was demonstrated two decades ago.[108,109] Due to the presence of diacyl peroxide radical species, the insertion of alkenes onto Si–H groups at the surface provides high-quality alkyl monolayers at 100 °C in 1 h. Octadecene gives an octadecyl monolayer on

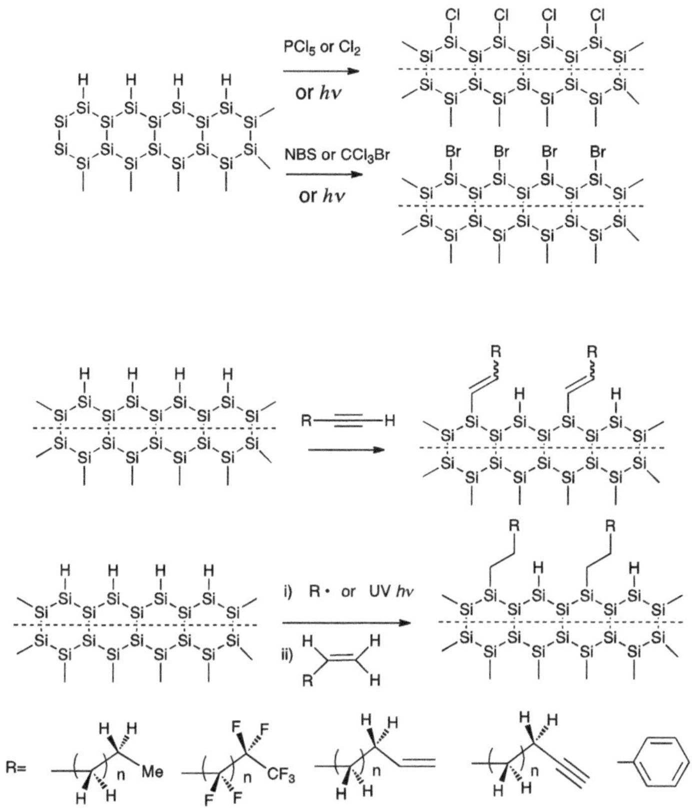

Chart 13.23

the surface with a high coverage, dense packing, and a tilt angle of approximately 30°. The *c*-Si surfaces showed excellent chemical resistance toward air, aerated organic solvents, water, inorganic acids, aqueous ammonia, and fluoride ion. Very minimal oxidation of *c*-Si wafer is possible, which is technologically important. Because many types of surface functionalization with various ordinary chemical reactions, such as C–H activation, ester hydrolysis, hydroboration, and grafting from polymerization, are possible.[110,111]

13.4 Experimental Procedures

13.4.1 Poly(diarylsilane)s[17,18]

Monomer synthesis. Dichloro[*p*-(S)-2-methylbutylphenyl](*p*-*n*-butylphenyl)silane (**1**) is prepared by lithiation of *p*-(S)-2-methylbutylphenylbromide in dry diethylether at 0 °C. The resulting aryl lithium ether solution is added to an equimolar amount of trichloro(*p*-*n*-butylphenyl)silane (prepared by addition of *p*-*n*-butylphenyl lithium to silicon tetrachloride) in dry *n*-hexane (Chart 13.24).

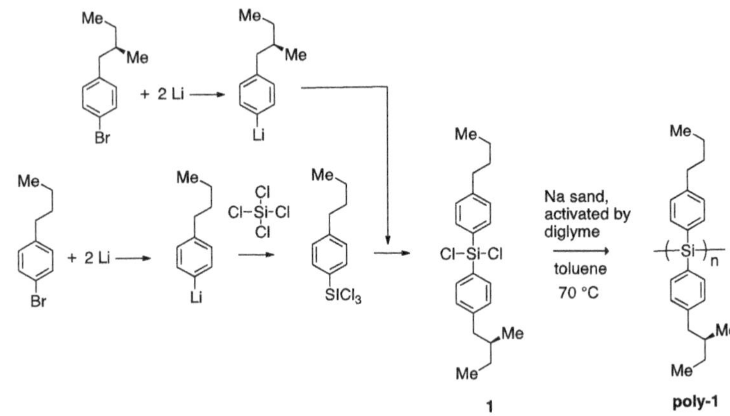

Chart 13.24

Table 13.3 Data for four diaryldichlorosilanes.

	bp (°*C/Torr*)	*[α]$_D$* (*neat, 20°C*)	*NMR δ (ppm)* ^{29}*Si*	^{13}*C*
1	172/1.2	8.10	6.27	11.44, 13.92, 18.92, 22.36, 29.19, 33.31, 35.77, 36.50, 43.40, 128.46, 129.00, 129.19, 133.97, 134.14, 145.78, 146.92
2	165/0.3	11.67	6.10	11.43, 18.82, 29.14, 36.58, 43.21, 128.08, 131.47, 131.73, 132.63, 134.75, 141.67
3	155/0.7	6.49	6.11	11.43, 13.92, 18.83, 22.36, 29.13, 33.32, 35.77, 36.56, 43.21, 128.05, 128.47, 128.99, 131.46, 131.74, 132.59, 134.14, 134.76, 141.66, 146.99
4	158/0.2	9.30	6.30	11.53, 19.01, 29.28, 36.58, 43.49, 129.12, 129.28, 134.07, 145.85

The solution is filtered and distilled in a vacuum to produce pure **1** with a 62% yield. Data for **1** and the other diaryldichlorosilanes (dichlorobis[*m*-(S)-2-methylbutylphenyl]silane (**2**), dichloro[*m*-(S)-2-methylbutylphenyl]-(*p*-*n*-butylphenyl)silane (**3**), and dichlorobis[*p*-(S)-2-methylbutylphenyl]silane) (**4**) are given in Table 13.3.

Polymer synthesis. Polymerization of **1**–**4** is achieved with a slightly modified Wurtz coupling. For example, sodium is stirred in toluene at 70 °C and initially activated by the addition of a few drops of diglyme to produce fine sodium particles. Compound **1** is added to a flask containing 2.5 equivalents of sodium in toluene and stirred at 70 °C, and the mixture is allowed to react for 2.5 h. The solution becomes a viscous purple mixture that is diluted with dry toluene. Further filtration through a 10-μm PTFE filter under N_2 pressure gives a clear toluene solution of crude polymer. The high M_w fraction ($M_w = 333\ 000$; yield 3.3%) is isolated by centrifugation after fractional precipitation in a mixed isopropanol–toluene solution and vacuum drying at 80 °C for 24 h.

13.4.2 Fluoroalkylpolysilane Homo- and Copolymers[56,57]

Monomer synthesis. The monomer *i*-butyl-*n*-decyldichlorosilane (**5**) is synthesized as follows. First, 1-bromodecane (TCI, 25.2 g, 114 mmol) is reacted with magnesium (Wako, 3.28 g, 135 mmol) in dry THF (Wako, 80.0 mL) for 1 h at 80 °C. The Grignard reagent is added to *i*-butyltrichlorosilane (Shin-Etsu, 20.0 g, 104 mmol) in a mixture of dry THF (80.0 mL) and diethyl ether (Wako, 300 mL) at 35 °C. After 1 h, the salts are removed by filtration from the reaction mixture. The crude product is purified by vacuum distillation to yield the desired monomer as a colourless liquid, with 54.0% yield (17.0 g). The boiling point is determined to be 117–120 °C (8.0 Torr) and the ^{1}H NMR results (399.00 MHz, $CDCl_3$) are as follows (ppm): 0.88 (t, -CH_2-CH_3), 1.02 (d,-CH-CH_3), 1.09 (m, -Si-CH_2-), 1.26 (m, -CH_2-), 1.50 (m, -Si-CH_2-CH_2-CH_2-) and 2.00 (m, -*CH*-CH_3). The ^{13}C NMR results (100.00 MHz, $CDCl_3$) are as follows (ppm): 14.1, 21.3, 22.4, 22.7, 24.2, 25.6, 29.1, 29.3, 29.4, 29.6, 30.4, 31.9, and 32.5. The ^{29}Si NMR data (79.00 MHz, $CDCl_3$) show a single peak at 32.6 ppm.

Methyl-3,3,3-trifluoropropyldichlorosilane (**6**) and methyl-*n*-propyldichlorosilane (**7**) are purchased from Shin-Etsu and purified by vacuum distillation prior to polymerization.

Polymer synthesis. All reactions are carried out in glassware that is dried under a dry nitrogen gas atmosphere. Details of the homopolymer synthesis (**poly-5**, **poly-6** and **poly-7**) have been described in previous papers.[57] Addition of a catalytic amount of 18-crown-6 and 15-crown-5 to most reactions of dialkyldichlorosilanes with sodium in hot toluene leads to reproducible high yields of HMW polysilanes with a monomodal distribution. However, several alkyl-3,3,3-trifluoropropyldichlorosilanes are exceptions and do not produce the corresponding homopolymers when crown ethers are present in the reaction mixture (Chart 13.25).

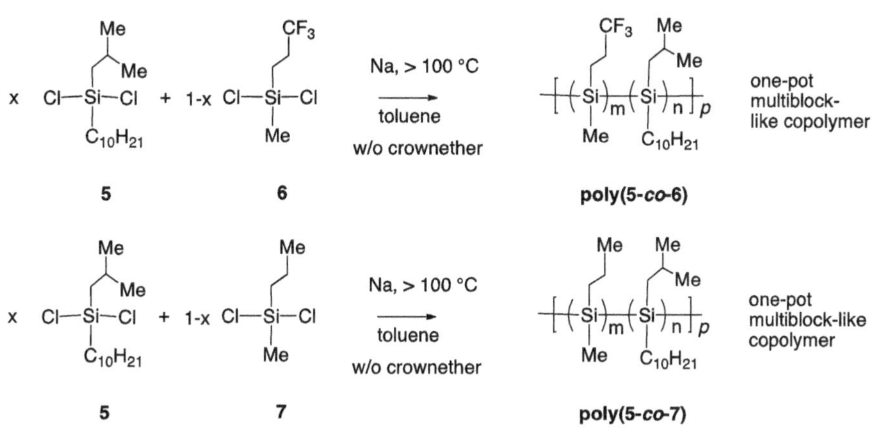

Chart 13.25

Poly[(*i*-butyl-*n*-decylsilane)$_m$-*co*-(methyl-3,3,3-trifluoropropylsilane)$_n$] (**poly(5-*co*-6)**) is prepared as follows. A freshly distilled mixture of **5** (2.70 g, 9.10 mmol) and **6** (1.90 g, 9.10 mmol) is added dropwise to 0.84 g (36.0 mmol) of sodium (Wako) suspended in 10.0 mL of refluxing toluene (Wako) and vigorously stirred under atmospheric nitrogen at 120 °C. After 2 h, 90 mL of dry toluene (Wako) are added to reduce the solution viscosity, and stirring is continued for an additional 1 h at 100 °C. The hot reaction mixture is immediately passed through a PTFE filter (2 μm pore size) under N_2 gas pressure. The solvents for fractional precipitation, 2-propanol, ethanol, and methanol, are carefully added to the clear filtrate. Several portions of white precipitate are collected by centrifugation and dried under vacuum at 80 °C. Data from ^{1}H NMR (399.00 MHz, $CDCl_3$, ppm): 0.37 (br, -Si-CH_3), 0.87 (t, -CH_2-CH_3), 0.98 (br, -Si-CH_2-, -CH-CH_3), 1.10 (br, -Si-CH_2-CH_2-CF_3), 1.26 (br, -CH_2-), 1.73 (br, -*CH*-CH_3), 2.06 (br, -Si-CH_2-CH_2-CF_3). ^{29}Si NMR (79.00 MHz, $CDCl_3$, ppm): –23.1 (*Si*-CH_2-CH(CH_3)-CH_3), –31.7 (*Si*-CH_2-CH_2-CF_3). ^{19}F NMR (564.69 MHz, $CDCl_3$, ppm): –69.9; FT-IR (cm^{-1}) 2955 ($\nu_{as}(CH_3)$), 2925 ($\nu_{as}(CH_2)$), 2872 ($\nu_s(CH_3)$), 2855 ($\nu_s(CH_2)$), 1262 ($\delta_s(SiCH_3)$), 1214 ($\nu_{as}(CF)$), 1198 ($\nu(CC)$), and 1130 ($\nu_s(CF)$). For PSi**1**, the composition fractions are determined by ^{1}H NMR.

Poly[(*i*-butyl-*n*-decylsilane)$_n$-*co*-(methyl-*n*-propylsilane)$_m$] (**poly(5-*b*-7)**) is prepared as follows. A mixture of 0.92 g (40.4 mmol) of sodium metal and 0.05 g (0.20 mmol) of 18-crown-6 (Wako) in 10.0 mL of dry toluene is added dropwise to a mixture of **5** (3.02 g, 10.1 mmol) and **7** (1.58 g, 10.1 mmol) in a nitrogen gas atmosphere at 120 °C. After 1 h, 100 mL of dry toluene are added to reduce the solution viscosity, and stirring is continued for an additional 1 h at 100 °C. The hot reaction mixture is passed through a 2-μm PTFE filter under nitrogen gas pressure. The precipitating solvents, 2-propanol, ethanol, and methanol, are carefully added to the clear filtrate. Several portions of white precipitate are collected by centrifugation and dried under vacuum at 80 °C. The ^{1}H NMR (399.00 MHz, $CDCl_3$, ppm) data are: 0.25 (br, -Si-CH_3), 0.85 (br, -Si-CH_2-CH_2-CH_3), 0.87 (t, -CH_2-CH_3), 0.98 (br, -Si-CH_2-CH_2-CH_3, -Si-CH_2-, -CH-CH_3), 1.26 (br, -CH_2-), 1.37 (br, -Si-CH_2-CH_2-CH_3), 1.78 (br, -*CH*-CH_3). ^{29}Si NMR (79.00 MHz, $CDCl_3$, ppm): –23.0 (*Si*-CH_2-CH(CH_3)-CH_3) and –32.5 (*Si*-CH_2-CH_2-CH_3).

13.4.3 Poly(alkylarylsilane) Homo- and Copolymers[112,113]

Materials: The molecular structures of (*S*)-2-methylbutylphenyl dichlorosilane (**8**), methylphenyldichlorosilane (**9**) and *n*-hexyl(*m*-tolyl)dichlorosilane (**10**) are shown in Chart 13.26. A typical method for synthesizing poly[((*S*)-2-methylbutyl-phenylsilane)$_{0.25}$-*co*-(methylphenylsilane)$_{0.75}$] (**poly(8-*co*-9)**) and poly[((*S*)-2-methylbutyl-phenylsilane)$_{0.20}$-*co*-(*n*-hexyl-*m*-tolylsilane)$_{0.80}$] (**poly-(8-*co*-10)**) is given. The same procedure is applied to synthesize copolymers with different molar fractions of chiral **8** ($x = 0.05$, 0.10, 0.25, 0.30, and 0.50), and achiral **9** and **10**.

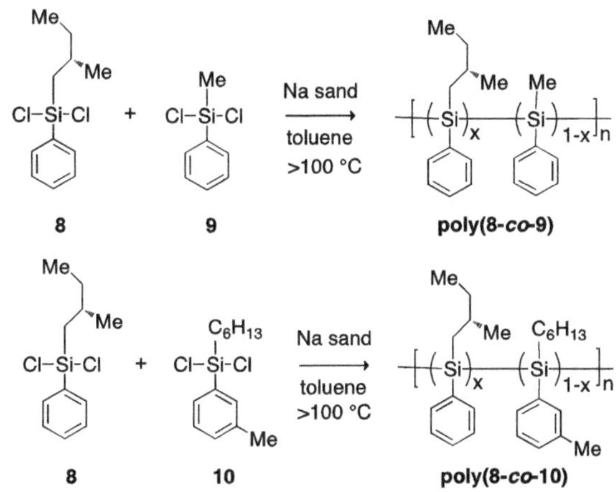

Chart 13.26

A mixture of **8** (1.6g) and **9** (5.0g) is added dropwise into toluene with 1.3 g sodium at 110 °C under an Ar gas atmosphere. After refluxing the mixture by stirring it for 5 min, sodium is crushed into small pieces and the reaction mixture turns bright purple. After the solution is allowed to cool at 65 °C, the rest of the starting materials are added at one time. By adding 100 mL of toluene, the solution is filtered using a 5-μm pore PTFE membrane filter under Ar gas. By carefully adding isopropylalcohol to the filtrate, the precipitate is isolated using a Kubota 5100 centrifuge apparatus at 4000 rpm for 30 min. The precipitate is dried at 100 °C overnight under a vacuum and the purified **poly(8-*co*-9)** (0. 22 g) is obtained as a white solid with a 7% yield. The M_w value is 9.8×10^5 and the polydispersity index (PDI) is 1.8. The ^{13}C NMR data in $CDCl_3$ are: –6.28, 10.90, 22.38, 25.35, 32.91, 127.19, 133.27, and 135.00. The ^{29}Si NMR data in $CDCl_3$ are –41.41 and –39.42.

A mixture of **8** (1.5 g) and **10** (5.0 g) is added dropwise by stirring into toluene with 1.1 g sodium at 110 °C under Ar gas for 5 min. The color of the reaction mixture changes to bright purple. After addition of 100 mL toluene, the solution is filtrated using a 5-μm pore PTFE disc under Ar gas pressure. A small amount of isopropanol is added carefully to the filtrate and the precipitate is isolated by centrifugation at 4000 rpm for 30 min. After drying the precipitate at 100 °C overnight under a vacuum, purified **4** (0.04 g) is obtained as a white solid with a yield of 1%. The M_w value is 2.4×10^5 and PDI is 2.2. The ^{13}C NMR data in $CDCl_3$ are: 14.07, 20.74, 21.19, 22.49, 31.31, 32.11, 128.26, 130.44, 132.39, 132.62, 133.90 and 137.95. The ^{29}Si NMR data in $CDCl_3$ are – 43.41 and –41.42.

The (*S*)-2-methylbutylphenylsilane contents in **poly(8-*co*-9)** and **poly(8-*co*-10)** are almost the same as the feed ratios based on the peak ratio from ^{1}H-NMR. All of the samples have high M_w of more than 10^5.

13.4.4 Poly(alkylsilyne)s and Vacuum Pyrolysis[79,114]

Synthesis. Soluble alkylpolysilynes are synthesized by Wurtz reduction of the corresponding alkyltrichlorosilanes (Shin-Etsu and Gelest) in the presence of 12-crown-4 as cocatalyst under N_2 atmosphere (Chart 13.26). All alkyltrichlorosilanes are distilled prior to use. To prevent any contact of alkylpolysilynes with air and moisture during all procedures, including preparation, isolation, and sample sealing in a glass tube, the polymerization of *n*-butyltrichlorosilane (**11**) is achieved in the following manner as shown in Chart 13.27.

Methylcyclohexane (TCI, dried over 4A molecular sieves or CaH_2, 4 mL) containing Na (Wako, 0.43 g, 19 mmol) and 12-crown-4 (Aldrich, 0.02 g, 0.11 mmol) is placed in a four-necked 100-mL flask and refluxed at 120 °C with mechanical stirring. To this mixture, **11** (Shin-Etsu, 0.98 g, 5.1 mmol) dissolved in methylcyclohexane (4 mL) is added dropwise. After the addition is complete, the reaction mixture is stirred for 1 h and allowed to cool to room temperature. The reaction vessel is moved to a glove box filled with 99.9% N_2 gas.

A clear yellow solution containing *n*-butylpolysilyne is obtained by a pressurized filtration of the reaction mixture using a fluorinated membrane filter (Advantec, Labodisc, 0.50 μm pore) to remove excess Na and NaCl. The polymer is isolated by precipitating the solution in dry acetone (Aldrich) and dried in the box that is connected to an external vacuum pump. No detectable Si–O–Si vibrations at 1000–1100 cm^{-1} are observed for the fresh *n*-butylpolysilyne. Ethyl, *n*-propyl, *i*-butyl, *n*-pentyl, *n*-hexyl, *n*-octyl, *n*-decyl, and *n*-dodecyl derivatives are prepared in a similar manner, except the 3,3,3-trifluoropropyl derivative. The polymerization of 3,3,3-trifluoropropyltrichlorosilane and alkyltrichlorosilane does not require the use of a crown ether. Soluble poly(methylsilyne) is not isolated due to its extreme insolubility. The isolated yields of the polysilynes are typically 40–50% when the crude polysilynes are precipitated into methanol outside the glove box with no precautions taken against air/moisture. These values are almost identical to those given in a previous report.[75,76]

Preparation of Polysilyne Film and Intube Vacuum Pyrolysis. A methylcyclohexane solution of *n*-butylpolysilyne is placed into a glass tube (ID 5 mm, OD 7 mm), followed by manual coating of the tube inner wall and drying by blowing with N_2 gas. The SNP film deposited in a glass tube is connected to a two-way vacuum bulb. The glass tube coated with the polysilyne

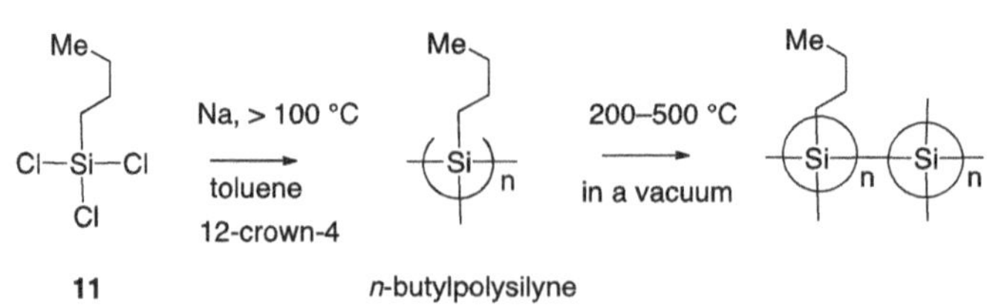

Chart 13.27

film is removed from the glove box and sealed with a hand burner using vacuum techniques (0.3 Torr by rotary pump or 5×10^{-5} Torr by turbo molecular pump). In pyrolysis experiments and optical measurements, the glass tube is placed in a housing made of an aluminum block that is on a digitally controlled hotplate (Thermolyne), and the temperature of the housing is monitored with a chromel-alumel thermocouple.

References

1. F. S. Kipping and J. E. Sands, *J. Chem. Soc. Trans.*, 1921, **119**, 830–847.
2. F. S. Kipping, *J. Chem. Soc.*, 1924, **125**, 2291–2297.
3. F. S. Kipping, *Proc. R. Soc. Lond. Ser. A*, 1937, **159**, 139–148.
4. C. A. Burkhard, *J. Am. Chem. Soc.*, 1949, **71**, 963–964.
5. W. P. Jarvie, H. J. Winkler, D. J. Peterson and H. Gilman, *J. Am. Chem. Soc.*, 1961, **83**, 1921–1924.
6. H. Gilman and G. L. Schwebke, *J. Organomet. Chem.*, 1965, **3**, 382–392.
7. H. Gilman and G. L. Schwebke, *J. Am. Chem. Soc.*, 1964, 2693–2699.
8. H. Gilman and R. A. Tomasi, *J. Org. Chem.*, 1963, **28**, 1651–1653.
9. R. West, *J. Organomet. Chem.*, 1986, **300**, 327–346.
10. R. D. Miller and J. Michl, *Chem. Rev.*, 1989, **89**, 1539–1410.
11. K. Tamao, E. West and K. Tatsumi (eds), *J. Organomet. Chem.*, 2003, **685**, 1–288.
12. N. Auner and J. Weis (Eds), *Organosilicon Chemistry I: From Molecules to Materials*, Wiley-VCH, Weinheim, Germany, 2005.
13. R. West, M. J. Fink and J. Michl, *Science*, 1981, **214**, 1343–1344.
14. T. Shimoda, Y. Matsuki, M. Furusawa, T. Aoki, I. Yudasaka, H. Tanaka, H. Iwasawa, D. Wang, M. Miyasaka and Y. Takeuchi, *Nature*, 2006, **440**, 783–786.
15. R. D. Miller and R. Sooriyakumaran, *Macromolecules*, 1988, **21**, 3120–3122.
16. R. D. Miller and R. Sooriyakumaran, *J . Polym. Sci., Part C. Polym. Lett. Ed.*, 1987, **25**, 321–326.
17. J. R. Koe, M. Fujiki and H. Nakashima, *J. Am. Chem. Soc.*, 1999, **121**, 9734–9735.
18. J. R. Koe, M. Fujiki, M. Motonaga and H. Nakashima, *Macromolecules*, 2001, **34**, 1082–1089.
19. J. R. Koe, M. Fujiki, M. Motonaga and H. Nakashima, *Chem. Commun.*, 2000, 389–390.
20. R. H. Cragg, R. G. Jones, A. C. Swain and S. J. Webb, *J. Chem. Soc., Chem. Commun.*, 1990, 1147–1148.
21. M. Fujino and H. Isaka, *Chem. Commun.*, 1989, 466–467.
22. M. Fujiki, M., *Macromol. Rapid Commun.*, 2001, **22**, 539–563.
23. M. Fujiki, J. R. Koe, K. Terao, T. Sato, A. Teramoto and J. Watanabe, *Polym. J.*, 2003, **35**, 297–344.
24. M. Fujiki, *J. Organometal. Chem.*, 2003, **685**, 15–34.

25. K. Sakamoto, K. Obata, H. Hirata, M. Nakajima and H. Sakurai, *J. Am. Chem. Soc.*, 1989, **111**, 7641–7642.
26. K. Sakamoto, M. Yoshida and H. Sakurai, *Macromolecules*, 1990, **23**, 4494–4496.
27. T. Sanji, K. Takase and H. Sakurai, *J. Am. Chem. Soc.*, 2001, **123**, 12690–12691.
28. H. Sakurai, R. Honbori and T. Sanji, *Organometallics*, 2005, **24**, 4119–4121.
29. R. E. Trujillo, *J. Organomet. Chem.*, 1980, **198**, C27–C28.
30. J. P. Wesson and T. C. Williams, *J. Polym. Sci., Polym. Chem. Ed.*, 1980, **18**, 959–965.
31. J. P. Wesson and T. C. Williams, *J. Polym. Sci., Polym. Chem. Ed.*, 1981, **19**, 65–72.
32. R. West, L. D. David, P. I. Djurovich, K. L. Stearley, K. S. V. Srinivasan and H. Yu., *J. Am. Chem. Soc.*, 1981, **103**, 7352–7354.
33. C. Aitken, J. F. Harrod and Samuel, *J. Organomet. Chem.*, 1985, **279**, C11–C13.
34. C. T. Aitkin, J. F. Harrod and E. Samuel, *J. Am. Chem. Soc.*, 1986, **108**, 4059–466.
35. L. S. Chang and J. Y. Corey, *Organometallics*, 1989, **8**, 1885–1893.
36. B. J. Grimmond and J. Y. Corey, *Organometallics*, 1999, **18**, 2223–2229.
37. J. Y. Corey, *Adv. Organomet. Chem.*, 2004, **51**, 1–52.
38. J. Y. Corey, *Chem. Rev.*, 2011, **111**, 863–1071.
39. H. G. Woo and T. D. Tilley, *J. Am. Chem. Soc.*, 1989, **111**, 8043–8044.
40. T. D. Tilley, *Acc. Chem. Res.*, 1993, **26**, 22–29.
41. J. P. Banovetz, K. M. Stein and R. M. Waymouth, *Organometallics*, 1991, **10**, 3430–3432.
42. J. P. Banovetz, Y.-L. Hsiao and R. M. Waymouth, *J. Am. Chem. Soc.*, 1993, **115**, 2540–2541.
43. Y.-L. Hsiao and R. M. Waymouth, *J. Am. Chem. Soc.*, 1994, **116**, 9779–9780.
44. B.-H. Kim, M.-S. Shik, M. A. Kim and H. G. Woo, *J. Organomet. Chem.*, 2003, **685**, 93–98.
45. L. Rosenberg, C. W. Davis and J. Yao, *J. Am. Chem. Soc.*, 2001, **123**, 5120–5121.
46. L. Rosenberg, C. W. Davis and J. Yao, *J. Organomet. Chem.*, 2003, **685**, 107–112.
47. A. E. Wetherby, N. T. Mucha and R. Waterman, *ACS Catal.*, 2012, **2**, 1404–1407.
48. J. R. Koe, D. R. Powell, S. Hayase, J. J. Buffy and R. West, *Angew. Chem. Int. Ed.*, 1998, **37**, 1441–1442.
49. J. R. Koe, M. Motonaga, M. Fujiki and R. West, *Macromolecules*, 2001, **34**, 706–712.
50. M. Fujino, T. Hisaki, M. Fujiki and N. Matsumoto, *Macromolecules*, 1992, **25**, 1079–1083.

51. M. Fujino, T. Hisaki and N. Matsumoto, *J. Polym. Sci. Part A, Polym. Chem.*, 1995, **33**, 2279–2283.
52. A. Saxena, M. Fujiki, M. Naito, K. Okoshi and G. Kwak, *Macromolecules*, 2004, **37**, 5873–5879.
53. S.-Y. Kim, A. Saxena, G. Kwak, M. Fujiki and Y. Kawakami, *Chem. Commun.*, 2004, 538–539.
54. A. Saxena, M. Fujiki, R. Rai, S.-Y. Kim and G. Kwak, *Macromol. Rapid Commun.*, 2004, **25**, 1771–1775.
55. M. Naito, M. Nakamura, K. Terao, T. Kawabe and M. Fujiki, *Macromolecules*, 2010, **43**, 7919–7923.
56. T. Kawabe, M. Naito and M. Fujiki, *Macromolecules*, 2008, **41**, 1952–1960.
57. T. Kawabe, M. Naito and M. Fujiki, *Polym. J.*, 2008, **40**, 317–326.
58. H. Sohn, R. R. Huddleston, D. R. Powell, R. West, K. Oka and X. Yonghua, *J. Am. Chem. Soc.*, 1999, **121**, 2935–2936.
59. H. Sohn, M. J. Sailor, D. Magde and W. C. Trogler, *J. Am. Chem. Soc.*, 2003, **125**, 3821–3830.
60. J. Chen, C. C. W. Law, J. W. Y. Lam, Y. Dong, S. M. F. Lo, I. D. Williams, D. Zhu and B. Z. Tang, *Chem. Mater.*, 2003, **15**, 1535–1546.
61. S. J. Toal, D. Magde and W. C. Trögler, *Chem. Commun.*, 2005, 5465–5467.
62. L. T. Canham, *Appl. Phys. Lett.*, 1990, **57**, 1046–1048.
63. G. Cullis, L. T. Canham and P. D. J. Calcott, *J. Appl. Phys.*, 1997, **82**, 909–965.
64. T. Shono, S. Kashimura, M. Ishifune and R. Nishida, *J. Chem. Soc. Chem. Commun.*, 1991, 1160–1161.
65. S. Kashimura, M. Ishifune, N. Yamashita, H.-B. Bu, M. Takebayashi, S. Kitajima, D. Yoshihara, Y. Kataoka, R. Nishida, S. Kawasaki, H. Murase and T. Shono, *J. Org. Chem.*, 1999, **64**, 6615–6621.
66. M. Ishifune, S. Kashimura, Y. Kogai, Y. Fukuhara, T. Kato, H.-B. Bu, N. Yamashita, Y. Murai, H. Murase and R. Nishida, *J. Organomet. Chem.*, 2000, **611**, 26–31.
67. M. Okano, H. Fukai, M. Arakawa and H. Hamano, *Electrochem. Commun.*, 1999, **1**, 223.
68. K. Yamada and M. Okano, *Electrochemistry*, 2006, **74**, 668–671.
69. A. Watanabe, *J. Organomet. Chem.*, 2003, **685**, 122–133.
70. R. West, in G. Wilkinson, G. A. Stone and E. W. Abel (eds.), *Comprehensive Organometallic Chemistry*, Vol. 2, Pergamon, New York, 1982 (Chapter 9).
71. R. West, in P. Patai and Z. Rappoport (eds.), *The Chemistry of Organic Silicon Compounds*, Part 2, Wiley, Chichester, UK, 1989 (Chapter 19).
72. T. Tsumuraya, S. A. Batcheller and S. Masamune, *Angew. Chem. Int. Ed. Engl.*, 1991, **30**, 902–930.
73. E. Hengge and R. Janoschek, *Chem. Rev.*, 1995, **95**, 1495–1526.

74. H. Matsumoto, S. Kyushin, M. Unno and R. Tanaka, *J. Organomet. Chem.*, 2000, **611**, 52–63.
75. P. A. Bianconi and T. W. Weidman, *J. Am. Chem. Soc.*, 1988, **110**, 2342–2344.
76. P. A. Bianconi, F. C. Schilling and T. W. Weidman, *Macromolecules*, 1989, **22**, 1697–1704.
77. K. Furukawa, M. Fujino and N. Matsumoto, *Macromolecules*, 1990, **23**, 3423–3426.
78. T. J. Cleij, S. K. Tsang and L. W. Jenneskens, *Macromolecules*, 1999, **32**, 3286–3294.
79. M. Fujiki, Y. Fujimoto, A. Saxena, T. Kawabe and G. Kwak, *Polym. Chem.*, 2012, **3**, 3256–3265.
80. H. Okamoto, Y. Kumai, Y. Sugiyama, T. Mitsuoka, K. Nakanishi, T. Ohta, H. Nozaki, S. Yamaguchi, S. Shirai and H. Nakano, *J. Am. Chem. Soc.*, 2010, **132**, 2710–2718.
81. Y. Sugiyama, H. Okamoto, T. Mitsuoka, T. Morikawa, K. Nakanishi, T. Ohta and H. Nakano, *J. Am. Chem. Soc.*, 2010, **132**, 5946–5947.
82. H. Okamoto, Y. Sugiyama and H. Nakano, *Chem. Eur. J.*, 2011, **17**, 9864–9887.
83. M. S. Brandt, G. Vogg and M. Stutzmann, in *Silicon Chemistry* P. Jutzi and U. Schubert (eds.), Wiley-VCH, Weinheim, 2003, Chapter 15, 194–213.
84. P. E. Eaton and T. W. Cole, *J. Am. Chem. Soc.*, 1964, **86**, 962–964.
85. T. J. Kats and N. Acton, *J. Am. Chem. Soc.*, 1973, **95**, 2738–2739.
86. P. E. Eaton, Y. S. Or and S. J. Branca, *J. Am. Chem. Soc.*, 1981, **103**, 2134–2136.
87. G. Maier, S. Pfriem, U. Schäfer and R. Matusch, *Angew. Chem. Int. Ed. Engl.*, 1978, **17**, 520–521.
88. S. Nagase, *Angew. Chem. Int. Ed. Engl.*, 1989, **28**, 329–330.
89. N. Wiberg, C. M. Fünger and K. Polborn, *Angew. Chem. Int. Ed. Engl.*, 1993, **32**, 1054–51056.
90. K. Matsumoto, Y. Higuchi, H. Hoshino, Y. Koike, Y. Naoi and N. Nagai, *J. Chem. Soc. Chem. Commun.*, 1988, 1083–1084.
91. H. Matsumoto, K. Higuchi, S. Kyushin and M. Goto, *Angew. Chem. Int. Ed. Engl.*, 1992, **31**, 1354–1356.
92. K. Furukawa, M. Fujino and N. Matsumoto, *Appl. Phys. Lett.*, 1992, **60**, 2744–2745.
93. G. Maier, J. Neudert, O. Wolf, D. Pappusch, A. Sekiguchi, M. Tanaka and T. Matsuo, *J. Am. Chem. Soc.*, 2002, **124**, 13819–13826.
94. T. Iwamoto and M. Kira, *Adv. Organomet. Chem.*, 2006, **54**, 73–148.
95. H. Ottosson and P. G. Steel, *Chem. Eur. J.*, 2006, **12**, 1576–1585.
96. T. Iwamoto, M. Tamura, C. Kabuto and M. Kira, *Science*, 2000, **290**, 504–506.
97. M. Kira, T. Iwamoto and C. Kahuto, *J. Am. Chem. Soc.*, 1996, **118**, 10303–10304.

98. M. Weidenbruch, S. Willms, W. Saak and G. Henkel, *Angew. Chem. Int. Ed. Engl.*, 1997, **36**, 2503–2504.
99. S. Ishida, T. Iwamoto, C. Kabuto and M. Kira, *Nature*, 2003, **421**, 725–727.
100. M. Kira, *J. Organomet. Chem.*, 2004, **689**, 4475–4488.
101. A. Sekiguchi, R. Kinjo and M. Ichinohe, *Science*, 2004, **305**, 1755–1757.
102. N. Wiberg, S. K. Vasisht, G. Fischer and P. Mayer, *Z. Anorg. Allg. Chem.*, 2004, **630**, 1823–1828.
103. T. Sasamori, K. Hironaka, Y. Sugiyama, N. Takagi, S. Nagase, Y. Hosoi, Y. Furukawa and N. Tokitoh, *J. Am. Chem. Soc.*, 2008, **130**, 13856–13857.
104. T. Sasamori, J. S. Han, K. Hironaka, N. Takagi, S. Nagase and N. Tokitoh, *Pure Appl. Chem.*, 2010, **82**, 603–612.
105. K. Abersfelder, A. J. P. White, H. S. Rzepa and D. Scheschkewitz, *Science*, 2010, **327**, 564–566.
106. K. Suzuki, T. Matsuo, D. Hashizume, H. Fueno, K. Tanaka and K. Tamao, *Science*, 2011, **331**, 1306–1309.
107. Y. Matsuo and M. Maruyama, *Chem. Commun.*, 2012, **48**, 9334–9342.
108. M. R. Linford and C. E. D. Chidsey, *J. Am. Chem. Soc.*, 1993, **115**, 12631–12632.
109. M. R. Linford, P. Fenter, P. M. Eisenberger and C. E. D. Chidsey, *J. Am. Chem. Soc.*, 1995, **117**, 3145–3155.
110. J. M. Buriak, *Chem. Commun.*, 1999, 1051–1060.
111. J. M. Buriak, *Chem. Rev.*, 2002, **102**, 1271–1308.
112. S. Toyoda and M. Fujiki, *Macromolecules*, 2001, **34**, 640–644.
113. S. Toyoda and M. Fujiki, *Macromolecules*, 2001, **34**, 2630–2634.
114. M. Fujiki, Y. Kawamoto, M. Kato, Y. Fujimoto, T. Saito, S. Hososhima and G. Kwak, *Chem. Mater.*, 2009, **21**, 2459–2466.

CHAPTER 14

Alternating Polyheterocycles

KAZUO TANAKA AND YOSHIKI CHUJO*

Department of Polymer Chemistry, Graduate School of Engineering, Kyoto University, Katsura, Nishikyo-ku, Kyoto 615-8510, Japan
*Email: chujo@chujo.synchem.kyoto-u.ac.jp

14.1 Introduction

Alternating polymers are defined as a type of polymer consisting of two different repeating units in which the two units appear alternately in the main chain (Chart 14.1). Particularly, by connecting directly or *via* multibonds, various intriguing functions can be accomplished. The dual properties originating from each component are readily obtained. For example, the solubility for various solvents involving water and the film processability of the product can be readily tuned by modulating the side chains in one of the monomer units without changing other electronic functions.[1] This advantage is useful for fabricating devices. In addition, extended π-conjugation systems often show unique characteristics originating from the combination of each component. For instance, the alternating polymers containing electron-withdrawing and -donating groups show strong emission and narrow bandgaps between HOMO and LUMO because of charge transfer (CT) between the two units.[2] In particular, the degree of the electronic interaction can be modulated, leading to large color changes.[3–7] These materials are versatile for solar cells and emitting devices.

RSC Polymer Chemistry Series No. 9
Conjugated Polymers: A Practical Guide to Synthesis
Edited by Klaus Müllen, John R. Reynolds and Toshio Masuda

Published by the Royal Society of Chemistry, www.rsc.org

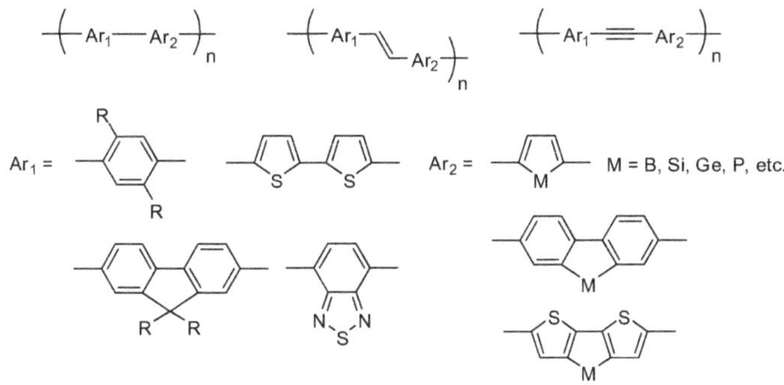

Chart 14.1

Heteroatom-containing compounds have attracted attention as building blocks to construct advanced optical, electronic, and optoelectronic materials based on the conjugated polymers. The synergetic effects based on heteroatoms and conjugation systems have abundant possibilities to induce significant characteristics that are hardly accomplished with only C, H, O, and N elements. Thus, the fusion of the characteristics of the heteroatom and conjugated polymers is expected to show superior properties distinctly different from those of each component. In this chapter, a series of alternating polyheterocycles are reviewed. Heteroles (heterocyclopentadienes) and bridged-type heterocycles with boron, silicon, germanium, phosphorus, and transition metals are mainly discussed. The fundamental issues of heterocycle-containing polymers, such as electronic properties and synthetic methods, and applications, are explained. The experimental procedures of the synthesis of alternating polymers are described in the latter part.

14.1.1 Alternating Polymers of Boron

Boron belongs to the 13th group and has one less electron than carbon. Many organoboron-containing polymers including the alternating type have been synthesized *via* the crosscoupling reaction in the presence of metal catalysts.[8] The introduction of boron atoms into the conjugation system through the polymer main chain induces the lowering of LUMO levels and strong electron affinities. The vacant p-orbital would be generated in organoboron complexes containing a tricoordinate boron atom (eqn (14.1)). The large absorption and emission bands can be induced from the donor–acceptor interaction with a tricoordinate boron atom.[9] Furthermore, based on the strong electron affinity with fluoride anion, boron-containing polymers can work as sensors.[10] However, because of instability, the construction of borole rings with a tricoordinate boron atom is still challenging although *via* theoretical and monomer studies, the unique properties originated from antiaromaticity of tricoordinate boron-containing polyboroles have been predicted.[11]

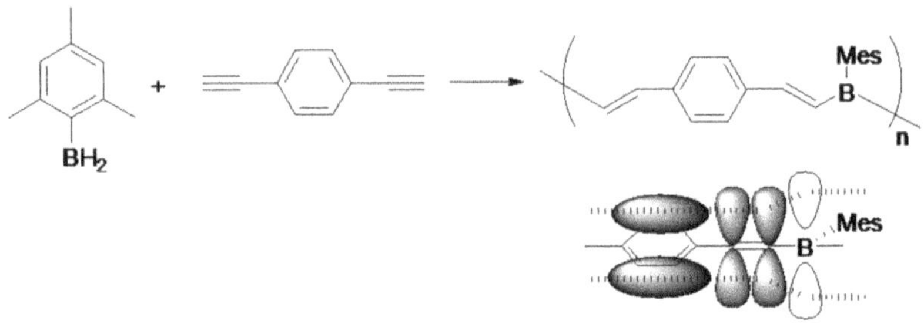

(14.1)

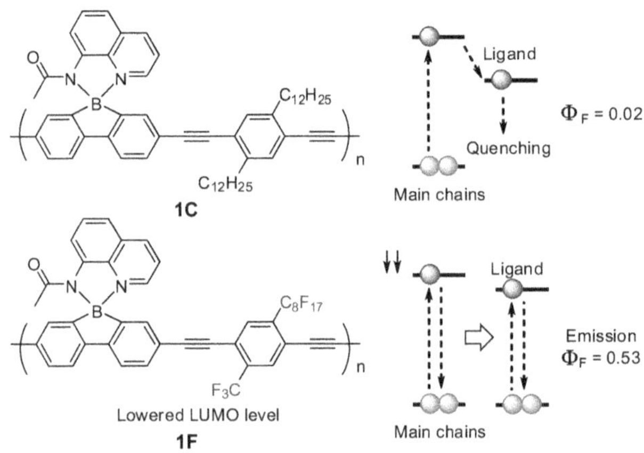

Chart 14.2

A wide variety of highly emissive polymers involving tetracoordinate boron have been synthesized, and their electronic structures investigated.[8] Most organoboron-containing alternating polymers showed strong emission. Moreover, by the electronic interaction between the organoboron complex and the comonomer, the function can be modulated. For example, distinct responses after the excitation depending on the energy levels of the main chains from the experiments with π-conjugated polymers **1 F** and **1C** containing aminoquinoline-borafluorene complexes in the main chains (Chart 14.2).[12] The energy gaps between the polymer main chains and the aminoquinolate moiety can be tuned by the substituent effect of the comonomers. Electron-donating groups in *p*-phenylene-ethynylene (PPE) units reduced the energy levels of the polymer main chains and suppressed the emission from the polymers *via* electron transferring from the main chain to the aminoquinolate moiety. This behavior is of significance for inducing the efficient charge separation in solar cells. In contrast, an electron-withdrawing chain enhanced fluorescence from the aminoquinolate moiety *via* energy transfer. This emission property is applicable in displays. Accordingly, the aminoquinoline-borafluorene polymers promise to be a versatile platform for fabricating efficient optoelectronic devices.

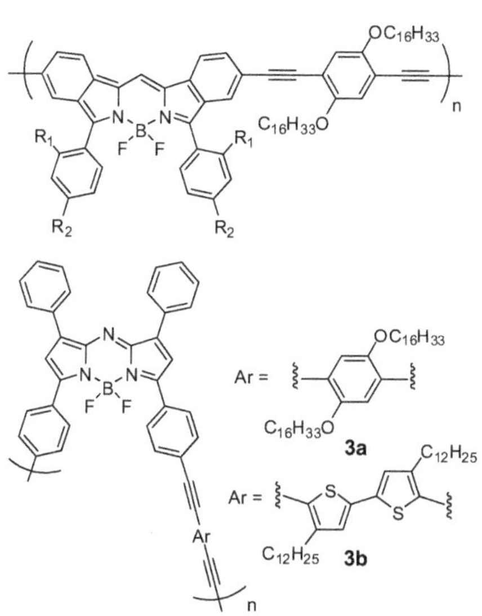

Chart 14.3

The deep-red emissive polymers **2** and **3** were obtained from the alternating polymers with the tetracoordinate boron-containing organoboron complexes.[13,14] The development of the near-infrared (NIR) light sources is a recent topic with high relevance because the light in the NIR region shows relatively higher permeability through air and vital organs. Thus, NIR-emissive polymers are of the one versatile and practical candidates for fabricating the NIR-emissive solid materials. The alternating polymers **2** and **3** with the conjugation-extended boron dipyrromethene (BODIPY) derivatives were also synthesized (Chart 14.3).[13,14] The resulting polymers showed good solubility in common organic solvents including THF, benzene, toluene, chloroform, and dichloromethane, *etc.* Surprisingly, the conjugation of the BODIPY derivatives with PPE led to red shifts in photoluminescence spectra, originating from the extension of π-conjugation through the polymer main chain in comparison with the monomers. Finally, the alternating polymers emitted in the range from the deep-red to near-infrared region with emission spectral maxima at around 691–720 nm and 753–777 nm, and exhibited high quantum yields (Φ_F = 5–49%), respectively.

Because of the vacant p orbital and the lowering effect of LUMO of the boron atom, organoboron-containing polymers are expected to be an efficient n-type polymer that has superior ability to transport electrons. Boron quinolate-containing polymers were prepared, and their charge-carrier abilities were evaluated in the multilayered devices (Chart 14.4).[15] The mobilities for **4a** and **4b** were determined to be the close value observed from the conventional device, which was prepared *via* the vapor deposition of Alq_3. This is the first study to prove that the organoboron-containing conjugated polymers are a superior electron-transporting material in practical devices.

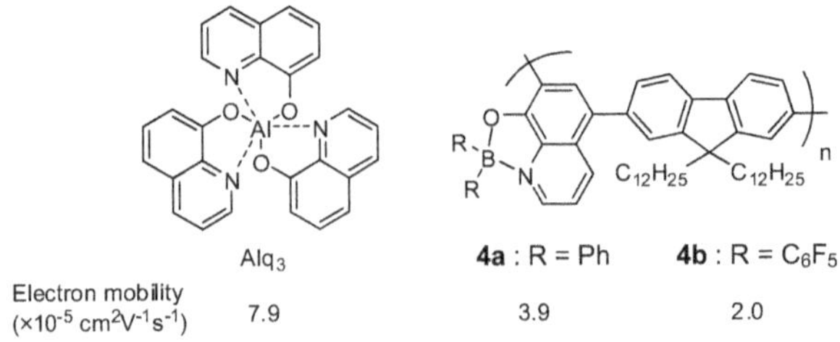

Chart 14.4

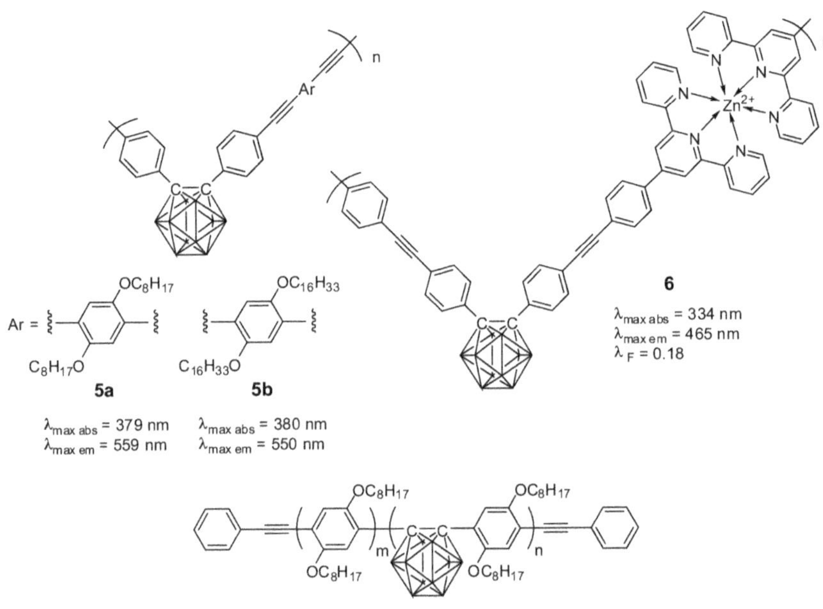

polymers	n	$\lambda_{max\ abs}$ (nm)	$\lambda_{max\ em}$ (nm)
7a	0	431	471, 496
7b	0.05	428	471, 497
7c	0.08	425	471, 497
7d	0.14	419	469, 497

Chart 14.5

The carborane-containing alternating polymers **5**–**7** have been synthesized *via* metal-catalyzed coupling, the carborane construction after the polymerization, and the coordination reactions (Chart 14.5).[16–18] A variety of unique optical properties were found from the carborane-containing polymers. For instance, the *o*-carborane-containing polymers **5a** and **5b** showed the aggregation-induced emission (AIE)[19,20] in water dispersion; *i.e.* nonemissive polymers in the solution state were induced to emit intensely by aggregation or film formation.[16] In addition, the regulation of the emission properties was presented by modulating the electronic structures and the molecular motion of the polymers. These polymers are promising as highly emissive solid materials.

The preparation of polymers was generally executed *via* the metal-catalyzed crosscoupling reactions with di-substituted organoboron monomers. Typical examples are described in the Section 14.2.

14.1.2 Alternating Polymers of Silicon

Siloles and silacyclopentadienes are composed of five-membered silacycles. In this section, the properties and applications of silole-containing alternating polymers are illustrated. In particular, their interesting optical and electrical characteristics are focused upon. Several chemosensors for detecting the explosive compounds have been developed based on silole-containing alternating polymers (Chart 14.6).[20,21] Lai and coworkers reported the silole-containing alternating polymers.[22] The emission colors significantly depended on the type of comonomers. In addition, the 2,5-tetraphenylsilole-ethyne polymer **8a** exhibited a stronger emission from the aggregation than that from the solution. Furthermore, the emission properties of **8a** can be changed by adding nitro compounds. Therefore, the polymer **8a** is applicable as a chemosensor for detecting explosive materials. Hyperbranched alternating polymers containing silole units (**9**, HPS) were also found to show the AIE in the green region.[23] HPS shows good solubility in acetonitrile, THF, and chloroform, slight solubility in methanol, but no solubility in water. HPS solutions in its good solvents hardly possess emission properties. In contrast, by adding water to the acetonitrile solution, the strong emission can be observed caused by the aggregation of silole molecules. Similar emission is also observed on the TLC plate after casting and drying. Furthermore, a recent report presents the chemical sensor for explosive compounds using silole-containing

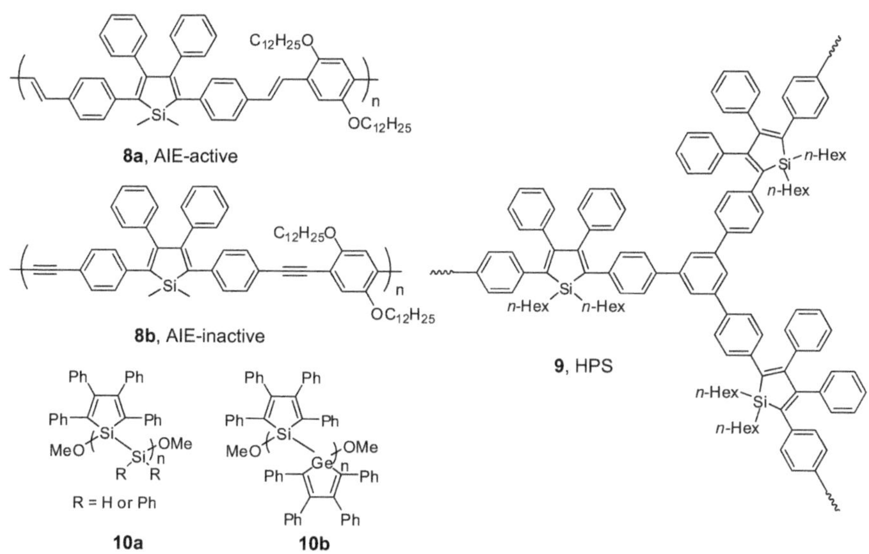

Chart 14.6

linear polymers.[24] Based on the quenching effects caused by the presence of the targets, a small amount of nitro compounds can also be detected. Silole–silane **10a** and –germole **10b** (germanium-containing cyclopentadiene) copolymers also work as chemosensors based on the quenching effect by nitro compounds.

Recently, the silicon-bridged fused ring systems have been reported for the application of optoelectronic devices.[25–28] The introduction of silicon as a bridging atom between adjacent aromatic units brings several benefits such as an increase of the effective conjugation length and reduction of the conformational disorder. The reduction of the LUMO energy level owing to the antibonding lobes of silicon contributes to increasing the open-circuit voltage and a potential increase in efficiency in solar cells.[29,30] In addition, the tetravalent silicon offers the possibility to attach alkyl side chains, resulting in improvements in the solubility and processability of the polymers. Yang and coworkers and Brabec and coworkers independently reported that the incorporation of silicon into the carbon bridge in cyclopentadithiophene can increase the crystallinity of the polymers,[31,32] resulting in the enhancement of charge-carrier mobilities.[33] Moreover, it was also demonstrated that the silicon bridge possessed high environmental stability.[34] Heeney and coworkers synthesized the donor–acceptor-type polymers **11** and demonstrated as a solar cell material (Chart 14.7).[35] Thus, fused silole and silafluorene have large possibility to be applied as materials with low bandgaps in solar cells.

Several reactions for the formation of silole rings are outlined in Scheme 14.1.[36] Cyclization from diiodide dienes can be applied for the preparation of the bridged systems. As mentioned below, siloles were obtained *via* the transmetallation from zirconocyclopentadiene. Most siloles have enough stability in the presence of metal catalysts. Similarly to the organoboron

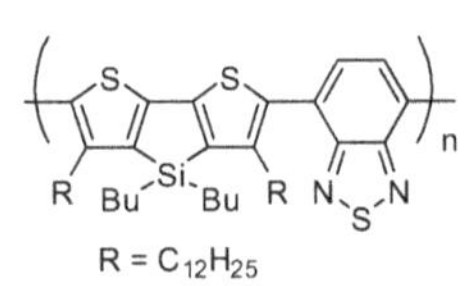

11

Chart 14.7

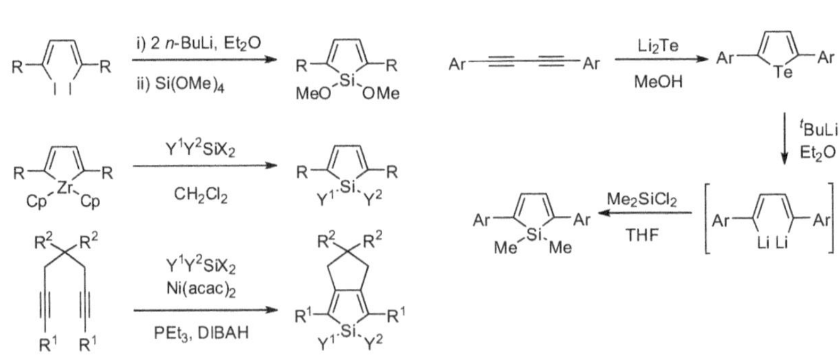

Scheme 14.1

complexes, a variety of crosscoupling reactions yield the desired alternating polymers.

14.1.3 Alternating Polymers of Germanium

Germanium is also a metalloid element bearing both metallic and nonmetallic characteristics. In particular, the alternating polymers with the germanium bridged system have been focused on as a material for highly efficient solar cells, similarly to silicon.[35,37–43] Germanium is located below silicon in the periodic table, but due to the d-block contraction, both elements exhibit similar covalent radii, with an average C–Ge bond being only slightly longer than a C–Si bond (1.96 *vs.* 1.88, respectively). From recent computer modeling studies, it is suggested that the relatively larger size of germanium than that of silicon could play a significant role in enhancing in ordering and stronger π-stacking. Indeed, Leclerc and coworkers demonstrated that the germole-containing alternating polymer **12a** showed a power conversion efficiency of 4.1% in photovoltaic cells (Chart 14.8).[37] Further experimental findings on the electronic properties and the structure–property relationships of germanium-containing polymers are needed to understand the role of germanium in the conjugated systems in the polymer chains for receiving highly efficient optoelectronic devices.

Two examples of the formation of the germole structure are explained here for preparing the monomers. According to the previous report, the synthesis of **14** is described (eqn (14.2)).[44] As an intermediate, the zirconocyclopentadiene was prepared *via* the intramolecular zirconocene coupling of **13** in a deep red solution of THF. The product was transformed to the germole dibromide by reacting with $GeBr_4$. Followed by the treatment with MeLi, the germole **14** can be obtained as a bright yellow oil after purification by column chromatography. When alternating procedures were employed with the transmetallation reagents such as $GeCl_4$ and Me_2GeBr_2, the decreases of overall reaction yields were reported.

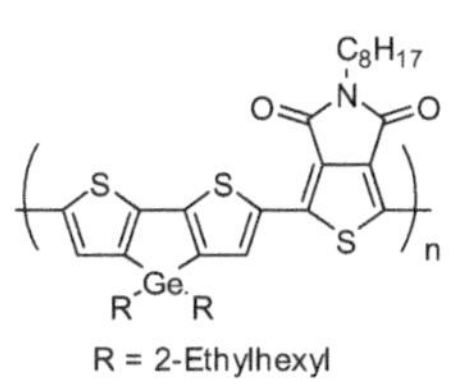

12a

Chart 14.8

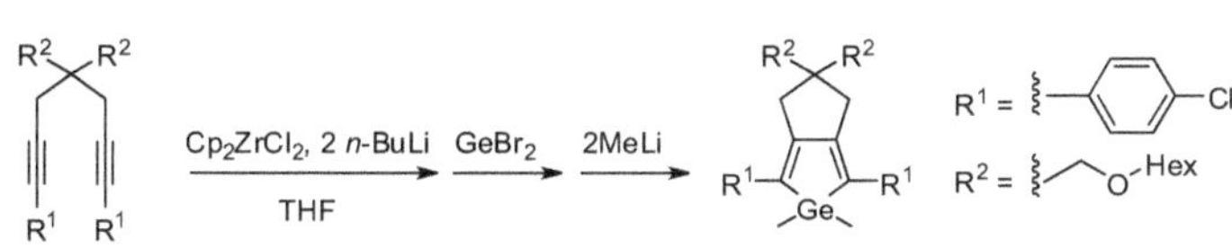

13 **14** (14.2)

Next, as illustrated in eqn (14.3), under strict Schlenk conditions, a reaction with dibromodithiophene **15** and *n*-BuLi (1.6 M) at –78 °C followed by the addition of the germane derivative GeBr(*n*-octyl)$_2$ generated dithienogermole **16** in 75% yield.[37] After the replacement of bromo groups to Ge, the monomer for Stille coupling was obtained.

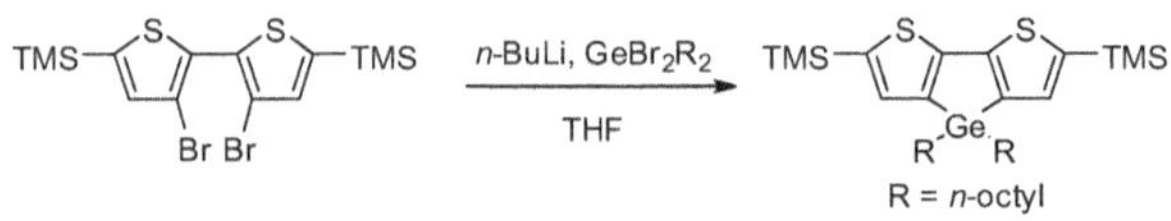

15 **16**

(14.3)

Germoles can be also obtained *via* titanacyclopentadiene. The preparation protocol is described in Example 13.[45]

14.1.4 Alternating Polymers of Phosphorus

Phospholes also have the low-lying exocyclic P–R σ* orbitals that can effectively overlap with the π system.[46–48] Similarly as silicon, this results in a low-energy LUMO and consequently, a high electron affinity (suitable for n-type conductivity). Furthermore, a reduced HOMO–LUMO separation of π-conjugated system of polyheteroles containing second-row elements like silicon or phosphorus is observed since these heteroles exhibit a low level of aromaticity, which results in a better electronic interaction of their endocyclic π electrons with other aromatic subunits at the 2,5-positions. As described in Chart 14.9, phosphole rings have three further advantages: 1) the role as a ligand for complexation with heavy metals, 2) the possibility to readily tune the electronic property of phosphole-based π-conjugated systems by chemical modifications of the reactive P atoms, and 3) the increased effectiveness of the hyperconjugative interaction in 1,1′-biphosphole moieties leading to further reduction of the HOMO–LUMO gaps. Because of these characteristics, phosphole-containing materials are expected to be an advanced material for optoelectronic devices.

Compared to the silole-containing polymers, phosphole has received little attention as a building block for extended π-conjugated systems, despite its unique characteristics and numerous applications in organic and metal coordination chemistry. The first practically applicable synthesis was independently presented in 1959 by two groups.[49,50] The first successful synthesis of phosphole-containing π-conjugated polymers was established by Tilley and

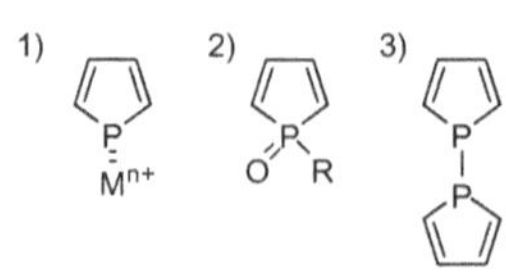

Chart 14.9

coworkers (eqn (14.4)).[51] Based on the zirconocene-mediated oxidative cyclization of dialkyne **17** and the transmetallation to yield phosphole **18**, the subsequent reaction of the zirconacyclopentadiene-containing polymer with $PhPCl_2$ by the polymer reaction gave both 2,4- and 2,5-substituted phosphole-containing polymers. The resulting polymers exhibited bluish-green emission with a large Stokes shift. Réau and coworkers also prepared phosphole-containing polymers through the electrolytic polymerization by using cyclic voltammetry.[52,53] They demonstrated the possibility of applying them to optoelectronic devices. The alternating polymers containing phospholes have been already synthesized.[54–57] In view of the abundant possibility of phosphole-containing conjugated systems, the establishment of the synthesis and the exploration of further properties of novel conjugated polymers having a phosphole unit in the main chain are still desired.

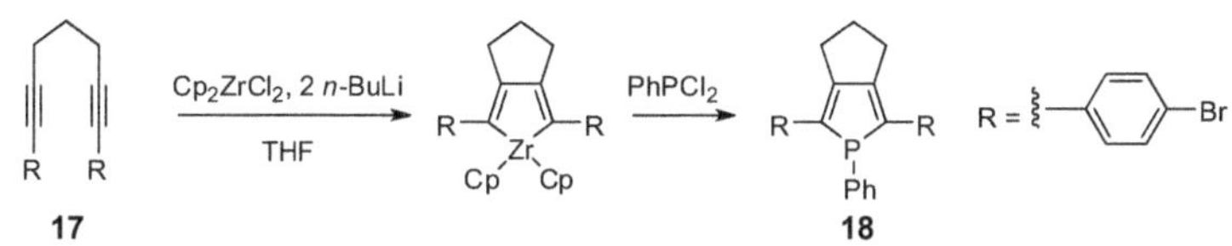

(14.4)

14.1.5 Alternating Polymers of Transition Metals

Transition-metal-containing heterocyclopentadienes are versatile intermediates for preparing various kinds of heteroatom-containing polymers *via* polymer reactions (Scheme 14.2). For example, the titanacyclopentadiene-unit containing polymer **19** was yielded in the reaction of diethynylbenzene and a titanium complex. Without isolation of the intermediate polymer, the following treatments with various reagents presented alternating polymers **20a–d** composed of 1,3-conjugated diene units with the modification groups.[58] The resulting polymers showed unique emission properties originating from the conjugation system through the polymer main chains. Notably, the titanacyclopentadiene unit can be also transformed not only to benzene **21a**[59] but also a variety of heterocyclopentadienes **21b–e**.[60,61] Cobalt, zirconium, and other titanium complexes can be a key material for producing other functional units including heteroatoms. These methods are feasible for parallel synthesis of a wide variety of heterocycle-containing polymers.

14.2 Experimental Procedures

14.2.1 Materials and General Polymerization Procedures

When the monomer complexes have enough stability, the metal-catalyzed coupling reactions are general procedures for polymerization. Suzuki–Miyaura, Sonogashira–Hagihara, Mizorogi–Heck, Negishi, and Stille crosscoupling reactions are applicable for preparing alternating copolymers. The typical protocols for synthesis and characterization are described here to obtain the

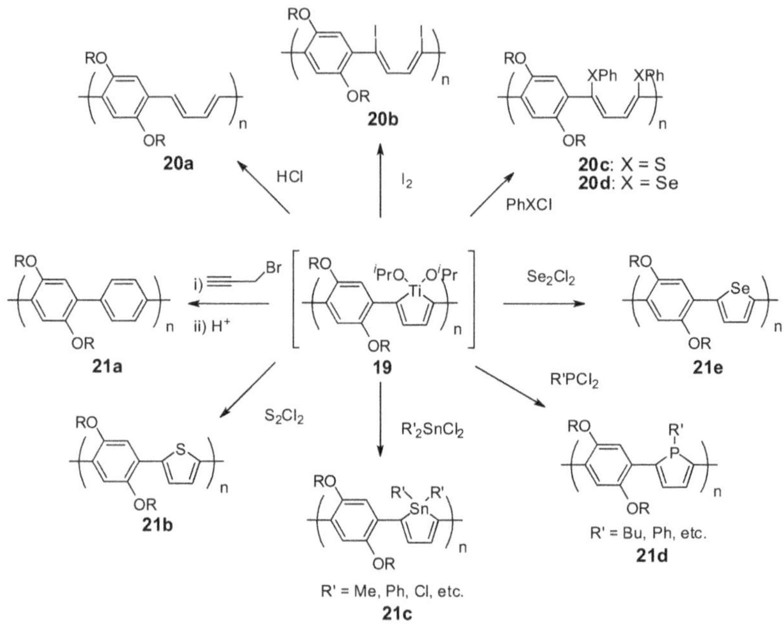

Scheme 14.2

polymers for evaluating their optical properties: The monomer unit was designed and synthesized with the dual polymerizable functional groups such as halogen or ethynyl groups. The crosscoupling reactions were executed in the presence of the metal catalysts with the comonomers (PPE, fluorene, dithiophene, *etc.*). The monomer consumption and the chain growth were evaluated from the profiles of size-exclusion chromatography (SEC). In our experiences, the amounts of the oligomers or the intramolecularly cyclized products tend to increase in the product when the monomer concentration is below 0.1 M in the initial reaction solution. After finishing the increase of the molecular weight of the product, the reaction solution was poured into the poor solvents (methanol, and so on). To remove the metal species thoroughly, the products were dissolved and reprecipitated several times and, if needed, treated with Soxhlet extraction. After drying, the polymers were obtained as powders. The number-average molecular weights (M_n) and the polydispersity indices (M_w/M_n) of the polymers were measured by SEC with the polystyrene standards. In the case of the π-conjugated polymers, the total shape of the polymer chain could be different from those of the standards in the column of SEC. In addition, when the Sonogashira–Hagihara coupling was applied in the polymerization reaction, the defects of the monomer unit is possible because of the Gracer-type side reaction. To check the chemical structures, the evaluation of the integration rates of the peaks in the ^{1}H NMR spectra between the complex and the comonomer is necessary. The mass spectroscopic measurements with the matrix-assisted laser desorption and ionization method are also a convenient tool to confirm the chain length. The formation of the heteroatom-containing complexes was confirmed by comparing the chemical shifts of the peaks in the multinuclear NMR (^{11}B, ^{29}Si, ^{31}P, *etc.*) to those of the model compounds. To ignore the effects on the optical properties by the chain ends, the polymers that possessed at least five repeat units should be used. The obtained polymers were dissolved into the organic solvents or cast onto the substrates to measure the optical properties. The quantum yields of emission were determined as an absolute value using the integration sphere.

14.2.2 Alternating Polymers of Boron

Example 1: Example of Borafluorene-Containing Alternating Polymers[12]

The representative reaction *via* metal-catalyzed crosscoupling for preparing **1F** is outlined in eqn (14.5). The solution containing monomer **22** (0.108 g, 0.180 mmol), 1.4-diethynyl-2-perfluorooctyl-5-trifluoromethylbenzene (0.110 g, 0.180 mmol), CuI (1.7 mg, 9.0 μmol), $Pd(PPh_3)_4$ (10.4 mg, 9.0 μmol) and 0.9 mL of triethylamine in 1.8 mL of THF was incubated at 40 °C for 48 h. After the reaction, a small amount of chloroform was added and poured into a large excess of methanol to precipitate the polymer. The polymer was purified by repeated precipitations from a small amount of chloroform into a large excess of methanol and hexane, respectively. The desired polymer **1F** was obtained a yellow solid in 84% yield. The M_n and the M_w/M_n of the polymers, measured by SEC in THF toward polystyrene standards, were 10 400 and 2.5 for **1F**, and 20 300 and 2.8 for **1C**, respectively.

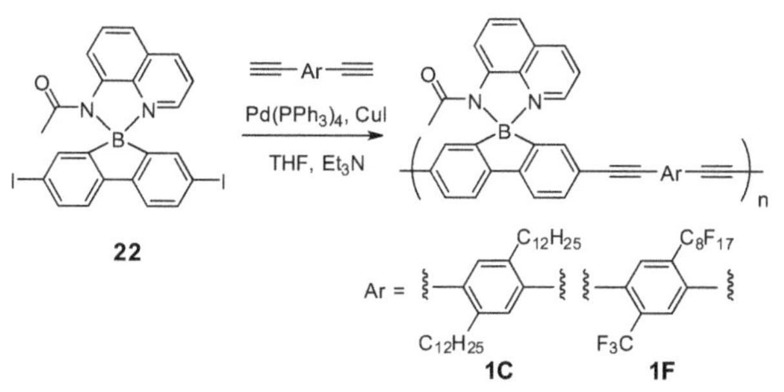

(14.5)

Example 2: Preparation of Organoboron Complexes Containing Alternating Polymers[13]

The typical procedure is explained as the preparation of **2a** (eqn (14.6)). After the mixture containing **23a** (52.2 mg, 0.075 mmol), 1,4-diethynyl-2,5-dihexadecyloxybenzene (45.5 mg, 0.075 mmol), $Pd(PPh_3)_4$ (4.33 mg, 3.75 μmol), and CuI (0.71 mg, 3.75 μmol) in THF (1.5 mL) and triethylamine (0.75 mL) was stirred at 50 °C for 48 h, the solvent was evaporated out *in vacuo*. The residue was extracted with chloroform, washed with 10% aqueous NH_3 solution and then water, and dried over $MgSO_4$. After removing solvents, the residue was dissolved in a small amount of chloroform and poured into a large excess of methanol to precipitate the polymer. The product was collected by filtration with suction, further washed with acetone, and then dried under vacuum at 60 °C for 12 h. Polymer **2a** was obtained as a deep green blue solid. The M_n and the M_w/M_n were 8200 and 1.94 for **2a**, 7900 and 2.12 for **2b**, and 8100 and 2.32 for **2c**, respectively. The chemical structures of the polymers were confirmed by ^1H NMR, ^{11}B NMR, and IR spectroscopies, besides elemental analysis. Aza-BODIPY-containing polymers were also synthesized *via* Sonogashira reaction in a similar manner.

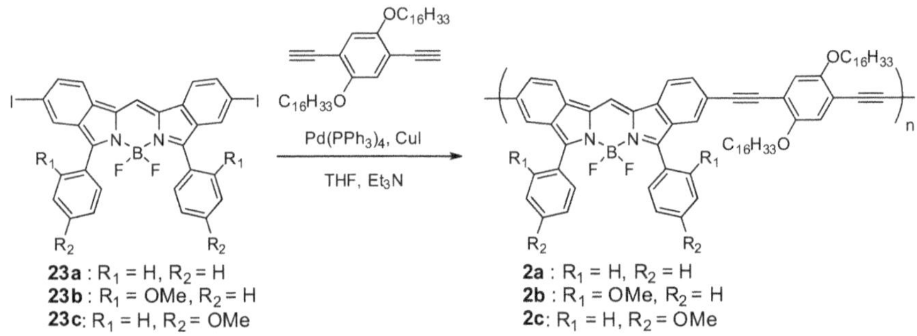

(14.6)

Example 3: Preparation of Boron-Quinolate-Containing Polymers[15]

The polymers **4a** and **4b** were obtained *via* the boron complexation after polymerization (eqn (14.7)). The reaction mixture containing 5,7-dibromo-8-benzylquinoline (1.50 g, 3.82 mmol), 9,9-didodecylfluorene-2,7-diboronic acid (1.83 g, 3.82 mmol), $Pd(OAc)_2$ (0.172 g, 0.76 mmol), S-Phos (0.313 g, 0.76 mmol) and K_3PO_4 (3.24 g, 15.3 mmol) in THF (80 mL) and water (20 mL) was refluxed for 48 h. After filtration to remove precipitates, the filtrate was condensed. The residue was dissolved in chloroform, and washed with aqueous NH_3 solution, water, and brine, dried over $MgSO_4$, and volatile products were evaporated. The residue was dissolved again in a small amount of chloroform and poured into a large excess of methanol to precipitate the polymer. The polymer was collected by filtration with suction and dried under vacuum at room temperature for 5 h. The polymer **24** was obtained as a pale yellow solid in 92% yield and used directly for the next step.

Boron tribromide in dichloromethane (1.0 M, 10 mL) was added to a solution of **24** (1.50 g, 2.00 mmol) in dichloromethane (30 mL) at room temperature under nitrogen. The resulting mixture was refluxed for 12 h. After reaction, the mixture was washed with saturated $NaHCO_3$ solution, 5% Na_2SO_3 solution, and brine. The organic layer was dried over $MgSO_4$, and the solvent was removed by evaporation. The residue was dissolved in a small amount of chloroform and poured into a large excess of hexane to precipitate the polymer. The polymer was collected by filtration with suction and dried under vacuum at r.t. for 6 h. The precursor polymer **25** was obtained as a yellow solid.

Triphenylborane (0.75 g, 3.10 mmol) was added to a solution of **25** (0.40 g, 0.62 mmol) in toluene (50 mL) at room temperature under nitrogen. The resulting mixture was refluxed for 12 h, and the solvent was evaporated. The residue was dissolved in a small amount of chloroform and poured into a large excess of methanol to precipitate the polymer. The polymer was collected by filtration with suction and dried under vacuum at 60 °C for 5 h. The desired polymer **4a** was obtained as an orange solid in 96% yield. To receive **4b**, $(C_6F5)_2BF \cdot OEt_2$ was instead used in the complexation. The M_n and M_w/M_n of polymer **4a** were 28 100 and 2.03, respectively.

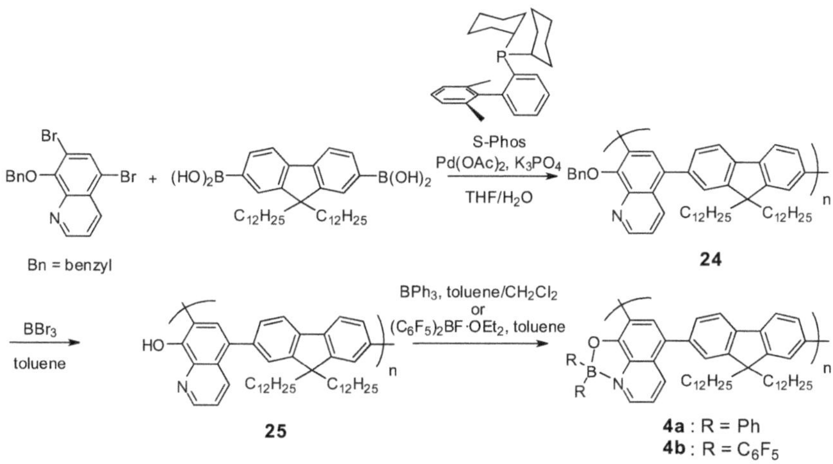

(14.7)

Example 4: Preparation of Carborane-Containing Polymers[16]

Carborane-containing polymers can be obtained according to eqn (14.8).[16] A typical procedure is as follows: The mixture containing Bis(4-iodophenyl)-*o*-carborane (**26**) (55.0 mg, 0.10 mmol), 1,4-diethynyl-2,5-bisoctyloxybenzene (34.0 mg, 0.10 mmol), $Pd(PPh_3)_4$ (5.80 mg, 0.010 mmol) and CuI (0.95 mg, 0.005 mmol) in THF (1.0 mL) and triethylamine (0.5 mL) was stirred at room temperature for 48 h. After reaction, the mixture was diluted with chloroform and washed with aqueous NH_3 solution, water, and brine. The organic layer was dried over $MgSO_4$. After evaporation and reprecipitation in methanol, the polymer **5a** was obtained as a yellow powder in 89% yield. The M_n and M_w/M_n of the polymer **5a** were 2800 and 3.5, respectively.

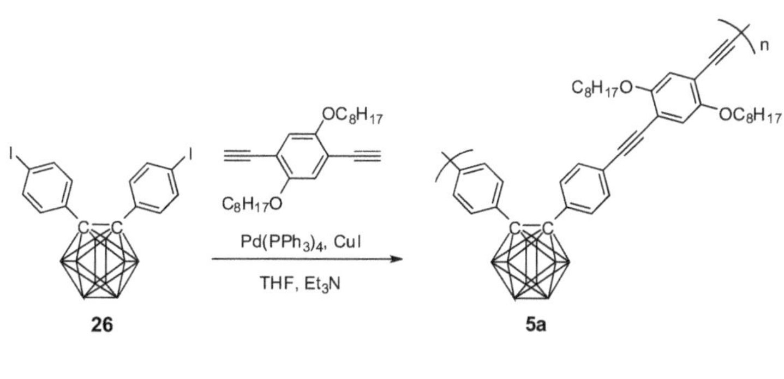

(14.8)

Example 5: Preparation of Coordination Polymers Containing Carborane[17]

Polymer **6** was prepared according to eqn (14.9).[17] After stirring a solution containing ditopic ligand **27** (24 mg, 25 mmol) and a methanol solution of $Zn(OAc)_2 \cdot 2H_2O$ (10 mM, 2.5 mL, 25 mmol) in chloroform (10 mL) for 1 h at room temperature, excess NH_4PF_6 was added into the solution. The resulting solution was poured into ether, and the precipitate **6** was collected as a light yellow solid in 83% yield.

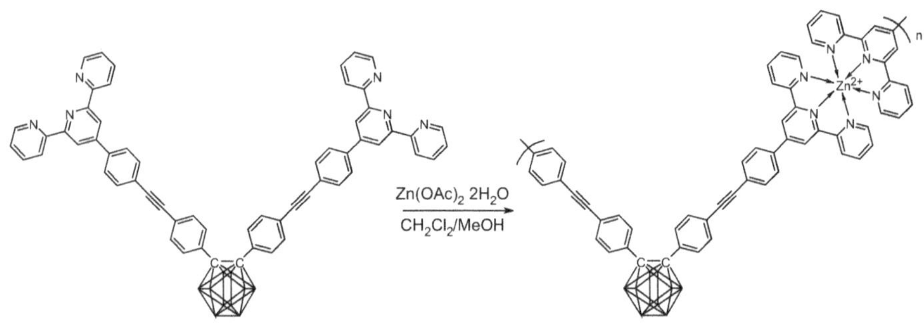

27 **6**

(14.9)

Example 6: Polymer Reaction for Carborane Formation[18]

The polymers **7a**–**7d** were prepared *via* polymer reactions as shown in eqn (14.10). To the precursor polymer **28** (35.7 mg, 100 mmol; $M_n = 4900$, $M_w/M_n = 2.40$) were added decaborane (610 mg, 5 mmol) and *N,N′*-dimethylaniline (1200 mg, 10 mmol) dissolved in toluene (2.5 mL) and THF (2.5 mL). The mixture was heated at 110 °C for 24 h. The reaction was quenched by rapid cooling with liquid nitrogen. The reaction mixture was diluted in chloroform, and the insoluble product was filtered off. The filtrate was evaporated and reprecipitated in methanol. The product was isolated by filtration and dried under vacuum. The resulting polymer **7a** was obtained as a brown powder in 90% yield. The M_n and M_w/M_n of the polymer **7a** were 10 700 and 2.33, respectively. By changing the feed ratio of decaborane to the triple bonds (0.05, 0.10, 0.20 and 0.40) in the polymer **28**, the introduction rates were modulated to obtain the polymers **7b**–**d**.

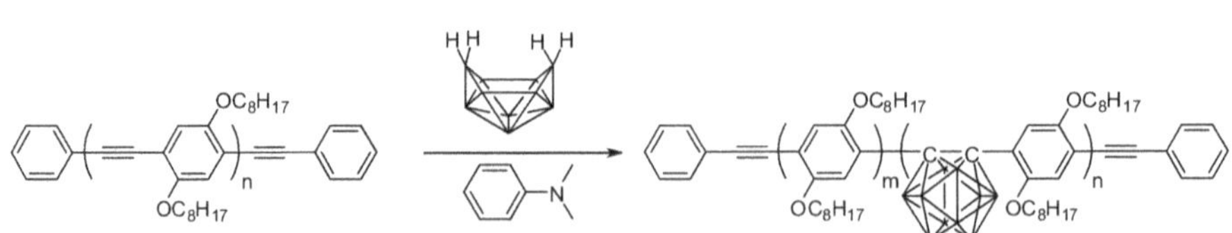

28 **7a–d**

(14.10)

14.2.3 Alternating Polymers of Silicon

Example 7: Preparation of Silole-Containing Polymers[22]

According to the previous report, the synthetic protocol for the polymer **8a** is described (eqn (14.11)).[22] The reaction mixture containing **29**, diiode comonomer, palladium acetate (4 mol%), tri-*o*-tolylphosphene (20 mol%), tributylamine (2.5 equiv) in DMF (5 mL per mmol of **29**) was heated at 110 °C overnight under an argon atmosphere. After reaction, the mixture was poured into methanol. The precipitate was collected by filtration and washed with methanol. The polymer **8a** was obtained after a Soxhlet extractor with methanol for 2 days and then dried under vacuum (56%). The M_n and M_w/M_n of polymer **8a** were 16 200 and 1.51, respectively.

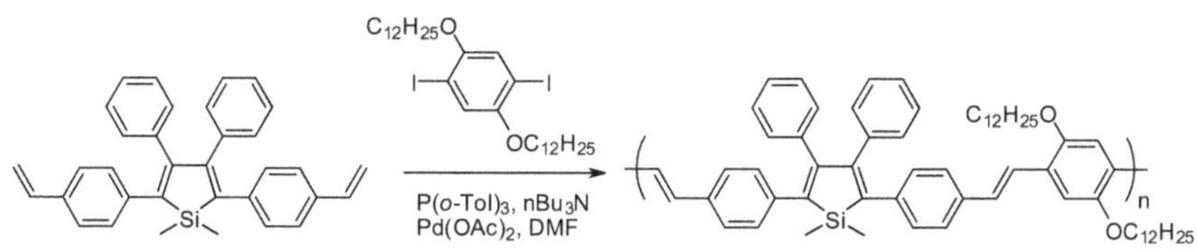

29 **8a**

(14.11)

Example 8: Preparation of Silole-Germole Alternating Polymers[24]

The representative procedure for the preparation of silole–germole alternating polymers is shown in eqn (14.12).[24] A dark yellow solution of silole dianion was obtained from the mixture containing 1,1-dichloro-2,3,4,5-tetraphenylsilole **30** (5.0 g, 11.0 mmol) and lithium (0.9 g, 129.7 mmol) in THF (120 mL) after stirring for 8 h at room temperature. After removal of excess lithium, 1 mol equiv of Ph_2SiCl_2 (11 mmol) was slowly added to a solution of tetraphenylsilole dianion, and stirred at room temperature for 2 h. After refluxing for 3 d the reaction mixture was cooled to room temperature and quenched with methanol. After removing volatiles under reduced pressure, THF (20 mL) was added to the residue, and then the polymer was precipitated by slow addition of methanol (700 mL). From the third cycle of dissolving–reprecipitation followed by freeze drying, the polymer **10a** was obtained as a yellow powder in 47%. The M_w and M_w/M_n of polymer **10a** were 4800 and 1.04, respectively. The germole-containing alternating polymers can be obtained from the similar method with 1,1-dichloro-2,3,4,5-tetraphenylgermole used instead of Ph_2SiCl_2.

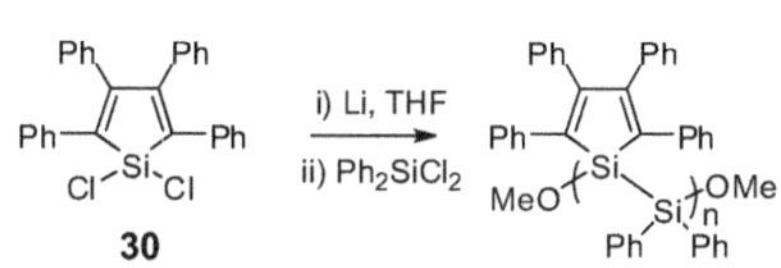

10a

(14.12)

Example 9: Preparation of Conjugated Polymers Containing Silicon-Bridged Dithiophene Units[35]

According to the previous report, the synthetic procedure of the polymer **11** is illustrated (eqn (14.13)).[35] The reaction mixture containing **31** (289.8 mg, 0.36 mmol), 2,1,3-benzothiadiazole-4,7-bis(boronic acid pinacol ester) (140.0 mg, 0.36 mmol), $Pd(PPh_3)_4$ (4.2 mg, 0.0036 mmol), degassed aqueous 1 M Na_2CO_3 (0.7 mL), and 2 drops of Aliquat 336 in degassed toluene (3 mL) in a high-pressure microwave reactor tube was thoroughly degassed under argon and heated for 3 days at 120 °C. After cooling to room temperature, the precipitate was obtained in methanol (100 mL). After Soxhlet extraction with methanol, acetone, hexane and chloroform, the chloroform solution of the product was condensed. After reprecipitation in methanol, a purple solid of 11 was obtained *via* filtration and drying under vacuum in 66% yield. The M_n and M_w/M_n of the polymer **11** were 53 600 and 1.97, respectively.

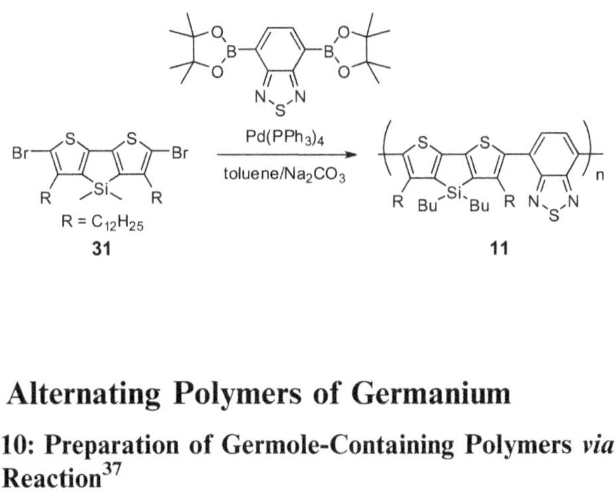

(14.13)

14.2.4 Alternating Polymers of Germanium

Example 10: Preparation of Germole-Containing Polymers *via* Suzuki-Miyaura Coupling Reaction[37]

According to the previous report, the synthetic procedure of the polymer **12b** is described (eqn (14.14)).[37] The reaction mixture containing 0.284 g (0.457 mmol) of **31**, 0.1775 g (0.457 mmol) of 2,1,3-benzothiadiazole-4,7-bis(boronic acid pinacol ester), 18.0 mg (0.020 mmol) of $Pd_2(dba)_3$, 50.0 mg (0.122 mmol) of S-Phos, and 0.5 mL of an aqueous Na_2CO_3 (2 M) in 5 mL of degassed toluene was vigorously stirred for 72 h at 90–100 °C. After reaction completion, 5.3 mL (0.05 mmol) of bromobenzene was added. After stirring for 1 h, 6 mg (0.05 mmol) of phenylboronic acid was added and the reaction was heated for three more hours. After cooling to room temperature, the precipitate was collected with a 0.45-μm nylon filter in methanol/water (9 : 1) and washed in a Soxhlet apparatus with acetone, hexanes and then chloroform. The chloroform fraction was condensed to 20–30 mL and then poured in methanol/water (9 : 1). After filtration through 0.45 mm nylon filter, the polymer **12b** was obtained in 26% yield. The M_n and M_w/M_n of polymer **12b** were 8900 and 1.9, respectively.

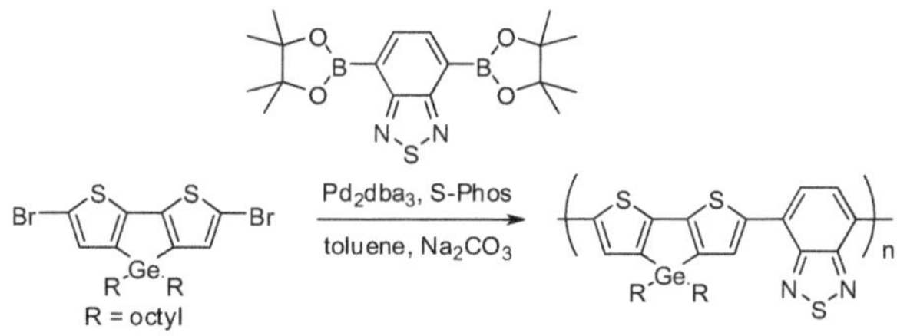

(14.14)

Example 11: Preparation of Germole-Containing Polymers *via* Stille Coupling Reaction[37]

The polymer **12c** can be prepared with the similar method to Example 10 (eqn (14.15)).[37] The mixture containing monomer **32** (0.1154 g, 0.1463 mmol), 4,7-dibromo-2,1,3-benzothiadiazole (0.409 g, 0.1393 mmol), $AsPh_3$ (16% mol) and $Pd_2(dba)_3$ (4% mol) in 1.2 mL of degassed chlorobenzene was heated by microwaves at 120 °C for 5 min, 150 °C for 5 min and finally 180 °C for 2 h. After cooling to room temperature, the precipitate was collected with 0.45 μm nylon filter in cold methanol (100 mL) and washed in a Soxhlet apparatus with acetone, hexanes and then chloroform. The chloroform fraction was condensed and then poured in methanol. After filtration through 0.45 μm nylon filter, the polymer **12c** was obtained in 73% yield. The M_n and M_w/M_n of polymer **12c** were 12 400 and 1.7, respectively.

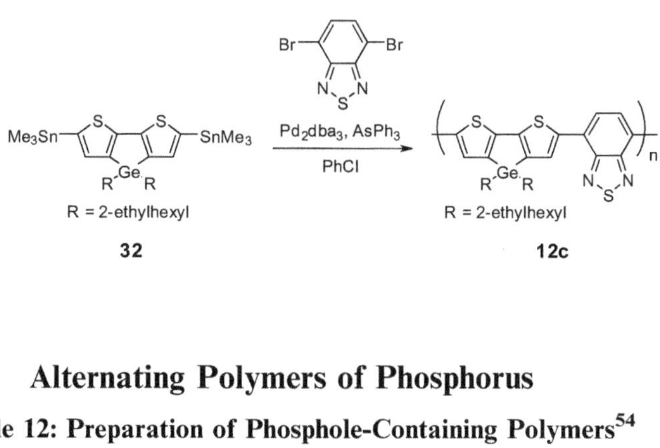

(14.15)

14.2.5 Alternating Polymers of Phosphorus

Example 12: Preparation of Phosphole-Containing Polymers[54]

A typical procedure is shown in eqn (14.16). A mixture containing **18** (0.77 g, 1.50 mmol), diethynyl monomer (0.49 g, 1.50 mmol), $PdCl_2(PPh_3)_2$ (53 mg, 0.075 mmol), CuI (14 mg, 0.075 mmol), triethylamine (10 mL), and THF (30 mL) was refluxed for 48 h. After cooling to room temperature, precipitated ammonium salts were filtered and washed with THF. The filtrate was concentrated and poured into degassed methanol (50 mL) to precipitate the polymer. These procedures were performed under an argon atmosphere. The polymer **33** was obtained in 67% yield after filtration, washing with methanol, and drying *in vacuo* as a dark yellow powder. The M_n and M_w/M_n values were

3500 and 1.41, respectively. From the ^{31}P NMR measurements, it was found that the upfield shift (δ +27.2 ppm) was observed from the polymer **33** in comparison with that of monomer **18** (δ +35.6 ppm) after the formation of the conjugated polymer. Réau and coworkers explained these phenomena by the fact that the size of the fused carbocycle on the phosphole ring influenced the angles around the phosphorus atom, resulting in the shift of ^{31}P signals.

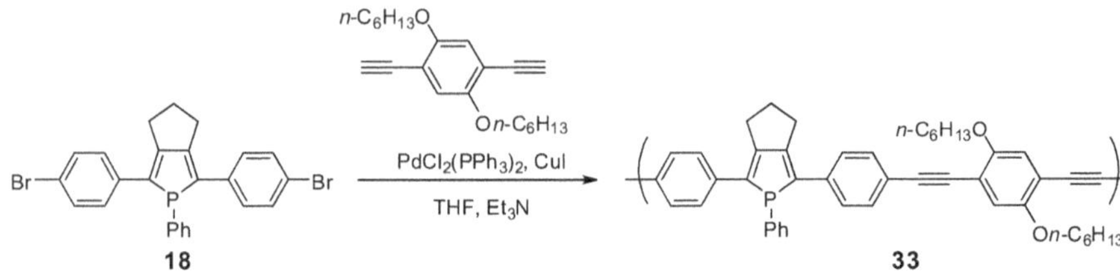

(14.16)

14.2.6 Alternating Polymers of Transition Metals

Example 13: Preparation of Germole-Containing Polymers *via* Titanacyclopentadiene[45]

The germole units were formed *via* titanacyclopentadiene intermediates (eqn (14.17)).[45] The mixture containing Cp_2TiCl_2 (0.27 g, 1.10 mmol), dry THF (20 mL), 1,4-bis(1-hexynyl)benzene (0.24 g, 1.00 mmol), and *n*-BuLi (1.60 M in hexane, 1.40 mL, 2.24 mmol) was stirred at −78 °C under argon for 15 min. Then, the mixture was warmed to room temperature and kept stirring overnight. After removing THF by evaporation, diethyl ether (30 mL) and the $GeCl_4$ (0.24 g, 1.10 mmol) were added at ambient temperature. The product containing the titanacyclopentadiene polymer **34** was transformed to **35** during stirring overnight. After removing precipitations, MeLi (1.04 M in diethyl ether, 7.0 mL, 7.28 mmol) was added to the filtrate at −78 °C. After warming to room temperature and stirring overnight, 5.0 mL of methanol was added into the reaction mixture. After removing volatiles by evaporation, the residue was dissolved with 3.0 mL of dichloromethane. The polymer **36** was collected as a yellow solid in 50% yield *via* precipitation in methanol from the dichloromethane-soluble fraction. The M_n and M_w/M_n values were estimated as 3400 and 2.12, respectively.

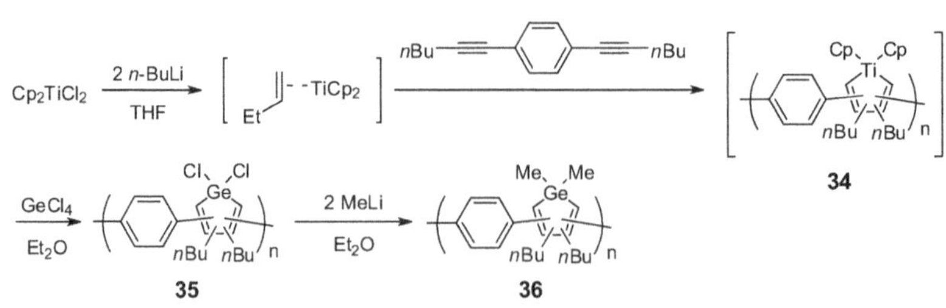

(14.17)

Example 14: Preparation of Stannole-Containing Polymers *via* Titanacyclopentadiene[60]

The polymer **38** can be also obtained *via* a titanacyclopentadiene intermediate (eqn (14.18)). After preparation of **34** as described above, a diethyl ether (30 mL) solution of $SnCl_4$ (0.29 g, 1.1 mmol) was added at ambient temperature. To transform to 1,1-dichlorostannole units (**37**), the mixture was stirred overnight. After the same operations shown in Example 13, the polymer **38** was collected as a yellow solid in 41% yield *via* precipitation in methanol from the dichloromethane-soluble fraction. The M_n and M_w/M_n values were estimated as 2800 and 2.1, respectively.

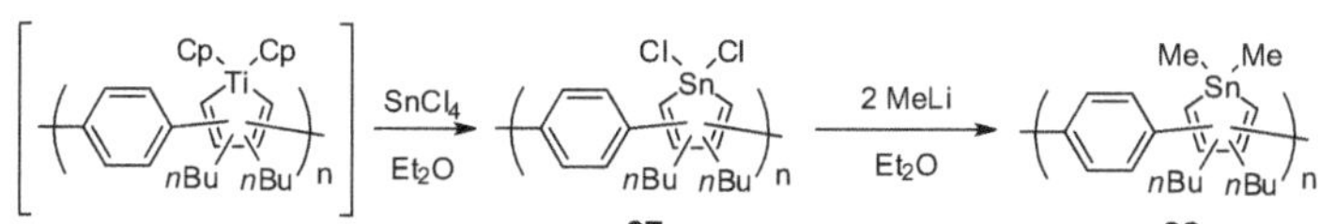

(14.18)

Example 15: Preparation of Conjugated Polymer Containing 1,4-Bismercapto-Substituted 1,3-Diene Units *via* Titanacyclopentadiene[58]

The polymer **40** was prepared *via* titanacyclopentadiene (eqn (14.19)).[58] The mixture of a diethyl ether (20 mL) solution of 1,4-diethynyl-2,5-di(2-ethylhexyloxy)benzene (0.191 g, 0.500 mmol) with $Ti(OPr^i)_4$ (0.202 g, 0.700 mmol) and ^{i}PrMgCl (1.0 M, 1.40 mL, 1.40 mmol) was stirred at –78 °C under argon for 30 min. After addition of PhSCl (0.173 g, 1.20 mmol) at –50 °C, the mixture was warmed to room temperature and kept stirring for 3 h. After quenching the reaction by adding hydrochloric acid (1.0 M, 50 mL), dichloromethane (50 mL) was added and the mixture was kept stirring for 12 h. The aqueous layer was extracted three times with dichloromethane (total 150 mL), and all the organic layers were collected and dried over $MgSO_4$. After condensation, the solid was dissolved in a small account of dichloromethane (ca. 5 mL). The polymer **40** was collected as an orange solid in 86% yield *via* precipitation in methanol from the dichloromethane-soluble fraction. The M_n and M_w/M_n values were estimated as 5,700 and 1.6, respectively.

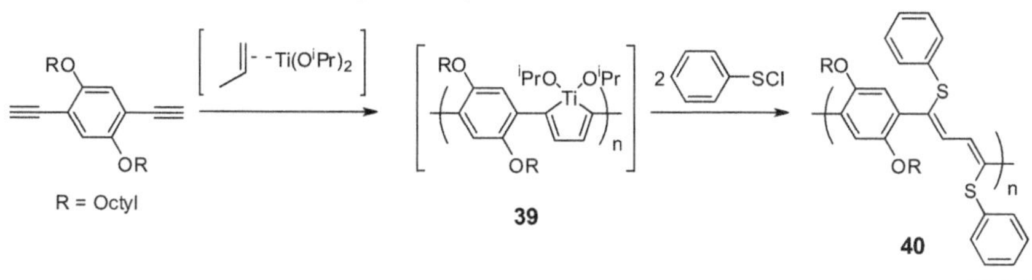

(14.19)

Example 16: Preparation of Group 16 Heterole-Containing Polymers[61]

From the titanacyclopentadiene **39**, the polymer **41** can be prepared (eqn (14.20)).[61] A cyclopentyl methyl ether solution of **39** (0.19 g, 0.50 mmol) was added to Se_2Cl_2 (0.16 g, 0.70 mmol) at –50 °C, and the resulting mixture was warmed to room temperature. After stirring for 3 h, the resulting mixture was poured into hydrochloric acid (1.0 M, 50 mL). The aqueous layer was extracted three times with dichloromethane (total 50 mL), and all the organic layers were collected and dried over $MgSO_4$. After condensation, the solid was dissolved in a small account of dichloromethane (*ca.* 5 mL). The polymer **41** was collected as an orange solid in 86% yield *via* precipitation in methanol from the dichloromethane-soluble fraction.

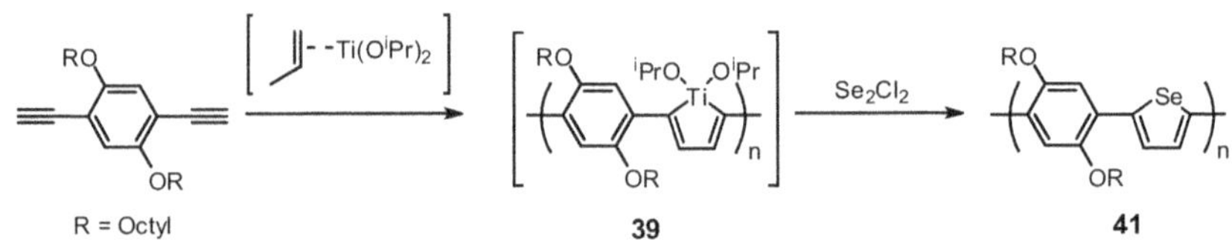

(14.20)

References

1. B. Liu, W. L. Yu, Y. H. Lai and W. Huang, *Macromolecules*, 2002, **35**, 4975.
2. A. Nagai, K. Kokado, Y. Nagata and Y. Chujo, *Macromolecules*, 2008, **41**, 8295.
3. Y. Nagata and Y. Chujo, *Macromolecules*, 2007, **40**, 6.
4. Y. Nagata and Y. Chujo, *Macromolecules*, 2008, **41**, 2809.
5. Y. Tokoro, A. Nagai, K. Kokado and Y. Chujo, *Macromolecules*, 2009, **42**, 2988.
6. Y. Tokoro, A. Nagai and Y. Chujo, *Macromolecules*, 2010, **43**, 6229.
7. Y. Nagata and Y. Chujo, *Macromolecules*, 2008, **41**, 3488.
8. K. Tanaka and Y. Chujo, *Macromol. Rapid Commun.*, 2012, **33**, 1235.
9. N. Matsumi, K. Naka and Y. Chujo, *J. Am. Chem. Soc.*, 1998, **120**, 5112.
10. M. Miyata and Y. Chujo, *Polym. J.*, 2002, **34**, 967.
11. A. Iida and S. Yamaguchi, *J. Am. Chem. Soc.*, 2011, **133**, 6952.
12. Y. Tokoro, A. Nagai, K. Tanaka and Y. Chujo, *Macromol. Rapid Commun.*, 2012, **33**, 550.
13. A. Nagai and Y. Chujo, *Macromolecules*, 2010, **43**, 193.
14. R. Yoshii, A. Nagai and Y. Chujo, *J. Polym. Sci., Part A: Polym. Chem.*, 2010, **48**, 5348.
15. A. Nagai, S. Kobayashi, Y. Nagata, K. Kokado, H. Taka, H. Kita, Y. Suzuri and Y. Chujo, *J. Mater. Chem.*, 2010, **20**, 5196.
16. K. Kokado and Y. Chujo, *Macromolecules*, 2009, **42**, 1418.
17. K. Kokado and Y. Chujo, *Dalton Trans.*, 2011, **40**, 1919.
18. K. Kokado and Y. Chujo, *Polym. J.*, 2010, **42**, 363.
19. A. Qin, J. W. Y. Lam and B. Z. Tang, *Prog. Polym. Sci.*, 2012, **37**, 182.

20. J. Liu, J. W. Y. Lam and B. Z. Tang, *J. Inorg. Organomet. Polym.*, 2009, **19**, 249.
21. S. J. Toal and W. C. Trogler, *J. Mater. Chem.*, 2006, **16**, 2871.
22. L. H. Zhang, T. Jiang, L. B. Wu, J. H. Wan, C. H. Chen, Y. B. Pei, H. Lu, Y. Deng, G. F. Bian, H. Y. Qiu and G. Q. Lai, *Chem. Asian J.*, 2012, **7**, 1583.
23. M. Haeussler, A. J. Qin and B. Z. Tang, *Polymer*, 2007, **48**, 6181.
24. H. Sohn, M. J. Sailor, D. Magde and W. C. Trogler, *J. Am. Chem. Soc.*, 2003, **125**, 3821.
25. M. Yuan, P. Yang, M. M. Durban and C. K. Luscombe, *Macromolecules*, 2012, **45**, 5934.
26. B. C. Schroeder, R. S. Sshraf, S. Thomas, A. J. P. White, L. Biniek, C. B. Nielsen, W. Zhang, Z. Huang, P. S. Tuladhar, S. E. Watkins, T. D. Anthopoulos, J. R. Durrant and I. McCulloch, *Chem. Commun.*, 2012, **48**, 7699.
27. J. Y. Wang, S. K. Hau, H. L. Yip, J. A. Davies, K. S. Chen, Y. Zhang, Y. Sun and A. K. Y. Jen, *Chem. Mater.*, 2011, **23**, 765.
28. R. S. Ashraf, Z. Chen, D. S. Leem, H. Bronstein, W. Zhang, B. Schroeder, Y. Geerts, J. Smith, S. Watkins, T. D. Anthopoulos, H. Sirringhaus, J. C. de Mello, M. Heeney and I. McCulloch, *Chem. Mater.*, 2011, **23**, 768.
29. S. Yamaguchi and K. Tamao, *J. Chem. Soc., Dalton Trans.*, 1998, 3693.
30. J. Ohshita, *Macromol. Chem. Phys.*, 2009, **210**, 1360.
31. H. Y. Chen, J. Hou, A. E. Hayden, H. Yang, K. N. Houk and Y. Yang, *Adv. Mater.*, 2010, **22**, 371.
32. M. Morana, H. Azimi, G. Dennler, H. J. Egelhaaf, M. Scharber, K. Forberich, J. Hauch, R. Gaudiana, D. Waller, Z. Zhu, K. Hingerl, S. S. van Bavel, J. Loos and C. J. Brabec, *Adv. Funct. Mater.*, 2010, **20**, 1180.
33. H. Usta, G. Lu, A. Facchetti and T. J. Marks, *J. Am. Chem. Soc.*, 2006, **128**, 9034.
34. J. Hou, H. Y. Chen, S. Zhang, G. Li and Y. Yang, *J. Am. Chem. Soc.*, 2008, **130**, 16144.
35. J. S. Kim, Z. Fei, D. T. James, M. Heeney and J. S. Kim, *J. Mater. Chem.*, 2012, **22**, 9975.
36. M. Hissler, P. W. Dyer and R. Réau, *Coord. Chem. Rev.*, 2003, **244**, 1.
37. D. Gendron, P. O. Morin, P. Berrouard, N. Allard, B. R. Aïch, C. N. Garon, Y. Tao and M. Leclerc, *Macromolecules*, 2011, **44**, 7188.
38. N. Allard, R. B. Aich, D. Gendron, P. L. T. Boudreault, C. Tessier, S. Alem, S. C. Tse, Y. Tao and M. Leclerc, *Macromolecules*, 2010, **43**, 2328.
39. C. M. Amb, S. Chen, K. R. Graham, J. Subbiah, C. E. Small, F. So and J. R. Reynolds, *J. Am. Chem. Soc.*, 2011, **133**, 10062.
40. Z. Fei, J. S. Kim, J. Smith, E. B. Domingo, T. D. Anthopoulos, N. Stingelin, S. E. Watkins, J. S. Kim and M. Heeney, *J. Mater. Chem.*, 2011, **21**, 16257.

41. Z. Fei, Y. Kim, J. Smith, E. Buchaca Domingo, N. Stingelin, M. A. McLachlan, K. Song, T. D. Anthopoulos and M. Heeney, *Macromolecules*, 2012, **45**, 735.
42. J. Ohshita, Y.-M. Hwang, T. Mizumo, H. Yoshida, Y. Ooyama, Y. Harima and Y. Kunugi, *Organometallics*, 2011, **30**, 3233.
43. Y. M. Hwang, J. Ohshita, Y. Harima, T. Mizumo, Y. Ooyama, Y. Morihara, T. Izawa, T. Sugioka and A. Fujita, *Polymer*, 2011, **52**, 3912.
44. B. L. Lucht, M. A. Buretea and T. D. Tilley, *Organometallics*, 2000, **19**, 3469.
45. W. M. Zhou and I. Tomita, *Polym. Bull.*, 2008, **61**, 603.
46. O. Fadhel, Z. Benkö, M. Gras, V. Deborde, D. Joly, C. Lescop, L. Nyulászi, M. Hissler and R. Réau, *Chem. Eur. J.*, 2010, **16**, 11340.
47. J. Crassous and R. Réau, *Dalton Trans.*, 2008, 6865.
48. M. G. Hobbs and T. Baumgartner, *Eur. J. Inorg. Chem.*, 2007, 3611.
49. F. C. Leavitt, T. A. Manuel and F. Johnson, *J. Am. Chem. Soc.*, 1959, **81**, 3163.
50. E. H. Braye and W. Hübel, *Chem. Ind.*, 1959, 1250.
51. S. S. H. Mao and T. D. Tilley, *Macromolecules*, 1997, **30**, 5566.
52. C. Hay, C. Fischmeister, M. Hissler, L. Toupet and R. Réau, *Angew. Chem. Int. Ed.*, 2000, **10**, 1812.
53. C. Hay, M. Hissler, C. Fischmeister, J. Rault-Berthelot, L. Toupet, L. Nyulászi and R. Réau, *Chem. Eur. J.*, 2001, **7**, 4222.
54. H. S. Na, Y. Morisaki, Y. Aiki and Y. Chujo, *Polym. Bull.*, 2007, **58**, 645.
55. Y. Morisaki, H. S. Na, Y. Aiki and Y. Chujo, *Polym. Bull.*, 2007, **58**, 777.
56. H. S. Na, Y. Morisaki, Y. Aiki and Y. Chujo, *J. Polym. Sci. A: Polym. Chem.*, 2007, **45**, 2867.
57. Y. Morisaki, Y. Aiki and Y. Chujo, *Macromolecules*, 2003, **36**, 2594.
58. H. Nishiyama and I. Tomita, *Macromol. Chem. Phys.*, 2010, **211**, 2248.
59. I. Tomita and M. Ueda, *Macromol. Symp.*, 2004, **209**, 217.
60. W. M. Zhou and I. Tomita, *J. Inorg. Organomet. Polym.*, 2009, **19**, 113.
61. H. Nishiyama, T. Kino and I. Tomita, *Macromol. Rapid Commun.*, 2012, **33**, 545.

CHAPTER 15

Donor–Acceptor Alternating Copolymers

WENTAO LI AND WEI YOU*

University of North Carolina at Chapel Hill, Department of Chemistry, Chapel Hill, North Carolina, 27599-3290, United States
*Email: wyou@unc.edu

15.1 Introduction

Donor–acceptor (D–A) alternating copolymers have significantly contributed to the performance improvement of key organic electronic applications, such as efficiency of bulk-heterojunction organic photovoltaics (OPVs)[1–11] and charge-carrier mobility of organic field effect transistors (OFET).[22–24] A D–A alternating copolymer usually contains two structurally distinct moieties of opposite electronic properties: an electron-rich moiety, *i.e.*, the 'donor', and an electron-deficient moiety, *i.e.*, the 'acceptor'. This unique 'donor' – 'acceptor' alternating structure offers a significant benefit for the design of conjugated polymers: both HOMO and LUMO energy levels (and bandgaps) can be individually fine tuned.

We will start this chapter with a brief review of the origin of the donor–acceptor concept, followed by a short summary of the key chemistries that are used for the synthesis of D–A copolymers (*e.g.*, Suzuki coupling and Stille coupling), with a focus on their application towards the polymer synthesis. The classical step-growth polymerization will be next discussed within the context of Pd-catalyzed polycondensation. Then we will briefly discuss two recent, interesting advances in the synthesis of D–A copolymers: Pd-free polycondensation and chain-growth Suzuki polymerization. Since the synthesis of D–A

RSC Polymer Chemistry Series No. 9
Conjugated Polymers: A Practical Guide to Synthesis
Edited by Klaus Müllen, John R. Reynolds and Toshio Masuda

Published by the Royal Society of Chemistry, www.rsc.org

copolymers involves, to a large extent, the design and synthesis of monomers/structural moieties, we will present the synthetic schemes and discuss the synthetic details of a few representative 'donors' and 'acceptors' that have contributed to D–A copolymers of superior performance in the OPV field. Finally, we will present detailed synthetic procedures of three high-performance polymers, followed by a short discussion on typical characterization methods, to conclude this chapter.

15.1.1 Inception of the Concept of D–A Alternating Copolymers

Prior to the year of 1993, the bandgap of conjugated polymers had already been decreased to around 1.0 eV by stabilizing the quinoid structure.[12] However, in 1993, ten Hoeve and coworkers proposed a completely different yet more versatile approach towards small bandgap semiconducting polymers: using regularly alternating conjugated 'donor' and 'acceptor' moieties to construct the conjugated backbone.[13] Schematically illustrated in Figure 15.1, the essence of this approach resides in the fact that the LUMO and HOMO energy levels of such alternating 'donor–acceptor' copolymers are largely determined by the LUMO of the acceptor moiety and the HOMO of the donor moiety, respectively. To experimentally verify their hypothesis, the same authors selected squaric acid and croconic acid as strong acceptors and copolymerized them with several donor moieties. Though only polymers of low molecular weight were obtained, the bandgaps of such copolymers were indeed decreased significantly to around 0.8 eV with two polymers showing extremely low bandgaps below 0.5 eV.

15.1.2 Advantages of D–A Alternating Copolymers and the State-of-the-Art

Although the original motivation of building D–A alternating copolymers was to decrease the bandgap of such conjugated polymers, the most noteworthy feature of this D–A alternating approach resides in its fine control of *both* bandgap and energy levels. Unlike inorganic materials and homopolymers of (usually) fixed bandgap and energy levels with little or no tunability, the D–A alternating copolymers allow one to adjust both the bandgap and the

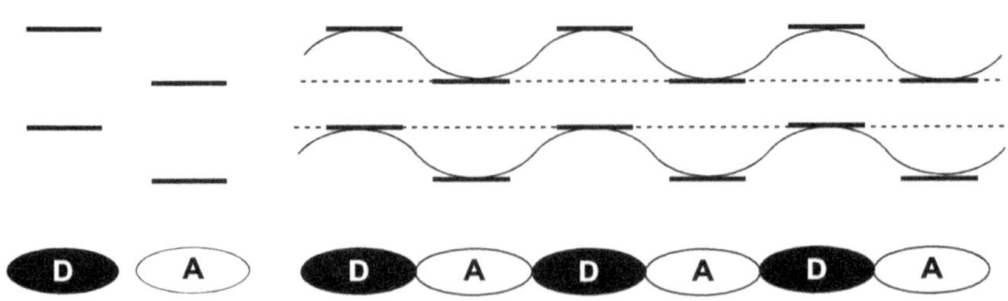

Figure 15.1 Schematic illustration of the bandgap engineering of D–A copolymer.

HOMO/LUMO levels of conjugated polymers with unprecedented ease.[2,14] As stated earlier, the LUMO of the D–A copolymer is largely localized on the acceptor moiety, while the HOMO level is mainly determined by the donor moiety. Therefore, one can select donor and acceptor structural moieties with appropriate HOMO and LUMO levels, respectively, to approach the target energy levels and the bandgap.[15,16] Further modification by substitution on donor and acceptor unit with electron-rich or -withdrawing functional groups allows additional fine tuning of energy levels (and the bandgap). Given the fact that a huge library of D–A copolymers with experimentally determined energy levels and bandgap is readily available in the literature, the design of a new D–A copolymers with specific energy levels is not much of a challenge.

This unique feature of D–A alternating copolymers – fine control of energy levels and bandgap – has attracted a lot of attention in various research areas within organic electronics, especially in organic photovoltaics where the precise engineering of energy levels and bandgap is of paramount importance. Almost all record-high efficiencies of organic solar cells were achieved with D–A alternating copolymers,[8,17–21] some of which are listed in Chart 15.1. All these excellent materials are great testimonies to the success of D–A alternating copolymers of these unique features. For example, in PBnDT-DTBT, BnDT (benzodithiophene) was selected as the 'donor' moiety and DTBT (4,7-dithiophenyl benzothiadiazole) as the 'acceptor' moiety because each of them had a suitable energy level and BnDT-based polymers had demonstrated high charge-carrier mobility.[22] Further substitution of the conjugated backbone by fluorine decreased the HOMO and LUMO levels of the copolymer simultaneously, resulting in an increased open-circuit voltage and enhanced efficiency, over 7%, in PBnDT-DTffBT-based solar cells. Besides

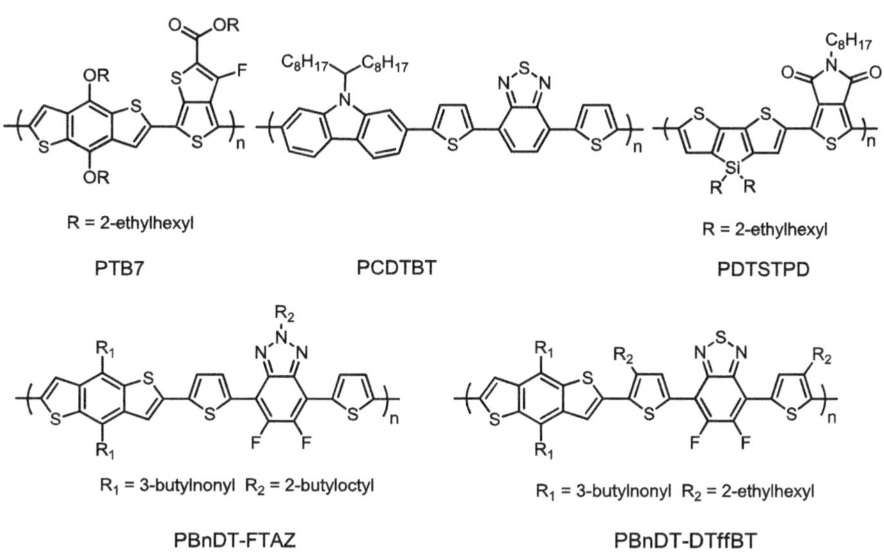

Chart 15.1 Representatives of high performance D–A polymers in OPVs.

their dominance in OPVs, D–A copolymers have also found great success in OFET with noteworthy mobility values and coveted ambipolar characteristics.[23–25]

15.2 General Methods for D–A Copolymerization

Fully conjugated polymers are typically synthesized by polycondensation *via* transition-metal-catalyzed cross coupling reactions, since these coupling reactions provide a powerful tool to connect sp^2-hybridized carbons. In practice, however, a number of factors – including the reactivity of monomers and their stoichiometry, solvent effect, and end-group effect – need to be optimized for any specific polymerization in order to achieve appreciable molecular weight. Because these factors vary case-by-case, we will summarize general principles one needs to consider when preparing these D–A alternating copolymers before we elaborate on specific procedures as examples.

15.2.1 Suzuki Reaction *vs.* Stille Reaction

Among the various coupling reactions, the Suzuki reaction and the Stille reaction are the most commonly employed ones to construct D–A copolymers. The Suzuki reaction was discovered in 1979 by Suzuki and coworkers,[26] where the authors demonstrated the coupling of two sp^2-hybridized carbons by eliminating the boronic acid of one alkenyl and the halogen of the other alkenyl. The scope of the Suzuki reaction has since been expanded to include many aromatic units,[27] and this methodology has become very popular in constructing aryl–aryl linkages in organic molecules and conjugated polymers. Typically, the boronic acid can be introduced by reaction of an aryl–lithium[28] or Grignard species with borate ester, or *via* Suzuki coupling between halogenated aromatics and diboronyl ester/diboronic acid.[29,30] One major advantage of the Suzuki reaction is its tolerance to a wide range of functional groups. Another significant benefit is the low toxicity of its reagents and byproducts (especially when compared with the tin-containing monomers and byproducts from the Stille reaction). On the other hand, the tin-containing monomers, key components in the Stille reaction, can usually be prepared ultrapure *via* recrystallization, while it is often difficult to purify monomers containing boronic acids. Any unquantified impurity in the monomers can upset the stoichiometry of a polymerization, rendering precise molecular weight control more difficult (see Section 15.2.4).

The Kosugi–Migita–Stille reaction, commonly known as the Stille reaction, was discovered by Stille and coworkers[31] and Migita and coworkers[32] in the late 1970s almost simultaneously. In this chapter, we will use the more commonly referred 'Stille reaction' for brevity. The Stille reaction boasts milder reaction conditions, lower oxygen sensitivity and a broader tolerance over most functional groups. Instead of using organoboron to activate the sp^2 carbon, the Stille reaction employs organotin to serve the same function. The organotin compound can be conveniently attached to the aromatic compounds by butyl lithium and trimethyltin chloride. Though easier to purify, the toxicity of stannylated monomers (and the byproducts) is a major concern.

Though both Suzuki and Stille reactions have been widely used to prepare conjugated polymers (including D–A copolymers), there are some subtle issues to consider when it comes to choose which reaction to use. For example, it is worth noting that the electron richness of stannyl aromatics decides whether these monomers are suitable for Stille-based polymerization or not. Mechanistically, relatively electron-rich thiophenes undergo the transmetalation step more readily than stannylbenzenes. Thus, stannylbenzenes experience low reactivity under Stille reaction conditions.[33] Correspondingly, most thiophene-based aromatics are polymerized *via* Stille reactions, whereas a Suzuki reaction is a better option for benzene-based compounds. For example, fluorene and carbazole based polymers are usually prepared by Suzuki reaction, whereas polymers with cyclopenta[2,1-*b*:3,4-*b*′]dithiophene, silolo[3,2-*b*:4,5-*b*′]dithiophene or benzo[1,2-*b*:4,5-*b*′]dithiophene are often polymerized *via* Stille reaction. Due to its broader utilization over the Suzuki reaction in preparing D–A copolymers, Stille reaction-based polymerization will be the focus of this chapter, with a brief discussion on the Suzuki-based polymerization also included (Section 15.2.3).

15.2.2 Factors to Consider from the Perspective of Stille Polymerization

Polymerizations utilizing the Stille reaction have been thoroughly reviewed by Yu and coworkers.[33,34] Here, we only briefly summarize some important aspects relevant to the synthesis of D–A copolymers, including monomer reactivity, solvent effect and types of catalyst and loading. The readers should refer to the original reviews for more information and further details.

Monomers: Mechanistically, coupling an electron-rich organotin molecule with an electron- deficient halide/triflate molecule promotes the desired C–C formation. Therefore, in order to obtain D–A copolymers of high molecular weight, electron-rich 'donor' moieties are usually di-stannylated, whereas the electron-deficient 'acceptor' moieties are typically halogenated. Iodinated acceptors are generally more reactive due to the labile carbon–iodine bond, which also lowers the stability of the iodinated acceptors. On the other hand, chlorinated acceptors are relatively rare because of their low reactivity. Therefore, with a good balance of reactivity and stability, brominated acceptors are the most common ones for polymerization.

Solvents: Solvents serve multiple roles in the polymerization. They need to not only solubilize monomers and stabilize the catalyst, but also dissolve as-formed conjugated polymers. Poor solubility of synthesized polymers in the solvent system leads to premature precipitation of polymers before they reach the desirable high molecular weight. Solvents such as DMF and NMP with high polarity and good coordination ability are desirable since they can function as coordination ligands to the palladium and thereby help maintain the catalyst in the solution. However, conjugated polymers often suffer from low solubility in these polar solvents. In this scenario, systems of mixed solvents, such as toluene/DMF, are employed to achieve high molecular weight. Additionally, high boiling point solvents, such as chlorobenzene, are preferred when the polymerization is carried out at high temperature (*e.g.*, under microwave conditions).

Catalysts: Various catalyst systems have been developed for Stille coupling. Although $Pd(PPh_3)_4$ is the most commonly employed one, PPh_3 can be easily oxidized by trace amount of oxygen. In addition, the excess PPh_3 ligands would retard the Stille reaction.[35,36] Another common catalyst, $Pd(PPh_3)_2Cl_2$, with a reduced amount of PPh_3, is even more prone to oxygen attack. Furthermore, the Pd (II) in $Pd(PPh_3)_2Cl_2$ needs to be reduced to Pd(0) by the distannyl compound before entering the catalytic cycle.[37] Therefore, excess distannyl monomers need to be added to maintain the desired stoichiometric balance of monomers in order to achieve high molecular weight (see Section 15.2.4 for further discussion). For these reasons, catalyst systems based upon a more air-stable source of Pd(0), *e.g.*, $Pd_2(dba)_3$,[38] become prevalent in the synthesis of conjugated copolymers. $Pd_2(dba)_3$ is typically paired with weak donating ligands such as P(*o*-tolyl)$_3$.[39] Unlike PPh_3, excess P(*o*-tolyl)$_3$ does not inhibit the Stille reaction.[36] Usually, four equivalents of P(*o*-tolyl)$_3$ (for each Pd atom) are required to ensure the stability of the catalyst and to facilitate the formation of high molecular weight polymers.[33]

15.2.3 Factors to Consider from the Perspective of Suzuki Polymerization

The Suzuki reaction shares many common similarities and features with aforementioned Stille reaction, such as similar catalytic cycles and Pd-based catalysts, and wide tolerance of functionalities.[27] Highlighted below are a few notable factors one needs to consider when choosing Suzuki polymerization to prepare D–A polymers. Interested readers are referred to a more general review for details on Suzuki polycondensation.[40]

Monomers: For Suzuki polymerization, the stability of the monomers is an issue when deciding what monomers to prepare. These boronic acids tend to dehydrate and form cyclic trimeric anhydride (boroxines). This unavoidable dehydration makes it difficult to purify the diboronic acid monomers and can easily upset the stoichiometric balance, a prerequisite to the desirable high molecular weight (see Section 15.2.4 for a more detailed discussion). For these reasons, boronic esters, especially pinacol boronic ester, are often employed instead, for their better solubility and stability, although these boronic esters are considered less reactive than the boronic acids. Unlike Stille polymerization, Suzuki polymerization does not necessarily require the dihalogenated species to be the electron-deficient 'acceptor'. For example, Müllen *et al.* synthesized several D–A polymers by such an 'inverse' Suzuki reaction where electron-deficient benzothiadiazole was functionalized with pinacol boronic ester.[41,42]

Solvents: Though organic-soluble bases have been used in Suzuki-type aryl–aryl coupling reactions,[43] such an anhydrous Suzuki reaction with organic-soluble base has not been applied yet for polymerizations. Typically, Suzuki polymerizations are carried out in biphasic, mixed solvents with toluene or xylene as the organic phase and potassium carbonate solution as the aqueous phase. Other solvent systems, especially these homogeneous organic solvents, are much less reported.

15.2.4 Factors to Consider from the Perspective of Step Growth

The Stille reaction-based polymerization, by its polycondensation nature, falls into the major category of step-growth polymerization. Thus, characteristics concerning step-growth polymerization still exist and related fundamental principles apply, such as reaction kinetics, molecular-weight distribution and control, and end-group modifications. We will briefly discuss the latter two in the context of Stille polymerizations.

In general, a high molecular weight is required to achieve the promised properties of any given polymer.[44] This characteristic holds true for conjugated polymers as well. For example, longer polymer chain/higher molecular weight has been shown to produce higher mobilities in OFETs and higher efficiencies in OPVs.[45–47] In order to achieve high molecular weight, we still need to pay attention to the classical Carothers equation, in addition to the aforementioned factors that are specific to the Stille polymerization (*e.g.*, monomers, solvents, and catalysts).

$$\overline{X_n} = \frac{1+r}{1+r-2rp} \tag{15.1}$$

$$\overline{X_n} = \frac{1}{1-p} (\text{when } r=1) \tag{15.2}$$

$$\lim_{p\to 1} \overline{X_n} \to \frac{1+r}{1-r} \tag{15.3}$$

According to the Carothers equation (eqn (15.1)), which describes the relationship among molecular weight, monomer feed ratio (r, $0<r\leq 1$) and monomer conversion (p, $0<p\leq 1$), high molecular weight can only be obtained at very high conversion (p approaching unity) and with stoichiometrically balanced monomers (eqn (15.2)). Therefore, typically a long reaction time is required to ensure a high conversion, and equal moles of the two monomers are used. The required stoichiometric balance of monomers has important implications for the Stille polymerization in particular. For example, the Stille reaction is often accompanied by the homocoupling between stannylated compounds.[37] This undesired homocopouling would consume the stannlyated monomers and upset the original stoichiometry, thereby leading to a low molecular weight according to eqn (15.3). Similarly, any un-quantified

contamination in the monomers could also change the stoichiometric balance and result in low molecular weight. One such example comes from the copolymerization of distannyl-dithienosilole and dibromo-benzothiadiazole, where only the polymerization with a feed ratio of 1.05 : 1.00 (distannyl : dibromo) reached the highest molecular weight in that study.[48] It is plausible that undetectable contaminants in the organo-tin monomer could be the culprit. One can compensate for the monomers consumed in nonpolymerization processes (or for the contaminants, if their quantity is known). For example, when $Pd(PPh_3)_2Cl_2$ is used, Pd(II) must be first reduced to Pd(0) by distannyl compound prior to entering the catalytic cycle. As a result, $a\%$ loading of $Pd(PPh_3)_2Cl_2$ would consume at least $a\%$ distannyl monomers. In this scenario, maintaining a stoichiometric balance of two monomers requires $1+a\%$ of distannyl monomers. This excess amount of distannyle monomers, however, is not necessary for catalyst systems with $Pd_2(dba)_3$ or $Pd(PPh_3)_4$.

On the other hand, the simplified Carothers eqn (15.3) also offers a theoretical basis to control the molecular weight by intentionally varying the stoichiometric ratio of the monomers (r). Table 15.1 summarizes the data from our own investigation of the impact of molecular weight on the performance of a high-performance D–A copolymer, PBnDT-FTAZ. The experimentally determined molecular weight (number average, M_n) matched quite well the calculated one based on the monomer feed ratio (r). Please note that the molecular-weight values measured by GPC are not the *absolute* values, rather only *relative* ones (to the standard used in GPC). Thus, one should pay attention to the trend of varying molecular weight by the feed ratio, which was consistent with the trend predicted by the Carothers eqn (15.3). From a practical standpoint, finding the appropriate molecular weight could be crucial to maximizing the device performance of any given polymer in solar cells. In our own study (Table 15.1), we discovered that only the polymer with a theoretical M_n of 40 kg/mol (experimental value of 55 kg/mol) achieved device efficiencies above 7%. Interestingly, a higher molecular weight (*e.g.*, 71 kg/mol) adversely impacted the efficiency of the devices. Similar effects of molecular weight on device efficiencies have also been observed in other polymers.[49–51] It is believed

Table 15.1 Controlling molecular weight of PBnDT-FTAZ and its impact on efficiency of BHJ devices.[a]

ratio	r	$M_n^{theor.}$ *(kg/mol)*	$M_n^{exper.\ b}$ *(kg/mol)*	$M_w^{exper.\ b}$ *(kg/mol)*	*PDI*[b]	J_{sc} *(mA/cm²)*	η^c(%)
1.105	0.9048	10	18.3	34.4	1.89	3.61	1.42
1.050	0.9524	20	36.5	76.6	2.10	11.56	4.96
1.025	0.9756	40	55.2	119	2.15	14.63	7.83
1.000	1.0000	60	71.4	156	2.19	12.69	6.41

[a]Polymerized with 0.1 mmol of dibromo-FTAZ and corresponding amount of distannyl-BnDT (*e.g.*, 0.1105 mmol) with 2% $Pd_2(dba)_3$ and 16% $P(o\text{-}tol)_3$ as the catalyst in 0.6 mL of o-xylene in a microwave reactor at 200 °C for 10 min.

[b]Estimated by GPC (trichlorobenzene as eluent, polystyrene standard, 135 °C).

[c]Devices were fabricated in normal architecture: ITO/PEDOT:PSS/polymer:PCBM/Ca/Al and tested under 1 Sun, AM 1.5 G irradiation.

that while polymers of low molecular weight cannot build effective charge-transport pathways, polymers with too high a molecular weight tend to aggregate excessively and inhibit the formation of an effective bulk heterojunction.

Another interesting fact of step-growth polymerization is that typically the chain ends of formed polymers contain unreacted functional groups. On one hand, these functional groups offer great opportunities for further functionalization, *e.g.*, introducing additional segments of other polymers. On the other hand, these reactive species at chain ends, *e.g.*, organo-tin in the case of Stille polymerization, could introduce defects into bulk polymer films, interact with the interfaces in the BHJ solar cells, and lower the stability of the device. For these reasons, capping these reactive chain ends with monofunctionalized thiophene or benzene is quite popular. For example, Bazan *et al.* end-capped a D–A copolymer by subsequently adding 2-bromothiophene and 2-stannylthiophene *in situ* after the planned polymerization time. With these two polymers (as-prepared copolymer vs. the end-capped one), the authors conducted a detailed study of the effect of end capping a D–A copolymer on BHJ solar cells.[52] It was discovered that the end-capped polymer exhibited better thermal stability, fewer defects and slightly higher photovoltaic efficiency in devices.

15.2.5 New Development: Direct Arylation and Chain-Growth Polymerization

As previously mentioned (Section 15.2.1), the Stille reaction/polymerization suffers from the toxicity of trialkyltin and the instability of stannylated compounds. Furthermore, if the donor molecule contains nucleophile-sensitive groups such as carbonyl, the common approach of adding trimethyltin *via* treating the substrate with *n*-BuLi and Me_3SnCl becomes impractical, and only few alternative solutions are available. To avoid these drawbacks associated with the Stille polymerization, the research community has made notable progress in exploring 'tin-free' approaches. For example, a new type of reaction called 'direct arylation' has recently been employed to synthesize conjugated small molecules and polymers. The first D–A copolymer synthesized by direct arylation was reported by Leclerc and coworkers in 2011.[53] In this study, a strong acceptor, 5-alkyl-thieno[3,4–c]pyrrole-4,6-dione, was copolymerized with brominated bithiophene to afford a polymer with very high molecular weight, even higher than that obtained *via* the conventional Stille polymerization (Scheme 15.1). For details on the mechanism and synthetic protocols of the direct arylation-based polymerization, the readers are referred to Chapter 19 in this book.

Another reaction worth noting as a potentially game-changing advance in the synthesis of D–A copolymers is the 'controlled' chain growth Suzuki polymerization. Compared with the step growth, the chain-growth polymerization possesses a faster reaction rate and potentially offers fine control over the molecular weight, polydispersity (PDI), and end groups if the propagating chain end can be made 'controlled' or 'living'. Although the synthesis of conjugated polymers *via* the chain growth Suzuki coupling has progressed since

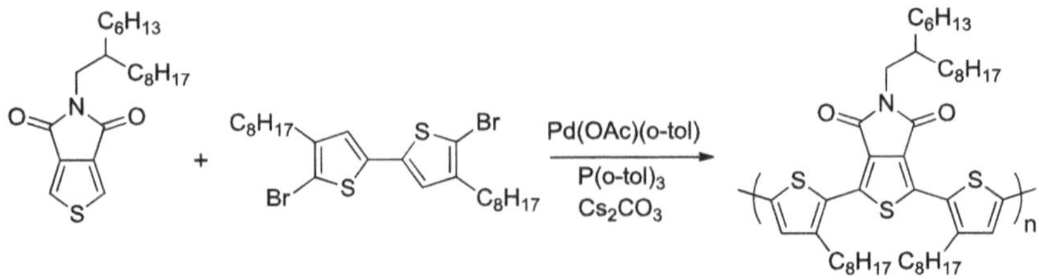

Scheme 15.1 Polymerization of TPD *via* direct arylation.

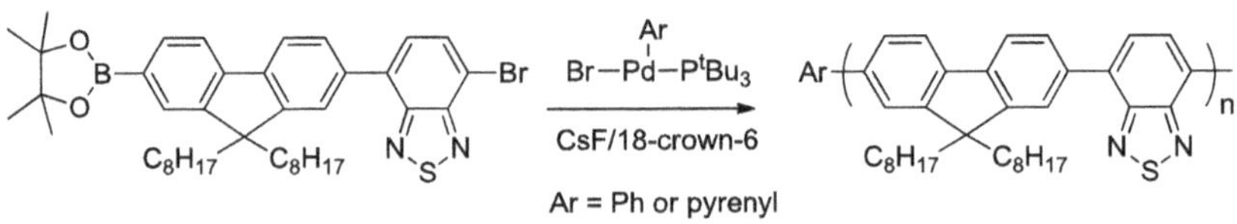

Scheme 15.2 Chain-growth Suzuki polymerization of PFDTBT in ref. 58.

2007,[54–57] its application to the synthesis of D–A alternating copolymers had not been reported until recently, when an AB-type monomer with the structure of 'boronic ester-D–A-bromo' underwent a chain-growth self-polymerization to yield a D–A copolymer, poly(fluorene-benzothiadiazole), of relatively low molecular weight (Scheme 15.2).[58] Although the synthesis of conjugated polymers *via* a chain growth mechanism is still in its early development, it is certainly plausible to achieve 'controlled' syntheses of conjugated polymers with more complex structures after further understanding of the mechanism and developing better catalyst systems.

15.3 Synthesis of Selected Monomers and Typical Polymerizations

Over the past twenty years numerous D–A copolymers with a great variety of donor/acceptor moieties have been designed and synthesized. It is practically impossible and unnecessary to include all related chemistries in the limited space of this chapter. In the next section, we will select a few commonly employed monomers (including both donors and acceptors) with broad applications and excellent properties (especially for OPV devices), and describe their syntheses in detail. Then we will discuss the specifics of synthesizing D–A copolymers *via* Stille polymerization. To conclude this chapter, three representative syntheses of D–A copolymers will be provided as examples.

15.3.1 Synthesis of Donor Monomers

Fluorene[10] and carbazole[11] have long been used as donor units in constructing D–A copolymers since they are commercially available and easy to functionalize. To ensure the solubility of the resulting polymers, alkylation of fluorene and carbazole is necessary. Alkylation can be achieved by employing a strong

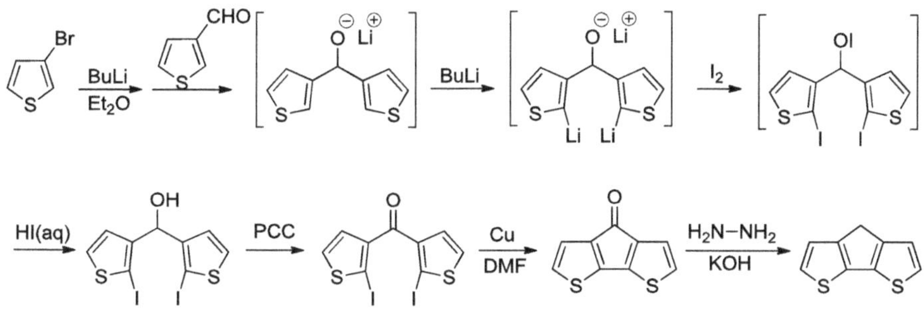

Scheme 15.3 Synthesis of CPDT.

base, such as NaO^{t}Bu or BuLi, to deprotonate the centered five-membered ring so that the formed anion can work as a nucleophile to undergo a substitution reaction with alkyl bromide.[59–61]

Replacing the two flanking benzene units of the fluorene with thiophene units gives cyclopenta[2,1-*b*:3,4-*b*′]dithiophene (CPDT), which is popular because CPDT can mitigate the steric hindrance from the flanking benzenes of the fluorene. The synthesis of CPDT had been quite tedious until Brzezinski and Reynolds demonstrated an efficient three-step route, as shown in Scheme 15.3.[62] Specifically, the commercially available 3-bromothiophene is first lithiated in anhydrous ethyl ether under inert atmosphere. This lithiated thiophene anion undergoes the nucleophilic attack to the 3-thiophenecarboxaldehyde to form a bithiophene adduct with an alkoxide. Adding 2 eq. of BuLi to deprotonate at the 2,2′ positions of the bisthiophene, followed by quenching with iodine and neutralizing with HI, offers an intermediate with all functional groups/handles in place. This intermediate is then subjected to the oxidation of the alcohol, followed by an Ullmann coupling to close the central ring to afford the cyclopentadithiophen-4-one in high yield. Finally, a Wolff-Kishner reduction is employed to yield the unsubstituted CPDT, which can be alkylated similarly to the alkylation of the fluorene.

Substituting the sp^3 carbon in the center of CPDT with silicon leads to an interesting analog, silolo[3,2-*b*:4,5-*b*′]dithiophene, whose synthesis requires two constructing fragments (Scheme 15.4).[48] One of them, 3,3′-dibromo-2,2′-bithiophene, is prepared by the reduction of 3,3′,5,5′-tetrabromo-2,2′-bithiophene with zinc dust and acetic acid. The other one, dichlorodialkylsilane, is synthesized *via* platinum-catalyzed hydrosilylation of dichlorosilane with two molecules of alkenes of selected alkyl chain length. The dianion formed after lithium-halogen exchange of the dibromobithiophene is then cyclized with the dichlorodialkylsilane, yielding the target molecule, silolodithiophene.

With a suitable HOMO level, good charge transport and versatility in further functionalization, benzo[1,2-*b*:4,5-*b*′]dithiophene (BnDT) has emerged to be the most popular donor unit to construct the D–A copolymers used in solar cells.[22] Many BnDT derivatives and related polymers have been synthesized and carefully studied.[5,7] Generally, the synthesis of BnDT-based monomers starts with the preparation of benzo[1,2-*b*:4,5-*b*′]dithiophene-4,8-dione, which

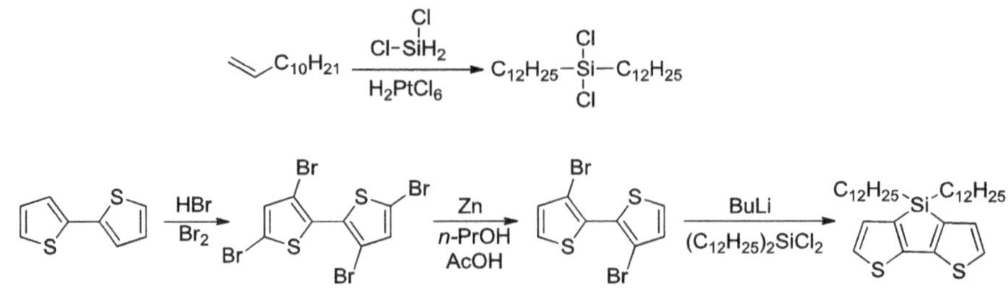

Scheme 15.4 Synthesis of silolodithiophene.

can be constructed following a two-step reaction scheme with high yield (Scheme 15.5).[63] The subsequent alkylation of the central 1,4-benzoquinone can be done in a number of ways. For example, aliphatic side chains can be attached by subjecting the central two carbonyls to nucleophilic attack by either an alkyl Grignard reagent[64] or deprotonated alkyne.[22] Removing the hydroxyl by tin(II) chloride under acidic conditions affords the rearomatized and alkylated BnDT. An important variation is to apply the 2-alkylated thiophene (after deprotonation) as the nucleophile, which can result in an interesting product with extended conjugation perpendicular to the BnDT.[65] Alternatively, if the central 1,4-benzoquinone is reduced to the hydroquinone, one can create the BnDT with alkoxy side chains by treating the deprotonated hydroquinone with an alkyl bromide of desirable chain length.[66]

Prior to the Stille polymerization, all these thiophene-based donor molecules need to be converted into stannylated monomers, typically *via* BuLi in anhydrous THF followed by treating the lithiated anion with trialkyltin chloride. Usually Me_3SnCl is preferred albeit its stronger toxicity than the butyl analog (*i.e.*, Bu_3SnCl) because the latter can render it much more difficult to purify monomers through recrystallization.

15.3.2 Synthesis of Acceptor Monomers

While donor units are primarily used to maintain appropriate HOMO levels in D–A copolymers, acceptor units are largely responsible for the control of the LUMO level. This is particularly important for D–A copolymers used for organic solar cells, since the more desirable narrow bandgap can be achieved by carefully selecting strong acceptor units. Due to the electron-deficient nature of these acceptor moieties, their syntheses often involve harsh conditions and quite delicate chemistry. Below we will briefly describe the synthesis of several important acceptor units that have contributed to the discovery of D–A copolymers of exceptional performance in organic electronics.

Commonly known as DTBT, the structural unit of 4,7-di(thiophene-2-yl)benzo[c][1,2,5]thiadiazole has been extensively used to construct D–A copolymers of small bandgaps. Shown in Scheme 15.6, the synthesis of DTBT starts with the cyclization of 1,2-diaminobenzene under thionyl chloride to form the central unit, benzothiadiazole (BT). This BT unit is then brominated

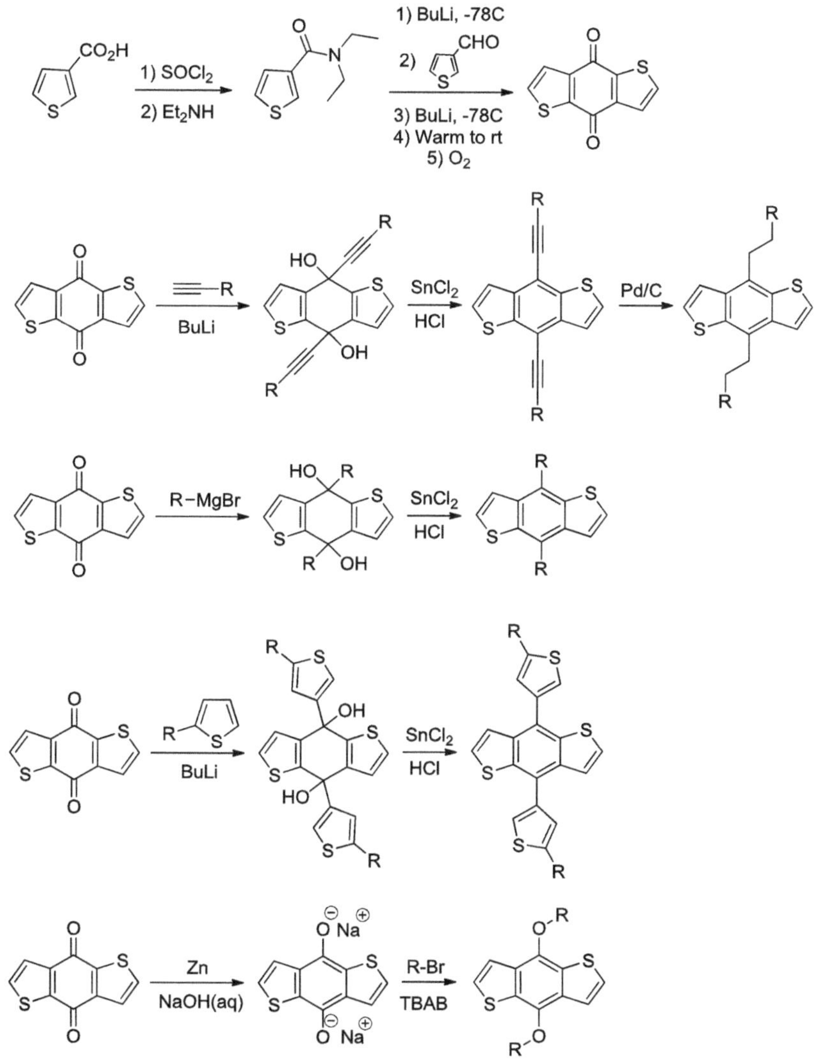

Scheme 15.5 Synthesis of BnDT derivatives.

in hot hydrobromic acid together with bromine. The product of this step, dibrominated BT, can directly serve as a strong acceptor monomer in synthesizing D–A polymers. However, if DTBT is desired, the dibrominated BT needs to undergo additional Stille coupling reactions with stannylated 3-alkylated thiophenes in order to accomplish the DTBT with two flanking thiophenes.[67]

The fluorinated version of DTBT, 5,6-difluoro-4,7-di(thiophen-2-yl)benzo[c][1,2,5]thiadiazole, *i.e.*, DTffBT, can be synthesized similarly to the preparation of the DTBT. However, due to its strong electron deficiency, ffBT can only be brominated in HBr/Br_2 after 72 h. Instead, a harsh condition of iodination in fuming sulfuric acid is able to halogenate this strong acceptor

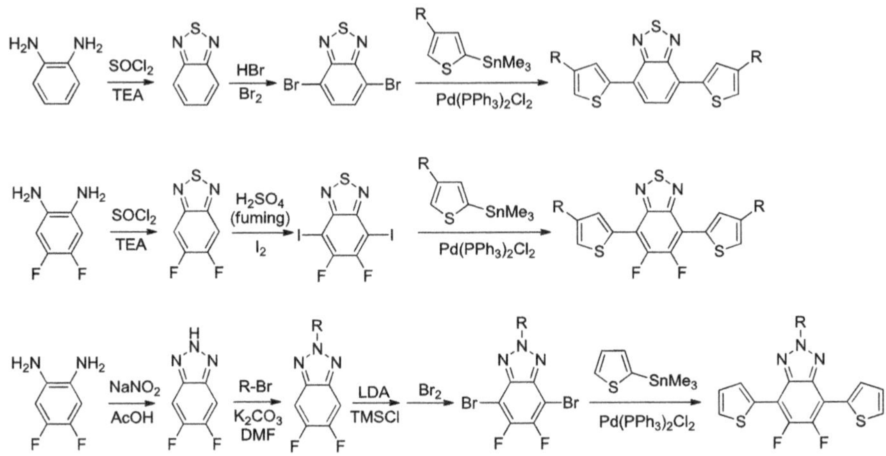

Scheme 15.6 Synthesis of DTBT, DTffBT and FTAZ.

overnight (Scheme 15.6), which, as discussed above, gives a diiodo-ffBT intermediate with a lower stability than dibromo-BT.[19]

A unique acceptor, 5,6-difluoro-2-alkyl-4,7-di(thiophen-2-yl)-2H-benzo-[d][1,2,3]triazole, *i.e.*, FTAZ, structurally similar to the DTffBT, has distinctly different electrochemical properties from those of DTffBT. The alkylation of the central nitrogen in fluorinated benzotriazole is suggested to be carried out in polar aprotic solvents such as DMF to promote the S_N2 reaction. However, typical brominations of the alkylated FTAZ with bromine ended with low yields and a new route was thus developed. The 4 and 7 positions of fluorinated benzotriazole are first deprotonated by a non-nucleophilic base, LDA, and then quenched by trimethylsilyl chloride (TMSCl). The as-formed C–Si bonds are more 'activated' than the original C–H bonds so that the alkyl FTAZ benzotriazole can then be successfully brominated (Scheme 15.6).[21]

The first high-performance polymer with over 7% efficiency in its photovoltaic devices, namely PTB7, was constructed with BnDT as the donor and 2-ethylhexyl-3-fluorothieno[3,4–b]thiophene-2-carboxylate as the acceptor.[68,69] The synthetic route is outlined in Scheme 15.7. Fluorine is first attached onto 4,6-dihydrothieno[3,4–*b*]thiophene-2-carboxylic acid by $PhSO_2NF$ *via* electrophilic fluorination. This fluorinated intermediate then undergoes Steglich esterification to attach a solubilizing side chain. Finally, an oxidation by mCPBA, followed by the dehydration by acetic anhydride, affords the fluorinated thienothiophene.

Another acceptor unit that has achieved high-performance D–A copolymers with a number of donors is thieno[3,4–*c*]pyrrole-4,6-dione, called TPD.[17,70–73] The synthesis of TPD can start with un-substituted thiophene, which can be converted into 3,4-dibromothiophene *via* selective debromination of the tetrabrominated thiophene. After the cyano substitution and hydrolysis, as-formed thiophene-3,4-dicarboxylic acid is brominated to provide the halogens

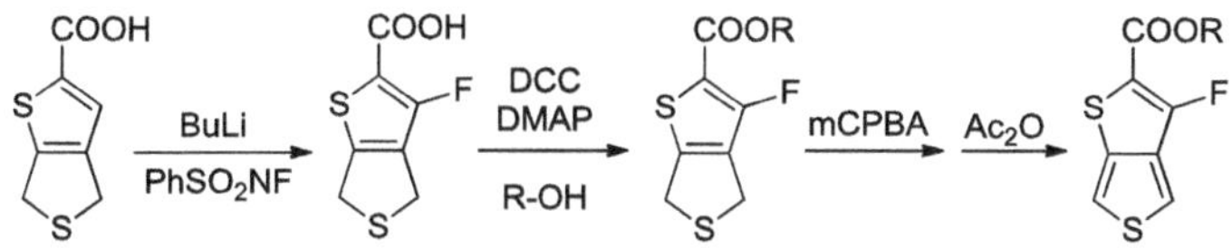

Scheme 15.7 Synthesis of 3-fluorothienothiophene. For PTB7, R = 2-ethylhexyl.

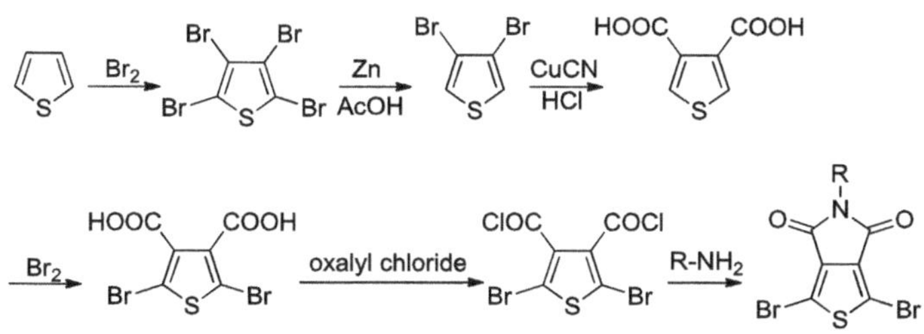

Scheme 15.8 Synthesis of TPD.

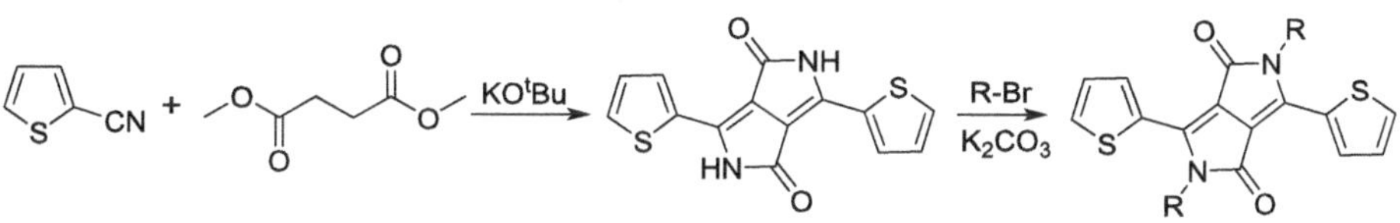

Scheme 15.9 Synthesis of DPP.

necessary for Stille polymerization. To complete the synthesis of TPD, these two carboxylic groups are converted to more reactive acyl chlorides and further react with an aliphatic amine for alkylation, finished by the ring closure (Scheme 15.8).

Recently, another interesting acceptor, 2,5-dialkyl-3,6-di(thiophen-2-yl)-tetrahydropyrrolo[3,4–c]pyrrole-1,4(2 H,5 H)-dione (*i.e.*, DPP), has gained popularity due to its attractive features. DPP based D–A copolymers usually have a much smaller bandgap (*e.g.*, 1.3 eV) which is very useful in OPVs, and exhibit bipolar transport that is desirable for OFETs.[38,74–78] Furthermore, the synthesis of DPP is very simple and straightforward. A solution of 2-thiophenecarbonitrile (3.0 eq.) in *t*-amyl alcohol is added to KO^{t}Bu (3.5 eq.) under argon atmosphere. Another solution of dimethyl succinate (1.0 eq.) in *t*-amyl alcohol is added slowly at 100 °C. After distilling off the byproduct of methanol, the remaining solution is further refluxed for 2 h and then neutralized. A simple filtration yields the second molecule in Scheme 15.9 in high yield, which undergoes normal alkylation and bromination to afford DPP monomers (Scheme 15.9).

As an easily accessible acceptor, isoindigo features convenient two-step synthesis and relatively low LUMO level. An aldol condensation between commercially available 6-bromoisatin and 6-bromooxindole in acid forms 6,6′-dibromoisoindigo in high yield, followed by typical *N*-alkylations to give the final product in an overall yield about 80% (Scheme 15.10).[79]

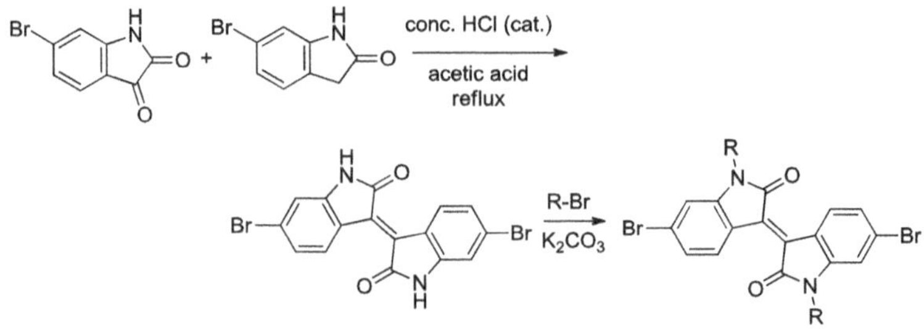

Scheme 15.10 Synthesis of isoindigo.

Except TPD and isoindigo, which were brominated during its synthesis, all other acceptor units introduced above can be converted into monomers for Stille polymerization, usually by subjecting them to simple brominations with NBS.

15.3.3 Purification of Reagents

The purity of both monomers for the Stille polymerization is crucial. This will not only ensure an accurate stoichiometric control (see Section 15.2.4 for details), but also minimize possible negative impact to the device properties by these impurities. Typically, both distannylated donor monomer and dibromo acceptor monomer need to be recrystallized or purified by prep HPLC prior to the polymerization, in order to ensure a high level of purity. An excellent case study demonstrating monomer purity's effect on the molecular weight of the D–A copolymer was provided by Osaka *et al.* In this study,[80] the authors used three different procedures to purify both monomers and observed essentially no difference in the NMR spectra of these monomers after the different purification processes. However, the purity of the monomers significantly affected the observed molecular weight of their polymers (from 13 kg/mol to 73 kg/mol), which noticeably influenced the photovoltaic properties of devices based on these polymers.

The purity of the catalyst is equally important. This simple fact, however, is often ignored because usually these catalysts are purchased from reputable vendors and regarded as 'pure'. However, Zalesskiy and Ananikov recently cautioned that commercially available $Pd_2(dba)_3$ could contain up to 40% Pd nanoparticles, which would not participate in the catalytic cycle.[81] For D–A copolymerization, such a high level of impurity in the catalyst could lead to unexpectedly low molecular weight and even worse, more metal residue in the finished polymers. Therefore, it is strongly recommended that $Pd_2(dba)_3$ be carefully purified prior to the Stille polymerization. Following is a specific procedure to purify $Pd_2(dba)_3$, adapted from ref. 81.

The $Pd_2(dba)_3$ from a commercial source is dissolved into chloroform and the insoluble Pd nanoparticles are removed by filtration. The filtered solution is evaporated on a rotary evaporator and the remaining solid is dissolved in a minimum amount of chloroform. A 4-fold excess of acetone is slowly added to form a 'biphasic' system, which is refrigerated at −18 °C overnight. The precipitates are filtered and washed by cold acetone to yield $Pd_2(dba)_3 \cdot CHCl_3$, which can retain 95% purity in air within one month.

15.3.4 Typical Procedure of Polymerization and Purification of Product

The preparation of D–A alternating copolymer *via* the Stille polymerization is usually carried out in anhydrous solvent under inert atmosphere. After adding all starting materials in flame-dried glassware, three pump-refill cycles are usually required to completely remove the oxygen. Then, anhydrous solvent is injected and purged with argon. For the polymerization with conventional heating, the amount of solvent must be sufficient to dissolve not only the starting materials, but also the as-formed polymers, so that the step polymerization can proceed smoothly, leading to a high molecular weight product. However, too much solvent could result in a low monomer concentration with low reaction rate and extended reaction time.

On the other hand, microwave-assisted Stille polymerization provides an efficient tool to obtain polymers of high molecular weight under increased pressure and temperature, with less solvent and shorter reaction time.[82,83] This is because the dipole moment of a solvent molecule is excited in the oscillating electromagnetic field and the reaction rate is significantly accelerated by internal heating. For microwave-assisted Stille polymerization of D–A polymers, the ratio of solvent to the projected polymer can be as low as 0.6 mL/mmol, depending upon the solubility of the polymer. The upper limit of the temperature should be set below 300 °C and the reaction time cannot be too long to avoid undesired side reactions or decomposition of the polymers. In some reported cases, the temperature was raised stepwise to improve the conversion at the last stage of polymerization, in order to achieve a high molecular weight.[48,52,84]

After polymerization, the crude product is dissolved in a minimal amount of solvent such as chlorobenzene and then precipitated in methanol. To remove the low molecular weight oligomers, the crude polymer after precipitation is washed in a Soxhlet extractor sequentially from poor solvents to good solvents. A common sequence of solvents is methanol (MeOH), ethyl acetate (EtOAc)/acetone, hexane, tetrohydrofuran (THF) and chloroform ($CHCl_3$). Depending on the polymer's solubility, such a sequence of extracting solvents can be varied. The ultimate goal of such practice is to obtain the fraction of the highest molecular weight, which is usually the least-soluble portion in the crude polymer. If removing the residual palladium is desirable, one can stir the polymer solution with chelating reagents such as EDTA or sodium diethyldithiocarbamate[85] to form stable palladium coordinates, or simply filter such a polymer solution through Celite or syringe filter. The purified polymer solution is then concentrated and precipitated in MeOH again to afford the final product. In some cases where the target polymer is hardly soluble, the residual solid in the thimble after Soxhlet extraction is collected as product without further purification, for example, in the case of PBnDT-DTffBT.

15.3.5 Representative Syntheses of D–A Alternating Copolymers

Example 15.1: Synthesis of PCDTBT *via* Suzuki Coupling

Diboronic ester carbazole (263.0 mg, 0.4 mmol) and dibromo-DTBT (174.1 mg, 0.38 mmol) are dissolved in 20 mL toluene in a 50 mL round-bottomed flask equipped with condenser. $Pd(PPh_3)_4$ (2%, 10 mg) is added after the solution is purged with N_2 for 30 min. After heating at 110 °C for 10 min, tetraethylammnium hydroxide (1.8 mL, 20% by weight in water) is added and the reaction is heated to reflux for overnight. Phenylboronic acid (1 mL) and bromobenzene (1 mL) are added sequentially and the mixture is allowed to react for 2 h each. After cooling to r.t., the mixture is precipitated in MeOH, filtered and Soxhlet extracted with ether and $CHCl_3$. The Pd residue is removed by passing $CHCl_3$ portion through a column packed with alumina, Celite and silica gel. The precipitation of the concentrated filtrate (30 mL) in 300 mL of MeOH is dried under vacuum overnight to afford the PCDTBT polymer (Scheme 15.11).[86]

Example 15.2: Synthesis of PBnDT-FTAZ *via* Stille Coupling by Conventional Heating

A flame-dried 25 mL 2-neck round-bottomed flask equipped with a condenser is charged with distannyl-BnDT (179.6 mg, 0.204 mmol) and dibromo-FTAZ (129.1 mg, 0.200 mmol). The flask is placed under vacuum and refilled with argon before $Pd_2(dba)_3 \cdot CHCl_3$ (4.1 mg, 0.004 mmol, 0.02 eq.) and P(o-tol)$_3$ (9.7 mg, 0.032 mmol, 0.16 eq.) are added. After three vacuum–refill cycles, anhydrous toluene (12 mL) is added *via* syringe. The solution is purged with Argon for 20 min and then heated on an oil bath at 110 °C for 72 h. The mixture is precipitated in 200 mL of MeOH, filtered and Soxhlet extracted sequentially by EtOAc, hexane, THF and chloroform. The chloroform portion is concentrated and precipitated again in 200 mL of MeOH, affording the PBnDT-FTAZ polymer after drying under vacuum (Scheme 15.12).[21]

Example 15.3: Synthesis of SilDT-BT by Stepwise Microwave Heating

Distannyl-SilDT (250.0 mg, 0.233 mmol), dibromo-BT (65.0 mg, 0.221 mmol), $Pd(PPh_3)_4$ (13.5 mg, 0.0117 mmol, 0.05 eq.) are added to a dry 5 mL microwave vial. *o*-Xylene (1 mL) is added to the vial and the solution is degassed with argon for 30 min. The mixture is subject to the following conditions in microwave reactor: 100 °C for 2 min, 120 °C for 2 min, 140 °C for 5 min, 160 °C for 5 min and 170 °C for 40 min. The vial is resubmitted to the microwave reactor after adding bromobenzene (0.1 eq.) and heat at 100 °C for 1 min, 120 °C for 1 min, 140 °C for 2 min and 160 °C for 5 min. Another 0.1 eq. of trimethyl(phenyl) stannane is added and the vial is subjected to the same heating procedure as above. The product is precipitated in MeOH and Soxhlet extracted with acetone, hexane and chloroform. The polymer solution in chloroform is vigorously stirred with aqueous sodium diethyldithiocarbamate at 50 °C for 2 h. The organic phase is dried, concentrated and precipitated to yield the polymer SilDT-BT after drying under vacuum (Scheme 15.13).[84]

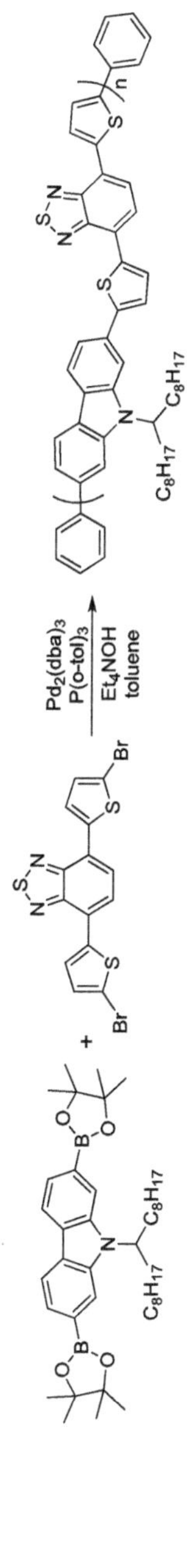

Scheme 15.11 Synthesis of PCDTBT *via* Suzuki coupling.

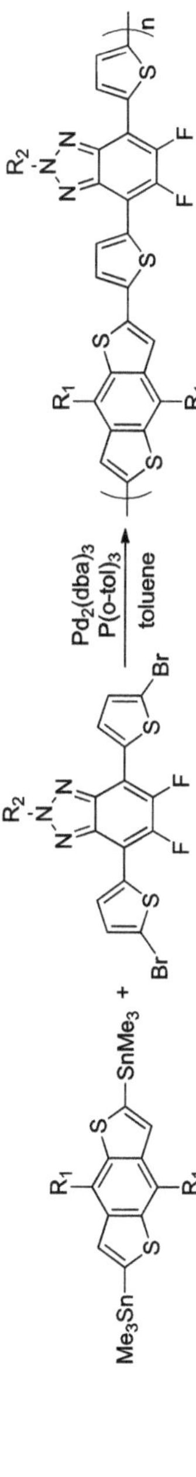

Scheme 15.12 Synthesis of PBnDT-FTAZ *via* Stille coupling.

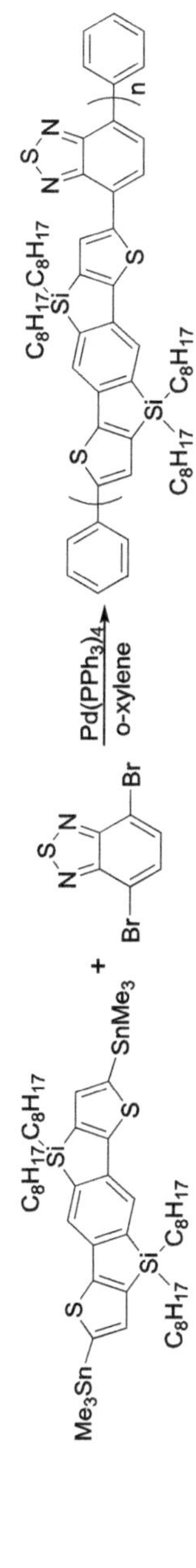

Scheme 15.13 Synthesis of SilDT-BT *via* Stille coupling.

15.3.6 Characterization and Remarks

The molecular weight and polydispersity index (PDI) of D–A alternating copolymers are measured by gel permeation chromatography (GPC). Since these polymers typically have rigid conjugated backbones and strong interchain stacking, a high-temperature GPC with good solvents for these polymers (*e.g.*, 1,2,4-trichlorobenzene at 135 °C) is preferred. Such conditions should be able to dissolve these polymers into single chains, enabling the accurate assessment of the molecular weight and PDI.

The chemical structures of D–A alternating copolymers are routinely characterized by nuclear magnetic resonance (NMR) spectroscopy. Again, since the polymer chains need to be properly solvated and dispersed to expose all of the characteristic protons, higher temperatures near 100 °C are necessary. Therefore, deuterated solvents with high boiling points, such as 1,1,2,2-tetrachloroethane-D2 ($C_2D_2Cl_4$), are typically used.

The optical properties of D–A copolymers are usually assessed by UV-vis absorption spectroscopy. The extinction coefficient of a polymer film is typically of the order of 10^4 cm^{-1}. The optical bandgap can be estimated from the onset of the absorption band of polymer thin films. Additionally, the shift between the absorption maximum of polymer solution spectrum and that of the polymer thin-film spectrum is used to estimate the degree of interchain interaction. Stronger stacking and aggregation of polymers in film is usually accompanied by a larger red shift of the absorption maximum from the solution to the film.

Cyclic voltammetry (CV) is often used to determine the electrochemical properties of D–A copolymers. Usually, the CV setup consists of a Ag/$AgNO_3$ reference electrode, a platinum wire counterelectrode and a glassy carbon electrode with a drop-cast polymer film. First, the CV curve of ferrocene/ferrocenium redox couple (Fc/Fc^+) should be measured to calibrate the reference electrode. The polymer's HOMO energy level in electron volts is calculated from the onset of the oxidation peak (E_{ox}) according to eqn (15.4). If the reduction peak of polymer is not clearly visible, the LUMO level can be approximated from the HOMO measured by CV and the bandgap measured by UV-vis.

$$\mathrm{HOMO} = -[4.8\ \mathrm{eV} + \mathrm{e}(E_{\mathrm{ox}} - E_{\mathrm{F_c/F_c^+}})] \tag{15.4}$$

References

1. Y. F. Li, *Acc. Chem. Res.*, 2012, **45**, 723–733.
2. H. X. Zhou, L. Q. Yang and W. You, *Macromolecules*, 2012, **45**, 607–632.
3. L. Y. Bian, E. W. Zhu, J. Tang, W. H. Tang and F. J. Zhang, *Prog. Polym. Sci.*, 2012, **37**, 1292–1331.
4. C. B. Nielsen, M. Turbiez and I. McCulloch, *Adv. Mater.*, 2013, **25**, 1859–1880.

5. P. Sista, M. C. Biewer and M. C. Stefan, *Macromol. Rapid Commun.*, 2012, **33**, 9–20.
6. I. McCulloch, R. S. Ashraf, L. Biniek, H. Bronstein, C. Combe, J. E. Donaghey, D. I. James, C. B. Nielsen, B. C. Schroeder and W. M. Zhang, *Acc. Chem. Res.*, 2012, **45**, 714–722.
7. L. J. Huo and J. H. Hou, *Polym. Chem.*, 2011, **2**, 2453–2461.
8. Y. Y. Liang and L. P. Yu, *Acc. Chem. Res.*, 2010, **43**, 1227–1236.
9. J. W. Chen and Y. Cao, *Acc. Chem. Res.*, 2009, **42**, 1709–1718.
10. O. Inganas, F. L. Zhang and M. R. Andersson, *Acc. Chem. Res.*, 2009, **42**, 1731–1739.
11. N. Blouin and M. Leclerc, *Acc. Chem. Res.*, 2008, **41**, 1110–1119.
12. Y. Ikenoue, F. Wudl and A. J. Heeger, *Synth. Met.*, 1991, **40**, 1–12.
13. E. E. Havinga, W. ten Hoeve and H. Wynberg, *Synth. Met.*, 1993, **55**, 299–306.
14. H. X. Zhou, L. Q. Yang, S. Stoneking and W. You, *ACS Appl. Mater. Interfaces*, 2010, **2**, 1377–1383.
15. B. G. Kim, X. Ma, C. Chen, Y. Ie, E. W. Coir, H. Hashemi, Y. Aso, P. F. Green, J. Kieffer and J. Kim, *Adv. Funct. Mater.*, 2013, **23**, 439–445.
16. P. M. Beaujuge, C. M. Amb and J. R. Reynolds, *Acc. Chem. Res.*, 2010, **43**, 1396–1407.
17. T. Y. Chu, J. P. Lu, S. Beaupre, Y. G. Zhang, J. R. Pouliot, S. Wakim, J. Y. Zhou, M. Leclerc, Z. Li, J. F. Ding and Y. Tao, *J. Am. Chem. Soc.*, 2011, **133**, 4250–4253.
18. S. H. Park, A. Roy, S. Beaupre, S. Cho, N. Coates, J. S. Moon, D. Moses, M. Leclerc, K. Lee and A. J. Heeger, *Nat. Photonics*, 2009, **3**, 297–302.
19. H. X. Zhou, L. Q. Yang, A. C. Stuart, S. C. Price, S. B. Liu and W. You, *Angew. Chem., Int. Ed.*, 2011, **50**, 2995–2998.
20. Y. M. Sun, C. J. Takacs, S. R. Cowan, J. H. Seo, X. Gong, A. Roy and A. J. Heeger, *Adv. Mater.*, 2011, **23**, 2226–2230.
21. S. C. Price, A. C. Stuart, L. Q. Yang, H. X. Zhou and W. You, *J. Am. Chem. Soc.*, 2011, **133**, 4625–4631.
22. H. L. Pan, Y. N. Li, Y. L. Wu, P. Liu, B. S. Ong, S. P. Zhu and G. Xu, *Chem. Mater.*, 2006, **18**, 3237–3241.
23. J. S. Ha, K. H. Kim and D. H. Choi, *J. Am. Chem. Soc.*, 2011, **133**, 10364–10367.
24. Y. N. Li, S. P. Singh and P. Sonar, *Adv. Mater.*, 2010, **22**, 4862–4866.
25. H. N. Tsao, D. M. Cho, I. Park, M. R. Hansen, A. Mavrinskiy, D. Y. Yoon, R. Graf, W. Pisula, H. W. Spiess and K. Mullen, *J. Am. Chem. Soc.*, 2011, **133**, 2605–2612.
26. N. Miyaura, K. Yamada and A. Suzuki, *Tetrahedron Lett.*, 1979, **20**, 3437–3440.
27. N. Miyaura and A. Suzuki, *Chem. Rev.*, 1995, **95**, 2457–2483.
28. J. Kristensen, M. Lysen, P. Vedso and M. Begtrup, *Org. Lett.*, 2001, **3**, 1435–1437.
29. G. A. Molander, S. L. J. Trice and S. D. Dreher, *J. Am. Chem. Soc.*, 2010, **132**, 17701–17703.

30. L. T. Pilarski and K. J. Szabo, *Angew. Chem., Int. Ed.*, 2011, **50**, 8230–8232.
31. D. Milstein and J. K. Stille, *J. Am. Chem. Soc.*, 1978, **100**, 3636–3638.
32. M. Kosugi, K. Sasazawa, Y. Shimizu and T. Migita, *Chem. Lett.*, 1977, **6**, 301–302.
33. Z. N. Bao, W. K. Chan and L. P. Yu, *J. Am. Chem. Soc.*, 1995, **117**, 12426–12435.
34. B. Carsten, F. He, H. J. Son, T. Xu and L. P. Yu, *Chem. Rev.*, 2011, **111**, 1493–1528.
35. W. J. Scott and J. K. Stille, *J. Am. Chem. Soc.*, 1986, **108**, 3033–3040.
36. V. Farina and B. Krishnan, *J. Am. Chem. Soc.*, 1991, **113**, 9585–9595.
37. J. S. Moore, *Makromol. Chem., Rapid Commun.*, 1992, **13**, 91–96.
38. Y. Ishii, *Ann. N.Y. Acad. Sci.*, 1974, **239**, 114–128.
39. U. Asawapirom, R. Guntner, M. Forster, T. Farrell and U. Scherf, *Synthesis*, 2002, 1136–1142.
40. J. Sakamoto, M. Rehahn, G. Wegner and A. D. Schluter, *Macromol. Rapid Commun.*, 2009, **30**, 653–687.
41. M. Zhang, H. N. Tsao, W. Pisula, C. D. Yang, A. K. Mishra and K. Mullen, *J. Am. Chem. Soc.*, 2007, **129**, 3472–3473.
42. X. Guo, S. R. Puniredd, M. Baumgarten, W. Pisula and K. Mullen, *J. Am. Chem. Soc.*, 2012, **134**, 8404–8407.
43. D. A. Widdowson and R. E. Wilhelm, *Chem. Commun.*, 1999, 2211–2212.
44. G. Odian, *Principles of Polymerization*, Wiley, Hoboken, 4th edn, 2004, Chapter 1, 19.
45. P. Schilinsky, U. Asawapirom, U. Scherf, M. Biele and C. J. Brabec, *Chem. Mater.*, 2005, **17**, 2175–2180.
46. C. Goh, R. J. Kline, M. D. McGehee, E. N. Kadnikova and J. M. J. Frechet, *Appl. Phys. Lett.*, 2005, **86**.
47. R. J. Kline, M. D. McGehee, E. N. Kadnikova, J. S. Liu, J. M. J. Frechet and M. F. Toney, *Macromolecules*, 2005, **38**, 3312–3319.
48. R. C. Coffin, J. Peet, J. Rogers and G. C. Bazan, *Nat. Chem.*, 2009, **1**, 657–661.
49. W. Ma, J. Y. Kim, K. Lee and A. J. Heeger, *Macromol. Rapid Commun.*, 2007, **28**, 1776–1780.
50. C. Muller, E. G. Wang, L. M. Andersson, K. Tvingstedt, Y. Zhou, M. R. Andersson and O. Inganas, *Adv. Funct. Mater.*, 2010, **20**, 2124–2131.
51. T. Y. Chu, J. P. Lu, S. Beaupre, Y. G. Zhang, J. R. Pouliot, J. Y. Zhou, A. Najari, M. Leclerc and Y. Tao, *Adv. Funct. Mater.*, 2012, **22**, 2345–2351.
52. J. K. Park, J. Jo, J. H. Seo, J. S. Moon, Y. D. Park, K. Lee, A. J. Heeger and G. C. Bazan, *Adv. Mater.*, 2011, **23**, 2430–2435.
53. P. Berrouard, A. Najari, A. Pron, D. Gendron, P. O. Morin, J. R. Pouliot, J. Veilleux and M. Leclerc, *Angew. Chem., Int. Ed.*, 2012, **51**, 2068–2071.
54. H. H. Zhang, C. H. Xing and Q. S. Hu, *J. Am. Chem. Soc.*, 2012, **134**, 13156–13159.

55. A. Yokoyama, H. Suzuki, Y. Kubota, K. Ohuchi, H. Higashimura and T. Yokozawa, *J. Am. Chem. Soc.*, 2007, **129**, 7236–7237.
56. T. Beryozkina, K. Boyko, N. Khanduyeva, V. Senkovskyy, M. Horecha, U. Oertel, F. Simon, M. Stamm and A. Kiriy, *Angew. Chem., Int. Ed.*, 2009, **48**, 2695–2698.
57. E. Elmalem, F. Biedermann, K. Johnson, R. H. Friend and W. T. Huck, *J. Am. Chem. Soc.*, 2012, **134**, 17769–17777.
58. E. Elmalem, A. Kiriy and W. T. S. Huck, *Macromolecules*, 2011, **44**, 9057–9061.
59. F. Brisse, G. Durocher, S. Gauthier, D. Gravel, R. Marques, C. Vergelati and B. Zelent, *J. Am. Chem. Soc.*, 1986, **108**, 6579–6586.
60. Y. Koizumi, S. Seki, A. Acharya, A. Saeki and S. Tagawa, *Chem. Lett.*, 2004, **33**, 1290–1291.
61. S. Maruyama, H. Hokari, T. Wada and H. Sasabe, *Synthesis*, 2001, 1794–1799.
62. J. Z. Brzezinski and J. R. Reynolds, *Synthesis*, 2002, 1053–1056.
63. D. W. Slocum and P. L. Gierer, *J. Org. Chem.*, 1976, **41**, 3668–3673.
64. T. M. Pappenfus, D. T. Seidenkranz and E. W. Reinheimer, *Heterocycles*, 2012, **85**, 355–364.
65. L. J. Huo, J. H. Hou, S. Q. Zhang, H. Y. Chen and Y. Yang, *Angew. Chem., Int. Ed.*, 2010, **49**, 1500–1503.
66. P. Beimling and G. Kossmehl, *Chem. Ber.*, 1986, **119**, 3198–3203.
67. A. J. Moule, A. Tsami, T. W. Buennagel, M. Forster, N. M. Kronenberg, M. Scharber, M. Koppe, M. Morana, C. J. Brabec, K. Meerholz and U. Scherf, *Chem. Mater.*, 2008, **20**, 4045–4050.
68. Y. Y. Liang, D. Q. Feng, Y. Wu, S. T. Tsai, G. Li, C. Ray and L. P. Yu, *J. Am. Chem. Soc.*, 2009, **131**, 7792–7799.
69. Y. Y. Liang, Z. Xu, J. B. Xia, S. T. Tsai, Y. Wu, G. Li, C. Ray and L. P. Yu, *Adv. Mater.*, 2010, **22**, E135–E138.
70. Y. P. Zou, A. Najari, P. Berrouard, S. Beaupre, B. R. Aich, Y. Tao and M. Leclerc, *J. Am. Chem. Soc.*, 2010, **132**, 5330–5331.
71. M. S. Su, C. Y. Kuo, M. C. Yuan, U. S. Jeng, C. J. Su and K. H. Wei, *Adv. Mater.*, 2011, **23**, 3315–3319.
72. C. M. Amb, S. Chen, K. R. Graham, J. Subbiah, C. E. Small, F. So and J. R. Reynolds, *J. Am. Chem. Soc.*, 2011, **133**, 10062–10065.
73. C. E. Small, S. Chen, J. Subbiah, C. M. Amb, S. W. Tsang, T. H. Lai, J. R. Reynolds and F. So, *Nat. Photonics*, 2012, **6**, 115–120.
74. L. J. Huo, J. H. Hou, H. Y. Chen, S. Q. Zhang, Y. Jiang, T. L. Chen and Y. Yang, *Macromolecules*, 2009, **42**, 6564–6571.
75. A. B. Tamayo, X. D. Dang, B. Walker, J. Seo, T. Kent and T. Q. Nguyen, *Appl. Phys. Lett.*, 2009, **94**, 103301.
76. S. Loser, C. J. Bruns, H. Miyauchi, R. P. Ortiz, A. Facchetti, S. I. Stupp and T. J. Marks, *J. Am. Chem. Soc.*, 2011, **133**, 8142–8145.
77. O. P. Lee, A. T. Yiu, P. M. Beaujuge, C. H. Woo and T. W. Holcombe, J. E. Millstone, J. D. Douglas, M. S. Chen and J. M. J. Frechet, *Adv. Mater.*, 2011, **23**, 5359–5363.

78. H. Bronstein, Z. Y. Chen, R. S. Ashraf, W. M. Zhang, J. P. Du, J. R. Durrant, P. S. Tuladhar, K. Song, S. E. Watkins, Y. Geerts, M. M. Wienk, R. A. J. Janssen, T. Anthopoulos, H. Sirringhaus, M. Heeney and I. McCulloch, *J. Am. Chem. Soc.*, 2011, **133**, 3272–3275.
79. J. G. Mei, K. R. Graham, R. Stalder and J. R. Reynolds, *Org. Lett.*, 2010, **12**, 660–663.
80. I. Osaka, M. Saito, H. Mori, T. Koganezawa and K. Takimiya, *Adv. Mater.*, 2012, **24**, 425–430.
81. S. S. Zalesskiy and V. P. Ananikov, *Organometallics*, 2012, **31**, 2302–2309.
82. R. Hoogenboom and U. S. Schubert, *Macromol. Rapid Commun.*, 2007, **28**, 368–386.
83. K. Kempe, C. R. Becer and U. S. Schubert, *Macromolecules*, 2011, **44**, 5825–5842.
84. B. C. Schroeder, Z. G. Huang, R. S. Ashraf, J. Smith, P. D'Angelo, S. E. Watkins, T. D. Anthopoulos, J. R. Durrant and I. McCulloch, *Adv. Funct. Mater.*, 2012, **22**, 1663–1670.
85. K. T. Nielsen, K. Bechgaard and F. C. Krebs, *Macromolecules*, 2005, **38**, 658–659.
86. N. Blouin, A. Michaud and M. Leclerc, *Adv. Mater.*, 2007, **19**, 2295–2300.

CHAPTER 16

Conjugated Polyelectrolytes

ANAND PARTHASARATHY, XUZHI ZHU AND KIRK S. SCHANZE*

University of Florida, Center for Macromolecular Science and Engineering, Gainesville, FL- 32611-7200, USA
*Email: kschanze@chem.ufl.edu

16.1 Introduction

Conjugated polyelectrolytes (CPEs) have been the focus of considerable research since the report in 1999 of their ability to serve as a fluorescent transducer for an avidin-biotin recognition biosensor.[1] Since this early study that focused on the application of an anionic sulfonated poly(phenylene vinylene) (PPV), there have been a considerable number of other CPE systems that have been synthesized, characterized and used in studies of materials and/or biosensor applications.[2–6] Conjugated polyelectrolytes feature two primary structural characteristics: 1) a π-conjugated backbone that imparts the polymer with its optical and electronic characteristics; 2) pendant ionic solubilizing groups, such as sulfonate ($-SO_3^-$), carboxylate ($-CO_2^-$) or tetraalkylammonium ($-NR_3^+$), which provide the polymers with solubility in polar solvents, including water (Charts 16.1 and 16.2). In addition, the interplay between the hydrophilic and polar ionic side groups and the hydrophobic conjugated polymer backbone affords CPEs with amphiphilic character, and many studies have explored the self-assembly of CPEs in solution to form nanoscale colloidal aggregates (polymer micelles) as well as at solid/liquid interfaces to form self-assembled nanostructured layer-by-layer (LbL) films.[2]

The purpose of this chapter is to provide detailed synthetic procedures for preparation of important categories of conjugated polyelectrolytes. In the

RSC Polymer Chemistry Series No. 9
Conjugated Polymers: A Practical Guide to Synthesis
Edited by Klaus Müllen, John R. Reynolds and Toshio Masuda

Published by the Royal Society of Chemistry, www.rsc.org

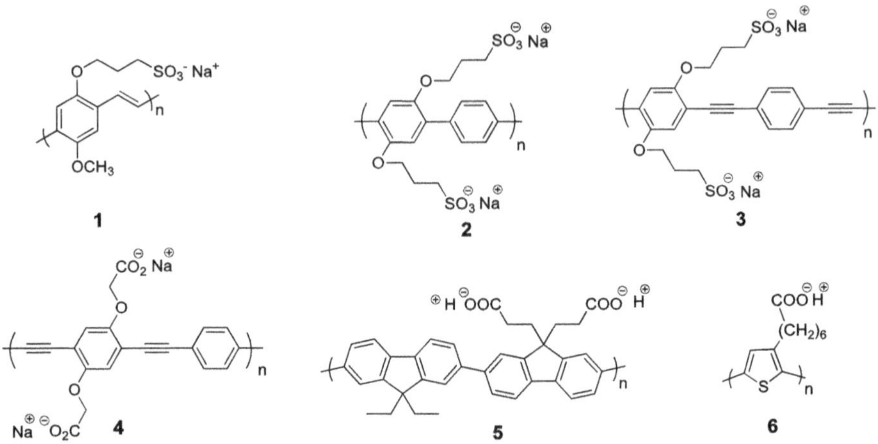

Chart 16.1

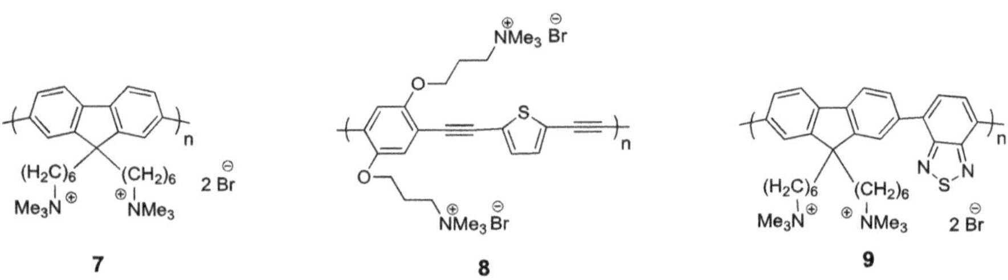

Chart 16.2

introduction, we will provide a brief overview of the history and important applications of CPEs, followed by an overview of the primary methods by which this important class of conjugated polymers can be synthesized. The chapter concludes with a set of detailed procedures for synthesis of specific CPE structures. The interested reader may wish to refer to a number of recent comprehensive review articles that describe CPE properties and applications studies.[2–7]

16.1.1 A Brief History of Conjugated Polyelectrolytes

Shortly after discovery of methods for preparation of organic soluble derivatives of poly(phenylene vinylene) and polythiophene, Wudl and coworkers reported preparation of each of these polymers with sulfonated side groups.[8,9] The specific objective of this work was based on the idea that the anionic sulfonate groups could make the polymers "self-doping", wherein the covalently linked anionic units could serve to provide charge compensation when the polymer backbone is in the oxidized (semiconducting) state. These first studies focused on the synthetic monomer and polymer chemistry. However, the authors reported in their papers that their newly synthesized conjugated polyelectrolytes were water soluble, and in the case of the PPV derivative, the solutions exhibited weak fluorescence characteristic of the PPV backbone. Over the next few years, there were a few other reports describing the synthesis of water-soluble conjugated polyelectrolytes, most notably poly(para-phenylene) derivatives prepared by metal-catalyzed coupling reactions.[10,11]

In 1999, Whitten and coworkers reported that the anionic PPV-type CPE (**1**, Chart 16.1) gave rise to the unusual property of fluorescence "superquenching" by an oppositely charged electron-acceptor quencher ion (dimethylviologen).[1] They further used the superquenching phenomenon to develop a fluorescent biosensor that relied on the avidin–biotin interaction to give rise to a fluorescence signal from CPE **1**. By taking advantage of the superquenching interaction between a covalently linked viologen–biotin quencher ligand and **1** they were able to detect avidin in solutions with low nanomolar concentration. This interesting and important study opened up a floodgate of research activity in the area of biosensor applications of CPEs, and a number of groups rapidly reported the development of fluorescent and colorimetric CPE-based assays for proteins, enzymes, nucleic acids and a variety of small molecules and metal ions. In parallel, a number of groups also began to explore the use of CPEs in optoelectronic device applications, where the ionic and surface active properties of CPEs, combined with the fluorescent and semiconductive properties of the backbone were leveraged.[2–3]

16.1.2 Synthesis of Conjugated Polyelectrolytes – General Considerations

Conjugated polyelectrolytes are generally prepared by the same methods used to synthesize their nonionic congeners. For example, poly(para-phenylene) and poly(phenylene ethynylene) based CPEs are prepared by Pd-catalyzed coupling (*e.g.*, Suzuki, Stille or Sonagashira) reactions, while polythiophene CPEs are prepared *via* oxidative polymerization, Negishi coupling or Grignard metathesis polymerization. The important distinction in CPE synthesis is the timing and method for incorporation of the ionic units. In general, two different broad approaches have been used in the literature. The first, which we refer to as the "direct approach" involves the direct polymerization of ionic monomers to afford the conjugated polyelectrolyte, eqn (16.1).

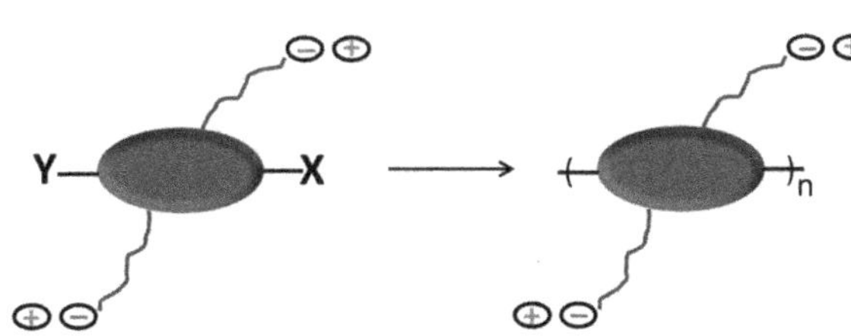

(16.1)

This method has some advantages, in that the polyelectrolyte is prepared directly from the ionic monomers. However, a disadvantage of the approach is that it is much more difficult to apply gel permeation chromatography (GPC) to determine the relative molecular weight of the resulting CPE sample. GPC molecular weight determination of water soluble, amphiphilic polymers is difficult to do because of the need for special columns and instrumentation compatible with the aqueous mobile phase. In addition, the tendency of CPEs to aggregate in aqueous solution can further complicate the molecular weight and polydispersity analysis.

A second approach which has been widely used we refer to as the "precursor approach", eqn (16.2).

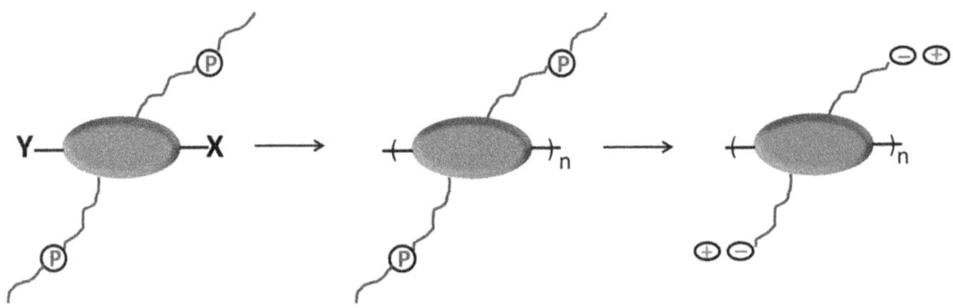

(16.2)

In this approach the monomers used in the polymerization are uncharged because the ionic units are protected or masked (*e.g.*, as esters or tertiary amines); thus, polymerization leads to a precursor conjugated polymer that is uncharged. The ionic groups are unmasked in a subsequent reaction that is typically carried out under mild conditions and in quantitative yield to afford the CPE (*e.g.*, base hydrolysis of the ester, or quaternization of the tertiary amine by methyl iodide). Despite the extra synthetic step, the precursor approach has several distinct advantages. First, the precursor conjugated polymer is uncharged and soluble in organic solvents, so its relative molecular weight and polydispersity can be obtained by standard, organic-phase GPC. Secondly, because of the good solubility of the uncharged precursor in organic solvents, high-resolution ^{1}H and ^{13}C NMR spectroscopy can be applied for structural and end-group analysis of the polymer. Experience shows that NMR spectroscopy of CPEs in aqueous (or polar organic) solvents typically gives rise to very broad lines, due to the slow rotational and translational diffusion of the chains in the viscous solvents (and also due to possible chain aggregation).

16.1.3 Conjugated Polyelectrolyte Synthesis by the Direct Approach

Most early studies of conjugated polyelectrolytes and their synthesis relied on preparations that followed the direct method. Here we provide some summary examples of polymers that are prepared by the direct route, with a brief discussion of the properties of the resulting polymers.

In 1994, Child and Reynolds reported synthesis of the anionic, sulfonate-substituted poly(para-phenylene) CPE **2** *via* an AA/BB polymerization

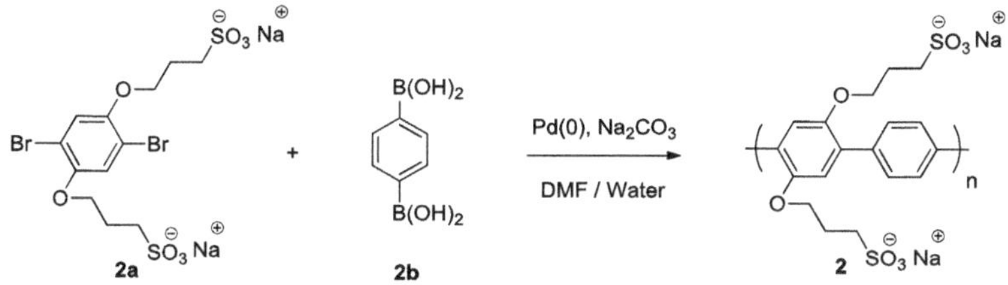

Scheme 16.1

using palladium-catalyzed Suzuki coupling chemistry.[10] Thus, by reacting an anionic, bis-sulfonate substituted 1,4-dibromobenzene with 1,4-benzenediboronic acid with a Pd(0) catalyst under in basic aqueous/DMF solution **2** was obtained (Scheme 16.1).

The resulting polymer was purified by dialysis against water using a 3500 g mol^{-1} cutoff membrane. As noted above, GPC analysis of the polymer was not feasible due to its propensity to strongly adsorb to the column from the aqueous mobile phase. The UV absorption of the polymer was consistent with the poly(para-phenylene) backbone structure, exhibiting a maximum at 344 nm. In a later, more detailed study, Reynolds and coworkers[11] explored more carefully the conditions used for synthesis of **2**, and by using a 4-tert-butylbenzene boronic acid end-capping agent they were able to use ^{1}H NMR end-group analysis to show that samples of **2** with a degree of polymerization, X_n, up to 50 were obtained.

Pinto and coworkers were the first to report the synthesis of an anionic CPE featuring a poly(phenylene ethynylene) (PPE) backbone.[12,13] The PPE-type CPEs are of considerable interest because of their comparatively high fluorescence quantum yields, which make them suitable for application as fluorescent ion and biosensors. By using a direct approach, the anionic alternating copolymer **3** was prepared in a Sonogashira coupling reaction between a bis-sulfonate 1,4-diiodobenzene monomer and 1,4-diethynylbenzene. The reaction was carried out in aqueous/DMF solution, with a Pd(0)/Cu(I) catalyst system in the presence of an amine base (Scheme 16.2):

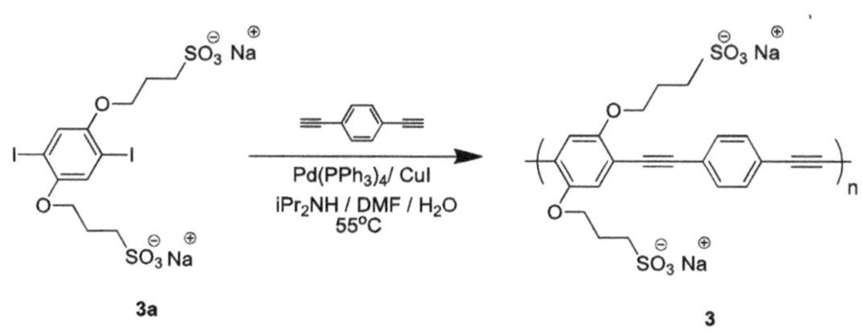

Scheme 16.2

The resulting anionic PPE **3** was purified by aqueous dialysis, and its structure elucidated by ^{1}H NMR spectroscopy. The relative molecular weight of a sample of **3** was determined by intrinsic viscometry, and it was found that

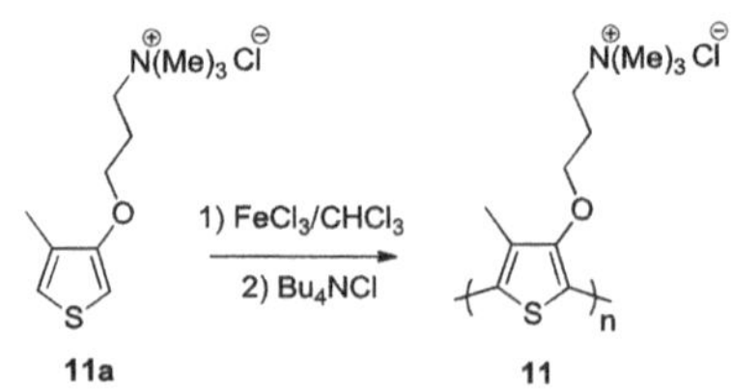

Scheme 16.3

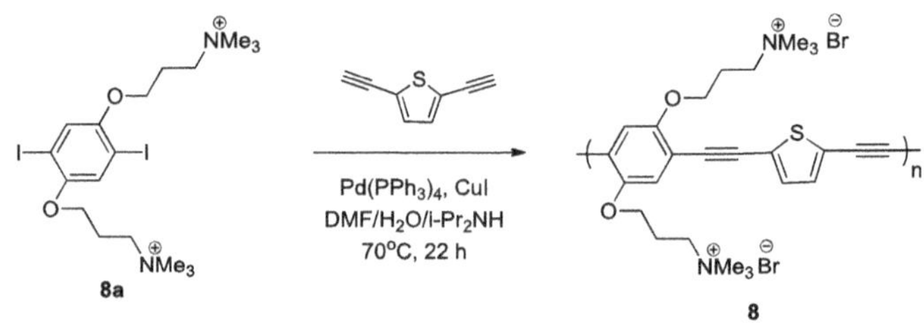

Scheme 16.4

the behavior of the polymer is consistent with an $X_n \sim 100$, corresponding to ~ 200 phenylene–ethynylene repeat units.[13]

Most cationic CPEs gain their charge due to the ammonium unit, R-NR_3^+. Several examples exist in the literature of ammonium-substituted CPEs that are prepared by the direct method.

Shinkai and coworkers reported cationic polythiophene **11** obtained by the direct oxidative polymerization of the tetraalkylammonium salt of a thiophene monomer (Scheme 16.3).[14] The molecular weight of the polymer was not determined; however, its absorption and fluorescence spectral properties were consistent with the polythiophene backbone structure. Schanze and coworkers reported the use of a direct polymerization method to prepare the cationic poly(arylene ethynylene) **8** that features a backbone that alternates thienylene-ethynylene and phenylene–ethynylene repeat units (Scheme 16.4).[15] Thus, polymerization of the cationic tetraalkylammonium substituted 1,4-diiodobenzene with 2,5-diethynylthiophene *via* a Sonogashira coupling protocol catalyzed by Pd(0)/Cu(I) catalyst in aqueous/DMF afforded **8**. The resulting polymer was purified by aqueous dialysis using an 8 kD molecular weight cutoff membrane. As with the previous examples, the molecular weight of the cationic CPE was not determined.

16.1.4 Conjugated Polyelectrolyte Synthesis by the Precursor Method

A variety of CPEs have been prepared by the precursor method, including those that feature polythiophene, polyfluorene and poly(phenylene ethynylene) backbones. In most cases the precursors feature ester (for anionic) and free amine (for cationic) functionality that can be easily converted to the ionic form in a high yield follow-up reaction.

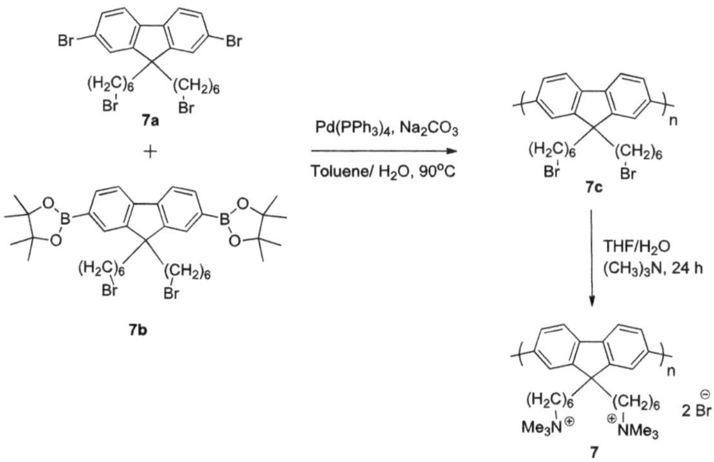

Scheme 16.5

Cationic polyfluorenes have been used in a variety of applications, including biosensors for DNA sequence detection and in optoelectronic devices, for example to serve as an electron-injection/transporting (ETL) layer in polymer light-emitting diodes.[3] A precursor approach is typically used in the preparation of cationic polyfluorenes, where as shown in Scheme 16.5 neutral fluorene monomers are subjected to Suzuki polymerization to afford a precursor polymer. The neutral precursor polymer **7c** is soluble in organic solvents (THF, $CHCl_3$) and therefore can be characterized by GPC and high-resolution NMR. Following the polymerization, the CPE is produced by in a high yield S_N2 reaction by treatment of the dibromohexyl polyfluorene **7c** with a solution of trimethylamine in THF.

As noted above, poly(phenylene ethynylene)s have been widely investigated as fluorescent sensors. The precursor approach has been used to synthesize a number of anionic PPEs that feature carboxylate pendant groups. A prototypical example of the use of the precursor approach to synthesize PPE-type CPEs is provided by the synthesis of the carboxylate-substituted alternating copolymer, **4**. Thus, as shown in Scheme 16.6, **4** is prepared by Sonogashira polymerization of an ester protected 1,4-diiodobenzene monomer **4a** with 1,4-diethynyl benzene to afford precursor polymer **4b** in which the carboxyl groups are protected as dodecyl esters. Owing to the presence of the long-chain esters, precursor **4b** is very soluble in organic solvents, and it can be characterized by organic phase GPC; typical Sonogashira polymerizations give rise to samples with M_n>100 kD and PDI ~2.0. Subsequent saponification of the ester groups in **4b** by using hydroxide base in a mixed organic/aqueous solvent gives rise to conjugated polyelectrolyte **4**, which is finally purified by aqueous dialysis. In a detailed study, Zhao *et al.* reported the synthesis of a series of samples of **4** in which an end-capping strategy was used to control the degree of polymerization.[16] The series of end-capped

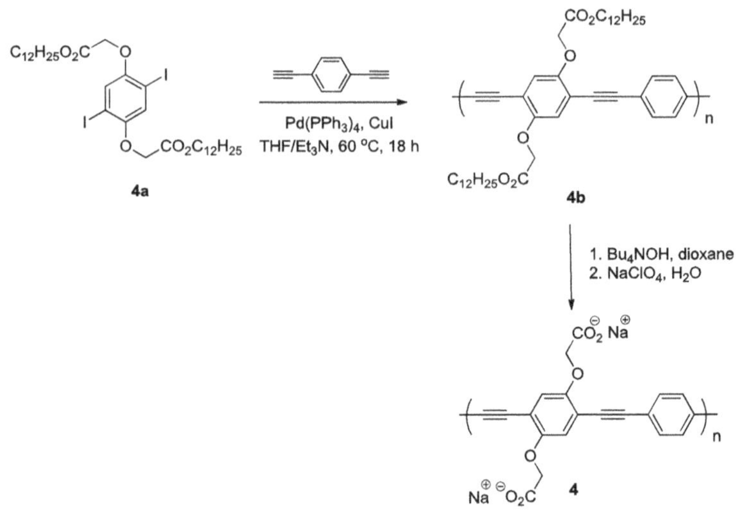

Scheme 16.6

precursor ester polymers **4b** were analyzed by GPC, and the molecular weight distribution obtained from the GPC analysis (using polystyrene standards) was compared to that obtained by analysis of ^{1}H NMR spectra using peaks characteristic of the end groups. This comparison showed that the GPC systematically overestimates the degree of polymerization by a factor of ~1.5.

Finally, it is important to note that an unusual feature of poly(phenylene ethynylene)-type CPEs such as **4** (or **2**) is that they can be difficult to redissolve if the solutions are taken to dryness without taking any special precaution. However, we have found empirically that solids of the CPEs obtained by lyophilization of the aqueous solutions are able to be redissolved.

16.2 Experimental Procedures

16.2.1 Materials and General Considerations

The monomers are synthesized from commercially available starting materials. The number of steps for synthesis varies depending on the structure and functionality that needs to be introduced. The most common methods of purifications include column chromatography, crystallization and distillation.

Palladium-catalyzed coupling reactions play a vital role in the synthesis of CPEs. Similar to most transition-metal-catalyzed reactions, it is important to eliminate oxygen by degassing the solution with argon or nitrogen. Typically, the polymerization of CPEs is carried out in polar solvents such as DMF, THF and MeOH; water can be used as a cosolvent to improve the solubility of the ionic monomers and the solubility of the polymers formed during the course of the polymerization. The polymerization is quenched by pouring the polymer solution over an appropriate solvent or solvent mixture that dissolves the unreacted monomers and low molecular weight polymers. Solvents such as acetone, methanol or solvent mixture comprising of methanol/acetone/ether are commonly used for precipitation of CPEs. The precipitated polymer is redissolved in a suitable solvent and reprecipitated several times to remove the starting materials and the low molecular weight polymers. Finally, the aqueous polymer solution is transferred to a dialysis membrane (choice depends on the molecular weight of the polymer) and dialyzed against Millipore water for 2 days to ensure the removal of oligomers and any inorganic byproducts formed during the reaction. The aqueous solution is then lyophilized to afford the solid polymer. Specific monomers are prepared according to citation shown in the particular section below.

16.2.2 Example Procedures

Example 1: Polymerization Procedure for Substituted PPV, 1[8]

To a solution of monomer[8] **1a** (1.075 g, 2 mmol) in 5 mL of MeOH is added 0.48 mL of 25% NaOMe in MeOH at 0 °C under nitrogen. Immediately, a viscous gum of **1b** is formed. After 15 min, the solvent is decanted and 8 mL of water and 2 mL of DMF are added. The resulting mixture is heated to reflux for 5–10 min, which gives a clear, colorless solution. The solution is dialyzed against deionized water with Spectropor membrane tubing of 6000 MWCO for 24 h to remove the low molecular weight reactants and products. The obtained precursor polymer **1c** is cast into films from aqueous solution (Scheme 16.7).

An aqueous solution of the precursor polymer **1c** (0.2 N) is mixed with the same volume of ethylene glycol and excess of sodium methoxide in methanol. The resulting mixture is heated to the boiling point. Upon removal of all the

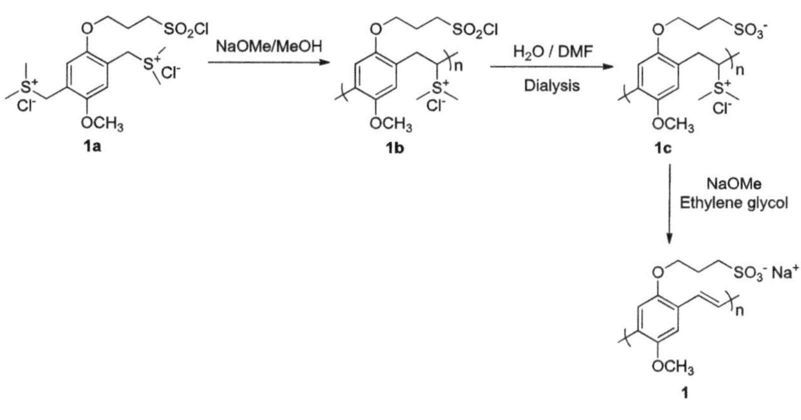

Scheme 16.7

water, the temperature of the solution is increased to 190 °C and the solution turned red. The solution is maintained at reflux for 26 h under nitrogen. Subsequently, the reaction mixture is cooled, and dialyzed against deionized water for 24 h to afford a red aqueous solution of substituted PPV, **1**. From this solution, a hygroscopic black film can be cast on a plastic weighing bowl at room temperature *in vacuo*.

Example 2: Polymerization Procedure for Poly[2,5-bis(3-sulfonatopropoxy)-1,4-phenylene-*alt*-1,4-phenylene, 2[10] (Refer to Scheme 16.1)

In an argon dry box, tris[(sulfonatophenyl)phosphine]palladium(0) trisodium salt[17,18] (45 mg, 0.079 mmol) is transferred to a Schlenk flask with a magnetic stir bar. In a separate flask, phenylenebis(boronic acid) **2b** (0.23 g,1.39 mmol), 1,4-dibromo-2,5-bis-(3-sulfonatopropoxy)benzene **2a** (0.78 g, 1.39 mmol) and sodium carbonate (0.99 g, 9.3 mmol) are dissolved in a mixture of 47 mL of doubly distilled water and 20 mL of DMF. This solution is heated to 85 °C to dissolve all the solids, degassed with an argon stream, and transferred *via* cannula to the flask containing the catalyst. The resulting solution is allowed to stir under argon at 85° C for 10 h and the reaction mixture is evaporated by boiling until the solution volume is approximately 25 mL. The black precipitate is removed by filtration, and the remaining liquid is gradually added to 250 mL of cold acetone. This results in the formation of a brown/tan precipitate that is collected by filtration. The polymer is dissolved in a minimum of hot distilled water, reprecipitated by cooling, and collected by filtration. Finally, the polymer is dissolved in distilled water, dialyzed for 72 h in a 3500 g mol^{-1} cutoff membrane and vacuum dried. The polymer has a good solubility in water, DMSO, aqueous base, dilute aqueous acid, and is slightly soluble in strong aqueous acid. The polymer is insoluble in acetone, ethanol, methylene chloride, and acetonitrile. Yield 64%.

Example 3: Polymerization Procedure for Polythiophene Derivatives 6 and 10[19]

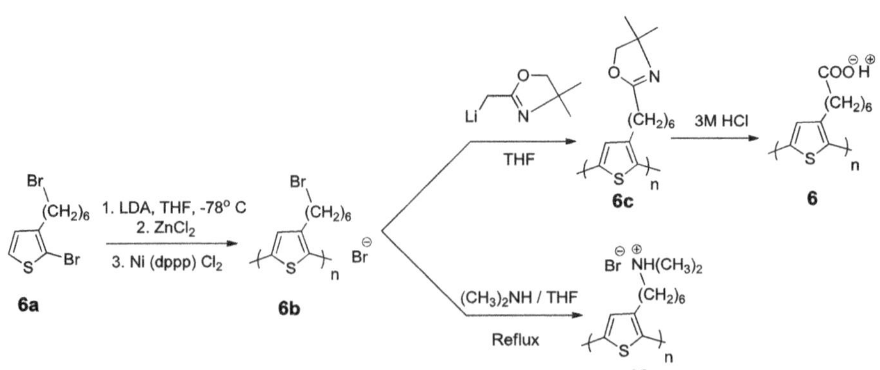

Scheme 16.8

Synthesis of HT-poly(3-bromohexylthiophene), 6b

Dry diisopropylamine (12 mmol) is dissolved in 25 mL of freshly distilled THF and the solution is cooled to –70 °C. To the above solution, *n*-BuLi (10 mmol) is added and the solution is stirred for 1.5 h. Subsequently, 2-bromo-3-hexylbromothiophene **6a** (3.26 g, 10 mmol) is added in one portion and the mixture is stirred for 2 h at –78 °C. To the above reaction mixture, $ZnCl_2$ (1.37 g, 10 mmol) is added in one portion and the reaction is stirred until the $ZnCl_2$ is dissolved. The reaction mixture is allowed to warm to –5 °C and Ni(dppp)Cl_2 (dppp = 1,3-diphenylphosphinopropane) (27 mg, 0.05 mmol) is added. The mixture is allowed to warm to room temperature, stirred for 20 min and quenched with methanol. The solid polymer is washed with methanol by using a Soxhlet extractor. The polymer is then dissolved by Soxhlet extraction with chloroform. Finally, chloroform is removed and the residue is dried under vacuum to yield **6b** (61% yield, $M_n = 14\ 630$ (GPC)) (Scheme 16.8).

Synthesis of 6c

A solution of 2,4,4-trimethyloxazoline (2.26 g, 20 mmol) in 30 mL dry THF is cooled to –70 °C and *n*-BuLi (18 mmol) is added dropwise. After stirring it for 2 h, a solution of **6b** (0.49 g, 2 mmol) in 30 mL dry THF is added *via* cannula and the addition is completed in 30 s. After removal of the cooling bath, the mixture is stirred for 1 h and quenched with hexane. The solid polymer **6c** is further purified by washing with hexane by using Soxhlet extractor (93% yield, $M_n = 16\ 610$ (GPC)).

Synthesis of 6

A sample of **6c** (278 mg, 1 mmol) is dissolved in 3M HCl (60 mL) in a 300-mL one-necked round-bottomed flask and refluxed for 12 h. The solid is filtered, rinsed with H_2O, and dried to afford the product **6** in 86% yield. $M_n = 15\ 200$ (GPC).

Synthesis of 10

To a solution of **6b** (490 mg, 2 mmol) in 30 mL dry THF is added dimethylamine in THF solution (30 mmol) and the mixture is stirred at 60 °C overnight. The mixture is dissolved in methanol/water (1 : 1 v/v) and K_2CO_3 is added. The precipitate is dissolved in chloroform and the solution is washed with water. Finally, chloroform was evaporated to yield the final polymer **10** (313 mg, 73% yield). $M_n = 15\ 630$ (GPC).

Example 4: Polymerization Procedure for Polythiophene Derivative, 11[14] (Refer to Scheme 16.3)

Anhydrous $FeCl_3$ (0.590 g, 3.64 mmol) is added to a solution of monomer **11a**[14] (255 mg, 0.91 mmol) in 25 mL of dry $CHCl_3$, under N_2 atmosphere. The resulting mixture is stirred at room temperature for 24 h. The solvent is removed under vacuum and the solid is washed with methanol. The crude product is filtered and further purified by Soxhlet extraction for 24 h against acetone. The insoluble fraction is dried and redissolved in methanol by adding a few drops of hydrazine. Subsequently, a saturated solution of tetrabutylammonium chloride in acetone is added to the polymer solution. The precipitate is collected by filtration and washed with acetone. Finally, the polymer is dried under reduced pressure.

Example 5: Polymerization Procedure for Polyfluorene Derivative, 9[20]

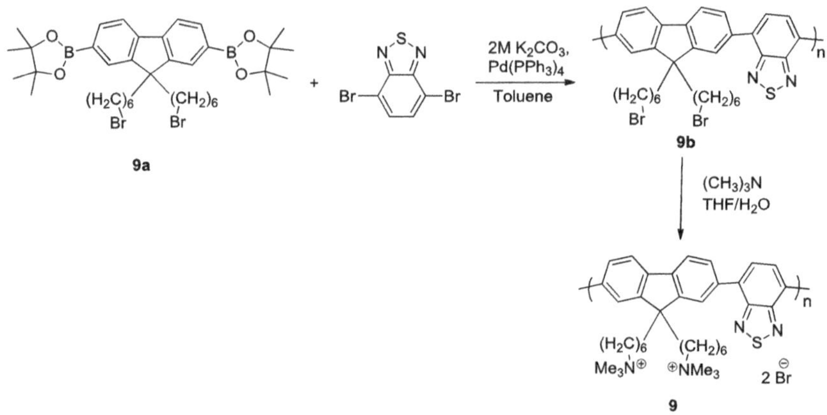

Scheme 16.9

Synthesis of 9

To a solution of 4,7-dibromo-2,1,3-benzothiadiazole (73.5 mg, 0.25 mmol) in a mixture of water (3 mL) and toluene (6 mL), potassium carbonate (828 mg, 6 mmol) and monomer **9a**[20] (182 mg, 0.25 mmol) are added. The mixture is degassed for 30 min and $Pd(PPh_3)_4$ (5 μmol) is added. The resulting mixture is heated at 85 °C for 20 h and then poured into a large amount of acetone. The precipitate is filtered, washed with methanol and acetone and then dried under vacuum to afford **9b** as a bright yellow powder. Yield 65%.

GPC (THF, polystyrene standard): $M_n = 12\ 000$; PDI 1.7.

To a solution of the precursor polymer **9b** (70 mg, 1.1 mmol) in THF (10 mL) at –78 °C, trimethylamine (2 mL) is added dropwise. The mixture is allowed to warm to room temperature and stirred for 24 h. After the removal of most of the solvent, the residue is poured into acetone to afford **9** as a light brown fiber. Yield 89% (Scheme 16.9).

Example 6: Polymerization Procedure for Polyfluorene Derivative, 5[21]

To a solution of monomers **5a**[21] (474 mg, 1 mmol) and **5b**[21] (580 mg, 1 mmol) in dimethoxyethane (10 mL), tetrabutylammonium fluoride (4 mmol) is added. The flask is subjected to five vacuum–argon cyclings and $Pd(PPh_3)_4$ (0.01 mmol) is added. The resulting mixture is heated at 100 °C for 48 h. The mixture is then cooled to room temperature and poured over a large amount of

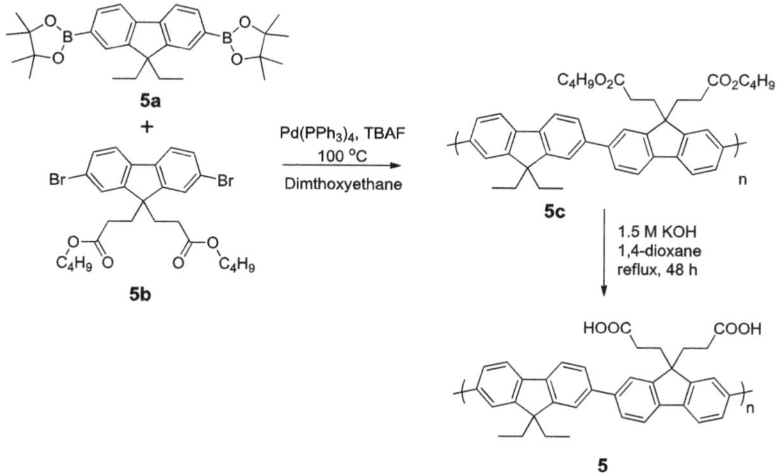

Scheme 16.10

methanol to precipitate the polymer, **5c**. GPC (THF, polystyrene standard): $M_n = 12\,700$; PDI 1.8.

For hydrolysis, the precursor polymer **5c** (6.2 mg, 0.97 mmol) is dissolved in 1,4-dioxane (4 mL) with 2 mL of 1.5 M KOH solution and the mixture is heated at 95 °C overnight. The solvent is removed under vacuum and 1 M HCl solution is added to the aqueous solution until the pH = 4. The precipitate is collected by filtration, washed with copious amounts of water and dried under vacuum to afford **5** (Scheme 16.10).

Example 7: Polymerization Procedure for Polyfluorene Derivative, 7[22] (Refer to Scheme 16.5)

In a 25-mL round-bottomed flask, 9,9-bis(6-bromohexyl)-2,7-dibromofluorene **7a**[22] (325 mg, 0.5 mmol), 2,2′-(9,9-bis(6-bromohexyl)-9H-fluorene-2,7-diyl)-bis(4,4,5,5-tetramethyl-1,3,2-dioxaborolane) **7b**[22] (372 mg, 0.5 mmol) and potassium carbonate (6 mmol) are placed. A mixture of toluene (5 mL) and water (3 mL) is added to the flask. The resulting mixture is degassed for 30 min before $Pd(PPh_3)_4$ (5 μmol) is added. The mixture is stirred at 85 °C for 24 h and then poured into a large amount of methanol. The precipitate is filtered and washed with methanol and acetone. The polymer is dried under vacuum for 24 h to afford the neutral polymer **7c**. GPC (THF, polystyrene standard): $M_n = 33\,000$; PDI 1.8.

Cationic polymer 7. To a solution of precursor polymer **7c** (50 mg) in THF (10 mL) at −78 °C, trimethylamine (2 mL) is added dropwise. The precipitate is redissolved by adding 10 mL of water. After the mixture is cooled to −78 °C, additional trimethylamine (2 mL) is added and the resulting mixture is allowed to warm to room temperature and stirred for another 24 h. The solvent is removed under vacuum and the residue is poured into a large amount of acetone to afford a light yellow precipitate. The polymer is collected by filtration and dried under vacuum.

Example 8: Polymerization Procedure for PPE derivative, 3[12] (Refer to Scheme 16.2)

In a Schlenk flask, monomer **3a**[12] (1.008 g, 1.55 mmol) and 1,4-diethynylbenzene (0.189 g, 1.50 mmol) are dissolved in 40 mL of a 1 : 1 mixture of DMF and water at 60 °C under argon. The resulting clear solution is deoxygenated by vacuum/argon cycling (3 times). In a separate round-bottomed flask, $Pd(PPh_3)_4$ (45.0 μmol) and CuI (45 μmol) are suspended in a mixture of diisopropylamine/DMF (20 mL of 1 : 1 mixture). The mixture is deoxygenated as described above and subsequently added to the solution containing the monomers by means of a cannula. The final mixture is again deoxygenated by vacuum–argon cycling and is then warmed to 50–55 °C and stirred under a positive pressure of argon for 14 h. The resulting solution is viscous, brown in color and exhibits an intense blue fluorescence when illuminated with a near-UV lamp. The solution is cooled and then slowly added to 1 L of a methanol/acetone/ether mixture (10 : 40 : 50 v : v : v). The isolated polymer is dissolved in DMSO/water and reprecipitated from methanol/acetone/ether mixture (3 times). The polymer is then dissolved in 150 mL water and the solution is filtered through quantitative filter paper, followed by a 10–20-μm fritted glass filter, and finally through a 0.8-μm nylon membrane. Finally, 0.05 g of sodium cyanide is added, and the resulting solution is dialyzed against water (Millipore Nanopure) using a 6–8 kD MWCO cellulose membrane. Yield 68%.

Example 9: Polymerization Procedure for PPE derivative, 8[15] (Refer to Scheme 16.4)

A solution of 3,3′-[(2,5-diiodo-1,4-phenylene)bis(oxy)]bis(N,N,N-trimethylpropan-1-aminium) bromide **8a**[15] (108 mg, 0.15 mmol) in a mixture of DMF (4.5 mL), H_2O (3 mL) and $^iPr_2NH$ (1 mL) is deoxygenated with argon for 30 min. To this solution, CuI (0.02 mmol), $Pd(PPh_3)_4$ (0.01 mmol) and 2,5-diethynylthiophene (20 mg, 0.15 mmol) are added under argon. The resulting mixture is heated at 70 °C for 22 h and then poured into a large amount of acetone. The precipitated polymer is collected by filtration and redissolved in a small amount of water and treated with NaCN. The polymer solution is filtered through a 0.22 μm cellulose membrane and further purified by dialysis against deionized water using regenerated cellulose membrane (6–8 kD MWCO). Finally, the polymer solution is lyophilized to afford a yellow-tan solid. Yield 51%.

Example 10: Polymerization Procedure for PPE derivative, 4[23] (Refer to Scheme 16.6)

Synthesis of precursor polymer, 4b

A solution of 2,5-bis(dodecyloxycarbonylmethoxy)-1,4-diiodobenzene **4a**[23] (204 mg, 0.25 mmol) and 1,4-diethynylbenzene (32 mg, 0.25 mmol) in 20 mL of dry THF/Et_3N (v/v = 2/1) is degassed with argon for 40 min. Then $Pd(PPh_3)_4$ (8.7 mg, 7.5 μmol) and CuI (4 mg, 7.5 μmol) are added under argon.

The reaction is stirred at 60 °C for 18 h. The obtained viscous suspension is then poured into 150 mL of methanol. The light yellow fibers are collected by centrifugation. The polymer is further purified by two repeated cycles of dissolution in THF and precipitation into methanol. A small amount (~30 mg) of the polymer is dried completely under vacuum and used for NMR study immediately. The rest of the polymer is kept as a THF solution for the next step.

Hydrolysis. To a solution of **4b** (117 mg, 0.20 mmol) in 30 mL of dioxane/THF (5/1) is added 1.5 mL of 1 M $(n\text{-}Bu)_4NOH$ in methanol. The resulting mixture is stirred at room temperature for 24 h. During the hydrolysis reaction, 2 mL of water is systematically added in order to maintain a clear solution (without polymer precipitation). Then, a solution of 0.20 g of $NaClO_4$ (1.6 mmol) in 3 mL of water is added to the hydrolyzed polymer solution. The solution is allowed to stir for another 6 h and then poured into 400 mL of cold acetone, resulting in precipitation of **4** as a yellow powder. The polymer is then dissolved in 50 mL of deionized water (several drops of 1 N NaOH solution are added) and filtered through a 0.22-μm cellulose filter membrane (Millipore brand). After filtration the polymer is further purified by dialysis using a regenerated cellulose membrane (12 kD molecular weight cutoff) against water (pH = 10). The CPE is obtained as a light yellow solid after removal of water by lyophilization.

References

1. L. H. Chen, D. W. McBranch, H. L. Wang, R. Helgeson, F. Wudl and D. G. Whitten, *Proc. Natl. Acad. Sci. USA*, 1999, **96**, 12287–12292.
2. H. Jiang, P. Taranekar, J. R. Reynolds and K. S. Schanze, *Angew. Chem. Int. Ed.*, 2009, **48**, 4300–4316.
3. A. Duarte, K.-Y. Pu, B. Liu and G. C. Bazan, *Chem. Mater.*, 2011, **23**, 501–515.
4. S. W. Thomas, G. D. Joly and T. M. Swager, *Chem. Rev.*, 2007, **107**, 1339–1386.
5. M. R. Pinto and K. S. Schanze, *Synthesis*, 2002, 1293–1309.
6. Y. Liu, K. Ogawa and K. S. Schanze, *J. Photochem. Photobiol., C: Photochem. Rev.*, 2009, **10**, 173–190.
7. B. Liu and G. Bazan, ed., *Conjugated Polyelectrolytes*, Wiley-VCH, Weinheim, 2013.
8. S. Q. Shi and F. Wudl, *Macromolecules*, 1990, **23**, 2119–2124.
9. A. O. Patil, Y. Ikenoue, F. Wudl and A. J. Heeger, *J. Am. Chem. Soc.*, 1987, **109**, 1858–1859.
10. A. D. Child and J. R. Reynolds, *Macromolecules*, 1994, **27**, 1975–1977.
11. S. Kim, J. Jackiw, E. Robinson, K. S. Schanze, J. R. Reynolds, J. Baur, M. F. Rubner and D. Boils, *Macromolecules*, 1998, **31**, 964–974.
12. C. Y. Tan, M. R. Pinto and K. S. Schanze, *Chem. Commun.*, 2002, 446–447.
13. X. Zhao, M. R. Pinto, L. M. Hardison, J. Mwaura, J. Müller, H. Jiang, D. Witker, V. D. Kleiman, J. R. Reynolds and K. S. Schanze, *Macromolecules*, 2006, **39**, 6355–6366.

14. C. Li, M. Numata, A. H. Bae, K. Sakurai and S. Shinkai, *J. Am. Chem. Soc.*, 2005, **127**, 4548–4549.
15. S. Chemburu, E. Ji, Y. Casana, Y. Wu, T. Buranda, K. S. Schanze, G. P. Lopez and D. G. Whitten, *J. Phys. Chem. B*, 2008, **112**, 14492–14499.
16. X. Zhao, H. Jiang and K. S. Schanze, *Macromolecules*, 2008, **41**, 3422–3428.
17. T. Bartik, B. Bartik, B. E. Hanson, T. Glass and W. Bebout, *Inorg. Chem.*, 1992, **31**, 2667–2670.
18. D. R. Coulson, *Inorg. Synth.*, 1972, **13**, 121.
19. L. Zhai and R. D. McCullough, *Adv. Mater.*, 2002, **14**, 901–905.
20. B. Liu and G. C. Bazan, *Proc. Natl. Acad. Sci. USA*, 2005, **102**, 589–593.
21. R. N. Brookins, K. S. Schanze and J. R. Reynolds, *Macromolecules*, 2007, **40**, 3524–3526.
22. K. Y. Pu, Z. Fang and B. Liu, *Adv. Funct. Mater.*, 2008, **18**, 1321–1328.
23. H. Jiang, X. Y. Zhao and K. S. Schanze, *Langmuir*, 2006, **22**, 5541–5543.

CHAPTER 17

Self-Doped Polymers

M. RAMESH KUMAR AND MICHAEL S. FREUND*

Department of Chemistry, University of Manitoba, Winnipeg, Manitoba, R3T 2N2, Canada
*Email: michael.freund@ad.umanitoba.ca

17.1 Introduction

Advances in the synthesis of self-doped conducting polymers have been motivated by the need for increased solubility, electrochromic switching speeds and enhanced charge-storage performance in polymer-based batteries and supercapacitors. The first report of "self-doping" in conducting polymers was published by Wudl *et al.*[1,2] in 1987. In their reports, a conjugated polymer was described in which a significant fraction of monomer units contained covalently attached, negatively charged functional groups acting as a stable/immobile dopant ions. The sodium salts and acid forms of poly(3-thiophene ethanesulfonate) and poly(3-thiophene butanesulfonate) were the first self-doped conducting polymers reported to show water solubility in neutral (insulating) and doped (conducting) states.[2] Following these initial reports, there have been significant advances in the synthesis and properties of this class of polymers, and they have found use in a wide range of technologies.[3]

17.1.1 Self-Doped Conducting Polymers

The concept of doping in conducting polymers is significantly different from conventional doping in inorganic semiconductors such as silicon. For example,

RSC Polymer Chemistry Series No. 9
Conjugated Polymers: A Practical Guide to Synthesis
Edited by Klaus Müllen, John R. Reynolds and Toshio Masuda

Published by the Royal Society of Chemistry, www.rsc.org

conducting polymers are doped to a level of tens of parts-per-hundred as opposed to parts-per-thousand or less in the case of silicon. This high level of doping results in significant changes in the polymer structure and carrier interactions that produced new states in the bandgap of the polymer. The evolution of these new states produce significant changes in optical and electrical properties between the undoped and doped states.[4] Conjugated polymers are typically p-doped, where charge carriers are delocalized positive charges. While there are reports of n-doped forms of conjugated polymers, they are typically less stable due to the significantly negative potentials required to reduce the backbone. In the case of p-doped polymers, charge neutrality is maintained by the presence of anions in the doped state (*e.g.*, doped polythiophene, Chart 17.1). Typically, these anions are sufficiently small to be mobile within the bulk polymer and therefore move in and out of the polymer as the charge on the backbone changes either due to redox or acid–base chemistry.

In its simplest form, a self-doped conducting polymer contains a covalently bound ionic species, which acts as an immobile dopant (*e.g.*, self-doped polythiophene in Chart 17.1). Under these conditions, charge neutrality must be maintained by cations. In situations where the cations are small, their high mobility results in significantly faster redox switching speeds compared to larger anions. In cases where the ionic state of the polymer backbone is governed by acid–base chemistry (*e.g.*, polyaniline), the presence of a covalently linked strong acid will substantially increase the pKa of the oxidized polymer backbone, thereby maintaining the protonated conducting form of the polymer under higher pH conditions compared to regularly doped forms of the polymer.

The level of covalently attached ionizable functional groups can vary from complete, where every monomer unit is modified (100%), to a fraction of the level required for complete doping (<30–40%). Synthetically, this can be achieved by copolymerization with varying concentrations of both the modified and unmodified monomers or by controlling the conditions for post-polymerization modification. The resulting level of modification can be used to tune the solubility, molecular weight and pH-dependent conductivity of the resulting polymer. In cases where the level of modification is above that required for complete doping (>30–40%), the additional ionizable groups can increase the ionic conductivity of the polymer, further enhancing the performance of the material.

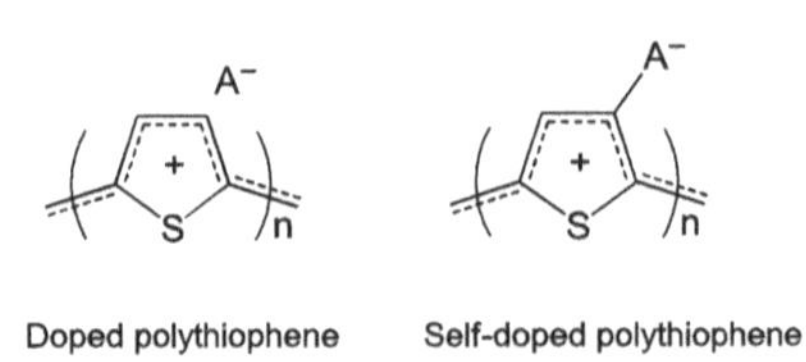

Chart 17.1

17.1.2 Types and Classes

Self-doped conducting polymers can be synthesized by functionalizing the monomer (prepolymerization modification) or the polymer (postpolymerization modification) with charged groups. These self-doped conducting polymers have been prepared with various repeat units including acetylene, pyrrole, aniline and thiophene, using various functional groups including sulfonic acid, alkyl and alkoxysulfonic acid, carboxylic acid, quaternary ammonium salts, phosphonic acid, acetic acid and boronic acid. For the most part, self-doped conducting polymer derivatives have been prepared either with ionizable functional groups (usually sulfonate) attached directly to the aromatic polymer backbone or bound through a spacer consisting of alkane or aromatic groups. Some examples of polyaniline derivatives are shown in Chart 17.2.

Pre- and postpolymerization modification can present significant challenges. For example, it is often difficult to control molecular weight, degree of modification and the distribution of monomeric forms within copolymers. For example, during copolymerization with modified and unmodified monomers,

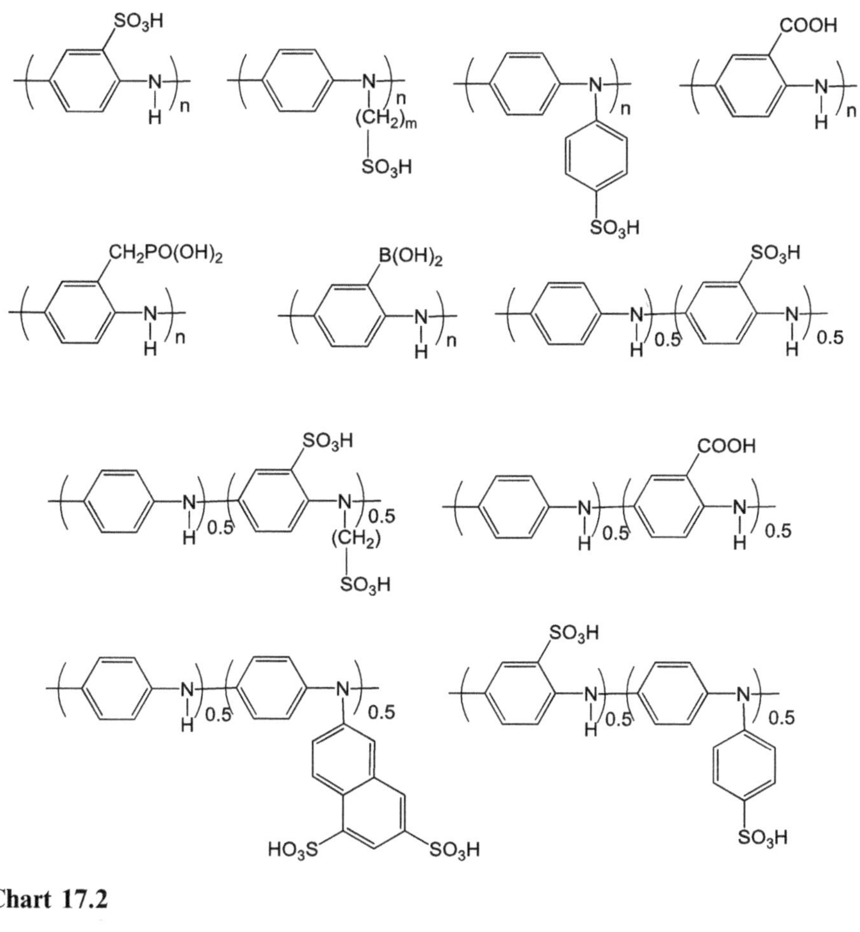

Chart 17.2

steric effects associated with added functional groups impact the coupling of radical cations. This leads to termination, resulting in lower molecular weight copolymers as well as nonstatistical distributions of the monomers in the copolymer produced. Depending on the harshness of the method used for post-polymerization modification, this approach can result in the lowering of molecular weight of the base polymer and variable degrees of self-doping. As a result, careful characterization of the polymers produced by any method is required to assure a reasonable level of confidence in the synthesis of self-doped polymers. Sometimes, postpolymerization modification needs the use of toxic and corrosive chemicals (*e.g.*, concentrated H_2SO_4 in sulfonation of polyanilines), which would make this method less suitable for scaling up. In addition, most of these postmodification methods are highly sensitive to process conditions, which leads to different backbone structures and random degrees of modification. On the other hand, the main drawbacks in prepolymerization modification are; that the substituted monomers are less reactive, which makes it hard for them to polymerize under normal conditions; if electron-withdrawing groups are present in the monomer as substituents, it will be challenging to polymerize it alone and in such cases these monomers need only to be copolymerized with its parent monomer or any other monomer that does not have any electron withdrawing group as substituent.[3,5,6]

17.1.3 Doping Mechanisms and Properties

Considerable work has been performed to characterize the doping mechanism of self-doped conducting polymers. In contrast to electrochemical cycling for most p-type conducting polymers, where anions move in and out of the bulk polymer to maintain charge neutrality, in the case of self-doped conducting polymers, cations must move. This process has been termed "cation popping" and has been established as a method for voltage-controlled cation exchange.[7]

As illustrated in Scheme 17.1, oxidation of the conjugated backbone results in expulsion of cations from the bulk polymer whereas reduction results in cations from the contacting electrolyte being driven into the bulk polymer. The substantially higher mobility of cations results in an order of magnitude faster redox switching speeds and better long-term stability.[8]

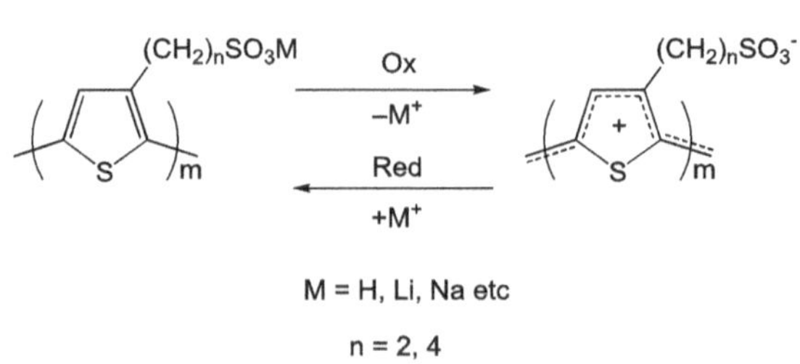

Scheme 17.1 Reproduced by permission of Elsevier (1987).[1]

In addition to positively impacting issues of dedoping of pH and thermally sensitive polymers such as polyaniline, self-doped forms enable the combination of p- and n-type conducting polymers without their spontaneous discharging by the concerted recombination of carriers and counterions. This has resulted in the formation of conjugated polymer p/n junctions, not possible with standard conjugated polymers.[9]

Self-doping can be further exploited by utilizing functional groups, where ionization is a result of complexation reactions. For example, boronic acid groups react reversibly with diols and Lewis-base donors to form boronates. As a result, this chemistry can be used to reversibly switch between self-doped and nonself-doped forms of conjugated molecules to impact solubility, reactivity and conductivity.[10]

17.2 Experimental Procedures

17.2.1 Common Synthetic Mechanisms and Outcomes

Self-doped conducting polymers are conjugated polymers, in which at least a fraction of monomer units contain covalently attached, ionizable functional groups that can act as immobile dopants. These polymers can be synthesized in a number of ways including chemical or electrochemical modification of the parent polymer as well as chemical or electrochemical polymerization of substituted monomers to form homo- or copolymers.[3,11]

17.2.1.1 Postpolymerization Functionalization

Though it is possible to obtain self-doped substituted polymers through oxidative polymerization of appropriate monomer, the presence of an electron-withdrawing dopant group (*e.g.*, sulfonate) on the monomer can make oxidation and subsequent polymerization difficult, typically resulting in a lower molecular weight and lower conductivity in the obtained polymer.[12] Therefore, modification of the backbone structure of the polymer has been exploited by many groups. For example, postpolymerization functionalization of polyaniline has been performed to synthesize self-doped polyanilines by adding various nucleophilies such as cyanide, carbanions, thiols, amines and sulfite under oxidative conditions.[13–16]

The very first report on the synthesis of self-doped, water-soluble polyaniline by Yue and Epstein[17] was based on direct electrophilic substitution of negatively charged sulfonic acid functional groups onto the backbone of polyaniline. The resulting polymer had nearly half of the aniline units sulfonated, and the presence of anionic SO_3^- groups makes the polymer soluble in water and in many organic solvents. In a similar way, Sotzing *et al.*[18] carried out ring sulfonation of poly(thieno[3,4-b]thiophene) using fuming sulfuric acid to synthesize a self-doped, low-bandgap, water-soluble sulfonated poly(thienothiophene). Though sulfonation of polyaniline is performed primarily through postpolymerization modification methods, it is rare in the case of other polymers such as polypyrrole.[19]

17.2.1.2 Chemical Polymerization of Monomer Derivatives

In order to overcome issues such as inductive and steric effects in polymerizing substituted monomers, less-reactive substituted monomers can be either copolymerized with unsubstituted monomers or homopolymerized under more controlled conditions.[13] In addition, it is important for the substitution to avoid locations that will impede polymer growth. For example, in the case of aniline, the substitution should only be in the meta and/or ortho positions and in the case of aromatic heterocyclics such as thiophene and pyrrole, substitution should only occur in the β position.[6]

Homopolymerization

Although postpolymerization methods are predominantly used for synthesizing sulfonated polyanilines, a major drawback is the lack of control over the level of sulfonation. Oxidative polymerization of substituted monomers can be used to increase the level of sulfonation. In this approach a relatively strong oxidizing agent is used to yield polymerizable radical cations in solution.[11] Efforts to directly polymerize aniline-3-sulfonic acid resulted in very low molecular weight polymer (MW <2000) with an unimpressive conductivity of 3×10^{-6} S cm^{-1}. However, adding an electron-donating group (-OCH_3) in the form of 2-methoxyaniline-5-sulfonic acid, and polymerizing under basic conditions (using 2.5 M aqueous pyridine) in the presence of ammonium persulfate oxidant significantly improves the reactivity and solubility of the monomer, resulting in a higher molecular weight polymer (10 000) with better conductivity (0.04 S cm^{-1}).[20] Similar to polyaniline, several sulfonate and carboxy-containing monomer derivatives of thiophene have also been chemically polymerized to form the homopolymer using various oxidizing agents.[21,22] However, chemical polymerization of sulfonated pyrrole is not commonly employed due to the problems with overoxidation.[3] Suzuki coupling has also been employed by many to synthesize a number of water-soluble, alkoxy-substituted,[23] carboxy-substituted,[24] in which every two phenylene ring has one carboxy group, and phenoxycarboxymethylene-substituted poly(p-phenylene)s.[25] Ring-opening metathesis polymerization (ROMP) of functionalized cyclo-octatetraenes (RCOT) is widely executed in order to synthesize self-doped, alkyl and phenyl-substituted polyacetylenes.[9,26–29]

Copolymerization

While there are reports of making homopolymers of self-doped polymers (*e.g.*, 2-methoxyaniline-5-sulfonic acid), many derivatized monomers do not polymerize alone to yield their respective homopolymers. For example, if the substituents are bulky (*e.g.*, 1-(2-amino)-pronane-3-sulfonic acid or 1-(2-amino)-butane-3-sulfonic acid), the homopolymerization is not possible.[30] In these cases, bulky substituent containing monomers need to be copolymerized with an unsubstituted monomer such as aniline, thiophene, *etc*. An advantage of performing a copolymerization is that it can introduce an additional variable (monomer ratios) through which the properties of the polymer including conductivity, solubility and molecular weight can be tuned. For example, Jo *et al.* have synthesized water-soluble, highly conductive, graft copolymers of polyaniline[31] and polypyrrole[32] with polystyrene sulfonic acid using ammonium persulfate as an oxidant.

17.2.1.3 Electrochemical Polymerization of Monomer Derivatives

Similar to chemical synthesis, homo- and copolymer forms of self-doped polymers can also be synthesized electrochemically using either aqueous or nonaqueous media. Typically, electropolymerization is carried out using a three-electrode configuration in the presence of a supporting electrolyte and monomer dissolved in an appropriate solvent. When sufficiently high positive potentials (anodic) are applied, oxidation of the monomer at the electrode surface results in electropolymerization. As cation radicals react to form oligomers and polymer they typically deposit in the form of a thin film on the electrode surface or dissolve in the electrolyte depending on its molecular weight and solubility.[11] Since deposited polymer is in an oxidized conducting state, polymerization continues at the surface of the electrodeposited polymer until the potential is stopped.

Homopolymerization

Fabre *et al.*[33] and Freund *et al.*[10,34,35] have homopolymerized aniline-boronic acid electrochemically in the presence of fluoride. Hur *et al.*[36] electrochemically deposited the homopolymer of sulfonated polypyrrole on stainless steel electrode plates. Highly conducting, soluble sulfonated homopolymers of aniline and thiophene have been polymerized electrochemically at high potentials (+1.70 V *vs.* Ag/AgCl) by Yildiz *et al.*[37] using FSO_3H as supporting electrolyte. They report conductivities for sulfonated polyaniline and polythiophene of 0.058 and 21 S cm^{-1} respectively. The higher conductivities are attributed to the presence of the SO_3^- counterions from the electrolyte in addition to the substituents. Many substituted monomers including metanilic acid[38] do not polymerize alone under electrochemical conditions. In these cases, typically only soluble oligomers form in the electrolyte during oxidation. However, the addition of very low concentrations of parent aniline in the electrolyte can lead to the deposition of thin films on the electrode surface. There are several reports in which different derivatives of sulfonated polythiophenes have been electro-polymerized under various conditions.[2,37,39]

Copolymerization

The majority of the sulfonated self-doped polymers are synthesized as copolymers since the presence of the electron-withdrawing sulfonic acid groups in the substituted monomers make it less suitable for oxidative polymerization.

Electrochemical copolymerization of aniline-disulfonic acid (ADSA) with parent aniline monomer has been demonstrated by many groups,[6,40–42] and it has been demonstrated that increasing the percent of substituted monomer in the mixture greatly reduces the polymer growth due to steric hindrance and decreasing nucleophilicity. In some cases, electrochemical copolymerization begins with the polymerization of unsubstituted monomer followed by the polymerization of the substituted monomer.[38] A soluble, sulfonated, self-doped copolymer of aniline-thiophene has been prepared potentiostatically (+1.70 V *vs.* Ag/AgCl) using FSO_3H/acetonitirle as supporting electrolyte by Yıldız *et al.*[43] The results suggest that increasing the concentration of thiophene in the electrolyte increases the conductivity since the conductivity contribution is much higher for polythiophene relative to polyaniline. The sulfur-to-nitrogen ratios of the copolymers were measured in their reduced form in order to avoid the contribution from SO_3^- counterions (apart from self-dopants) that could be possibly present in the oxidized copolymers.

17.2.2 Common Methods of Characterization

17.2.2.1 Elemental Analysis

Elemental analyzers are widely used to determine the elemental composition or the empirical formula of materials, especially organic compounds. In this method, the compound (*e.g.*, polymer) is combusted with excess oxygen in a controlled environment, and various traps collect the gases formed including carbon dioxide, water, nitric oxide and sulfur dioxide. By analyzing the quantity of all trapped gases, the mole ratios of C, H, N and S are determined.

In the case of sulfonated polyanilines, most applications and properties including, conductivity and solubility depend on the degree of modification, which can be determined from the sulfur-to-nitrogen (S/N) ratio. Several groups have used elemental analyzers to determine the S/N ratio and doping level of self-doped, homo- and copolymers of sulfonated polyanilines.[17,43–47] Chan *et al.*[48,49] have used two different elemental analyses to determine the N/P ratio and the doping level of poly(*o*-aminobenzylphosphonic acid); combustion elemental analysis to detect mol% of nitrogen and inductively coupled plasma (ICP) elemental analysis to determine the mol% of phosphorous. In a similar manner, the elemental ratios and dopant levels of several self-doped polymers, including polythiophene-phosphonic acid,[50] sulfonated polythiophenes,[37] copolymers of polypyrrole with sulfonated polypyrroles[51,52] and self-doped 3-substituted polypyrroles[53] have been determined using elemental analysis techniques.

17.2.2.2 X-Ray Photoelectron Spectroscopy

X-ray photoelectron spectroscopy (XPS), sometimes referred to as electron spectroscopy for chemical analysis (ESCA), is another technique used for determining the quantitative elemental composition, empirical formula, chemical and electronic states of a materials. Tang *et al.*[41] used XPS to track the changes in S/N and C/N ratios of poly(aniline-co-aniline-2,5-disulfonic acid) as the concentration of aniline-2,5-disulfonic acid in the monomer electrolyte varied. Yue and Epstein[54] used XPS peak areas to calculate the S/N ratios of sulfonated polyanilines and found that the XPS determined S/N ratios are in good agreement with those of chemical elemental analysis. In addition, they also used XPS to calculate the level of doping, since the amine and imine subunits of polyaniline result in chemical shifts in N(1 s) peaks related to doped and undoped states.[55] The oxidation level of the sulfonated polyaniline is generally obtained from the XPS spectrum by deconvoluting peaks into imine and amine nitrogen components.[56]

The electronic structure, protonation level and elemental composition of the aniline homopolymer and aniline–anthranilic acid copolymers have been studied thoroughly using XPS by Chan *et al.*[5] The XPS results indicated that the Cl^- : N^+ ratio is 1 for HCl-doped polyaniline and decreases with increasing anthranilic acid in the monomer solution, indicating that the protons on the carboxylate group self-dope some of the nitrogen atoms in order to maintain the charge neutrality of the polymer. There are several other reports in which XPS has been used to determine the doping level of various self-doped polymers, including poly[n-(3'-thienyl)alkanesulfonic acid],[57,58] poly(aniline boronic acid),[59] carboxylated polyaniline,[60] and poly(o-aminobenzylphosphonic acid).[48]

17.2.2.3 Fourier Transform-Infrared (FT-IR) Spectroscopy

FT-IR spectroscopy is a chemically specific technique and is widely used to identify chemical compounds and their substituent groups. Chan *et al.*[48] have used FT-IR to monitor structural changes in poly(*o*-aminobenzylphosphonic acid) as a result of exposure to different concentrations of NaOH. They found that increasing the concentration of base shifts the peak at 1580 cm^{-1} to higher frequencies, indicating the decline in NH^+ concentration, required for conductivity in the bulk polymer. In another report,[61] nanofibers of poly(*o*-aminobenzylphosphonic acid) were synthesized using a self-assembly process, and FT-IR analyses of these samples indicated that the monomer feed ratios between aniline and *o*-aminobenzylphosphonic acid do not affect the chemical structure of the polymer obtained. Sahin *et al.*[62] synthesized sulfonated polypyrrole electrochemically using pyrrole as monomer and FSO_3H as electrolyte, and used FT-IR to determine the level of sulfonation. The polymer obtained exhibited additional peaks at 1040, 700 and 630 cm^{-1} relative to the polymer prepared without FSO_3H, resulting in the assignment of the new peaks to O=S=O symmetric stretching, S–O stretching and C–S stretching vibrational mode respectively. This confirmed the presence of sulfonate groups in the polymer, prepared with the presence of FSO_3H. In a similar way, FT-IR has also been used to confirm the presence of sulfonate groups using the same assignments in electrochemically polymerized sulfonated, self-doped polythiophene.[37]

17.2.2.4 Spectroelectrochemistry

In spectroelectrochemical measurements, *in situ* spectroscopic measurements (commonly ultraviolet, visible, near-infrared and Raman absorption spectroscopes) are carried out while the system is under electrochemical control in order to observe changes in doping, chemical modification or electropolymerization.[63] Electrochemical homo- or copolymerization of monomer derivatives initially yields intermediate species that quickly react, rendering standard analytical tools ineffective for studying electropolymerization reactions at electrode surfaces. As a result *in situ* spectroelectrochemical studies have been employed to investigate the early stages of electropolymerization.[64] A fast scan UV-Vis spectrophotometer was used by Wen *et al.*[64] to determine the detailed mechanism behind the homopolymerization of diphenylamine-4-sulfonic acid and its copolymerization with unsubstituted aniline. In this work, spectra were recorded simultaneously with the slow scan potential sweep polymerization of monomer. At sufficiently positive potentials, a broad band appeared at 425 nm, attributed to the diphenylamine cation radicals. At higher potentials (0.9 V *vs.* Ag/AgCl), the 425-nm band shifted to 400 nm and a new broad peak appeared at approximately 550–740 nm. The 400-nm peak was attributed to the reaction between the cation radicals and monomers to form oligomers, and the peak at 550–740 was assigned to the transition of oligomers into polymer. The same group has also used fast-scan UV-Vis spectroscopy to investigate the early stages of electrochemical copolymerization of diaminobenzenesulfonic acid (DABSA) with several derivatives of aniline including pure aniline (ANI), o-toluidine (OT) and N-methyl aniline (NMA).[65] In this work, the copolymers were grown at constant potential (0.8 V *vs.* Ag/AgCl) and UV-Vis spectra were recorded continuously during the polymerization process. During the growth of copolymer, an additional band appeared at 672, 633 and 688 nm for the copolymers of DABSA with ANI, OT and NMA, respectively. This band did not appear during their homopolymerization, and is attributed to excitonic transition of monomer units in the copolymer. This exitonic transition is largely influenced by the presence of sulfonic acid groups, and as a result dedoping the copolymer decreases this band intensity. By analyzing the intensities of this band (doped *vs.* undoped), the level of doping (or the presence of DABSA units) in the copolymer could be estimated, and the obtained trend in doping was ANI > NMA > OT.

Malinauskas *et al.*[66] have used an *in situ* Raman spectroelectrochemical technique to study the degradation kinetics of self-doped sulfonated polyaniline. When a positive potential (+0.9 V *vs.* Ag/AgCl) was applied to the polymer in a solution (pH ranging from 0.5 to 9), the intensity of Raman bands decreases with time due to electrochemical degradation of polymer, and the rate of which increased with decreasing pH.

17.2.2.5 Electrochemical Quartz Crystal Microbalance (EQCM)

Quartz crystal microbalance (QCM) is a technique that monitors changes in mass of thin films by measuring the frequency change of an oscillating quartz crystal resonator. While QCM is used extensively for monitoring metal deposition in vacuum, it has also been developed as an *in situ* electrochemical technique (EQCM) to study changes in the mass at electrode surfaces during electrochemical growth of thin films/metals, dissolution of metals/corrosion, adsorption of material and mass transfer of ions/solvent across the film/electrolyte interface during redox reactions.[67,68] In the case of self-doped conducting polymers, EQCM can be used to determine the quantity and direction of the mass flux between the polymer and electrolyte. For example, Reynolds *et al.* have used EQCM to track the cation flux in and out of self-doped polypyrrole, poly{pyrrole-co[3-(pyrrole-1-yl)propanesulfonate]}[52] and poly[3,6-(carbaz-9-yl)propanesulfonate][69] at various redox potentials. In addition, they found that for polypyrrole, the anion (tosylate) also diffuses in and out of the film upon addition/expulsion of cations in order to maintain charge neutrality. However, in the case of poly{pyrrole-co[3-(pyrrole-1-yl)propanesulfonate]}, the cations are the only mobile species and the anions are immobile. In another example,[70] oxidative potentials applied to self-doped sulfonated polyaniline in HCl, resulted in mass increases, which the authors attributed to ion or solvent insertion into the polymer. The authors were able to calculate the equivalent molar mass of the exchanged species from the slope of charge-frequency plot. Lonergan[27] has carried out EQCM studies on n- and p-doped polyacetylenes (P_C and P_A in Scheme 17.10) in Bu_4NBF_4/CH_3CN electrolyte. The data showed that during p-doping of P_A and n-doping of P_C, the polymers lose their cations and anions, respectively, to reach their internally compensated states.

17.2.3 Polyanilines

Example 1: Polyaniline-Sulfonic Acid – Synthesis of Aniline/Sulfonated Aniline Copolymer

For the synthesis of sulfonated polyaniline, 0.5 g of emeraldine base is added into 40 mL of fuming sulfuric acid at room temperature. During sulfonation, the color of the polyaniline turns from dark-purple to dark-blue, and then the solution is left at room temperature for approximately 2 h with constant stirring. Afterwards, the addition of 200 mL of methanol at 10–20 °C results in a precipitation of green powder. The green precipitate is collected in a Buchner funnel and washed several times using methanol until the pH of the filtrate reaches 7. The green color of the filtrate indicates that the polymer is slightly soluble in water. Finally, the filter cake is dried under vacuum for 24 h, and the elemental analysis indicates the potential structures shown in Chart 17.3. Though the polymer obtained is only slightly soluble in water, it became highly soluble once it is converted into its sodium salt form by exposure to an aqueous basic (0.1 M NaOH) solution. The conductivity of this polymer is 0.1 S cm^{-1} without external doping.[17]

Though initially a sulfur to nitrogen (S/N) ratio of only 0.5 was obtained using various synthetic methods, it is later reported that a ratio of 0.75 can be obtained (with 70% yield) by reducing the emerandine base prior to the addition of fuming sulfuric acid.[44] This polymer has a higher conductivity (1 S cm^{-1}), which is also stable over the pH range of 0 to 14.

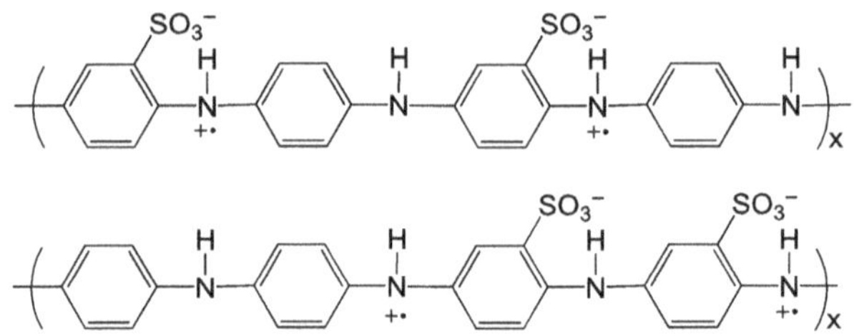

Chart 17.3 Reproduced by permission of the American Chemical Society (1990)[17].

Example 2: Homopolymer of Sulfonated Aniline – Electrochemical Polymerization of Aniline-2,5-Disulfonic Acid (ADSA)

The electrochemical polymerization of aniline-2,5-disulfonic acid (ADSA) is performed by applying a constant current density of 0.3 mA cm^{-2} on a Pt plate electrode in 0.1 M monomer containing 0.5 M HCl electrolyte. The polymerization is carried out at room temperature under inert atmosphere. After polymerization, ethanol is added to precipitate the polymer from solution, and the polymer is vacuum dried at 70 °C for 24 h. The molecular weight of the poly(ADSA) obtained varies from 5000 to 12 000 depending on synthetic conditions, and the conductivity ranges in between 0.04 and 0.3 S cm^{-1}, which is also independent up to pH 10.[42]

Example 3: Polyaniline-Carboxylic Acid (PACA) – Copolymerization of Aniline with Anthranilic Acid (AA), *m*-Aminobenzoic Acid (ABA), *m*-Aminobenzenesulfonic Acid (ABS), and Aniline-2, 5-Disulfonic Acid (ADSA)

Aniline can be copolymerized electrochemically with three different substituted anilines, anthranilic acid (AA, **3a** in Chart 17.4), *m*-aminobenzoic acid (ABA, **3b**), and *m*-aminobenzenesulfonic acid (ABS, **3c**) using 0.05 M of aniline with 0.05 M of substituted aniline in 0.2 M of H_2SO_4 as electrolyte. A constant potential of 0.9 V *vs.* Ag/AgCl is applied for the growth of the copolymer on a glassy carbon disk electrode up to the charge density of 100 mC cm^{-2}. Cyclic voltammetric studies show that the addition of 0.05 M substituted aniline to 0.05 M aniline in the deposition solution results in a drop in current and the peak potential shifts approximately +150 mV, indicating that the addition of substituted aniline hinders the polymer deposition. Among the three substituted anilines, ABS forms the most stable copolymer with better electrochemical properties.[6] The homopolymerization of aniline derivatives yield polymers with very little electroactivity.

As another example,[5] aniline is copolymerized with anthranilic acid by adding a 50-mL solution containing 16 mmol sodium persulfate (oxidant) dropwise into a 100-mL solution that contains aniline and anthranilic acid (with varying mole ratios). A solution pH of 0.1 is maintained by adding 1.0 M HCl and the solution is left at room temperature with constant stirring for 48 h for the polymerization to proceed. Afterwards, the powder produced is filtered and dried in a desiccator containing phosphorus pentoxide (P_2O_5). The

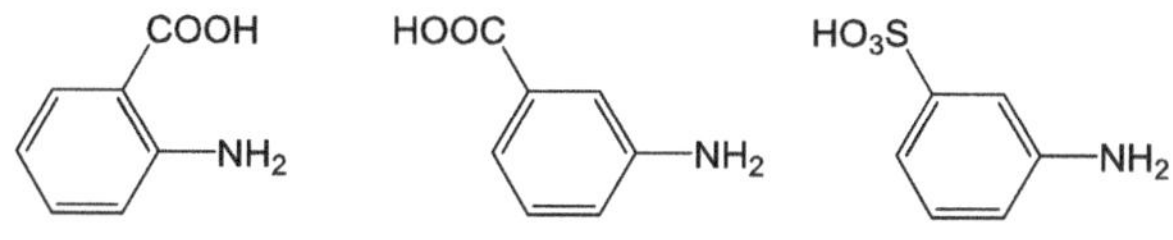

3a **3b** **3c**

Chart 17.4

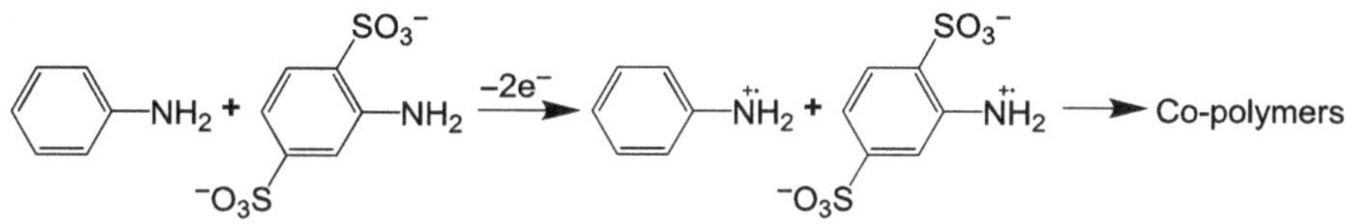

Scheme 17.2 Reproduced by permission of the Elsevier (1994).[41]

polymer obtained is soluble in *N*-methyl-2-pyrrolidone (NMP) and dimethyl sulfoxide (DMSO), and the solubility increases with the percent of anthranilic acid present in the monomer mixture. However, increasing the percent of anthranilic acid in the monomer solution lowers the yield as well as the conductivity of the polymer. For example, a ratio of anthranilic acid to aniline of 1 : 2 results in a yield of 50% and a conductivity of 4 S cm^{-1}, and increasing the ratio to 2 : 1 lowers the yield and conductivity to 32% and 2×10^{-5} S cm^{-1}, respectively. Though the yield and conductivity of the polymer decrease with anthranilic acid content, the solubility increases.

The electrochemical copolymerization of aniline (Scheme 17.2) and aniline-2, 5-disulfonic acid (ADSA) is carried out under potentiodynamic conditions using 0.5 M HCl as solvent, and 0.2 M aniline with 0.2 M ADSA at various monomer ratios. Pt plates are used as working and counterelectrodes, and the cycling is performed between –0.2 and 0.9 V (*vs.* SCE) at the rate of 50 mV s^{-1} in an inert atmosphere. As above, the addition of ADSA to the aniline-containing solution greatly decreases the polymer deposition even at low ADSA to aniline ratios (1 : 199). While no ADSA is observed in the polymer when concentrations are below 5 mM, a concentration of 150 mM results in the formation of copolymer with an S/N ratio of 0.46. The conductivities of the polymers obtained are largely independent on the concentration of ADSA and are in the range of 0.1 to 0.3 S cm^{-1} for concentrations in the range of 5 to 100 mM of ADSA.[41]

Example 4: Polyaniline-Phosphonic Acid (PANI-PA) – Synthesis of Homopolymers of PANI-Bu/Bn/DcPA and Poly(*o*-aminobenzylphosphonic Acid)

The homopolymers of PANI-Bu/Bn/DcPA (where Bu = *n*-butyl, Bn = benzyl, Dc = *n*-decyl, and PA = phosphonic acid) are synthesized by treating polyaniline with substituted phosphonic acids. PANI-Bu/EtPA and PANI-BnPA derivatives are prepared by adding 0.4 M of acid to a finely ground 3.0 mmol PANI base. The dark-green precipitate obtained is filtered and dried under vacuum at 60 °C for approximately 15 h. To synthesize PANI-DcPA, 3.0 mmol of PANI base is dissolved in 1 : 1 cyclohexane-petroleum ether solution containing 0.3 M of

Table 17.1 Atomic composition and dc conductivity. Reproduced by permission of the American Chemical Society (1994).[49]

Sample	*bulk stoichiometry*[a]	*P/N*	*O/P*	*surface stoichiometry (XPS)*[b]	*P/N*	*O/P*	*conductivity (S cm^{-1})*
PANI-BnPA	$C_{12.1}H_{12.6}N_{1.00}P_{0.87}O_{4.22}$	0.87	4.9	$C_{14.7}N_{1.00}P_{1.20}O_{3.5}$	1.2	2.9	2.4
PANI-DcPA	$C_{9.7}H_{11.3}N_{1.00}P_{0.24}O_{1.27}$	0.24	5.3	$C_{29.5}N_{1.00}P_{1.90}O_{5.7}$	1.9	3	2.3×10^{-4} [b] 4.0×10^{-8} [c]
PANI-Bu/EtPA	$C_{8.1}H_{11.5}N_{1.00}P_{0.58}O_{2.96}$	0.58	5.1	$C_{10.3}N_{1.00}P_{8.8}O_{2.8}$	8.3	3.4	11.0

[a]Normalized to N = 100.
[b]Measurement made on freshly prepared dull green sample.
[c]Measurement made on black sample after several weeks.

DcPA at ~100 °C under N_2. The precipitate obtained is light-green in color, which is later filtered and dried under vacuum at 60°C for 24 h.

Table 17.1 shows the conductivities and phosphorous to nitrogen ratio (P/N) for the homopolymers of polyaniline self-doped with various derivatives of phosphonic acids. Among these, PANI-BnPA has the highest P/N ratio and PANI-Bu/EtPA has the highest conductivity (11.0 S cm^{-1}). In addition, the conductivity of these polyanilines (PANI-BnPA and PANI-Bu/EtPA) are stable at elevated temperatures; decreasing by less than an order of magnitude when exposed at 150 °C for 7 h. In contrast, the conductivity of PANI-HCl decreases more than three order of magnitudes for the same treatment.[49]

To synthesize poly(*o*-aminobenzylphosphonic acid), the monomer, *o*-aminobenzylphosphonic acid (**4a** in Chart 17.5) is prepared using a three-step chemical reaction, and is used as a starting material to synthesize several phosphonic acid-doped polyanilines (**4b–4e**). To synthesize polymer **4b,** initially a pale-yellow solution of 9.9 mmol monomer **4a** is added to 3.5 M HCl at 0 °C. Then, 9.9 mmol of a 20-mL sodium persulfate solution is added dropwise, and the solution is maintained in an ice bath at 0–5 °C for 16 h. The polymer generated is rinsed using 2 M HCl and deionized water, followed by a vacuum drying at room temperature for about 16 h to yield a dark-green powder of polymer **4b**. The addition of 0.50, 1.0, and 2.0 mol per PO_3 of NaOH with polymer **4b** results in hygroscopic polymers of **4c, 4d** and **4e,** respectively. Polymer **4f** can be obtained by adding 2 M HCl to polymer **4b**. Polymers **4b** and **4f** exhibit higher conductivities (10^{-3} S cm^{-1}), and the conductivity of polymer **4b** is independent of pH up to 6. The conductivities of polymers **4c–4e** are lower (see Table 17.2) because the addition of NaOH deprotonates the polyaniline backbone and forms the sodium salt of the phosphonic acid groups.[48]

Example 5: Polyaniline-Boronic Acid – Synthesis of Poly(aniline-co-3-aminophenylboronic Acid)-without Fluoride (5a)

Enzymatic copolymerization is employed to synthesize copolymers of aniline and 3-aminobenzeneboronic acid (with different mole ratios). The electrolyte is

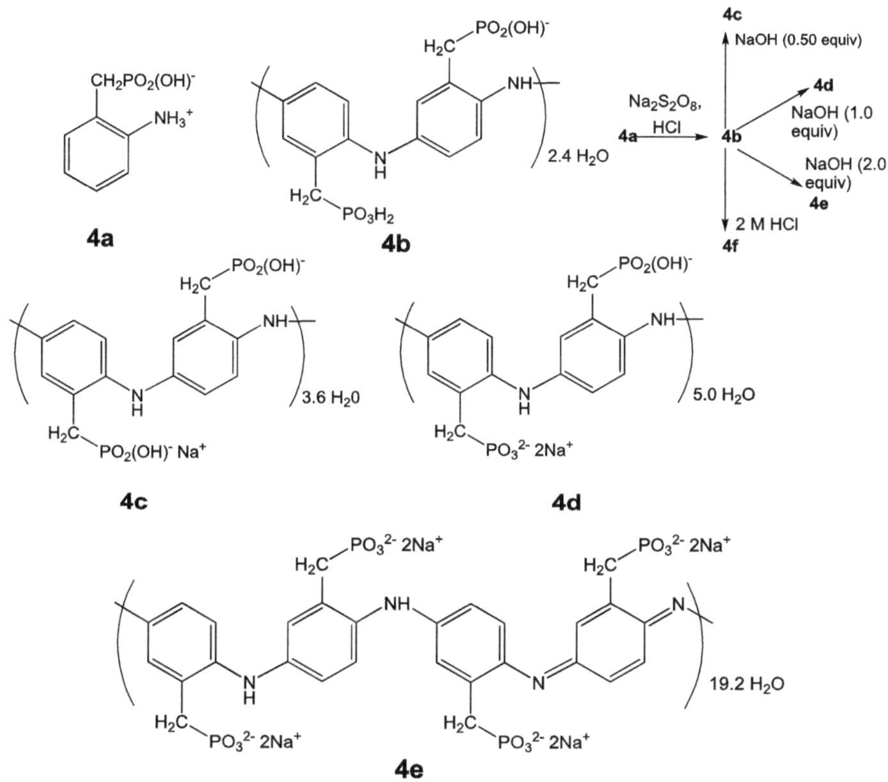

Chart 17.5 Reproduced by permission of the American Chemical Society (1992).[5]

Table 17.2 Bulk atomic composition, surface atomic composition, and dc conductivity (σ) of polymers **4b–4 f**. Reproduced by permission of the American Chemical Society (1995).[48]

Polymer	*Bulk atomic ratio*	*Surface atomic ratio*	*σ, (S. cm^{-1})*
4b	$C_{7.0}H_{10.0}N_{1.0}P_{0.9}O_{4.2}Cl_{0.05}$	$C_{7.7}N_{1.0}P_{1.0}O_{3.2}$	1.5×10^{-3}
4c	$C_{7.0}H_{10.7}N_{1.0}P_{0.9}O_{4.9}Na_{0.5}Cl_{0.05}$	$C_{7.6}N_{1.0}P_{1.0}O_{3.5}$	0.6×10^{-3}
4d	$C_{7.0}H_{12.3}N_{1.0}P_{1.0}O_{5.3}Na_{0.9}Cl_{0.05}$	$C_{8.1}N_{1.0}P_{1.0}O_{4.6}$	0.08×10^{-3}
4e	$C_{7.0}H_{15.2}N_{1.0}P_{0.9}O_{8.3}Na_{1.8}Cl_{0.05}$	$C_{9.8}N_{1.0}P_{0.8}O_{4.7}$	1.0×10^{-7}
4f	$C_{7.0}H_{9.9}N_{1.0}P_{0.9}O_{4.0}Cl_{0.2}$	$C_{9.0}N_{1.0}P_{1.0}O_{3.5}Cl_{0.15}$	1.0×10^{-3}

prepared by dissolving 20 mM of aniline (**5b** in Scheme 17.3), 10 mM 3-aminobenzeneboronic acid (AB, **5a**) and 30 mM of polyelectrolyte template (sulfonated polystyrene, **5c**) together in a 10-mL of sodium phosphate buffer solution of pH ~4.5. Afterwards, 10 mg of enzyme-containing biocatalyst, horseradish peroxidase (HRP) is added. A stoichiometric quantity of H_2O_2 is further added in the solution while stirring to initiate the reaction. Further dropwise addition of diluted stoichiometric amount of H_2O_2 helps to avoid inhibition of HRP due to the presence of excess H_2O_2. The reaction mixture is

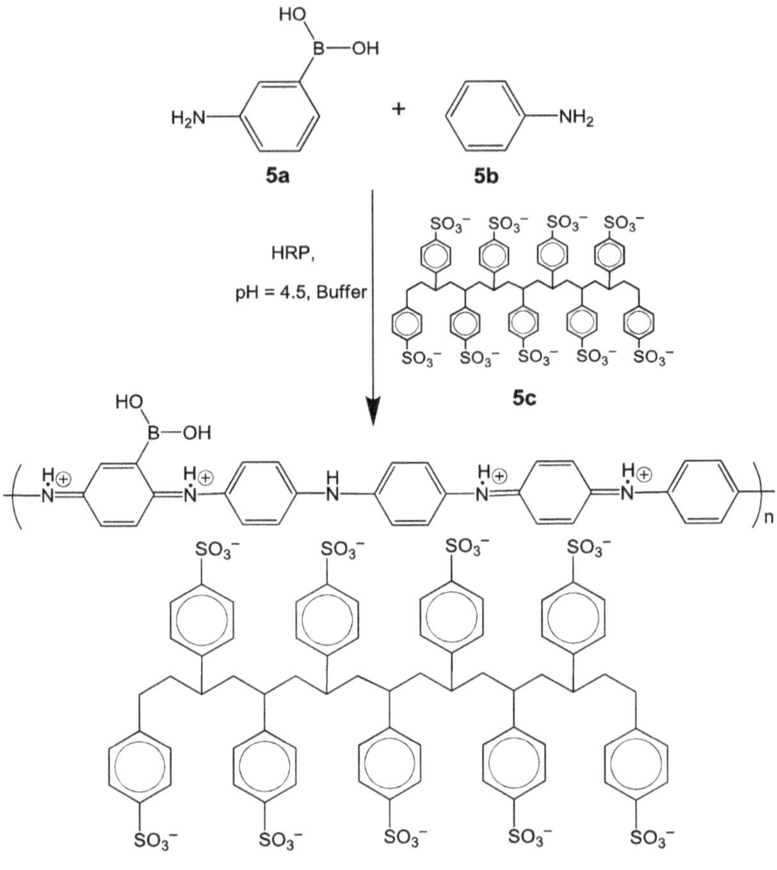

Scheme 17.3 Reproduced by permission of the American Chemical Society (2004).[10]

left at room temperature with constant stirring for 2 h for the polymerization to proceed. The polymer (**5d**) is precipitated by adding concentrated NaOH, and rinsed several times using an acetone–water mixture to remove oligomers, unreacted monomers, free sulfonated polystyrene templates, buffer salts and HRP from the polymer.

Increasing the mole ratio of aniline with respect to AB in the reaction mixture increases the conductivity of the poly(aniline-co-aminobenzeneboronic acid) initially up to 10^{-6} S cm^{-1}. However, when the mole ratio of [anilie]/[AB] exceeds 2, the conductivity declines until a ratio of 8 is reached.[71]

Example 6: Electrochemical Synthesis of Polyaniline-Boronic Acid (with Fluoride)

The monomer, aniline boronic acid (**6a** in Scheme 17.4) is prepared in a pH 7.4 phosphate buffer saline (PBS) solution, and later 0.5 M HCl is added to bring its pH to 5. Potentiodynamic polymerization of the monomer is carried out on a glassy carbon disk electrode between –0.1 to 1.0 V at the rate of 100 mV s^{-1} until

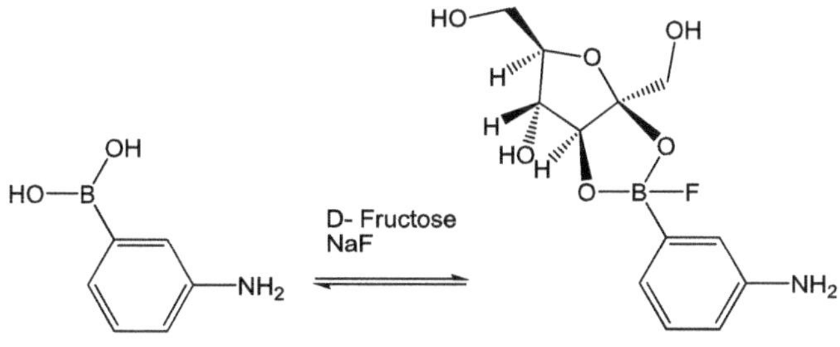

6a **6b**

Scheme 17.4 Reproduced by permission of the American Chemical Society (1997).[73]

the charge under the oxidation peak reaches 0.9 mC cm^{-2}. The cycling under acidic condition or with low fluoride content (40 mM) results in a redox-active film but incomplete polymerization. Increasing the fluoride content further (300 mM) produces continuous polymer films with higher cathodic current. This self-doped polyaniline conserves its higher conductivity at elevated pH, which shows that even the electropolymerization can be performed at upper pH values.[34]

The self-doped, water-soluble poly(**6b**) is synthesized through the complexation reaction of 3-aminophenylboronic acid **(6a)** with D-fructose in the presence of fluoride. For the polymerization reaction, an oxidizing agent, ammonium persulfate (40 mM, 0.5 mL) is slowly added in a 19.5-mL solution containing 40 mM of **6a**, 10 M of D-fructose and 40 mM of soldium fluoride. The solution is left at room temperature with constant stirring overnight and diluted further in order to reduce the concentration of fluoride. The resulting precipitate is redissolved in water and centrifuged several times, which eventually results in a precipitation of poly(**6b**). The switching from poly(**6b**) to poly(**6a**) is achieved by removing the D-fructose group in the presence of phosphate-buffered saline solution.[10]

17.2.4 Polypyrroles

Example 7: Polypyrrole-Sulfonic Acid – Copolymerization of Poly{pyrrole-co-[3-(pyrrol-1-yl)propanesulfonate]} (7a)

For the synthesis of poly{pyrrole-co-[3-(pyrrol-1-yl)propanesulfonate]} (**7a** in Chart 17.6), initially the monomer, potassium 3-(pyrrol-1-yl)propanesulfonate is prepared by reacting the pyrrole with potassium and 1,3-propane sulfonate, followed by a vacuum drying at 60 °C for 24 h. Homopolymerization of potassium 3-(pyrrol-1-yl)propanesulfonate and copolymerization of it with pyrrole (1 : 1 ratio) are carried out on a Pt electrode by cycling between 0.0 to 1.5 V (*vs.* Ag/AgCl) at a rate of 100 mV s^{-1} in a Bu_4NBF_4-containing acetonitrile solution. Elemental analysis shows that one unit of substituted pyrrole is present per three units of pyrrole rings.[51] Although the copolymer is considered as self-doped, the expected conductivity is lower due to the presence of nitrogen-bound substituents that generally reduces the conductivity, similar to N-alkyl substituted polypyrroles.

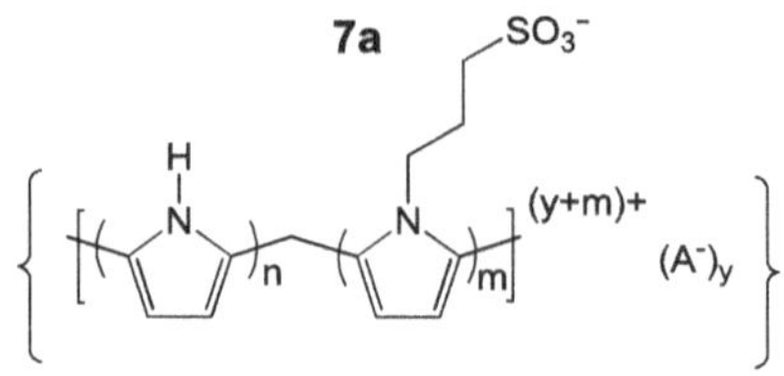

Chart 17.6 Reproduced by permission of the Royal Society of Chemistry (1987).[48]

Example 8: Polypyrrole-Carboxyl Acid – Electrochemical Synthesis of Poly(3-methylpyrrole-4-carboxylic Acid)

3-Methylpyrrole-4-carboxylic acid monomer (**8a**) has been polymerized electrochemically using 7 mM of solution **8a** as monomer, 0.1 M Et_4NClO_4 as supporting electrolyte and acetonitrile as solvent. The polymerization is carried out on a Pt electrode at a constant potential of 1.1 V (*vs.* SSCE) for approximately 20 min in an unstirred electrolyte solution. This process yields a golden-brown polymer film (poly(**8a**)), coated with a thin powdery deposit on the electrode and soluble oligomers in the electrolyte. The polymer samples prepared by cycling the potentials between 0.0 to 1.5 V (*vs.* SSCE) at the rate of 100 mV s^{-1} do not show the presence of the powdery coating over the polymer film. The measured conductivity of poly(**8a**) in its reduced and oxidative states are 10^{-7} and 10^{-5} S cm^{-1}, respectively.[72]

17.2.5 Polythiophenes

Example 9: Polythiophene-Sulfonic Acid – Synthesis of Sodium Poly(3-thiophene-β-ethanesulfonate) (P3-ETSNa) and Sodium Poly(3-(thiophene-δ-butanesulfonate)) (P3-BTSNa)

Upon oxidation, the water-soluble polymers of sodium poly(3-thiophene-β-ethanesulfonate) (P3-ETSNa) and sodium poly(3-(thiophene-δ-butanesulfonate)) (P3-BTSNa) yield self-doped polymers by losing a proton concomitant (Scheme 17.1). Since these monomers do not polymerize directly, their methyl ester forms (**9a** and **9b** in Chart 17.7) are polymerized first and later converted into polyelectrolytes using sodium iodide in acetone. When dipped in sodium iodide containing acetone, these methyl ester films of P3-ETS and P3-BTS rapidly dissolve, dedope and yield orange precipitates. The precipitates are soluble in Me_2SO, and produce tacky, orange films after evaporation. The conductivity of poly(**9a**) is in the range of 10^{-7} to 10^{-2} S cm^{-1} depending on relative humidity, and exposing the polymer to bromine vapor further dopes the polymer and increases the conductivity to ~10 S cm^{-1}.[2]

Example 10: Polythiophene-Carboxylic Acid – Synthesis of Water-Soluble 2,5-Poly(thiophene-3-propionic Acid) and Poly(3-thiophene Methyl Acetate) (P3TMA)

The polymerization of monomer **10a** (in Scheme 17.5) by a CuO-modified Stille coupling route produce regioregular oxazoline polythiophene (**10b**) with a

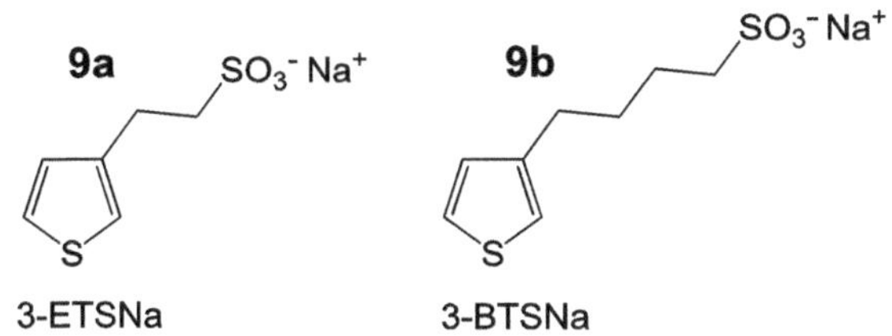

Chart 17.7 Reproduced by permission of the American Chemical Society (1987).[51]

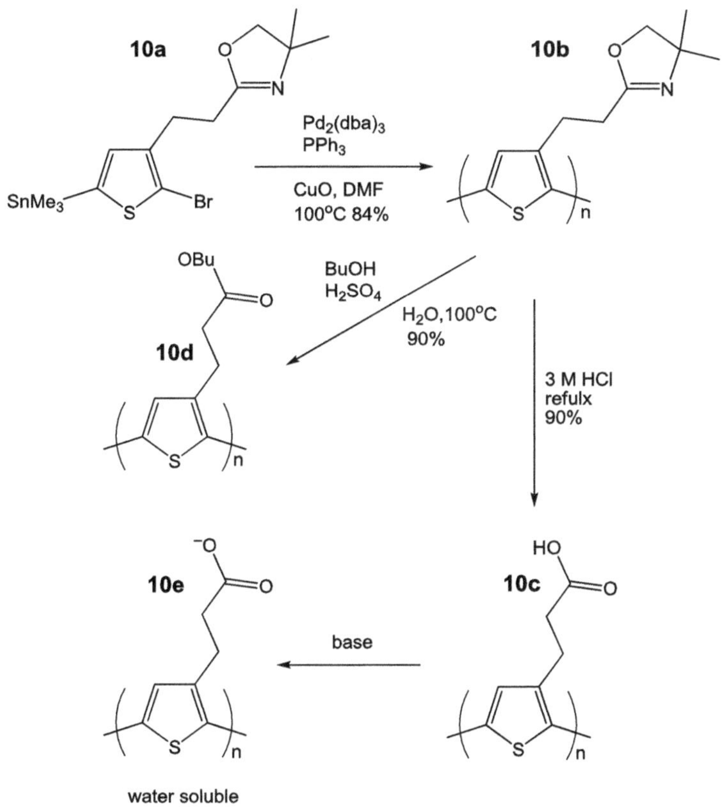

Scheme 17.5 Reproduced by permission of the American Chemical Society (1999).[74]

molecular weight of 8000 (yield, 84%). The carboxylic group that is protected as oxazoline in polymer **10b** can be easily converted into its propionic acid form by a hydrolysis reaction to give a dark-purple solid of 2,5-poly(thiophene-3-propionic acid) (**10c**, PTPA) with 90% yield. Though **10c** is not very soluble in CD_3OD (to perform ^1H NMR), it becomes highly soluble as soon as it is treated with Cs_2CO_3 to obtain its cesium salt form. Dissolving **10c** in aqueous base results in a purple solution of water-soluble ammonium salt of **10e**. In addition, dissolving **10c** in a bulky cation (Bu_4N^+) containing solution yields a

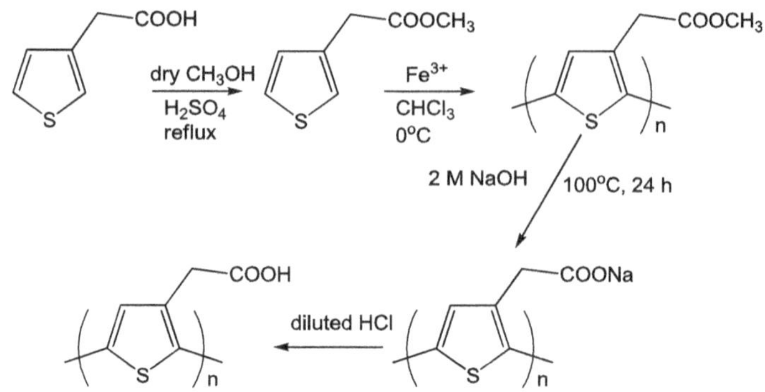

Scheme 17.6 Reproduced by permission of the Electrochemical Society (1990).[69]

yellow product, in which the polymer chains are twisted and solvated presumably due to the bulky cations reducing aggregation.[73]

For the synthesis of P3TMA, 10.0 g of 3-thiophene acetic acid (3TAA) monomer is refluxed in 50 mL dry methanol with a drop of concentrated sulfuric acid for 24 h to protect from oxidative decomposition of its carboxylic acid group during the oxidative-coupling polymerization process. Diethyl ester is used to extract the methanol-evaporated residue, and after rinsing with deionized water, the extract is dried using anhydrous $MgSO_4$. The protected monomer 3TAA is polymerized chemically in 20 mL chloroform using 40 mmol of ferric chloride as an oxidant. After the polymerization, a large amount of methanol is added in the solution to precipitate the brown-red colored poly(3-thiophene methyl acetate) (P3TMA) with a yield of 75% (Scheme 17.6).[74]

Example 11: Polythiophene-Phosphonic Acid – Synthesis of Poly(3-(3′-thienyloxy)propanephosphonate) (P3TOPP)

The direct polymerization of 3-(3′-thienyloxy) propanephosphonic acid (3TOPP) and its sodium salt do not yield polymer, however, an ethyl-ester form of 3TOPP, 3TOPPEt, is polymerized electrochemically using 0.1 M of $LiClO_4$/acetonitrile:dichloromethane (1 : 1) as supporting electrolyte. The monomer is prepared by adding 21.3 mmol of 3TOPPEt in 25.5 mmol (4.2 mL) of triethylphosphite solution, followed by an overnight heat at 150 °C. Pt and carbon rods are used as working and auxiliary electrodes, respectively, and a constant potential of 1.8 V (*vs.* sodium SCE) is applied for the electropolymerization. The progress of polymerization is followed using *in situ* spectrophotometry by monitoring the $\pi \rightarrow \pi^*$ transition. Once the polymerization is completed, the poly(3TOPPEt) is precipitated using diethyl ether (41% yield), and the silyl dealkylation of poly(3TOPPEt) is carried out in bromotrimethylsilane (2.9 mmol, 0.38 mL) containing CH_2Cl_2 in a dry box. Further hydrolysis using distilled water yields acidic form of poly(3TOPPEt) (yield 48%).[75]

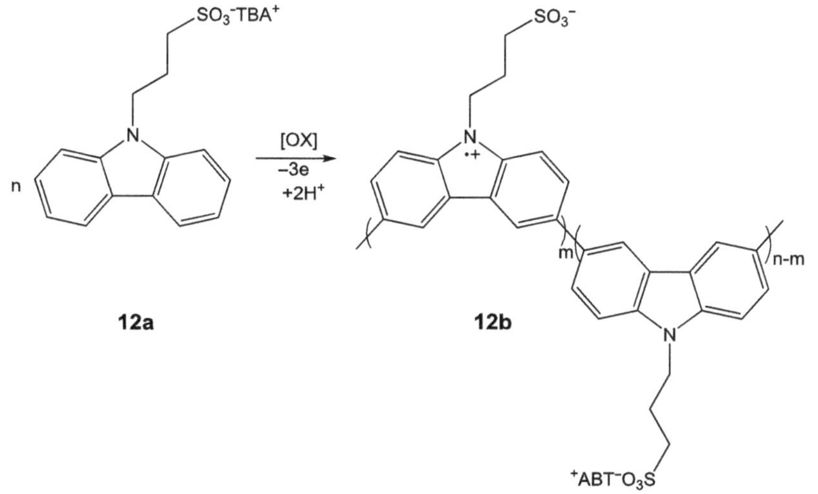

Scheme 17.7 Reproduced by permission of the American Chemical Society (1994).[23]

17.2.6 Polycarbazoles

Example 12: Poly(3,6-carbaz-9-yl)propanesulfonate

Synthesis of the monomer, tetrabutylammonium (carbaz-9-yl)propanesulfonate (**12a** in Scheme 17.7) is achieved by initially refluxing carbazole (0.012 M) with potassium (0.012 M) in THF. Once the traces of potassium have disappeared, the solution is cooled to room temperature and 0.012 M of propane sultone is added. The resulting slurry is washed with THF, vacuum dried, and equimolar tetrabutylammonium hydroxide (40% solution) is added. Afterwards, methylene chloride is added to the solution to extract the TBA salt, yielding a white solid of monomer **12a**. The monomer **12a** (10 mM) is electrochemically polymerized in 0.1 M $TBAClO_4$/MeCN solution using Pt as working electrode. Applying a constant current of 2 mA cm^{-2} results in a partially oxidized mixture of polymer (PCPS-TBA, **12b**) and monomer, in which sulfonate anion groups act as dopants. After removing the excess monomer by thoroughly washing with MeCN and vacuum drying for 24 h, a brown, water-soluble polymer is isolated. The measured conductivity of **12b** is 2×10^{-7} S cm^{-1} and could not be further increased by exposing it to Br_2 or I_2 vapor.[69]

17.2.7 Poly(*p*-phenylene)s

Example 13: Sulfonatealkoxy-Substituted Poly(*p*-phenylene) – Synthesis of Poly[2,5-bis(3-sulfonatopropoxy)-1,4-phenylene-*alt*-1,4-phenylene] ($PPP\text{-}ORSO_3$)

A homogeneous Suzuki coupling method has been used to synthesize poly[2,5-bis(3-sulfonatopropoxy)-1,4-phenylene-*alt*-1,4-phenylene] (**13b** in Scheme 17.8). To synthesize **13b**, 1.39 mmol of phenylenebis-(boronic acid), 1.39 mmol of 1,4-dibromo-2,5-bis-(3-sulfonatopropoxy)benzene (**13a**) and

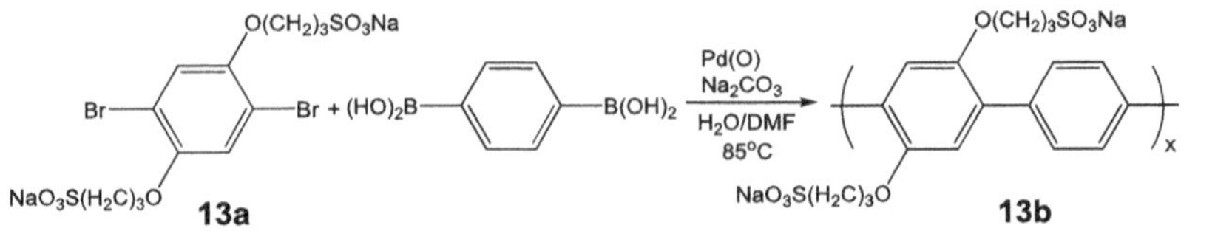

Scheme 17.8 Reproduced by permission of the American Chemical Society (1990).[76]

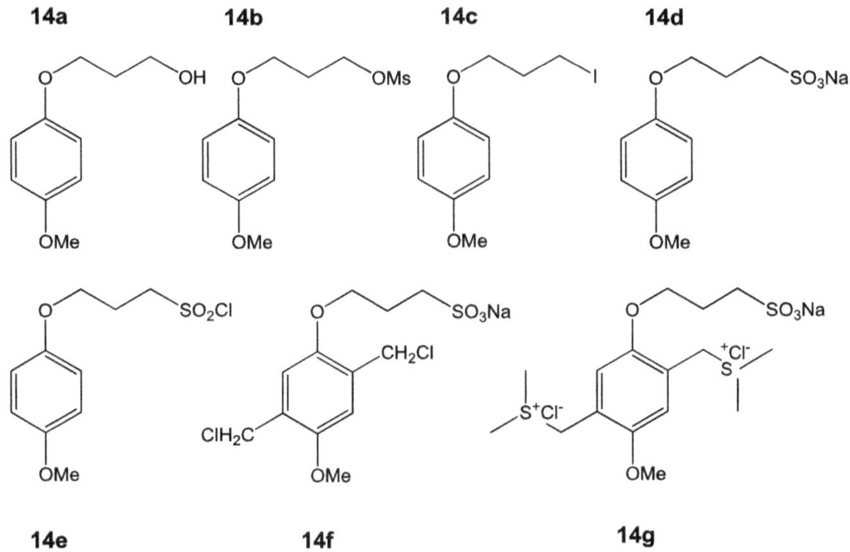

Chart 17.8 Reproduced by permission of the American Chemical Society (1987).[2]

0.99 g of sodium carbonate are dissolved in a mixture containing 40 mL of water and 20 mL of dimethylformamide at 85 °C. The tan precipitate obtained is filtered after performing several cycles of dissolving in hot water and reprecipitating by cooling. The filtrate is dissolved again in distilled water, and dialyzed using 3500 g mol^{-1} cutoff membrane for 72 h. The resulted polymer is soluble in water, aqueous bases, diluted aqueous acids and DMSO, but not in acetone and acetonitrile. The terminal sulfonate group of the polymer enhances the water solubility of the polymer's rigid backbone and also acts as a compensating dopant during the redox cycling.[23]

17.2.8 Poly(phenylenevinylene)s

Example 14: Water-Soluble Poly(phenylenevinylene) Derivative

A water-soluble poly(phenylenevinylene) derivative is synthesized using 4-methoxyphenol as a starting material. 4-Methoxyphenol (180 mmol) is initially reacted with 3-chloropropanol (456.5 mmol) in the presence of sodium ethoxide in order to alkylate it to give compound **14a** (in Chart 17.8) with the yield of 66%. Then, 3.25 mmol of freshly diluted pyridine is added with 28 mmol of compound **14a**. The reaction mixture is stirred overnight at room temperature, extracted twice using ether and 10% HCl, and finally dried to give compound **14b** (yield 65%). Then, 50 mmol of compound **14b** is added in 100 mL acetone that contains 45 mmol of NaI. After 24 h stirring at room temperature, the resulting precipitate is filtered, extracted using chloroform, and finally solvent evaporated to obtain 93% of oily **14c**. A concentrated solution of **14d** is obtained by reacting 61 mmol of **14c** with 122 mmol of sodium sulfite, followed by washing with anhydrous acetone and drying under vacuum at 80 °C (yield 88%). To synthesize **14e**, 80 mL of DMF containing 20.5 g of **14d** is added dropwise on to 20 mL of thionyl chloride. Subsequent washing using water and drying with $MgSO_4$ yield 13.3 g of yellow crystalline solid powder of **14e**. Addition of 56.7 mmol of **14e** (in 80 mL dioxane) to a mixture of 150 mL of 37% formaldehyde and 100 mL of HCl produces a white solid of **14f** (92% yield). Further reaction of 2.55 mmol of **14f** (in 5 mL acetone) with 5 mL of methanol, followed by solvent evaporation and drying yield 76% of compound **14g**.

Polymerization of **14g** (2 mmol) in 5 mL of methanol in the presence of 0.48 mL of 25% sodium methoxide produces highly viscous gel-like polymer **14h** (in Scheme 17.9). Refluxing polymer **14h** using water or DMF yields colorless polymer **14i,** which is later converted into its zwitterionic form (**14j**) by dialyzing against deionized water for 24 h. A fully conjugated derivative of PPV, polymer **14k,** is obtained by heating the precursor film **14j** at 200 °C under vacuum for 4 h. In another method, the precursor **14j** is mixed with ethylene glycol and excess NaOMe/methanol solution. Then, the reaction mixture is heated to its boiling point in order to remove the water, and the temperature of the bath further increased to 190 °C. Once the solution turns red, it is refluxed under N_2 for 26 h. Further cooling and dialyzing against deionized water yield red aqueous solution, using which a black hygroscopic film of **14l** can be cast under vacuum at room temperature. The conductivity of **14k** is 10^{-6} S cm^{-1}, and for **14l,** it ranges from 10^{-4} to 10^{-2} S cm^{-1} depending on humidity.[76]

17.2.9 Polyindoles

Example 15: Poly(indole-5-carboxylic Acid)

Poly(5-carboxyindole) is electropolymerized from a solution containing 5 mM of 5-carboxyindole monomer and 0.1 M of tetraethylammonium tetrafluoroborate (TEATFB) in acetonitrile. Electropolymerization is carried out on a Pt rotating-working electrode (4 Hz) at 1.4 V (*vs.* SCE) for 15 to 360 s. It is reported that better quality films are obtained on rotating-disk electrodes. Once the polymerization is completed, the films are washed in stages using mixtures of acetonitrile and water, starting with greater quantities of acetonitrile and finishing with pure water (80/20; 60/40; 40/60; 20/80; and finally pure water). The polymer attained is conductive in its partially oxidized state,

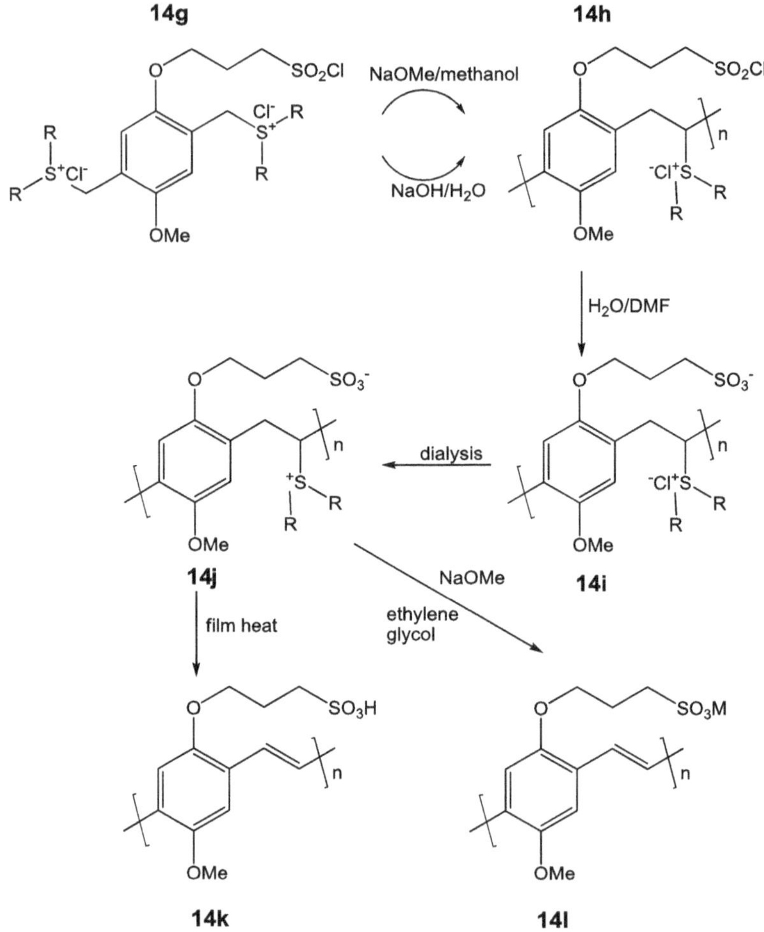

R= CH_3 or R, R=$(CH_2)_4$; M=H^+, NH_4, NA^+

Scheme 17.9 Reproduced by permission of the American Chemical Society (1999).[29]

however, it is nonconducting in its completely reduced and completely oxidized states.[77]

17.2.10 Polyacetylenes

Example 16: p- and n-Type Self-Doped Polyacetylenes

Cationic (**P_C**) and anionic (**P_A**) ionomers of polyacetylene have been synthesized[9,29] using ring-opening metathesis polymerization (ROMP).[26,78] For the ROMP process, tungsten alkylidene (**16a** in Scheme 17.10) is used as a catalyst, and alkylammonium and sulfonate functionalized cyclooctatetraenes (**16b**) are used as starting monomer materials to obtain **P_C** and **P_A**, respectively. The monomer is prepared by adding 150 mg (0.04 mmol) of

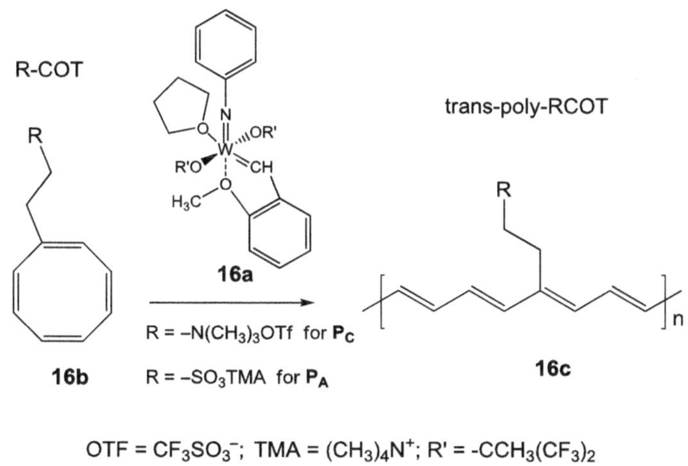

Scheme 17.10 Reproduced by permission of the American Chemical Society (1999).[29]

16b to 25 mg (0.02 mmol) of catalyst **16a** in 200 μL of CH_2Cl_2. The solution is mixed thoroughly and allowed to stand for 12 h before using it further. The dark-solid precipitate obtained (**16c**) is vacuum filtered and rinsed with toluene several times until the filtrate becomes colorless. The calculated yields for both $\mathbf{P_C}$ and $\mathbf{P_A}$ are in the range of 60–70%. To measure the conductivity, the solid polymer is dissolved in DMSO (dimethyl sulfoxide) and the solution is cast dropwise onto a glass slide. Once the solvent evaporates, the film is dried under vacuum for 12 h. Oxidative doping of the polymer can be achieved by exposing the polymer film into iodine vapor for 1 h and removal of trace I_2 under vacuum. The molecular weight (M_n) and conductivity (iodine doped) for the $\mathbf{P_A}$ are 35 000 and 0.7–0.4 S cm^{-1}, respectively.

References

1. A. O. Patil, Y. Ikenoue, N. Basescu, N. Colaneri, J. Chen, F. Wudl and A. J. Heeger, *Synth. Met.*, 1987, **20**, 151–159.
2. A. O. Patil, Y. Ikenoue, F. Wudl and A. J. Heeger, *J. Am. Chem. Soc.*, 1987, **109**, 1858.
3. M. S. Freund and B. A. Deore, *Self-Doped Conducting Polymers*, Wiley and Sons, West Sussex, England, 2007.
4. J. L. Bredas and G. B. Street, *Acc. Chem. Res.*, 1985, **18**, 309.
5. H. S. O. Chan, S. C. Ng, W. S. Sim, K. L. Tan and B. T. G. Tan, *Macromolecules*, 1992, **25**, 6029.
6. A. A. Karyakin, A. K. Strakhova and A. K. Yatsimirsky, *J. Electroanal. Chem.*, 1994, **371**, 259.
7. Y. Ikenoue, J. Chiang, A. Patil, F. Wudl and A. J. Heeger, *J. Am. Chem. Soc.*, 1988, **110**, 2983.

8. Y. Ikenoue, H. Tomozawa, Y. Saida, M. Kira and H. Yashima, *Synth. Met.*, 1991, **40**, 333.
9. C. H. W. Cheng and M. C. Lonergan, *J. Am. Chem. Soc.*, 2004, **126**, 10536.
10. B. A. Deore, I. Yu and M. S. Freund, *J. Am. Chem. Soc.*, 2004, **126**, 52.
11. A. Malinauskas, *J. Power Sources*, 2004, **126**, 214.
12. G. Dáprano, E. Proynov, M. Lebuf, M. Leclerc and D. R. Salahub, *J . Am. Chem. Soc.*, 1996, **118**, 9736.
13. C. Barbero, H. J. Salavagione, D. F. Acevedo, D. E. Grumelli, F. Garay, G. A. Planes, G. M. Morales and M. C. Miras, *Electrochim. Acta*, 2004, **49**, 3671.
14. C. C. Han and R. C. Jeng, *Chem. Commun.*, 1997, 553.
15. C.-C. Han, S.-P. Hong, K.-F. Yang, M.-Y. Bai, C.-H. Lu and C.-S. Huang, *Macromolecules*, 2001, **34**, 587.
16. C.-C. Han, W.-D. Hseih, J.-Y. Yeh and S.-P. Hong, *Chem. Mater.*, 1999, **11**, 480.
17. J. Yue and A. J. Epstein, *J. Am. Chem. Soc.*, 1990, **112**, 2800.
18. B. Lee, V. Seshadri, H. Palko and G. A. Sotzing, *Adv. Mater.*, 2005, **17**, 1792.
19. E. T. Kang, K. G. Neoh, Y. L. Woo and K. L. Tan, *Polym. Commun.*, 1991, **32**, 412.
20. S. Shimizu, T. Saitoh, M. Uzawa, M. Yuasa, K. Yano, T. Maruyama and K. Watanabe, *Synth. Met.*, 1997, **85**, 1337.
21. Y. lkenoue, Y. Saida, M.-A. Kira, H. Tomozawa, H. Yashima and M. Kobayashi, *J. Chem. Soc., Chem. Commun.*, 1990, 1694.
22. P. G. Baeuerle, K. U. Gaudl, F. Wuerthner, N. S. Sariciftci, H. Neugebauer, M. Mehring, C. Zhong and K. Doblhofer, *Adv. Mater.*, 1990, **2**, 490.
23. A. D. Child and J. R. Reynolds, *Macromolecules*, 1994, **27**, 1975.
24. T. I. Wallow and B. M. Novak, *J. Am. Chem. Soc.*, 1991, **113**, 7411–7412.
25. I. U. Rau and M. Rehahn, *Polymer*, 1993, **34**, 2889–2893.
26. C. B. Gorman, E. J. Ginsburg, S. R. Marder and R. H. Grubbs, *Angew. Chem. Int. Ed.*, 1989, **28**, 1571.
27. M. C. Lonergan, C. H. Cheng, B. L. Langsdorf and X. Zhou, *J . Am. Chem. Soc*, 2002, **124**, 690.
28. B. L. Langsdorf, X. Zhou and M. C. Lonergan, *Macromolecules*, 2001, **34**, 2450.
29. B. L. Langsdorf, D. H. Xin Zhou, Adler and M. C. Lonergan, *Macromolecules*, 1999, **32**, 2796.
30. V. Prevost, A. Petit and F. Pla, *Synth. Met.*, 1999, **104**, 79.
31. W. J. Bae, K. H. Kim, Y. H. Park and W. H. Jo, *Chem. Commun.*, 2003, 2768.
32. W. J. Bae, K. H. Kim and W. H. Jo, *Macromolecules*, 2005, **38**, 1044.
33. M. Nicolas, B. Fabre, G. Marchand and J. Simonet, *Eur. J. Org. Chem*, 2000, 1703.
34. B. A. Deore and M. S. Freund, *Analyst*, 2003, **128**, 803.
35. B. A. Deore and M. S. Freund, *Chem. Mater.*, 2005, **17**, 2918.

36. E. Hur, G. Bereket and Y. Sahin, *Mater. Chem. Phys.*, 2006, **100**, 19.
37. Y. A. Udum, K. Pekmez and A. Yıldız, *Synth. Met.*, 2004, **142**, 7.
38. J. Y. Lee and C. Q. Cui, *J. Electroanal. Chem.*, 1996, **403**, 109.
39. E. Turac, R. Varol, M. Ak, E. Sahmetlioglu and L. Toppare, *Des. Monomers Polym.*, 2008, **11**, 309.
40. R. Guo, J. N. Barisci, P. C. Innis, C. O. Too, G. G. Wallace and D. Zhou, *Synth. Met.*, 2000, **114**, 267.
41. H. Tang, A. Kitani and S. Ito, *Electrochim. Acta*, 1997, **42**, 3421.
42. H. Tang, A. Kitani, T. Yamashitab and S. Itob, *Synth. Met.*, 1998, **96**, 43.
43. Y. A. Udum, K. Pekmez and A. Yıldız, *Eur. Polym. J.*, 2005, **2005**, 1136.
44. X. L. Wei, Y. Z. Wang, S. M. Long, C. Bobeczko and A. J. Epstein, *J. Am. Chem. Soc.*, 1996, **118**, 2545.
45. Y. A. Udum, K. Pekmez and A. Yıldız, *J Solid State Electrochem.*, 2006, **10**, 110.
46. L. V. Lukachova, E. A. Shkerin, E. A. Puganova, E. E. Karyakina, S. G. Kiseleva, A. V. Orlov, G. P. Karpacheva and A. A. Karyakin, *J. Electroanal. Chem.*, 2003, **544**, 59.
47. S.-A. Chen and G.-W. Hwang, *J. Am. Chem. Soc.*, 1995, **117**, 10055.
48. H. S. O. Chan, P. K. H. Ho, S. C. Ng, B. T. G. Tan and K. L. Tan, *J. Am. Chem. Soc.*, 1995, **117**, 8517.
49. H. S. O. Chan, S. C. Ng and P. K. H. Ho, *Macromolecules*, 1994, **27**, 2159.
50. K. K. Stokes, K. Heuze and R. D. McCullough, *Macromolecules*, 2003, **36**, 7114.
51. N. S. Sundaresan, S. Basak, M. Pomerantz and J. R. Reynolds, *J. Chem. Soc., Chem. Commun.*, 1987, 621.
52. J. R. Reynolds, N. S. Sundaresan, M. Pomerantz, S. Basak and C. K. Baker, *J. Electroanal. Chem. Interfacial Electrochem.*, 1988, **250**, 355.
53. E. E. Havinga, W. T. Hoeve, E. W. Meijer and H. Wynberg, *Chem. Mater.*, 1989, **1**, 650.
54. J. Yue and A. J. Epstein, *Macromolecules*, 1991, **24**, 4441.
55. X.-L. Wei, M. Fahlman and A. J. Epstein, *Macromolecules*, 1999, **32**, 3114.
56. S.-A. Chen and G.-W. Hwang, *J. Am. Chem. Soc.*, 1994, **116**, 7939.
57. S. A. Chen and M. Y. Hua, *Macromolecules*, 1993, **26**, 7108.
58. M.-Y. Hua, S.-W. Yang and S.-A. Chen, *Chem. Mater.*, 1997, **9**, 2750.
59. E. Shoji and M. S. Freund, *Langmuir*, 2001, **17**, 7183.
60. S.-C. Kim, J. Whitten, J. Kumar, F. F. Bruno and L. A. Samuelson, *Macromol. Res.*, 2009, **17**, 631.
61. C.-H. Yang, Y.-K. Chih, H.-E. Cheng and C.-H. Chen, *Polymer*, 2005, **46**, 10688.
62. Y. Sahin, A. Aydın, Y. A. Udum, K. Pekmez and A. Yıldız, *J. Appl. Polym. Sci.*, 2004, **93**, 526.
63. W. Kaim and J. Fiedler, *Chem. Soc. Rev.*, 2009, **38**, 3373.
64. T.-C. Wen, C. Sivakumar and A. Gopalan, *Electrochim. Acta*, 2001, **46**, 1071.
65. T.-C. Wen, L.-M. Huang and A. Gopalan, *Electrochim. Acta*, 2001, **46**, 2463.

66. R. Mazeikiene, A. Statino, Z. Kuodis, G. Niaura and A. Malinauskas, *Electrochem. Commun.*, 2006, **8**, 1082.
67. H. Varela, M. Malta and R. M. Torresi, *Quimica Nova*, 2000, **23**, 664.
68. D. A. Buttry and M. D. Ward, *Chem. Rev.*, 1992, **92**, 1355.
69. Y.-J. Qiu and J. R. Reynolds, *J. Electrochem. Soc.*, 1990, **137**, 900.
70. C. Barbero, M. C. Miras, R. Kötz and O. Haas, *J. Electroanal. Chem.*, 1997, **437**, 191.
71. P. Huh, S.-C. Kim, Y. Kim, Y. Wang, J. Singh, J. Kumar, L. A. Samuelson, B.-S. Kim, N.-J. Jo and J.-O. Lee, *Biomacromolecules*, 2007, **8**, 3602.
72. P. G. Pickup, *J. Electroanal. Chem.*, 1987, **225**, 273.
73. R. D. McCullough, P. C. Ewbank and R. S. Loewe, *J. Am. Chem. Soc.*, 1997, **119**, 633.
74. B. Kim, L. Chen, J. Gong and Y. Osada, *Macromolecules*, 1999, **32**, 3964.
75. A. Viinikanoja, J. Lukkari, T. Aaritalo, T. Laiho and J. Kankare, *Langmuir*, 2003, **19**, 2768.
76. S. Shi and F. Wudl, *Macromolecules*, 1990, **23**, 2119.
77. P. N. Bartlett, D. H. Dawson and J. Farrington, *J. Chem. Soc., Faraday Trans.*, 1992, **88**, 2685.
78. C. B. Gorman, E. J. Ginsburg and R. H. Grubbs, *J. Am. Chem. Soc.*, 1993, **115**, 1397.

CHAPTER 18

Fused Heterocycle Polymers

SANDEEP KAUR, ALEXANDER L. KANIBOLOTSKY AND PETER J. SKABARA*

WestCHEM, Department of Pure and Applied Chemistry, University of Strathclyde, 295 Cathedral Street, Glasgow G1 1XL, UK
*Email: peter.skabara@strath.ac.uk

18.1 Introduction

The synthesis of conjugated polymers comprising fused heterocyclic units remains a subject of extensive research in the field of materials science, which provides a wide range of possibilities for altering the electronic properties of organic semiconductor materials. Among the various methods of polymer synthesis there are those that involve a heterocyclization reaction during the polymerization and those that use monomers with already prepared fused heterocyclic units. The latter are more widely used due to the possibility for more flexible design of heterocyclic systems within a monomeric building block.

As a polymerization technique for creating polyarene systems, a wide range of transition-metal-catalyzed coupling procedures by direct arylation method or Kumada, Stille, Suzuki, Negishi, Yamamoto protocols are available. For the synthesis of polymers where the conjugation along the polymeric backbone is retained by unsaturated acyclics unit between the monomers, a series of condensation techniques such as Wittig–Horner–Emmons, Knoevenagel or azomethine condensation as well as palladium catalyzed Heck and Sonogashira coupling can be used.

If the monomer forms a stable radical cation at the aryl peripheral edge upon oxidation, then an oxidative coupling becomes an option as a polymerization

RSC Polymer Chemistry Series No. 9
Conjugated Polymers: A Practical Guide to Synthesis
Edited by Klaus Müllen, John R. Reynolds and Toshio Masuda

Published by the Royal Society of Chemistry, www.rsc.org

procedure. Various oxidizing reagents ($FeCl_3$, $NO^+BF_4^-$, *etc.*) can be used. An alternative kind of oxidative polymerization technique known as electropolymerization is also worth discussing. This method requires a minute amount of monomeric material (a few micrograms), forms a polymeric film immediately on a conducting substrate and provides a facile strategy of evaluating its electronic and electrochemical properties from the as-made polymer film. The morphology of the electropolymerized film can be easily controlled by the electrodeposition conditions.

The present chapter deals with the synthesis of conjugated polymers with various kinds of fused heterocyclic units, focusing on the experimental details of their synthesis. Taking into account that the structure of the monomeric building block affects the properties of the final polymer and defines the methods that can be used for its synthesis, the discussion of the available polymerization methods will be closely related to the type of heterocyclic unit incorporated in its backbone (Table 18.1).

Table 18.1 Heterocycles units, their abbreviations and structures mentioned in the chapter.

Heterocyclic Units	*Abbreviation*	*Structure*
Dithiin	–	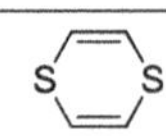
Tetrathiafulvalene	**TTF**	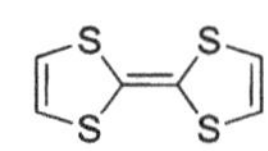
Diketopyrrolopyrrole	**DPP**	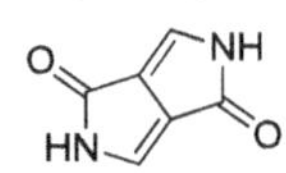
Thieno[3,4-*c*]pyrrole-4,6-dione	**TDP**	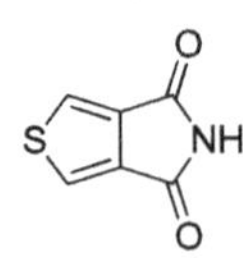
4,4-difluoro-4-bora-3a-4a-diaza-*s*-indacene	**BODIPY**	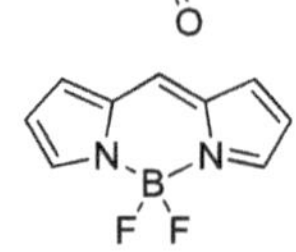
Isoindigo	**iI**	
Azoles: i) Benzobisthiazole ii) Benzo[1,2-d:4,5-d′]bisoxazoles iii) Benzo[1,2-d:5,4-d′]bisoxazole iv) Benzothiadiazole v) Benzoselenodiazole vi) Benzotriazole vii) thieno[3,4-*b*]pyrazine	 i) **BBT**: X = S ii) **BBO**:X = O iii) **BBO** iv) **BT**: X = S v) **BSe**: X = Se vi) **BTA**: X = N **TP**	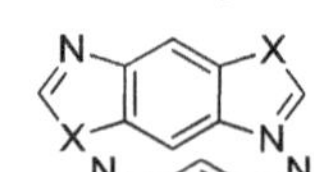 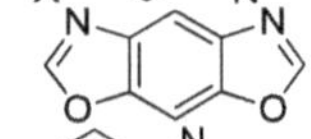 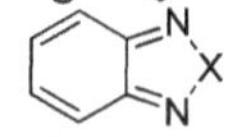 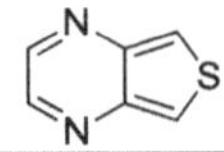

18.1.1 Dithiin-Based Polymers

The successful studies of poly(3,4-ethylenedioxythiophene) PEDOT as a material for electronic applications created a wide interest in the chalcogen effects of replacing the oxygen with either sulfur or selenium atoms. Due to a quite low oxidation potential of ethylenedichalcogenothiophenes and stability of their cation radicals, the main method for preparing the polymer is oxidative polymerization. With the eagerness to resolve the insolubility of PEDOT, Kanitzidis and co-workers[1] were the first to describe the synthesis of the sulfur analog of EDOT, 3,4-ethylenedithiothiophene (EDTT) from thieno[3,4-*d*][1,3]-dithiole-2-thione (**1a**). The monomer was successfully polymerized electrochemically and by chemical oxidation yielding a material that was effectively soluble in *N*-methyl-2-pyrrolidone (NMP) and partially in tetrahydrofuran (THF). Meijer and co-workers[2] provided an alternative synthesis of EDTT from dimethoxythiophene and 1,2-ethanedithiol. This procedure was later optimized by Bendikov and co-workers.[3] The electronic characteristics of PEDTT (**P1**) were found to be inferior compared to those of PEDOT due to the effects of the chalcogen S—S non-covalent interactions, causing a twist in the chain conformation, hindering electron delocalization and therefore increasing the bandgap and the material's oxidation potential.[4]

In contrast, the conformation of the polymer chain in PEDOT is highly planar due to extensive S–O contacts between repeat units that provide a pseudoribbon-type structure in the polymer. The electronic properties of copolymers of EDOT-EDTT provide excellent proof of the twisting effect of the EDTT unit, compared to the planarizing effect of EDOT, in polythiophene systems.[5] However, the lack of the intense intermolecular chalcogen–chalcogen interactions, as seen in PEDOT, was found to improve the solubility of its sulfur analog PEDTT. Skabara and co-workers have reported a simple and cleaner route towards PEDTT and other analogs *via* the spontaneous polymerization of the crystalline forms of the 2,5-dibromothiophene derivatives (**P2-3**) (Chart 18.1).[6]

Among other analogs of PEDOT, poly(3,4-ethylenediselenothiophene) **P4** (PEDST), prepared electrochemically, was found to adopt a more planar structure than PEDTT, with its bandgap being closer to that of PEDOT. PEDST possesses a considerably lower LUMO level than either of its analogs due to the increased polarizibility of the selenium atom compared to oxygen

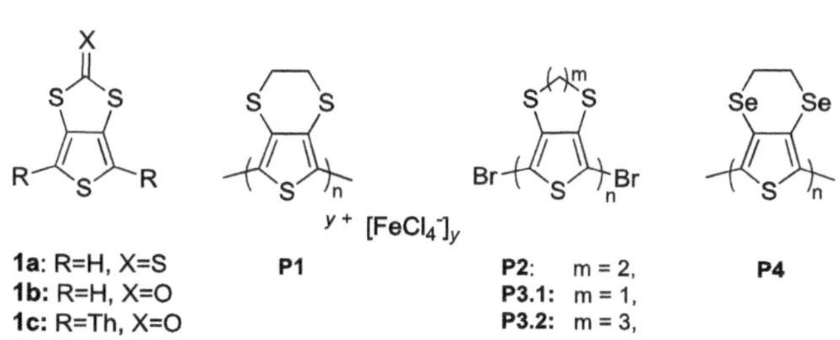

Chart 18.1

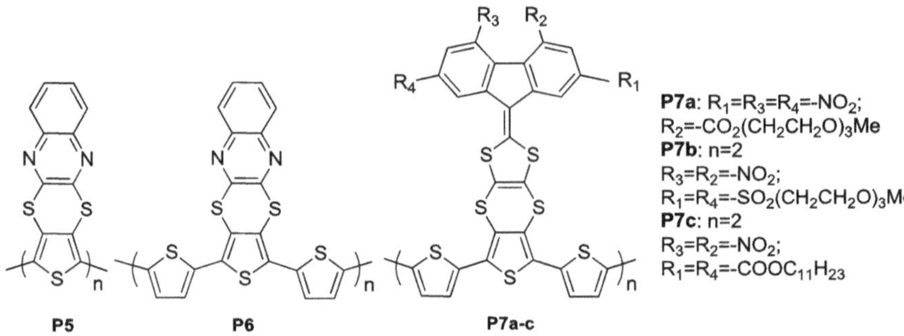

Chart 18.2

and sulfur, indicating that PEDST should be a more efficient electron-transport material in the series.[7]

The thieno[3,4-d][1,3]dithiole-2-one **1b** and its 4,6-dithienyl derivative **1c** were converted to the corresponding thieno[3′,4′:5,6][1,4]dithiino[2,3-*b*]quinoxaline fused heterocycle monomers that were easily electropolymerized forming **P5**[8] and **P6** (Chart 18.2).[9] The polymer **P5** exhibited a transformation in voltammetric response upon complexation to metal cations in solution, elucidating its suitability as a sensor.[8] Polymer **P6,** with one dithiinoquinoxaline per terthiophene repeat unit showed a fast electrochromic switch from an orange color, in neutral form, to a blue color in the oxidized state. The switching characteristics followed at 650 nm revealed a change in transmittance of 44% over cycles of 1 s switching rate *cf.* PEDOT (44% at 2.2 s). This fast switching performance is presumed to be attributed to the bent conformation of the 1,4-dithiin ring, which creates a less tightly packed material, facilitating the ingression and egression of anions throughout the film upon redox switching.[9]

A push–pull fluorenylidenedithiole unit was incorporated into a conjugated polythiophene architecture through the fusion to a dithiin ring. The polymer, prepared by oxidative polymerization, featured a charge-transfer component orthogonal to the main conjugated backbone. Polymer **P7a** turned out not to be soluble in common organic solvents.[10] The oxidative polymerization of other monomers with push–pull units possessing two solubilizing ester groups led to the isolation of dimers **P7b** and **P7c** only. The latter was used for fabrication of bulk heterojunction organic solar cells (BHJSCs) (Chart 18.2).[11]

Other versatile building blocks with fused dithiol-2-one, dithiin and thiophene or furan heterocyclic units (**2a** and **2b**) were used to synthesize a series of monomers bearing the redox active 1,1,2,2-tetrathioethene unit. The teraryl monomers were electrochemically polymerized, and the resulting polymers were studied as potential electrochromic materials **P8**[12] and **P9–11** (Chart 18.3).[13] The di(thienyl)furan polymer **P8** exhibited a lower electrochemical bandgap and superior switching speeds compared to its corresponding terthiophene analog **P9** (*e.g.*, 1.97 eV *cf.* 2.11 eV). The narrower bandgap of **P8** is explained by the lower aromaticity of furan over that of thiophene, which provides more efficient charge delocalization across the backbone of the polymer **P8.**[12,13] Notably, polyfuran

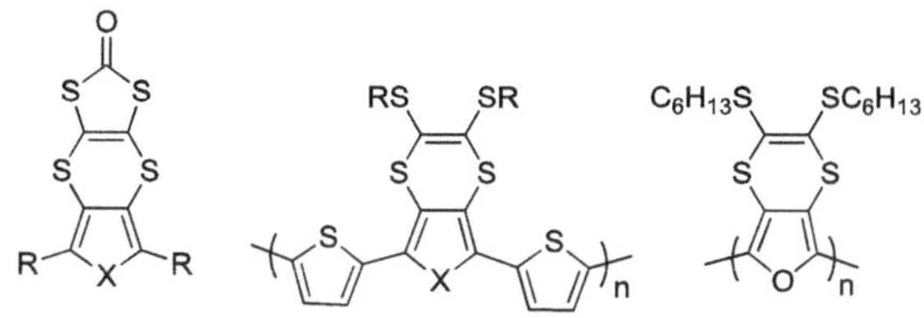

Chart 18.3

P12 with a dithiin ring incorporated onto the polymer backbone was recently synthesized chemically *via* Yamamoto polymerization.[14] Being an inverse analog of PEDOT and with a vinylene bridge instead of an ethylene unit, this polymer features planarizing chalcogen–chalcogen (S-O) interactions along the polymer backbone.

18.1.2 Tetrathiafulvalene (TTF)-Based Polymers

The fusion of independent electroactive units and subsequent incorporation of the fused system into a polymer can create materials with interesting electrochemical properties. Among them are the conjugated polymers containing the tetrathiafulvalene (TTF) unit. TTF has attracted vast attention as an electron donor.[14,15] The strong electron-donor properties of TTF are provided by the gain in aromaticity accompanying the formation of the dithiolium unit upon oxidation. Structural modifications of TTF derivatives enable the fine tuning of the system's electronic properties. Additionally, the presence of sulfur atoms attached to the TTF unit can increase the dimensionality of intermolecular interactions.

Several groups have succeeded in attaching TTFs to poly(thiophene), in which the fulvalene units are held by non-conjugated links and render the TTFs redox independent.[16,17]

Using building blocks **1c**, **1d** and **2b** a series of monomers comprising fused TTF thiophene heterocycle systems were synthesized *via* phospite-mediated crosscoupling reactions. Polymer **P13** was the only one that was successfully prepared by electrodeposition and showed truly hybrid electrochemical behavior.[18,19] The preparation of **P14** and **P15** *via* electrodeposition was prevented due to the TTF moiety being the site of the radical cation, limiting aryl-coupling at the peripheral thiophenes. The polymers **P14**, **P15**, **P16** were synthesized using chemical polymerization techniques; Yamamoto homocoupling for **P14**[17] and Stille cross-coupling for **P15** and **P16**.[18] An organic solar cell was fabricated from **P16**, and its photovoltaic performance has been reported (Chart 18.4)[18,19].

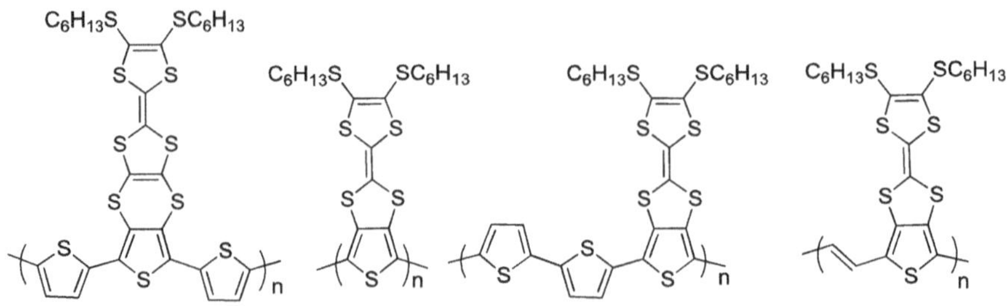

P13 P14 P15 P16

Chart 18.4

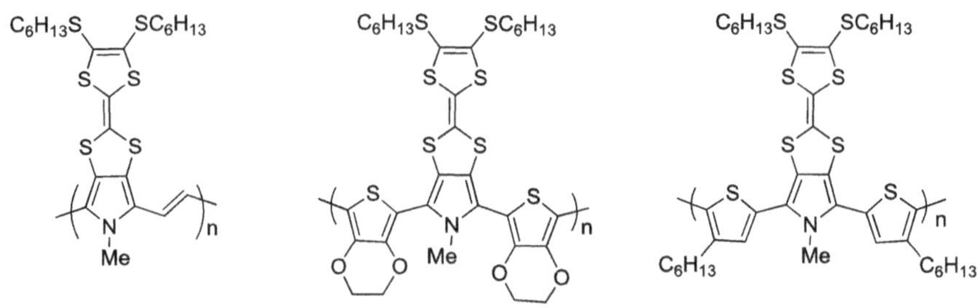

P17 P18 P19

Chart 18.5

Fused pyrrolo-TTF units were also incorporated into conjugated polymer architectures using Stille coupling for polymer **P17**, and electropolymerization for **P18**, **P19**. UV-vis absorption spectroscopy revealed a wider bandgap for **P17** in comparison to its thieno analog **P16** due to the decrease in polarizibility of the heteroatom (nitrogen *vs.* sulfur) and steric interactions between repeat units.[20] Despite the pyrrolo-TTF units being stronger electron donors than that of thieno-TTF, the two redox-active elements of the new polymers function independently upon oxidation (Chart 18.5).[20]

18.1.3 Diketopyrrolopyrrole (DPP)-and Thieno[3,4-*c*]-pyrrole-4,6-dione (TDP)-Based Polymers

A useful strategy in the design of new polymeric materials for electronic applications is the incorporation of a chromophore unit into the conjugated backbone of the polymer. One such class of chromophoric compounds that have gained interest over the past decade is represented by diketopyrrolopyrrole (DPP) derivatives. The chromophoric properties are attributed to the intramolecular charge transfer (ICT) between the aryl substituents at the 3- and 6- positions to the electron deficient lactam units. Tieke *et al.*[21] have reported an

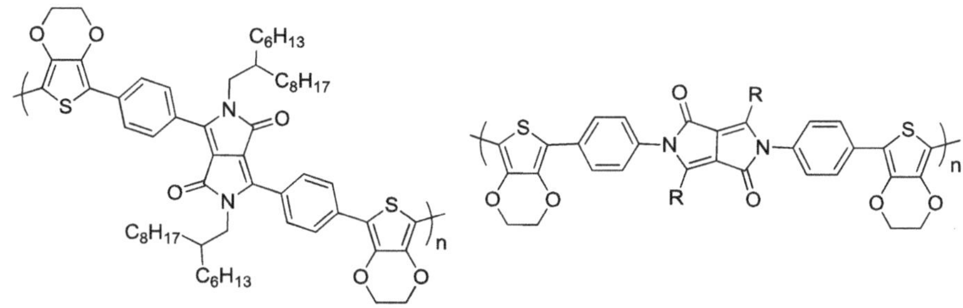

P20 P21

Chart 18.6

extensive review of the development and progression of DPP derivatives, which underlines the high level of current interest in these materials.

DPP monomers can be polymerized either through 3,6-aryl substituents (*e.g.* **P20**) or *via* those at the 2,5-position (*e.g.* **P21**).[22,23] The effect of the substitution pattern of the DPP chromophore in the polymer backbone was investigated using both electropolymerization[22] of EDOT functionalized derivatives and Suzuki polycondensation[23] products. The incorporation of the DPP unit into polymeric backbones *via* the 3- and 6-positions leads to a significant change in the optical and electrochemical properties of the polymer compared to those of the DPP monomer, whereas polymerization through the 2- and 5-positions provides a polymer that essentially retains the properties of the basic DPP-monomer unit. In this case, the disruption of π-conjugation at the lactam N-atoms is responsible for the independent behavior of the monomeric unit within the polymer. Taking this into account, the most promising DPP co-monomers for creating low bandgap polymers are those that contain polymerizable units at the 3- and 6- positions of DPP[22,23] (Chart 18.6).

Skabara *et al.* recently prepared the polymer **P22**, using the high yielding co-polymerization of a dibromothieno-TTF compound with a dithienyl-DPP derivative *via* Suzuki polycondensation.[24] This polymer showed good organic field effect transistor OFET device characteristics (hole mobility of 4×10^{-2} cm^2 V^{-1} s^{-1}, with an on/off ratio of 10^4), with excellent air stability that is believed to be assisted by the presence of the strong electron-donating TTF unit within the polymer (Chart 18.7).

Stille coupling has been used by Li *et al.*[25] to synthesize the donor–acceptor co-polymer **P23** consisting of dithienyl-DPP and fused thienothiophene (TT) units. OFET devices fabricated from this polymer showed a high hole mobility of 0.94 cm^2 V^{-1} s^{-1} due to significant intermolecular π–π donor–acceptor interactions between TT and DPP fused heterocyclic units.[25] Additionally, a stannyl derivative of dithienothiophene has been recently co-polymerized by Shahid *et al.* with bis(5-bromo-2-thiophenyl)- and bis(5-bromo-2-selenophenyl)-DPP comonomers.[26] Both polymers exhibited ambipolar characteristics in OFET devices. The thiophene-based co-polymer analog showed a higher

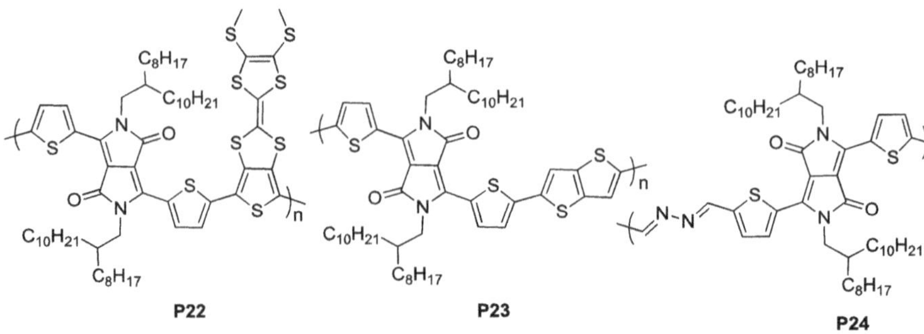

Chart 18.7

photovoltaic performance than its selenophene analog (power-conversion efficiency, PCE = 5.10% *vs.* 4.05%, at 1:2 polymer:$PC_{71}BM$ ratio), due to a higher open-circuit voltage.[26]

Li and co-workers[27] described the polycondensation of a dithienyl-DPP dialdehyde derivative with hydrazine yielding the polymer **P24** that showed ambipolar charge-transport characteristics (electron and hole mobilities of 0.41 $cm^2\ V^{-1}\ s^{-1}$ and 0.36 $cm^2\ V^{-1}\ s^{-1}$, respectively). The presence of strongly electron-withdrawing azine linkages alongside the DPP unit in the conjugated backbone of the polymer assists in lowering the LUMO level and provides the high value of electron mobility (Chart 18.7)[27].

The donor–acceptor (D–A) approach realized in alternating electron-rich and electron-deficient heterocycles has been the main strategy in creating low bandgap donor materials for bulk heterojunction solar cells. Similar to DPP, thieno[3,4-*c*]pyrrole-4,6-dione (TDP) is another electron-deficient fused heterocyclic system.[28–30] On the other hand, bridging the bithiophene at 3,3′-positions by Si and Ge atoms increases the conjugation of the two thiophenes and creates electron-rich fused heterocycles. A co-polymer of alternating TDP and dithienosilole units (**P25**) synthesized *via* Stille crosscoupling polymerization, exhibited impressive performance as a donor material in a BHJSC (PCE of 7.3% for **P25**: $PC_{71}BM$ (1:2) and an active layer of 1 cm^2).[28] In comparison, using dithienogermole (DTG) **(P26)** instead of dithienosilole as an electron-rich heterocyclic unit increases the short-circuit current, fill factor and PCE of the solar cell.[29] It was rationalized that the better performance of the Ge analog in the BHJSC is attributed to the longer Ge–C bond compared to that of Si–C, which elongates the alky chains further from the fused thiophene rings, thereby reducing steric hindrance and fostering close π-intermolecular interactions. TDP has been used in direct arylation polycondensation with bromo-substituted bithiophene by Leclerc and co-workers[30] The activating effect of the imide group on the protons of the fused thiophene unit makes this coupling procedure efficient, clean and provides the polymer **P27** with a higher molecular weight in comparison with the corresponding Stille polycondensation method (Chart 18.8).[30]

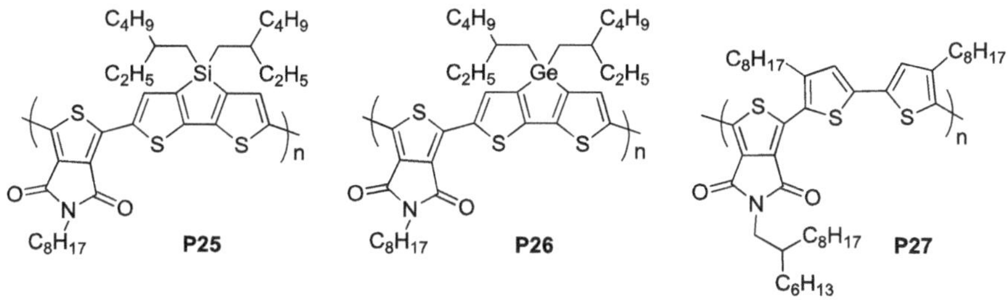

Chart 18.8

18.1.4 4,4-Difluoro-4-boro-3a-4a-diaza-*s*-indacene (BODIPY)- and Isoindigo (iI)-Based Polymers

Over the past two decades, several research disciplines have taken a keen interest in the dye molecule 4,4-difluoro-4-bora-3a-4a-diaza-*s*-indacene known as BODIPY. This fused heterocycle consists of two pyrrole rings linked by methine and difluoroboron bridges and is considered as a cyclic cyanine dye with the positive charge stabilized by the BF_2^- substituent. BODIPY core compounds can be prepared using relatively short synthetic routes in multi-gram scale by applying approaches that have been developed in the field of porphyrin chemistry.[31] Dehaen and co-workers have recently published a comprehensive review on the synthesis and application of BODIPY derivatives.[31] Due to their strong absorption and emission properties, BODIPY-based materials have been used for various applications such as biological labeling reagents, tunable laser dyes, electroluminescent devices, electrogenerated chemiluminescence, chemosensing and solar cells.[31,32]

The BODIPY moiety has been incorporated into conjugated polymer architectures both by electrochemical polymerization and Stille cross-coupling procedures. In the former case, the core was functionalized at the α-position with EDOT units and the final monomer was electropolymerized to provide the polymer **P28**.[33] Cyclic voltammetry revealed remarkable reversibility of both n- and p-doping processes with a very narrow electrochemical bandgap ($E_g = 0.8$ eV) of the polymer that is highly attractive for application to organic photovoltaics (Chart 18.9).[33]

The coupling of α,α′-dichloro BODIPY derivatives with distannylated bis-EDOT and bis-EDTT provided the polymers **P29a** and **P29b**, respectively, in moderate yields.[32] These polymers exhibited low optical bandgaps (1.18 and 1.35 eV for **P29a** and **P29b**, respectively). The pristine organic photovoltaic (OPV) devices fabricated from these polymers demonstrated decent short-circuit current implying ambipolar behavior of the materials, which was confirmed by time-of-flight measurements. When blended with $PC_{71}BM$, PCEs of 0.95% and 0.45% were observed for **P29a** and **P29b**, respectively. The broad spectral response observed from these materials highlights their potential as candidates for future broadband light-sensing and -harvesting applications.[32]

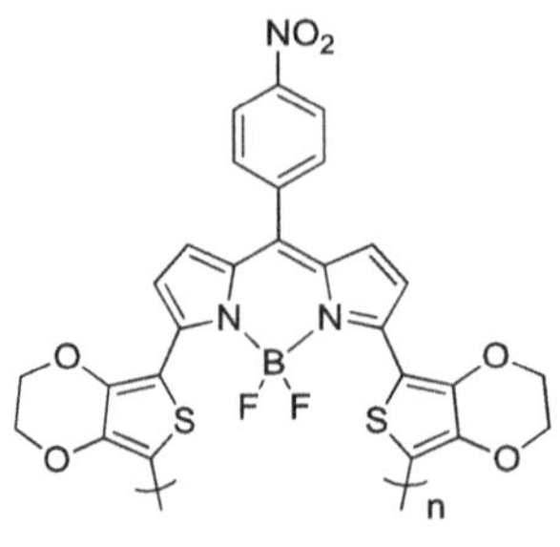

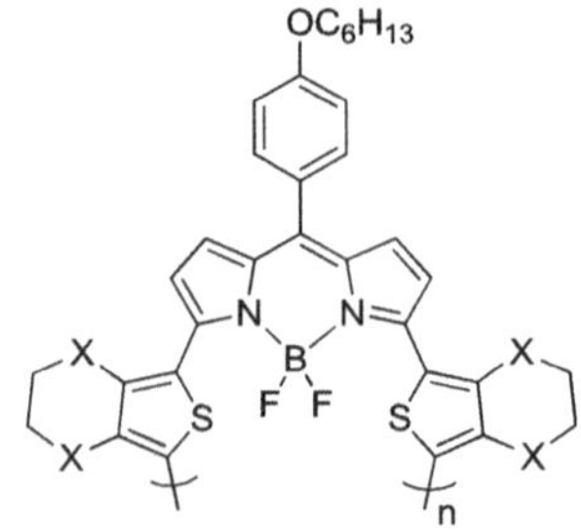

Chart 18.9

Another new conjugated dye known as isoindigo (**iI**) has been of recent interest. Isoindigo-based materials are already known in the dye industry and are readily available from renewable resources. This **iI** heterocycle possesses two lactam units, which results in strong electron-withdrawing characteristics.[34] Reynolds' group has studied several **iI**-based materials which have been found to possess promising optical, electrochemical and photovoltaic properties.[35–37] It was noted that when different donor units were combined with isoindigo, the resulting polymers exhibited totally different optoelectronics properties. The photovoltaic performance of these isoindigo-based polymers have been independently reported by Zhang *et al.*,[38] Liu *et al.*[39] and Wang *et al.*[40] Despite the moderate OPV performances, they possess broad absorption spectra, high extinction coefficients and favorable energy levels, which encourage further synthetic manipulation to be explored in order to improve device performances.[34,35,40]

Extending the conjugation of the donor unit in the donor–acceptor chain would increase the delocalization of the HOMO along the polymer backbone and enhance p-type behavior.[34] Wang *et al.*[34] recently performed Stille cross-coupling reactions of 6,6′-bis(5-bromo-4-alkyl-2-thienyl)-isoindigo with commercially available bis(trimethylstannyl)thiophene yielding high molecular weight polymers **P30** that are thermally stable T_d ~380 °C. The terthiophene moiety is quite planar and encourages π–π stacking within the polymer backbone, giving excellent PV performance of PCE = 6.3%, from a **P30**:$PC_{71}BM$ (1:1.5) blend processed with 2.5% v/v 1,8-diiodooctane (DIO) in *o*-dichlorobenzene (DCB) (Chart 18.10).[34]

Recently, Reynolds and co-workers[35] investigated the incorporation of a different electron-rich donor, dithienosilole, due to the aforementioned benefits. The successful co-polymerization of 6,6′-dibromo-N,N′-(2-ethylhexyl)-isoindigo with 2,6-bistrimethylstannyl-4,4-bis-(2-ethylhexyl)-dithieno[3,2-b:2′,3′-d]silole was achieved by using conventional Stille cross-coupling, affording **P31** in 94% yield. The catalyst used was $Pd_2(dba)_3$ as Pd(0) source and $P(o\text{-tolyl})_3$ as the ligand. The polymerization was terminated by end-capping with tributyltin thiophene, yielding high molecular weight polymers with thermal stability up to ~410 °C and a PCE of 4% in a BHJ organic solar cell comprising a donor–acceptor blend ratio of 1:4 (**P31**: $PC_{70}BM$).[35]

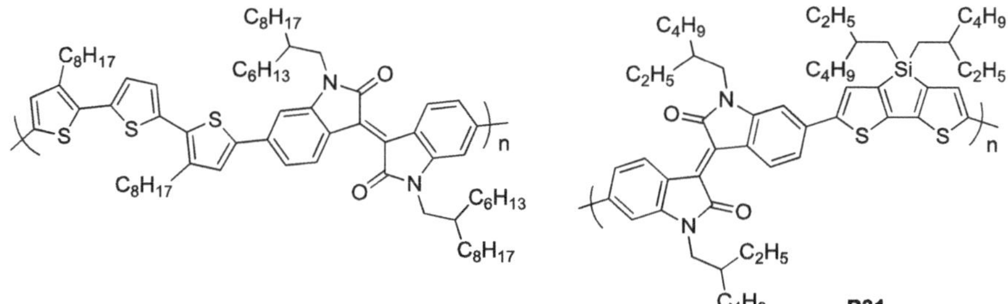

Chart 18.10

18.1.5 Azole- and Pyrazine-Based Polymers

Fused azole units are widely used for their incorporation into polymer conjugated architectures. The electron-withdrawing >C=N- imino units provide an electron deficiency to the whole fused aromatic system yielding a significant degree of electron affinity for the parent polymer. Moreover, the presence of an additional heteroatom in the fused heterocycle creates the possibility for multidimensional self-assembly due to non-covalent interchain interactions, as was shown for benzobisthiazole (BBT) monomeric units.[41] Polymers with fused azole units have attracted interest since the 1980 s due to their rod-like rigidity, stability, strong intermolecular contacts and self-assembling behavior in the solid state.[42] There are two different ways the fused benzoazole heterocycle can be incorporated into a polymer: Type 1) by polymerization through the 2,2′-position, so that the complete unit is located within the conjugated backbone of the polymer, or Type 2) by connection exclusively through the benzene ring.

In the case of the BBT unit the first type of the polymers (**P32**, **P33**) is more developed than the type 2 (**P34**) (Chart 18.11). The morphology of the polymers provided by non-covalent interchain interactions brought in by the BBT unit allows the fabrication of stable and durable devices (Chart 18.11).[43]

The early synthesis of a series of conjugated polymers **P35** with BBT units was carried out by Osaheni and Jenekhe *via* high-temperature condensation reactions of diacid and diaminobenzenedithiol monomer building blocks and forming the fused thiazole rings in the conjugated backbone upon polymerization.[44,45] This methodology requires strong acids, inert conditions and extensive work-up procedures. However, in terms of functional group compatibility, a more flexible strategy has been adopted where the fused heterocyclic system is built up during monomer synthesis. Using this approach and subsequent Stille polymerization, the BBT unit was incorporated into a series of DA polymers with dithienopyrrole **P32**, cyclopentadithiophene **P33a** and dithienosilole **P33b**.[43] Using Stille cross-coupling, the BBT unit was also incorporated into a polymeric backbone *via* its 4- and 8-positions **P34** (the aforementioned type-2 polymer). The high value of conductivities upon doping and a 78 nm red shift of absorption in the solid compared to that in $CHCl_3$ solution was rationalized on the basis of planarization of the 4,8-bis(2-thienyl)benzo[1,2-*d*:4,5-*d*′]bisthiazole unit and aggregation due to π–π stacking.[46]

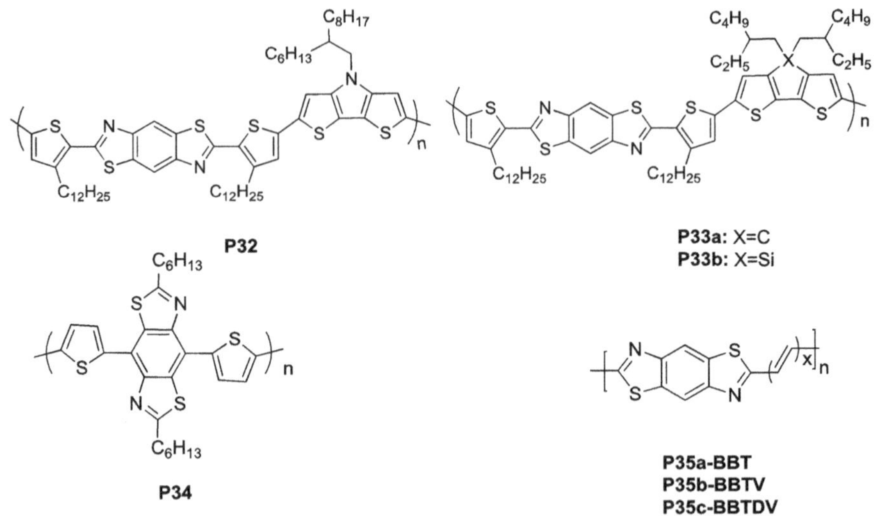

Chart 18.11

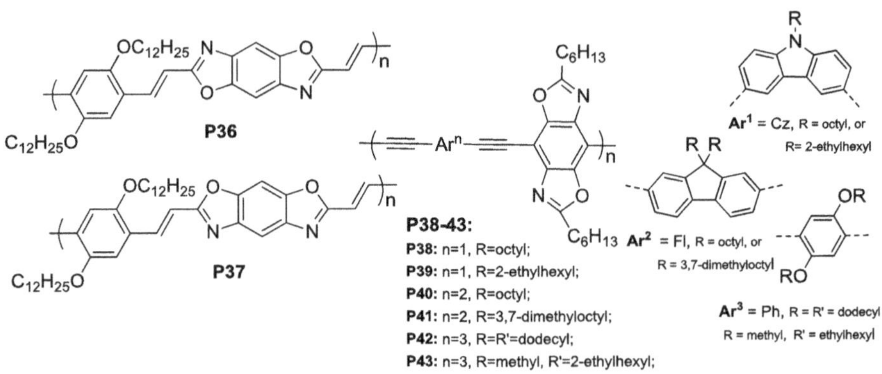

Chart 18.12

The synthesis of conjugated polymers with an analogous fused azole system – benzobisoxazole (BBO) was initially performed using a simple condensation of 4,6-diamino-l,3-benzenediol and terephthalic acid.[42] This synthetic strategy required harsh conditions similar to those used for the preparation of BBT-containing polymers that limits the types of substituents that may be successfully incorporated into the polymer backbone. Jefferies-El *et al.*[47] circumvented these issues by applying Horner–Wadsworth–Emmons methodology to bis-(diethylphosphonate) derivatives of 2,6-dimethyl BBTs and 2,5-didodecyloxyterephthalaldehyde, which provided the soluble polymers **P36** and **P37** that are thermally stable (Chart 18.12).[47]

Conjugated BBO-containing polymers with type-2 positioning of fused oxazole units were synthesized using the Sonogashira coupling protocol.[48]

Alkyl substitution at the 2- and 6- positions of BBO gave an increase in solubility for the final polymers (**P38–43**) by disrupting the π-stacking between polymer chains.[48] Integrating BBO into polymeric backbones *via* its benzene ring improved Förster resonance energy transfer (FRET) between the poly(*N*-vinylcarbazole) host and BBO co-polymer guest materials, highlighting the importance of effective conjugation pathways in the design of emissive materials for guest–host polymer light-emitting diodes (PLEDs).[48]

Fused azole systems with three heteroatoms can be represented by benzothiadiazole (BT), benzoselenodiazole (BSe), benzotriazole (BTA).[49] Among them the heterocycle BT is most widely used in conjugated co-polymers. Derivatives of BT have been widely used in organic electronic devices, especially in terms of electrochromic applications. While incorporated into the conjugated backbone these BT moieties bring to the polymeric material such properties as strong ICT transition dipole, multicolored electrochromism and efficient photoluminescence.

The most illustrative examples of the donor–acceptor BT-containing polymers are those with fused germole and silole units as donor moieties, consisting of thiophene and benzene aromatic units that have been forced into co-planarity by bridging atoms (C, Si, Ge). For the synthesis of the germanium analog (**P44**), Heeney and co-workers used Suzuki coupling of dibromogermaindacenodithiophene with 2,1,3-benzothiadiazole-4,7-bis(boronic acid pinacol ester) in high yield.[50] Due to the instability of silaindacenodithiophene under basic conditions, Stille polycondensation was used to synthesize the silicon analog **P45**.[51] Interestingly, while the C-bridged co-polymer **P46** is an amorphous material, the Si and Ge analogs display a certain degree of crystallinity. Crystalline polymers exhibit higher charge carrier mobilities that is suitable for a number of applications, such as OFETs and OPVs. The featured low bandgap polymers displayed OPV power conversion efficiencies of 5.02% in blends of **P44**: $PC_{71}BM$ (1:4), and 4.3% for **P45**: $PC_{71}BM$ (Chart 18.13).[50]

Another set of BT containing polymers (**P47–49**) featuring the 4,7-di(thiophen-2-yl)-2,1,3-benzothiadiazole chromophore unit were synthesized using Suzuki coupling protocol from 4,7-bis(5-bromothiopene-2-yl)-2,1,3-benzothiadiazole. The blends of PCBM with polymers **P47, P48** and **P49** were used for fabrication of OPV cells and showed PCE of 2.67%, 3.05% and 1.47%, respectively (Scheme 18.1).[52,53]

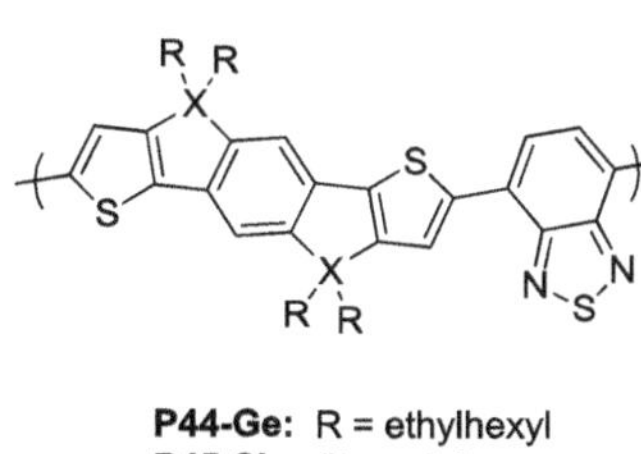

P44-Ge: R = ethylhexyl
P45-Si: R = octyl
P46-C: R = octyl

Chart 18.13

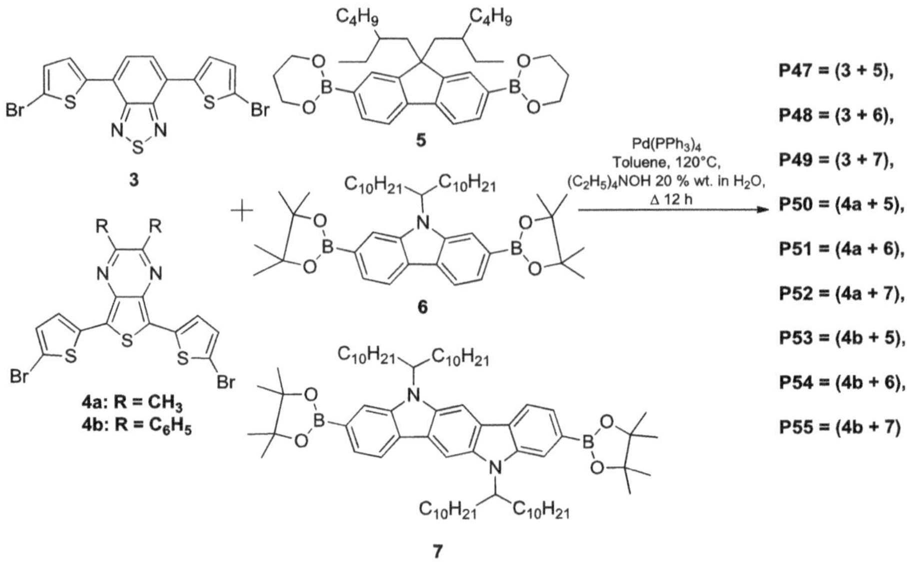

Scheme 18.1

The isomer of BT, thieno[3,4-*b*]pyrazine (TP) is an another example of an electron-deficient fused heterocyclic unit, which has been introduced extensively into polymer conjugated backbones, creating a host of new low bandgap polymer (**P50–55**).[52,53] The low bandgap characteristics for these materials is influenced by the enhanced quinoidal nature of the polymer backbone as a result of the fused pyrazine ring, the lack of significant steric interactions between the pyrazine and neighbouring thiophenes allowing a fully planar backbone geometry.[52,53]

18.2 Experimental Procedures

18.2.1 Materials and General Polymerization Procedures

The monomers presented here have been prepared *via* multistep syntheses, each having been derivatized with desirable functional groups as per device/application requirements. Cited papers for the polymerizations presented here will refer the reader to the appropriate journals for the corresponding monomer synthesis. For our own work, glassware consists mainly of three-necked round-bottomed flasks, Schlenk tubes, microwave vials, that have been baked, dried effectively and kept under vacuum to remove any residuals of water or air. All polymerization reactions are performed under Schlenk conditions with care taken to eliminate the presence of air and water. The crude reaction mixture is conventionally precipitated into excess methanol (MeOH) followed by filtration of the solid, which is then subjected to successive Soxhlet extraction with acetone, MeOH, dichloromethane (CH_2Cl_2) and chloroform ($CHCl_3$) as required. Soxhlet extraction is an effective purification technique to remove undesired small molecules, oligomers and residual catalyst by solubilizing them in the more polar solvents, leaving relatively pure polymer in the final extraction. Transition-metal-based catalysts such as bis(cyclooctadiene)nickel(0) ($Ni(COD)_2$), palladium(II) acetate $Pd(OAc)_2$, and tris(dibenzylideneacetone)-dipalladium(0) ($Pd_2(dba)_3$) were purchased from commercial sources; tetrakis(triphenylphosphine)palladium(0) ($Pd(PPh_3)_4$) was prepared following the standard reported procedure.[54]

For electropolymerization an anode made of glassy carbon (CWE), gold (Au), platinum disk or plate (Pt), or tin-doped indium-oxide-coated glass (ITO) are commonly used as working electrodes. A platinum wire or mesh is used as the counterelectrode. The potential is measured *vs* silver/silver chloride (Ag/AgCl) or saturated calomel (SCE) electrodes. Alternatively, a silver wire can be used as a pseudoreference with calibration to the half-wave potential ($E^{1/2}$) of the ferrocene/ferrocenium (Fc/Fc^+) redox couple. Organic solvents that have a large potential window, stability, high relative permittivity with good dissociation of the electrolyte, thus creating an ionic conductive medium are preferred (*e.g.* acetonitrile (MeCN), dichloromethane (CH_2Cl_2), propylene carbonate ($C_4H_6O_3$)). Organic solvents are either anhydrous (HPLC) commercial grade or degassed by Ar/N_2 prior to use. The supporting electrolyte typically used is *n*-tetrabutylammonium hexafluorophosphate ($TBAPF_6$, 0.1 M). The monomer concentration is normally in the range of 10^{-4}–10^{-3} M.

18.2.2 Dithiin-Based Polymers

Example 1: Polymerization of 3,4-Ethylenedithiathiophene (8) with Ferric Chloride to give P1[1]

Anhydrous $FeCl_3$ (3.5 mmol, 0.57 g) is dissolved in 80 mL of CH_3CN and stirred for 10 min. To this red-orange solution, a solution of 3,4-ethylenedithiathiophene **8** (0.86 mmol, 0.15 g) in CH_3CN (20 ml) is added dropwise. A dark-green precipitate immediately forms and the reaction is stirred for 24 h under N_2 at room temperature. The dark-green solid is filtered through a glass frit and washed with fresh CH_3CN (3×20 mL) and dry under vacuum for 24 h. An amount of 0.076 g of product is obtained in the doped form (~0.25 $FeCl_4^-$ per repeat unit). To obtain the neutral form of the polymer, first perform Soxhlet extraction of the green solid with MeOH, followed by acetone, and then treated with 20 mL of $N_2H_4 \cdot H_2O$ for 12 h. A brown neutral poly(3,4-ethylenedithiathiophene) **P1** is achieved in 43% yield (0.065 g). Elemental analysis (EA): Energy-dispersive spectroscopy (ESD) showed almost no impurity of Fe and Cl (<0.5%). Calcd (%) for $C_6H_4S_5$: C 41.9; H 2.30. Found (%): C 40.38; H 2.34. GPC: $M_n = 3.03 \times 10^3$, $M_w = 5.75 \times 10^3$, PDI = 1.9, using NMP/0.5 wt% LiCl as the eluent (Scheme 18.2).[1]

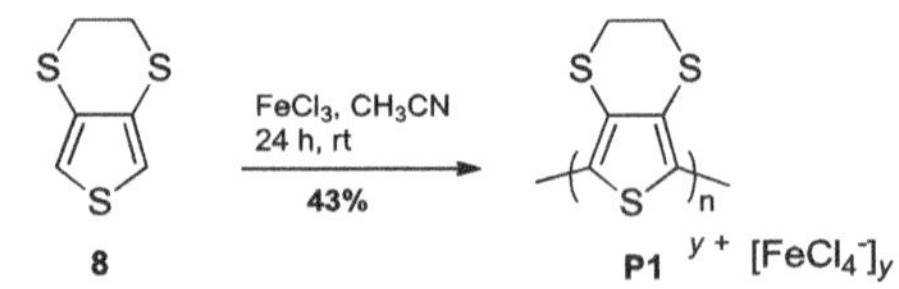

Scheme 18.2

Example 2: Electropolymerization of 3,4-Ethylenedithiathiophene (8) using Bulk Electrolysis[1]

Bulk doped polymer is obtained under potentiostatic electrochemical conditions, holding the oxidation potential of an electrolytic solution of EDTT **8** at +1.15 V *vs* SCE at ambient temperature for 45 min. The three electrode single compartment cell contains a 20 mM solution of the monomer (0.40 mmol, 70 mg) with 0.1 M $(Bu_4N)ClO_4$ (2.00 mmol, 648 mg) as electrolyte in acetonitrile (20 mL, HPLC Aldrich grade), platinum plate as a working electrode with an area of 1.8 cm^2, a platinum wire as counterelectrode and SCE as a reference electrode. The solution is degassed by bubbling with argon for 20 min prior to use and maintained under an argon blanket whilst performing bulk electrolysis. After deposition, the film is rinsed with neat CH_3CN and dried under vacuum overnight. The doped polymer has ~0.33 equiv. of ClO_4^- per repeat unit. Neutral (undoped) polymer, is obtained by electrochemically reducing the doped film at −0.4 V *vs.* SCE until the residual cathodic current reached a constant value, then rinsed, and further chemically reduced by hydrazine hydrate under nitrogen for 12 h. The polymer is then washed with MeOH and dried under vacuum overnight, yielding a brown powder of neutral poly-(3,4-ethylenedithiathiophene) **P1** (9 mg, 13%). EA/ESD shows no chloride impurity. Calcd. (%) for $C_6H_4S_5$: C 41.90; H 2.30. Found (%): C 39.80; H 2.75. GPC: $M_n = 4.75 \times 10^3$, $M_w = 8.08 \times 10^3$, PDI = 1.7, using NMP/0.5 wt% LiCl as an eluent.[1]

Example 3: Oxidative Polymerization of 9-(5,7-Di(thiophen-2-yl)-[1,3]-dithiolo[4,5-b]thieno[3,4-e][1,4]dithiin-2-ylidene)-2,5,7-trinitro-9 H-fluorene-4-carboxylic Acid (Triethylene Glycol Monomethyl Ether) Ester (9) to give P7a[10]

To a solution of **9** (0.15 mmol, 131 mg) in nitrobenzene (30 mL) at 60 °C, anhydrous iron(III)chloride (1.23 mmol, 200 mg) is added. After stirring the reaction mixture at 60 °C for 20 h, the volume of the solvent is reduced to *ca.* 15 mL under vacuum. The precipitate collected after the addition of methanol (50 mL) is washed with MeOH and ether to give **P7a** as a black-violet solid (110 mg, 84%). EA: Calcd. (%) For $C_{36}H_{23}$ $N_3O_{11}S_7$: C, 48.16; H, 2.56; N, 4.68. Found (%): C, 47.41; H, 2.40; N, 3.81 (Scheme 18.3).[10]

Example 4: Yamamoto Polymerization of 5,7-Dibromo-2,3-bis(hexylthio)-[1,4]dithiino[2,3-c]furan (10) to give P12[14]

Bis(1,5-cyclooctadienyl) nickel(0) (1.27 mmol, 350 mg) is weighed out in a glovebox and sealed in a reaction flask equipped with a condenser. Toluene (5 mL) and DMF (5 mL) were added and the solution stirred under Ar.

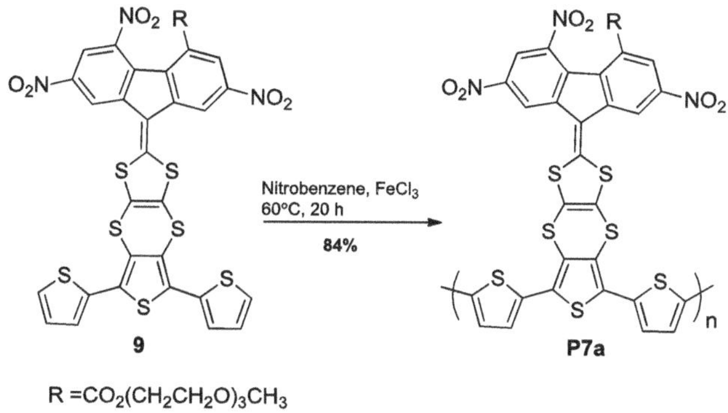

Scheme 18.3

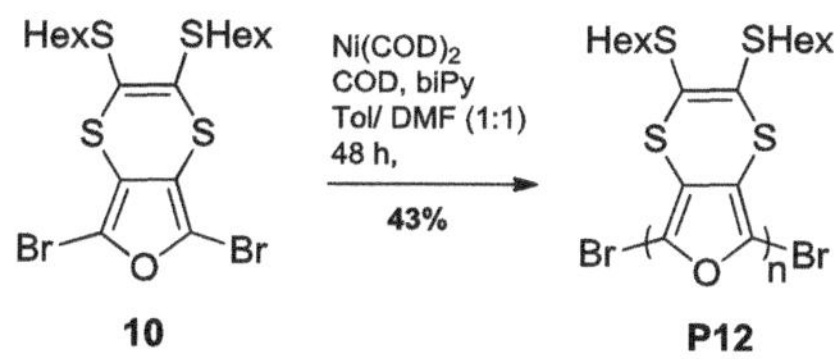

Scheme 18.4

A solution of 5,7-dibromo-2,3-bis(hexylthio)-[1,4]dithiino[2,3-c]furan (monomer **10**) (0.91 mmol, 1.0 eq., 496 mg), 1,5-cyclooctadiene (0.91 mmol, 0.11 mL) and 2,2′-bipyridyl (1.27 mmol, 1.40 eq., 199 mg) in toluene (5 mL) and DMF (5 mL) is added, causing an immediate formation of a dark brown solution. The reaction mixture is heated to 80 °C and stirred under Ar for 48 h. After this time, the reaction mixture is cooled to room temperature and concentrated to a dark brown residue. Cold MeOH (150 mL) is added, causing a precipitate to form. The precipitate is filtered and dried, then subjected to Soxhlet extraction with MeOH for 18 h, acetone for 24 h and CH_2Cl_2 for 24 h. The organic layer is concentrated to afford **P12** as a dark brown, shiny solid (214 mg, 43%); ^1H NMR (500.13 MHz, $CDCl_3$): δ_H 0.70–0.97 (6 H, bm), 1.10–1.65 (16 H, bm), 2.90–2.96 (4 H, bm); ν_{max} ATR-IR/cm^{-1}: 2954, 2924, 2853, 1444, 1412, 1282, 1256; GPC: $M_w = 6\ 200$, $M_n = 3\ 800$, PDI = 1.63, using chlorobenzene as eluent. TGA: heating from 40–250 °C at 5 °C/min gave a 11% mass loss, with an onset at 170 °C. A second run heating from 40–550 °C at 10 °C/min gave an overall loss of 60% mass with an onset at 170 °C. DSC, N_2: (T_g) at 86 °C (heating cycle), (T_c) at 69 °C (cooling cycle), polymer degradation above 150 °C (heating cycle) (Scheme 18.4).[14]

18.2.3 TTF-Based Polymers

Example 5: Yamamoto Polymerization of 4,6-Dibromo-2-(4,5-bis(hexylsulfanyl)-1,3-dithiol-2-ylidene)-thieno[3,4-d][1,3]dithiole (11) to give P14[55]

Compound **11** (0.86 mmol, 560 mg) is dissolved in anhydrous 1:1 DMF-toluene (10 mL) and added dropwise to a solution of $Ni(COD)_2$ (1.16 mmol, 320 mg),

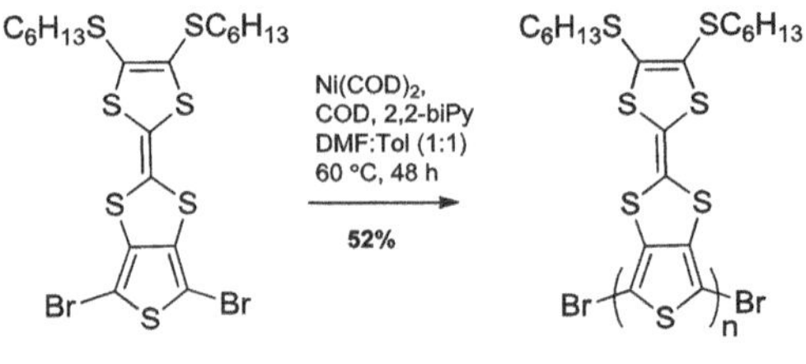

Scheme 18.5

COD (0.1 mL) and 2,2'-bipy (1.15 mmol, 180 mg) in DMF (10 mL). The reaction is stirred under nitrogen in the dark for 48 h at 60 °C. The mixture is then poured into MeOH (200 mL) and the resulting precipitate is washed sequentially with MeOH, acetone and CH_2Cl_2 (each for 16 h) using a Soxhlet apparatus. The CH_2Cl_2 fraction is evaporated and the residue dried under vacuum to leave a dark red film. The polymer is redissolved in a small amount of CH_2Cl_2 and then added to MeOH (30 mL); the precipitate formed is collected by filtration to afford the product as a deep red powder **P14** (220 mg, 52% yield); EA for $C_{500}H_{650}Br_2S_{175}$: Calcd. (%) C, 48.35; H, 5.24; Br, 1.29; Found (%) C, 47.36; H, 5.16; Br, 1.25; corresponding to DP = 25 units; ^1H NMR (300 MHz, $CDCl_3$): δ_H 0.96 (6 H, br s), 1.41 (12 H, br s), 1.72 (4 H, br s), and 2.90 (4 H, br s); ν_{max} (KBr)/cm^{-1} 2924.5, 2851.2, 1458.6, 1440.3, 1260.2, 760.5 and 722.9 (Scheme 18.5).[55]

Example 6: Stille Crosscoupling Polymerization of Dibromothieno-TTF (11) and Distannyl Derivative of Bithiophene (12) to give P15[18]

To 20 mL of freshly distilled anhydrous toluene under a N_2 atmosphere, monomer **11** (0.77 mmol, 500 mg) and 5,5′-bis(trimethylstannyl)-2,2′-bithiophene **12**[56] (0.77 mmol, 380 mg) is added and the resultant mixture is stirred at room temperature for 10 min to yield a yellow solution. Maintaining the inert atmosphere, 5 mol% (0.0385 mmol, 45 mg) of tetrakis(triphenylphosphine)palladium[0] catalyst ($Pd(PPh_3)_4$) is added, and the subsequent mixture is refluxed for 24 h to yield a purple/black solution. After cooling to room temperature the volume is reduced to ~5 mL under vacuum and a large excess of MeOH is added (100 mL), precipitating a dark blue/black powder. Purification is achieved by Soxhlet extraction with MeOH, acetone and finally CH_2Cl_2. The CH_2Cl_2 fraction is reduced in volume to ~5 mL and added dropwise to 100 mL MeOH to precipitate the product. Filtration and drying yielded a black powder of **P15** (263 mg, 55%). GPC: $M_n = 4\ 886$ and $M_w = 11\ 740$ with $CHCl_3$ as eluent. TGA: 40% mass loss within the range of 227–400 °C (Scheme 18.6).[18]

Example 7: Stille Crosscoupling Polymerization of Dibromothieno-TTF (11) and 1,2-Bis(tributylstannyl)ethylene (13) to give P16[18]

The monomer **11** (0.825 mmol, 540 mg) and 1,2-bis(tributylstannyl)ethylene **13**[57] (0.825 mmol, 500 mg), are added to 10 mL of freshly distilled anhydrous

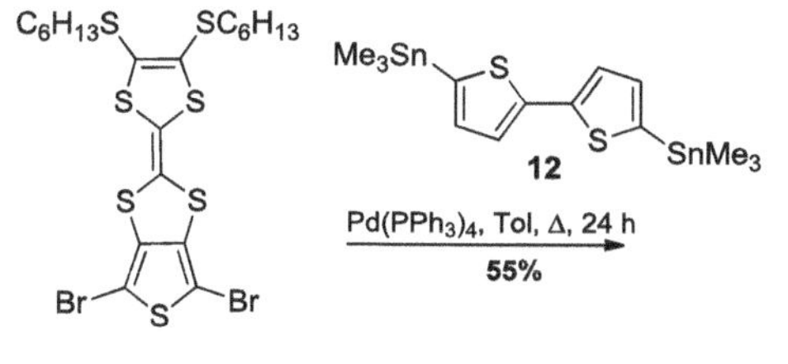

11 P15

Scheme 18.6

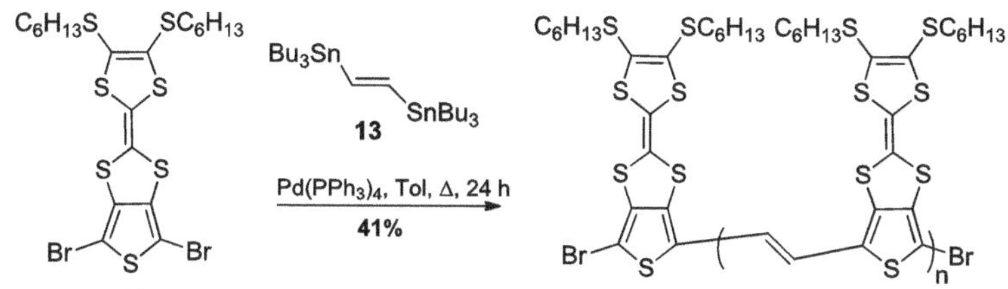

11 P16

Scheme 18.7

toluene under a nitrogen atmosphere. After stirring the mixture at room temperature for 10 min, a 5 mol% of tetrakis(triphenylphosphine)palladium[0] catalyst (0.041 mmol, 48 mg) is added and the mixture is refluxed for 18 h to yield a purple/black solution. After cooling to room temperature the volume is reduced to ~5 mL under vacuum, and the crude polymer is precipitated as a dark blue/black powder by a large excess of methanol (100 mL). Purification is achieved by Soxhlet extraction with methanol, acetone and finally dichloromethane. The dichloromethane fraction (dark blue) was reprecipitated from a ~5 mL of CH_2Cl_2 and 100 mL of methanol. Filtration and drying yielded a blue/black powder **P16** (316 mg, 41%). This fraction is used for complete characterization. EA: Calcd. (%) for $n=6$: C, 48.66; H, 5.21; Br, 4.26; S, 41.87. Found (%): C, 48.90; H, 5.11; Br, 4.09; S, 41.99. GPC: $M_n =$ 3 158 and $M_w = 3750$ using $CHCl_3$ as eluent. TGA: decomposition begins at 167 °C with 8% mass loss by 251 °C and further 40% mass loss by 312 °C (Scheme 18.7).[18]

Example 8: Stille Crosscoupling Polymerization of 1,2-Bis(tributylstannyl)ethylene (13) with 4,6-Diiodo-5-methyl-2-(4,5-bis(hexylthio)-1,3-dithiole-2-ylidene)-5 H-[1,3]dithiolo)[4,5-*c*]pyrrole (14) to give P17[20]

1,2-Bis(tributylstannyl)ethylene **13**[57] is distilled under vacuum immediately prior to use. Compound **13** (0.90 mmol, 340 mg), 4,6-diiodo-5-methyl-2-(4,5-bis(hexylthio)-1,3-dithiole-2- yliden)-1,3-dithiolo[4,5-*c*]pyrrole **14** (0.56 mmol, 410 mg), and Pd(PPh3)4 (0.11 mmol, 130 mg) are dissolved in toluene (5 mL), and the orange mixture is heated under reflux under N_2 for 24 h. The dark purple mixture is poured into MeOH (100 mL). The precipitate formed is

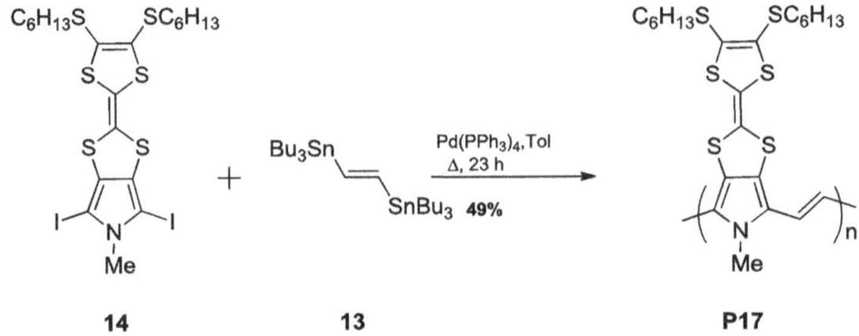

Scheme 18.8

filtered and the resulting solid subjected to Soxhlet extraction with MeOH for 24 h, then acetone for a further 24 h. Subsequent Soxhlet extraction of the remaining solid with CH_2Cl_2 for 24 h gave the main fraction **P17a** (140 mg, 49%), whilst a final extraction with $CHCl_3$ gave polymer **P17b** (7 mg, 2%) after evaporation of the solvent. Multiangle light-scattering MALS GPC analysis of **P17a** with THF as an eluent gave a weight-average molecular weight; $M_w = 317\,500$ with PDI = 1.02 (Scheme 18.8).[20]

18.2.4 DPP- and TDP-Based Polymers

Example 9: Suzuki–Miyaura Crosscoupling Polymerization of DPP and TTF Derivatives (15, 16) to give P22[24]

Compound **15** (0.7 mmol, 760 mg), compound **16** (0.7 mmol, 350 mg), tris-[dibenzylideneacetone]dipalladium(0) (50 mg) and tri-*tert*-butyl phosphonium tetrafluoroborate (60 mg) are dissolved in dry THF (20 mL) and tripotassium phosphate (1.0 mmol, 0.22 g) dissolved in water (3 mL) is added to the solution. The reaction is stirred at 80 °C under N_2 for 48 h, followed by the addition of 2-bromothiophene (0.01 mL, 16 mg, 0.1 mmol). The reaction is maintained at the same temperature and stirred for 6 h. 2-Thienylboronic acid (0.15 mmol, 19 mg) is then added, and the mixture is stirred for a further 6 h before being poured into MeOH (500 mL) and filtered. The solid is purified by Soxhlet extractions with MeOH, acetone and $CHCl_3$. The $CHCl_3$ fraction is redissolved in $CHCl_3$ and precipitated in MeOH to give a dark green–black powder of **P22** (620 mg, 73% yield). ^{1}H NMR ($CDCl_3$): δ_H 8.87–9.50 (2 H, br), 7.40–6.80 (2 H, br), 4.50–3.62 (4 H, br), 2.73–2.40 (6 H, br), 2.12–1.85 (2 H, br), 1.49–1.10 (64 H, br) and 0.94–0.80 (12 H, br); GPC (with chlorobenzene as an eluent): $M_n = 9.6$ kDa, PDI = 3.0. TGA: $T_d = 300$ °C. DSC: $T_c = 175$ °C (Scheme 18.9).[24]

Example 10: Stille Crosscoupling Polymerization of Distannylated Thienothiophene (18) with Dibromo-DPP Derivatives (17) to give P23[25]

3,6-Bis-(5-bromothiophen-2-yl)-2,5-bis(2-octyl-1-dodecyl)-1,4-dioxo-pyrrolo-[3,4-c]pyrrole **17** (0.23 mmol, 0.234 g), 2,5-bis(trimethylstannyl)-thieno[3,2-b]thiophene **18** (0.23 mmol, 0.107 g), bis(triphenylphosphine)palladium(II)

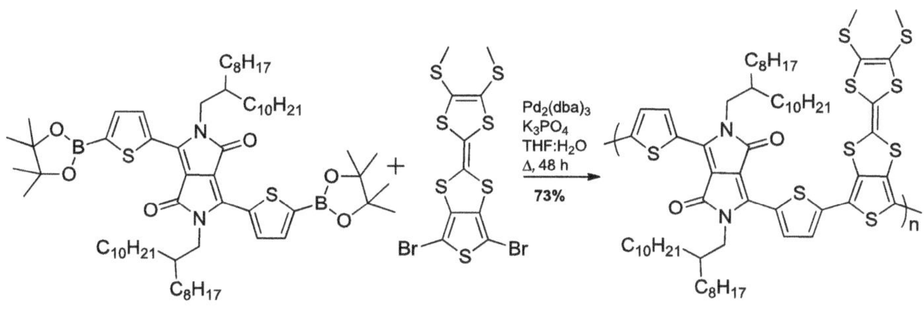

Scheme 18.9

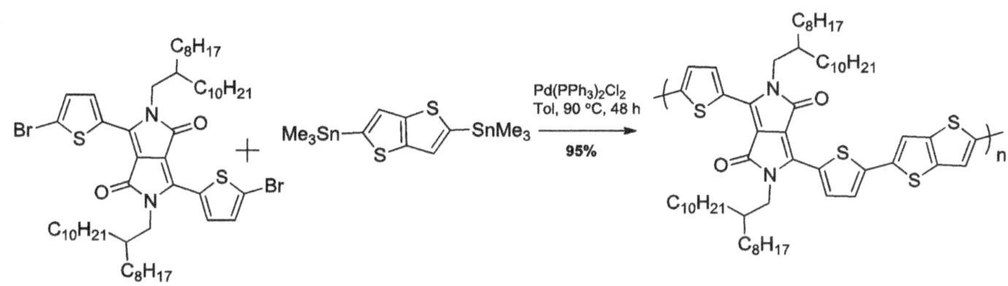

Scheme 18.10

dichloride (0.008 mmol, 5.3 mg), and anhydrous toluene (10 mL) are charged into a 25-mL flask. The mixture is heated to 90 °C and stirred for 48 h, followed by addition of bromobenzene (0.5 mL) to react with the unreacted trimethylstannyl end group. The mixture is further stirred at 90 °C for 6 h before cooling to room temperature, and poured into 200 mL of stirring MeOH. The solid is filtered off, washed with MeOH and dried. The solid is further purified by Soxhlet extraction using hexane, and redissolved with $CHCl_3$. Finally, a blue solid is obtained after removing the solvent **P23** (0.220 g, 95.4%). EA: Calcd. (%) for $C_{60}H_{88}N_2O_2S_4$: C 72.24, H 8.89, N 2.81; Found (%): C 71.98, H 9.05, N 2.83. GPC: $M_w = 211\ 300$, PDI = 2.36. DSC: $T_m = 338$, 357 °C. TGA: $T_d = \sim 350$ °C. UV-vis-near-IR: $\lambda_{max} = 810$ nm (in chloroform); 825 nm (thin film) (Scheme 18.10).[25]

Example 11: Condensation Polymerization of Hydrazine Monohydrate with DPP Derivative (19) to give a Polyazine-DPP Co-polymer P24[27]

To 0.322 g of hydrazine monohydrate is added ethanol to give a hydrazine solution (10.330 g, containing 31.2 mg of hydrazine monohydrate/g). To a 50-mL flask is added the above hydrazine solution (1.075 g, containing 0.67 mmol, 33.5 mg of hydrazine monohydrate) and 5,5′-(2,5-bis(2-octyldodecyl)-3,6-dioxo-2,3,5,6-tetrahydropyrrolo[3,4-c]pyrrole-1,4-diyl)bis(thiophene-2-carbaldehyde) **19** (0.67 mmol, 614.3 mg), ethanol (11 mL), and chloroform

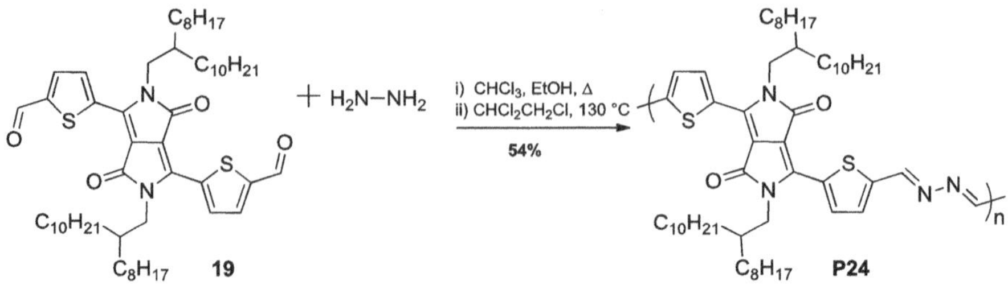

Scheme 18.11

(10 mL). The mixture is refluxed for 40 h and then the solvent is distilled out before 20 mL of 1,1,2-trichloroethane is added. The mixture is further heated at 130 °C for 12 h. The reaction mixture is then cooled to room temperature, diluted with 30 mL of $CHCl_3$, and added dropwise into stirring MeOH (200 mL). After filtration, the solid residue is subjected to Soxhlet extraction using hexane for 24 h to remove oligomers and other impurities. The remaining residue is further extracted with $CHCl_3$ for 24 h. Upon removal of the solvent, bluish-purple films obtained for **P24** (0.330 g, 54.0%). HT-GPC (1,2,4-trichlorobenzene as eluent, 140 °C): $M_n = 102\,016$, $M_w = 438\,410$ and PDI = 4.30. TGA: $T_d \sim 300$ °C. DSC (cooling scan): $T_g/T_c = \sim 125$ °C (Scheme 18.11).[27]

Example 12: Stille Crosscoupling of Thieno[3,4-*c*]pyrrole-4,6-dione (20) with Dithieno[3,2-*b*:2′,3′-*d*]silole (21) using Standard Pd(PPh$_3$)$_4$ Catalyst to give P25_1[28]

In a 50-mL oven-dried one-necked flask, 4,4-bis(2-ethylhexyl)-2,6-bis-(trimethyltin)-dithieno[3,2-*b*:2′,3′-*d*]silole **21** (0.534 mmol, 397.5 mg) and 1,3-dibromo-5-octylthieno[3,4-*c*]pyrrole-4,6-dione **20** (0.507 mmol, 214.7 mg), is added. The flask is purged 3 times with Ar, and then $Pd(PPh_3)_4$ (0.035 mmol, 40 mg), is added in a N_2-filled glove-box. Degassed toluene/DMF (8 mL, 10:1, v/v) is added under Ar atmosphere, and the resulting solution is stirred at 115 °C under Ar. After 42 h, trimethylphenyltin (0.08 mmol, 19 mg) in 0.75 mL of degassed toluene/DMF (10:1, v/v) is added to the reaction flask and the reaction is kept at 115 °C for an additional 5 h. Bromobenzene (0.32 mmol, 50 mg) in 0.75 mL of degassed toluene/DMF (10:1, v/v) is then added to the reaction flask, and the reaction is kept at 110 °C for an additional 8 h to complete the end-capping reaction. The dark-blue solution is cooled to room temperature and precipitated into MeOH. The resultant polymer is collected by filtration, dried and extracted successively with hexanes and CH_2Cl_2 using a Soxhlet extraction apparatus. The remaining solid is extracted with 200 mL of $CHCl_3$. After evaporation of the $CHCl_3$ under reduced pressure, MeOH is added and the polymer **P25_1** is collected by filtration (130 mg, 38% yield). ^{1}H NMR (400 MHz, 1,2-dichlorobenzene-d$_4$, 50 °C, ppm): δ_H 8.59 (s, 1 H), 7.52 (s, 1 H), 3.64 (br, 2 H), 2.10–0.60 (m, 49 H). EA: Calcd. (%) for $C_{38}H_{53}NO_2S_3Si$: C, 67.11; H, 7.86; N, 2.06, O, 4.70; S, 14.14; Si, 4.13. Found: C, 66.65; H, 7.85; N, 2.39; S, 13.97 (Scheme 18.12).[28]

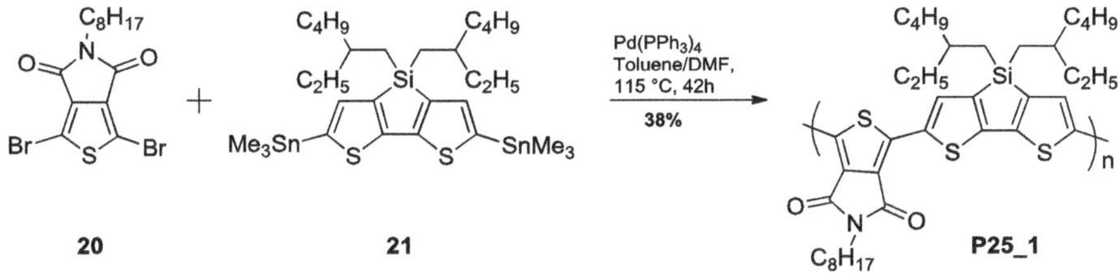

Scheme 18.12

Example 13: Stille Crosscoupling Polymerization of Distannylated-dithieno[3,2-*b*:2′,3′-*d*]silole/germole (21, 22) with Dibromo-thieno[3,4-*c*]pyrrole-4,6-dione (20) to give P25_2 and P26[29]

2,2′-Bistrimethylstannyl-4,4′-bis-(2-ethylhexyl)-dithieno[3,2-b:2′,3′-d]germole **22** (0.881 mmol, 695 mg), 1,3-dibromo-5-octyl-4*H*-thieno[3,4-*c*]pyrrole-4,6(5*H*)- dione **20** (0.881 mmol, 373 mg), $Pd_2(dba)_3$ (0.007 mmol, 6 mg), and $P(o\text{-tol})_3$ (0.042 mmol, 13 mg) are added to a 50-mL Schlenk tube, which is evacuated and backfilled with Ar four times. Toluene (12 mL, degassed by 4 freeze/pump/thaw cycles) is then added, and the mixture is heated to 115 °C and stirred for 5 days. The contents are then cooled to 60 °C, and a large spatula tip of diethylammonium diethyldithiocarbamate and 25 mL degassed toluene is added and stirred for 1 h. The mixture is then pipetted into 300 mL of MeOH where the polymer precipitates, and this is filtered into a cellulose thimble, and purified by Soxhlet extraction with MeOH (24 h), acetone (24 h) hexanes (24 h), CH_2Cl_2 (24 h), and $CHCl_3$ (3 h). The $CHCl_3$ soluble fraction is filtered through a 0.45-μm pore size PTFE syringe filter directly into acetone, and the solid is collected and residual solvent removed under vacuum (24 h) to give polymer **P26** as a dark solid (466 mg). ^1H NMR ($CDCl_3$, 500 MHz): δ_H 8.46 (br, 1 H), 7.4 (br, 1 H), 3.72 (br, 2 H), 2.04 (br, 2 H), 1.8–0.7 (br, 47 H). EA Calc. for $C_{38}H_{53}GeNO_2S_3$: C 62.98, H 7.37, N 1.93. Found C 63.26, H 7.55, N 1.85.[29]

Similar reaction and purification conditions are adopted for the silole polymer analog **P25_2**, (286 mg). GPC: $M_n = 31\ 300$, $M_w = 53\ 800$ and a PDI = 1.71 using $CHCl_3$ as eluent, bulk material results. ^1H NMR ($CDCl_3$, 500 MHz): δ_H 8.5 (bs, 1 H), 7.5 (bs, 1 H), 3.5 (bs, 2 H), 2.2–0.8 (m, 49 H). EA Calc. for $C_{38}H_{53}SiNO_2S_3$: C 67.11, H 7.85, N 2.06. Found: C 67.44, H 7.85, N 2.06 (Table 18.2 and Scheme 18.13).[28]

Example 14: Direct Heteroarylation Polycondensation of 5-Alkythieno[3,4-*c*]pyrrole-4,6-dione (23) with Dibromo-dialkyl-bithiophene (24) to give P27[30]

5-(2-Hexyldecyl)-5 H-thieno[3,4-c]pyrrole-4,6-dione **23** (0.25 mmol, 94.4 mg), 5,5′-dibromo-4,4′-dioctyl-2,2′bithiophene **24** (0.25 mmol, 128.1 mg), *trans*-di(μ-acetato)bis[*o*-(di-*o*-tolyl-phosphino)benzyl]dipalladium(II) (4 mol%, 4.71 mg) as the catalyst, (8 mol%, 7.04 mg) of tris(*o*-methoxyphenyl)-phosphine ligand and Cs_2CO_3 (0.50 mmol, 162.9 mg) are put in a Biotage microwave vial (size 2 to 5 mL) equipped with a magnetic stirring bar. The vial is sealed with a cap and then purged with nitrogen to remove oxygen. THF (1 mL) is added and the

Table 18.2 Polymerization yields and GPC characterizations of dithieno[3,2-b:20,30-d]silole/germole - TDP copolymers **P25–26**.

Polymer	*Yield (%)*	M_n *(kDa)*	M_w *(kDa)*	*PDI*
[28]**P25_1**[a]	38	28		1.6
[29]**P26**[b]	73	47.5	82.2	1.73
[29]**P25_2**[b]	58	31.3	53.8	1.71
	–	42,5*	65.3*	1.54*

GPC using $C_6H_5Cl^a/CHCl_3{}^b$ as eluent with polystyrene standards.
*After GPC purification.

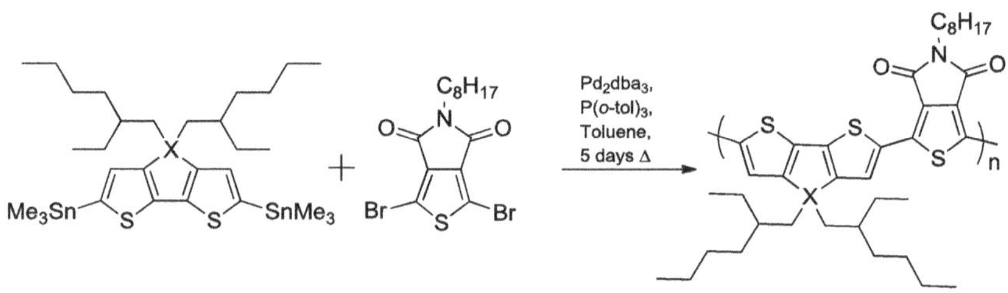

21: X = Si
22: X = Ge

20

P25_2: X = Si, **58%**
P26: X = Ge, **73%**

Scheme 18.13

reaction is heated with an oil bath at 120 °C (reaction under pressure). At the end of the reaction time, the reaction is cooled and the corresponding 5-alkylthieno[3,4-c]pyrrole-4,6-dione is added as a capping agent (50 mg in 1 mL). The solution is heated again at 120 °C for 1 h to complete the end-capping procedure. After an additional hour of reaction, the whole mixture was cooled to room temperature and poured into 500 mL of cold MeOH. The precipitate is filtered. Soxhlet extractions with acetone followed by hexanes removed catalytic residues and low molecular weight materials. Polymers are then extracted with $CHCl_3$. The solvent is reduced to about 10 mL and the mixture is poured into cold MeOH. The precipitate is filtered to provide **P27** with 96% yield. GPC = M_n of 56 kDa. ^{1}H NMR spectra are similar to those reported by Marks and co-workers.[58] DSC: $T_m = 225$ °C (heating cycle) and $T_c = 201$ °C (cooling cycle) (Scheme 18.14).[30]

18.2.5 BODIPY- and Isoindigo-Based Polymers

Example 15: Stille Crosscoupling Polymerization of 4,4-Difluoro-4-bora-3*a*,4*a*-diaza-*s*-indacene (BODIPY) Core Monomer (25) with Bis-EDOT and Bis-EDTT Derivatives (26, 27) to give P29a,b[32]

To a solution of **25** (0.7 mmol, 450 mg or 0.5 mmol, 360 mg) in dry toluene (50 mL) is added one equivalent of a solution either of bis-EDOT **26** (0.7 mmol, 332 mg) or bis-EDTT **27** (0.5 mmol, 233 mg) in dry toluene (25 mL). $Pd(PPh_3)_4$ catalyst (30 mg or 25 mg) is added and the reaction is stirred under nitrogen at

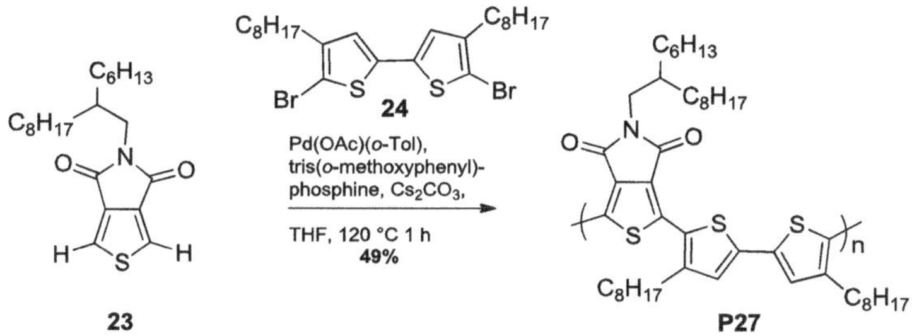

Scheme 18.14

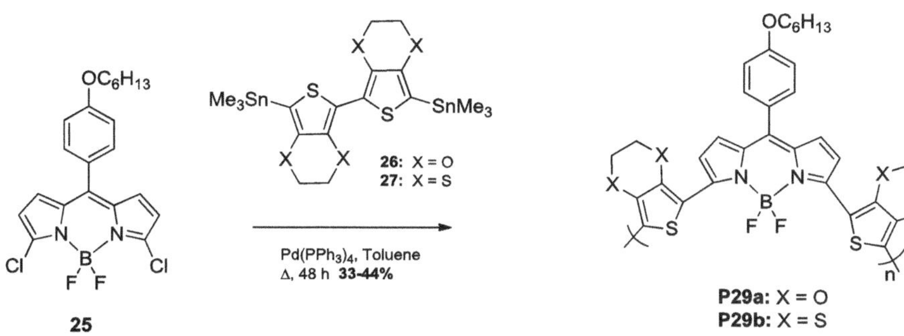

Scheme 18.15

110 °C for 48 h. The reaction mixture is cooled to room temperature and 2-bromothiophene (0.35 mmol, 60 mg or 0.25 mmol, 50 mg) and $Pd(PPh_3)_4$ (10 mg) were added. The reaction is heated to 110 °C for 5 h and cooled again to room temperature to add trimethyl stannylated EDOT (0.35 mmol, 100 mg) or trimethyl stannylated EDTT (0.35 mmol, 125 mg) and stirred at 110 °C again overnight. The crude product is precipitated into methanol and collected by filtration. Soxhlet purification is carried out using methanol, acetone and the main fractions are extracted from CH_2Cl_2 to give black polymers of **P29a-BDP-bis-EDOT** (200 mg, 44%) or **P29b-BDP-bis-EDTT** (220 mg, 33%). ^{1}H NMR ($CDCl_3$): **P29a-BDP-bis-EDOT**: δ_H 7.48 (2 H, br), 7.28 (2 H, br), 7.01 (2 H, br), 6.80 (2 H, br), 4.28 (8 H, br), 4.06 (2 H, br), 1.85 (2 H, br), 1.60–1.20 (6 H, br) and 0.96 (3 H, br). **P29b-BDP-bis-EDTT**: δ_H 7.70–7.34 (4 H, br), 7.13–6.73 (4 H, br), 4.07 (2 H, br), 3.21(8 H, br), 1.86 (2 H, br), 1.63–1.26 (6 H, br) and 0.96 (3 H, br) (Scheme 18.15).

Example 16: Stille Crosscoupling of Distannylated-dithieno-silole (21) with Dibromo-Isoindigo Monomers (28) to give P31[35]

The faint yellow oil 2,2′-bistrimethylstannyl-4,4′-bis-(2-ethylhexyl)-dithieno[3,2-*b*:2′,3′-*d*]silole (monomer **21**) is added to a tared clean glass vial, which is kept under vacuum. The oil, weighing (0.510 mmol, 381.6 mg) once dry, is then diluted in 1 mL of hexanes and transferred to a dry Schlenk flask equipped with a stirrer bar. The dilution/transfer of **21** is repeated four times so that the compound is completely transferred to the reaction Schlenk flask, as a total of 5 mL of hexanes solution. The hexane is then evaporated carefully under vacuum

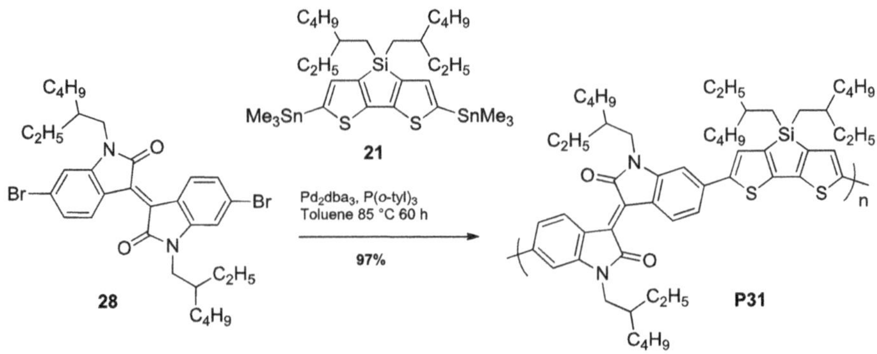

Scheme 18.16

for 2 h at room temperature. After refilling the reaction flask with argon, 6,6′-dibromo-N,N′-(2-ethylhexyl)-isoindigo **28** (0.510 mmol, 330.4 mg), $Pd_2(dba)_3$ ($CHCl_3$ adduct, 0.015 mmol, 15 mg) and P(*o*-tolyl)$_3$ (0.03 mmol, 10 mg) are added to the flask and purged with 3 vacuum/Ar refill cycles. Degassed toluene (5 mL, 5 freeze–pump–thaw cycles prior to addition) is then added to the flask and the reaction medium is stirred and heated to 85 °C under Ar. The reaction medium viscosity had notably increased after 12 h at 85 °C, and is left to stir for an additional 2 days at 85 °C, after which the temperature was increased to 100 °C for 2 h. After the reaction medium is cooled back to 85 °C, a solution of 2-bromothiophene (0.2 mmol) in degassed toluene (2 mL) is added to the flask along with a small amount of the catalytic system and allowed to react for 6 h. A solution of 2-(tributylstannyl)-thiophene (0.2 mmol) in degassed toluene (2 mL) is subsequently added to the flask and allowed to react for 12 h. At this point, the reaction medium is then cooled to 60 °C and a spatula tip of diethylammonium diethyldithiocarbamate is added to the flask. After 2 h of stirring at 60 °C, the reaction medium is precipitated into MeOH (200 mL). The precipitates are filtered and collected into a cellulose thimble, then purified in a Soxhlet apparatus using MeOH (24 h), hexanes (12 h) and $CHCl_3$ (2 h). The MeOH and hexanes fractions are discarded, while the $CHCl_3$ fraction is precipitated into MeOH (200 mL). The precipitates are collected and dried to afford a dark purple solid **P31** (0.494 mg, 97%). ^{1}H NMR ($CDCl_3$, 500 MHz): δ_H 9.4–8.8 ppm (br, 2 H), 7.8–7.0 (br, 4 H), 6.8–6.2 (br, 2 H), 4.0–3.6 (br, 4 H), 2.0–1.0 (br, 40 H), 1.0–0.6 (br, 24 H). EA Calc. for $C_{56}H_{76}O_2S_2Si$: C, 74.61; H, 8.50; N, 3.11. Found: C, 74.22; H, 8.61; N, 3.02.GPC: $M_n = 36\,000$, $M_w = 99\,600$ with a PDI = 2.77, using THF as eluent. TGA: $T_d =$ 410 °C (5% mass loss) (Scheme 18.16).[35]

18.2.6 Azole- and Pyrazine-Based Polymers

Example 17: Condensation Polymerization of Diacids (30–32) with Diaminobenzenedithiol (29) to give P35a–c having Benzobisthiazole Units in the Polymer Backbone[44]

Poly(benzobisthiazole) (P35a-BBT). Compound **29** (15.78 mmol, 3.87 g) is dissolved in 14.66 g of deaerated 77% polyphosphoric acid (PPA). After dehydrochlorination at 70 °C under vacuum, oxalic acid **30** (15.78 mmol, 1.42 g) is added together with 12 g of fresh P_2O_5 under positive pressure. The reaction temperature is raised gradually to 120 °C over 10 h, then to 140 °C, and finally to 180–200 °C. The reaction is allowed to proceed at elevated temperatures (180–200 °C) for 24 h. The mixture is precipitated into water and purified by extraction of PPA with water for 2 days to provide polymer **P35a**. ^{1}H NMR (CD_3NO_2 containing $AlCl_3$): δ_H 8.90–9.20 (2 H, Ar–H). ν_{max} (cm^{-1}) 1467, 1406, 1313, 889, 860, 685.

Poly(benzobisthiazole-2,6-diylvinylene) (P35b-BBTV). In the first attempt to synthesize this polymer, compound **29** (13.87 mmol, 3.4 g) is dehydrochlorinated in 15 g of deaerated 77% PPA. After complete dehydrochlorination at 70 °C, fumaric acid **31** (13.87 mmol, 1.61 g) is added together with ~11 g of fresh P_2O_5 to compensate for the water of condensation. The temperature is raised to 120 °C over 6 h. Within a few hours of raising the temperature to 120 °C, there is a rapid increase in the viscosity of the reaction mixture that eventually balled-up the mechanical stirrer. After ~6 h, the highly viscous reaction mixture is precipitated in water, shredded into small pieces with a blender and purified by refluxing in water for 2–3 days. A second polymerization is carried out using a dilute polymer concentration (5 wt%). Compound **29** (6.8 mmol, 1.667 g), is dehydrochlorinated in 18.5 g of de-aerated 77% PPA. Fumaric acid **31** (6.8 mmol, 0.789 g) and P_2O_5 (9 g) are added. The temperature is gradually raised to 120 °C over 6 h, then 160 °C, and finally to 180 °C. The mixture becomes shiny dark green and is allowed to proceed for 8 h at 180 °C before precipitating the polymer in water. The polymer is shredded into small pieces with a blender to facilitate purification, which consisted of extraction for 2 days with a large volume of water to provide polymer **P35b**. ^{1}H NMR (CD_3NO_2 containing $AlCl_3$): δ_H 8.65 (2 H, Ar-H), 9.2–9.33 (2 H, -HC=CH-). ν_{max} (KBr)/cm^{-1} 1663, 1600, 1515, 1400, 1313, 1052, 944, 856, 689.

Poly(benzobisthiazole-2,6-diyldivinylene) (P35c-BBTDV). Compound **29** (17.57 mmol, 4.31 g) is reacted with *trans,trans*-muconic acid **32** (17.57 mmol, 2.5 g), following the same procedure outlined for PBTV synthesis. The polymer concentration is 5.5 wt%. The reaction temperature is raised to 80 °C and held for 6 h at this temperature before finally increasing it to 120 °C. The reaction is allowed to proceed overnight for a total of 16 h. The polymer is precipitated in water and purified in a similar way to PBTV purification to provide **P35c**. δ_H 9.02 (2 H, Ar–H), 8.21 (2 H, -HC=CH-), 7.82 (2 H, -HC=CH-). ν_{max} (KBr)/cm^{-1} 1601, 1480, 1400, 1315, 1053, 974, 845, 680 (Scheme 18.17).

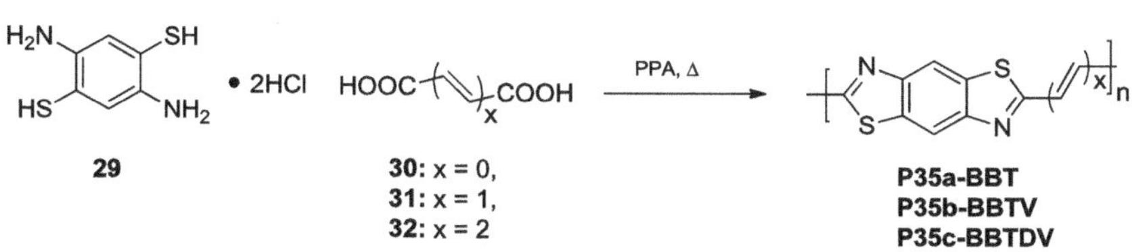

Scheme 18.17

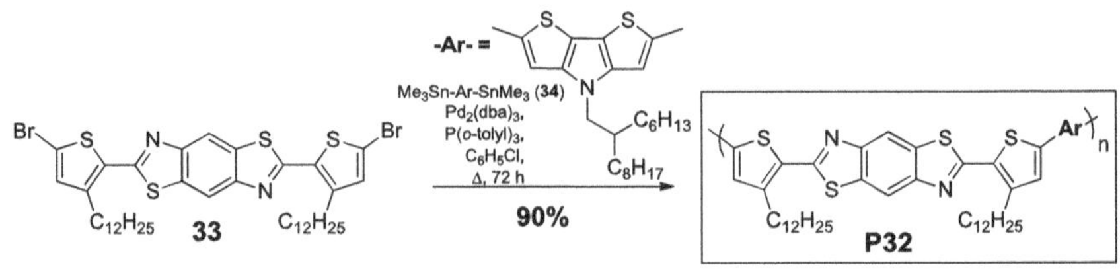

Scheme 18.18

Example 18: Stille Cross-coupling Co-polymerization of 2,6-Bis(5-bromo-3-dodecylthiophen-2-yl)benzo[1,2-*d*:4,5-*d'*]bisthiazole (33) and 4-(2-Hexyldecyl)-2,6-bis-(trimethylstannyl)-4 H-dithieno[3,2-*b*:2',3'-*d'*]pyrrole (34) to give P32-PBTDTP[43]

A mixture of **33** (0.246 mmol, 210 mg), 4-(2-hexyldodecyl)-2, 6-bis-(trimethyltin)-4*H*-dithieno[3,2-*b*:2',3'-*d'*]pyrrole **34** (0.246 mmol, 180 mg), tris(dibenzylidene-acetone)dipalladium(0) (0.0049 mmol, 4.5 mg), and tris-*o*-tolylphosphine (0.0197 mmol, 6.0 mg) in 5.4 mL of chlorobenzene is refluxed for 72 h. The reaction solution is poured into 200 mL of 5% hydrochloric acid/methanol solution and stirred for 5 h. The filtered solid is subjected to Soxhlet extraction with MeOH and hexane for 24 h each. **P32-PBTDTP-1** is extracted with $CHCl_3$ to give a shiny green solid (68.8 mg), whereas the insoluble fraction **P32-PBTDTP-2** (172 mg) remains on the filter paper (overall yield was 241 mg, 90%). ^{1}H NMR ($C_6D_4Cl_2$, 60 °C): δ_H 8.43–8.17 (m, 4 H), 7.30–7.00 (m, 2 H), 4.10 (s, br, 2 H), 3.18 (s, br, 4 H), 1.91–0.98 (m, 77 H). DSC: no thermal transitions observed up to 390 °C (Scheme 18.18).[43]

Example 19: Stille Crosscoupling Polymerization using Bis-trimethylstannyl derivative of Bisthienylbenzobisthiazole (35) to give P33a,b

Poly[(4,4'-bis(2-ethylhexyl)-4*H*-cyclopenta[2,1-*b*:3,4-*b'*]dithiophene)-2,6-diyl-*alt*-(2,5-bis(3-dodecylthiophen-2-yl)benzo[1,2-*d*;4,5-*d'*]bisthiazole)] (P33a-PBTPDT)[43]

A mixture of **35** (414 mg, 0.406 mmol), 2,6-dibromo-4,4'-bis(2-ethylhexyl)-4*H*-cyclopenta[2,1-*b*:3,4-*b'*]dithiophene **36** (0.406 mmol, 228 mg), tris(dibenzylidene-acetone)dipalladium(0) (0.008 08 mmol, 7.4 mg), and tris-*o*-tolylphosphine (0.0315 mmol, 9.9 mg) in 6 mL of chlorobenzene is refluxed for 72 h. A Similar precipitation and purification procedure to the one described for **P32** is followed. The product is dried in a vacuum oven for 12 h and collected as a dark solid **P33a-PBTPDT** (381 mg). ^{1}H NMR ($C_6D_4Cl_2$, 60 °C): δ_H 8.48 (s, 2 H), 7.44–7.27 (m, 4 H), 3.16 (s, 4 H), 2.12–0.82 (m, 80 H) (Scheme 18.19).[43]

Poly(4,4'-bis(2-ethylhexyl)dithieno[3,2-*b*:2',3'-*d*]silole)-2,6-diyl-*alt*-(2,5-bis(3-dodecylthiophen-2-yl)benzo[1,2-*d*;4,5-d']bisthiazole) (P33b-PBTEHS)[43]

The protocol follows a similar synthetic procedure as detailed for **P33a-PBTPDT**, this time using one equivalent of 2,6-dibromo-4,4-bis(2-ethylhexyl)-dithieno[3,2-*b*:4,5-*b'*]silole **37** (0.374 mmol, 216 mg) in 20 mL of chlorobenzene.

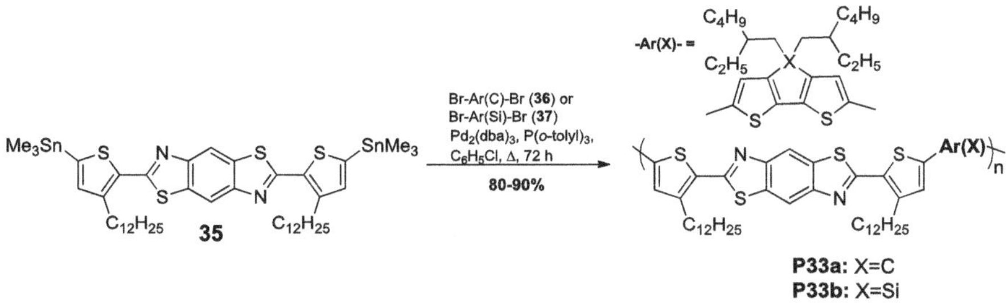

Scheme 18.19

Table 18.3 Polymerization yields and GPC characterization of BBT polymers **P32–33**.

Polymer	*Yield (%)*	$M_n{}^a$ *(kDa)*	$M_w{}^a$ *(kDa)*	*PDI*^{a}
P32-PBTDTP-1	26	10.3	32.1	3.12
P32-PBTDTP-2	64	52.4	147.6	2.82
P33a-PBTPDT	86	9.87	13.9	1.41
P33b-PBTEHS	92	7.8	19.2	2.46

^{a}GPC with chlorobenzene as an eluent at 60 °C using polystyrene standards.[43]

The isolation and purification procedure is similar to that for **P33a-PBTPDT**. The product **P33b-PBTEHS** is collected as a dark solid (380 mg). ^{1}H NMR ($CDCl_3$): δ_H 8.48 (br, 2 H), 7.34–7.31 (br, 4 H), 3.06 (m, 8 H), 1.81 (br, 8 H), 1.30 (br, 50 H), 0.91 (br, 18 H). DSC analyses of the polymers **P33a,b** showed no thermal transitions observed up to 390 °C (Table 18.3).[43]

Example 20: Synthesis of Dialkoxy-Substituted Poly(phenylenevinylene) Benzobisthiazoles (P36 and P37) by Horner–Wadsworth–Emmons Reaction[47]

Benzobisoxazole monomer (**38** or **39**) (1.00 mmol, 460 mg) and 2,5-didodecyloxy-1,4-benzene dicarboxaldehyde **40** (1.00 mmol, 503 mg) are dissolved in THF (15 mL). Then, a 1.0 M solution of potassium *tert*-butoxide in THF (2.5 mmol, 2.5 mL) is added to the solution and the mixture stirred at room temperature for 48 h to obtain an opaque, dark red solution. An additional amount of THF (10 mL) is added and the reaction stirred for a further 72 h. The solution is then precipitated by pouring into MeOH and filtered to obtain a red-orange solid. The resultant polymer is purified *via* Soxhlet extraction (MeOH, hexane and THF). Evaporation of the THF layer yields the polymers **P36-*cis*-DDO-PPVBBO** and **P37-*trans*-DDO-PPVBBO**. ^{1}H NMR (400 MHz, THF-d_8): **P36-*cis*-DDO-PPVBBO**: δ_H 0.886 (m, CH_3-), 1.29 (br m, -$(CH_2)_{10}$-), 4.16 (br m, -OCH_2-), 7.72–8.10 (br m, Ar-H and vinylic peaks), 10.41 terminal aldehyde peak, 10.68 terminal P(=O)(OH)2. **P37-*trans*-DDO-PPVBBO**: δ_H 0.866 (m, CH_3-), 1.29 (br m,-$(CH_2)_{10}$-), 4.17 (br m, -OCH_2-), 7.62–8.079 (br m, Ar-H and vinylic peaks). TGA shows both polymers to be thermally stable with mass loss onset occurring only above 300 °C under N_2[47] (Table 18.4 and Scheme 18.20).

Table 18.4 Polymerization yields, GPC characterization and thermal properties of BBO polymers **P36–37**.

Polymer	*Yield (%)*	$M_n{}^a$ *(kDa)*	$M_w{}^a$ *(kDa)*	*PDI*	$T_d{}^b$ *(°C)*	$T_g{}^c$ *(°C)*
P36-*cis*-DDO-PPVBBO	40	5.9	14.4	2.45	315	118
P37-*trans*-DDO-PPVBBO	53	6.8	18.1	2.64	304	99

[a]GPC using THF as eluent at 35 °C with polystyrene standards.
[b]5% weight loss temperature by TGA under N_2.
[c]Glass-transition temperature reported from 2nd DSC scan, heating rate 15 °C min^{-1} under N_2.[47]

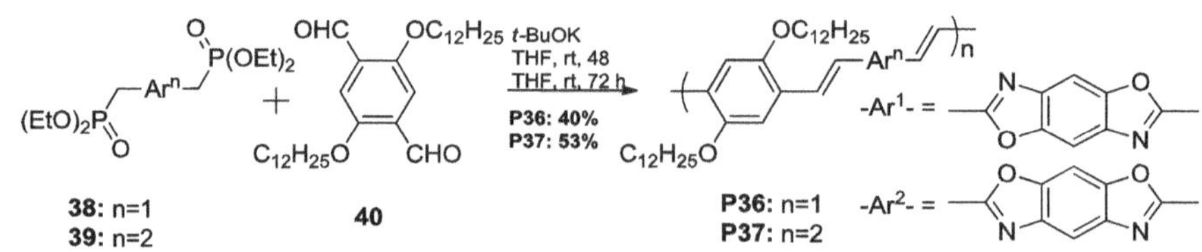

Scheme 18.20

Example 21: Sonogashira Crosscoupling of 4,8-Dibromo-2,6-dihexylbenzo[1,2-*d*;4,5-*d*′]bisoxazole (41) with Diethynyl Comonomers to give P38-43[48]

To a flame-dried 50-mL Schlenk flask is added 4,8-dibromo-2,6-dihexyl-benzo[1,2-*d*;4,5-*d*′]bisoxazole **41** (0.5 mmol, 243.1 mg), diethynyl comonomer (0.5 mmol), N,N-diisopropylamine (6 mL), and toluene (12 mL). The mixture is degassed by bubbling Ar through it for 30 min followed by addition of Pd(PPh3)4 (2 mol%, 11.6 mg) and copper iodide (5 mol%, 4.8 mg). The reaction is heated to 70 °C and stirred under Ar for 24 h. The polymer is then precipitated into MeOH and the solid is filtered and subjected to a Soxhlet extraction with MeOH followed by acetone to remove residual catalyst and low molecular weight material and finally with THF to dissolve the polymer. Evaporation of the THF solvent yields the polymers **P38–43**, as solids of varying colors (Scheme 18.21).[48]

^{1}H NMR (400 MHz, $CDCl_3$): **P38-PBOCz-O**: δ_H 0.91 (9 H), 1.27–1.49 (22 H), 1.92–2.05 (6 H), 3.10 (4 H), 4.34 (2 H), 7.42 (2 H), 7.85 (2 H), 8.49 (2 H). **P39-PBOCz-EH**: 0.86–0.92 (12 H), 1.27–1.54 (20 H), 2.00–2.11 (5 H), 3.10 (4 H), 4.21 (2 H), 7.42 (2 H), 7.85 (2 H), 8.49 (2 H). **P40-PBOF-O**: 0.68 (4 H), 0.84 (6 H), 0.95 (6 H), 1.11–1.27 (20 H), 1.41 (8 H), 1.53 (4 H), 2.01 (8 H), 3.09 (4 H), 7.68–7.78 (6 H). **P41-PBOF-DMO**: 0.51–1.44 (56 H), 1.92 (8 H), 3.00 (4 H), 7.62–7.66 (6 H). **P42-PBOP-D**: 0.86–0.95 (12 H), 1.23–1.28 (32 H), 1.39 (8 H), 1.48-1.59 (8 H), 1.98 (8 H), 3.05 (4 H), 4.16 (4 H), 7.24 (2 H). **P43-PBOP-MEH**: 0.81–0.98 (12 H), 1.27–1.52 (20 H), 1.99 (5 H), 3.04 (4 H), 4.02 (5 H), 7.25 (2 H) (Table 18.5).[48]

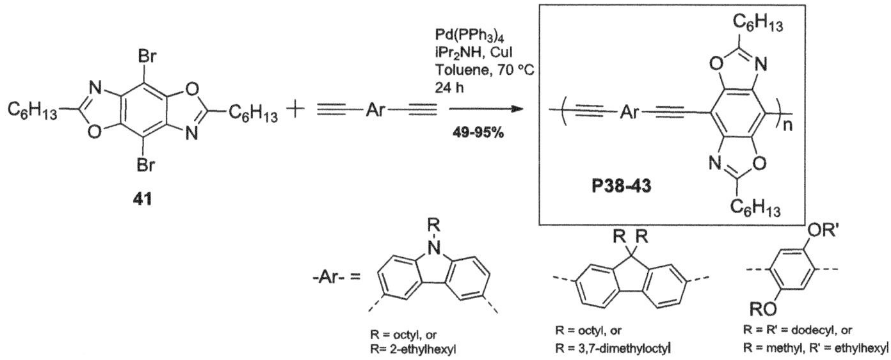

Scheme 18.21

Table 18.5 Polymerization yields, color, GPC characterization and thermal properties of BBO polymers **P38–43**.

Polymer	*Yield (%)*	*Color*	M_n[a] *(kDa)*	M_w[a] *(kDa)*	*PDI*	T_d[b] *(°C)*	T_g[c] *(°C)*
P38-PBOC$_z$-O	74	Yellow	13.0	23.8	1.8	354	139
P39-PBOC$_z$-EH	95	Yellow	47.9	74.9	1.6	353	140
P40-PBOF-O	73	Yellow	38.2	145.0	3.8	352	123
P41-PNOF-DMO	49	Yellow	51.9	188.3	3.6	298	110
P42-PBOP-D	59	Orange	33.1	45.8	1.4	320	-
P43-PBOP-MEH	51	Orange	26.2	87.2	3.3	310	-

[a]Determined by GPC in $CHCl_3$ using polystyrene standards.
[b]5% mass loss temperature by TGA in air.
[c]Glass-transition temperature reported from 2nd scan, heating rate 15 °C/min under N_2.[48]

Example 22: Suzuki Cross-coupling Polymerization of 2,7-Dibromo-4,4,9,9-tetraalkyl-4,9-digerma-4,9-dihydro-*s*-indaceno[1,2-*b*:5,6-*b*′]dithiophene (dibromodigermaindacenodithiophene) (42) with 2,1,3-Benzothiadiazole-4,7-bis(boronic Acid Pinacol Ester) (43) to give P44-PGeTPTBT[50]

In a 20-mL high-pressure microwave reactor tube, equipped with a sealed septum is added compound **42** (0.26 mmol, 260.0 mg), 2,1,3-benzothiadiazole-4,7-bis(boronic acid pinacol ester) **43** (0.26 mmol, 101.5 mg), and $Pd(PPh_3)_4$ (0.005 mmol, 6.0 mg). The tube is sealed and flushed with Ar, and then degassed toluene (3.5 mL), and degassed aqueous 1.0 M Na_2CO_3 (0.7 mL) and 2 drops of Aliquat 336 are added. The solution is thoroughly degassed under Ar, and then the Ar inlet is removed and the reaction is heated for 3 days at 120 °C. After cooling to room temperature, the polymer is precipitated into MeOH (100 mL), and filtered through a Soxhlet thimble. The polymer is extracted (Soxhlet) with MeOH, acetone, hexane and $CHCl_3$. The $CHCl_3$ solution is concentrated and precipitated into MeOH and the precipitant is filtered and dried under vacuum to obtain polymer **P44-PGeTPTBT**, as a brown solid (182 mg, yield: 82%).

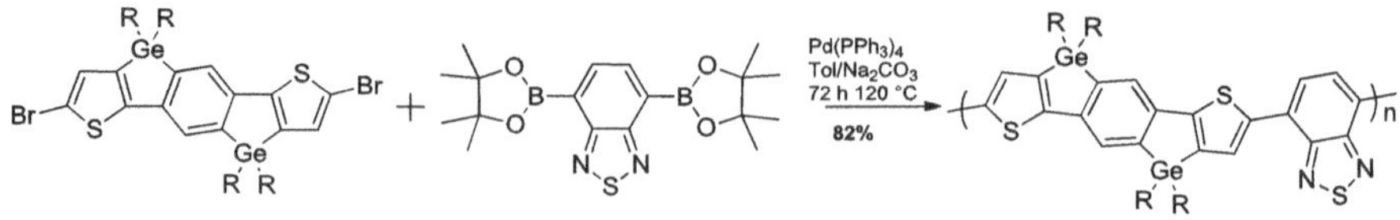

42 43 P44: R= ethylhexyl

Scheme 18.22

^{1}H NMR ($CDCl_3$, 400 MHz): δ_H 8.26 (broad, 2 H), 7.94 (broad, 2 H), 7.79 (broad, 2 H), 1.63–1.61 (broad, 4 H), 1.47–1.25 (broad, 40 H), 0.88–0.80 (broad, 24 H). Anal. Calcd. for $C_{52}H_{74}Ge_2N_2S_3$)n: C, 64.48; H, 7.70; N, 2.89. Found: C, 64.47; H, 7.83; N, 2.99. GPC (in chlorobenzene at 80 °C): M_n = 32 000, M_w = 75 000; PDI = 2.3 (Scheme 18.22).[50]

18.2.7 Characterization and Remarks

The extent of conjugation can be estimated by UV-Vis-NIR spectroscopy. Usually conjugated polymers are deeply colored compounds. Fused heterocyclic systems that favor the quinoid structure within the polymeric backbone (BT, TP) create an intense charge-transfer electronic transition in the visible spectrum. Incorporation of a BODIPY structural unit into the polymer architecture provides a strong bathochromic shift into the near-IR region and increased absorption compared to the isolated dye moecule. A special kind of characterization for conjugated polymers in general and for the system with incorporated electroactive fused heterocycle units in particular is provided by electrochemical methods which complement the UV-Vis-NIR measurement. Optical and electrochemical properties provide valuable information about potential applications of the conjugated polymers.[32,33] However, in some complex polymers that bear electron-rich or electron-deficient heterocycles, care must be taken to recognize the possibility of localized redox activity. In some cases (*e.g.* in the tetrathiafulvalene materials presented above), the bandgaps determined by optical and electrochemical methods are very different. In such situations, the HOMO (or LUMO) is not involved in the π–π* transition, in which the latter represents the linear conjugated portion of the polymer.

In summary, fused heterocycles provide several atrributes for the parent polymer that include the following: a propensity for aromatization and localization of electron density on heteroatoms; increased ability for extensive conjugation within the whole polymeric backbone; multiple possibilities for interchain interactions through short contacts between heteroatoms; increased electroactivity and stabilization of multiple polarons/bipolarons. These features provide complexity in electronic properties that are manifested entirely through the incorporation of fused heterocycles into the conjugated backbone of a polymer.

References

1. C. Wang, J. L. Schindler, C. R. Kannewurf and M. G. Kanatzidis, *Chem. Mater.*, 1995, **7**, 58.
2. F. Goldoni, B. M. W. Langeveld-Voss and E. W. Meijer, *Synth. Commun.*, 1998, **28**, 2237.
3. Y. H. Wijsboom, A. Patra, S. S. Zade, Y. Sheynin, M. Li, L. J. W. Shimon and M. Bendikov, *Angew. Chem Int. Ed.*, 2009, **48**, 5443.
4. A. Cravino, H. Neugebauer, A. Petr, P. J. Skabara, H. J. Spencer, J. J. W. McDouall, L. Dunsch and N. S. Sariciftci, *J. Phys. Chem. B*, 2006, **110**, 2662.
5. H. J. Spencer, P. J. Skabara, M. Giles, I. McCulloch, S. J. Coles and M. B. Hursthouse, *J. Mater. Chem.*, 2005, **15**, 4783.
6. H. J. Spencer, R. Berridge, D. J. Crouch, S. P. Wright, M. Giles, I. McCulloch, S. J. Coles, M. B. Hursthouse and P. J. Skabara, *J. Mater. Chem.*, 2003, **13**, 2075.
7. H. Pang, P. J. Skabara, S. Gordeyev, J. J. W. McDouall, S. J. Coles and M. B. Hursthouse, *Chem. Mater.*, 2006, **19**, 301.
8. L. M. Goldenberg, P. J. Skabara, D. M. Roberts, R. Berridge, E. Orti, P. M. Viruela and R. Pou-Amerigo, *J. Mater. Chem.*, 2000, **10**, 2458.
9. R. Berridge, S. P. Wright, P. J. Skabara, A. Dyer, T. Steckler, A. A. Argun, J. R. Reynolds, R. W. Harrington and W. Clegg, *J. Mater. Chem.*, 2007, **17**, 225.
10. P. J. Skabara, I. M. Serebryakov, I. F. Perepichka, N. S. Sariciftci, H. Neugebauer and A. Cravino, *Macromolecules*, 2001, **34**, 2232.
11. P. J. Skabara, R. Berridge, I. M. Serebryakov, A. L. Kanibolotsky, L. Kanibolotskaya, S. Gordeyev, I. F. Perepichka, N. S. Sariciftci and C. Winder, *J. Mater. Chem.*, 2007, **17**, 1055.
12. S. Kaur, N. J. Findlay, A. L. Kanibolotsky, S. E. T. Elmasly, P. J. Skabara, R. Berridge, C. Wilson and S. J. Coles, *Polym. Chem.*, 2012, **3**, 2277.
13. J. C. Forgie, A. L. Kanibolotsky, P. J. Skabara, S. J. Coles, M. B. Hursthouse, R. W. Harrington and W. Clegg, *Macromolecules*, 2009, **42**, 2570.
14. S. Kaur, N. J. Findlay, F. C. Coomer, R. Berridge and P. J. Skabara, *Macromol. Rapid. Commun.*, 2013, DOI: 10.1002/marc.201300357.
15. M. Bendikov, F. Wudl and D. F. Perepichka, *Chem. Rev.*, 2004, **104**, 4891.
16. M. R. Bryce, A. Chissel, J. Gopal, P. Kathirgamanathan and D. Parker, *Synth. Met.*, 1991, **39**, 397–400.
17. J. L. Segura and N. Martín, *Angew. Chem. Int. Ed.*, 2001, **40**, 1372–1409.
18. R. Berridge, P. J. Skabara, C. Pozo-Gonzalo, A. Kanibolotsky, J. Lohr, J. J. W. McDouall, E. J. L. McInnes, J. Wolowska, C. Winder, N. S. Sariciftci, R. W. Harrington and W. Clegg, *J. Phys. Chem. B*, 2006, **110**, 3140.
19. P. J. Skabara, D. M. Roberts, I. M. Serebryakov and C. Pozo-Gonzalo, *Chem. Commun.*, 2000, 1005.

20. A. L. Kanibolotsky, J. C. Forgie, S. Gordeyev, F. Vilela, P. J. Skabara, J. E. Lohr, B. M. Petersen and J. O. Jeppesen, *Macromol. Rapid Commun.*, 2008, **29**, 1226.
21. B. Tieke, A. R. Rabindranath, K. Zhang and Y. Zhu, *Beilstein J. Org. Chem.*, 2010, **6**, 830–845.
22. K. Zhang, B. Tieke, J. C. Forgie and P. J. Skabara, *Macromol. Rapid Commun.*, 2009, **30**, 1834.
23. K. Zhang and B. Tieke, *Macromolecules*, 2008, **41**, 7287.
24. D. Cortizo-Lacalle, S. Arumugam, S. E. T. Elmasly, A. L. Kanibolotsky, N. J. Findlay, A. R. Inigo and P. J. Skabara, *J. Mater. Chem.*, 2012, **22**, 11310.
25. Y. Li, S. P. Singh and P. Sonar, *Adv. Mater.*, 2010, **22**, 4862.
26. M. Shahid, R. S. Ashraf, Z. Huang, A. J. Kronemeijer, T. McCarthy-Ward, I. McCulloch, J. R. Durrant, H. Sirringhaus and M. Heeney, *J. Mater. Chem.*, 2012, **22**, 12817–12823.
27. W. Hong, B. Sun, H. Aziz, W.-T. Park, Y.-Y. Noh and Y. Li, *Chem. Commun.*, 2012, **48**, 8413.
28. T.-Y. Chu, J. Lu, S. Beaupré, Y. Zhang, J.-R. M. Pouliot, S. Wakim, J. Zhou, M. Leclerc, Z. Li, J. Ding and Y. Tao, *J. Am. Chem. Soc.*, 2011, **133**, 4250.
29. C. M. Amb, S. Chen, K. R. Graham, J. Subbiah, C. E. Small, F. So and J. R. Reynolds, *J. Am. Chem. Soc.*, 2011, **133**, 10062.
30. P. Berrouard, A. Najari, A. Pron, D. Gendron, P.-O. Morin, J.-R. Pouliot, J. Veilleux and M. Leclerc, *Angew. Chem. Int. Ed.*, 2012, **51**, 2068.
31. N. Boens, V. Leen and W. Dehaen, *Chem. Soc. Rev.*, 2012, **41**, 1130.
32. D. Cortizo-Lacalle, C. T. Howells, S. Gambino, F. Vilela, Z. Vobecka, N. J. Findlay, A. R. Inigo, S. A. J. Thomson, P. J. Skabara and I. D. W. Samuel, *J. Mater. Chem.*, 2012, **22**, 14119.
33. J. C. Forgie, P. J. Skabara, I. Stibor, F. Vilela and Z. Vobecka, *Chem. Mater.*, 2009, **21**, 1784.
34. E. Wang, Z. Ma, Z. Zhang, K. Vandewal, P. Henriksson, O. Inganäs,, F. Zhang and M. R. Andersson, *J. Am. Chem. Soc.*, 2011, **133**, 14244.
35. R. Stalder, C. Grand, J. Subbiah, F. So and J. R. Reynolds, *Polym. Chem.*, 2012, **3**, 89.
36. L. A. Estrada, D. Y. Liu, D. H. Salazar, A. L. Dyer and J. R. Reynolds, *Macromolecules*, 2012.
37. R. Stalder, J. Mei and J. R. Reynolds, *Macromolecules*, 2010, **43**, 8348–8352.
38. G. Zhang, Y. Fu, Z. Xie and Q. Zhang, *Macromolecules*, 2011, **44**, 1414.
39. B. Liu, Y. Zou, B. Peng, B. Zhao, K. Huang, Y. He and C. Pan, *Polym.Chem.*, 2011, **2**, 1156.
40. E. Wang, Z. Ma, Z. Zhang, P. Henriksson, O. Inganas, F. Zhang and M. R. Andersson, *Chem. Commun.*, 2011, **47**, 4908.

41. H. Pang, P. J. Skabara, D. J. Crouch, W. Duffy, M. Heeney, I. McCulloch, S. J. Coles, P. N. Horton and M. B. Hursthouse, *Macromolecules*, 2007, **40**, 6585–6593.
42. R. C. Evers, F. E. Arnold and T. E. Helminiak, *Macromolecules*, 1981, **14**, 925–930.
43. E. Ahmed, S. Subramaniyan, F. S. Kim, H. Xin and S. A. Jenekhe, *Macromolecules*, 2011, **44**, 7207.
44. J. A. Osaheni and S. A. Jenekhe, *Chem. Mater.*, 1992, **4**, 1282.
45. J. A. Osaheni and S. A. Jenekhe, *Macromolecules*, 1993, **26**, 4726.
46. I. T. Kim, S. W. Lee, S. Y. Kim, J. S. Lee, G. B. Park, S. H. Lee, S. K. Kang, J.-G. Kang, C. Park and S.-H. Jin, *Synth. Met.*, 2006, **156**, 38.
47. J. F. Mike, A. J. Makowski, T. C. Mauldin and M. Jeffries-El, *J. Polym. Sci. Part A: Polym. Chem.*, 2010, **48**, 1456.
48. J. J. Intemann, E. S. Hellerich, B. C. Tlach, M. D. Ewan, C. A. Barnes, A. Bhuwalka, M. Cai, J. Shinar, R. Shinar and M. Jeffries-El, *Macromolecules*, 2012, **45**, 6888.
49. A. Balan, D. Baran and L. Toppare, *Polym. Chem.*, 2011, **2**, 1029–1043.
50. Z. Fei, R. S. Ashraf, Z. Huang, J. Smith, R. J. Kline, P. D'Angelo, T. D. Anthopoulos, J. R. Durrant, I. McCulloch and M. Heeney, *Chem. Commun.*, 2012, **48**, 2955.
51. R. S. Ashraf, Z. Chen, D. S. Leem, H. Bronstein, W. Zhang, B. Schroeder, Y. Geerts, J. Smith, S. Watkins, T. D. Anthopoulos, H. Sirringhaus, J. C. de Mello, M. Heeney and I. McCulloch, *Chem. Mater.*, 2010, **23**, 768–770.
52. S. C. Rasmussen, R. L. Schwiderski and M. E. Mulholland, *Chem. Commun.*, 2011, **47**, 11394–11410.
53. E. Zhou, J. Cong, S. Yamakawa, Q. Wei, M. Nakamura, K. Tajima, C. Yang and K. Hashimoto, *Macromolecules*, 2010, **43**, 2873.
54. D. R. Coulson, L. C. Satek and S. O. Grim, in *Inorganic Synthesis*, McGraw-Hill, Inc., USA, Vol XIII, 1972, p. 121.
55. P. J. Skabara, R. Berridge, E. J. L. McInnes, D. P. West, S. J. Coles, M. B. Hursthouse and K. Mullen, *J. Mater. Chem.*, 2004, **14**, 1964.
56. S. Kotani, K. Shiina and K. Sonogashira, *J. Organomet. Chem.*, 1992, **429**, 403.
57. A. F. Renaldo, J. W. Labadie and J. K. Stille, *Org. Synth.*, 1989, **67**, 89.
58. X. Guo, R. P. Ortiz, Y. Zheng, M.-G. Kim, S. Zhang, Y. Hu, G. Lu, A. Facchetti and T. J. Marks, *J. Am. Chem. Soc.*, 2011, **133**, 13685.

CHAPTER 19

Direct Arylation/Heteroarylation Polycondensation Reactions

LAUREN G. MERCIER, AGNIESZKA PRON AND MARIO LECLERC*

Department of Chemistry, Université Laval, Québec City, QC, Canada G1V 0A6
*Email: mario.leclerc@chm.ulaval.ca

19.1 Introduction

Aryl C–C bond formation is universal in many aspects of synthetic chemistry - from pharmaceuticals, biologically active compounds, conjugated small molecules, polymers, *etc.* This process can be carried out catalytically under Stille, Suzuki, Negishi, *etc.* conditions using a transition-metal catalyst that reacts with "preactivated" groups such as halogens (Br, I) and organometallic species (SnR_3, $B(OR)_2$, ZnX, *etc.*) as shown in Scheme 19.1a. Although these reactions are usually high yielding, preparation of the aryl reagents requires extra steps to install the "preactivated" groups, which are often unstable and ultimately produce stoichiometric amounts of waste. An alternative and promising methodology to these reactions is direct (hetero)arylation where a (hetero)aryl bromide is directly coupled with a (hetero)arene (Scheme 19.1b).[1–3]

Indeed this is an interesting reaction since it is dependent on the energy required to deprotonate the (hetero)arene, which can be overcome by activation of the C–H bonds through directing groups, acidity or the catalytic system. This methodology has been used to prepare materials for organic electronics, specifically those containing thiophene units, and has more recently been applied to the

RSC Polymer Chemistry Series No. 9
Conjugated Polymers: A Practical Guide to Synthesis
Edited by Klaus Müllen, John R. Reynolds and Toshio Masuda

Published by the Royal Society of Chemistry, www.rsc.org

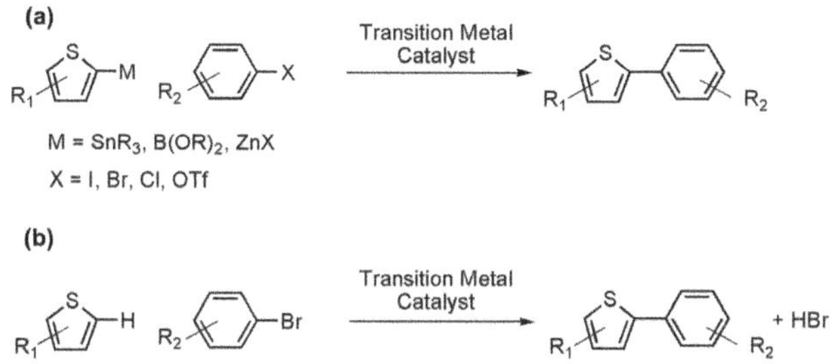

Scheme 19.1 C–C bond formation *via* (a) Stille, Suzuki, Negishi, *etc.* coupling reaction conditions; (b) direct arylation shown with a substituted thiophene and a substituted bromobenzene.

synthesis of conjugated polymers – termed direct (hetero)arylation polymerization (DHAP). Even though the application of this methodology to polymer synthesis is relatively new, it has indeed made a great impact on the field.[4] This chapter will focus on direct (hetero)arylation reactions in the context of thiophene-based aromatics given their ubiquitous use in organic electronics.

19.2 Reaction Conditions

19.2.1 General Comments

Direct arylation of thiophene C–H bonds is typically carried out with aryl bromides, but aryl iodides and aryl triflates have been used due to their higher reactivity. Late transition-metal catalysts are employed including Pd(II) pre-catalysts such as $Pd(OAc)_2$, Herrmann–Beller catalyst (*trans*-di(μ-acetato)bis[*o*-(di-*o*-tolylphosphino)benzyl]dipalladium(II)) or palladium *N*-heterocyclic carbene complexes.[5] There are also some examples of catalysis by rhodium complexes but under these reaction conditions aryl iodides give much superior reactivity than aryl bromides and triflates.[6,7] Electron-rich phosphine ligands may also be utilized, although there are many examples of phosphine-free conditions.[8] Triaryl, bidentate and ionic phosphines have all been investigated. The base plays a pivotal role in the reaction since it is believed to assist in C–H bond activation,[1,9] and one equivalent of acid is produced for every C–H bond activated. Typically, carbonate or carboxylate bases are used (Cs_2CO_3, K_2CO_3, CsOPiv and KOAc). Polar aprotic solvents (*N*,*N*-dimethylacetamide = DMAc, *N*,*N*-dimethylformamide = DMF, NMP = *N*-methylpyrrolidine) as well as non-polar aprotic solvents (toluene, xylenes) are used and reactions are carried out at temperatures greater than 100 °C, or with the assistance of microwave heating.[10] Polar aprotic solvents may also participate in the catalytic cycle by coordination to the metal, especially in the absence of phosphine ligands.[11] Other additives such as pivalic acid,[9,12,13] and oxidants such as metals[14] and molecular oxygen[15] have also been explored. Reactions are performed under a dry inert atmosphere and reagents are stored in this way as well.

Reaction conditions for C–H activation of thiophenes can be divided into three groups: 1) Heck (Jeffery) conditions,[16] although these are used infrequently; 2) with carboxylic acid additives; 3) without carboxylic acid additives. The last two types of conditions have been used in polar and aprotic solvents and may be carried out with a phosphine ligand.

19.2.2 Heck (Jeffery) Conditions

Palladium (II) acetate, tetrabutyl ammonium bromide (a phase-transfer catalyst) and potassium carbonate in a mixture of acetonitrile and water, which were originally published by Jeffery in 1996 for Heck reactions,[16] have also been applied to C–H activation of thiophenes. Lemaire and coworkers have coupled a number of 2- and 3-substituted thiophenes (-CHO, $-NO_2$, -CN) with iodobenzenes,[17] albeit in low yields (Scheme 19.2).

They then applied these conditions to the preparation of poly-3-alkylthiophene (**1**).[18,19] Low molecular weight material was obtained ($M_n \sim 3.0$ kDa, polydispersity index (PDI) = 1.1–2.1) with bromo- and iodo-substituted materials, but this was, however, the first example of application to polymeric materials (Chart 19.1).

Abdo *et al.* used similar conditions ($Pd(OAc)_2$, KOAc, Bu_4NBr, DMF, microwave) for the polymerization of thieno[3,4-*b*]pyrazines with dibromoalkylthiophenes and dibromo(dialkyl)bithiophenes.[20] With monomer **2-H** (Chart 19.2), partially soluble polymers (**3**, **4**) were obtained after only 15 min and increasing the reaction time or temperature further diminished the solubility of these polymers. The ^{1}H NMR spectra of the polymers synthesized by C–H activation were very different from those measured for traditional Stille polymerization methods and the authors attributed this to crosslinking in the 2-position of **2-H**.[20] However, blocking of this position with a methyl group (**2-CH_3**) enabled the formation of polymers with high molecular weights (**5**: $M_n = 18$ kDa, PDI = 1.7; **6**: $M_n = 23$ kDa, PDI = 1.6) and ^{1}H NMR

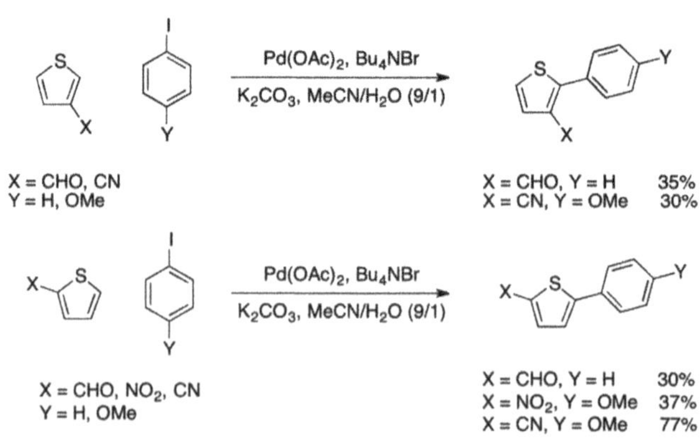

Scheme 19.2 Coupling of 2-substituted thiophenes (*top*), and 3-substituted thiophenes (*bottom*).

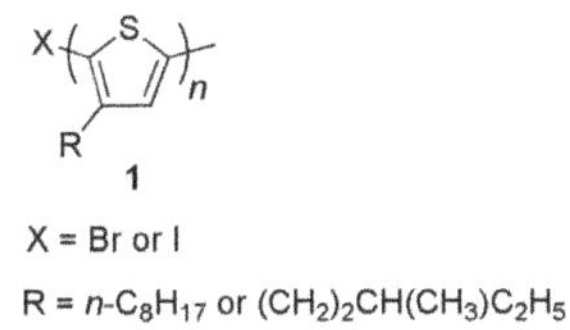

Chart 19.1

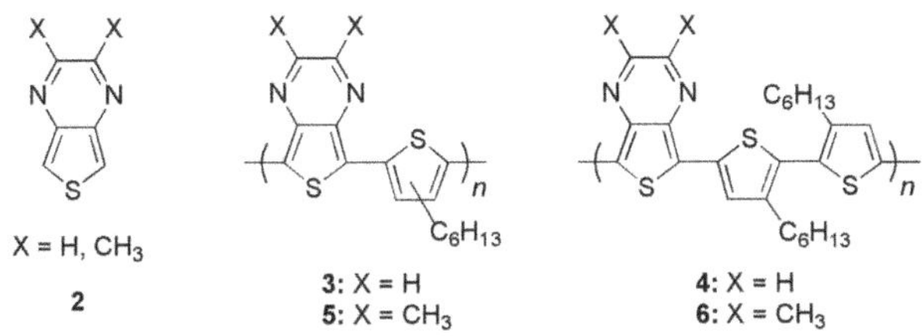

Chart 19.2

spectra identical to those of polymers prepared by Stille methods.[20] These results show that selectivity of C–H bonds has to be considered in the choice of monomers utilized.

19.2.3 Carboxylic Acid Additives

Carboxylic acid additives have been shown to assist in the deprotonation process (*vide infra*). Fagnou and coworkers have used palladium (II) acetate, a phosphine ligand, potassium carbonate and pivalic acid (PivOH) in DMAc or toluene to prepare several small molecules through the coupling of thiophenes and 2,2′-bithiophenes with arylbromides.[21,22] An example of a biscoupling of 2,2′-bithiophene with two equivalents of 1-bromo-3,5-dimethylbenzene to give **7** is shown in Scheme 19.3.

Thompson and coworkers polymerized 2-bromo-3-hexylthiophene with $Pd(OAc)_2$, PivOH and K_2CO_3 in DMAc at 95 °C and obtained **1-C$_6$H$_{13}$** with relatively high molecular weight ($M_n = 15.5$ kDa, PDI = 2.9).[23] Using the same conditions as Thompson and coworkers, Fujinami *et al.* polymerized bithiophene with a fluorene monomer, which gave rise to insoluble material, most likely resulting from crosslinking.[24] Protection of the 3,3′,4,4′-positions of bithiophene with methyl groups (3,3′,4,4′-tetramethylbithiophene) and phosphine-free conditions resulted in the formation of **8** with high molecular weight (Yield = 91%, $M_n = 32$ kDa, PDI = 2.5).[24] They later applied this methodology to the polymerization of 4,4-dialkyl-2,2′-bithiazole monomers with fluorene (**9**) and other phenylene (**10-13**) monomers and achieved molecular weights as high as 46 kDa (PDI = 2.3) for **9**[25] (Chart 19.3). This is much more efficient than what was obtained for traditional Suzuki–Miyura

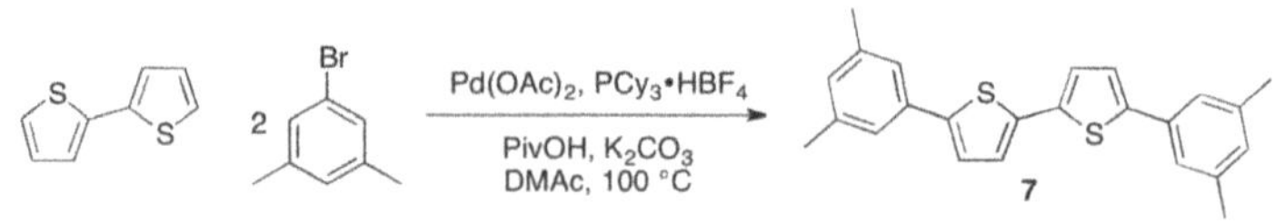

Scheme 19.3 Coupling of 2,2′-bithiophene.

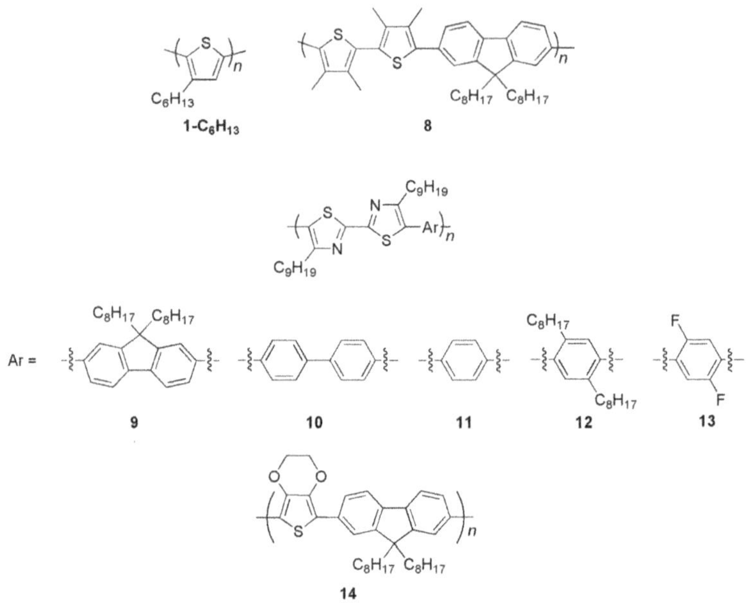

Chart 19.3

polymerization methods ($M_n = 18$ kDa). The monomers were rapidly consumed (10 min), however, some insoluble material was obtained due to C–H activation of other bonds on the arylene units. This was most pronounced for **11** and could be suppressed for **12** and **13** due to steric effects.[25,26] Recently Kanbara and coworkers used similar phosphine-free conditions to prepare high molecular weight EDOT-fluorene copolymer **14**.[26]

Leclerc's group prepared **15** and **16** (Chart 19.4) from brominated heteroarenes and thieno[4,3-*d*]thiazole with high molecular weights that were comparable to those obtained by Stille polymerization ($M_n = 32$ kDa, PDI = 3.4 for DHAP, $M_n = 32$ kDa, PDI = 2.8 for Stille).[27] They utilized the Herrmann–Beller catalyst, P(*o*-anisole)$_3$ and Cs_2CO_3 in toluene heated to 120 °C.[27] These conditions were also applied to the co- and homopolymerization of several thieno[3,4-*c*]pyrrole-4,6-dione (TPD) monomers substituted with branched and linear alkyl chains (**17**, **18**).[28] They found that higher molecular weights were achieved for TPD monomers bearing branched alkyl chains.[28]

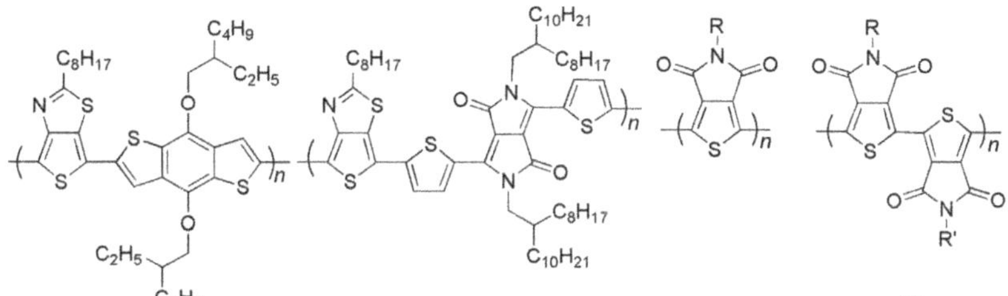

15 16 17 18

R = $CH_2CH(C_{10}H_{21})(C_8H_{17})$
R = $CH(C_8H_{17})_2$
R = $CH(C_8H_{17})_2$, R' = n-C_8H_{17}
R = $CH_2CH(C_{10}H_{21})(C_8H_{17})$, R' = n-C_8H_{17}

Chart 19.4

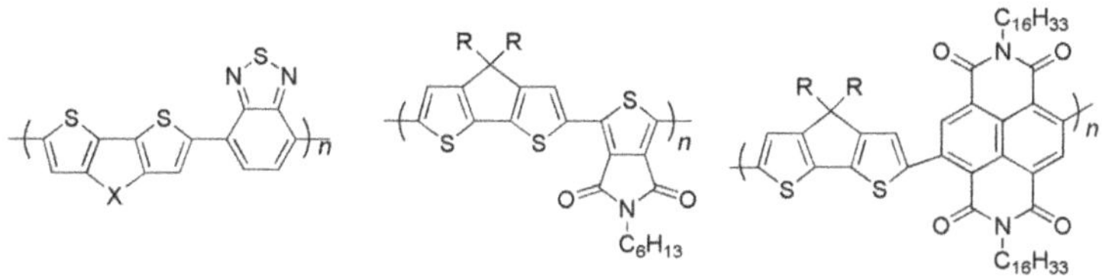

X = $C(CH_2CH(C_2H_5)(C_4H_9))_2$ **19**
$Si(CH_2CH(C_2H_5)(C_4H_9))_2$ **20**
$NCH(C_8H_{17})_2$ **21**

R = $CH_2CH(C_2H_5)(C_4H_9)$
22

R = $CH_2CH(C_2H_5)(C_4H_9)$
23

Chart 19.5

Horie and coworkers used phosphine-free conditions in NMP to couple 4,4-bis(2-alkyl)-4*H*-cyclopenta[2,1-*b*:3,4-*b*′]dithiophene (CPDT), *N*-(1-alkyl)dithieno[3,2-b:2′,3′-*d*]pyrrole and 4,4′-bis(2-alkyl)dithieno[3,2-*b*:2′,3′-*d*]-silole with dibrominated (hetero)arenes to give polymers **19–23** (Chart 19.5).[8] Polymers with high molecular weights were only achieved for **19** ($M_n = 72$ kDa, PDI = 4.5) due to side reactions and insolubility of the other polymers.[8]

19.2.4 Without Carboxylic Acid Additives

These are the most ubiquitous type of reaction conditions published in C–H activation literature and use a palladium catalyst, phosphine ligand and a base. Many examples of small molecules have been published and these conditions are tolerant of a number of functional groups.[29–33] In 2010, Ozawa and coworkers prepared **1-C_6H_{13}** using the Herrmann–Beller catalyst, different *ortho*-substituted aryl phosphines (P(*o*-anisole)$_3$, P(*o*-$NMe_2(C_6H_4)_3$) and Cs_2CO_3 in superheated THF (125 °C).[34] Under these conditions, high molecular weight materials were obtained in good yield with a large degree of regioregularity ($M_n = 30$ kDa, PDI = 1.6, yield = 99%, regioregularity = 98%).[34] C–H activation of a more electron-rich thiophene, 3,4-ethylenedioxythiophene (EDOT), with several arylbromides, has been investigated by Yu and coworkers who screened various catalysts, ligands, bases and solvents.[35] They found that

$Pd(OAc)_2$, a monodentate phosphine ($P(m\text{-tol})_3$ (m-tol = m-toluene)) and Cs_2CO_3 in toluene gave high yields of dicoupled products (**24–27**, Scheme 19.4).[35] This catalyst system was also applied to the biscoupling of EDOT dimers with arylbromides containing electron-rich and electron-poor groups in moderate yields (**28**, Chart 19.6).[35]

They later applied similar conditions to the polymerization of alkoxy-substituted EDOT and 3,4-propylenedioxythiophene (PProDOT) starting from a dibrominated and an unhalogentated monomer (Chart 19.7).[36] For **29**, high molecular weights and yields were obtained using $Pd(OAc)_2$ and $PCy_3 \cdot HBF_4$ (where Cy = cyclohexyl) in DMF or the Herrmann–Beller catalyst and P(o-anisole)$_3$ in THF (M_n = 10 kDa, PDI = 1.6; M_n = 8 kDa, PDI = 1.5, respectively). However, for **30** the latter conditions were optimal (M_n = 5 kDa, PDI = 1.1).[36]

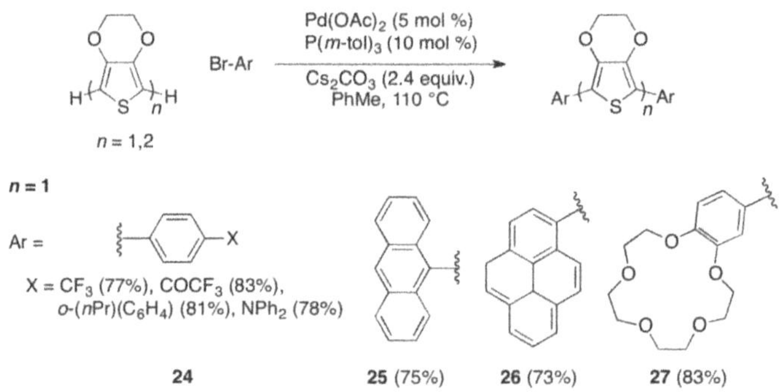

Scheme 19.4 Optimized conditions for the double C–H activation of EDOT.

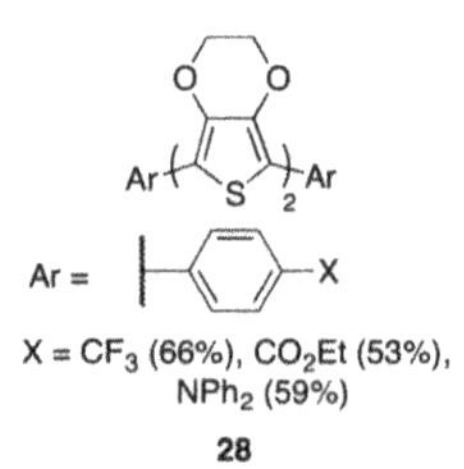

Chart 19.6

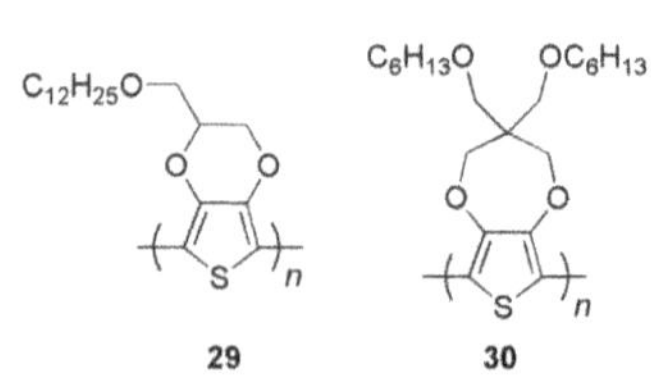

Chart 19.7

Kanbara and coworkers used similar conditions to copolymerize a number of electron-rich dibrominated arenes with perfluorinated arenes (**31**–**34**, Chart 19.8) using $Pd(OAc)_2$, $Pd(tBu)_2Me \cdot HBF_4$, K_2CO_3 in DMAc.[24,37,38] They found that addition of PivOH diminished polymer growth and observed some crosslinked material. They also were the first to publish the fabrication of devices prepared with polymers prepared by C–H activation. Polymer **31** was incorporated into organic light-emitting diodes (OLEDs) as a hole blocking layer and improved current efficiency compared to devices fabricated without this layer.[37] Scherf and coworkers applied similar conditions ($PCy_3 \cdot HBF_4$ was used) for the preparation of polymer **19** and found that higher molecular weights were obtained under phosphine-free conditions ($M_n = 40$ kDa, $M_n = 20$ kDa, respectively).[39] Although lower than what was achieved by Horie and coworkers (*vide supra*), these molecular weights were still larger than those previously reported for Suzuki ($M_n = 15$ kDa, PDI = 2.0)[40] and Stille ($M_n = 28$ kDa, PDI = 1.5)[41] methods. However, ^1H NMR spectroscopic analyses revealed the presence of the irregular coupling.

Leclerc and coworkers investigated the C–H activation of 5-alkyl-5*H*-thienopyrrole-4,6-dione (TPD)[42] for copolymerization of brominated bithiophene and terthiophene using similar conditions published by Ozawa (Chart 19.9).[34]

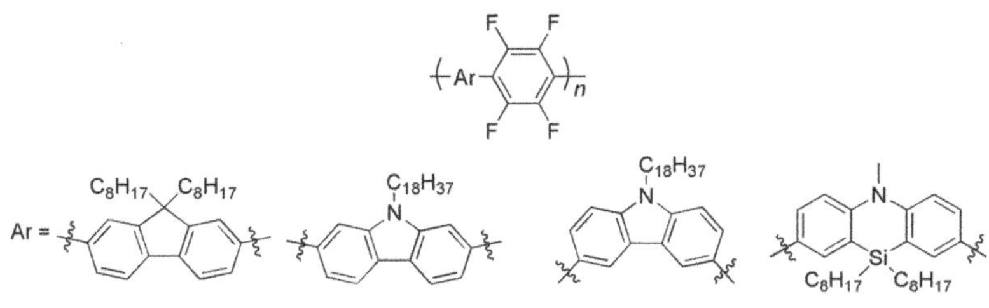

31 32 33 34

Chart 19.8

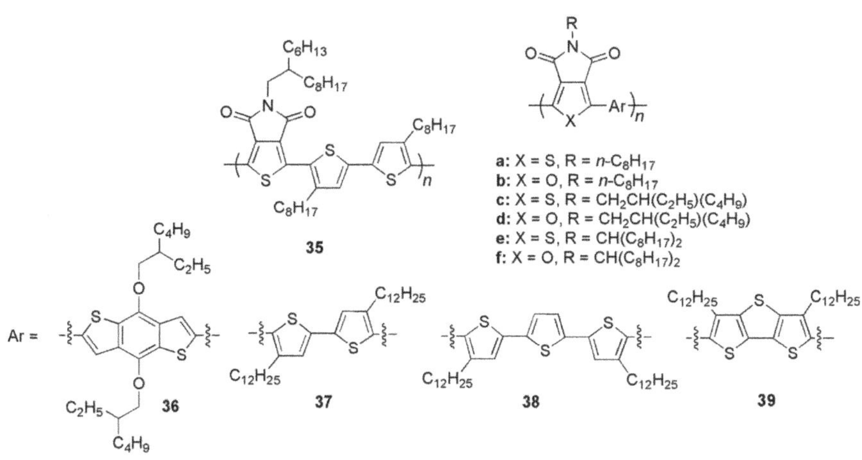

35 36 37 38 39

Chart 19.9

For **35**, they achieved molecular weights five times as large as those obtained for Stille polymerization ($M_n = 50$ kDa vs. $M_n = 9$ kDa). This can be attributed to the improved stability of the monomers during polymerization process. They also published the synthesis of a number of thiophene-based copolymers with TPD and the furan congener (5-alkyl-5*H*-furopyrrole-4,6-dione (FPD), **36a,b**–**39a,b**), using the same conditions with the addition of pivalic acid and switching to toluene as solvent.[43] High molecular weights were achieved as well.[43] It is worth noting that **38e** ($M_n = 44$ kDa) was successfully incorporated into BHJ solar cells. A power conversion efficiency (PCE) of 6.1% was achieved for **38e**/$PC_{71}BM$ blend with 3% of chloronaphthalene as an additive.[44]

19.3 Mechanistic Investigations

The mechanism of direct arylation has been studied experimentally and computationally and possible pathways include electrophilic aromatic substitution, Heck-type coupling and concerted metalation-deprotonation (CMD).[1,9,45–48] The reaction pathway is dependent on the substrate and the catalytic system employed,[7,49,50] however, most electron-rich (hetero)arenes seem to follow a base-assisted CMD pathway. Two catalytic cycles for the coupling of bromobenzene and thiophene are shown in Schemes 19.5 and 19.6. Scheme 19.5 depicts a carboxylate-mediated process[47,48,51] where C–H activation occurs

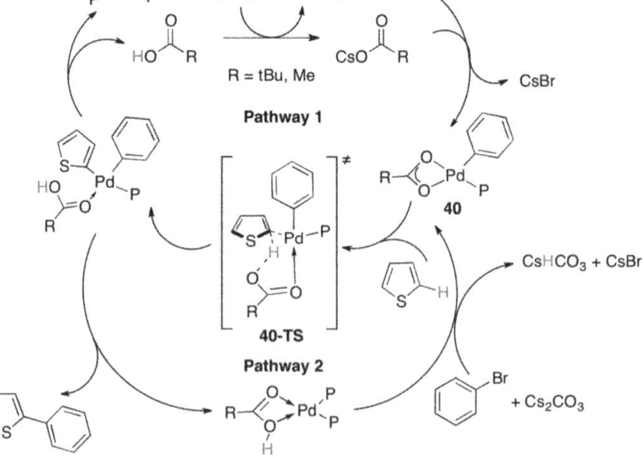

Scheme 19.5 Catalytic cycle for crosscoupling between thiophene and bromobenzene using a carboxylate additive.

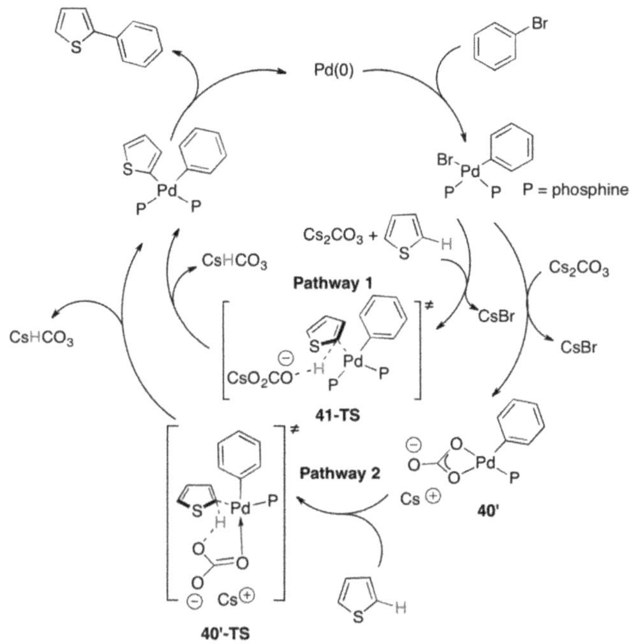

Scheme 19.6 Catalytic cycle for crosscoupling between thiophene and bromobenzene without a carboxylate additive using Cs_2CO_3 as a base.

intramolecularly, while Scheme 19.6 represents the reaction process in the absence of a carboxylate additive.[1,9]

Under carboxylate-mediated conditions (Scheme 19.5), oxidative addition of the carbon–halogen bond is followed by exchange of the halogen ligand for the carboxylate anion to form complex **40**. If a donating solvent, such as DMAc or DMF, is used, the solvent may coordinate to the metal center instead of the phosphine ligand.[11] With the assistance of the carboxylate ligand, this complex then deprotonates the thiophene substrate while simultaneously forming a metal–carbon bond through transition state **40-TS**.[47–49] The phosphine ligands, or solvent, can recoordinate to the metal center following Pathway 1, or the carboxylate group can remain coordinated throughout the entire process (Pathway 2).[13] Finally, reductive elimination renders the aryl-coupled product. In the absence of a carboxylate additive, after oxidative addition of the aryl bromide, the reaction follows one of the two pathways shown in Scheme 19.6. If a bidentate phosphine is employed, C–H activation of thiophene can follow Pathway 1, where deprotonation is assisted intermolecularly (**41-TS**).[1,9] When a monodentate phosphine is used, the reaction may follow Pathway 1 or Pathway 2.[1] The latter mechanism most closely resembles Pathway 2 in Scheme 19.5 where the carbonate coordinates to the metal center to give the Zwitterionic species **40′**.[1] From here, deprotonation occurs intramolecularly through transition state **40′-TS**. Reductive elimination then renders 2-phenylthiophene.

Density functional theory (DFT) calculations (B3LYP//DZVP/TZVP) of the CMD process reveal that the energy required to form transition state **40-TS** can be broken into two components - the energy needed to distort the aryl C–H bond and the energy to be overcome when the distorted aryl fragment interacts with the metal center.[48] The authors have computed the activation energy of a number of C–H bonds for various (hetero)arenes and compared the regioselectivity for those containing more than one reactive site. These arenes were then divided into three classes based on the factors that control the regioselectivity in the CMD reaction: 1) Class I (hetero)arenes depend on the energy required to distort each C–H bond (*i.e.* benzenes, perfluorinated benzenes and benzothiophenes), 2) Class II (hetero)arenes rely on the interaction energy between the metal center and the distorted aryl fragment (*i.e.* furans and thiazoles), and 3) Class III (hetero)arenes, where the site of reactivity depends on both the distortion and interaction energy (*i.e.* thiophenes and pyrrolles).[48] A small activation energy difference of 4.3 kcal/mol between the 2- and 3 – positions in thiophene was calculated according to these criteria.[48] This explains the common appearance of crosslinked materials during DHAP, and these side reactions should be considered during monomer selection. This study also investigated C5-functionalization of thiophenes with electron-donating and withdrawing groups in the 2-position. Thiophenes substituted with electron-donating groups are strong π-nucleophiles and have a strong interaction with the palladium center, but require more energy to distort the C–H bond.[48] On the contrary, electron-withdrawing groups on thiophenes decrease the π-nucleophilicity of the system but seem to lower the energy required to distort the C–H bond.[48] This implies that the acidity of the thiophene proton does not seem to affect the reaction since electron-rich thiophene units, such as EDOT, can undergo C–H activation. However, it is difficult to find trends based on the electronic effects of the substituents since the energy of distortion

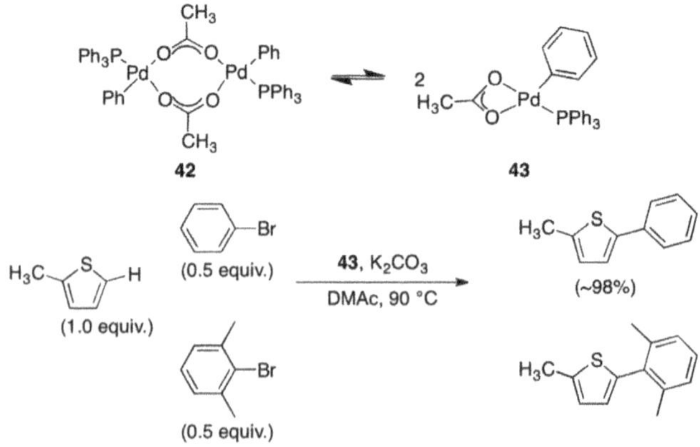

Scheme 19.7 Disproportionation of **42** into the active catalytic species **43** and competition experiment between bromobenzene and 2,6-dimethyl-1-bromobenzene.

and interaction is different for each substrate, but these calculations did correlate well with experimental results.[48]

Experimental verification of the CMD mechanism was published in 2012 by Ozawa and coworkers, who isolated a single crystal of complex **42** suitable for X-ray analysis and studied the catalytic activity towards C–H activation.[52] In solution, this dimeric species disproportionates into two molecules of **43**, which is the active catalytic species. They used this species under catalytic conditions to compare the rates of C–H activation of 2-methylthiophene in a competition experiment with half an equivalent of bromobenzene and 2-bromo-1,3-dimethylbenzene (Scheme 19.7). Oxidative addition, not C–H activation, was determined to be the rate-limiting step since coupling with bromobenzene took place preferentially.[52]

19.4 Experimental Procedures

19.4.1 Materials and General Polymerization Procedures

All C–H activation procedures for polymers reported thus far have been carried out under dry inert atmosphere using sealed vessels (Schlenk glassware). Similar to small-molecule procedures, polar (DMF, DMAc) and nonpolar (toluene, THF) aprotic solvents have been used and are degassed prior to use. Most thiophene-based monomers are not commercially available and must be synthesized according to literature procedures. It is very important that these monomers be extremely pure and free of all aryl impurities since other aryl bonds may undergo C–H activation and be incorporated into the polymer. All examples have employed palladium (II) acetate or the Herrmann–Beller catalyst. The latter can be prepared from $Pd(OAc)_2$ and tris-(*o*-tolyl)phosphine.[53] All phosphine ligands, anhydrous bases and pivalic acid are commercially available and are stored under inert atmosphere.

The reaction procedures reported in the literature were relatively similar and therefore only a few examples will be shown for each type of protocol. However, the work-up procedures for the reactions differed between research groups. Typical work-up procedures for polymerizations in polar solvents involved precipitation in aqueous solution (neutral, acidic or basic) while reactions carried out in nonpolar solvents were precipitated in aqueous methanol solutions. The crude polymer was washed with various solvents (acetone, methylvinylketone (MVK), hexanes, dichloromethane), typically through Soxhlet extraction, to remove lower molecular weight materials. The polymer was then extracted in chloroform or *o*-dichlorobenzene and reprecipitated in methanol. Often, the polymer was washed with a metal scavenger, such as ethylenediaminetetracetic acid disodium salt, to remove residual metals, and this can be done before or after removal of low molecular weight materials. Filtration of the chloroform solution through celite or silica gel has also been used for this purpose.

Throughout the experimental procedures, the amount of reagents used will be expressed with relation to the monomer(s). For copolymerizations, a one-to-one ratio of monomers is always utilized. All molecular weights and PDIs are estimated from size-exclusion chromatography (SEC) using polystyrene as a standard and regioregularity is estimated from ^{1}H NMR spectroscopy.

19.4.2 Polymerization Using Heck (Jeffery) Conditions

Example 1: Polymerization of 2-Bromo-3-alkylthiophenes

Several different examples of preparation of this polymer using DHAP have been published. Under Jeffrey conditions, potassium carbonate (2.5 equiv.) and tetrabutylammonium bromide (1.0 equiv.) are added to a solution of either 2-iodo-3-alkylthiophene or 2-bromo-3-alkylthiophene in DMF (0.8 M) (Scheme 19.8).[18,19] The solution is heated to 80 °C and then $Pd(OAc)_2$ (5 mol%) is added and the reaction is stirred for 48 h. The resulting polymers are filtered, washed with water and methanol. Then, the resulting polymer is dissolved in chloroform, washed again with water, dried with $MgSO_4$, filtered and the solvent removed *in situ*. Under these conditions, low molecular weight materials ($M_n \sim 3.0$ kDa, PDI = 1.1–2.1) are obtained with a high degree of regioregularity and a blue-shifted absorption maxima in UV-vis spectroscopy (Table 19.1).

19.4.3 Polymerization with Carboxylic Acid Additives

Example 2: Polymerization of 2-Bromo-3-alkylthiophenes

The same polymerization can be carried out with $Pd(OAc)_2$ (2 mol%), pivalic acid (0.3 equiv.) and K_2CO_3 (1.5 equiv.) in DMAc (0.05 M) at 95 °C to give

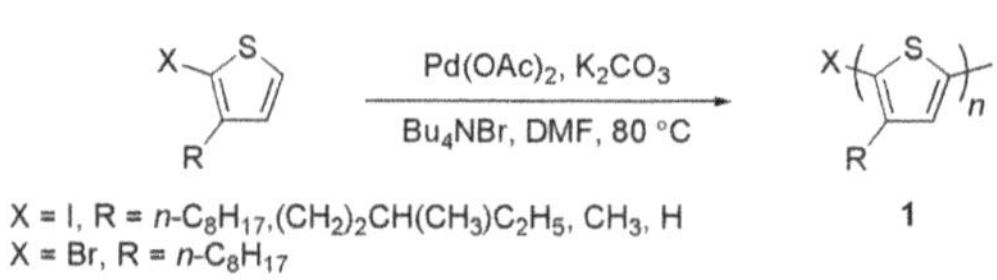

Scheme 19.8 Preparation of poly-3-alkylthiophenes from 2-halo-3-alkylthiophenes using Jeffrey conditions.

Table 19.1 Solvent effects for the polymerization of 2-iodo-3-octylthiophene using the conditions shown in Scheme 19.1.

Solvent	*Regioregularity [%]*[a]	λ_{max} *[nm]*[b]
PhMe	78	427
Dioxane	40	405
1-methyl-2-pyrrolidinone	70	429
$CHCl_3$	70	412
THF	88	440
DMF	80	438
–	98–99[c]	450[c]

[a]Determined by ^{1}H NMR spectroscopy.
[b]Measured in $CHCl_3$ solution.
[c]Prepared by the McCullough method.[54]

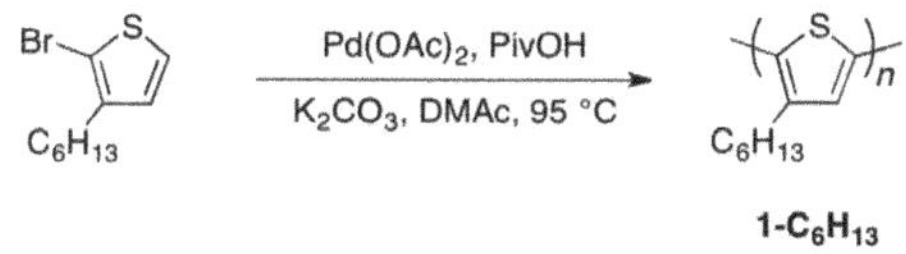

Scheme 19.9 Preparation of poly-3-hexylthiophene from 2-bromo-3-hexylthiophene using $Pd(OAc)_2$.

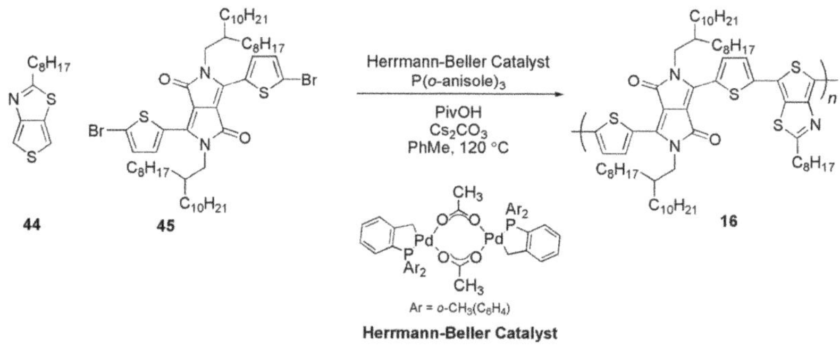

Scheme 19.10 Preparation of **15** using the Herrmann–Beller catalyst and pivalic acid.

1-C_6H_{13} in 39% yield (Scheme 19.9).[23] An $M_n = 15.5$ kDa (PDI = 2.9) and a regioregularity of 88% are obtained.[23] The lower yields and regioregularity in this case are most likely due to the decomposition of the palladium catalyst.

These conditions are also used for the copolymerization of 3,3′,4,4′-tetramethylbithiophene with several brominated monomers[24] and 2,2′-bithiazole monomers with fluorene (**9**) and other phenylene (**10–13**) monomers (Chart 19.3).[25] *N*-methylpyrrolidine (85–90 °C) is used as a solvent for the preparation of **19-23** (Chart 19.5).[8] Lower molecular weight materials are obtained in other solvents (DMF, DMAc, xylene, toluene, THF) due to the poor solubility of the polymer or the base. Without pivalic acid or the addition of phosphine ligands ($PCy_3 \cdot HBF_4$, XPhos, SPhos, *etc.*) also decreases the molecular weight of the resulting polymers. The polymers are first washed with ethanol and methylethylketone (MEK) to remove low molecular weight materials, and then extracted by Soxhlet using chloroform.[8]

Example 3: Polymerization of Thieno[4,3-*d*]thiazole

Leclerc and coworkers copolymerize monomer **44** with brominated monomers to prepare polymers **15** and **16** (Chart 19.4).[27] The reactions are carried out in a sealed microwave vial, which is charged with 3,6-bis(5-bromothiophen-2-yl)-2,5-bis(2-octyldodecyl)pyrrolo[3,4-*c*]pyrrole-1,4(2*H*,5*H*)-dione **45**, 2-octyl-thieno[3,4-*d*]-thiazole **44**, Herrmann–Beller catalyst (2 mol%), pivalic acid (0.3 equiv.), tri(*o*-anisole)phosphine (4 mol%) and cesium carbonate (3 equiv.) (Scheme 19.10). THF [0.1 M] is added, and the reaction mixture heated to 120 °C under pressure for 24 h. The mixture is cooled to room temperature and the polymer precipitated in methanol/water and filtered. The polymer is then washed *via* Soxhlet extraction with acetone, hexanes and extracted with chloroform. The chloroform fraction is concentrated and precipitated in methanol, filtered and dried and **15** obtained in 60 yield ($M_n = 17$ kDa, $M_w = 51$ kDa, PDI = 2.9).[27] The same conditions are also used for the preparation of **16**.[27]

Example 4: Polymerization of 5-Alkylthieno-[3,4-*c*]pyrrole-4,6-dione (TPD) and 5-Alkylfuro[3,4-*c*]pyrrole-4,6-dione (FPD)

The Leclerc group synthesis used DHAP to prepare copolymers containing bithiophene and terthiophene with TPD and FPD monomers.[42,43] The polymers are prepared by charging a microwave vial with the dibrominated thiophene, 5-alkylthieno-[3,4-*c*]pyrrole-4,6-dione (TPD) or 5-alkylfuro[3,4-*c*]pyrrole-4,6-dione (FPD), Herrmann–Beller catalyst (4 mol%), tri(*o*-anisole)-phosphine (8 mol%), Cs_2CO_3 (2.3 equiv), and pivalic acid (0.3 equiv.) (Scheme 19.11). Toluene (0.2 M) is added and the reaction mixture heated to 120 °C under pressure for 24–36 h. The workup for the reaction is the same as that outlined in Example 2. Polymers **36–39** are prepared in moderate to high yields (38–94%).[43]

Example 5: Polymerization of 3,4-Ethylenedioxythiophene (EDOT) with 2,7-Dibromo-9,9-dioctylfluorene

The group of Kanbara[26] uses DHAP to synthesize a high molecular weight EDOT–fluorene-based copolymer **14** (Chart 19.3). The polymer is prepared by stirring a mixture of EDOT (**46**), 2,7-dibromo-9,9-dioctylfluorene (**47**), $Pd(OAc)_2$ (1 mol%), 1-adamantanecarboxylic acid (0.3 eq), and K_2CO_3 (2.6 eq), in DMAc [0.3 M] for 6 h at 100 °C under an atmosphere of nitrogen

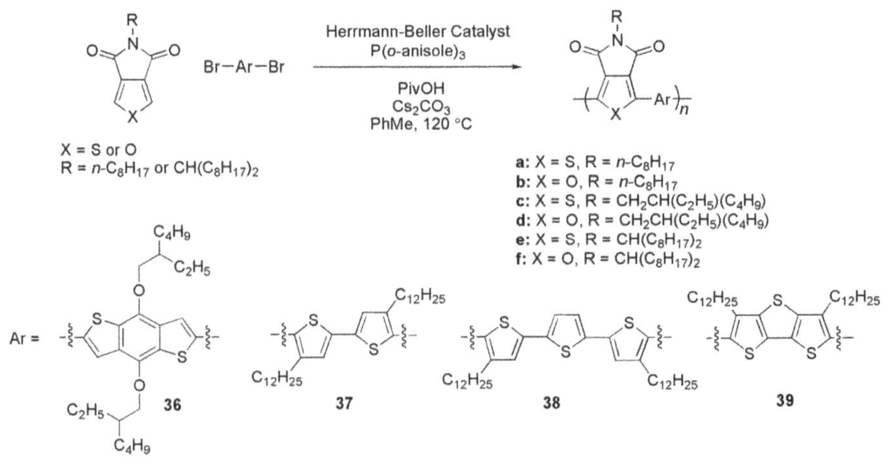

Scheme 19.11 Preparation of polymers using the Herrmann–Beller catalyst and pivalic acid in toluene.

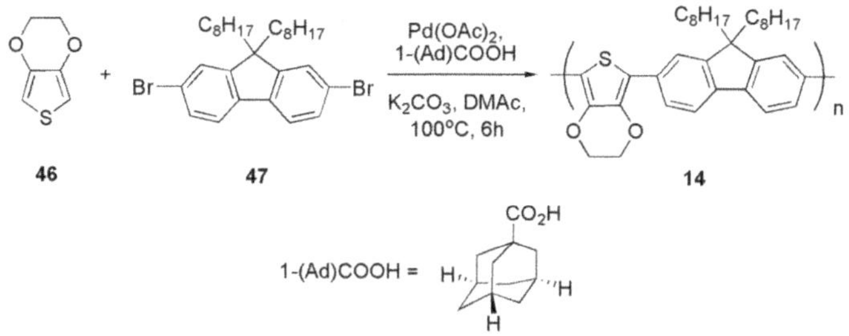

Scheme 19.12 Preparation of EDOT–fluorene-based polymer **14**.

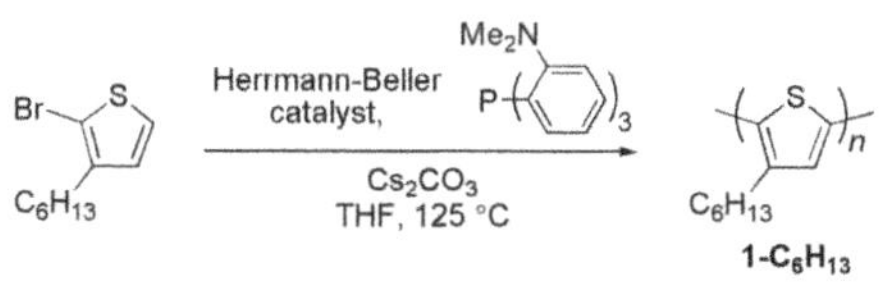

Scheme 19.13 Preparation of poly-3-hexylthiophene from 2-bromo-3-hexylthiophene using Herrmann–Beller catalyst.

(Scheme 19.12). To endcap the resulting polymer, a solution of EDOT (0.5 eq) and $Pd(OAc)_2$ (1 mol%) in DMAc is added to the reaction and the mixture is stirred at 100 °C for an additional 6 h. In contrast to the work up presented in Example 4, the mixture is poured immediately into an aqueous solution of ethylenediaminetetraacetic acid disodium salt and stirred overnight at room temperature in order to remove traces of catalyst. The precipitate is isolated by filtration and washed with HCl (0.1 M), distilled water, MeOH, and then hexanes. The remaining solid is dissolved in $CHCl_3$ and reprecipitated from MeOH to yield polymer **14** (98% yield, $M_n = 42$ kDa, PDI = 2.7).[26]

19.4.4 Polymerization without Carboxylic Acid Additives

Example 6: Polymerization of 2-Bromo-5-alkylthiophene

Ozawa and coworkers carried out DHAP of 2-bromo-3-hexylthiophene using similar conditions to those reported by Leclerc's group.[34] The monomer, Herrmann–Beller catalyst (1 mol%), $P(o\text{-}NMe_2(C_6H_4))_3$ (2 mol%), Cs_2CO_3 (1 equiv.) in THF (0.5 M) are heated to 125 °C for 24 h in a sealed Schlenk flask (Scheme 19.13).[34] Polymers with molecular weights of 30 kDa (PDI = 1.6) and regioregularity of 98% are obtained.[34] The mixture is cooled to room temperature, diluted with dichloromethane and washed with water several times. The polymer is then precipitated in methanol and isolated by centrifugation. The methanol is decanted off and the solid polymer is redissolved in dichloromethane, filtered through a membrane filter and the solvent removed *in vacuo* to give **1-C_6H_{13}** (99% yield).[34] The same conditions were used by Leclerc and

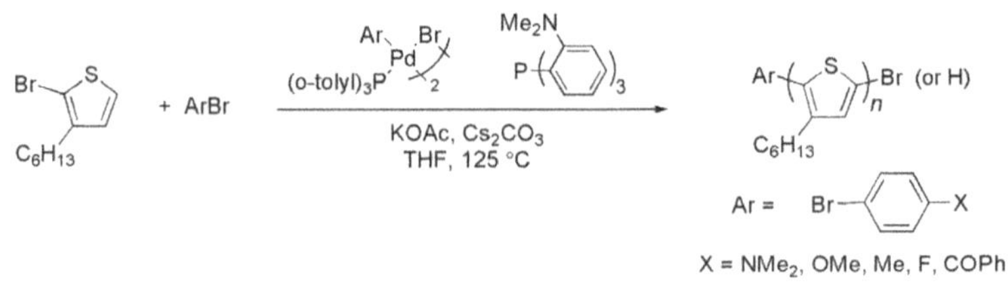

Scheme 19.14 End capping of poly-3-hexylthiophene with *p*-substituted arylbromides using C–H activation.

Table 19.2 Results for end-capped **1-C_6H_{13}** prepared by C–H activation.

X	*End cap [%]*[a]	*Regioregularity [%]*[a]	M_n *[kDa] (PDI)*[b]
H	95	99	7.1 (1.88)
NMe_2	98	99	11.8 (1.70)
OMe	86	–	6.5 (1.86)
Me	98	99	9.0 (1.94)
F	96	–	7.6 (1.96)
COPh	94	98	4.7 (1.70)
CO_2Et	90	99	6.8 (1.86)

[a]Estimated by ^{1}H NMR spectroscopy.
[b]Estimated by SEC calibrated to polystyrene standards.

coworkers for the preparation of **35**, although, they found that P(*o*-anisole)$_3$ gives the highest molecular weight (yield = 96%, M_n = 50 kDa, PDI = 2.6).[42]

Ozawa and coworkers are also able to end-cap **1-C_6H_{13}** *via* C–H activation using [PdAr(μ-Br)(P(*o*-tolyl)$_3$)]$_2$ (2 mol%), which is generated by heating Pd[P(*o*-tolyl)$_3$]$_2$ (2 mol%) and a variety of *p*-substituted aryl bromides (1 equiv.) in THF (1 M) at 60 °C for 3 h. To this solution is then added P(*o*-NMe_2(C_6H_4))$_3$ (2 mol%), Cs_2CO_3 (1 equiv.), KOAc (0.1 equiv.) and 2-bromo-3-hexylthiophene (Scheme 19.14).[55] A large percentage of the polymer is end-capped (86–98%) based on ^{1}H NMR spectroscopy, however, much lower molecular weights are achieved for these polymers compared to the uncapped analog (M_n up to 11.8 kDa, PDI = 1.70). A high degree of regioregularity is maintained (98–99%) (Table 19.2).[55]

Example 7: Polymerization of 3,4-Ethylenedioxythiophene (EDOT) and 3,4-Propylenedioxythiophene (ProDOT) Monomers

Yu's group recently published DHAP methodology with EDOT monomers,[8] which is an expansion of the work they published several years ago involving the coupling of EDOT dimers with several aryl bromides.[35] Optimized conditions for the polymerization of one equivalent of **48** with one equivalent of **49** use Pd(OAc)$_2$ (10 mol%), phosphine (20 mol%), Cs_2CO_3 (2.0 equiv.) and DMF (0.25 M) (Scheme 19.15).[36] Utilization of the Herrmann–Beller catalyst does not have a large effect on the degree of polymerization and reactions carried out in DMAc result in no polymerization. The reaction is heated to 100 °C for 48 h and upon completion is quenched with HCl (1 M) and the polymer precipitated

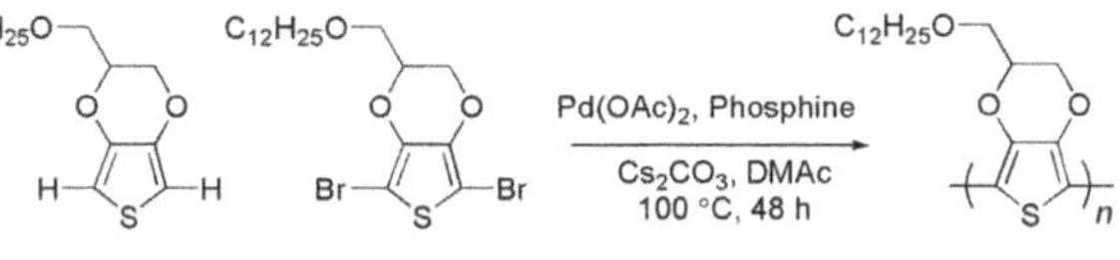

Scheme 19.15 Polymerization of EDOT monomers.

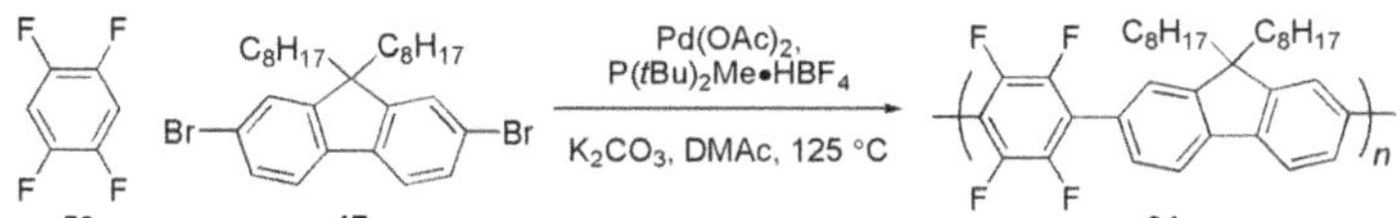

Scheme 19.16 Polymerization of 1,2,4,5-tetrafluorobenzene.

in methanol and water. The polymer is collected by filtration and dried by vacuum. For **29**, polymers with the highest molecular weight are obtained under phosphine-free conditions and with the alkyl phosphine $PCy_3 \cdot HBF_4$ ($M_n = 8.8$ kDa and 9.6 kDa, respectively).[36] However, much lower molecular weight materials are obtained using aryl phosphines (PPh_3: $M_n = 3.9$ kDa, P(*o*-anisole)$_3$: $M_n = 4.8$ kDa).[36]

Example 8: Polymerization of Fluorinated Arenes

Kanbara and coworkers copolymerize 1,2,4,5-tetrafluorbenzene with a number of brominated monomers.[24,37,38] For example, **31** is synthesized by combining the two monomers (**47** and **50**) with $Pd(OAc)_2$ (5 mol%), $PtBu_2Me \cdot HBF_4$ (10 mol%) and K_2CO_3 (2.0 equiv.) in anhydrous DMAc (1 M) for 24 h at 100 °C under an atmosphere of nitrogen (Scheme 19.16). The mixture is cooled to room temperature and poured into an aqueous solution of ethylene diamine tetraaceticacid disodium salt and the suspension is stirred for 1 h. The precipitated polymer is filtered and washed with HCl (1 M), water, methanol and finally hexanes. The solid is then dissolved in chloroform, filtered over Celite and then reprecipitated in methanol (79% yield, $M_n = 27$ kDa, PDI = 3.53).[37,38] Scherf and coworkers use similar conditions to those shown in Scheme 19.16 ($PCy_3 \cdot HBF_4$ was utilized) to prepare **19** (Chart 19.5) from 4,4-di(2-ethylhexyl)-cyclopenta[2,1-*b*:3,4-*b*′]dithiophene and 4,7-dibromo-2,1,3-benzothiadiazole, but found that phosphine-free media give the highest molecular weight material (70% yield, $M_n = 40$ kDa, PDI = 3.5).[39] After a similar work-up to that of the Kanbara group, the polymers are purified by Soxhlet extraction (methanol, ethyl acetate, hexanes, dichloromethane and finally chloroform) to isolate the high molecular weight fractions.[39]

19.5 Conclusions

Direct (hetero)arylation has recently been applied to the synthesis of regioregular polymers as well as alternating copolymers. This methodology utilizes monomers with aryl C–Br and C–H bonds to form new C–C bonds and in doing so 1) reduces the number of steps for monomer preparation since organometallic reagents are no longer necessary; 2) utilizes monomers that have long-term stability and are easily manipulated; 3) produces only HBr as a byproduct, unlike typical polymerization reactions that give rise to stoichiometric amounts of waste. However, DHAP is not without shortcomings. The reactions are, in some cases, not easily controlled and can give rise to other products. Many of the examples in the literature reported crosslinking in the polymer chains, which resulted from activation of more than one C–H bond on the aryl monomers. The energy required to activate the α- and β-aryl C–H bonds of thiophene favor the former by only 4.3 kcal/mol.[48] This has been proven experimentally (^{1}H NMR spectroscopy, mass spectrometry, *etc.*) and these pieces of evidence show that some polymers prepared by DHAP are clearly different from those prepared by Stille or Suzuki coupling reactions. These side reactions can be limited by rational choice of monomers and blocking of other active positions.[24] Prediction of what type of reaction conditions will polymerize a monomer, or set of comonomers, is difficult to predict since there does not seem to be a trend in literature reports. Optimization of specific reaction conditions has been reported by most research groups, but there is little variation between palladium sources, phosphine ligands, bases and solvent. Some groups have had success using DMAc as a solvent, and since this is a donating solvent, a phosphine ligand is often not necessary.

Direct (hetero)arylation polymerization is a methodology that can be added to a polymer chemists' toolbox. This reaction may be applied to the preparation of novel polymers synthesized from copolymers that are unstable as organometallic reagents. It can provide a more facile entry into known polymers if the monomers can be selectively activated, however careful optimization and control is required.

References

1. L. Ackermann, *Chem. Rev.*, 2011, **111**, 1315.
2. D. Alberico, M. E. Scott and M. Lautens, *Chem. Rev.*, 2007, **107**, 174.
3. T. W. Lyons and M. S. Sanford, *Chem. Rev.*, 2010, **110**, 1147.
4. A. Facchetti, L. Vaccaro and A. Marrocchi, *Angew. Chem. Int. Ed.*, 2012, **51**, 3520.
5. I. Ozdemir, Y. Gök, Ö. Özeroğlu, M. Kaloğlu, H. Doucet and C. Bruneau, *Eur. J. Inorg. Chem.*, 2010, **2010**, 1798.
6. S. Yanagisawa, T. Sudo, R. Noyori and K. Itami, *J. Am. Chem. Soc.*, 2006, **128**, 11748.
7. S. Yanagisawa, K. Ueda, H. Sekizawa and K. Itami, *J. Am. Chem. Soc.*, 2009, **131**, 14622.
8. S.-W. Chang, H. Waters, J. Kettle, Z.-R. Kuo, C.-H. Li, C.-Y. Yu and M. Horie, *Macromol. Rapid Commun.*, 2012, **33**, 1927.
9. D. Lapointe and K. Fagnou, *Chem. Lett.*, 2010, **39**, 1118.
10. M. Baghbanzadeh, C. Pilger and C. O. Kappe, *J. Org. Chem.*, 2011, **76**, 8138.

11. Y. Tan and J. F. Hartwig, *J. Am. Chem. Soc.*, 2011, **133**, 3308.
12. B. Liégault, I. Petrov, S. I. Gorelsky and K. Fagnou, *J. Org. Chem.*, 2010, **75**, 1047.
13. M. Lafrance and K. Fagnou, *J. Am. Chem. Soc.*, 2006, **128**, 16496.
14. S. Pivsa-Art, T. Satoh, Y. Kawamura, M. Miura and M. Nomura, *Bull. Chem. Soc. Jpn.*, 1998, **71**, 467.
15. C.-Y. He, Q.-Q. Min and X. Zhang, *Organometallics*, 2011, **31**, 1335.
16. T. Jeffery, *Tetrahedron*, 1996, **52**, 10113.
17. C. Gozzi, L. Lavenot, K. Ilg, V. Penalva and M. Lemaire, *Tetrahedron Lett.*, 1997, **38**, 8867.
18. J. Hassan, E. Schulz, C. Gozzi and M. Lemaire, *J. Mol. Catal. A: Chem.*, 2003, **195**, 125.
19. M. Sévignon, J. Papillon, E. Schulz and M. Lemaire, *Tetrahedron Lett.*, 1999, **40**, 5873.
20. N. I. Abdo, A. A. El-Shehawy, A. A. El-Barbary and J.-S. Lee, *Eur. J. Org. Chem.*, 2012, **2012**, 5540.
21. B. Liégault, D. Lapointe, L. Caron, A. Vlassova and K. Fagnou, *J. Org. Chem.*, 2009, **74**, 1826.
22. D. J. Schipper and K. Fagnou, *Chem. Mater.*, 2011, **23**, 1594.
23. A. E. Rudenko, C. A. Wiley, S. M. Stone, J. F. Tannaci and B. C. Thompson, *J. Polym. Sci., Part A: Polym. Chem.*, 2012, **50**, 3691.
24. Y. Fujinami, J. Kuwabara, W. Lu, H. Hayashi and T. Kanbara, *ACS Macro Lett.*, 2011, **1**, 67.
25. W. Lu, J. Kuwabara and T. Kanbara, *Polym. Chem.*, 2012, **3**, 3217.
26. K. Yamazaki, J. Kuwabara and T. Kanbara, *Macromol. Rapid Commun.*, 2012, **34**, 69.
27. N. Allard, A. Najari, J.-R. Pouliot, A. Pron, F. Grenier and M. Leclerc, *Polym. Chem.*, 2012, **3**, 2875.
28. P. Berrouard, S. Dufresne, A. Pron, J. Veilleux and M. Leclerc, *J. Org. Chem.*, 2012, **77**, 8167.
29. L. Chen, J. Roger, C. Bruneau, P. H. Dixneuf and H. Doucet, *Chem. Commun.*, 2011, **47**, 1872.
30. N. Laidaoui, J. Roger, A. Miloudi, D. El Abed and H. Doucet, *Eur. J. Org. Chem.*, 2011, **2011**, 4373.
31. A. Ohta, Y. Akita, T. Ohkuwa, M. Chiba, R. Fukunaga, A. Miyafuji, T. Nakata, N. Tani and Y. Aoyagi, *Heterocycles*, 1990, **31**, 1951.
32. T. Okazawa, T. Satoh, M. Miura and M. Nomura, *J. Am. Chem. Soc.*, 2002, **124**, 5286.
33. K. Si Larbi, H. Y. Fu, N. Laidaoui, K. Beydoun, A. Miloudi, D. El Abed, S. Djabbar and H. Doucet, *ChemCatChem*, 2012, **4**, 815.
34. Q. Wang, R. Takita, Y. Kikuzaki and F. Ozawa, *J. Am. Chem. Soc.*, 2010, **132**, 11420.
35. C.-Y. Liu, H. Zhao and H.-h. Yu, *Org. Lett.*, 2011, **13**, 4068.
36. H. Zhao, C.-Y. Liu, S.-C. Luo, B. Zhu, T.-H. Wang, H.-F. Hsu and H.-h. Yu, *Macromolecules*, 2012, **1**, 7783.

37. W. Lu, J. Kuwabara, T. Iijima, H. Higashimura, H. Hayashi and T. Kanbara, *Macromolecules*, 2012, **45**, 4128.
38. W. Lu, J. Kuwabara and T. Kanbara, *Macromolecules*, 2011, **44**, 1252.
39. S. Kowalski, S. Allard and U. Scherf, *ACS Macro Lett.*, 2012, **1**, 465.
40. J. Kettle, M. Horie, L. A. Majewski, B. R. Saunders, S. Tuladhar, J. Nelson and M. L. Turner, *Sol. Energy Mater. Sol. Cells*, 2011, **95**, 2186.
41. Z. Zhu, D. Waller, R. Gaudiana, M. Morana, D. Mühlbacher, M. Scharber and C. Brabec, *Macromolecules*, 2007, **40**, 1981.
42. P. Berrouard, A. Najari, A. Pron, D. Gendron, P.-O. Morin, J.-R. Pouliot, J. Veilleux and M. Leclerc, *Angew. Chem., Int. Ed.*, 2012, **51**, 2068.
43. S. Beaupré, A. Pron, S. H. Drouin, A. Najari, L. G. Mercier, A. Robitaille and M. Leclerc, *Macromolecules*, 2012, **45**, 6906.
44. J. Jo, A. Pron, P. Berrouard, W. L. Leong, J. D. Yuen, J. S. Moon, M. Leclerc and A. J. Heeger, *Adv. Energy Mater.*, 2012, **2**, 1397.
45. D. L. Davies, S. M. A. Donald and S. A. Macgregor, *J. Am. Chem. Soc.*, 2005, **127**, 13754.
46. S. I. Gorelsky, *Coord. Chem. Rev.*, 2013, **257**, 153.
47. S. I. Gorelsky, D. Lapointe and K. Fagnou, *J. Am. Chem. Soc.*, 2008, **130**, 10848.
48. S. I. Gorelsky, D. Lapointe and K. Fagnou, *J. Org. Chem.*, 2011, **77**, 658.
49. S.-Y. Tang, Q.-X. Guo and Y. Fu, *Chem. Eur. J.*, 2011, **17**, 13866.
50. K. Ueda, S. Yanagisawa, J. Yamaguchi and K. Itami, *Angew. Chem. Int. Ed.*, 2010, **49**, 8946.
51. S. I. Gorelsky, *Organometallics*, 2012, **31**, 4631.
52. M. Wakioka, Y. Nakamura, Q. Wang and F. Ozawa, *Organometallics*, 2012, **31**, 4810.
53. W. A. Herrmann, C. Brossmer, K. Öfele, C.-P. Reisinger, T. Priermeier, M. Beller and H. Fischer, *Angew. Chem. Int. Ed.*, 1995, **34**, 1844.
54. R. D. McCullough, *Adv. Mater.*, 1998, **10**, 93.
55. Q. Wang, M. Wakioka and F. Ozawa, *Macromol. Rapid Commun.*, 2012, **33**, 1203.

CHAPTER 20

Chain-Growth Catalyst-Transfer Polycondensations

ANTON KIRIY*[a] AND VOLODYMYR SENKOVSKYY[b]

[a] Leibniz-Institut für Polymerforschung Dresden e.V. Hohe Straße 6, 01069 Dresden, Germany; [b] NOVALED AG, Tatzberg 49, 01307 Dresden, Germany
*Email: kiriy@ipfdd.de

20.1 Catalyst-Transfer Polycondensation: Mechanism, Scope and Limitations

20.1.1 Introduction

π-Conjugated polymers (CPs) are commonly prepared *via* established transition-metal-catalyzed crosscouplings (*e.g.*, Stille and Suzuki couplings) or oxidative polymerizations, which usually follow a step-growth polycondensation mechanism.[1] CPs prepared *via* step-growth polycondensations frequently suffer from structural irregularity, a low degree of control over molecular weight (MW), MW distribution and end-group functionality. This often results in batch-to-batch variations and altered material properties, which is undesirable for applications. Simultaneous control over all of these factors is required to achieve reproducible material properties, and in turn, reproducible device parameters.

Poly(3-hexylthiophene), P3HT, is a good example to illustrate the impact of structural perfectness on fundamental properties of a polymer and performance of devices fabricated on its basis. In general, the asymmetry of 3-substituted thiophenes results in three possible coupling modes when two monomers are

RSC Polymer Chemistry Series No. 9
Conjugated Polymers: A Practical Guide to Synthesis
Edited by Klaus Müllen, John R. Reynolds and Toshio Masuda

Published by the Royal Society of Chemistry, www.rsc.org

linked between the 2- and the 5-positions: 2,5′- (or head–tail, HT), 2,2′- (or head–head, HH) and 5,5′- (or tail–tail, TT) couplings (Figure 9.1 in Chapter 9).[2]

P3HT produced by step-growth polymerizations usually has irregular structure (with HT, HH and TT dyads more or less randomly distributed along the backbone). Regiorandom P3HT is essentially amorphous in thin-film and bulk forms and has relatively poor electrical transport properties unattractive for optoelectronic devices. In sharp contrast, the physical properties and performance of highly regioregular head-to-tail P3HT (rrP3HT) prepared *via* controlled chain-growth polymerizations differ significantly from the performance of its regioirregular counterpart. rrP3HT has a great propensity to form semicrystalline lamellar microstructures.[3,4] Films of pristine rrP3HT were used in organic field effect transistors, which typically yield appreciably high charge carrier mobilities of 10^{-2} to 10^{-1} $cm^2/(V\ s)$.[5] In a highly doped state, rrP3HT exhibits electrical conductivity up to 1000 S/cm, which is 2 to 4 orders of magnitude higher than the conductivity for regioirregular P3HT.[2] The power-conversion efficiency of organic photovoltaic devices incorporating blends of rrP3HT and fullerene derivatives are in the 4–5% range,[6] whereas blends of irregular P3HT exhibit negligibly small photoeffect. Furthermore, a strong molecular weight (MW) dependence of properties and device performance was found for rrP3HT and other conjugated polymers.[7]

Another important issue is the polymer end groups. It is known that the end groups not only strongly influence charge-carrier properties of the polymer but also determine the photophysical and photochemical properties of the polymer backbone. For example, bromine end groups can induce intersystem crossing, which can significantly reduce the photoluminescence and adversely influence photochemical stability. In addition, new CP architectures are needed that predictably self-assemble into desirable nanomorphologies, thus solving a longstanding problem with improper morphologies of active layers in optoelectronic devices.[8] Synthesis of well-defined conjugated homopolymers, conjugated block copolymers and other complex architectures of conjugated polymers remain difficult to date *via* conventional step-growth polycondensations.[9]

Chain-growth catalyst-transfer polycondensation (CTP) is a rapidly developing polymerization method, as it allows, in many cases, the abovementioned limitations of step-growth polymerizations to be overcome.[10] CTP provides a straightforward access to well-defined conjugated homopolymers (*e.g.*, polythiophenes (**1**), polyfluorenes (**2**), polyphenylenes (**3**), *etc.*), alternate donor–acceptor copolymer (*e.g.*, **4**) and all-conjugated block polymers (*e.g.*, **5**), Chart 20.1.

20.1.2 Mechanism

In 2004/2005, McCullough and coworkers[11,12] and Yokozawa and coworkers[13,14] discovered that the Ni-catalyzed Kumada catalyst-transfer polymerization (KCTP), also referred to as Grignard metathesis polymerization (GRIM), of rrP3HT proceeds *via* chain growth, but not *via* a step-growth mechanism. An important prerequisite of the chain-growth mechanism in metal-catalyzed polycondensations is the use of AB-type monomers, having

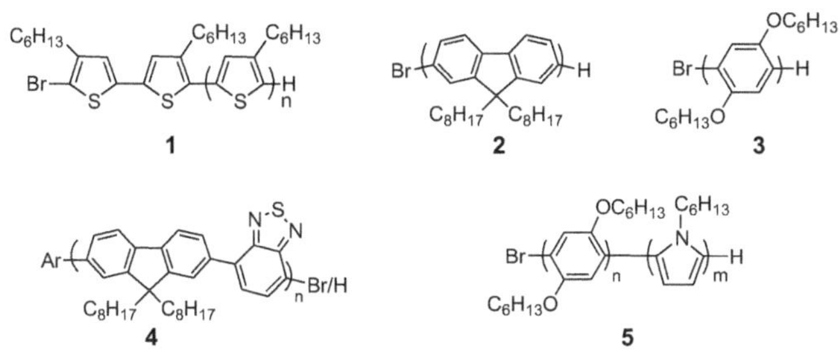

Chart 20.1

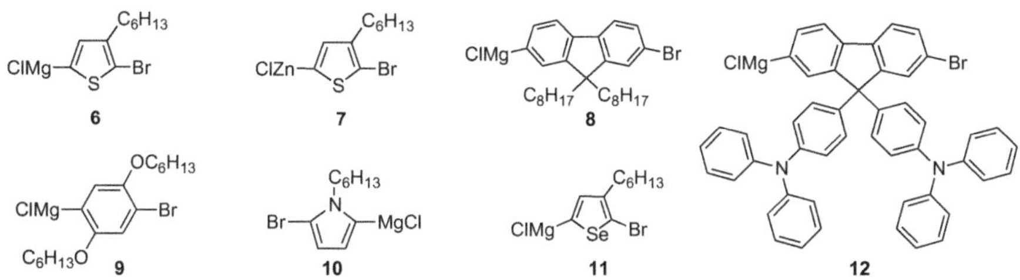

Chart 20.2

two potentially self-reactive groups in the same molecule (*e.g.*, **6–12**, Chart 20.2).

It is now generally accepted that KCTP proceeds by extension of tail-to-tail dimers (**14**) formed in the first step from Ni(dppp)Cl_2 (dppp = diphenylphosphinopropane or similar catalyst precursors) and two Grignard-type monomer molecules (20.1). As for the parent Kumada crosscoupling reaction, the catalytic cycle of the polycondensation involves trans-metalation (TM), reductive elimination (RE), and oxidative addition (OA) elementary steps (20.1). The Ni(0) species formed at the reductive elimination step associates with the nearest thiophene ring, forming an π-complex **13**, then "walks" toward a growing polymer chain end, and finally inserts into the terminal C–Br bond.[12,15] Thus, the rare for polycondensations chain-growth mechanism in this case is due to a unique propensity of coordination–unsaturated Ni(0) species to form π-complexes with the polymerized chain and transfer intramolecularly *via* a "ring-walking" process[16] rather than to diffuse to another chains or monomer molecules, in such a way contributing to the step-growth mechanism.

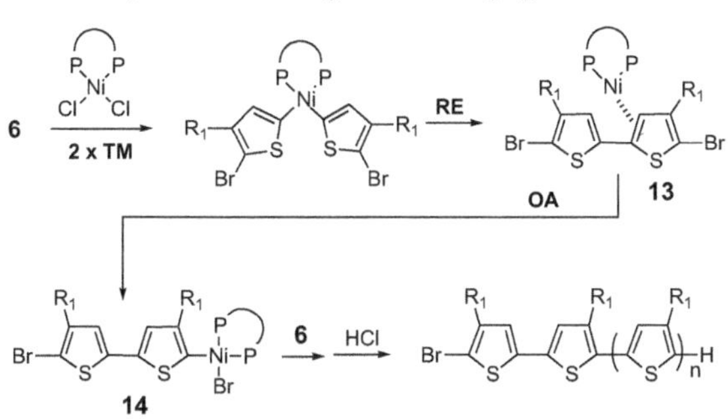

(20.1)

20.1.3 End-Functionalized Polymers and Brushes

KCTP conducted in the presence of $Ni(dppp)Cl_2$ or the like catalysts (*in situ* initiation) leads to polymers with the starting group derived from the monomer (20.1).[12] This approach restricts the applicability of KCTP in the synthesis of complex architectures of CPs such as polymer brushes and well-defined block copolymers. To overcome drawbacks of *in situ* initiation, *ex situ* initiation strategies were developed that use Ar-Ni(L_2)-X initiators (L = dppe or other bidentate ligands) prepared in an independent step. In the first approach, model *ex situ* initiators Ar-Ni(PPh_3)$_2$-X were synthesized by oxidative addition of $Ni(PPh_3)_4$[17] or $Ni(COD)_2/PPh_3$[18] to aryl halides (20.2). The monodentate ligand (*i.e.* PPh_3) can be further replaced by various bidentate ligands to give initiators **15** and **16**, which performs better in KCTP than initiators ligated by monodentate ligands.[19]

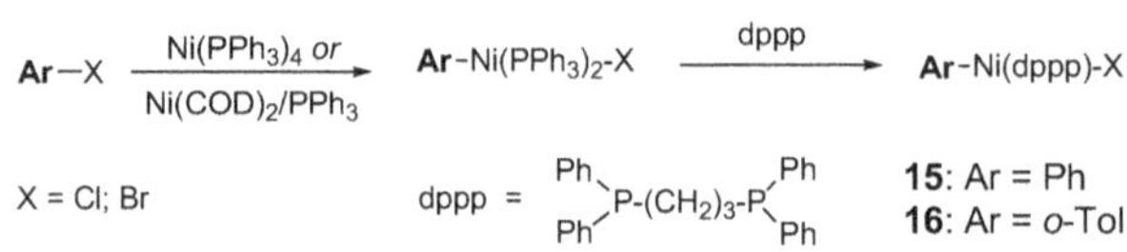

(20.2)

This approach, however, suffers from moderate selectivity of the OA process and low stability of the intermediate Ar–Ni(PPh_3)$_2$–X complex against disproportionation (resulting in homocoupling products, Ar–Ar) if *para*- or *meta*-substituted aryl halides are used as starting materials. This problem can be solved *via* introduction of alkyl substituents at the *ortho*-position relative to the Ni center (*e.g.*, in **16**), which significantly increase stability of the Ni complexes.[20]

Alternatively, *ortho*-substituted aryl halides can be activated in a two-step process that includes their transformation into corresponding Grignard or zinc-organic compounds followed by transmetalation with commercial and readily accessible $Ni(dppp)Cl_2$ (20.3).[21] A quintessence of this approach is that the reaction between $Ni(dppp)Cl_2$ and Ar-MgCl stops after the first transmetalation step leading to stable initiators **16–18** or similar complexes, because of steric hindrances that occur at the second transmetalation step caused by *ortho*-substituents.

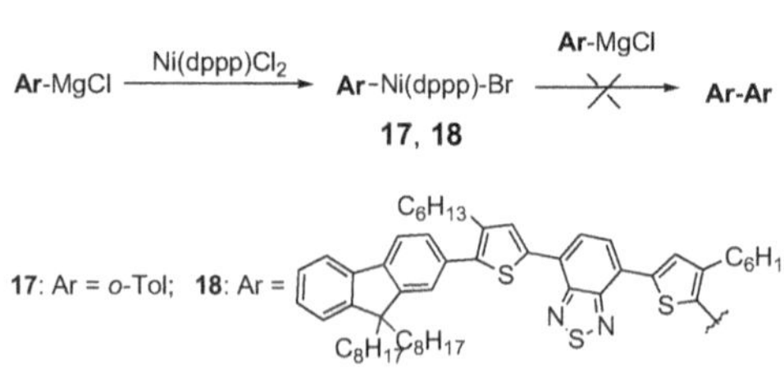

(20.3)

The most universal and selective *ex situ* initiation approach, among those reported to date, utilizes $Et_2Ni(bipy)$ as the activator.[22] According to the method, a library of initiators can be synthesized by treatment of readily available aryl halides with $Ni(bipy)Et_2$ followed by exchange of the bipy ligand by other ligands (*e.g.*, dppe, dppp, *etc.*, (20.4)).

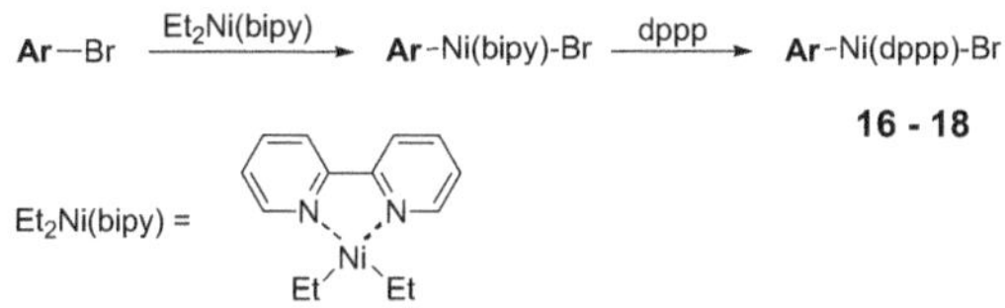

(20.4)

Importantly, with this method in hand, high activation selectivity and initiation efficiency can be achieved without incorporation of stabilizing groups into aryl halides and this method works efficiently with wide variety of aryl halides, including electron-deficient[23] and electron-rich ones. The broad applicability and high functional group tolerance of the Et_2Ni(bipy) approach was illustrated by the synthesis of well-defined conjugated polymer architectures, *e.g.*, star polymers,[24] and rod–coil block copolymers[25,26] and polymer brushes (20.5).[22,27–31]

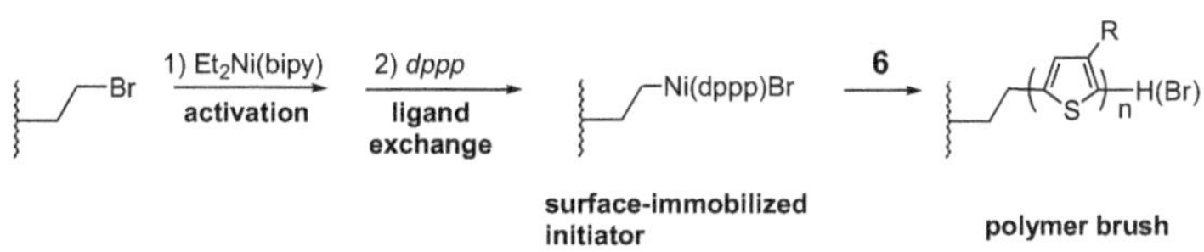

(20.5)

20.1.4 Fully Conjugated Block Copolymers

Block copolymers are able to self-organize into microphase-separated nanostructures driven by the immiscibility or the crystallinity difference between the two blocks.[32] This property is valuable for various optoelectronic devices, such as solar cells or ambipolar transistors, therefore block copolymers containing conjugated polymer blocks with a complementary electronic functions is an important target. All-conjugated P3HT-based block copolymers were recently prepared using conventional step-growth methods.[8,9] This approach relies on coupling of prepolymerized P3HT during the step-growth synthesis of other CPs.[33,34] While this approach is simple, the obtained raw products are usually contaminated with homopolymers and exhibit a broad polydispersity. A more straightforward and general strategy for preparation of block copolymers implies sequential polymerization of different monomers. This approach, however, requires both blocks to be prepared by living/controlled polymerizations, yet with the same catalyst and under the same reaction conditions. This method has become possible with the discovery of the chain-growth KCTP. In 2005, McCullough and coworkers first demonstrated the synthesis of a polythiophene-based all-conjugated block copolymer, poly(3-hexylthiophene)-*b*-poly(3-dodecylthiophene).[12] Later, different all-conjugated block copolymers with polythiophene backbone but different side chains were synthesized.[35–38] At the same time, all-conjugated block copolymers comprising the blocks with different conjugated backbones are more attractive for various applications. The first such block copolymer was prepared by Yokoyama *et al.*[39] They found common conditions for the quasiliving polymerization of phenylene- and pyrrole-based monomers **9** and **10**, respectively, which enabled the synthesis of the corresponding macroinitiator **19** and block copolymer **5** (20.6).

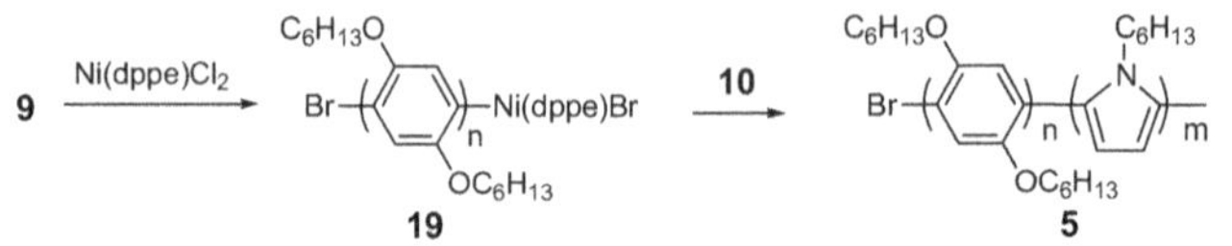

(20.6)

More recently, poly(3-hexylthiophene)-polyfluorene[40–42] and poly(3-hexylthiophene)-poly(3-hexylselenophene)[43] block copolymers were prepared. Interestingly, these materials showed a phase separation, even so they have the same side chain (n-hexyl). Recently, KCTP was used to prepare hydrophilic – hydrophobic block copolymer ionomers having polyphenylene backbone.[44] In this case, block copolymers with controlled block lengths were synthesized from a dibromo phenylene derivative having a neopentyl ester protected sulfonic acid group, followed by the polycondensation of hydrophobic dibromo hexyloxybenzene (20.7). The diblock copolymer ionomers were obtained by the removal of neopentyl groups, resulting in clear phase separation dependent on the block lengths. The well-developed microphase separation provided controlled water uptake and sufficiently high proton conductivity, especially at low relative humidity conditions. As such, the fine block copolymerization by using KCTP is a promising strategy for the development of hydrocarbon ionomers having well-defined ordered structures with reasonable proton conductivity for fuel-cell applications.[44]

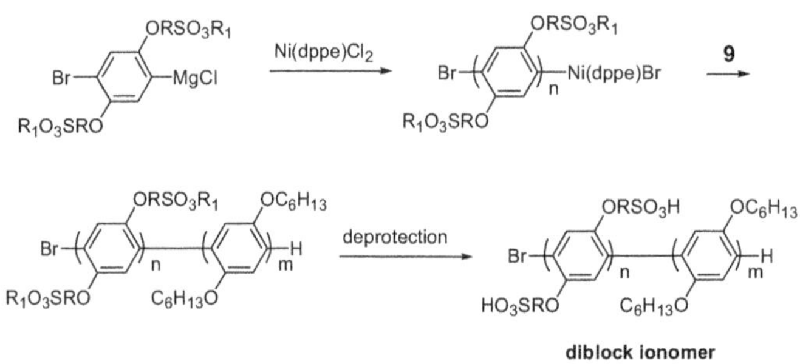

(20.7)

It is noteworthy that the order of monomer addition is an important parameter in the synthesis of all-conjugated block copolymers. For example, in the synthesis of block copolymers comprising polythiophene and poly(phenylene) blocks, efficient initiation and growth of the second block was only achieved when the poly(phenylene)-block was polymerized first followed by polythiophene, while the reverse case yielded the target block copolymer largely contaminated with poly(phenylene) and polythiophene homopolymers.[40–42] The observation of inefficient initiation of the second polythiophene block can be well understood with the mechanism of random catalyst walking.[16] According to this mechanism, the complexation ability of the conjugated monomer (and the resulting polymer) is likely to decide on the order of monomer addition. Apparently, it is energetically more favorable for Ni(0) to interact with the thiophene backbone than with the polyphenylene backbone.

20.1.5 Chain-Growth Polymerization of Electron-Deficient Monomers

Until recently, catalyst-transfer polycondensations were limited to the preparation of p-type π-conjugated polymers and block copolymers comprising relatively simple electron-rich building blocks (such as thiophene). On the other hand, well-defined conjugated donor–acceptor alternate copolymers and donor–acceptor all-conjugated block copolymers are needed for practical applications.[1] Very recently, controlled polymerizations of a few electron-deficient monomers were also reported.[45–47]

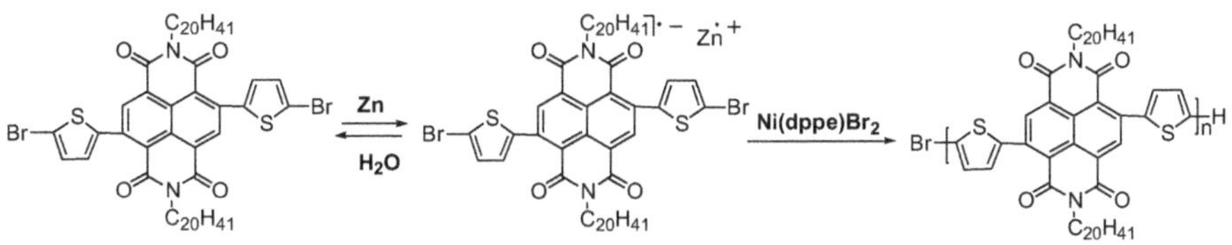

Br-TNDIT-Br **20** **21**

(20.8)

In one case, a nickel-catalyzed chain-growth polymerization of a highly unusual paramagnetic, anion-radical monomer **20** was performed (20.8). The monomer was prepared by reaction of activated Zn powder and naphthalene-diimide-based dibromo-precursor, Br-TNDIT-Br.[45,46] Interestingly, reaction of Br-TNDIT-Br with activated Zn did not lead to a usual zinc-organic compound Br-TNDIT-ZnBr, in which Zn is inserted between C and Br, but rather to a charge-transfer complex. This structural feature follows from the fact that the acidic work-up of **20** resulted in a quantitative recovering of Br-TNDIT-Br but not of Br-TNDIT-H, protonation product of Br-TNDIT-ZnBr. It is also noteworthy that preparation of usual Grignard-type monomers (*i.e.* Br-TNDIT-MgBr) or other conventional metalorganic compounds (*e.g.*, Br-TNDIT-Li) *via* halogen-metal exchange was not possible, presumably due to a high electron affinity of Br-TNDIT-Br, which promoted the electron-transfer rather than the atom exchange. Despite unusual monomer structure, the polymerization of **20** proceeded quickly at room temperature and followed the chain-growth mechanism leading to donor–acceptor alternating copolymer **21** with controlled MW and relatively low PDIs of 1.4–1.7.[46]

20.1.6 Chain-Growth Suzuki Polycondensation

Although KCTP is currently a dominating method for controlled synthesis of conjugated polymers and block copolymers, it suffers from a relatively high reactivity of organometallic monomers toward electrophilic groups (*e.g.*, carbonyl, cyano, ester, *etc.*). Suzuki coupling of aryl halides with organoboron compounds[48,49] is another general route to conjugated polymers.[50] Due to a high functional group tolerance, Suzuki coupling is potentially applicable to a polymerization of a wide range of electron-rich and electron-deficient monomers having in their structure functional groups incompatible with KCTP conditions.[50] Recently, several examples of chain-growth Suzuki polycondensation of aromatic monomers have been reported.[51–54]

Chain-growth Suzuki polycondensation proceeds with AB-type monomers (*e.g.*, **22**) having organoboron and halogen groups in the same monomer molecule. To ensure the (intramolecular) catalyst-transfer mechanism and, hence, the chain-growth propagation, a "stickiness" of the catalyst to the polymerized π-conjugated chain must be provided. Pd complex ligated by bulky and electron-rich $t\mathrm{Bu_3P}$ ligand was found to be the best, to date, catalytic system for chain-growth Suzuki polycondensation. Because of the high bulkiness of the ligand, it efficiently supports coordination-unsaturated ($t\mathrm{Bu_3P}$)Pd(0) species formed in the reductive elimination step, which further associates with the polymer backbone and then transfers intramolecularly to the carbon–halogen bond rather than diffuses into the bulk solution. A variety of three-coordinated ($t\mathrm{Bu_3}$)Pd($\mathrm{Ar_1}$)Br complexes can be routinely prepared from aryl halides and $(t\mathrm{Bu_3P})_2$Pd and they serve as external initiators. These catalysts have been used in controlled synthesis of polyfluorenes, polyphenylenes,[51–53] and polythiophenes[54] with narrow PDI, well-defined MW, having an Ar starting group.

Very recently, externally initiated chain-growth Suzuki polycondensation was applied for preparation of polyfluorenes having two different functional end groups on either side of the conjugated polymer (*e.g.*, shown in 20.9).[55] While the starting group (*i.e.* $\mathrm{Ar_1}$) transfers from the initiator, the end group ($\mathrm{Ar_2}$) was introduced by termination of the "living" polymer chain with aryls having boronic ester group (*e.g.*, $\mathrm{Ar_2}$-B(OR)$_2$).

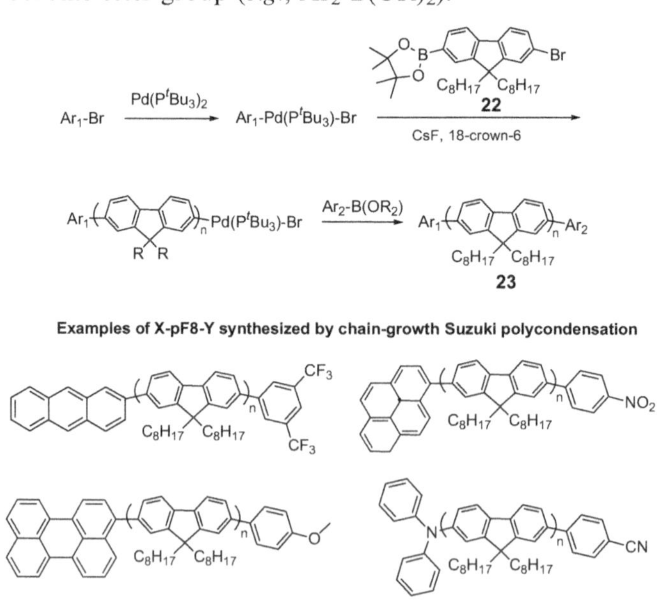

(20.9)

Furthermore, the living performance of the chain-growth Suzuki polycondensation was recently illustrated by the synthesis of polyfluorene–polyphenylene and polyfluorene-polythiophene block copolymers *via* sequential polymerization of two different monomers.[53,54]

A significant progress was recently achieved in preparation of more complex conjugated polymers (*e.g.*, donor–acceptor alternate copolymers) attractive for fabrication of solar cells and field-effect transistors. Thus, Huck and coworkers reported the controlled chain-growth Suzuki polymerization of a complex monomer comprising electron-deficient benzothiadiazole and relatively electron-rich fluorene units.[47] The polymerization yielded a well-defined n-type polymer, poly(9,9′-dioctylfluorene-co-benzothiadiazole) with controlled MW, relatively low PDI (<1.3) and specific end functions. In this work, a fluorescent pyrene (Pyr) end group was transferred from the initiator ((tBu$_3$P)Pd(Pyr)Br) prepared by OA of Pd(0) to 1-bromopyrene. In agreement with the catalyst-transfer mechanism, propagation occurred by subsequent transmetalation, reductive elimination, intramolecular migration of the Pd species to the C–Br chain end, followed by oxidative addition. MALDI-TOF revealed that every polymer chain carries the initiator aromatic moiety and either Br or H on the other side, which is in agreement with the proposed chain-growth mechanism.

20.1.7 Perspective

Controlled methods for preparation of conjugated polymers, *e.g.*, the nickel-catalyzed Kumada catalyst-transfer polycondensation and chain-growth Suzuki polymerization, are developing extremely fast. Since the discovery of the chain-growth mechanism of the Ni(dppp)Cl$_2$-initiated polymerization of poly(3-hexylthiophene) only less than a decade ago, a plethora of conjugated chain architectures have been prepared, and more and more different monomers can be polymerized. However, most of polymers prepared to date in a controlled manner have a relatively simple structure, electron-rich nature and wide bandgaps; so these materials are not very useful for practical applications. For example, alternate donor–acceptor copolymers with easily modulating electronic levels are needed for fabrication of field-effect transistors and solar cells. In this regard, step-growth polymerizations still remain the most universal tool in the synthesis of a variety of CPs for the above-mentioned applications. Indeed, a great number of complex building blocks, including relatively large π-conjugated oligomers and dye molecules, can be routinely polymerized by using step-growth Stille or Suzuki coupling polymerizations. At the same time, there are only a few examples in the literature that report controlled polymerization of alternating donor–acceptor building blocks. Therefore, in order for controlled polymerizations to have a practical importance, their scope should be significantly extended, so that a variety of donor and acceptor building blocks to be involved into controlled catalyst-transfer polymerizations. Among the most promising building blocks are, for example, cyclopentadithiophene-, dithienosilole-, benzodithiophene-based monomers, *etc.*, as well as a variety of dyes, such as diketopyrrolopyrroles, indigos and isoindigos, perylene bisimide and squaraine dyes, *etc*. Having such diverse controlled polymerization method in hand, one can target extremely complex and, at the same time, well-defined structures, such as all-conjugated donor–acceptor block copolymers. Without doubt, the prospect of preparing all-conjugated block copolymers with suitable p-type and n-type blocks in a controlled one-pot procedure is intriguing.

20.2 Experimental Procedures

20.2.1 Polythiophenes

General information. rrP3ATs can be prepared by polymerization of AB-type thiophenic monomers (*i.e*. 3-alkylthiophenes **6** and **7** having halogen function in the 2-position and metalorganic function in the 5-position) under anhydrous and air-free conditions in the presence of Ni complexes (typically, Ni(dppp)Cl_2 or Ni(dppe)Cl_2). Several protocols were developed, such as McCullough method,[56] Rieke method,[57] GRIM polymerization,[58] which differ mostly by monomer preparation procedures (see Chapter 9, Examples 9.8–9.10). All reactions must be performed under anhydrous and oxygen-free nitrogen or argon, using dry glassware. Solvents, (usually, tetrahydrofuran, THF) must be dried over Na benzophenone ketyl radical and freshly distilled prior to use. As these polymerizations follow the chain-growth mechanism; the DPs of resulting polymer products may be controlled by a [monomer]/[catalyst] ratio (corrected by monomer conversion). It should, however, be emphasized that satisfactory MW control, low PDI and end-group fidelity are achievable at mild polymerization conditions and in the absence of chain-termination impurities. Thus, a long-term reflux of reaction mixtures (when monomer is already fully consumed) leads to step-growth coupling of prepolymerized chains and, consequently, to an increase of MW and PDI increase. On the other hand, other alkylmagnesium halides or alkyl lithium (used for monomer preparation) react efficiently with propagating chain ends and cause chain-termination and chain reinitiation. It is therefore important to ensure the absence of such metalorganic impurities in the polymerization mixture. In view of this, in many cases, it is helpful to keep a slight excess of dihalo-monomer precursors relative to the activating metalorganics and to check the conversion degree of precursors into the active monomer prior the polymerization (*e.g*., by recording a ^{1}H NMR spectrum of the acidified monomer). If these precautions are taken into account, one can precisely control the MW of the resulting P3ATs by setting the corresponding monomer-to-initiator ratio.

Example 1: KCTP[13,14]

A superior control over MW, PDI, regioselectivity and end-group fidelity (compared to methods that utilize dibromo-precursors, Examples 9.9 and 9.10) can be achieved using 2-bromo-5-iodo-3-alkylthiophenes as monomer precursors (20.10) because: i) halogen–magnesium exchange leads to a single monomer (I/MgCl exchange occurs predominantly), which contributes to a higher head-to-tail regioselectivity; ii) I/MgCl exchange proceeds faster and cleaner than a Br/MgCl one, which minimizes the amount of chain-terminating alkyl-magnesium chlorides impurities.

Monomer 6. A round-bottomed flask equipped with a three-way stopcock is heated under reduced pressure and then cooled to room temperature under an argon atmosphere. 2-Bromo-3-hexyl-5-iodothiophene (373 mg, 1.00 mmol) is placed in the flask, and the atmosphere is replaced with argon. THF (5.0 mL) is added into the dry flask *via* a syringe and the mixture is stirred at 0 °C. Isopropylmagnesium chloride, *i*-PrMgCl, (2.0 M solution in THF, 0.50 mL, 1.0 mmol) is added *via* a syringe to the mixture and stirred at 0 °C for 1 h.

Afterwards, a suspension of Ni(dppp)Cl_2 (8.7 mg, 0.016 mmol, 1.60 mol%) in dry THF (5.0 mL) is added to the solution of monomer **6** *via* a syringe at 0 °C, and then the mixture is stirred at room temperature for 6 h.

Polymer separation procedure. The polymerization is stopped by quenching with HCl aqueous solution (5 M, 5 mL) and then precipitated with MeOH (300 mL). The resulting red precipitate is then filtered into a Soxhlet thimble. Soxhlet extractions are performed with methanol (to remove monomer and salts), hexane (to remove catalyst and oligomers), and chloroform. Yield: 0.12 g (72%), $M_n = 9300$, $M_w/M_n = 1.1$.

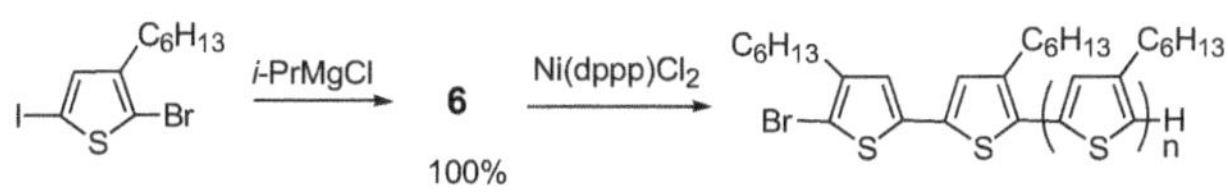

(20.10)

20.2.2 Polyfluorenes, Polycarbazoles, Polyphenylenes and Alternating Copolymers[59]

Example 2: Synthesis of Polyfluorene using Turbo-Grignards as Monomer Activators (20.11)

9,9-Dioctyl-2,7-dibromofluorene (**24**) (2.7 g, 5 mmol) in 15 mL of anhydrous THF is added under nitrogen to the previously prepared solution of *i*-PrMgCl · LiCl (5 mmol). The reaction mixture is stirred at room temperature for 12 h and monitored by GC-MS. Analysis of the reaction mixture confirmed the formation of mono-Grignard reagent in at least 85% yield. Ni(dppp)Cl_2 (0.02 g, 0.04 mmol) is added to the reaction mixture dissolved in 10 mL anhydrous THF. The polymerization is allowed to proceed for 20 min at room temperature followed by quenching of the reaction mixture with acidified methanol. The precipitated polymer is filtered through a thimble and subjected to methanol extraction. The polymer is dried under vacuum. $M_n = 15\ 000$; PDI = 1.45. Alternatively, **24** is activated by *t*-BuMgCl · LiCl in the presence of crown ether 15-c-5 ([*t*-BuMgCl] : [LiCl] : [15-c-5])1 : 1 : 1) under otherwise same reaction conditions. Almost double molecular weight of the resulting polymer is obtained in this case, $M_n = 29\ 000$; PDI = 1.5.

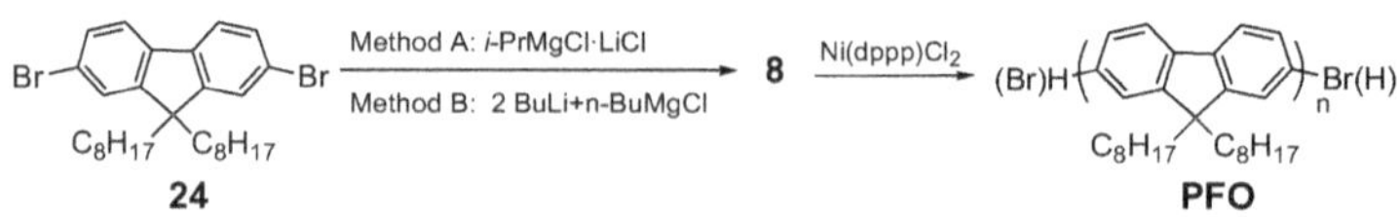

(20.11)

Example 3: Synthesis of Polyfluorene using Magnesium ATE Complex as Monomer Activator[59]

A dry 25-mL three-neck flask is flushed with nitrogen and charged with butyl lithium (1.6 M in hexanes, 1.1 mmol) and a solution of *n*-BuMgCl (2 M in THF, 0.55 mmol) in dry toluene (2 mL) at −10 °C. After stirring for 1 h at −10 °C, a solution of **24** (0.9 g, 1.65 mmol) in 5 mL of anhydrous THF is added under nitrogen. The reaction mixture is stirred for 1–5 h at 40 °C and monitored by GC-MS until such time that >80% of the monomer is converted into the mono-Grignard reagent. Ni(dppp)Cl_2 (0.01 g, 0.02 mmol) is added to the reaction mixture. The polymerization is allowed to proceed for 20 min at room temperature followed by quenching of the reaction mixture with methanol. The precipitated white polymer is filtered, washed with cold methanol and dried under vacuum for 12 h. Monomer conversion of 92% is observed in this case and polyfluorene with $M_n = 9700$ and PDI = 1.55 is obtained.

The same procedure is applicable but less efficient for preparation of polycarbazoles. The magnesium halogen exchange of 2,7-dibromo-*N*-octylcarbazole with *i*-PrMgCl · LiCl is significantly slower compared to the fluorene monomer. Almost no conversion is observed even after 24 h. In contrast, when lithium tributylmagnesate (*n*-Bu_3MgLi) is used for magnesium halogen exchange, the reaction is completed in 5 h at room temperature. The yield of polymerization is 50% after 5 h, and a polymer with $M_n = 2600$ and PDI = 1.23 is obtained.

Example 4: Improved Synthesis of Polyfluorene using Ni(acac)$_2$/dppp Catalyst (20.12)[60]

In a typical experiment, 10 mL THF is added into a mixture of Ni(acac)$_2$ (51.4 mg, 0.20 mmol) and dppp (84.1 mg, 0.204 mmol) in a-25 mL Schlenk tube flushed with argon. Then the solution is stirred at 25 °C for 20 min and directly used the polymerization. A solution of *i*-PrMgCl in THF (0.49 mmol) is added to a mixture of monomer precursor **25** (0.429 g, 0.50 mmol), LiCl (21.2 mg, 0.50 mmol) and THF (10 mL), under argon in a 25-mL Schlenk tube and stirred for 1 h at −20 °C (conversion of **25** was 92% as measured by ^{1}H NMR). Afterwards, Ni(acac)$_2$/dppp THF solution (1.5 mL, 6.5 mol% is added) is added in one portion and the mixture is stirred for 30 min at 0 °C. The polymerization mixture is quenched with HCl aqueous solution (5 M, 5 mL) and poured into water and extracted with toluene. The organic extracts are washed with brine and dried over $MgSO_4$. After the solvent is removed under reduced pressure, the GPC measurement is conducted ($M_n = 8.8$ kDa, PDI = 1.28). The polymer is further purified by the precipitation into acetone (191.9 mg, 59%).

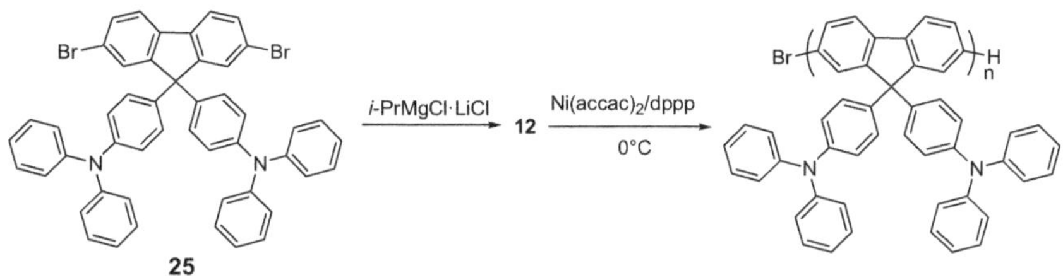

(20.12)

Example 5: Synthesis of Naphthalene-Diimide-Based Alternate donor–Acceptor Copolymer (20.13)[45]

The charge-transfer monomer **20** (20.8) is prepared in the argon glove box as follows: Br-TNDIT-Br (1.15 g, 1 mmol) is placed in a round-bottomed flask equipped with a magnetic stirrer and a septum. Dry THF (200 mL) and active Zn (0.2 g, 3 mmol) are added *via* a syringe (equipped with a thick enough needle to ensure an easy passage of Zn suspension), the mixture is stirred at room temperature for 1 h and then excess Zn is filtered out (20.13). Afterwards, a catalytic amount of Ni(dppe)Br_2 (0.154 g, 0.025 mmol) is added. Polymerization is conducted at room temperature for 5 h. Afterwards, the mixture is quenched with 5 M HCl and extracted with $CHCl_3$. The organic layers are washed with water, dried over anhydrous $MgSO_4$ and concentrated under reduced pressure to give the crude polymers. Acetone is added and the precipitates are washed with warm acetone, collected by filtration and dried in vacuum at 40 °C to give pure **21**. Yield: 0.85 g, 86%. $M_n = 23100$, $M_w/M_n = 1.6$ (oligo-(TNDIT) standard); $M_n =$ 94 700, $M_w/M_n = 1.8$ (PS standard).

(20.13)

20.2.3 End-Functionalized Polymers, Stars and Brushes

Example 6: External Initiation using Aryl Halides and Et_2Ni(bipy) as Activator[23]

In a glove box, 190 mg (0.203 mmol) of F8TBT-Br, 55 mg (0.20 mmol) of Et_2Ni(bipy),[22] and 89 mg (0.22 mmol) of dppe are dissolved in 3, 2, and 2 mL of THF, respectively, and the solutions are backtransferred outside the glove box *via* syringe (20.14). A 100-mL single-necked flask equipped with a magnetic stirrer and a septum is flame-dried under vacuum, cooled, and backfilled with argon. The F8TBT-Br solution is injected and stirred at room temperature. Et_2Ni(bipy) is added slowly *via* syringe, and the completion of reaction is spotted by a characteristic color change from the immediately appearing brown color to deep purple. The dppe solution is then added *via* syringe, and the solution is stirred for 1 min. To start polymerization, a solution of monomer **6** (4 mmol) in 40 mL of THF is added, and the mixture is stirred for 20 min. The resulting polymerization mixture is quenched by adding HCl (1 M in methanol) and precipitated by adding methanol. Soxhlet extraction is performed with methanol, hexanes and chloroform, and the chloroform fraction is passed through a plug of silica gel. The chloroform is removed under reduced pressure, methanol is added, and the solids are collected *via* filtration and dried at 40 °C overnight.

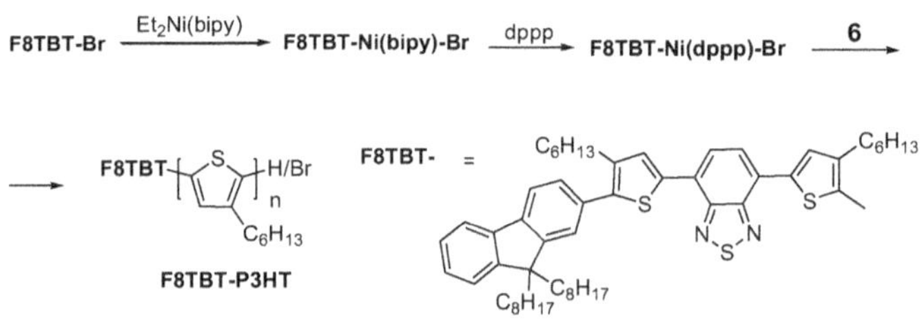

(20.14)

Example 7: External Initiation using Starting from Sterically Hindered Grignards and $NidpppCl_2$ (Synthesis of Defect-Free P3HT)[21]

Magnesium turnings (242 mg, 10 mmol) are placed into a double-neck round-bottomed flask equipped with a reflux condenser, a teflon-coated magnetic stirring bar and a rubber septum. The flask is evacuated and backfilled with dry argon. Dry THF (1 mL) and 1,2-dibromoethane (0.005 mL) are added and the mixture is stirred until evolution of ethylene gas is no longer observed. Then, dry THF (17 ml) is added, followed by the dropwise addition of 2-bromo-3-hexylthiophene (2.47 g, 10 mmol). Finally, the mixture is stirred under reflux for 1 h. After cooling to room temperature, the Grignard solution is withdrawn *via* syringe and filtered into an argon filled storage flask. $NidpppCl_2$ (1 mmol) is placed into a 25-mL round-bottomed flask equipped with a magnetic stirring bar, sealed with a rubber septum and the atmosphere is replaced with argon. THF (5 ml) is add *via* syringe and the mixture is cooled down to 0 °C. Grignard solution (1 mmol) is added dropwise under stirring, resulting in dissolution of the initially present solid and in formation of orange solution of initiator **26**, which is then used for the initiation of the polymerization (20.15).

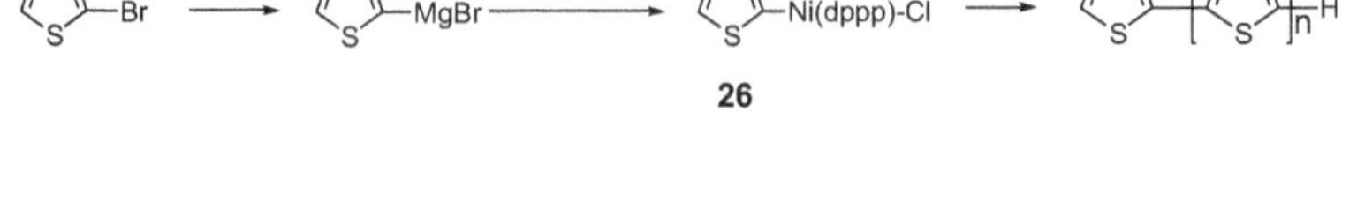

26

(20.15)

Example 8: "Hairy" P3HT Particles Prepared by Surface-Initiated KCTP (SI-KCTP)[22]

The monodispersed spherical ~460 nm in the diameter, silica-core organosilica-shell particles are prepared by a modified Stöber method *via* sequential sol-gel hydrolysis of tetramethyl orthosilicates (TMOS) and [2-(4-bromo-phenyl)-ethyl]-triethoxi-silane (**27**) under controlled conditions, according to the following procedure. Three solutions, I (containing 11.25 ml of TMOS in 67.5 ml of ethanol), II (containing 1.875 ml of TMOS, 2.92 g of **27** and 22.5 ml of ethanol), and III (containing 28.5 ml of 25% ammonia in water and 150 ml of ethanol), are prepared separately. Solutions I and II are consecutively added to solution III, with the rate 0.5 mL/min under an argon blanket into a round-bottomed flask under stirring at 440 min^{-1} and stirred overnight at r.t. The dispersion is then transferred out of the reactor, and the powders are washed with ethanol and ultrapure water by repeated centrifugation (at 5000 rpm for 15 min) and further dried at 70 °C for 12 h.

To develop |–Si–$(CH_2)_2$–C_6H_4–Ni(dppp)–Br initiator sites on the particles surface, Et_2Ni(bipy) solution in THF (0.070 g in 7 ml) is added at room temperature to the dispersion of the particles in THF (0.4 g in 5 ml) in several portions, adding the new portion after a clear indication of the consumption of a previous portion of Et_2Ni(bipy), *i.e.* after disappearance of the green color inherent to Et_2Ni(bipy). Normally, the reaction requires some induction period (0.5–2 h), but afterwards, it proceeds very rapidly. This "titration" procedure permits efficient conversion of all accessible |–Si–$(CH_2)_2$–C_6H_4–Br surface groups, thus avoiding contamination of the particles by unreacted Et_2Ni(bipy). Afterwards, the solution of dppp in THF (105.6 mg in 4 ml) is added to the suspension of 0.4 g particles in 4 ml of THF and allowed to react overnight at room temperature. Upon ligand exchange, the particles change their color from reddish-brown to orange. Prior the SI-KCTP, the particles are carefully purified from physiosorbed Ni compounds by 3 centrifugation/redispersion cycles (at 5000 rpm for 15 min) using anhydrous and degassed THF and CH_2Cl_2 in an argon atmosphere.

To perform SI-KCTP (20.5), the monomer **6** prepared as described in Example 1, is injected into the dispersion of the activated particles and the polymerization is performed at room temperature for 0.5 h. After the polymerization the resulting composite particles are subjected onto careful purifications to separate them from any ungrafted polymers, catalyst and byproducts. A typical purification procedure included several centrifugation/redispersion cycles using the following solvents and additives in the following consequence: 5 N HCl, water, aqueous hydrasine hydrate, water, methanol, $CHCl_3$, THF). Typically, less than 10% of unbound P3ATs is formed for the SI-KCTP from |–Si–$(CH_2)_2$–C_6H_4–Ni(dppp)–Br terminated particles. The grafting procedure leads to "hairy" P3HT particles having 20-nm thick P3HT layer (in the dry state).

Example 9: External Initiation using $Ni(COD)_2/PPh_3$ (Exemplified by the "Core-First" Synthesis of the Y-Shaped Three-Armed P3HT Star)[61] (20.16)

Initiator precursor **28** (405 mg, 0.5 mmol) is dissolved in anhydrous toluene (2 mL), then $Ni(COD)_2$ (455 mg, 1.65 mmol, 3.3 eq) and PPh_3 (1080 mg, 4.13 mmol, 8.3 eq) are added to the solution. The solution is heated to 50 °C and is left overnight. The color of the solution is changed from brown to light yellow. The resulting mixture is poured into anhydrous hexane. The precipitate is collected by filtration and washed with cold ether and hexane several times. Dppp (743 mg, 1.8 mmol) is added to the crude solution of the above Ni complex in anhydrous toluene. The solution is stirred at room temperature for 2 h. Then, the mixture is added dropwise into anhydrous hexane and filtrated carefully. The precipitate is collected by filtration and washed with cold ether and hexane several times. The crude **29** solution is added rapidly *via* syringe under nitrogen to the solution of monomer **6** in THF (1 mmol in 10 ml) prepared as described in Example 1. The reaction is allowed to warm to room temperature. After 4 h, 1 mL of 5 M HCl and 50 mL of methanol are added to quench the polymerization. The material is isolated by filtration, washed with methanol (2×10 mL) and cold hexane (5 mL), and dried in vacuo to afford pure regioregular Y-shaped P3HT (93.5 mg, 55% $M_n = 7.9$ PDI = 1.15).

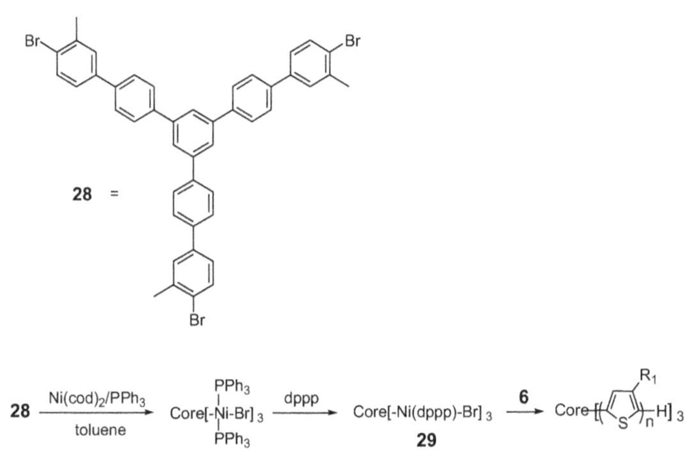

(20.16)

Example 10: Synthesis of Boc-Protected $P3HT-NH_2$ (20.17)[19]

Initiator preparation. A solution of Boc-protected initiator precursor **30**, prepared as shown in (20.20) (2.00 mmol, 0.861 g) in dry toluene (20 ml) is added under an argon atmosphere to $Ni(PPh_3)_4$ (2.00 mmol, 2.21 g). The reaction is allowed to proceed overnight, after which toluene (50 mL) is added and the solution is filtered. The supernatant is concentrated *in vacuo* and the product precipitated in MeOH at 0 °C. Finally, washing with *i*-PrOH yielded the **30**, ligated by monodentate as a yellow powder. Exchange of the monodentate PPh_3 ligand, by bidentate dppp one to give initiator **31** is performed *in situ*, just before the polymerization.

Polymerization. Solution of monomer **6** (2.00 mmol) in dry THF (20 mL) is prepared as described in Example 1 and converted into zinc derivative **7** by stirring of **6** with $ZnBr_2$ (3.00 mmol, 0.676 g) for 30 min. At the same time, a mixture of **30** (0.133 mmol, 0.135 g) and dppp (0.270 mmol, 0.111 g) in dry THF (10 mL) is stirred for 30 min at room temperature. In the next step, the two mixtures are combined and stirred for 3 h. The polymerization is then quenched by adding a few drops of NH_4Cl aqueous solution. The polymer is precipitated in methanol and filtered off. The polymer is fractionated by Soxhlet extraction with acetone and $CHCl_3$. The $CHCl_3$-soluble fraction is precipitated in methanol, filtered off and dried. Yield 238 mg (72%).

Deprotection. To a solution of P3HT-$NBoc_2$ (0.139 mmol, 23.0 mg) in $CHCl_3$ (80 mL), *p*-toluenesulfonic acid (0.5 mmol, 95.1 mg) is added, the mixture is heated to reflux and allowed to react during 48 h. Then, a $NaHCO_3$ solution is added and the mixture is extracted with $CHCl_3$. The organic layer is dried over K_2CO_3 and concentrated *in vacuo*. The deprotected amino-terminated P3HT is then precipitated in MeOH, filtered off and dried in vacuo. Yield = 18.6 mg (81%).

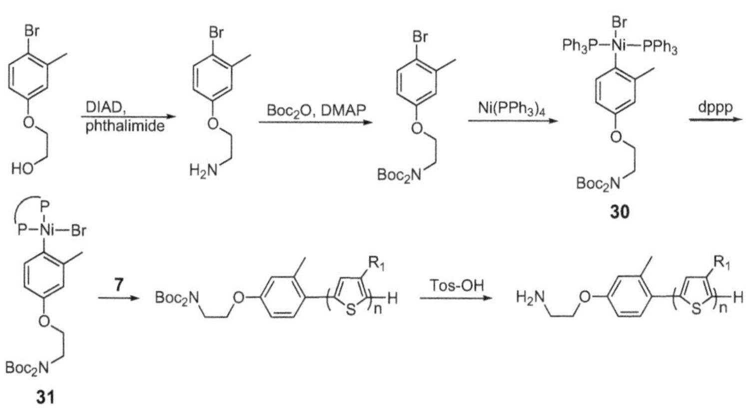

(20.17)

20.2.4 Rod–Coil Block Copolymers[62,63]

Example 11: Bifunctional Initiator for KCTP and NMRP (5)[62]

Bifunctional initiator **33** is prepared according to a two-step protocol that includes the reaction of aryl halides with $Et_2Ni(bipy)$, followed by the addition of an excess of the dppp ligand (20.18). A few drops of $Et_2Ni(bipy)$ solution in dry degassed toluene (total amount of $Et_2Ni(bipy)$ is 0.136 g, 0.5 mmol, in 5 mL of toluene) under an argon atmosphere are added to 0.248 g (0.6 mmol) of **32** dissolved in toluene (5 mL), and the mixture is warmed up to 40 °C until the disappearance of the green color inherent to $Et_2Ni(bipy)$. After the induction period that typically takes about 30 min, the reaction mixture is cooled to room temperature, and the remaining quantity of $Et_2Ni(bipy)$ solution is added dropwise; from this point, new added portions of $Et_2Ni(bipy)$ discolor immediately, and a deep-red color develops instead that indicates the formation of 4. Afterward, 0.272 g (0.66 mmol) of dppp in 5 mL toluene is added in one portion, and the solution is stirred for 30 min, during which the color of the reaction mixture changes from red to yellow-brown. This solution can be either used for further experiments without purification or **33** can be purified. In the latter case, 10 mL of hexane is added to the reaction mixture and is cooled to −40 °C for 1 h. Precipitated oil is separated from the solution by decantation and then used for further experiments.

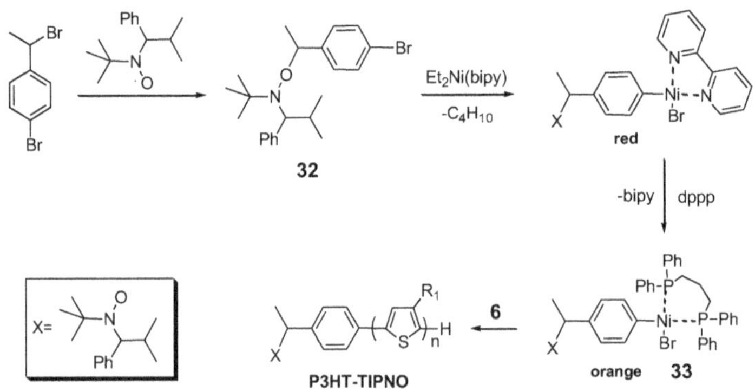

(20.18)

Example 12: Macroinitiator P3HT-TIPNO[62]

10 mmol of monomer **6** is prepared as described in Example 1. To prepare P3HT-TIPNO, 2.5 mL of freshly prepared solution of initiator **33** in toluene (0.1 mmol) is added in one portion to a solution of monomer **6** (10 mmol) (20.19). The mixture is stirred for 15 min at r.t. and then quenched by 5 M HCl. The reaction mixture is poured in water and extracted with $CHCl_3$. Combined organic layers are washed with brine, dried over anhydrous $MgSO_4$, and concentrated under reduced pressure. Finally, MeOH is added to the residue, and the insoluble material is collected by filtration and washed with MeOH, acetone, and hexane to give pure P3HT-TIPNO as a purple solid (1.16 g, 70%).

Example 13: Rod-Coil Block Copolymer Containing P3HT- and Fullerene-Based Blocks[26]

A glass pressure vessel is charged with 0.01 mmol of 0.04 mmol of P3HT-TIPNO, 43.2 mmol of styrene, 4.8 mmol of acetoxystyrene and a small portion of acetic anhydride (20.22). Afterwards, the mixture is degassed *via* three freeze-pump-thaw cycles. The reaction mixture is heated to 125–135 °C in a thermostated oil bath for 2 h to yield 2.183 g of **34**. For deprotection of **34**, it is dissolved in dioxane and heated to 40 °C. For each gram of **34**, 5 mL of hydrazine monohydrate is added, and the solution is stirred overnight. After evaporation of the solvent, the polymer is twice precipitated from water and dried, and the product is isolated as a purple powder. To perform etherification (attachment of fullerene), 0.1 g (0.094 mmol of –OH) of deprotected block copolymer is dissolved in 20 mL of chlorobenzene, then 200 mL of DMF is added slowly, and it is purged with nitrogen. To the solution are added 0.112 g (0.094 mmol) of the functionalized fullerene **35** and 49 mg (0.094 mmol) of Cs_2CO_3 and the reaction is stirred for 11 days at 50 °C. The filtered solution is dried; the product dissolved in THF and precipitated from hot hexane until the supernatant solution is colorless. The brownish block copolymer **36** is dried in vacuum (yield 0.161 g, 88%).

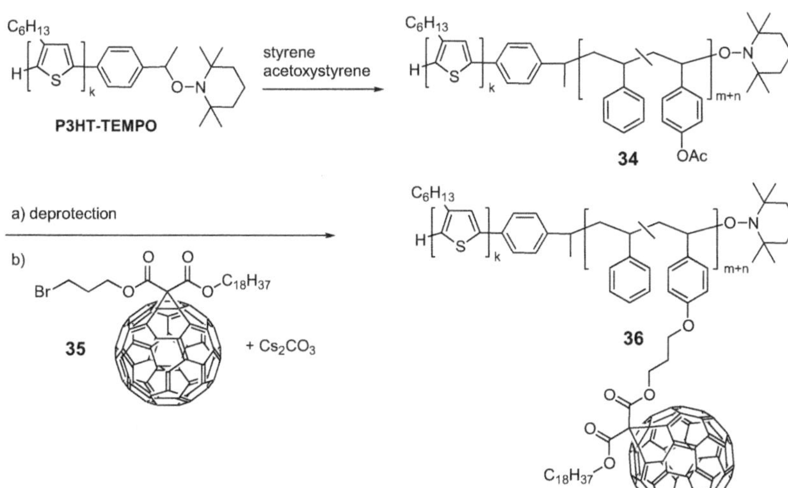

(20.19)

20.2.5 All-Conjugated Block Copolymers[39]

Example 14: Poly(*N*-hexylpyrrole)-*block*-(2,5-dihexyloxy-*p*-phenylene) (20.6)

A round-bottomed flask equipped with a three-way stopcock containing lithium chloride (45 mg, 1.1 mmol) is heated under reduced pressure, and then cooled to RT under a nitrogen atmosphere. To prepare monomer **9**, 1-bromo-2,5-dihexyloxy-4-iodobenzene (**37**) (496 mg, 1.03 mmol) is placed in the flask, and the atmosphere in the flask is replaced with nitrogen. Into the flask is added THF (5.0 mL) *via* a syringe, and the mixture is stirred at 0 °C. *i*-PrMgCl (2.0 M solution in THF, 0.48 mL, 0.96 mmol) is added *via* a syringe, and the mixture is stirred at room temperature for 1 h. To the mixture is added a suspension of $Ni(dppe)Cl_2$ (13.2 mg, 0.025 mmol, 2.5 mol%) in THF (5.0 mL) *via* a syringe, and then the mixture is stirred at room temperature for 0.5 h to give macro-initiator **19** ($M_n = 11300$ and $M_w/M_n = 1.13$). After the reaction mixture is cooled to 0 °C, a THF solution of monomer **10**, which had been prepared by treatment of 2,5-dibromo-1-hexylpyrrole (310 mg, 1.00 mmol) in THF (5.0 mL) with *i*-PrMgCl (2.0 M solution in THF, 0.48 mL, 0.96 mmol) in the presence of dppe (10.0 mg, 0.0251 mmol) at room temperature for 24 h, is added to the solution of **19** and the mixture is stirred at 0 °C for 9 h. To stop polymerization, 5 mol/L HCl is added, and the mixture is extracted with $CHCl_3$. The combined organic layer is washed with saturated aqueous $NaHCO_3$ and water, dried over anhydrous Na_2SO_4, and concentrated under reduced pressure to give a crude product **5** (770 mg, $M_n = 11\ 300$, $M_w/M_n = 1.13$). The crude **5** is dissolved in a small amount of $CHCl_3$, and the solution is poured into CH_3OH with vigorous stirring. The precipitated polymer is collected and dried under reduced pressure to give pure **5** as an orange viscous solid (223 mg, 51%, $M_n = 16\ 400$, $M_w/M_n = 1.16$).

Example 15: Polyphenylene-Based Diblock Ionomer (20.20)[44]

i-PrMgCl–LiCl complex (2.33 ml of 1.3 M solution in THF, 3.03 mmol) is added *via* a syringe to a solution of **38** (2.78 g, 3.33 mmol) in THF (13 ml) at room temperature in an argon-filled gloved box, and the reaction mixture is stirred at 40 °C for 5 h. The mixture is added to a suspension of $Ni(dppe)Cl_2$ (0.04 g, 0.076 mmol) in THF at room temperature, and stirred for 5 h to give macroinitiator **39** ($M_n = 14\ 900$, $M_w/M_n = 1.09$). To this reaction solution, the second monomer precursor, **37** (6.18 g, 14.2 mmol treated with *i*-PrMgCl–LiCl complex (9.91 ml of 1.3 M solution in THF, 12.9 mmol) dissolved in THF, 300 ml) is added and stirred for 5 h. After addition of 20 vol% MeOH solution of HCl, the insoluble material is washed well with MeOH and collected by suction filtration to give block copolymer **40** (Yield 2.89 g, 51.6% $M_n = 25\ 900$, $M_w/M_n = 1.11$).

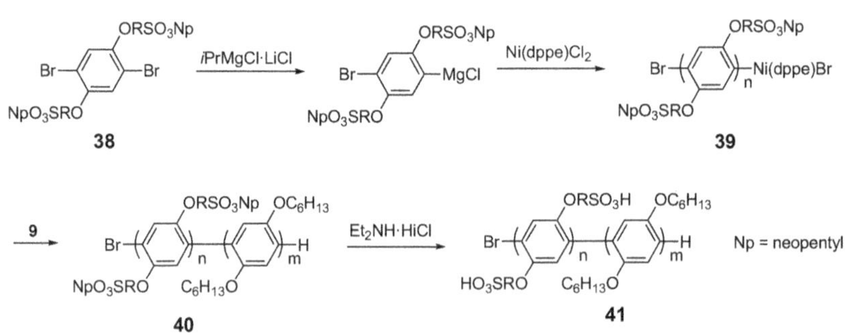

(20.20)

Deprotection. The neopentyl protected sulfonyl group in **40** (4.57 g) is cleaved with an *N*-methylpyrrolidone solution (40 ml) of diethylamine hydrobromide (5.45 g, 35.4 mmol) at 120 °C for 48 h to obtain **41**, the acid form of the copolymer **40**. After cooling to room temperature, the dark brown viscous reaction solution is poured into diethyl ether. Insoluble material is collected by suction filtration. The reddish brown product is stirred in 1 M HCl aq. for 48 h, and in water for 24 h to give **41** as a brown solid. Yield 3.49 g, 72.1%.

Example 16: Poly(3-hexylselenophene-*co*-3-hexylthiophenes)[43,63]

*Synthesis of the initiator **17** (20.21).* $Ni(COD)_2$ (517 mg, 1.88 mmol, 1.00 equiv) and PPh_3 (986 mg, 3.76 mmol, 2.00 equiv) are dissolved in toluene (10 mL) in an oven-dried 25-mL round-bottomed flask equipped with a stirring bar, under a nitrogen atmosphere. 2-Bromotoluene (354 mg, 2.07 mmol, 1.10 equiv) is added in a single portion and the mixture is stirred at r.t. for 1 h. Hexane (20 mL) is added and the resulting precipitate is collected by filtration. The solid is purified by recrystallization from THF/hexanes at −35 °C to afford **42** as an orange powder (345 mg, 24%). In a 25 mL oven-dried round-bottomed flask equipped with a stir bar, under a nitrogen atmosphere, **42** (340 mg, 0.45 mmol, 1.00 equiv) is dissolved in THF (5 mL). Then, dppe (198 mg, 0.50 mmol, 1.10 equiv) is added and the mixture is stirred at r.t. for 1 h. The volume of THF is reduced to ~1 mL under reduced pressure and hexanes (10 mL) are added. The resultant precipitate is isolated by filtration, washed with hexanes (10 mL) and diethyl ether (10 mL), and recrystallized from THF/hexanes (20 mL) at −35 °C to afford **17** as a fine yellow powder (215 mg, 76%).

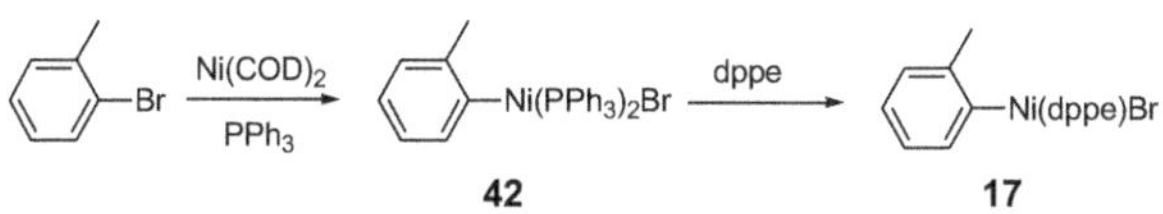

(20.21)

Random copolymer, P(3HT-co-3HS) (20.22).[63] Under a nitrogen atmosphere, solutions of **6** (3.50 mL, 0.145 M) and **11** (3.32 mL, 0.153 M) are combined in an oven-dried 100 mL Schlenk flask and diluted with THF to give an initial monomer concentration of 0.02 M. A stock solution of the nickel catalyst **17** in THF (5 mM, 2.0 mL, 10.1 μmol) is transferred to the flask *via* syringe. After stirring at r.t. for 30 min, the polymerization is quenched with 5 M aq HCl (5 mL) and extracted with $CHCl_3$ (3×5 mL) and dried *in vacuo*. The resultant blue solid is dissolved in a minimum amount of $CHCl_3$ (~1 mL) upon heating and precipitated into MeOH (20 mL). The precipitate is collected on Whatman filter paper and purified by Soxhlet extraction using MeOH, acetone, hexanes, dichloromethane, and chlorobenzene. The chlorobenzene fraction is dried in vacuo to afford the polymer as a blue solid (170 mg, 88%). Gel permeation chromatography relative to polystyrene standards (THF, 40 °C): $M_n = 32.6$ kDa, PDI = 1.11.

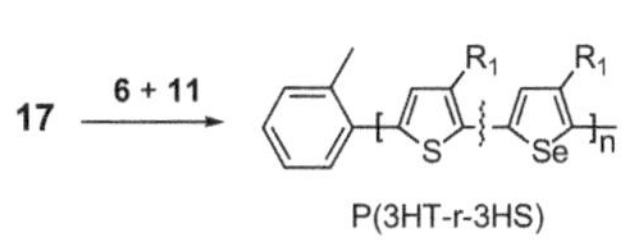

(20.22)

Block copolymerization, P(3HT-b-3HS) (20.23). Under a nitrogen atmosphere, a stock solution of **6** (3.90 mL, 0.153 M) is transferred to an oven-dried 100 mL round-bottomed flask and diluted with THF to give an initial monomer concentration of 0.01 M. A stock solution of the nickel catalyst **17** in THF (5 mM, 2.34 mL, 11.7 μmol) is transferred to the flask *via* syringe. After stirring at r.t. for 1 h, nearly complete conversion of **6** is confirmed by gas chromatography (>98%). GPC of the macroinitiator block indicated $M_n = 12.5$ kDa, PDI = 1.09. Subsequently, the stock solution of **11** (3.55 mL, 0.162 M) is injected. After an additional 1 h at r.t., the polymerization is quenched with 5 M aq HCl (5 mL) and extracted with $CHCl_3$ (3×5 mL) and dried *in vacuo*. The resultant blue solid is dissolved in a minimum amount of $CHCl_3$ upon heating and precipitated into MeOH. The precipitate is collected on Whatman filter paper and purified by Soxhlet extraction using MeOH, acetone, hexanes, CH_2Cl_2, and chlorobenzene. The chlorobenzene fraction is dried *in vacuo* to afford the polymer as a blue solid (182 mg, 81%). Gel permeation chromatography relative to polystyrene standards (THF, 40 °C): $M_n = 28.7$ kDa, PDI = 1.15.

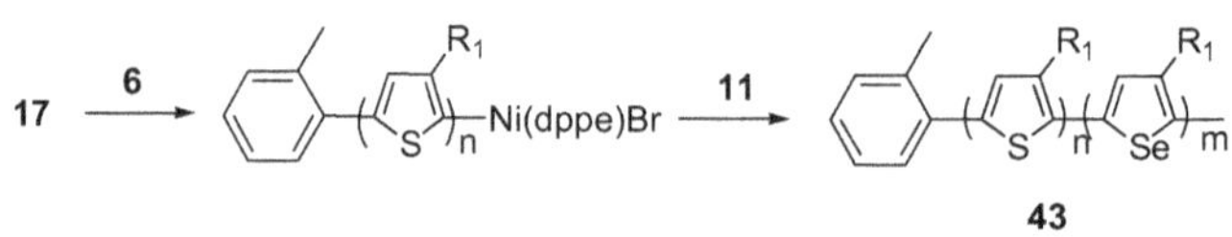

(20.23)

Gradient copolymerization, P(3HT-g-3HS) (20.24). Under a nitrogen atmosphere, a solution of **6** (15 mL, 0.01 M) is transferred to an oven-dried 25-mL Schlenk flask. A solution of **11** (3 mL, 0.05 M) is loaded into a 5-mL plastic syringe that attached to a syringe pump apparatus. With constant stirring, a stock solution of the nickel catalyst **17** in THF (5 mM, 0.5 mL, 3.0 μmol) is transferred to the flask *via* syringe, and the syringe pump is activated to deliver **11** at a rate of 0.24 mL/min. After 1 h at r.t., the polymerization is quenched with 5 M aq HCl (5 mL) and extracted with $CHCl_3$ (3×5 mL) and dried *in vacuo*. The resultant blue solid is dissolved in a minimum amount of $CHCl_3$ (~1 mL) upon heating and precipitated into MeOH (20 mL). The precipitate is collected on Whatman filter paper and purified by Soxhlet extraction using MeOH, acetone, hexanes, dichloromethane, and chlorobenzene. The chlorobenzene fraction is dried *in vacuo* to afford the polymer as a blue solid (54 mg, 84%). Gel permeation chromatography relative to polystyrene standards (THF, 40 °C): $M_n = 31.9$ kDa, PDI = 1.13.

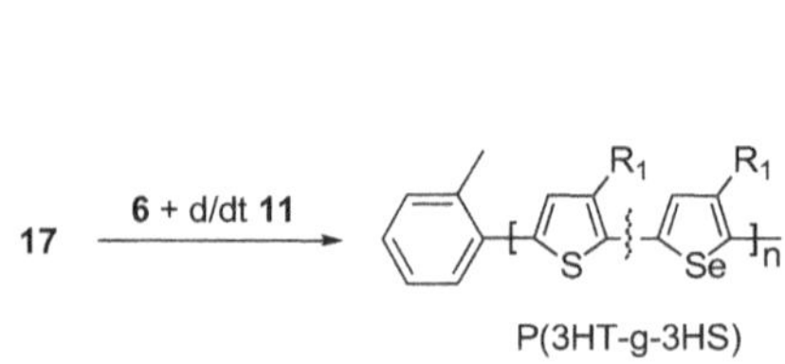

(20.24)

20.2.6 Suzuki Polymerization[51–55]

Example 17: Pd Initiators[55]

All the following operations are performed under glove-box atmosphere. To obtain initiators **44–47,** Pd(PtBu$_3$)$_2$ (1 equiv) and 1.1 equiv of the relevant aryl-bromides (3-perylene-, 2-anthracene-, 1-pyrene- or phenyl-bromide) are mixed together with toluene (3 mL) under an argon atmosphere at 70 °C and allowed to react for 2, 3, 4 and 5 h, respectively (20.9). For the diphenyl-amine (DPA) bromo initiator, the reaction time is 9 h. Later, the reaction mixture is poured into cold pentane (20 mL) and left to stand in the fridge for another 2 h until a bright yellow/orange precipitate is formed. The products are stored inside a glove-box fridge (–40 °C). The latter is collected by filtration, washed 5 times with cold pentane (5×4 mL) and dried under reduced pressure overnight. For example, 146.6 mg (70%) of the initiator **46** (yellow powder) is obtained starting from 110 mg (0.45 mmol) of Pd(PtBu$_3$)$_2$ and 137 mg (0.5 mmol) of bromo-pyrene.

Example 18: Pyrene-End-Functionalized Poly(9,9′-dioctylfluorene-co-benzothiadiazole) (20.25)

Under an argon atmosphere, monomer **48** (30 mg, 0.50 mmol), 18-crown-6 ether (350 mg, 1.31 mmol) and CsF (84 mg, 0.55 mmol) are dissolved in anhydrous THF (stabilizer free, 5.2 mL) and 0.78 mL water. The reaction flask is cooled to 0 °C and the polymerization is initiated by fast injection of solution of **48** into the initiator flask (2.5 μmol, 1 mM), which is also cooled to 0 °C. After 24 h the reaction is quenched with a mixture of 1 : 1 HCl (2 M):methanol. The precipitate is collected by filtration, washed with water and dried under a vacuum oven for 24 h to give **49** as a dark orange/red powder: M_n = 7 380, PDI = 1.30 (crude GPC, PS standards). Yield: 87%.

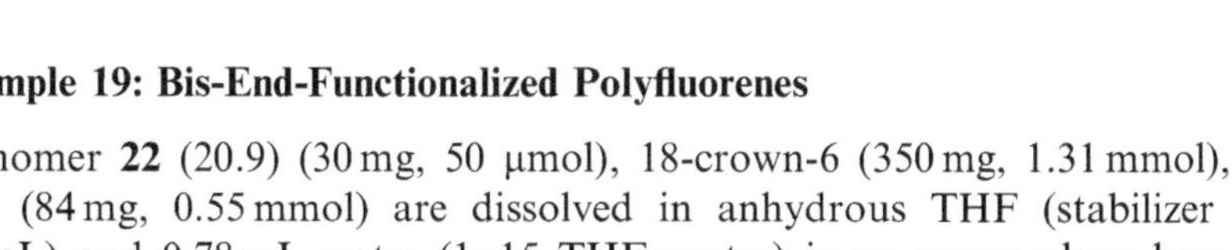

(20.25)

Example 19: Bis-End-Functionalized Polyfluorenes

Monomer **22** (20.9) (30 mg, 50 μmol), 18-crown-6 (350 mg, 1.31 mmol), and CsF (84 mg, 0.55 mmol) are dissolved in anhydrous THF (stabilizer free, 5.2 mL) and 0.78 mL water (1 : 15 THF : water) in an argon glove box. The reaction flask is carefully sealed with a septum and 2 syringes are attached to it; one contained a mixture of the monomer with CsF and crown ether and the second syringe with the end-capper solution. The reaction flask is taken out of the glove box and cooled to 0 °C and the polymerization is initiated by fast

Examples of bis-endfunctionalyzed polyfluorenes

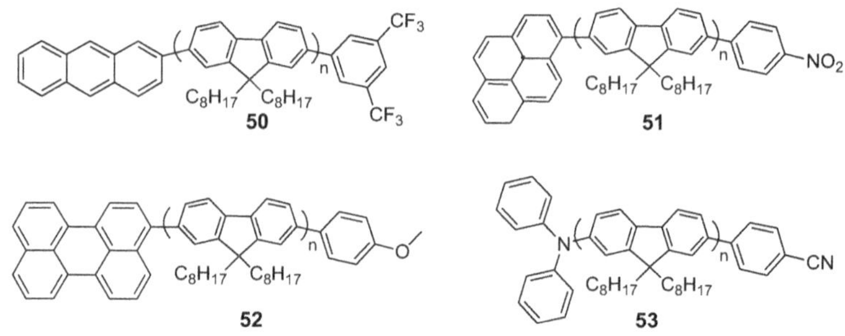

Chart 20.3

injection of the monomer solution into the initiator flask (5 mol%, 2.5 μmol, 1 mM), which is also cooled to 0 °C. After 15 min the reaction is quenched with one of the monoboronic-ester (or monoboronic-acid) end-capper (50 μmol) and stirred for additional 5 min before it is fully quenched with water. The precipitate is redissolved with $CHCl_3$, dried over $MgSO_4$, and the solvent removed under reduced pressure. In order to purify the polymer from the remaining monomer and initiator, the residue is redispersed with a small amount of $CHCl_3$ and reprecipitated with MeOH, centrifuged, collected and dried in a vacuum oven for 24 h to give bis-end-functionalized polyfluorenes **50–53** (Chart 20.3) as a light yellow color powder. In order to achieve high end-group fidelity, the polymerizations are quenched relatively early, *i.e.* after 15 min. Yield = 50–58%; M_n about 5000 g/mol, PDI = 1.1–1.2 (GPC, PS standards), DP = 10–11 (NMR end-group analysis). However, if longer polymeric molecular wires are desirable, different monomer-to-initiator ratios (*i.e.* feed ratios) and longer polymerization times can be used. In this case, polyfluorenes with M_n = 10 000 g/mol and PDI = 1.2–1.3 but a bit worse end-group fidelity are achievable.

Example 20: Polymerization with $Pd_2(dba)_3$/*t*-Bu_3P/*p*-BrC_6H_4I as the Initiator (20.9)[64]

In a drybox under a N_2 atmosphere, to 5-mL vial containing $Pd_2(dba)_3$ (5.5 mg, 0.006 mmol), THF (0.4 mL), *t*Bu3P (48 μL, 0.5 M solution in THF) is added *p*-bromoiodobenzene (40 μL, 0.25 M solution in THF). The mixture is stirred for 1 h at room temperature to generate the initiator. In another vial containing monomer **22** (0.1 mmol) and THF (5.4 mL) is added 0.5 mL of K_3PO_4 solution (0.5 mL, 2 M solution in water) and the mixture is cooled to 0 °C. The solution of the *in situ* generated initiator is quickly injected into the solution of monomer and the resulting mixture is stirred for 30 min at 0 °C. The reaction is quenched by pouring the mixture into 12 N HCl solution. The product is extracted with CH_2Cl_2 (3–20 mL). The organic layer is combined, washed by brine and dried with Na_2SO_4. The mixture is filtered and the filtrate is evaporated under reduced pressure. The residue is dissolved in a minimum amount of THF and the solution is added dropwise to methanol with stirring. The precipitate is collected by filtration, washed with methanol and water and dried under vacuum. The resulting poly(9,9-dioctyl-2,7-fluorenylene) is obtained in 80% yield (GPC: $M_n = 14\ 600$; PDI = 1.19).

Example 21: Suzuki Block Copolymerization (20.26)[53]

Initiator **47** prepared as described in Example 17 (1.0 mg, 2.1 μmol) is placed in a round-bottomed flask, and a vacuum is applied. The flask is filled with argon, THF (2 mL) is added and then the solution is degassed with argon by using a diaphragm pump. In a pear-shaped flask, monomer **22** (23 mg, 0.042 mmol), CsF (70 mg, 0.46 mmol), 18-crown-6 (0.29 g, 1.1 mmol), THF (4 mL) and water (0.6 mL) are added. The solution is similarly degassed. The polymerization is initiated by addition of the solution in the pear-shaped flask into the solution containing **47** in the round-bottomed flask *via* a cannula at 0 °C. The mixture is stirred at 0 °C for 4 h (macromonomer **54**: $M_n = 7300$, $M_w/M_n = 1.19$). A solution of monomer **55** (19 mg, 0.043 mmol) in THF (3 mL), which had been degassed, in a pear-shaped flask is added to the solution of **54**. The reaction mixture is stirred at 0 ° C for 1 h, and then the reaction is quenched with 12 mol/L hydrochloric acid. The solution is extracted with dichloromethane, and the organic layer is dried over $MgSO_4$, followed by concentration under reduced pressure to give 18.2 mg (68%) of **56** ($M_n = 13\ 000$, $M_w/M_n = 1.29$).

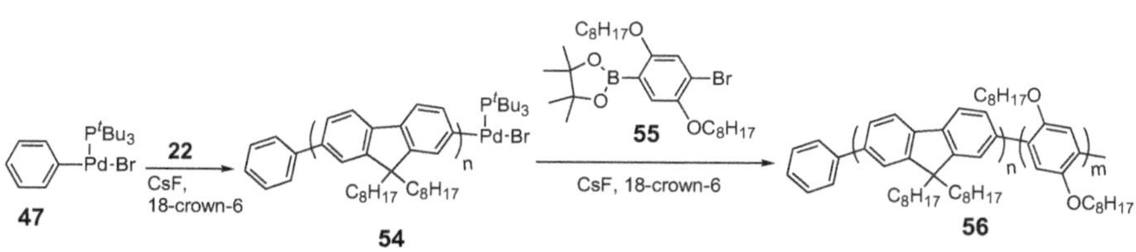

(20.26)

References

1. Y. J. Cheng, S. H. Yang and C. S. Hsu, *Chem. Rev.*, 2009, **109**, 5868.
2. I. Osaka and R. D. McCullough, *Acc. Chem. Res.*, 2008, **41**, 1202.
3. R. D. McCullough, S. Tristram-Nagle, S. P. Williams, R. D. Lowe and M. Jayaraman, *J. Am. Chem. Soc.*, 1993, **115**, 4910.
4. L. H. Jimison, M. F. Toney, I. McCulloch, M. Heeney and A. Salleo, *Adv. Mater.*, 2009, **21**, 1568.
5. H. Sirringhaus, P. J. Brown, R. H. Friend, M. M. Nielsen, K. Bechgaard, B. M. W. Langeveld-Voss, A. J. H. Spiering, R. A. J. Janssen, E. W. Meijer and P. Herwig, D. M. de Leeuw, *Nature*, 1999, **401**, 685.
6. G. Zhao, Y. He and Y. Li, *Adv. Mater.*, 2010, **22**, 4355.
7. P. Schilinsky, U. Asawapirom, U. Scherf, M. Biele and C. Brabec, *J. Chem. Mater.*, 2005, **17**, 2175.

8. R. C. Mulherin, S. Jung, S. Huettner, K. Johnson, P. Kohn, M. Sommer, S. Allard, U. Scherf and N. C. Greenham, *Nano Lett.*, 2011, **11**, 4846.
9. U. Scherf, A. Gutacker and N. Koenen, *Acc. Chem. Res.*, 2008, **41**, 1086.
10. A. Kiriy, V. Senkovskyy and M. Sommer, *Macromol. Rapid Commun.*, 2011, **32**, 1503.
11. E. E. Sheina, J. Liu, M. C. Iovu, D. W. Laird and R. D. McCullough, *Macromolecules*, 2004, **37**, 3526.
12. M. C. Iovu, E. E. Sheina, R. R. Gil and R. D. McCullough, *Macromolecules*, 2005, **38**, 8649.
13. A. Yokoyama, R. Miyakoshi and T. Yokozawa, *Macromolecules*, 2004, **37**, 1169.
14. R. Miyakoshi, A. Yokoyama and T. Yokozawa, *J. Am. Chem. Soc.*, 2005, **127**, 17542.
15. T. Beryozkina, V. Senkovskyy, E. Kaul and A. Kiriy, *Macromolecules*, 2008, **41**, 7817.
16. R. Tkachov, V. Senkovskyy, H. Komber, J.-U. Sommer and A. Kiriy, *J. Am. Chem. Soc.*, 2010, **132**, 7803.
17. V. Senkovskyy, N. Khanduyeva, H. Komber, U. Oertel, M. Stamm, D. Kuckling and A. Kiriy, *J. Am. Chem. Soc.*, 2007, **129**, 6626.
18. N. Doubina, M. Stoddard, H. A. Bronstein, A. K.-Y. Jen and C. K. Luscombe, *Macromol. Chem. Phys.*, 2009, **210**, 1966.
19. A. Smeets, P. Willot, J. Winter, P. Gerbaux, T. Verbiest and G. Koeckelberghs, *Macromolecules*, 2011, **44**, 6017.
20. H. A. Bronstein and C. K. Luscombe, *J. Am. Chem. Soc.*, 2009, **131**, 12894.
21. V. Senkovskyy, M. Sommer, H. Komber, R. Tkachov, W. T. S. Huck and A. Kiriy, *Macromolecules*, 2010, **43**, 10157.
22. V. Senkovskyy, R. Tkachov, T. Beryozkina, H. Komber, U. Oertel, M. Horecha, V. Bocharova, M. Stamm, S. A. Gevorgyan, F. C. Krebs and A. Kiriy, *J. Am. Chem. Soc.*, 2009, **131**, 16445.
23. H. Komber, V. Senkovskyy, R. Tkachov, K. Johnson, A. Kiriy, W. T. S. Huck and M. Sommer, *Macromolecules*, 2011, **44**, 9164.
24. V. Senkovskyy, T. Beryozkina, V. Bocharova, R. Tkachov, H. Komber, A. Lederer, M. Stamm, N. Severin, J. P. Rabe and A. Kiriy, *Macromol. Symp.*, 2010, **17**, 291.
25. E. Kaul, V. Senkovskyy, R. Tkachov, V. Bocharova, H. Komber, M. Stamm and A. Kiriy, *Macromolecules*, 2010, **43**, 77.
26. M. Heuken, H. Komber, T. Erdmann, V. Senkoyskyy, A. Kiriy and B. Voit, *Macromolecules*, 2012, **45**, 4101.
27. R. Tkachov, V. Senkovskyy, M. Horecha, U. Oertel, M. Stamm and A. Kiriy, *Chem. Commun.*, 2010, **46**, 1425.
28. R. Tkachov, V. Senkovskyy, U. Oertel, A. Synytska, M. Horecha and A. Kiriy, *Macromol. Rapid Commun.*, 2010, **31**, 2146.
29. N. Khanduyeva, V. Senkovskyy, T. Beryozkina, V. Bocharova, F. Simon, M. Nitschke, M. Stamm, R. Groetzschel and A. Kiriy, *Macromolecules*, 2008, **41**, 7383.

30. N. Khanduyeva, V. Senkovskyy, T. Beryozkina, M. Horecha, M. Stamm, C. Uhrich, M. Riede, K. Leo and A. Kiriy, *J. Am. Chem. Soc.*, 2009, **131**, 153.
31. N. Marshall, S. K. Sontag and J. Locklin, *Chem. Commun.*, 2011, **47**, 5681.
32. T. P. Lodge, *Macromol. Chem. Phys.*, 2003, **204**, 265.
33. R. Verduzco, I. Botiz, D. L. Pickel, S. M. Kilbey, K. Hong, E. Dimasi and S. B. Darling, *Macromolecules*, 2011, **44**, 530.
34. D. Izuhara and T. M. Swager, *Macromolecules*, 2011, **44**, 2678.
35. P.-T. Wu, G. Ren, C. Li, R. Mezzenga and S. A. Jenekhe, *Macromolecules*, 2009, **42**, 2317.
36. K. van den Bergh, I. Cosemans, T. Verbiest and G. Koeckelberghs, *Macromolecules*, 2010, **43**, 3794.
37. Y. Zhang, K. Tajima and K. Hashimoto, *Macromolecules*, 2009, **42**, 7008.
38. K. Ohshimitsu and M. Ueda, *Macromolecules*, 2008, **41**, 5289.
39. A. Yokoyama, A. Kato, R. Miyakoshi and T. Yokozawa, *Macromolecules*, 2008, **41**, 7271.
40. R. Miyakoshi, A. Yokoyama and T. Yokozawa, *Chem. Lett.*, 2008, **37**, 1022.
41. S. Wu, L. Bu, L. Huang, X. Yu, Y. Han, Y. Geng and F. Wang, *Polymer*, 2009, **50**, 6245.
42. A. E. Javier, S. R. Varshney and R. D. McCullough, *Macromolecules*, 2010, **43**, 3233.
43. J. Hollinger, A. A. Jahnke, N. Coombs and D. S. Seferos, *J. Am. Chem. Soc.*, 2010, **132**, 8546.
44. K. Umezawa, T. Oshima, M. Yoshizawa-Fujita, Y. Takeoka and M. Rikukawa, *ACS Macro Lett.*, 2012, **1**, 969.
45. V. Senkovskyy, R. Tkachov, H. Komber, M. Sommer, M. Heuken, B. Voit, W. T. S. Huck, V. Kataev, A. Petr and A. Kiriy, *J. Am. Chem. Soc.*, 2011, **133**, 19966.
46. V. Senkovskyy, R. Tkachov, H. Komber, A. John, J.-U. Sommer and A. Kiriy, *Macromolecules*, 2012, **45**, 7770.
47. E. Elmalem, A. Kiriy and W. T. S. Huck, *Macromolecules*, 2011, **44**, 9057.
48. N. Miyaura and A. Suzuki, *J. Chem. Soc. Chem. Commun.*, 1979, 866.
49. A. Suzuki, *Angew. Chem. Int. Ed.*, 2011, **50**, 6723.
50. J. Sakamoto, M. Rehahn, G. Wegner and A. D. Schlüter, *Macromol. Rapid Commun.*, 2009, **30**, 653.
51. A. Yokoyama, H. Suzuki, Y. Kubota, K. Ohuchi, H. Higashimura and T. Yokozawa, *J. Am. Chem. Soc.*, 2007, **129**, 7236.
52. T. Beryozkina, K. Boyko, N. Khanduyeva, V. Senkovskyy, M. Horecha, U. Oertel, F. Simon, M. Stamm and A. Kiriy, *Angew. Chem., Int. Ed.*, 2009, **48**, 2695.
53. T. Yokozawa, H. Kohno, Y. Ohta and A. Yokoyama, *Macromolecules*, 2010, **43**, 7095.
54. T. Yokozawa, R. Suzuki, M. Nojima, Y. Ohta and A. Yokoyama, *Macromol. Rapid Commun.*, 2011, **32**, 801.

55. E. Elmalem, F. Biedermann, K. Johnson, R. H. Friend and W. T. S. Huck, *J. Am. Chem. Soc.*, 2012, **134**, 17769.
56. R. D. McCullough, R. D. Lowe, M. Jayaraman and D. L. Anderson, *J. Org. Chem.*, 1993, **58**, 904.
57. T.-A. Chen, X. Wu and R. D. Rieke, *J. Am. Chem. Soc.*, 1995, **117**, 233.
58. R. S. Loewe, S. M. Khersonsky and R. D. McCullough, *Adv. Mater.*, 1999, **11**, 250.
59. M. C. Stefan, A. E. Javier, I. Osaka and R. D. McCullough, *Macromolecules*, 2009, **42**, 30.
60. A. Sui, X. Shi, S. Wu, H. Tian, Y. Geng and F. Wang, *Macromolecules*, 2012, **45**, 5436.
61. M. Yuan, K. Okamoto, H. A. Bronstein and C. K. Luscombe, *ACS Macro Lett.*, 2012, **1**, 392.
62. E. Kaul, V. Senkovskyy, R. Tkachov, V. Bocharova, H. Komber, M. Stamm and A. Kiriy, *Macromolecules*, 2010, **43**, 77.
63. E. F. Palermo and A. J. McNeil, *Macromolecules*, 2012, **45**, 5948.
64. H.-H. Zhang, C.-H. Xing and Q.-S. Hu, *J. Am. Chem. Soc.*, 2012, **134**, 13156.

Subject Index

References to figures are given in italic type. References to tables are given in bold type.